THE ELEMENTS

				3A	4A	5A	6A	1 **H** 1.00797 ±0.00001	2 **He** 4.0026 ±0.00005
				5 **B** 10.811 ±0.003	6 **C** 12.01115 ±0.00005	7 **N** 14.0067 ±0.00005	8 **O** 15.9994 ±0.0001	9 **F** 18.9984 ±0.00005	10 **Ne** 20.183 ±0.0005
1B		2B		13 **Al** 26.9815 ±0.00005	14 **Si** 28.086 ±0.001	15 **P** 30.9738 ±0.00005	16 **S** 32.064 ±0.003	17 **Cl** 35.453 ±0.001	18 **Ar** 39.948 ±0.0005
28 **Ni** 58.71 ±0.005	29 **Cu** 63.54 ±0.005	30 **Zn** 65.37 ±0.005	31 **Ga** 69.72 ±0.005	32 **Ge** 72.59 ±0.005	33 **As** 74.9216 ±0.00005	34 **Se** 78.96 ±0.005	35 **Br** 79.909 ±0.002	36 **Kr** 83.80 ±0.005
46 **Pd** 106.4 ±0.05	47 **Ag** 107.870 ±0.003	48 **Cd** 112.40 ±0.005	49 **In** 114.82 ±0.005	50 **Sn** 118.69 ±0.005	51 **Sb** 121.75 ±0.005	52 **Te** 127.60 ±0.005	53 **I** 126.9044 ±0.00005	54 **Xe** 131.30 ±0.005
78 **Pt** 195.09 ±0.005	79 **Au** 196.967 ±0.0005	80 **Hg** 200.59 ±0.005	81 **Tl** 204.37 ±0.005	82 **Pb** 207.19 ±0.005	83 **Bi** 208.980 ±0.0005	84 **Po** (210)	85 **At** (210)	86 **Rn** (222)

63 **Eu** 151.96 ±0.005	64 **Gd** 157.25 ±0.005	65 **Tb** 158.924 ±0.0005	66 **Dy** 162.50 ±0.005	67 **Ho** 164.930 ±0.0005	68 **Er** 167.26 ±0.005	69 **Tm** 168.934 ±0.0005	70 **Yb** 173.04 ±0.005	71 **Lu** 174.97 ±0.005

95 **Am** (243)	96 **Cm** (247)	97 **Bk** (247)	98 **Cf** (249)	99 **Es** (254)	100 **Fm** (253)	101 **Md** (256)	102 **No** (253)	103 **Lw** (257)

Atomic Weights are based on C¹²—12.0000 and Conform to the 1961 Values
of the Commission on Atomic Weights

Printed in U.S.A.

WILLIAM L. MASTERTON

PROFESSOR OF CHEMISTRY,
UNIVERSITY OF CONNECTICUT, STORRS, CONNECTICUT

EMIL J. SLOWINSKI

CHAIRMAN, DEPARTMENT OF CHEMISTRY
MACALESTER COLLEGE, ST. PAUL, MINNESOTA

SECOND EDITION

Chemical Principles

W. B. SAUNDERS COMPANY PHILADELPHIA / LONDON / TORONTO

W. B. Saunders Company: West Washington Square
                           Philadelphia, Pa. 19105

                           12 Dyott Street
                           London, WC1A 1DB
                           1835 Yonge Street
                           Toronto 7, Ontario

Chemical Principles

SBN 0-7216-6171-8

Print No.:   9   8   7   6   5

PREFACE

Of the many temptations which beset the authors of a general chemistry text, there is one which is to be avoided above all others. We refer to the tendency to produce an encyclopedic survey in which isolated facts are presented because they are "important" and elaborate theories are developed because they are "interesting." With this in mind, we have restricted ourselves to a body of material which can readily be covered in a year course in general chemistry. The student who is looking for a mathematical treatment of the Schrödinger equation or an extended discussion of the chemistry of the blast furnace will not find it here. He will, however, find a reasonably thorough treatment of the major principles of chemistry, illustrated by factual material which we hope is both relevant and interesting.

In presenting descriptive inorganic chemistry, we have departed from the traditional framework of the periodic table, in which successive families of elements are discussed ad nauseam. Instead, this material is organized around types of reactions in which substances participate. Using this approach, precipitation reactions are discussed in Chapter 16, acid-base reactions in Chapters 17 and 18, complex-ion formation in Chapter 19, and redox reactions in Chapters 20 to 22. Experience has convinced us that students find descriptive chemistry more meaningful and more interesting when it is presented in this context.

One of the major objectives of a student in general chemistry is to develop the ability to predict the spontaneity and extent of chemical reactions. To achieve this, he must become familiar with the principles of chemical thermodynamics. The challenge of presenting these principles and their applications at an elementary level is a difficult one; we frankly confess that our treatment in the first edition of this text left something to be desired. The principles of thermodynamics are introduced here in an early chapter (Chapter 4) which deals with thermochemistry, the First Law, and the enthalpy function. The more abstract concepts of free energy and entropy are presented in Chapter 13, after the student has had sufficient exposure to the principles of equilibrium to appreciate their relevance.

The area of chemical kinetics is one which has received far too little attention in the general chemistry course. Chapter 14, which is devoted to this subject, emphasizes the importance of rate studies in unravelling reaction mechanisms. This aspect of kinetics is further developed in interpreting substitution reactions involving complex ions (Chapter 19) and organic molecules (Chapter 24).

The chapters dealing with chemical bonding and structure have been extensively revised. After an expanded treatment of atomic structure in Chapter 7, the principles of ionic and covalent bonding are presented in Chapter 8. The difficulty that our students have in drawing simple Lewis structures has prompted us to include a separate section on this ancient but still fundamental topic. Subsequent sections of Chapter 8 deal with hybrid orbitals, molecular geometry, and an introduction to molecular orbital theory. In Chapter 9, the relationship between structure and physical properties for ionic, molecular, macromolecular, and metallic lattices is developed in some detail. Crystal geometry, types of packing, and crystal defects are discussed in Chapter 10.

Those who are familiar with the first edition will note two changes in format. The distinctive, two-color illustrations which replace the austere artwork of the previous edition were furnished by Mr. George Mass of the Versatron Corporation. The comments which appear in color along the margins of the text are our responsibility. We have used them to emphasize certain principles, to stimulate the student to think about what he is reading, or to provide a light touch where it seemed appropriate.

The many suggestions of the students and instructors who used the first edition have been extremely helpful. Among the comments, those of Professor Joseph Deck of the University of Louisville and Professor Charles E. Waring of the University of Connecticut were particularly pertinent. We are indebted to Professor Eugene Rochow of Harvard University for a detailed and thoughtful review of the manuscript. Finally, we appreciate the patience shown by the staff of the W. B. Saunders Co. toward two authors who could never resist the temptation to change galleys and page proof regardless of deadlines.

WILLIAM L. MASTERTON

EMIL J. SLOWINSKI

CONTENTS

1

SOME BASIC CONCEPTS 1

   1.1 WHAT CHEMISTRY IS 1
   1.2 MEASUREMENTS 2
       Length, 2. Volume, 4. Mass, 5. Temperature, 8.
       Other Measured Quantities, 11.

   1.3 RESOLUTION OF MATTER INTO PURE SUBSTANCES 12
       Distillation and Fractional Distillation, 12. Fractional
       Crystallization, 14. Chromatography, 15.

   1.4 IDENTIFICATION OF PURE SUBSTANCES 16
       Behavior of Pure Substances During Melting or Boiling,
       17. Physical Properties of Pure Substances, 18.

   1.5 KINDS OF SUBSTANCES — ELEMENTS AND COMPOUNDS 20
       The Discovery of the Elements, 21. Chemical Symbols
       of the Elements, 21.

       PROBLEMS 22

2

ATOMS, MOLECULES, AND IONS 24

   2.1 ATOMS 24
       Atomic Theory, 24. Structure of the Atom, 26.

   2.2 MOLECULES AND IONS 27
       Molecules, 27. Ions, 28.

2.3 RELATIVE MASSES OF ATOMS, ATOMIC WEIGHTS AND GRAM ATOMIC WEIGHTS 28

The Meaning of Atomic Weights. The Atomic Weight Scale, 30. Gram Atomic Weight, 31. Atomic Weights from Specific Heats. Gram Equivalent Weights, 31. Atomic Weights from Mass Spectra, 34.

2.4 ABSOLUTE MASSES OF ATOMS. AVOGADRO'S NUMBER ... 37
2.5 MASSES OF MOLECULES. MOLECULAR WEIGHT. GRAM MOLECULAR WEIGHT.. 38

PROBLEMS 39

3

CHEMICAL FORMULAS AND EQUATIONS.. 42

3.1 SIMPLEST (EMPIRICAL) FORMULAS FROM ANALYSIS 43
3.2 MOLECULAR FORMULAS 46
3.3 THE MOLE... 46
3.4 TRANSLATION OF REACTIONS INTO EQUATIONS 47
3.5 INTERPRETATION OF BALANCED EQUATIONS.................... 48
3.6 THEORETICAL YIELDS. PERCENTAGE YIELDS 51

PROBLEMS 53

4

ENERGY CHANGES IN CHEMICAL REACTIONS 56

4.1 SOME QUANTITATIVE RELATIONS INVOLVING HEAT FLOW 56
Laws of Thermochemistry, 59.

4.2 HEATS OF FORMATION OF PURE SUBSTANCES.................. 60
4.3 THE EXPERIMENTAL MEASUREMENT OF HEAT FLOW 63
4.4 THE FIRST LAW OF THERMODYNAMICS 65
4.5 A PHYSICAL INTERPRETATION OF THE ORIGIN OF HEAT FLOW 71
4.6 ENERGY CHANGES OTHER THAN HEAT FLOW 73
Mechanical Energy, 73. Electrical Energy, 73. Light, 74.

PROBLEMS 75

5

THE PHYSICAL BEHAVIOR OF GASES.. 79

5.1 SOME GENERAL PROPERTIES OF GASES 79

5.2 ATMOSPHERIC PRESSURE AND THE BAROMETER 80

The Manometer and the Measurement of Gas Pressure, 82.

5.3 BOYLE'S LAW 83
5.4 CHARLES' AND GAY-LUSSAC'S LAW 85
5.5 THE IDEAL GAS EQUATION... 87
5.6 OTHER APPLICATIONS OF THE IDEAL GAS LAW............... 91
5.7 MIXTURES OF GASES: DALTON'S LAW OF PARTIAL PRESSURES 95
5.8 REAL GASES 98
5.9 THE KINETIC THEORY OF GASES 99

Postulates of the Kinetic Theory of Gases, 100. The Ideal Gas Law, 100. Graham's Law, 102. Molecular Speeds, 105.

PROBLEMS 107

6

THE PERIODIC CLASSIFICATION OF THE ELEMENTS

THE PERIODIC CLASSIFICATION OF THE ELEMENTS 111

6.1 THE PERIODICITY OF PROPERTIES OF THE ELEMENTS...... 111
6.2 HISTORY OF THE PERIODIC LAW 114
6.3 THE PERIODIC TABLE 116
6.4 SOME OBSERVATIONS ON THE PERIODIC TABLE 119
6.5 PREDICTIONS ON THE BASIS OF THE PERIODIC TABLE 121

Correlation of Chemical Formulas, 123.

6.6 ANOMALIES AND LIMITATIONS OF THE PERIODIC CLASSIFICATION.. 124

PROBLEMS 125

7

THE STRUCTURE OF ATOMS

THE STRUCTURE OF ATOMS 127

7.1 ELECTRONS AND NUCLEI—THE COMPONENTS OF ATOMS 127

The Atomic Nucleus, 128. X-ray Spectra and Atomic Number, 129.

7.2 THE ELECTRONS IN ATOMS 132

The Quantum Theory, 133.

7.3 EXPERIMENTAL BASIS OF THE QUANTUM THEORY............ 134

Atomic Spectra, 134. Regularities in Atomic Spectra, 137.

CONTENTS

7.4   THEORIES OF THE ELECTRONIC STRUCTURE OF ATOMS ... 137
7.5   MODERN QUANTUM THEORY 140

Some Implications of the Wave Properties of Particles,
141.   Some Systems for which the One-dimensional
Particle in a Box Can Serve as a Model, 144.   The
Schrödinger Wave Equation, 146.

7.6   ELECTRON ARRANGEMENTS IN ATOMS............................ 147

Quantum Numbers of Electrons, 147.   Electron Configurations of Atoms, 150.   Geometric Representations
of Electron Charge Clouds, 158.

7.7   EXPERIMENTAL SUPPORT FOR ELECTRON CONFIGURATIONS 160

The Electronic Structure of Atoms and the Periodic
Table, 162.

7.8   THE CORRELATION OF ATOMIC AND PHYSICAL PROPERTIES
      WITH ELECTRON CONFIGURATIONS 163

      PROBLEMS 167

8

CHEMICAL BONDING 170

8.1   IONIC BONDING.   FORMATION OF SODIUM FLUORIDE FROM
      THE ELEMENTS... 170
8.2   CHARGES OF IONS FOUND IN IONIC COMPOUNDS 171

Monatomic Ions with Noble Gas Electronic Structures,
171.   Monatomic Cations with Non-Noble Gas Structures, 172.   Polyatomic Ions, 174.

8.3   SIZES OF MONATOMIC IONS................................. 174
8.4   THE COVALENT BOND.   FORMATION OF THE H_2 MOLECULE 176
8.5   POLAR AND NONPOLAR COVALENT BONDS.   ELECTRO-
      NEGATIVITY VALUES 179
8.6   COVALENT BOND DISTANCES AND BOND ENERGIES 182
8.7   LEWIS STRUCTURES.   THE OCTET RULE 185

Rules for Writing Lewis Structures, 187.   Resonance,
189.   Failure of the Octet Rule, 190.

8.8   MOLECULAR GEOMETRY...................................... 192

Electron Pair Repulsion, 192.   Molecules Containing
Multiple Bonds, 194.   Polarity of Molecules, 195.

8.9   HYBRID ATOMIC ORBITALS.................................. 197

sp Hybrid Bonds, 197.   sp² Hybrid Bonds, 198.   sp³
Hybrid Bonds, 199.   sp³d² Hybrid Bonds, 199.   Bonding Orbitals in Molecules Containing Multiple Bonds,
200.

8.10 MOLECULAR ORBITALS 201

First Period Elements (H, He). Combination of 1s
Atomic Orbitals, 203.   Second Period Elements.   Combination of 2s and 2p Orbitals, 204.   Other Molecules,
207.

PROBLEMS.. 207

9

PHYSICAL PROPERTIES IN RELATION TO STRUCTURE 212

9.1 IONIC SUBSTANCES 212

Melting Point.   Thermal Decomposition of Polyatomic
Ions, 212.   Electrical Conductivity, 214.   Water Solubility, 214.

9.2 MOLECULAR SUBSTANCES.. 214

Interatomic vs. Intermolecular Forces, 215.   Trends
in Melting and Boiling Points, 215.   Types of Intermolecular Forces, 216.

9.3 MACROMOLECULAR SUBSTANCES...................................... 221
9.4 METALS 222

General Properties, 223.   The Metallic Bond, 224.
Band Theory, 227.

9.5 ALLOTROPY... 230

Oxygen, 230.   Sulfur, 231.   Carbon, 233.   Phosphorus,
234.

PROBLEMS 235

10

LIQUIDS AND SOLIDS: CHANGES IN STATE... 238

10.1 NATURE OF THE LIQUID STATE 238
10.2 LIQUID-VAPOR EQUILIBRIUM... 240

Vapor Pressure, 241.   Critical Temperature, 245.   Boiling Point, 246.

10.3 NATURE OF THE SOLID STATE...................................... 248

Crystal Structure, 249.   Unit Cells, 251.   Types of
Packing, 252.   Crystal Defects, 255.   Macromolecular
Crystals, 257.

10.4 SOLID-VAPOR EQUILIBRIUM 259

Vapor Pressure of Solids, 259.   Heat of Sublimation, 260.

10.5 SOLID-LIQUID EQUILIBRIUM 260
10.6 PHASE DIAGRAMS 262
10.7 NONEQUILIBRIUM PHASE BEHAVIOR 264

       PROBLEMS.. 265

11

SOLUTIONS.. 269

11.1 SOLUTION PHASES.. 269
11.2 SOLUTION TERMINOLOGY 271

Solvent and Solute, 271.   Dilute and Concentrated Solutions, 271.   Saturated, Unsaturated, and Supersaturated Solutions, 272.

11.3 CONCENTRATION UNITS 272

Mole Fraction, 273.   Molality, 274.   Molarity, 275.

11.4 PRINCIPLES OF SOLUBILITY 276

Liquid-Liquid, 277.   Solid-Liquid, 278.   Gas-Liquid, 279.

11.5 EFFECT OF TEMPERATURE AND PRESSURE ON SOLUBILITY 280

Temperature, 280.   Pressure, 281.

11.6 ELECTRICAL CONDUCTIVITIES OF WATER SOLUTIONS...... 283
11.7 COLLIGATIVE PROPERTIES OF DILUTE SOLUTIONS 283

Nonelectrolytes, 284. Determination of Molecular Weights from Colligative Properties, 290. Electrolytes, 291.

       PROBLEMS.. 294

12

EQUILIBRIUM IN CHEMICAL SYSTEMS 299

12.1 AN EXAMPLE OF CHEMICAL EQUILIBRIUM. THE HI-H_2-I_2 SYSTEM... 300
12.2 THE GENERAL FORM OF THE EQUILIBRIUM CONSTANT EXPRESSION 304

12.3 APPLICATIONS OF THE EQUILIBRIUM CONSTANT EXPRESSION 305

Prediction of the Direction of Reaction, 306. Prediction of the Extent of Reaction, 308.

12.4 EFFECT OF CHANGES IN CONCENTRATION UPON THE POSITION OF AN EQUILIBRIUM 311

Changes in the Number of Moles of Reactants or Products, 311. Changes in Volume, 313. Le Chatelier's Principle, 316.

12.5 THE EFFECT OF TEMPERATURE ON CHEMICAL EQUILIBRIA 317
12.6 EQUILIBRIA INVOLVING SOLIDS OR LIQUIDS IN ADDITION TO GASES 318

PROBLEMS 320

13

THERMODYNAMICS AND THE SPONTANEITY AND EQUILIBRIUM PROPERTIES OF CHEMICAL REACTIONS 323

13.1 SPONTANEITY OF CHEMICAL REACTIONS: FREE ENERGY 323

Maximum Useful Work, 326. Free Energy Change, ΔG, 327.

13.2 DEPENDENCE OF ΔG ON TEMPERATURE. THE CONCEPT OF ENTROPY 330

Distinction Between ΔG and ΔH, 330. The Origin of Q'. Entropy Changes, 331. Gibbs-Helmholtz Equation, 335.

13.3 REACTION SPONTANEITY AND EQUILIBRIUM AS A FUNCTION OF PRESSURE: THE RELATION BETWEEN $\Delta G°$ AND THE EQUILIBRIUM CONSTANT FOR A REACTION 339

PROBLEMS 345

14

RATES OF REACTION 349

14.1 MEANING OF REACTION RATE 350
14.2 DEPENDENCE OF REACTION RATE UPON CONCENTRATION 352

First Order Reactions, 354. Reactions of Higher Order, 357.

14.3 DEPENDENCE OF REACTION RATE UPON TEMPERATURE... 359

14.4 CATALYSIS 362
14.5 THE COLLISION THEORY OF REACTION RATES 364
14.6 CHAIN REACTIONS.. 369

PROBLEMS... 371

15

REACTIONS OF ELEMENTS WITH EACH OTHER 376

15.1 OXIDATION AND REDUCTION....................................... 376

Loss and Gain of Electrons, 376. Oxidation Number, 377. Oxidation States of the Elements, 379. Oxidation and Reduction: General Definition, 381.

15.2 REACTIONS OF HYDROGEN.. 382

Saline Hydrides, 382. Metallic (Interstitial) Hydrides, 383. Molecular Hydrides, 384.

15.3 REACTIONS OF OXYGEN 389

Metals, 389. Nonmetals, 393.

PROBLEMS... 396

16

PRECIPITATION REACTIONS 400

16.1 NET IONIC EQUATIONS 400
16.2 SOLUBILITIES OF IONIC COMPOUNDS............................. 403
16.3 SOLUBILITY EQUILIBRIA 406

Qualitative Aspects: Common Ion Effect, 406. Quantitative Treatment: Solubility Product, 407.

16.4 PRECIPITATION REACTIONS IN ANALYTICAL CHEMISTRY... 411

Quantitative Analysis, 411. Qualitative Analysis, 414.

16.5 PRECIPITATION REACTIONS IN INORGANIC PREPARA-
TIONS 415
16.6 WATER SOFTENING 416

Lime-Soda Method, 417. Ion Exchange, 418.

PROBLEMS... 421

17

ACIDS AND BASES 425

   17.1 PROPERTIES OF ACIDIC AND BASIC WATER SOLUTIONS... 425
   17.2 IONS PRESENT IN ACIDIC AND BASIC WATER SOLUTIONS... 426

   Basic Solution: OH$^-$, 426. Acidic Solution: the Hydrated Proton or Hydronium Ion, 426. Equilibrium Between H$^+$ and OH$^-$: Concept of K$_w$, 427.

   17.3 pH 428
   17.4 FORMATION OF ACIDIC WATER SOLUTIONS 430

   Transfer of a Proton from a Neutral Solute Molecule, 430. Transfer of a Proton from an Anion (HSO$_4^-$, HSO$_3^-$, H$_2$PO$_4^-$), 431. Transfer of a Proton from a Cation (NH$_4^+$, Al^{3+}, Transition Metal Ions), 432.

   17.5 STRONG AND WEAK ACIDS: K$_a$ 433

   Expression for K$_a$, 434. Experimental Determination of K$_a$, 435. Interpretation of K$_a$, 436. Use of K$_a$ in Calculations, 437.

   17.6 FORMATION OF BASIC WATER SOLUTIONS 439
   17.7 WEAK BASE EQUILIBRIA: K$_b$ 441
   17.8 RELATIVE STRENGTHS OF ACIDS AND BASES 442

   Hydrides of the 7A, 6A, and 5A Elements, 443. Oxyacids, 444. Hydrolysis of Salts, 446.

   17.9 GENERAL CONCEPTS OF ACIDS AND BASES 447

   Brönsted-Lowry Concept, 447. The Lewis Concept, 451.

       PROBLEMS 453

18

ACID-BASE REACTIONS 457

   18.1 TYPES OF ACID-BASE REACTIONS 457
   18.2 ACID-BASE TITRATIONS 460

   Normality: Gram Equivalent Weights of Acids and Bases, 461. Acid-Base Indicators, 462.

   18.3 BUFFERS 466
   18.4 APPLICATION OF ACID-BASE REACTIONS IN INORGANIC SYNTHESIS 469

   Preparation of Salts, 469. Preparation of Volatile Acids or Bases, 470.

18.5 APPLICATION OF ACID-BASE REACTIONS IN QUALITATIVE ANAYLYSIS 471

Tests for Specific Ions, 471. Separation of Ions, 472.

18.6 AN INDUSTRIAL APPLICATION OF ACID-BASE REACTIONS: THE SOLVAY PROCESS.. 474

Preparation of $NaHCO_3$, 475. Preparation of Na_2CO_3 from $NaHCO_3$, 476. Preparation of CO_2 and Recovery of NH_3, 476. Summary of the Solvay Process, 476.

PROBLEMS... 477

19

COMPLEX IONS 481

19.1 CHARGES OF COMPLEX IONS: NEUTRAL COMPLEXES 482
19.2 COMPOSITION OF COMPLEX IONS 483

Central Atom, 483. Coordinating Group: Chelating Agents, 484. Coordination Number, 486.

19.3 GEOMETRY OF COMPLEX IONS...................................... 487

Coordination Number = 2, 487. Coordination Number = 4, 487. Coordination Number = 6, 489.

19.4 ELECTRONIC STRUCTURE OF COMPLEX IONS.................. 491

Valence Bond (Atomic Orbital) Approach, 493. Crystal Field Theory, 495.

19.5 RATE OF COMPLEX-ION FORMATION 498
19.6 COMPLEX-ION EQUILIBRIA 500
19.7 COMPLEX IONS IN ANALYTICAL CHEMISTRY 502

Qualitative Analysis, 502. Quantitative Analysis, 506.

PROBLEMS... 508

20

ELECTROLYTIC CELLS: BALANCING OXIDATION-REDUCTION EQUATIONS..................... 511

20.1 ELECTROLYSIS OF MOLTEN IONIC COMPOUNDS 512

Na from NaCl, 512. Al from Al_2O_3, 513. F_2 from KF-HF Mixture, 515.

20.2 ELECTROLYSIS OF WATER SOLUTIONS 515

Cathode Reactions (Reduction), 515. Anode Reactions (Oxidation), 516. Electrolysis of a Solution of $CuCl_2$, 516. Electrolysis of a Solution of NaCl, 516. Electrolysis of a Solution of $CuSO_4$, 517. Electroplating, 518.

20.3 FARADAY'S LAW OF ELECTROLYSIS 519
20.4 BALANCING OXIDATION-REDUCTION EQUATIONS 523

Oxidation Number Method of Balancing Equations, 526. Calculations Involving Balanced Equations, 527.

PROBLEMS 528

21

VOLTAIC CELLS: SPONTANEITY OF OXIDATION-REDUCTION REACTIONS 532

21.1 A SIMPLE VOLTAIC CELL: THE ZN-CU CELL 532
21.2 OTHER VOLTAIC CELLS: COMMERCIAL CELLS 535

Dry Cell (Leclanché Cell), 536. Lead Storage Battery, 537. Fuel Cells, 538.

21.3 STANDARD REDUCTION AND OXIDATION POTENTIALS 539

Assignment of Potentials, 540. Calculation of Cell Voltages from Standard Potentials, 541. Qualitative Interpretation of Standard Potentials, 543.

21.4 SPONTANEITY AND EXTENT OF OXIDATION-REDUCTION REACTIONS 545

Relation Between E° and ΔG°, 546. Relation Between E° and K, 547. Calculations Involving K, 549.

21.5 EFFECT OF CONCENTRATION ON VOLTAGE: THE NERNST EQUATION 550

Direction of Effect, 550. Magnitude of Effect, 552. Use of the Nernst Equation to Determine Concentrations of Ions in Solution, 553. Use of the Nernst Equation to Determine Reaction Spontaneity, 555.

PROBLEMS 556

22

OXIDIZING AGENTS IN WATER SOLUTION 560

22.1 H^+ ION 561

Reaction of Metals with Acids, 561. Reaction of Metals with Water, 562.

22.2   METAL CATIONS.. 562
22.3   CHLORINE.. 564
22.4   OXYGEN.. 566

Reaction with Ions in Solution, 567. Reaction with
Metals. Corrosion of Iron, 568.

22.5   OXYANIONS 572

Reactions of Metals with Nitric Acid, 575. Use of
MnO_4^- in Volumetric Analysis, 576. Use of Oxyanions
in Qualitative Analysis, 577.

22.6   THE PHOTOGRAPHIC PROCESS 578

Exposure, 578. Development, 579. Fixing, 580.
Preparation of the Positive, 580.

PROBLEMS.. 581

23

NUCLEAR REACTIONS... 585

23.1   NATURAL RADIOACTIVITY 586

Nature of Radiation, 587. Radioactive Series, 588.
Interaction of Radiation with Matter, 589.

23.2   RATE OF RADIOACTIVE DECAY..................................... 591

Logarithmic Rate Law, 591. Half-Life, 592. Age of
Rocks, 594. Age of Organic Material, 594.

23.3   BOMBARDMENT REACTIONS. ARTIFICIAL RADIOACTIVITY 595

Bombarding Particles, 596. Transuranium Elements,
597. Decay of Artificially Produced Radioactive Isotopes, 598. Uses of Radioactive Isotopes, 600.

23.4   NUCLEAR FISSION 601

Discovery, 601. Fissionable Isotopes, 602. Fission
Products, 602. Neutron Emission: Nuclear Chain Reactions, 603. Evolution of Energy: Nuclear Reactors,
604.

23.5   MASS-ENERGY RELATIONS 604

Calculations Involving Mass-Energy Conversions, 605.
Nuclear Stability: Binding Energy, 606.

23.6   NUCLEAR FUSION 610

PROBLEMS.. 611

24

ORGANIC CHEMISTRY.. 615

24.1 KINDS OF ORGANIC SUBSTANCES.................................... 616

Saturated Hydrocarbons: Paraffins and Cycloparaffins, 616. Unsaturated Hydrocarbons: Olefins and Acetylenes, 618. Aromatic Hydrocarbons, 620. Halogen-Containing Organic Compounds, 624. Oxygen-Containing Organic Compounds, 625.

24.2 SOME COMMON REACTIONS OF ORGANIC COMPOUNDS... 629

Substitution Reactions, 629. Elimination Reactions, 631.

24.3 PETROLEUM AND RUBBER: TWO MATERIALS OF IMPORTANCE IN THE CHEMICAL INDUSTRY............................ 633

Petroleum, 633. Rubber, 636.

24.4 NATURAL PRODUCTS... 638

Glucose: A Typical Carbohydrate and Sugar, 638. Amino Acids and Proteins, 641.

PROBLEMS.. 646

APPENDICES

1. MATHEMATICS... 651

   Exponential Notation... 651
   Common and Natural Logarithms 653

   Significant Figures.. 659
   Proportionality Constants... 661
   Linear Functions... 663
   Problems 666

2. NOMENCLATURE OF INORGANIC COMPOUNDS..................... 669

   Ionic Compounds 669
   Binary Compounds of the Nonmetals 671
   Oxyacids 671
   Coordination Compounds... 672

3. ATOMIC AND IONIC RADII 673

4. CONVERSION FACTORS AND CONSTANTS 674

5. ANSWERS TO SELECTED PROBLEMS 675

INDEX 687

Some Basic Concepts

1.1 WHAT CHEMISTRY IS

Although you may not have undertaken any formal study of chemistry, the importance of chemistry to your daily life is abundantly evident. Chemical changes in your body produce energy and growth from food and air; bodily functions themselves are controlled by sets of complex chemical substances called vitamins, hormones, and enzymes. The food you eat is produced by the chemical reaction called photosynthesis, and the clothes you wear consist in large part of synthetic fibers produced by chemical processes. The metals for automobiles and machines also are produced by chemical processes, and the operations by which glass, bricks, and pottery are made are further (and even older) examples of chemical processes. More recent triumphs of chemistry include the development of antibiotics and other drugs to combat disease and pain, the production of greatly improved gasoline from petroleum, and the synthesis of many new plastics. Entire industries have evolved from the development of these new materials, giving employment to many thousands of people. Chemistry operates in and around us every moment of our lives, and thus we need to know something about it. We particularly need to know how chemical changes take place and how they may be controlled to good use — which is what this book is about.

Chemistry is a science of substances, those materials of which the earth and the universe are composed. Specifically, the science of chemistry deals with the properties and structures of substances, and their preparation from and interaction with other substances. Since chemistry is a very broad subject, the boundaries which separate it from physics, geology, pharmacy, engineering, and the biological sciences are often indefinite. We find that these fields overlap extensively and that knowledge gained in one area is often applicable to another. Many of you, therefore, may be studying chemistry mainly because of its importance in other fields in which you have a specific interest.

Within chemistry itself there are several areas of study, which may be classified according to the types of matter that are of primary interest in that area. Thus we have *organic* chemistry, which deals with the many substances containing carbon, *inorganic* chemistry, whose province includes all other substances, and *nuclear* chemistry, in which we study reactions involving the nuclei of atoms. There are also areas in which the division is based on the purpose of the work in that area. Here we find *analytical* chemistry, which encompasses both qualitative and quantitative methods for investigating the composition of matter, and *physical* chemistry, which deals with the determination of the properties and structure of matter and the laws and theories of chemistry.

In this general course, we shall endeavor to introduce the principles that underlie all chemistry and to present applications of these principles so that we can obtain a clear understanding of chemical reactions. In discussing principles, we shall often use mathematical relations to calculate certain quantities. It is most important that you become familiar with these relations (there are many of them) and learn to use them intelligently.

1.2 MEASUREMENTS

Chemistry is an experimental science. Progress in the development of chemical principles is based upon carefully designed experiments carried out under controlled conditions. The person who performs an experiment has the responsibility of describing it in sufficient detail so that it can be repeated by another investigator.

A chemist who prepares the compound $Co(NH_3)_6Cl_3$ in the laboratory might describe his procedure as follows:

> Dissolve 10 g of NH_4Cl and 15 g of $CoCl_2 \cdot 6H_2O$ in 20 ml of boiling water. Add 1 g of charcoal, cool to room temperature, and add 32 ml of concentrated NH_3. Chill the solution to 0°C in an ice bath and add 30 ml of H_2O_2 while stirring. Heat at 60°C until the pink color has disappeared. Cool to 0°C and filter the crystals of $Co(NH_3)_6Cl_3$ through a Buchner funnel 5 cm in diameter.

$Co(NH_3)_6Cl_3$ is a coordination compound (Chapter 19); its preparation is described in greater detail in *Inorganic Syntheses*, Vol. II, p. 217.

Included in this description of experimental procedure are the measured values of several important quantities. These include the diameter of the Buchner funnel (a Buchner funnel **5 cm** in diameter), the **volume** of water (**20 ml** of boiling water), aqueous ammonia (**32 ml** of concentrated NH_3), and hydrogen peroxide (**30 ml** of H_2O_2), the **mass** of solid ammonium chloride (**10 g** of NH_4Cl) and cobalt chloride hexahydrate (**15 g** of $CoCl_2 \cdot 6H_2O$), and the **temperature** of the solution (chill the solution to **0°C** in an ice bath, heat at **60°C**). Anyone who wishes to repeat this experiment must know exactly what is meant by the units in which these quantities are expressed (**centimeters, milliliters, grams, degrees centigrade**). It is equally important that he know how length, volume, mass, and temperature are measured in the laboratory.

Length

The device which is most commonly used in the laboratory to measure lengths is the meter stick. The meter stick reproduces, with as much accuracy as possible, the fundamental unit of length in the metric system, the **meter**. The meter was originally intended to be 1/40,000,000 of the earth's meridian that passes through Paris. This distance was fixed by two marks on a platinum-iridium rod stored at the International Bureau of Weights and Measures at Sèvres, near Paris. In 1960 the standard international meter was defined as being 1650763.73 times the wavelength of a certain line in the visible spectrum of krypton.

Examination of a meter stick reveals that it is divided into one hundred equal parts, each one **centimeter** in length (1 cm = 10^{-2} m). A centimeter is in turn subdivided into ten equal parts, each one **millimeter** long (1 mm = 10^{-1} cm). A much smaller unit of length, which is widely used by chemists in reporting atomic dimensions, is the **angstrom** (1 Å = 10^{-8} cm). A much larger unit, familiar to track and field runners, is the **kilometer** (1 km = 10^3 m).

We frequently find it necessary to convert lengths reported in one unit (e.g., cm) to another unit (e.g., mm, Å). We can accomplish this by using conversion factors such as those given in the previous paragraph (cf. Example 1.1). Conversions within the metric system, in which the units are related to each other by powers of ten, are readily carried out. The arithmetic is somewhat more complicated when one makes conversions within the English system, in which

$$1 \text{ yard} = 3 \text{ feet; } 1 \text{ foot} = 12 \text{ inches}$$

Because of its simple, logical interrelations, the metric system is universally employed by scientists and is in common use in most countries. Conservatism, coupled with a certain amount of stubbornness, is responsible for the retention of the English system in the United States and the British Commonwealth.

Conversions between the metric and English systems can be carried out by using conversion factors such as

$$1 \text{ m} = 39.37 \text{ in; } 1 \text{ in} = 2.54 \text{ cm}$$

In equations of this kind, "1 m" can be taken to mean exactly one meter, and "1 in" to be exactly one inch. By the same token, in equations relating units in the same system, such as 1 yard = 3 feet, both numbers can be taken to be exact. On the other hand, the number 2.54 is not exact; it represents, to three significant figures, the length of an inch in centimeters. If a more precise conversion factor is required, we can write

$$1 \text{ in} = 2.540 \text{ cm}$$

or, if we wish a conversion factor accurate to seven significant figures,

$$1 \text{ in} = 2.540005 \text{ cm}$$

(See Appendix 1 for a discussion of significant figures.)

Example 1.1 illustrates the use of conversion factors to convert lengths from one unit to another. This operation is general and easy to learn. Indeed, you have undoubtedly used this approach, perhaps without realizing it, to convert dollars to cents, feet to inches, and so forth.

EXAMPLE 1.1. A student measures a piece of glass tubing to be 7.21 cm long. Express its length in

a. angstroms   b. inches   c. feet

SOLUTION

a. To convert centimeters to angstroms, we make use of the conversion factor

$$1 \text{ Å} = 1 \times 10^{-8} \text{ cm}$$

This equation can be expressed as: $\dfrac{1 \text{ Å}}{1 \times 10^{-8} \text{ cm}} = 1$

If we multiply 7.21 cm by the ratio 1 Å/10⁻⁸ cm, we do not change its magnitude, since we are, in effect, multiplying by unity. We do, however, change the units from cm to Å:

$$7.21 \ \cancel{\text{cm}} \times \frac{1 \text{ Å}}{1 \times 10^{-8} \ \cancel{\text{cm}}} = 7.21 \times 10^{8} \text{ Å}$$

Note that the ratio was written in such a way that "cm" appeared in both the numerator and the denominator of the product and hence cancelled out.

   b. A suitable conversion factor here is 1 in = 2.54 cm

The ratio becomes $\dfrac{1 \text{ in}}{2.54 \text{ cm}} = 1$

and we have $7.21 \ \cancel{\text{cm}} \times \dfrac{1 \text{ in}}{2.54 \ \cancel{\text{cm}}} = 2.84$ in

Note that here, as in part a, we retained three significant figures in the answer, since the length was known to that number of significant figures (Appendix 1).

   c. We do not have available a conversion factor immediately relating centimeters to feet. We do, however, know that

$$1 \text{ in} = 2.54 \text{ cm, and } 1 \text{ ft} = 12 \text{ in}$$

We may solve this problem by making use of these two conversion factors, first converting cm to in, and then in to ft.

$$7.21 \ \cancel{\text{cm}} \times \frac{1 \ \cancel{\text{in}}}{2.54 \ \cancel{\text{cm}}} \times \frac{1 \text{ ft}}{12 \ \cancel{\text{in}}} = 0.237 \text{ ft}$$

The conversion factor approach may seem awkward or artificial at first. As you become better acquainted with it, you will find it to be the simplest, most straightforward way of solving a wide variety of chemical problems.

Throughout this text we shall use the conversion factor approach illustrated in Example 1.1 to solve many problems. We should point out that this method is no substitute for analyzing a problem. It does, however, emphasize that whenever we are asked to solve a problem in chemistry, we must decide how we are to "convert" the information we are given to that which is required. Additional examples of the use of the **conversion factor method** will be found in this and succeeding chapters.

Volume

Units of volume in the metric system are very simply related to units of length. The basic unit of volume is the **cubic centimeter** (cc or cm³), which is the volume of a cube with an edge one centimeter long. Some of the conversion factors relating units of volume in the metric and English systems are listed in Table 1.1. You will note that a **liter** is equal to 1000 cc and is slightly larger than a quart (1 lit = 1.0567 qt).

In the laboratory we work with vessels that are designed to contain or deliver known volumes, within certain limits of accuracy. These include the **graduated cylinder**, which can be used to deliver volumes of liquids with an accuracy of 1 to 5 per cent, the **buret**, which delivers variable volumes with a limit of accuracy of about 0.1 per cent, and the **pipet**, which delivers a specified volume of liquid with an accuracy comparable to that of a buret.

TABLE 1.1 CONVERSION FACTORS RELATING VOLUME UNITS

Metric System	1 liter (lit) = 1000 milliliters (ml) = 1000 cubic centimeters (cc)
English System	1 gallon (gal) = 4 quarts (qt) = 8 pints (pt) = 231 in³
Metric to English	1 lit = 1.0567 qt = 61.025 in³

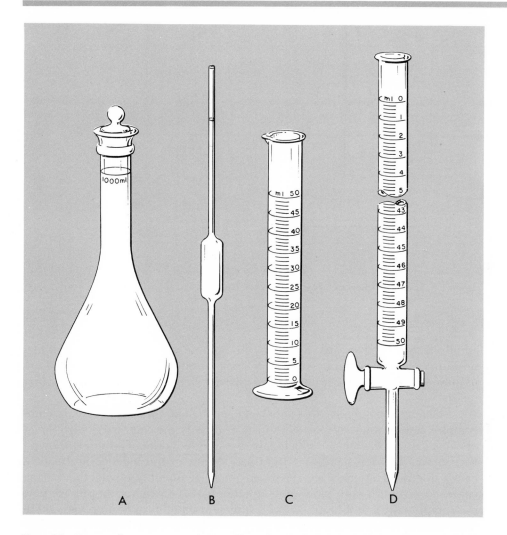

Figure 1.1 Devices for measuring volumes of liquids. *A*, Volumetric flask; *B*, pipet; *C*, graduated cylinder; *D*, buret.

Mass

In reporting the amount of a solid used in an experiment, we ordinarily specify its mass rather than its volume. There are many practical reasons for doing this. For one thing, crystal size can make a considerable difference in the amount of solid required to fill a container. A fine powder leaves much less unfilled air space in a vessel than do large chunks of solid. Moreover, the volumes of many solids are difficult to measure accurately. It would, for ex-

5

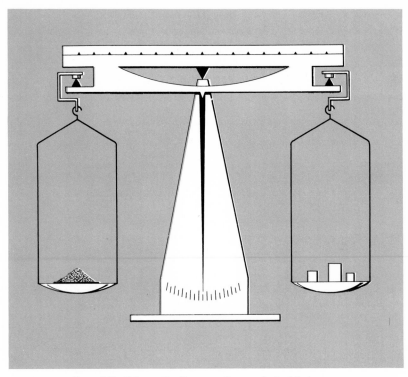

Figure 1.2 Schematic diagram of an analytical balance.

ample, be rather awkward to specify the volume of a one-gram strip of magnesium ribbon. Analogous situations arise frequently. When someone asks us how big our brother is, we never answer, "His volume is now about 17 quarts."

The amount of matter in a substance, or the **mass** of a substance, is a more complex notion than volume. Chemists ordinarily measure mass with the aid of a mechanical device called a balance, a simple model of which is shown in Figure 1.2. The left-hand and right-hand sides of the balance are made as nearly identical as possible. With no material on either pan, the balance comes to rest with both pans at an equal height. To **weigh** a sample, we place the sample on the left-hand pan and sufficient weights of known mass on the right-hand pan to restore balance, i.e., to bring the pans to the same height. At balance, the force attracting the sample to the earth is equal to the force attracting the metal pieces to the earth. This statement is equivalent to saying that, at balance, the weight of the sample is equal to the sum of the metal weights. The definition of weight is embodied in these two sentences.

Newton's Law states that under the conditions obtaining in the balance, the weight of the sample and its mass are proportional to each other through the same proportionality constant that relates the weight of the metal pieces to their mass. Mathematically, this statement can be expressed as

$$\text{weight of sample} = k \times \text{mass of sample}$$

$$\text{weight of metal} = k \times \text{mass of metal}$$

Dividing the two equations, one by the other, we obtain

$$\frac{\text{weight of sample}}{\text{weight of metal}} = \frac{\text{mass of sample}}{\text{mass of metal}}$$

Since, at balance,

$$\text{weight of sample} = \text{weight of metal}$$

it is true that, at balance,

$$\text{mass of sample} = \text{mass of metal}$$

At balance, then, not only are the weights of the sample and metal pieces equal, but their masses are also equal. That is, the balance detects both **equality of weight and equality of mass**.

Strictly speaking, in chemistry we are more interested in mass than in weight. The mass of an object is an inherent property of the object, since it is a measure of the amount of material in the object. By its very nature, the weight of an object does not remain constant over the surface of the earth, since weight depends on gravitational pull, which varies with altitude and latitude. The weight of an object, then, depends on where you weigh it. The mass of an object is presumably a constant throughout the universe. For this reason we record amounts of matter in terms of mass and set up standards on the basis of mass.

A piece of metal weighing 6 g on the surface of the earth should have a weight of about 1 g on the moon. How could you show experimentally that this is the case?

The standard unit of mass is the **gram** (g), which is defined as one-thousandth of the mass of a platinum-iridium block kept at the International Bureau of Weights and Measures. One gram is very nearly equal to the **mass of one cubic centimeter of water** under certain specified conditions of temperature and pressure. Thus the standard of mass is ultimately related to the standard of length in the metric system. Such a relation makes it very easy to estimate the mass of a given volume of water. For example, we can say that 250 ml or 250 cc of water will have a mass of just about 250 g.

Originally, the gram was thought to be exactly equal to the mass of one cubic centimeter of water at 4°C, the temperature at which the density of water is maximum. It was later found that the standard gram has a mass equal to that of 1.000027 cubic centimeters of water under the given conditions. Since the liter was originally defined as the volume of one kilogram of water under these conditions, this meant that one liter became equal to 1000.027 cubic centimeters. This small discrepancy was removed by an international agreement, reached in 1964, which re-defined a liter so as to make it exactly equal to 1000 cubic centimeters.

Some of the conversion factors relating units of mass in the metric and English systems are listed in Table 1.2. The use of these conversion factors is illustrated in Example 1.2.

TABLE 1.2 CONVERSION FACTORS RELATING MASS UNITS

Metric System	1 kilogram (kg)	$= 10^3$ grams (g)
	1 gram (g)	$= 10^3$ milligrams (mg)
English System	1 pound (lb)	$= 16$ ounces (oz)
	1 ton	$= 2000$ pounds (lb)
Metric to English	1 lb	$= 453.6$ g

EXAMPLE 1.2.   The directions for a certain experiment require that a student weigh out 526 g of a certain solid. How many mg are required? How many lb?

SOLUTION.   To convert g to mg, we note that $1 \text{ g} = 10^3 \text{ mg}$

$$526 \cancel{g} \times \frac{10^3 \text{ mg}}{1 \cancel{g}} = 526 \times 10^3 \text{ mg} = 5.26 \times 10^5 \text{ mg}$$

To convert g to lb, we use the conversion factor $1 \text{ lb} = 453.6 \text{ g}$

$$526 \cancel{g} \times \frac{1 \text{ lb}}{453.6 \cancel{g}} = 1.16 \text{ lb}$$

Note that three significant figures are retained in the answer.

Temperature

When we speak informally of temperature and temperature differences, we have confidence that our words are understood. Yet, if we were asked what temperature is, many of us would be at a loss for words. The concepts from which the idea of temperature arises are not obvious, and their development is somewhat more abstract than was the case with either volume or mass.

The human body is sensitive to contact with objects that we describe as being "hot" or "cold." If we dip a finger into a glass of ice water, we conclude that the mixture of ice and water is cold. If, by mistake, our finger comes in contact with boiling water, we may say among other things that the boiling water is hot. These sensations lead naturally to the concept of temperature; we say that hot objects have a higher temperature than cold objects.

To get a clearer idea as to what we mean by temperature and temperature differences, let us consider what really happens when we dip our finger into ice water. The sensation of coldness that we feel is caused by the flow of heat from our finger to the water. Indeed, heat is defined by an experiment of this sort. *Heat is that which flows spontaneously from a body at a higher temperature to one at lower temperature when they are placed in contact with each other.*

Our definition of heat may seem rather poor, since it does not endow heat with any sort of tangible nature. Early scientists considered heat to be a sort of fluid, which flowed from hot bodies to cold. They considered that bodies containing a large amount of the fluid, which was called caloric, were hot. It was never possible to demonstrate experimentally the existence of caloric, and the idea was finally dropped.

Fingers have certain characteristics which render them impractical as temperature-measuring devices. In the first place, they are relatively insensitive to small temperature differences. Furthermore, they tend to be overly sensitive to large temperature differences. A finger which has come in contact with molten metal or liquid air is no longer useful for estimating relative temperatures. A more practical device for measuring temperature is the mercury-in-glass thermometer shown in Figure 1.3. This is constructed from a piece of thick-walled capillary tubing which is blown out at one end to a thin bulb. Enough mercury is added to fill the bulb and part of the capillary, whereupon the upper end of the thermometer is pumped free of air and sealed.

Could you use a mercury thermometer to measure the temperature of molten iron or liquid air? How *would* you measure these temperatures?

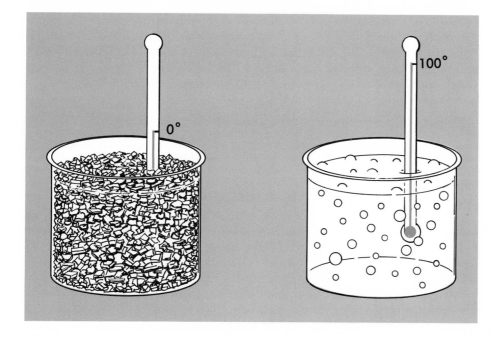

Figure 1.3 Centigrade thermometer in an ice-water mixture (0°C) and in water boiling at 1 atmosphere pressure (100°C).

The principle upon which the mercury-in-glass thermometer is based is a simple one. Mercury expands when it is heated. Thus the volume of a given amount of mercury is a function of its temperature: the higher the temperature, the greater will be the volume of the mercury. Relatively small temperature changes are easily detected by a change in the height of the mercury in the capillary column, because even a small change in the volume of mercury produces a noticeable change in the height of mercury within the fine capillary.

Using a thermometer of this type, we can establish a quantitative scale of temperatures. To do this, we first immerse the thermometer in a container which holds a mixture of ice and water. We mark the level of the mercury, arbitrarily assigning it a temperature of zero degrees (0°). We then place the thermometer in a container in which water is boiling at standard atmospheric pressure (760 mm Hg). We find that the mercury level remains constant so long as both water and steam are present in the container at one atmosphere pressure, and arbitrarily assign to this level a temperature value of 100°. The mercury level at 100° is higher than the level at 0° by a certain distance on the capillary. We measure this distance and divide it into 100 equal parts. Each division is marked on the capillary tube. Using the same size of division, we can engrave lines above 100° and below 0°.

If the thermometer is placed in a liquid of unknown temperature and the mercury level comes to rest 45 divisions above the zero line, we say that the temperature of the liquid is 45°. If the level comes to rest 130 divisions above the zero level we say that the temperature of the liquid is 130°. In this way we can measure any temperature within the range of the thermometer.

Since the scale of temperature we have set up is not the only one used

9

by scientists, we must always indicate the temperature scale used to express a specified temperature. The scale we have described is called the **centigrade** scale (or **Celsius** scale) since the boiling point of water was taken to be 100° and the freezing point of water to be 0°. In the examples given, the temperatures would properly be reported as 45° centigrade and 130° centigrade or, more commonly, 45°C and 130°C.

In setting up a scale of temperature as we have done, there are some disadvantages which are not at once apparent. Although mercury is ordinarily used in thermometers, other liquids, among them ethyl alcohol and pentane, are sometimes employed. We find that although thermometers made with different liquids agree with one another at 0°C and 100°C, at intermediate temperatures they may disagree by appreciable amounts. Obviously then, a scale of temperature should be set up in such a way that it is independent of the properties of any one substance. In our work with gases we shall present the results of efforts made in that direction.

The common temperature scale used in the United States is the Fahrenheit scale. On this scale 32° corresponds to the temperature of the freezing point of water and 212° to the temperature of its normal boiling point. The centigrade and Fahrenheit scales are related by a simple algebraic expression which can be derived easily (Figure 1.4).

From Figure 1.4 it is clear that there is a linear relationship between the two temperature scales. Mathematically, this relation must be of the form

$$y = a + b\, x$$

or
$$°F = a + b\ °C$$

Linear relationships are discussed in greater detail in Appendix 1.

where a and b are constants that can readily be evaluated. The constant a is the intercept on the vertical axis, 32; the constant b is the slope of the line:

$$b = \frac{\Delta y}{\Delta x} = \frac{212 - 32}{100 - 0} = 1.8$$

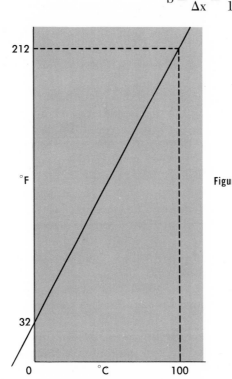

Figure 1.4   Relation between °F and °C.

Making these substitutions, we obtain

$$°F = 32 + 1.8 °C \qquad (1.1)$$

Another temperature scale, which is particularly useful in describing the behavior of gases, is the **absolute** or **Kelvin** scale. The origin and significance of this scale will be discussed in Chapter 5. The relation between °K and °C is

$$°K = 273 + °C \qquad (1.2)$$

Other Measured Quantities

Experiments carried out in chemical laboratories frequently involve the measurement of quantities other than length, volume, mass, and temperature. We may, for example, specify the **time** required to carry out a particular reaction. All of us are familiar with the units in which time is expressed.

1 day = 24 hours; 1 hour = 60 minutes; 1 minute = 60 seconds

In experiments involving gases, it is important to specify the **pressure** of the gas as well as its volume, mass, and temperature. The pressure of a gas is defined as the force per unit area of the container wall. Since the unit of force in the metric system is the dyne, the fundamental unit of pressure is the **dyne per square centimeter** (dyne/cm²). In the English system, pressure may be expressed in **pounds per square inch** (lb/in²).

Another unit commonly used to express pressure is **millimeters of mercury** (mm Hg). When we say that a gas exerts a pressure of 100 mm Hg, we mean that the pressure on the walls of the container is equal to the pressure that would be exerted by a column of mercury 100 mm high. The pressure equivalent to a millimeter of mercury is often referred to as a **torr**, after Evangelista Torricelli, a seventeenth century Italian scientist who was instrumental in developing an apparatus commonly used to measure pressures (see Chapter 5). Another unit used to express pressures is the **atmosphere**, which is defined to be exactly 760 mm Hg and is approximately the mean atmospheric pressure at sea level. Conversion factors relating these various units of pressure include:

1 atmosphere = 760 mm Hg = 14.7 lb/in² = 1.013×10^6 dynes/cm²

Associated with every chemical reaction is an **energy change**, whose magnitude is often specified in reporting the results of experiments (see Chapter 4). Chemists use a variety of energy units. Perhaps the most frequently used unit is the **calorie** (cal), which is defined as the heat required to raise the temperature of one gram of water from 14.5 to 15.5°C. Other energy units that we shall refer to in later chapters are the **kilocalorie** (kcal), the **erg**, the **liter atmosphere** (lit atm), and the **electron volt** (eV). Conversion factors relating these units include:

A liter atmosphere is the work done when the volume of a system increases by one liter against a pressure of one atmosphere.

1 kcal = 1000 cal
1 cal = 2.61×10^{19} eV = 4.184×10^7 erg = 4.129×10^{-2} lit atm

11

1.3 RESOLUTION OF MATTER INTO PURE SUBSTANCES

A chemist carrying out a reaction in the laboratory seldom, if ever, is fortunate enough to obtain a single, pure substance as a product. Instead, he obtains a mixture of two or more substances in varying amounts. He then is faced with the problem of separating from this mixture the particular substance in which he is interested.

The procedures used to separate matter into its components are not obvious. They depend upon our experimental observations about the behavior of matter when it is carried through changes in its state but is not changed into another kind of substance. Such changes in state are called **physical** changes. They must be distinguished from changes in which the substances present in the matter are converted to other substances; these are called **chemical** changes.

By experiment it has been determined that matter which is carried through physical changes in state has an inherent tendency to be resolved into its components. To resolve a sample of matter completely may require many properly conceived physical changes, but the separation is, in principle, considered possible. As a result of these changes, one ultimately obtains substances which cannot be further resolved by any physical changes. These materials are called **pure substances**.

We shall consider three different techniques which are used to separate pure substances from mixtures. Two of these, **distillation** and **fractional crystallization** have been in common use for many years; the third, **chromatography**, has become available more recently.

Distillation and Fractional Distillation

A mixture containing two components only one of which is volatile is readily separated into pure substances by **distillation**. The apparatus used and the techniques involved are illustrated in Figure 1.5, which shows the separation of sodium chloride from water. When the solution of sodium chloride in water is heated, the water boils off, leaving sodium chloride in the distillation flask. If desired, the water may be collected by passing the vapor down a condenser through which cold water is circulated.

If both constituents of a mixture are volatile, the simple technique illustrated in Figure 1.5 will not achieve a complete separation. Suppose, for example, that we wish to separate a mixture consisting of equal amounts of chloroform and benzene, two volatile organic liquids. We might attempt to do this by heating the mixture until approximately half of it has distilled over. If we examine the distillate, we will find that it contains predominantly, but not exclusively, the more volatile constituent, chloroform. Similarly, the residue in the distilling flask will be rich in the less volatile constituent, benzene; it will, however, contain an appreciable amount of chloroform. We could improve upon this partial separation by subjecting both distillate and residue to a second distillation. Each time we repeat the process, we will find that the distillate becomes richer in chloroform, while the benzene becomes concentrated in the residue. Eventually, we can separate the mixture into its pure components, by this process of **fractional distillation**. If we

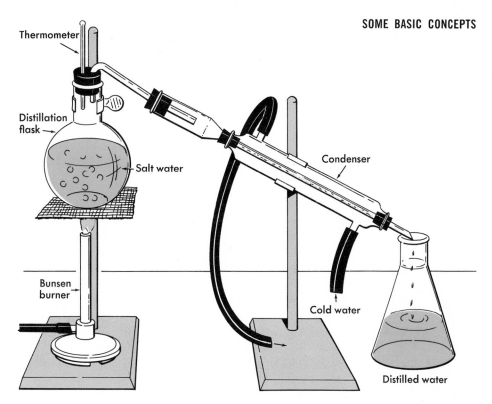

Figure 1.5 The distillation of salt water.

are willing to settle for 99, 95, or 90 per cent purity, we can cut down the number of distillations and the time accordingly.

A more efficient way to carry out a fractional distillation is to interpose between the distilling flask and the condenser a column of the type shown in Figure 1.6. The glass beads with which the column is packed offer a greatly increased surface upon which the less volatile constituents of the vapor will condense and then fall back into the distilling flask. The most volatile com-

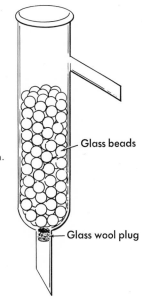

Figure 1.6 Column for fractional distillation.

ponent of the vapor escapes from the column and passes over into the distillate. To achieve a good separation, the distillation must be carried out quite slowly. Huge columns of a somewhat more sophisticated design than that shown in Figure 1.6 are used routinely to separate petroleum fractions that differ in volatility (e.g., gasoline, kerosene).

Fractional Crystallization

A common problem which arises in both inorganic and organic chemistry is that of purifying a solid substance contaminated by small amounts of other solids. The impure solid may be a reagent purchased from a chemical supply house or a product isolated from a reaction, or even a natural product.

An effective technique for removing small amounts of impurities from a solid sample is **fractional crystallization**. In its simplest form, fractional crystallization involves dissolving the solid in a minimal amount of a hot solvent, cooling the solution to room temperature, and filtering the purified solid that separates on cooling.

To appreciate the principle upon which this process is based, suppose that we wish to purify a sample of tartaric acid, an organic compound isolated from grape juice, contaminated with a small amount of another organic substance, succinic acid. Specifically, suppose that we are working with a sample containing about 5 g of succinic acid mixed with 95 g of tartaric acid. Referring to Figure 1.7, we deduce that this sample should dissolve completely in 100 g of water at 80°C (100 g of water at 80°C dissolves 98 g of tartaric acid and up to 71 g of succinic acid). Consider, now, what happens when this solution is cooled to 20°C. Since the solubility of tartaric acid at 20°C is only 18 g/100 g of water, we see that

$$95 \text{ g} - 18 \text{ g} = 77 \text{ g}$$

of tartaric acid will separate on cooling. On the other hand, all of the succinic acid (5 g) will stay in solution, since 100 g of water at 20°C will dissolve as much as 7 g of succinic acid. Consequently, by dissolving the sample in hot

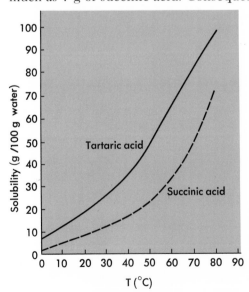

Figure 1.7 Solubilities of tartaric acid and succinic acid at various temperatures.

water at 80°C and cooling to 20°C, we should be able to separate about 77 g (81 per cent) of pure tartaric acid, uncontaminated with succinic acid.

In carrying out a fractional crystallization, it is important that we choose a solvent in which the desired product is much more soluble at high temperatures than at low temperatures. Otherwise, a large amount of the solid will be "lost" when the solution is cooled; that is, the solid will stay in solution at room temperature. The chemist who attempts to purify a solid product from a reaction mixture frequently has to use a trial-and-error method to find the best solvent for his purposes.

It is important to point out that fractional crystallization can be depended upon to yield a pure product only if the amount of impurities present is relatively small. In the example discussed, suppose the sample had contained 10 g of succinic acid instead of 5 g. Since the solubility of succinic acid at 20°C is 7 g/100 g water, it is evident that the tartaric acid isolated from the solution would have been contaminated by

$$10 \text{ g} - 7 \text{ g} = 3 \text{ g}$$

of succinic acid. A second recrystallization would have been required to give a pure product.

Two compounds, A and B, have identical solubilities at all temperatures. Can pure A be separated from a mixture of A and B? Why, or why not?

Chromatography

Both of the classical separation techniques which we have discussed are subject to certain limitations. Distillation or fractional distillation requires that the components of the mixture differ appreciably in volatility. Fractional crystallization becomes a tedious and inefficient process when large amounts of impurities are present. A recently developed technique which is more versatile than either of these methods is chromatography. In a typical experiment we use a tube of Pyrex glass, packed with a finely divided solid. The solid is often silica gel, small particles of quartz, or another chemically inert substance, which tends to hold or adsorb small amounts of other substances on its surface. The sample to be separated is put into solution and poured into the tube, which is called a **chromatography column**. The packing adsorbs the sample and holds it at the top of the column.

A solvent in which the sample is more soluble is then poured slowly through the column. The different components of the sample, which have somewhat different solubilities in this solvent, are gradually flushed down the column and are continuously dissolved in the solvent and readsorbed on the packing. As the experiment proceeds, the various components become separated into zones, or bands, in the column, and they finally pass from the bottom of the column as purified fractions, which may be pure substances. This procedure is called chromatography after the Greek word *chroma*, meaning color, since in some early experiments the separated bands were identified by their colors.

Column chromatography has the disadvantage that it is relatively slow, both in the separation itself and in the preliminary work of selection of suitable solvents and column materials. **Thin-layer chromatography**, (TLC) in which the packing is deposited as a thin layer on a flat plate, does not have these drawbacks. In this method the sample is applied, usually from solution, at the bottom of a plate, which is then placed in a tank in which there is a

15

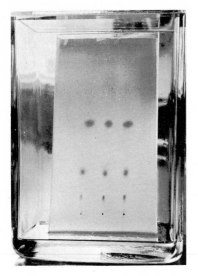

Figure 1.8 A thin-layer chromatogram showing the separation of three dyes (butter yellow, Sudan Red G, and indophenol) on activated Silica Gel G with benzene. Plate is shown in developing tank in solvent layer. (Photograph by L. A. Webb, courtesy of Dr. J. M. Bobbitt, University of Connecticut.)

shallow layer of solvent. By capillary action the solvent slowly rises on the coating, flowing through the sample and separating its components into spots, or zones. Detection is by selective staining or, if radioactive tracers are used, by use of a radiation counter. Since the experiment is rapid, easily performed, and requires only a few micrograms (1 microgram = 10^{-6} g) of sample, it has recently come into common use by organic chemists and biochemists (Figure 1.8).

In still another variation on the basic experiment, vapors rather than liquids are used. A column similar to that in the previously described experiment is employed, and the sample, either in solution or as a vapor, is injected into the top of the column where it is adsorbed. An inert gas, usually helium, is used as the eluting material, and the components in the sample gradually separate as they continuously vaporize in the helium and readsorb on the packing. Since the experiment is carried out at a temperature at which the sample has an appreciable volatility and the sample passes down the column as a vapor, the procedure is called **vapor phase chromatography** (VPC). The resolving power of this method far exceeds that of any other approach yet devised, and because of the exceedingly sensitive detectors that have been developed, it requires only micrograms of sample. Because of the importance of the method to chemical research, many commercial vapor chromatographs have been developed in the relatively few years since this kind of apparatus first became popular (Figure 1.9).

1.4 IDENTIFICATION OF PURE SUBSTANCES

In the preceding sections we have considered three examples of the many techniques that are available to a chemist who wishes to separate a mixture into its components. Two questions immediately arise: How do we know when we have achieved a separation? That is, what tests can we apply to decide whether the product separated from a mixture is indeed a pure

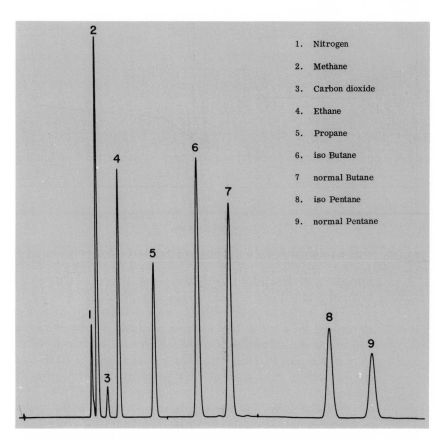

1. Nitrogen
2. Methane
3. Carbon dioxide
4. Ethane
5. Propane
6. iso Butane
7 normal Butane
8. iso Pentane
9. normal Pentane

Figure 1.9 Vapor chromatogram of a natural gas synthetic blend. (Courtesy of F & M Scientific Division of Hewlett-Packard, Avondale, Pa.)

substance? Once we are satisfied that we do indeed have a pure substance, how do we identify it? In other words, how do we decide precisely which of the many thousands of known substances we are working with? Let us now consider some of the methods that are available to the chemist to help him answer these questions.

Behavior of Pure Substances During Melting or Boiling

One of the most effective methods of checking the purity of a solid is to measure its melting point. A pure crystalline solid, if heated slowly, shows a sharp melting point, as indicated in the diagram at the left of Figure 1.10. The temperature of the solid increases steadily until the melting point is reached. At that point, the entire sample melts. The temperature stays constant as long as any solid remains; then and only then does the temperature rise, as the liquid absorbs heat.

The melting point behavior of an impure solid differs from that of a pure substance in one important respect. It is ordinarily found, for reasons which we shall discuss in Chapter 11, that the temperature rises steadily during the melting process (Figure 1.10). The last crystals melt at a temperature higher than the initial melting point. Any evidence of a deviation from a horizontal

If the melted sample is cooled, we would expect the curve of Figure 1.10 to be retraced in the reverse direction. Frequently, however, the liquid "supercools" before freezing, thereby obscuring the horizontal portion of the cooling curve.

17

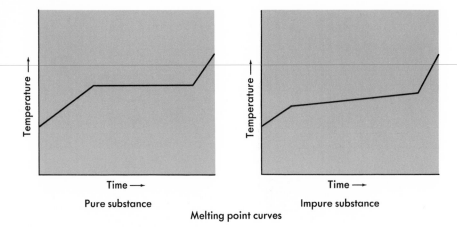

Figure 1.10 Melting point curves of a pure substance (left) and a mixture (right).

temperature-time plot during a melting process leads us to suspect the presence of impurities.

The purity of a liquid can be tested in a manner quite similar to that described for a solid. In this case, a constant temperature during the boiling process is the criterion of purity. The heating curve for a pure liquid looks very much like that shown at the left of Figure 1.10. If the liquid is impure, we ordinarily find that the temperature rises steadily during the boiling process.

We conclude that the purity of a sample can be tested by observing its behavior when it melts or boils. There are, of course, many other ways of checking the purity of a reagent or a product formed in a reaction. If the sample is volatile, for example, one might run a vapor-phase chromatogram. The appearance of a single, sharp peak (Figure 1.9) would imply a pure product. If we find more than one peak in the chromatogram, we conclude that significant amounts of impurities are present.

Physical Properties of Pure Substances

Let us suppose that a student has separated a product from a reaction mixture and has shown by one method or another that it is a pure substance. The simplest way to identify it is to compare its physical properties with those of known substances, which can be found in the literature. Suppose, for example, that after conducting a reaction of chlorine with benzene, a student isolates a pure liquid that boils at 80°C and freezes at 5°C. Looking in a handbook or other source of such information, he finds, to his chagrin, that these temperatures correspond to the boiling point and freezing point of benzene, the substance that he started with. Learning from experience, he carries out the reaction under more rigorous conditions, obtaining a liquid that boils at 132°C and refuses to freeze in an ice bath. The properties of his product are consistent with those of chlorobenzene (BP = 132°C, FP = −45°C).

At this stage, on the basis of the measurement of a single property, the student is in no position to state categorically that he has prepared chlorobenzene (there are approximately 50 known substances that boil between 130 and 135°C). To confirm that the compound he has prepared is indeed

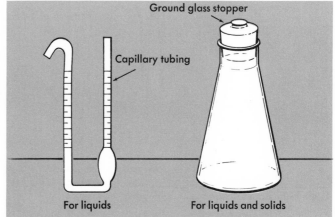

Figure 1.11 Pycnometers for measuring densities.

Ground glass stopper

Capillary tubing

For liquids   For liquids and solids

chlorobenzene, he may wish to determine another of its physical properties. One such property, which is relatively easy to measure, is its **density**.

The density of a substance can be determined by measuring the mass and volume of a sample.

$$d = M/V \qquad\qquad (1.3)$$

Density, like melting point or boiling point, is an **intensive** property; that is, it is independent of the amount of sample. If we measure out successive samples of pure water having volumes of 10 ml, 20 ml, and 50 ml, we find that they have masses of 10 g, 20 g, and 50 g respectively. The density of each of these samples, as defined by equation 1.3, is a fixed quantity, namely 1.0 g/ml, which is characteristic of the pure substance water. Mass and volume, taken separately, are **extensive** properties, in that they are directly proportional to the amount of sample.

The density of a liquid such as chlorobenzene is readily measured using a device known as a pycnometer (Figure 1.11). A small container of precisely known volume (about 5 ml) is first weighed empty and then weighed again filled with the liquid. The mass of the liquid is obtained by taking the difference of the masses obtained; the density is found by dividing this mass by the volume of the container. The pycnometer can be calibrated with a liquid of known density.

At a given temperature, the mass or volume of a pure substance can be varied, but not its density.

EXAMPLE 1.3. The student referred to determines the density of his liquid sample using a pycnometer known to have a volume of 5.000 ml. The empty pycnometer weighs 3.412 g; when filled with liquid at room temperature the total mass is 8.937 g. Calculate the density of the liquid.

SOLUTION. The mass of the liquid is 8.937 g − 3.412 g = 5.525 g. Since the volume of the sample is known to be 5.000 ml, its density must be

$$d = \frac{M}{V} = \frac{5.525 \text{ g}}{5.000 \text{ ml}} = 1.105 \text{ g/ml}$$

The density of chlorobenzene is listed as 1.106 g/ml at 20°C and 1.101 g/ml at 25°C. It would appear that the measured density agrees reasonably well with that of chlorobenzene; if a further check is desired, it would be well to carry out the density determination at a carefully controlled temperature.

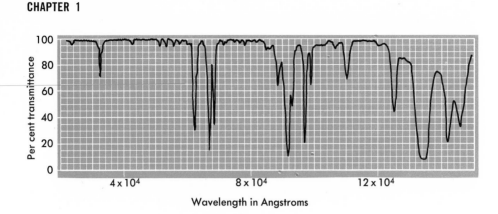

Figure 1.12 Infrared spectrum of chlorobenzene.

The density of a solid crystalline material is not so easily determined as that of a liquid. One approach is to weigh a sample of crystals and add them to a container of known mass and volume. The container is then filled with a nonsolvent liquid of known density and weighed. From these measurements the volume of the crystals can be found and their density calculated (Figure 1.11).

Many other physical properties can be used to identify a pure substance. One of these is its **infrared absorption spectrum**. Many substances absorb infrared radiation at well-defined, characteristic wavelengths. By exposing the substance to infrared radiation covering a wide range of wavelengths and measuring the absorption at each wavelength, it is possible to obtain an absorption spectrum of the type shown in Figure 1.12. The infrared spectrum of a pure substance, unlike its density or its boiling point, can be used to identify the substance unambiguously, provided comparison spectra are available.

1.5 KINDS OF SUBSTANCES—ELEMENTS AND COMPOUNDS

Of the many thousands of pure substances which have been isolated over the years, approximately 100 are unique in one important respect. No one has ever succeeded in resolving these substances, called **elements**, into two or more substances that differ in their chemical properties from the original. All other pure substances are classified as **compounds**. A compound, by definition, is a pure substance that can be resolved into two or more elements.

Many different methods have been used to bring about chemical changes in which compound substances are resolved into elements. The English chemist Joseph Priestley decomposed mercuric oxide into the elements mercury and oxygen by exposing it to an intense beam of sunlight focused by a powerful lens. Sir Humphry Davy showed that ordinary quicklime, long believed to be an elementary substance, was a compound, by passing an electric current through melted lime and demonstrating that two different substances (calcium and oxygen) were produced at the electrodes.

Many of you have probably seen a similar experiment in which the compound substance water is broken down into the elements hydrogen and oxygen by electrolysis.

The Discovery of the Elements

Some elements were discovered in ancient times, and new elements continue to be discovered even today. Some elementary substances are readily isolated, and ancient peoples probably noted their presence free in nature or in the remains of fires. Among the elements known since antiquity are gold, silver, copper, iron, lead, tin, mercury, sulfur, and carbon.

During the alchemical period, prior to 1700, a few more elements were found. Among them were arsenic, antimony, bismuth, phosphorus, and zinc. The alchemist, working purely empirically, could not do more than obtain a few more of the easily resolved elements.

During the period 1770 to 1800, with the explanation of combustion and the discovery of oxygen, nitrogen, and hydrogen by Lavoisier and Priestley, chemistry began as a science. In about 1800 the atomic theory, as proposed by Dalton, gave recognition to the existence of elements. Between 1770 and 1830 about 35 elements were discovered, twice as many as in all the preceding centuries. It is obvious that a few men, working on the basis of correct principles, can do far more than many who rely almost exclusively on chance.

A rather remarkable fact is that, although chemistry developed tremendously during the nineteenth century, 14 of the 103 elements have been discovered since about 1935. The reason for this is not that the early chemists were rather lackadaisical in their search for new elements, but lies in a completely unexpected area. The elements found since 1935 do not actually exist in nature, but were synthesized from known elements by chemists and physicists. These new elements are all unstable and spontaneously decompose to other elements, in some cases very quickly. Both the synthesis and decomposition of these elements involve far more energy than is observed in any physical or chemical changes, and so we do not consider that such reactions invalidate our criterion for the existence of elementary substances.

Chemical Symbols of the Elements

In dealing with the physical and chemical behavior of elementary and compound substances one can describe experimental observations with words, as we have done in this section. Although this is always possible, chemists make extensive use of a brief and informative shorthand notation to accomplish the same purpose.

Basic to the system of shorthand notation are the symbols for the elements. As you know, each element has a name, which in many cases was given to it by its discoverer. Thus we have the elements cobalt, boron, polonium, and gold. In addition each element is represented by a symbol. The symbol consists of one or two letters, taken frequently, but not always, from the first two letters in the name of the element. For the elements just mentioned, the symbols are Co, B, Po, and Au. Some symbols derive from the Latin name of the element or one of its compounds. The names and symbols of the elements are an important part of the language of chemistry, and the reader is encouraged to become familiar with them. A table giving the names and symbols of the elements may be found inside the front cover of this book.

PROBLEMS

1.1 In what areas of chemistry would you expect the following to fall?
   a. The synthesis of rubber.
   b. The percentage of the elements in the composition of a stainless steel.
   c. Spectroscopy.
   d. Radioactive fallout.
   e. The development of semiconductors.

1.2 The volume of a sample of gas is 21.8 ml. Express the volume in liters and cubic feet.

1.3 Distinguish between weight and mass.

1.4 A metal sample weighs 34.9 g. Express its mass in milligrams and in pounds.

1.5 When you say "The temperature of the room is 27°C," what do you really mean?

1.6 Express 22°C in °F and °K.

1.7 Suppose that in setting up the Fahrenheit scale of temperature, 20° had been chosen as the freezing point of water and 240° as its normal boiling point. What would have been the relationship between the Fahrenheit and centigrade scales?

1.8 In a certain reaction, 12.0 kcal of energy are evolved per gram of reactant. How many ergs are evolved per mg of reactant?

1.9 The density of water at a certain temperature is 0.997 g/ml. What would be the mass in pounds of one quart of water?

1.10 An empty pycnometer weighing 4.921 g has a volume of 5.000 ml. When filled with a liquid, it weighs 12.175 g. Calculate the density of the liquid in g/ml.

1.11 Suggest an appropriate method of separating
   a. The components of a water solution of potassium chloride.
   b. A gaseous mixture of ethane and propane.
   c. A mixture of pentane and decane, two organic liquids.
   d. A mixture of potassium nitrate and sodium chloride, two water soluble solids.
   e. The components of air.

1.12 Suppose you had a mixture containing 80 g of tartaric acid and 20 g of succinic acid. (Refer to Figure 1.7.)
   a. What is the smallest quantity of water required to dissolve this sample at 80° C?
   b. To what temperature could you cool this solution without precipitating any succinic acid?
   c. At the temperature referred to in b, how much tartaric acid would have crystallized from solution?

1.13 Describe some of the tests that a chemist might perform to demonstrate that a liquid sample, believed to be ethyl alcohol, is pure. How could he show that the sample is indeed ethyl alcohol? How could he show that ethyl alcohol is a compound rather than an element?

1.14 A metal sample has a mass of 12.0 g and a volume of 1.25 ml. Calculate its density in g/ml and lb/ft³.

1.15 A temperature scale is set up on the basis of the properties of benzene, a common organic liquid. On the benzene scale, 0°B is set at the freezing point of benzene, which occurs at 5°C; 100°B is set at the temperature at which benzene normally boils, which is also 80°C. Find the temperature on the benzene scale, in °B, at which water normally boils.

1.16 The pressure in a vessel containing oxygen gas is 624 mm Hg. What is the pressure in atmospheres?

1.17 To the pycnometer in Problem 1.10 is added 3.454 g of an unknown crystalline solid. A liquid having a density of 1.466 g/ml is then added to fill the pycnometer. The final mass of the pycnometer plus solid plus liquid is 13.099 g. Find the density of the solid.

1.18 A student has a solution containing 12.0 g of sodium chloride, 0.50 g of potassium chloride, and 80.0 g of water.
   a. How could he separate pure water from this solution? How could he check the purity of the water he obtained?
   b. Suggest a reasonable procedure for separating pure sodium chloride from this mixture. How could the student identify it as sodium chloride?

1.19 Suppose you have a mixture of 70 g of succinic acid and 15 g of tartaric acid. If this mixture is dissolved at 80°C in 100 g of water and then cooled to 20°C, how much succinic acid will come out of solution? How much tartaric acid?

°1.20 The volume of one gram of water at atmospheric pressure is given by the equation

$$V = V_0(1 - 4.5 \times 10^{-5}\,t + 6.7 \times 10^{-6}\,t^2 - 1.9 \times 10^{-8}\,t^3)$$

in which V_0 is its volume in ml at 0°C and t is its temperature in °C. A water-in-glass thermometer is calibrated at 0°C and 100°C. What temperature would it read in a bath that is actually at 50°C? What temperature would it indicate for a sample actually at 5°C? What can be said about using water as a thermometric fluid?

°1.21 A student has a mixture of 80 g of tartaric acid and 20 g of succinic acid. He dissolves this in the minimum quantity of pure water at 80°C and cools it to 20°C. The crystals that separate are impure, containing both tartaric acid and succinic acid. How many grams of each solid are present? If the student repeats this procedure a second time, how many grams of each solid will be present in the solid he obtains? How many times must he repeat this fractional crystallization to obtain a pure product?

°1.22 A manufacturer of thermometers uses a capillary column with a diameter of 0.10 mm. He wants the degree markings on the thermometer to be exactly 2 mm apart. The density of mercury decreases by 0.018 per cent per °C. What must be the volume of the bulb at the bottom of the thermometer?

---

* Those problems in this text which require extra effort and ability on the part of the student will be indicated by an asterisk. Answers to problems numbered in color can be found in Appendix 5.

2

Atoms, Molecules,
and Ions

We have discussed some of the tools that chemists use and some of the experimental procedures they employ to isolate and identify pure substances. In this chapter we shall examine some of the units of matter that form the building blocks of all pure substances. These include atoms, molecules, and ions.

2.1 ATOMS

Atomic Theory

The notion that matter ultimately consists of discrete particles is an old one. About 400 B.C. this idea was put forward in the writings of Democritus, a Greek philosopher, who apparently had been introduced to it by his teacher, a man named Leucippus. The idea was rejected by Plato and Aristotle, and it was not until about 1650 A.D. that it again was suggested, this time by the Italian physicist Gassendi. Sir Isaac Newton (1642–1727) supported Gassendi's arguments with these words:

> . . . it seems probable to me that God, in the Beginning, formed Matter in solid, massy, hard, impenetrable, movable Particles, of such Sizes and Figures, and with such other Properties, and in such Proportions to Space, as most conduced to the End for which he formed them. . . .

Until about 1800, the idea that matter is particulate in nature was based mainly on the intuition of its adherents, who included, in addition to Newton, the English scientists Boyle and Higgins. It remained for John Dalton, an English chemist and schoolteacher, to place this idea on a firm experimental basis. In about 1808, Dalton proposed a simple explanation of the then known laws of chemistry, which came to be known as the **atomic theory**.

Dalton's theory stated that all elements consist of tiny particles called **atoms**. He postulated that all the atoms in a given elementary substance are alike and that compound substances are formed when one or more atoms of one element combine in a definite proportion with one or more atoms of another element. This theory, simple though it was, was very convincing in its ability to explain the experimental facts and generalizations deduced from them.

While certain of Dalton's ideas proved untenable as chemists learned more about the structure of matter, the essentials of his theory have withstood the test of time. Three of Dalton's main postulates, which now comprise modern atomic theory, are given here with an example to illustrate the meaning of each.

1. *An element is composed of extremely small particles called atoms. The atoms of a given element all exhibit identical chemical properties.* The element oxygen is made up of oxygen atoms, all of which behave chemically in the same way.

2. *Atoms of different elements have different chemical properties. In the course of an ordinary chemical reaction, no atom of one element disappears or is changed into an atom of another element.* The chemical behavior of oxygen atoms is different from that of hydrogen atoms or any other kind of atom. When the elementary substances hydrogen and oxygen combine with each other, all the hydrogen atoms and all the oxygen atoms that react are present in the water formed, and no atoms of any other element are formed in the process.

3. *Compound substances are formed when atoms of more than one element combine. In a given pure compound the relative numbers of atoms of the elements present will be definite and constant. In general, these relative numbers can be expressed as integers or simple fractions.* In the compound substance water, hydrogen atoms and oxygen atoms are combined with each other. For every oxygen atom present, there are always two hydrogen atoms. Ammonia, a gaseous compound of nitrogen and hydrogen, always contains three hydrogen atoms for every nitrogen atom.

The third postulate offers a rational explanation for the **Law of Constant Composition**, which states that a **compound**, regardless of its origin or method of preparation, **always contains the same elements in the same proportions by weight**. Clearly, if the atom ratio of the elements in a compound is fixed, their proportions by weight must also be fixed. The validity of this law became generally recognized at about the same time that Dalton's theory appeared. Prior to 1808, many people agreed with the French chemist, Berthollet, who believed that the composition of a compound could vary over wide limits, depending on how it was prepared. Joseph Proust, a French expatriate working in Madrid, refuted Berthollet's ideas by showing that the "compounds" Berthollet had cited were actually mixtures.

It is now known that in many compounds, particularly metal oxides and sulfides, the atom ratio may vary slightly from a whole-number value. Careful analyses of apparently homogeneous samples of a certain oxide of tungsten give atom ratios of oxygen to tungsten varying from 2.88:1 to 2.92:1, rather than the 3:1 ratio that one might expect. In another case, it has been found that different samples of zinc oxide may contain up to 0.03 per cent of excess zinc atoms above the hypothetical 1:1 ratio. Interestingly enough, zinc oxide was one of the compounds cited by Berthollet in disputing the Law of Constant Composition. The data he gave showed variations of up to 3 per cent in zinc content, approximately 100 times that now accepted.

Compounds of this type are often referred to as "Berthollides" or nonstoichiometric compounds. The deviations from constant composition arise because of defects in the crystal structure. In later chapters, we shall examine the nature of these defects and their effect on certain of the properties of the solids in which they occur.

The same postulate led Dalton to formulate another of the basic quantitative laws of chemistry, the **Law of Multiple Proportions**, which states that

25

Can you think of a metal which forms two different oxides in which the weights of metal combining with 1 g of oxygen are in the ratio 2:1? 3:2?

when two elements combine to form two different compounds, the weights of one which combine with a fixed weight of another are in a simple ratio of whole numbers (2:1, 3:2, and so forth). Dalton reasoned that elements A and B might form two compounds, in one of which two atoms of A were combined with one of B, while in the other, one atom of A was combined with one atom of B. If this happened, the weight of A combined with a fixed weight of B in the first compound would be twice that in the second. The validity of this law was rapidly established, partly on the basis of experiments that Dalton himself carried out on the oxides of carbon, sulfur, and nitrogen.

Despite the successes of Dalton's atomic theory, it was not immediately accepted by all scientists. Physicists were reluctant to give up the concept that matter is continuous in nature. Many chemists felt that it was a waste of time to speculate on the particulate nature of matter, and that natural laws should be based exclusively upon experimentally measured quantities. As late as 1900, the well-known German chemist, Ostwald, in writing a basic textbook of general chemistry that was in common use for a generation, deliberately avoided all mention of atomic or other elementary particles.

Structure of the Atom

Like any useful scientific theory, atomic theory raised many new questions. Even before Dalton's ideas had been generally accepted, many philosophers and scientists were speculating as to whether atoms, tiny as they are, might in turn be broken down into still smaller particles. Nearly one hundred years were to pass before this question could be answered in the affirmative on the basis of experimental evidence. The first conclusive proof of the existence of subatomic particles was obtained by the English physicist J. J. Thomson in 1897. He was able to show that the so-called cathode rays given off when an electric discharge is passed through a gas at low pressures consist of a stream of negatively charged particles called **electrons**, whose masses are much less than those of the atoms from which they are derived. Fourteen years later, in 1911, another English physicist, Ernest Rutherford, showed that most of the mass of the atom was concentrated in a positively charged **nucleus**, having a diameter that is very small compared to that of the atom itself.

In Chapter 7, we shall consider in greater detail some of the experiments that led to our present-day picture of the structure of the atom. For the time being, it will be sufficient to give a brief description of the two principal structural features of the atom.

How might one determine the diameter of an electron? a hydrogen atom?

1. **Electrons** are present in all atoms and so are present in all matter. The mass of an electron is only about $\frac{1}{2000}$ of that of the lightest atom, the hydrogen atom. The diameter of an electron is only about $\frac{1}{10000}$ of that of the hydrogen atom. Probably the most important property of an electron is its electric charge. Each electron carries the same amount of charge; the magnitude of the charge, which is negative, is 1.60×10^{-19} coulomb.

All atoms contain an integral number of electrons. This number, which may vary from 1 to over 100, is characteristic of an atom of a particular element. All hydrogen atoms contain one electron; all atoms of the element lawrencium contain 103 electrons. We shall have more to say later about the way in which these electrons are distributed in relation to one another. For the time being, it is sufficient to point out that electrons are found in the outer regions of atoms and form a cloud of negative charge about the positively charged atomic nucleus.

2. A **nucleus** is located at the center of the atom. The nucleus has about the same diameter as an electron and a mass very nearly equal to that of the atom. It carries a positive charge, just equal in magnitude and opposite in sign to the total negative charge of the electrons in the atom, making the atom an electrically neutral particle.

The charge of the nucleus, which must be equal to the number of electrons, is characteristic of the atom of a particular element. All hydrogen atoms have nuclei which carry a unit positive charge equal in magnitude to that of the electron; all nuclei of the element lawrencium have a charge of +103 on the same scale.

Rather recently it has been discovered that nuclei of atoms of the same element may differ in mass. For example, two different kinds of hydrogen atoms are found in nature; in one of them the nucleus has a mass just twice that of the other. Carbon atoms are found to have three different masses with relative magnitudes 12, 13, and 14. When an element contains atoms with nuclei having "n" possible masses, we say that the element has n **isotopes**. Thus, carbon has three isotopes; tin, remarkably enough, has 10 isotopes.

A few elements, including sodium, fluorine, phosphorus, and gold, are anisotopic $(n = 1)$.

2.2 MOLECULES AND IONS

Individual atoms are rarely encountered in matter. Only in a very few elementary substances, the so-called noble gases (He, Ne, Ar, Kr, Xe, Rn), can we consider individual atoms to be the fundamental structural units of which the substance is composed. The physical and chemical properties of all other elementary and compound substances are most readily understood in terms of two other types of structural units, both derived from atoms.

Molecules

The fundamental structural unit in most volatile substances, including both elements and compounds, is the molecule, which is an aggregate of atoms held together by relatively strong forces called chemical bonds. In contrast, the forces between molecules are relatively weak. Molecules, therefore, do not usually interact strongly with one another but behave more or less as independent particles.

Carbon dioxide is one example of a molecular substance. The carbon dioxide molecule contains one carbon atom and two oxygen atoms. As shown in Figure 2.1, each of the oxygen atoms is bonded to the carbon atom. Carbon dioxide molecules are discrete particles which, under ordinary conditions, are stable; that is, they do not react chemically with one another or with most other molecules. The properties of carbon dioxide are determined by the properties of its molecules.

Many other pure substances with which you are familiar also contain simple molecules. Figure 2.1 shows schematic diagrams of the molecules that are present in oxygen, water, and ethylene. Oxygen, like several elementary substances, is molecular. Oxygen molecules are diatomic; that is, they

How could you show that oxygen molecules are diatomic?

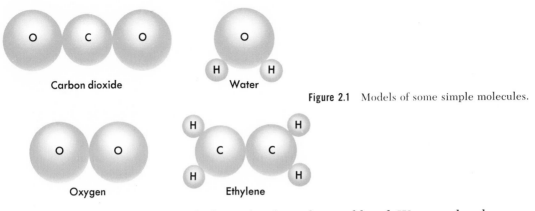

Carbon dioxide

Water

Oxygen

Ethylene

Figure 2.1 Models of some simple molecules.

contain two atoms linked together by a chemical bond. Water molecules contain one oxygen atom bonded to two hydrogen atoms. The water molecule has a bent structure, whereas the carbon dioxide molecule is linear. In ethylene (the gas which is the raw material for the production of polyethylene) the molecule contains two carbon atoms and four hydrogen atoms; each carbon atom is bound to two hydrogen atoms and to the other carbon atom.

Ions

Under certain circumstances, it is possible to remove electrons from a neutral atom, leaving behind a positively charged particle that is somewhat smaller than the original atom. Alternatively, electrons may be added to certain atoms to form particles that carry a negative charge and are somewhat larger than the original atom. These charged particles, whether positive or negative, are called **ions**. Examples of positive ions (**cations**) include the sodium ion Na^+, formed from a sodium atom by the loss of a single electron, and the calcium ion Ca^{2+}, derived from a calcium atom by extraction of two electrons. Among the more common negative ions (**anions**) are Cl^- and F^-, formed when atoms of chlorine and fluorine acquire an extra electron.

Many compounds contain positively and negatively charged ions as their fundamental structural units. The structures of two such ionic compounds, sodium chloride and calcium fluoride, are shown in Figure 2.2. In sodium chloride, there are an equal number of ions of opposite charge (Na^+, Cl^-). In calcium fluoride, there are two F^- ions for every Ca^{2+} ion, giving the electrical balance that must be maintained in any macroscopic sample of an ionic compound. It is important to point out that in these compounds and, indeed, in all ionic substances, there is a continuous, three-dimensional network of chemical bonds holding the oppositely charged ions together. These bonds, which owe their origin to the electrostatic attraction between oppositely charged ions, are ordinarily extremely strong.

Certain compounds contain neither ions nor small discrete molecules. An example is SiO_2, whose structure is discussed on p. 221.

2.3 RELATIVE MASSES OF ATOMS, ATOMIC WEIGHTS AND GRAM ATOMIC WEIGHTS

One of the postulates of atomic theory states that "An element is composed of extremely small particles called atoms." The question arises as to

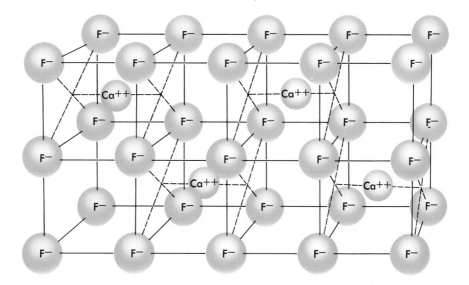

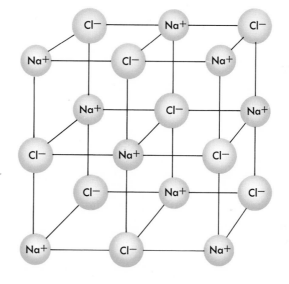

Figure 2.2 The structures of two ionic crystals, sodium chloride and calcium fluoride. The structure shown persists throughout a macroscopic sodium chloride crystal; Na+ and Cl- ions alternate along lines forming the edges of cubes in the lattice. In the calcium fluoride crystal, the fluoride ions are in a simple cubic lattice; the calcium ions are at the centers of alternate cubes along each of the three principal axes.

precisely *how* small (or how large) atoms of different elements are. What is more important, we should like to know the weights of atoms. We should like to know, for example, how much atoms of elements such as sodium and fluorine weigh.

From an experimental standpoint, it is much simpler to determine relative rather than absolute masses of atoms. That is, it is easier to obtain the ratio of the masses of sodium and fluorine atoms than it is to determine the masses of either of these atoms individually in grams or pounds. Fortunately, so far as the development of atomic theory is concerned, we find that we can obtain almost as much information from relative as from absolute atomic masses.

The problem of determining the relative masses of atoms of different elements has occupied the attention of physical scientists since the time of Dalton. The men and women who have worked in this field include many of

29

the best-known chemists of the nineteenth and twentieth centuries, and they have worked in laboratories in many different countries: Berzelius (1779–1848) at the University of Stockholm, Dumas (1800–1884) and later Marie Curie (1867–1934) at the Sorbonne in Paris, Cannizzaro (1826–1910) at Genoa, Palermo, and Rome, and T. W. Richards (1868–1928), the first American to receive a Nobel Prize in chemistry, at Harvard University.

We shall shortly consider some of the experimental methods that have been developed to determine the relative masses of atoms. Before doing so, it will be helpful if we define two quantities, atomic weight and gram atomic weight.

The Meaning of Atomic Weights. The Atomic Weight Scale

The word "average" is used here because of the existence of isotopes (Section 2.1).

Relative weights of different atoms are most often expressed in terms of atomic weights. The **atomic weight of an element is a number that indicates how heavy**, on the average, **an atom of an element is compared to an atom of another element**. For example, the fact that hydrogen has an atomic weight slightly greater than 1 and oxygen has an atomic weight of almost exactly 16 tells us that a hydrogen atom is about $\frac{1}{16}$ as heavy as an oxygen atom. In another case we conclude that since the atomic weight of sulfur is 32.064, an atom of sulfur must be a little more than twice as heavy as an oxygen atom. To be exact, a sulfur atom is 32.064/15.9994 = 2.0041 times as heavy as an oxygen atom. In general, for two elements X and Y,

$$\frac{\text{mass of an atom of X}}{\text{mass of an atom of Y}} = \frac{\text{atomic weight of X}}{\text{atomic weight of Y}}$$

When atomic weights were being determined in the first half of the nineteenth century, the atomic weight of the lightest element, hydrogen, was generally taken to be 1. It soon became apparent that oxygen would be a more convenient base for atomic weights since it forms stable compounds with many more elements than does hydrogen. The Belgian chemist J. S. Stas, in a series of experiments initiated in 1850, obtained precise values for the atomic weights of several elements relative to that of oxygen, which he took to be exactly 16. It was supposed at the time that these two standards, i.e., H = 1.000, O = 16.000, were equivalent to each other. However, in the period from 1882 to 1895, Professor E. W. Morley of Western Reserve University demonstrated by the synthesis and analysis of water that the ratio of the atomic weights of hydrogen and oxygen differed from 1 : 16 by nearly 1 per cent. Since most atomic weights had been established relative to oxygen, it was agreed in 1905 to take the atomic weight of oxygen to be exactly 16.

Can you explain *why* atomic weights on the physicists' scale should be slightly greater than those on the chemists' scale?

In the 1930's, physicists interested in nuclear reactions were faced with the problem of determining the relative masses of elementary particles. Using the mass spectrograph (see p. 34) to determine these quantities, they found it most convenient to use the mass of one specific isotope as the basis for an atomic weight scale. They chose the most common isotope of oxygen as a standard, assigning it an atomic weight of exactly 16. Since naturally occurring oxygen, on which the chemists' scale was based, contained traces of heavier isotopes, the two scales differed slightly from each other. To convert from the chemists' to the physicists' scale of atomic weights, it was necessary to multiply by 1.000275.

As the years passed, the existence of two atomic weight scales differing slightly from each other proved sufficiently annoying to cause scientists to work for the adoption of a single scale that would be acceptable to both physicists and chemists. This was accomplished in 1961 when it was agreed to set up a uniform scale in which the atomic weight 12 was assigned to the most common isotope of carbon. Such a compromise satisfied the physicists' requirement that a single isotope be used as a standard. For chemists, it resulted in a change of only about 0.004 per cent from the old chemical scale. The atomic weight of oxygen, for example, changed from 16.0000 to 15.9994 with the adoption of the new scale. The atomic weights used throughout this text are all based on this so-called "carbon-12" scale.

Gram Atomic Weight

The **gram atomic weight** of an element is the **weight in grams that contains the same number of atoms as twelve grams of carbon-12** (or 15.9994 g of oxygen, 1.00797 g of hydrogen, and so forth). If the atomic weight of an element is known, its gram atomic weight can be written down immediately. For example, consider sulfur: since the atomic weight of this element is 32, a sulfur atom must be $\frac{32}{16}$ as heavy as an oxygen atom. If the number of atoms in sixteen grams of oxygen is represented by N, then the weight of an equal number, N, of sulfur atoms must be

$$\tfrac{32}{16} \times \text{wt. of N oxygen atoms} = \tfrac{32}{16} \times 16 \text{ g} = 32 \text{ g}$$

In other words, the gram atomic weight of an element is numerically equal to its atomic weight. It must be kept in mind, however, that while the atomic weight of an element is dimensionless, its gram atomic weight has the unit of grams. It is often stated, somewhat loosely, that the gram atomic weight is the "atomic weight expressed in grams."

Atomic Weights from Specific Heats. Gram Equivalent Weights

In presenting his atomic theory in 1808, Dalton suggested that the atomic weights of elements could be determined if it were possible to select samples of different elements that contained the same number of atoms. The weights of these samples would then be in the same ratio as the atomic weights of the elements of which they are composed. Although Dalton himself rejected this approach as impractical, it served as the basis of an early method of estimating atomic weights of metallic elements. In 1819, two French chemists, Dulong and Petit, suggested that the amount of heat required to raise the temperature of a solid element by a given amount, such as 1°C, should depend only on the *number* of atoms in the sample and not upon the *type* of atom. Samples that require the same amount of heat to raise their temperature by a unit amount should then contain the same number of atoms. Stated another way, since 1 gram atomic weight of every element contains the same number of atoms, a constant amount of heat should be required to raise the temperature of 1 gram atomic weight of any solid element by 1°C.

The Law of Dulong and Petit is expressed most simply in terms of a physical property known as the **specific heat**, which is the amount of heat required to raise the temperature of one gram of a substance one degree centigrade. In modern form, the law states that the product of the specific heat of a metallic element multiplied by its gram atomic weight is approximately 6 cal/°C.

$$GAW \times C \approx 6 \text{ cal/°C} \qquad (2.1)$$

where GAW = gram atomic wt (g) and C = specific heat (cal/g°C).

TABLE 2.1   LAW OF DULONG AND PETIT

ELEMENT	GAW	C	GAW × C
Na	23.0 g	0.295 cal/g°C	6.79 cal/°C
Mg	24.3 g	0.246 cal/g°C	5.98 cal/°C
Ca	40.0 g	0.155 cal/g°C	6.21 cal/°C
Cu	63.5 g	0.0921 cal/g°C	5.85 cal/°C
Ag	108 g	0.0558 cal/g°C	6.02 cal/°C

If Equation 2.1 were exact, it would be possible to determine accurately the gram atomic weight of a metal by measuring only its specific heat. Unfortunately, as Table 2.1 indicates, the equation is by no means exact; for many metals, the product of the specific heat and the gram atomic weight differs from 6 cal/°C by 10 per cent or more.

We can obtain a very precise value for the gram atomic weight of a metal by measuring one other quantity in addition to its specific heat. This quantity is the **gram equivalent weight** of the metal, which is defined as **the weight that combines with or is chemically equivalent to eight grams of oxygen**. There is a simple relationship between the gram atomic weight (GAW) and the gram equivalent weight (GEW):

$$GAW = n (GEW), \text{ where } n = 1, 2, 3 \ldots \qquad (2.2)$$

16.0 g of S, 35.5 g of Cl are chemically equivalent to 8.00 g of oxygen in the sense that they combine with the same weight of a given metal (e.g., 12.16 g of Mg).

The gram equivalent weight of a metal can be determined very accurately in the laboratory. A simple apparatus, capable of producing data from which gram equivalent weights can be calculated to ±0.1 per cent, is shown in Figure 2.3; the calculations are illustrated in Example 2.1. An approximate value of the gram atomic weight, calculated from specific heat data, allows us to decide whether n in Equation 2.2 is 1, 2, 3, or some other integer. Knowing the exact value of the gram equivalent weight and the proper value of n, we can then use Equation 2.2 to obtain an accurate value for the gram atomic weight of the metal (Example 2.2).

EXAMPLE 2.1.   A sample of nickel oxide weighing 2.000 g is reduced with hydrogen to give a residue of pure nickel weighing 1.572 g. Calculate the gram equivalent weight of nickel.

SOLUTION.   In order to calculate the weight of nickel that is combined with eight grams of oxygen in nickel oxide, we need a "conversion factor" that relates grams of nickel to grams of oxygen. The data gives us this factor since we know that 1.572 g of nickel are combined with or are chemically equivalent to

$$2.000 \text{ g} - 1.572 \text{ g} = 0.428 \text{ g of oxygen}$$
$$1.572 \text{ g Ni} \simeq 0.428 \text{ g O}$$

Using this factor to "convert" g of oxygen to g of nickel, in much the same way that we might convert centimeters to inches, we obtain

$$\text{GEW Ni} = 8.000 \text{ g O} \times \frac{1.572 \text{ g Ni}}{0.428 \text{ g O}} = 29.4 \text{ g Ni}$$

EXAMPLE 2.2. The specific heat of nickel is 0.104 cal/g°C.
  a. Using Equation 2.1, calculate an approximate value for the gram atomic weight of nickel.

  b. Using Equation 2.2 in conjunction with the GEW of nickel calculated in Example 2.1, obtain an accurate value for the gram atomic weight of nickel.

SOLUTION.

  a. GAW Ni $\times$ 0.104 cal/g°C $\approx$ 6 cal/°C; GAW Ni $\approx \dfrac{6}{0.104}$ g $= 58$ g

  b. In Example 2.1 we calculated the GEW of nickel to be 29.4 g. Clearly, n in Equation 2.2 is 2, and

$$\text{GAW Ni} = 2 \times 29.4 \text{ g} = 58.8 \text{ g}$$

In the century and a half that has passed since the work of Dulong and Petit, chemists have worked out several other methods of determining the atomic weights of elements. One of these, suggested by Stanislao Cannizzaro in 1858, proved to be particularly useful for determining atomic weights of elements such as carbon, hydrogen, and nitrogen, which form a large number of gaseous compounds. The theory upon which this method is based and the

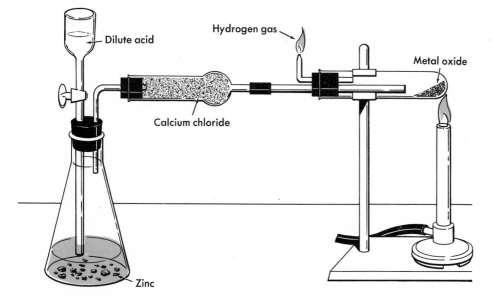

Figure 2.3 Reduction of oxides by hydrogen.

calculations involved are illustrated by Problem 2.28 at the end of this chapter. In 1870, the establishment of the periodic table by Mendeleev (Chapter 6) offered still another method of estimating atomic weights. Both these approaches, like that of Dulong and Petit, led to approximate atomic weights, from which more precise values could be calculated from Equation 2.2. Indeed, all chemical methods for determining the relative weights of atoms rely ultimately on the accurate determination of the gram equivalent weights of elements.

Atomic Weights from Mass Spectra

During the past few decades, a new and quite different approach to the determination of atomic weights has been developed, primarily by physicists. This technique makes use of an instrument known as a mass spectrometer, a simple model of which is shown in Figure 2.4. This instrument, which measures the charge to mass ratio of ions, can be used to investigate a wide variety of chemical problems. Before discussing how the mass spectrometer can be applied to atomic weight measurements, let us consider the principle upon which it is based.

In a mass spectrometer (Figure 2.4), a gaseous sample in the ionization chamber is bombarded by a stream of electrons emitted from a heated filament. A few of the gas particles are converted to positive ions, which pass through a narrow slit and are accelerated by a voltage of 500 to 2000 volts toward a magnetic field. The field deflects the ions from their straight-line path toward the collector.

For a given accelerating voltage and magnetic field strength, the extent to which the ion beam is deflected depends upon the charge and mass of the

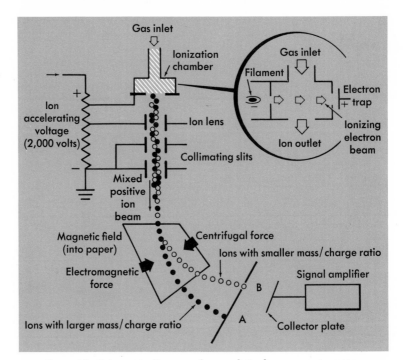

Figure 2.4 Schematic diagram of an analytical mass spectrometer.

ions. The greater the charge of the ion, the greater its deflection will be. If a +1 ion is bent to point A (Figure 2.4), a +2 ion of the same mass might appear at B. The deflection of an ion is inversely related to its mass; the lighter the ion, the more readily it is pulled off its course by the magnetic field. Of two ions which have the same charge but different masses, the heavier one might arrive at A, while the lighter ion is deflected to B. By moving the collector from A to B, we can detect these two ion beams separately. In practice, it is simpler to use a stationary collector and adjust the accelerating voltage or the magnetic field strength so that first one ion beam and then the other strikes the collector.

Depending upon the nature of the gas being analyzed, one may obtain either a relatively simple or an exceedingly complex mass spectrum. In the case of helium, which consists of individual atoms almost all of which have a mass of 4, the principal species formed is a +1 ion (mass to charge ratio = 4). A few ions of charge +2 (mass to charge ratio = 2) may also be formed. If the gas used is neon, the mass spectrum is more complex. Since ordinary neon consists of three different isotopes, a beam of +1 neon ions is split into three parts when it passes through the magnetic field of a mass spectrometer. The strongest emerging beam is one in which the ions have a mass to charge ratio of 20 (the most abundant isotope of neon has a mass of 20). Two weaker beams, consisting of ions of masses 21 and 22, can also be detected. The mass spectrum of neon as observed on a recorder attached to the collector is shown in Figure 2.5 (only +1 ions are considered).

As an illustration of how the mass spectrometer can be used to determine atomic weights, consider the element helium. In principle, it is possible to calculate the mass of this ion if we know the extent to which a +1 helium ion of mass 4 is deflected by the magnetic field. The appropriate equation is

$$m = \frac{qH^2r^2}{2E} \qquad (2.3)$$

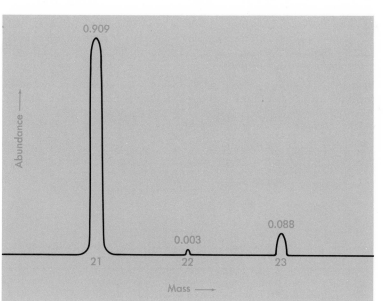

Figure 2.5 The mass spectrum of neon.

where m is the mass of the ion, q is its charge, H is the magnetic field strength, r is the radius of curvature of the ion beam as it passes from the magnetic field, and E is the accelerating voltage. In practice, both H and r are difficult to determine accurately. To circumvent this problem, we could measure the voltage necessary to bring a +1 ion of known mass such as that derived from carbon-12 to the same point, using the same magnetic field strength. Since q, H, and r are the same for the two ions, Equation 2.3 tells us that

For any two ions brought to the same point by accelerating voltages E_1 and E_2, $m_1E_1 = m_2E_2$.

$$m\ He \times E\ He = m\ C\text{-}12 \times E\ C\text{-}12 \qquad (2.4)$$

or

$$\frac{m\ He}{m\ C\text{-}12} = \frac{E\ C\text{-}12}{E\ He}$$

Experimentally, we find that the voltage required for the carbon-12 beam is 0.3336 times that required to bring helium-4 ions to the same point (i.e., E C-12/E He = 0.3336).

Hence $\qquad \dfrac{m\ He}{m\ C\text{-}12} = 0.3336; \ m\ He = 0.3336 \times 12.000 = 4.003$

But, since naturally occurring helium contains essentially 100 per cent of this isotope, the number 4.003 must also represent the atomic weight of the element helium.

The determination of the atomic weight of an element such as neon, in which more than one isotope is present, is a more complicated problem. Here it is necessary to determine not only the mass but also the relative abundance of each isotope. We can estimate the latter by measuring the relative heights of the peaks shown in Figure 2.5. Once the masses and abundances of the isotopes of an element have been determined, the element's average atomic weight is readily calculated (Example 2.3).

EXAMPLE 2.3. From mass spectra it can be determined that the element neon consists of three isotopes, whose masses on the carbon-12 scale are 19.99, 20.99, and 21.99. The abundances of these isotopes are, respectively, 90.92 per cent, 0.25 per cent, and 8.83 per cent. Calculate an accurate value for the atomic weight of neon.

SOLUTION. We can find the atomic weight of any element by adding the contributions of each of the isotopes. In the case of neon, we have:

	mass	$\times$	fraction		
neon-20	19.99	$\times$	0.9092	=	18.17
neon-21	20.99	$\times$	0.0025	=	0.05
neon-22	21.99	$\times$	0.0883	=	1.94
		atomic weight of neon =			20.16

Mass spectra, like infrared spectra, can be used to identify both elementary and compound substances. When introduced into the ionization chamber of a mass spectrometer, a compound substance is broken up into charged fragments that are characteristic of that substance. For example, the +1 ions formed from a sample of water vapor might include the species H_2O^+ (mass to charge ratio = 18) and HO^+ (mass to charge ratio = 17). At a given voltage, these species are formed in a definite ratio characteristic of the substance water.

The mass spectrometer can be used to identify rather complicated organic molecules containing many different atoms. Even though the mass spectra produced are exceedingly complex, reflecting the presence of many different ions, they can be deciphered, provided that comparison spectra are available. It is even possible, using a mass spectrometer, to identify the individual components present in a mixture of reaction products.

2.4 ABSOLUTE MASSES OF ATOMS. AVOGADRO'S NUMBER

A knowledge of the atomic weights of elements makes it possible for us to calculate the relative masses of different atoms. While such quantities are of great value in chemistry, we should like to go one step further and calculate the absolute masses of atoms. In order to do so, the number of atoms in one gram atomic weight of any element must be known. This number, known as Avogadro's number and given the symbol N, may be calculated in various ways; two different methods are illustrated by Problems 2.29 and 2.30 at the end of this chapter.

Avogadro's number, to four significant figures, has been found to be 6.023×10^{23} (i.e., 602,300,000,000,000,000,000,000). This is a number so large as to defy comprehension. Some idea of its magnitude may be gained when one realizes that if the entire population of the world were to be assigned to counting the number of atoms in one gram atomic weight of an element, each person counting one atom per second and working a 48-hour week, the task would require more than three billion years. From another point of view, the fact that Avogadro's number is so huge means that the atom itself is almost inconceivably small. Individual atoms are far too small to be seen with the most powerful microscope or to be weighed on the most sensitive analytical balance (Example 2.4).

In the head of a common pin there are more atoms than there are people on the earth.

EXAMPLE 2.4.

    a. Calculate the mass of a hydrogen atom.

    b. Calculate the number of magnesium atoms in a sample weighing 1.0×10^{-6} g. (This is the smallest sample of magnesium that can be weighed on the most sensitive analytical balance.)

SOLUTION.

    a. Knowing that one gram atomic weight of hydrogen weighs 1.008 g and contains 6.023×10^{23} atoms, we have

$$1.008 \text{ g H} = 6.023 \times 10^{23} \text{ atoms H}$$

$$\text{mass of 1 atom H} = 1 \text{ atom H} \times \frac{1.008 \text{ g H}}{6.023 \times 10^{23} \text{ atoms}}$$

$$= 1.674 \times 10^{-24} \text{ g}$$

    b. 1 GAW Mg = 24.31 g Mg = 6.023×10^{23} atoms Mg

$$\text{number of atoms Mg} = 1.0 \times 10^{-6} \text{ g} \times \frac{6.023 \times 10^{23} \text{ atoms}}{24.31 \text{ g}}$$

$$= 2.5 \times 10^{16} \text{ atoms } (25,000,000,000,000,000)$$

2.5 MASSES OF MOLECULES. MOLECULAR WEIGHT. GRAM MOLECULAR WEIGHT

In dealing with substances such as carbon dioxide, oxygen, or ethylene, in which the fundamental building block is the molecule, it is convenient to consider a number known as the molecular weight, which is defined in a manner analogous to atomic weight. The **molecular weight** is a number that tells us how heavy a molecule of a substance is compared to an atom of the most common isotope of carbon, carbon-12. When we learn, for example, that the molecular weight of carbon dioxide is 44, we deduce that a carbon dioxide molecule is $\frac{44}{12}$ as heavy as a carbon-12 atom.

Recall the analogous definition of atomic weight (Section 2.3).

Knowing the composition of a molecule, we can readily calculate the molecular weight of the corresponding substance. For example, once it is established that the oxygen molecule is diatomic, it follows that the molecular weight of oxygen must be exactly twice its atomic weight.

$$2 \times 16.00 = 32.00$$

Again, knowing that the ethylene molecule contains two carbon atoms and four hydrogen atoms, we can calculate the molecular weight of ethylene to be 28.05.

$$
\begin{aligned}
2 \times \text{AW of C} &= 2 \times 12.01 = 24.02 \\
4 \times \text{AW of H} &= 4 \times 1.008 = \underline{4.03} \\
& 28.05
\end{aligned}
$$

While this is a very convenient method of calculating molecular weights on paper, it requires that we know the atomic composition of the molecule. Often it is the other way around, and molecular compositions are ordinarily deduced from measured molecular weights. Two different experimental approaches to the determination of molecular weights will be discussed later (Chapters 5 and 11).

In dealing with molecular substances, we find it convenient to define a quantity known as the **gram molecular weight**, which bears the same relation to molecular weight as does gram atomic weight to atomic weight. We have seen that the gram atomic weight of an element is the weight in grams which contains Avogadro's number of atoms; the gram molecular weight of an elementary or compound substance is the weight in grams which contains Avogadro's number (6.023×10^{23}) of molecules. Thus 16.00 g of oxygen or 1.008 g of hydrogen contains 6.023×10^{23} *atoms*; 32.00 g of oxygen (2 atoms of oxygen per molecule) or 18.02 g of water (2 atoms of hydrogen, 1 atom of oxygen per molecule) contains 6.023×10^{23} *molecules*.

The gram molecular weight can be obtained directly from the molecular weight, just as the gram atomic weight is obtained from the atomic weight.

Element	AW	GAW	Molecular Substance	MW	GMW
Oxygen	16.00	16.00 g	Oxygen	32.00	32.00 g
Chlorine	35.45	35.45 g	Water	18.02	18.02 g
Uranium	238.0	238.0 g	Ethylene	28.05	28.05 g

Knowing the gram molecular weight of a substance, we can readily calculate the weight of an individual molecule (Example 2.5).

EXAMPLE 2.5.   Calculate the mass in grams of a water molecule.

SOLUTION.   We know that the gram molecular weight of water is 18.02 g; one gram molecular weight of any substance contains 6.023×10^{23} molecules. Therefore

$$18.02 \text{ g water} = 6.023 \times 10^{23} \text{ molecules water}$$

$$\text{mass} = 1 \text{ molecule} \times \frac{18.02 \text{ g}}{6.023 \times 10^{23} \text{ molecules}} = 2.992 \times 10^{-23} \text{ g}$$

Note the similarity to Example 2.4.

PROBLEMS

2.1   Differentiate clearly between
   a. An atom and a molecule.
   b. Gram equivalent weight and gram atomic weight.
   c. Atomic weight and gram atomic weight.
   d. The Law of Constant Composition and the Law of Multiple Proportions.
   e. An ionic and a molecular compound.

2.2   Using a table of atomic weights, give the total number of electrons in a
   a. Na atom
   b. F^- ion
   c. Carbon dioxide molecule
   d. Ca^{2+} ion

2.3   The element chromium forms three different oxides in which the percentages of chromium are 52.0, 68.4, and 76.5 respectively. Show how these data illustrate the Law of Multiple Proportions.

2.4   Consider the structure of sodium chloride shown in Figure 2.2, keeping in mind that this structure is extended indefinitely in three dimensions. How many Na^+ ions must there be around each Cl^- ion? In practice, the Na^+ and Cl^- ions are touching each other. Draw a two-dimensional section of the sodium chloride structure, showing the oppositely charged ions in contact.

2.5   When 1.260 g of a certain metal oxide is reduced by hydrogen, 1.019 g of metal is formed. Calculate the gram equivalent weight of the metal.

2.6   One gram equivalent weight of a metal combines not only with 8.00 g of oxygen but also with 16.0 g of sulfur.
   a. What mass of nickel sulfide is formed when one gram of nickel (GEW = 29.4 g) reacts with sulfur?
   b. What are the percentages by weight of nickel and sulfur in nickel sulfide?

2.7   Estimate the specific heat of manganese (AW = 55).

2.8   The specific heat of a certain metal is 0.079 cal/g°C. Its gram equivalent weight is 23.24 g. Calculate, to four significant figures, the gram atomic weight of this element.

2.9   In measuring the atomic weight of a certain isotope, it is found that, with a given magnetic field strength, the voltage required to bring a +1 ion to the collector is 502 volts; with a +1 ion derived from carbon-12, the corresponding voltage is 795 volts. Calculate the atomic weight of the isotope, using Equation 2.4.

2.10 The element chlorine consists of two isotopes of masses 34.97 and 36.97. The lighter isotope comprises 76.0 per cent of natural chlorine. Calculate the aver-

39

age atomic weight of chlorine. Sketch a mass spectrum of chlorine analogous to Figure 2.5 (consider only +1 ions).

2.11 Suppose the atomic weight of oxygen had been taken as 8.00 instead of 16.00.
a. What would be the atomic weight of nickel?
b. How heavy would a sulfur atom be relative to an oxygen atom?
c. What would be the numerical value of Avogadro's number?

2.12 Calculate the mass of the following in grams.
a. 1.24 gram atomic weights of iron.
b. 0.683 gram molecular weights of phosphorus trichloride (one phosphorus, three chlorine atoms per molecule).
c. One atom of iron.
d. One molecule of phosphorus trichloride.

2.13 Calculate the number of atoms in
a. Twelve molecules of carbon dioxide.
b. 1.62 gram atomic weights of sulfur.
c. 1.62 gram atomic weights of iron.
d. 1.62 grams of sulfur.
e. 1.62 grams of iron.

2.14 Describe briefly how each of the following quantities might be measured.
a. The gram equivalent weight of copper.
b. The atomic weight of calcium, using Equation 2.2.
c. The atomic weight of chlorine, using the mass spectrometer (cf. Problem 2.10).

2.15 The compound lead dioxide, which contains 13.4 per cent oxygen by weight, loses oxygen when heated to form compounds containing 9.34 and 7.17 per cent oxygen by weight.
a. Explain how data of this sort might have led Berthollet to dispute the Law of Constant Composition.
b. Show that these data are consistent with the Law of Multiple Proportions.

2.16 Consider the structure of calcium fluoride shown in Figure 2.2. Demonstrate, to your own satisfaction, that if this structure is repeated indefinitely in three dimensions there will be one calcium ion for every two fluoride ions.

2.17 When 2.24 g of a certain metal oxide is reduced with hydrogen, 1.76 g of metal is formed. What is the gram equivalent weight of the metal?

2.18 When a sample of a certain metal weighing 1.600 g is heated with sulfur, 1.960 g of metal sulfide is formed. Calculate the gram equivalent weight of the metal (cf. Problem 2.6).

2.19 The specific heat of a certain metal is 0.086 cal/g°C; its gram equivalent weight is 23.24 g. Calculate, to four significant figures, the gram atomic weight of the metal.

2.20 Draw a graph of specific heat vs. atomic weight for metals.

2.21 Draw a graph of accelerating voltage vs. atomic weight of ions (Equation 2.4).

2.22 The element magnesium consists of three isotopes of masses 24.0, 25.0, and 26.0. The percentages of these isotopes are 78.6, 10.1, and 11.3 respectively. Calculate the atomic weight of magnesium.

2.23 How many atoms of calcium are required to balance 20 atoms of oxygen? 15 atoms of argon? 2×10^4 atoms of neon?

2.24 Calculate
a. The number of grams in 4.12 gram atomic weights of phosphorus.
b. The number of gram atomic weights in 209 g of palladium.
c. The mass of the number of potassium atoms which is just equal to the number of atoms in 8.00 g of oxygen.

2.25 Calculate
   a. The mass in grams, of a carbon-12 atom.
   b. The number of atoms of copper in a piece of copper wire 1.00 mm in diameter and 1.0 m long (density Cu = 8.9 g/cc).
   c. The mass in grams of 2.14×10^{12} molecules of carbon dioxide.

°2.26 The chloride of a certain metal contains 54.7 per cent of chlorine by weight. If the atom ratio of metal to chlorine is 1:1, what is the atomic weight of the metal? Suppose the atom ratio is 1:2? 1:3? Using a table of atomic weights, suggest what the metal might be.

°2.27 When a sample of a certain metal weighing 12.0 g is heated to 50.0°C and is added to 10.0 g of water at 25.0°C, the final temperature is found to be 27.8°C. Analysis of an oxide of this metal shows that it contains 30.1 per cent oxygen. Calculate the gram atomic weight of the metal.

°2.28 Cannizzaro, in 1858, suggested a method of deducing the atomic weight of an element such as hydrogen, based on the following reasoning:
   It is known from Avogadro's Law (Chapter 5) that in one liter of any gas at 0°C and 1 atm, there is a certain definite number of molecules x. In at least one of the gaseous compounds of hydrogen, there should be one atom of hydrogen per molecule. Consequently, by studying the weights of hydrogen in one liter of several of its gaseous compounds, it should be possible to deduce the weight of x hydrogen atoms. A similar study applied to a series of gaseous compounds of oxygen should yield the weight of x oxygen atoms.
   Applying Cannizzaro's reasoning to the data given below, obtain the atomic weight of hydrogen.

Hydrogen Compounds	Wt 1 liter (0°C, 1 atm)	%H	Oxygen Compounds	Wt 1 liter (0°C, 1 atm)	%O
Hydrogen bromide	3.636 g	1.246	Nitrous oxide	1.975 g	36.35
Acetylene	1.170 g	7.743	Oxygen difluoride	2.422 g	29.63
Ammonia	0.765 g	17.76	Carbon dioxide	1.975 g	72.71
Hydrogen sulfide	1.532 g	5.915	Nitrogen dioxide	2.071 g	69.55

°2.29 When a direct electric current is passed through a water solution containing Cu^{2+} ions, the following reaction occurs:

$$Cu^{2+} + 2\ e^- \longrightarrow Cu$$

A total of 1.93×10^5 coulombs is required to produce one gram atomic weight of copper by this reaction. Given that an electron carries a charge of 1.60×10^{-19} coulomb, calculate the number of atoms in one gram atomic weight of copper (Avogadro's number).

°2.30 By a technique known as x-ray diffraction (Chapter 10), it is possible to determine the geometric pattern in which atoms are arranged in a crystal and the distances between atoms. In this way, it has been shown that in a crystal of sodium, two atoms effectively occupy the volume of a cube 4.29 Å on an edge. Knowing that the density of sodium is 0.970 g/cc, calculate the number of atoms in one gram atomic weight of sodium (Avogadro's number).

3

Chemical Formulas
and Equations

Chemists often summarize many of the essential features of a reaction which they have carried out in the laboratory by means of a balanced equation which expresses, among other things, the relationship between the amounts of reactants and products. Before we can write a chemical equation, we must know the formulas of reactants and products. In this chapter we shall consider two different types of chemical formulas – the simplest or empirical formula, and the molecular formula.

1. The **simplest** or **empirical** formula of a compound substance gives the simplest whole number ratio between the numbers of atoms of the different elements making up the compound. An example is the simplest formula of water, H_2O, which tells us that in this compound there are twice as many hydrogen atoms as oxygen atoms. In potassium chlorate, it has been established by experiment that the three elements potassium, chlorine, and oxygen are present in an atomic ratio of 1:1:3; the simplest formula of potassium chlorate is $KClO_3$.

Regardless of how many elements a compound contains or how complicated its structure may be, its simplest formula can be obtained by subjecting it to an experimental technique known as **quantitative analysis**, in which the proportions by weight of the constituent elements are determined. Whenever a new compound is prepared, it is analyzed and, on that basis, is assigned a simplest formula. Even if the compound has been prepared previously, it is often analyzed to determine its purity. The importance of analysis as a research tool may be inferred from the observation that in one recent (January 17, 1968) issue of the Journal of the American Chemical Society, elementary analyses were reported for some 111 different compounds.

2. The **molecular** formula indicates the number of atoms in a molecule of a molecular substance. The formula O_2 tells us that the elementary substance oxygen consists of diatomic molecules. The molecular formula of water, like its empirical formula, is H_2O; there are two atoms of hydrogen and one atom of oxygen per molecule of water. But the composition of the hydrogen peroxide molecule is indicated by its true molecular formula H_2O_2; the empirical formula of this compound would be HO.

3.1 SIMPLEST (EMPIRICAL) FORMULAS FROM ANALYSIS

In order to determine experimentally the simplest formula of a compound, we must establish by chemical analysis the proportions by weight of the elements making up the compound. Combining this information with the known atomic weights of the elements, we can readily calculate the simplest formula. The reasoning involved is indicated in Example 3.1.

EXAMPLE 3.1. In the course of his research a graduate student isolates a compound, purifies it, and sends it out for analysis. He receives from the analyst the following information:

$$\%C = 50.7; \ \%H = 4.25; \ \%O = 45.1$$

from these data, calculate the simplest formula of the compound.

SOLUTION. We wish to determine the relative numbers of atoms of C, H, and O in the compound. We can do this by calculating the relative number of gram atomic weights of C, H, and O in a given weight of the compound, for example, 100 g.

In 100 g of the compound, we have 50.7 g of C, 4.25 g of H, and 45.1 g of O. The numbers of gram atomic weights of C, H and O in 100 g of the compound are:

$$\text{no. GAW of C} = 50.7 \text{ g C} \times \frac{1 \text{ GAW C}}{12.0 \text{ g C}} = 4.23 \text{ GAW C}$$

$$\text{no. GAW of H} = 4.25 \text{ g H} \times \frac{1 \text{ GAW H}}{1.01 \text{ g H}} = 4.21 \text{ GAW H}$$

$$\text{no. GAW of O} = 45.1 \text{ g O} \times \frac{1 \text{ GAW O}}{16.0 \text{ g O}} = 2.82 \text{ GAW O}$$

The quantities we have just calculated represent the relative numbers of gram atomic weights of C, H, and O in any given weight of the compound. They also represent relative numbers of atoms of the three elements in any sample of the compound, *since the conversion factor relating the number of gram atomic weights to the number of atoms* (1 GAW = 6.02×10^{23} atoms) *is the same for all elements*. In other words, the atom ratio of C:H:O in this compound is 4.23 : 4.21 : 2.82.

To deduce the simplest formula, we need to know the simplest, whole number ratio between these three numbers. Perhaps the simplest way to obtain this is to divide each number by the smallest, 2.82:

$$\frac{4.23}{2.82} = 1.50; \ \frac{4.21}{2.82} = 1.50; \ \frac{2.82}{2.82} = 1.00$$

We deduce that for every O atom, there are 1.5 C atoms and 1.5 H atoms. Multiplying by two to obtain the simplest whole number ratio, we have

$$3\,C : 3\,H : 2\,O; \text{ simplest formula} = C_3H_3O_2$$

Make sure you understand why the simplest formula can be found in this way. Students often memorize the procedure for determining formulas without referring to the principle involved.

If we know the percentages by weight of the constituent elements, the calculation of the simplest formula of a compound requires little or no knowledge of chemistry. It is important, however, that a chemist be familiar with the experimental methods of analysis which are used to establish these percentages. Only then can he judge how much confidence to place in them. In this, an elementary course, we cannot hope to become familiar with more than a small fraction of the methods of elemental analysis that are employed by analytical chemists. It may, however, be instructive to consider some of the types of experiments involved in establishing empirical formulas of compounds.

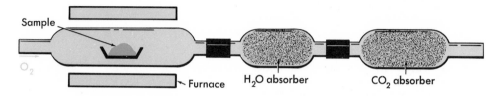

Figure 3.1 Schematic diagram of an absorption train for carbon-hydrogen analysis.

The percentages of carbon and hydrogen in an organic compound can be determined with the aid of the apparatus shown in Figure 3.1. A sample, which may weigh as little as a few milligrams, is burned in oxygen to form carbon dioxide and water. The amounts of CO_2 and H_2O produced are determined by measuring the increase in weight of the two absorption tubes. From these data, it is possible to calculate the percentages of carbon and hydrogen (Example 3.2). If only one other element (e.g., oxygen) is present, its percentage can be calculated by difference.

EXAMPLE 3.2. A sample of the compound referred to in Example 3.1, weighing 2.00 mg, is burned to form 3.72 mg of CO_2 and 0.753 mg of H_2O. Calculate the percentages of C, H, and O in this compound.

SOLUTION. Let us first calculate the weight of carbon in 3.72 mg of CO_2. Since one gram molecular weight of CO_2, 44.0 g, contains one gram atomic weight of carbon, 12.0 g, we deduce that 12.0/44.0 of the carbon dioxide is carbon

$$\text{no. mg C} = 3.72 \text{ mg} \times \frac{12.0}{44.0} = 1.01 \text{ mg C}$$

In a similar manner, since the fraction of hydrogen in water is 2.02/18.0:

$$\text{no. mg H} = 0.753 \text{ mg} \times \frac{2.02}{18.0} = 0.0845 \text{ mg H}$$

Noting that the sample weighed 2.00 mg, we have

$$\% \text{ C} = \frac{1.01}{2.00} \times 100 = 50.5\%; \ \% \text{ H} = \frac{0.0845}{2.00} \times 100 = 4.23\%$$

To obtain the percentage of oxygen (assuming that no other element is present), we subtract the combined percentages of carbon and hydrogen from 100:

$$\% \text{ O} = 100.0 - (50.5 + 4.2) = 45.3\%$$

Reflecting on the experimental procedure illustrated in Figure 3.1 and Example 3.2, we realize that the determination of the percentage of carbon depends upon converting a weighed sample of the unknown to a compound of known composition, CO_2. Similarly, the determination of hydrogen requires that the composition of the compound to which it is converted, H_2O, be known.

The same principle can be applied to determine the percentages by weight of the elements in inorganic compounds. For example, the percentage of chlorine in a water-soluble metal chloride can be determined by precipitating a weighed sample with silver nitrate, which forms the insoluble compound silver chloride, AgCl. The silver chloride can be filtered off and weighed; knowing the weight of the sample and that of the AgCl to which it is converted, we can calculate the percentage of chlorine. Once again the percentage of the metal can be obtained by difference if it is the only other element present (Example 3.3).

EXAMPLE 3.3. A sample of scandium chloride weighing 2.159 g is dissolved in water and treated with Ag^+ to precipitate quantitatively the Cl^- ions as AgCl. The precipitated silver chloride weighs 6.134 g. Calculate the percentage composition and the simplest formula of scandium chloride.

SOLUTION. Let us first calculate the weight of chlorine in the precipitated silver chloride. Since one gram formula weight of AgCl, 143.32 g, contains one gram atomic weight of Cl, 35.45 g, it follows that 35.45/143.32 of the silver chloride is chlorine. Hence,

$$\text{weight Cl in 6.134 g AgCl} = \frac{35.45}{143.32} \times 6.134 \text{ g} = 1.518 \text{ g}$$

This is the weight of chlorine not only in the precipitated silver chloride but also in the original sample of scandium chloride. The percentage composition of scandium chloride must then be:

$$\% \text{ Cl} = \frac{\text{wt. Cl}}{\text{wt. sample}} \times 100 = \frac{1.518 \text{ g}}{2.159 \text{ g}} \times 100 = 70.31\%$$

$$\% \text{ Sc} = 100.00 - 70.31 = 29.69\%$$

To calculate the simplest formula of scandium chloride

$$29.69 \text{ g Sc} \simeq 70.31 \text{ g Cl}$$

$$\text{no. of GAW Sc in 29.69 g} = 29.69 \text{ g Sc} \times \frac{1 \text{ GAW Sc}}{44.96 \text{ g Sc}} = 0.660$$

$$\text{no. of GAW Cl in 70.31 g} = 70.31 \text{ g Cl} \times \frac{1 \text{ GAW Cl}}{35.45 \text{ g Cl}} = 1.98$$

Hence $0.660 \text{ GAW Sc} \simeq 1.98 \text{ GAW Cl}$
$1 \text{ GAW Sc} \simeq 3 \text{ GAW Cl}$ Simplest formula $= ScCl_3$

We should point out that it is not necessary to calculate the percentage composition of a compound in order to determine its simplest formula. So long as we know the relative weights of the elements making up the compound, we can obtain the simplest formula. In the compound scandium chloride, referred to in Example 3.3, since we know that 1.518 g of Cl is combined with (2.159 − 1.518) g of scandium, we have

$$\text{no. GAW Sc} = 0.641 \text{ g Sc} \times \frac{1 \text{ GAW Sc}}{44.96 \text{ g Sc}} = 0.0143 \text{ GAW Sc}$$

$$\text{no. GAW Cl} = 1.518 \text{ g Cl} \times \frac{1 \text{ GAW Cl}}{35.45 \text{ g Cl}} = 0.0428 \text{ GAW Cl}$$

To make sure you understand what we are doing here, calculate the simplest formula of the compound referred to in Example 3.2 *without* working through the per cent composition.

45

Hence 0.0143 GAW Sc $\simeq$ 0.0428 GAW Cl

1 GAW Sc $\simeq$ 3 GAW Cl   Simplest formula = $ScCl_3$

3.2 MOLECULAR FORMULAS

The molecular formula of a substance can be derived from the simplest formula if the molecular weight of the substance is known. We write the molecular formula O_2 for oxygen because it has been shown experimentally that oxygen has a molecular weight of 32. The molecular weight of water vapor is known to be 18; consequently, water must have the molecular formula H_2O rather than H_4O_2, H_6O_3, or some other multiple of the simplest formula.

EXAMPLE 3.4. The molecular weight of the compound described in Example 3.1 has been found to be approximately 140. Determine the molecular formula of the compound.

SOLUTION. The simplest formula of this compound has been shown to be $C_3H_3O_2$. The corresponding formula weight is:

$$3 \times 12 + 3 \times 1 + 2 \times 16 = 71$$

Clearly, the molecular weight (140) is twice the formula weight. It follows that the simplest formula must be multiplied by two to obtain the molecular formula.

$$\text{Molecular formula} = C_6H_6O_4$$

Note that an approximate value of the molecular weight is sufficient to determine the molecular formula from the simplest formula. An experiment that establishes the molecular weight of this compound to be in the range 130 to 150 would be conclusive proof that its simplest formula should be multiplied by 2 rather than 1, 3, or some other integer to obtain the molecular formula.

3.3 THE MOLE

In considering mass relations in reactions, chemists make frequent use of a quantity known as the mole. Always associated with a formula, **a mole contains Avogadro's number of formula units**. To understand what this statement means, consider the substance water, whose molecular formula is H_2O. One mole of H_2O contains Avogadro's number of molecules; it weighs 18.0 g (the gram formula weight of H_2O). In a similar manner:

1 mole of C contains 6.02×10^{23} carbon atoms and weighs 12.0 g
1 mole of CO_2 contains 6.02×10^{23} CO_2 molecules and weighs 44.0 g
1 mole of Na^+ contains 6.02×10^{23} Na^+ ions and weighs 23.0 g

The mole is frequently used in connection with an empirical formula.

Consider, for example, the ionic compound sodium chloride in which there is one Na^+ ion for every Cl^- ion. We can, if we wish, say that one mole of NaCl contains Avogadro's number of formula units of NaCl. A "formula unit" of NaCl consists of one Na^+ ion and one Cl^- ion. A mole of NaCl must then contain Avogadro's number of Na^+ ions and Avogadro's number of Cl^- ions. The mass of one mole of NaCl is readily calculated.

How many ions are there in one mole of NaCl? $CaCl_2$?

$$1 \text{ mole of NaCl weighs } 23.0 \text{ g} + 35.5 \text{ g} = 58.5 \text{ g}$$

Similarly, $\quad\quad 1 \text{ mole of } CaCl_2 \text{ weighs } 40.0 \text{ g} + 71.0 \text{ g} = 111.0 \text{ g}$

EXAMPLE 3.5.
    a. Calculate the mass in grams of 2.40 moles of K_2CrO_4.
    b. How many moles of $CaBr_2$ are there in 312 g?
    c. Calculate the number of molecules in 0.0820 mole of CO_2.

SOLUTION.
    a. One mole of K_2CrO_4 weighs $2(39.1 \text{ g}) + 52.0 \text{ g} + 4(16.0 \text{ g}) = 194.2$ g

$$\text{mass } K_2CrO_4 = 2.40 \text{ moles} \times \frac{194.2 \text{ g}}{1 \text{ mole}} = 466 \text{ g}$$

    b. Mass of one mole of $CaBr_2 = 40.0 \text{ g} + 2(79.9 \text{ g}) = 199.8$ g

$$\text{no. of moles } CaBr_2 = 312 \text{ g} \times \frac{1 \text{ mole}}{199.8 \text{ g}} = 1.56 \text{ moles}$$

    c. 6.02×10^{23} molecules $= 1$ mole

$$\text{no. of molecules} = 0.0820 \text{ mole} \times \frac{6.02 \times 10^{23} \text{ molecules}}{1 \text{ mole}} =$$
4.94×10^{22} molecules

3.4 TRANSLATION OF REACTIONS INTO EQUATIONS

The process of deducing the balanced chemical equation that corresponds to a given reaction is by no means as simple as many general chemistry students appear to believe. To illustrate the process that is involved, let us consider a specific reaction, the combustion of ethane. It has been established experimentally that ethane, a gaseous hydrocarbon of molecular formula C_2H_6, reacts with elementary oxygen of the air to give two products. One of these products, carbon dioxide, is a gas; the other product, water, is a liquid at 25°C and atmospheric pressure. To write a balanced equation for this reaction, we proceed as follows:

1. We start by writing an unbalanced equation in which the formulas of the reactants appear on the left and those of the products on the right. For molecular substances, molecular formulas are preferred to simplest formulas. For the combustion of ethane, we would write

$$C_2H_6 + O_2 \longrightarrow CO_2 + H_2O$$

47

2. We balance the equation by making it conform to the Law of Conservation of Mass, which requires that we have the same number of atoms of each element on both sides of the equation. To accomplish this, we might begin by writing a coefficient of 2 for the CO_2 and 3 for H_2O. In this way, we obtain 2 carbon and 6 hydrogen atoms on both sides.

$$C_2H_6 + O_2 \longrightarrow 2\ CO_2 + 3\ H_2O$$

To complete the balancing process, we note that there are now 7 atoms of oxygen on the right; to obtain 7 oxygen atoms on the left, we could use a coefficient of 7/2 for O_2:

$$C_2H_6 + \frac{7}{2}\ O_2 \longrightarrow 2\ CO_2 + 3\ H_2O \tag{3.1}$$

Equation 3.1 is one of an infinite number of balanced equations that could be written to represent this reaction. Three others are

$$2\ C_2H_6 + 7\ O_2 \longrightarrow 4\ CO_2 + 6\ H_2O \tag{3.2}$$
$$4\ C_2H_6 + 14\ O_2 \longrightarrow 8\ CO_2 + 12\ H_2O$$
$$5.2\ C_2H_6 + 18.2\ O_2 \longrightarrow 10.4\ CO_2 + 15.6\ H_2O$$

There is nothing sacred about Equation 3.2. For certain purposes Equation 3.1 is more convenient.

Ordinarily, we prefer to work with the balanced equation which gives the simplest, whole number coefficients for reactants and products (Equation 3.2).

3. Throughout this text, the physical states of reactants and products will be indicated in the equation. We shall use the letters (g), (l), or (s) to represent gases, liquids, and solids respectively. Thus, for Equation 3.2, we write

$$2\ C_2H_6(g) + 7\ O_2(g) \longrightarrow 4\ CO_2(g) + 6\ H_2O(l) \tag{3.3}$$

Unless stated otherwise, it will be assumed that the reactants and products are at 25°C and one atmosphere pressure. The letters (g), (l), and (s) will be written to indicate the stable state of each substance under these conditions.

3.5 INTERPRETATION OF BALANCED EQUATIONS

The coefficients of a balanced equation tell us the relative numbers of formula units of reactants and products participating in a chemical reaction. For example, the equation

$$2\ C_2H_6(g) + 7\ O_2(g) \longrightarrow 4\ CO_2(g) + 6\ H_2O(l)$$

indicates that 2 formula units (2 molecules) of C_2H_6 react with 7 formula units (7 molecules) of O_2 to form 4 formula units (4 molecules) of CO_2 and 6 formula units (6 molecules) of H_2O. Similarly, the equation

$$4\ Li(s) + O_2(g) \longrightarrow 2\ Li_2O(s)$$

can be interpreted to mean that 4 formula units (4 atoms) of Li react with 1 formula unit (1 molecule) of O_2 to yield 2 formula units of Li_2O.

We must keep in mind that the coefficients of a balanced equation represent only the simplest whole number ratio of formula units. For example, Equation 3.3 tells us that not only do 2 molecules of C_2H_6 react with 7 molecules of O_2 to give 4 molecules of CO_2 and 6 molecules of H_2O, but also that

$$20 \text{ molecules } C_2H_6 + 70 \text{ molecules } O_2 \longrightarrow 40 \text{ molecules } CO_2 +$$
$$60 \text{ molecules } H_2O$$

$$2000 \text{ molecules } C_2H_6 + 7000 \text{ molecules } O_2 \longrightarrow 4000 \text{ molecules } CO_2 +$$
$$6000 \text{ molecules } H_2O$$

or, in general, that

$$2x \text{ molecules } C_2H_6 + 7x \text{ molecules } O_2 \longrightarrow 4x \text{ molecules } CO_2 + \tag{3.4}$$
$$6x \text{ molecules } H_2O$$

where **x can be any positive number**.

A situation which is of particular interest so far as mass relations are concerned is that where x in Equation 3.4 is Avogadro's number, N:

$$2N \text{ molecules } C_2H_6 + 7N \text{ molecules } O_2 \longrightarrow 4N \text{ molecules } CO_2 +$$
$$6N \text{ molecules } H_2O$$

Recalling from Section 3.3 that one mole contains Avogadro's number of molecules (N molecules), we see that

$$2 \text{ moles } C_2H_6 + 7 \text{ moles } O_2 \longrightarrow 4 \text{ moles } CO_2 + 6 \text{ moles } H_2O \tag{3.5}$$

The relationship implied by Equation 3.5 is a general one. **The coefficients of a balanced equation represent the relative numbers of moles of reactants and products,** as well as the relative number of formula units. For example, the equation

$$4 \text{ Li}(s) + O_2(g) \longrightarrow 2 \text{ Li}_2O(s)$$

tells us that in the reaction of lithium with oxygen, 4 moles of Li combine with 1 mole of O_2 to give 2 moles of Li_2O, or that, in this reaction,

$$4 \text{ moles Li} \simeq 1 \text{ mole } O_2 \simeq 2 \text{ moles } Li_2O$$

where the symbol $\simeq$ is taken, as usual, to mean "chemically equivalent to."

We saw in the previous section that a mole of a substance has a certain weight in grams, which can be calculated from the gram atomic weights of the elements in the formula of the substance. Thus, 1 mole of Li weighs 6.94 g (1 gram atomic weight of Li); 1 mole of O_2 weighs 32.0 g (2 × GAW of O); 1 mole of Li_2O weighs 29.88 g (2 × GAW Li + GAW O). Consequently, for the reaction of lithium with oxygen, not only is

$$4 \text{ moles Li} \simeq 1 \text{ mole } O_2 \simeq 2 \text{ moles } Li_2O$$

but also $\qquad 4(6.94 \text{ g Li}) \simeq 1(32.0 \text{ g } O_2) \simeq 2(29.88 \text{ g } Li_2O)$

or $\qquad 27.76 \text{ g Li} \simeq 32.0 \text{ g } O_2 \simeq 59.76 \text{ g } Li_2O$

Similarly, for the reaction represented by Equation 3.5,

$$2 \text{ moles } C_2H_6 \simeq 7 \text{ moles } O_2 \simeq 4 \text{ moles } CO_2 \simeq 6 \text{ moles } H_2O$$

$$2(30.0 \text{ g } C_2H_6) \simeq 7(32.0 \text{ g } O_2) \simeq 4(44.0 \text{ g } CO_2) \simeq 6(18.0 \text{ g } H_2O)$$

In summary, the coefficients of a balanced equation give us directly the relative numbers of **formula units** or the relative numbers of **moles** of reactants and products. They also make it possible to calculate the relative numbers of grams of reactants or products. Stated in a slightly different way, the coeffi-

49

cients of a balanced equation give us, directly or indirectly, the **conversion factors** which allow us to relate the numbers of formula units, moles, or grams of reactants and products in a chemical reaction. The use of these conversion factors in calculations involving balanced equations is illustrated in Examples 3.6 and 3.7.

EXAMPLE 3.6. Consider the reaction

$$2\ C_2H_6(g) + 7\ O_2(g) \longrightarrow 4\ CO_2(g) + 6\ H_2O(l)$$

a. Calculate the number of molecules of O_2 required to react with 26 molecules of C_2H_6.
b. How many moles of H_2O are produced from 2.52 moles of C_2H_6?
c. How many grams of H_2O are formed from 12.0 g of C_2H_6?

SOLUTION. In each case we obtain the required conversion factor from the coefficients of the balanced equation.

a. 2 molecules $C_2H_6 \simeq 7$ molecules O_2

$$\text{no. molecules } O_2 = 26 \text{ molecules } C_2H_6 \times \frac{7 \text{ molecules } O_2}{2 \text{ molecules } C_2H_6}$$

$$= 91 \text{ molecules } O_2$$

b. 2 moles $C_2H_6 \simeq 6$ moles H_2O

$$\text{no. moles } H_2O = 2.52 \text{ moles } C_2H_6 \times \frac{6 \text{ moles } H_2O}{2 \text{ moles } C_2H_6}$$

$$= 7.56 \text{ moles } H_2O$$

c. In this case we require a conversion factor relating g of H_2O to g of C_2H_6. This can be obtained from the conversion factor used in b.

$$2 \text{ moles } C_2H_6 \simeq 6 \text{ moles } H_2O$$

$$2(30.0 \text{ g } C_2H_6) \simeq 6(18.0 \text{ g } H_2O);$$
$$60.0 \text{ g } C_2H_6 \simeq 108 \text{ g } H_2O$$

$$\text{no. g } H_2O = 12.0 \text{ g } C_2H_6 \times \frac{108 \text{ g } H_2O}{60.0 \text{ g } C_2H_6} = 21.6 \text{ g } H_2O$$

21.6 g is the maximum amount of water that can be produced from 12.0 g of ethane. The actual amount of water formed might be considerably less (Section 3.6).

EXAMPLE 3.7. Consider the reaction

$$4\ Li(s) + O_2(g) \longrightarrow 2\ Li_2O(s)$$

a. Calculate the number of molecules of O_2 required to react with 1.26 moles of Li.
b. How many grams of Li_2O can be formed from 1.26 moles of Li?

SOLUTION.

a. From the coefficients of the balanced equation, we can relate moles of Li to moles of O_2. Knowing that 1 mole of O_2 contains 6.02×10^{23} molecules of O_2, we can then calculate the number of molecules of O_2.

$$\text{no. moles } O_2 = 1.26 \text{ moles } Li \times \frac{1 \text{ mole } O_2}{4 \text{ moles } Li} = 0.315 \text{ mole } O_2$$

$$\text{no. molecules } O_2 = 0.315 \text{ mole} \times \frac{6.02 \times 10^{23} \text{ molecules}}{1 \text{ mole}}$$
$$= 1.90 \times 10^{23} \text{ molecules } O_2$$

It was not really necessary to solve for the number of moles of O_2. The arithmetic could have been set up in a single expression involving two conversion factors:

$$\text{no. molecules } O_2 = 1.26 \text{ moles Li} \times \frac{1 \text{ mole } O_2}{4 \text{ moles Li}} \times$$
$$\frac{6.02 \times 10^{23} \text{ molecules } O_2}{1 \text{ mole } O_2} = 1.90 \times 10^{23} \text{ molecules } O_2$$

b. As in part a, we can proceed in two steps.
Calculate the no. of moles of Li_2O; 4 moles Li $\eqsim$ 2 moles Li_2O
Calculate the no. of grams of Li_2O; 1 mole $Li_2O = 29.88$ g Li_2O

$$\text{no. g } Li_2O = 1.26 \text{ moles Li} \times \frac{2 \text{ moles } Li_2O}{4 \text{ moles Li}} \times \frac{29.88 \text{ g } Li_2O}{1 \text{ mole } Li_2O}$$
$$= 18.8 \text{ g } Li_2O$$

3.6 THEORETICAL YIELDS. PERCENTAGE YIELDS

In carrying out a reaction in the laboratory, a chemist seldom uses exactly equivalent quantities of reactants. Instead, he usually works with an excess of one reactant, hoping in this way to quantitatively convert the other reactant, which may be more expensive or more difficult to obtain, to products. Consider, for example, the reaction of benzene with nitric acid:

$$C_6H_6(l) + HNO_3(l) \longrightarrow C_6H_5NO_2(l) + H_2O(l) \qquad (3.6)$$

Let us suppose that we wish to form one mole of nitrobenzene, $C_6H_5NO_2$, from one mole of benzene, C_6H_6. In principle, one mole of nitric acid could be used to accomplish this purpose. In practice, if we wish to convert as much of the benzene as possible to nitrobenzene and make the reaction go as fast as possible, it is advisable to use a considerable excess of HNO_3.

In calculating the **theoretical yield**, i.e., the maximum amount of product that can be produced in the reaction, we must be sure to base the calculation on the "critical" reagent, the one that is not in excess. If, for example, a student were to allow one mole of benzene to react with five moles of nitric acid, he could not hope to obtain more than one mole of nitrobenzene, for he would run out of benzene to be nitrated. The theoretical yield of nitrobenzene, using these quantities of reagents, would be one mole (123 g).

The **actual yield** of product in a reaction is invariably less than the theoretical yield. For example, if we react one mole of benzene with excess nitric acid, the actual yield of nitrobenzene will not be one mole. Instead, we may obtain 0.90, 0.80, or even as little as 0.50 mole of product. There are many reasons for this. In the first place, the reaction may not go to completion. Frequently, an equilibrium is set up whose position is such that significant amounts of starting material remain unreacted (Chapter 12). Another factor which reduces the yield of product below that to be expected theoretically is the possibility of side reactions. For example, in the reaction of benzene

51

with nitric acid, small quantities of dinitrobenzene, $C_6H_4(NO_2)_2$, are likely to be formed, thereby reducing the yield of the desired product, $C_6H_5NO_2$. Finally, even if the desired product is obtained in nearly the theoretical yield, part of it is likely to be lost in the separation and purification processes that follow the preparation.

Clearly, it is the goal of a chemist carrying out a reaction to make the **percentage yield**, defined as

$$\% \text{ yield} = \frac{\text{actual yield}}{\text{theoretical yield}} \times 100$$

as large as possible. To do this, the chemist tries to choose reaction conditions such that the desired reaction will occur to as great an extent as possible without significant interference from side reactions. He can often do this by changing the temperature at which the reaction is carried out or the time during which it is allowed to proceed. Once the reaction is over, the chemist attempts to isolate as much pure product as possible; to do this requires a knowledge of chemical principles and the practice of good experimental techniques.

EXAMPLE 3.8. A student attempting to prepare bromobenzene, C_6H_5Br, by the reaction of benzene, C_6H_6, with bromine

$$C_6H_6(l) + Br_2(l) \longrightarrow C_6H_5Br(l) + HBr(g)$$

is instructed to weigh out 20.0 g of benzene and 50.0 g of bromine. What is the theoretical yield of bromobenzene (i.e., the maximum amount of C_6H_5Br that could be formed) in this experiment? If the student actually obtains 28.0 g of bromobenzene, what is the per cent yield?

SOLUTION. In order to calculate the theoretical yield, we must first decide which reagent to base our calculations on. From the coefficients of the equation, it is clear that one mole of bromine is required for every mole of benzene. In this experiment, there are available

$$\frac{20.0}{78.1} \text{ moles} = 0.256 \text{ mole } C_6H_6 \qquad \frac{50.0}{159.8} \text{ moles} = 0.313 \text{ mole } Br_2$$

$$(1 \text{ mole } C_6H_6 = 78.1 \text{ g}, 1 \text{ mole } Br_2 = 159.8 \text{ g})$$

Clearly, the bromine is in excess; the theoretical yield must then be calculated by determining how much bromobenzene can be produced from the amount of benzene available.

Referring again to the equation, it is clear that one mole of C_6H_6 could yield one mole of C_6H_5Br. We have calculated that 0.256 mole of C_6H_6 is available; it follows that the theoretical yield of C_6H_5Br must be 0.256 mole. Or, since one mole of C_6H_5Br weighs 157.0 g,

$$\text{Theoretical yield} = 0.256 \text{ mole } C_6H_5Br \times \frac{157.0 \text{ g } C_6H_5Br}{1 \text{ mole } C_6H_5Br} = 40.2 \text{ g } C_6H_5Br$$

$$\% \text{ yield} = \frac{\text{actual yield}}{\text{theoretical yield}} \times 100 = \frac{28.0 \text{ g}}{40.2 \text{ g}} \times 100 = 69.7\%$$

(The remainder of the benzene is consumed in side reactions, producing dibromo- and tribromobenzenes, $C_6H_4Br_2$ and $C_6H_3Br_3$.)

Instead, we could calculate two different theoretical yields, one based on 20 g of C_0H_6, the other on 50 g of Br_2. The smaller calculated yield would, of course, be the correct answer.

PROBLEMS

3.1 Find the percentages by weight of the elements in
a. H_2SO_4 b. $Ca(ClO_3)_2$ c. $C_7H_5N_3O_6$

3.2 Calculate the simplest formulas of compounds with the following compositions.
a. 15.6% Ca, 40.6% Cr, 43.8% O
b. 17.7% N, 6.4% H, 15.2% C, 60.7% O
c. 22.0% Co, 31.4% N, 6.8% H, 39.8% Cl

3.3 An organic compound weighing 4.00 mg is burned to form 9.60 mg of CO_2 and 1.96 mg of H_2O. The sample is known to contain only three elements: C, H, and O. What is its simplest formula? What additional information would you need to establish its molecular formula?

3.4 On reaction with silver nitrate a sample of indium chloride weighing 0.500 g forms 0.972 g of AgCl. Determine the simplest formula of indium chloride.

3.5 A sample of barium chlorate (Ba, Cl, O) weighing 0.850 g is heated in air to form 0.582 g of barium chloride, which is then converted to 0.800 g of AgCl. Determine the simplest formula of barium chlorate.

3.6 Calculate
a. The mass in grams of 0.623 moles of $HClO_4$.
b. The number of moles in 24.5 g of methyl alcohol, CH_3OH.
c. The number of molecules in 1.67 moles of trinitrotoluene.

3.7 Write balanced equations for the reactions that occur when
a. A sample of ethylene, a gaseous hydrocarbon, is burned in air (cf. Chapter 2).
b. An oxide of iron of empirical formula Fe_3O_4 is reduced to the metal by hydrogen.

3.8 Consider the reaction

$$4\ NH_3(g) + 5\ O_2(g) \longrightarrow 4\ NO(g) + 6\ H_2O(l)$$

a. How many moles of NO can be formed from 1.48 moles of O_2?
b. Calculate the number of grams of H_2O formed from 6.16 moles of NH_3.
c. How many grams of O_2 will be required to react with 40.0 g of NH_3?
d. Calculate the number of moles of NO formed from 12.6 g of NH_3.

3.9 Carbon dioxide can be prepared by allowing concentrated sulfuric acid to drop on sodium hydrogen carbonate:

$$2\ NaHCO_3(s) + H_2SO_4(l) \longrightarrow 2\ CO_2(g) + Na_2SO_4(s) + 2\ H_2O(l)$$

a. If the $NaHCO_3$ used in this preparation is 94 per cent pure, how many grams of it will be required to form 10.0 g of CO_2?
b. If the sulfuric acid is used in the form of a water solution which is 52 per cent H_2SO_4 by weight and has a density of 1.26 g/ml, what volume of this solution must be used to produce 10.0 g of carbon dioxide?

3.10 Nitrobenzene, $C_6H_5NO_2$, can be prepared from benzene by the following reaction:

$$C_6H_6(l) + HNO_3(l) \longrightarrow C_6H_5NO_2(l) + H_2O(l)$$

a. If we start with 10.0 g of benzene and 15.0 g of nitric acid, what is the theoretical yield of nitrobenzene?
b. If 8.15 g of nitrobenzene are actually formed in this reaction, what is the percentage yield?

What weight of metal could be formed by decomposing
a. 12.0 g of $SrCO_3$?   b. 1.82 g of $Ni(NO_3)_2$?   c. 5.00 g of $[Co(NH_3)_5Cl]Cl_2$?

3.12 Calculate the simplest formulas of compounds that have the following compositions.
a. 19.3% Na, 26.8% S, 53.9% O   c. 19.2% P, 2.5% H, 78.3% I

b. 79.3% Tl, 9.9% V, 10.8% O   d. 14.2% Ni, 61.3% I, 20.2% N, 4.3% H

3.13 An impure sample of $CuCl_2$ weighing 1.600 g is dissolved in water and electrolyzed to give 0.345 g of copper. Assuming that the impurity contains no copper, calculate the percentage of $CuCl_2$ in the sample.

3.14 A sample of gallium bromide weighing 1.000 g is dissolved in water and treated with excess silver nitrate solution. A total of 1.820 g of silver bromide, AgBr, is formed. What is the simplest formula of gallium bromide?

3.15 An organic compound containing carbon, hydrogen, and chlorine is found to have a molecular weight of about 150. When 0.593 g of this compound is burned in an absorption train, 1.063 g of carbon dioxide and 0.145 g of water are formed. Find the simplest and molecular formulas of the compound.

3.16 When a certain compound containing cobalt, carbon, and oxygen is heated in air, it decomposes to form an oxide of cobalt. A sample of the compound weighing 2.500 g, when treated in this way, gives off 0.928 g of carbon dioxide and leaves an oxide residue weighing 1.686 g. Analysis shows the oxide to contain 73.4 per cent cobalt by weight. Determine the simplest formula of the original compound and the oxide.

3.17 Calculate
a. The mass in grams of 1.28 moles of $Ca(ClO_3)_2$.
b. The number of moles in 169 g of sugar, $C_{12}H_{22}O_{11}$.
c. The total number of ions in 12.6 g of $CaCl_2$.

3.18 Write a balanced equation for the combustion of the liquid hydrocarbon octane, which has the molecular formula C_8H_{18}.

3.19 Consider the combustion of propyl alcohol:

$$2\ C_3H_8O(l) + 9\ O_2(g) \longrightarrow 6\ CO_2(g) + 8\ H_2O(g)$$

a. How many moles of CO_2 can be formed from 3.68 moles of propyl alcohol?
b. How many moles of O_2 are required to react with 0.931 mole of propyl alcohol?
c. How many molecules of H_2O can be formed from 1.58×10^{-5} moles of O_2?
d. What is the maximum number of moles of CO_2 that can be formed from a mixture of 1.81 moles of propyl alcohol and 7.43 moles of O_2?

3.20 Consider the reaction

$$3\ NO_2(g) + H_2O(l) \longrightarrow 2\ HNO_3(l) + NO(g)$$

Calculate
a. The number of grams of HNO_3 that can be formed from 12.0 g of NO_2.
b. The number of grams of H_2O required to react with 34.6 g of NO_2.
c. The number of moles of NO that can be formed from 1.64 g of NO_2.
d. The number of grams of HNO_3 that can be formed from 2.94 moles of NO_2.

3.21 Ammonia can be produced by heating ammonium chloride with calcium oxide.

$$2\ NH_4Cl(s) + CaO(s) \longrightarrow 2\ NH_3(g) + CaCl_2(s) + H_2O(g)$$

a. How many grams of NH_4Cl are needed to produce 0.182 mole of NH_3?
b. If it is desired to use a 50 per cent excess of calcium oxide, how many grams of CaO should be used to produce 0.182 mole of NH_3?

3.22 A student in the organic chemistry laboratory prepares ethyl bromide, C_2H_5Br, by reacting ethyl alcohol with phosphorus tribromide:

$$3\ C_2H_5OH(l) + PBr_3(l) \longrightarrow 3\ C_2H_5Br(l) + H_3PO_3(s)$$

He is told to react 24.0 g of ethyl alcohol with 39.0 g of phosphorus tribromide.
a. What is the theoretical yield of ethyl bromide?
b. If the student actually obtains 36.0 g of C_2H_5Br, what is the percentage yield?

3.23 A student in the inorganic chemistry laboratory wishes to prepare 15 g of the compound $[Co(NH_3)_5SCN]Cl_2$, starting with $[Co(NH_3)_5Cl]Cl_2$:

$$[Co(NH_3)_5Cl]Cl_2(s) + KSCN(s) \longrightarrow [Co(NH_3)_5SCN]Cl_2(s) + KCl(s)$$

He is instructed to use a 75 per cent excess of potassium thiocyanate, KSCN, and is told that he can expect to get a 63 per cent yield in the reaction. How many grams of the two starting materials should he use?

°3.24 A mixture of SnO and SnO_2 weighing 1.000 g is heated with hydrogen; the tin formed weighs 0.850 g. Calculate the percentage of SnO in the mixture.

°3.25 A research chemist prepares a compound which he believes to be $[Co(NH_3)_5NO_2]Cl_2\cdot H_2O$. He sends it out for analysis and receives the following results:

	Sample 1	Sample 2
% N	30.12	30.22
% Co	21.2	21.2
% Cl	25.52	25.59

a. In your opinion, are these results consistent with the above formula?
b. Shortly before the analysis is received, the chemist performs a test which makes him suspect that the compound may actually be $[Co(NH_3)_5NO_3]Cl_2$. On the basis of the analytical results, can you distinguish between these two possibilities?
c. Suggest a way that one might be able to distinguish clearly between these two possibilities.

°3.26 A student determines the formula of copper sulfide by heating a piece of copper wire weighing 0.536 g with 0.438 g of sulfur. The excess sulfur burns off as SO_2. What weight of SO_2 is formed
a. If the formula of the sulfide formed is Cu_2S?
b. If the formula of the sulfide formed is CuS?

°3.27 A student is asked to prepare 0.250 mole of a pure compound C by the reaction sequence

$$2\ A \longrightarrow B \quad \text{and} \quad 3\ B \longrightarrow 2\ C$$

in which A and B represent two other compounds. He is told to expect a 71 per cent yield in the first step of this sequence and an 82 per cent yield in the second step. He is also required to purify compound C by recrystallizing it from hot water; it is estimated that 22 per cent of it will be lost in the recrystallization. How many moles of A should he start with?

4

Energy Changes in Chemical Reactions

During the course of chemical reactions we ordinarily observe effects which imply that energy changes are occurring simultaneously with the chemical changes. If a mixture of hydrogen and oxygen is ignited, the reaction produces water so rapidly and so violently as to also produce a flame and sometimes an explosion. When magnesium metal burns in air, forming magnesium oxide, a very bright light is emitted as well as a great deal of heat. When concentrated aqueous solutions of barium chloride and sodium sulfate are mixed, barium sulfate precipitates and the temperature of the mixture rises several degrees centigrade. We ordinarily find that the products of a spontaneous chemical reaction are warmer than the reactants were before the reaction began, but there are some processes, such as the solution of potassium nitrate in water, in which the products become cooler than the reactants. In this chapter we shall consider energy effects such as those mentioned, with a view toward developing a quantitative treatment of energy changes in chemical reactions and some understanding of the origin and interrelationships of such changes.

4.1 SOME QUANTITATIVE RELATIONS INVOLVING HEAT FLOW

In order to deal quantitatively with the energy changes that are associated with chemical reactions, we need to set up some well-defined system for measuring energy changes. In the reaction in which magnesium is burned in oxygen

$$Mg(s) + \tfrac{1}{2} O_2(g) \longrightarrow MgO(s)$$

the magnesium oxide forms at a very high temperature, being produced in a white hot flame. It then cools rapidly to a white powder. Heat and light are clearly evolved, but it is probably not obvious how we might determine the exact amount of energy evolved, or perhaps even what quantities might profitably be measured.

There are several useful approaches to the measurement of energy changes, each having its application in specific chemical problems. In Chapter 13 we will examine energy changes from the point of view of the work such changes can produce, and we will find that the spontaneity and equilibrium relations of chemical reactions are most directly connected with their capacity for doing useful work. Another approach is to inquire into the heat effects associated with chemical reactions, noting the magnitude and direction of heat flow and relating these to the pure substances taking part in the reactions. In this chapter we shall be concerned mainly with this latter approach.

You will recall that heat has been defined as that manifestation of energy which flows spontaneously from a warmer to a cooler body when the two are placed in contact. Heat flow is ordinarily measured in calories, one calorie being the amount of heat required to raise the temperature of one gram of water by one degree centigrade (strictly speaking, from 14.5 to 15.5°C). Since the calorie is in the metric system, 1 kilocalorie (kcal) = 1000 calories (cal). Experimental measurement of heat flow is readily accomplished by letting the sample under study exchange heat with a reference sample, often a measured amount of water.

In order to describe heat flow properly it is necessary to define several terms in a thermodynamic sense. The sample of material with respect to which the heat flow occurs is called the **system**. The system is separated from the rest of the universe, which is called the **surroundings**, by a real or defined **boundary**. The system will often consist of the sample under investigation, separated by its container wall, the boundary, from the rest of the laboratory, which effectively comprises the surroundings. Heat may flow into a system from the surroundings only by crossing the boundary. Heat flow is usually denoted by the symbol Q, and has both magnitude and direction. Heat flow *into* a system from its surroundings is taken to be positive ($Q = +$); heat flow from the system is taken to be negative ($Q = -$). In attempting to determine heat flow one must be careful to define the boundary inside of which the system exists, since the value of Q often depends on position, which may be arbitrarily chosen, of the boundary of the system.

A simple example of an experiment involving heat flow might be helpful. Let us assume that we have a metal sample weighing 80 grams at a temperature of 100°C and 50 grams of water at 25°C in a polystyrene coffee cup, fitted with a cover (Figure 4.1). The experiment consists of placing the hot metal in the water in the cup, replacing the cover, and measuring the temperature of the water, which we find goes up to 28°C. Depending on how we choose to define the system, the heat flow Q might have any one of several values:

I. System = coffee cup plus water plus metal. In this case the heat flow would equal the amount of heat that passed through the wall of the cup into the water and metal. Since the wall of the cup is a thermal insulator, essentially no heat flow would occur, and Q would be nearly zero.

II. System = metal sample. When the hot metal is put in the water, heat flows from the metal to the water, and the water is warmed from 25°C to 28°C.

57

By the definition of the calorie, 50 calories must be absorbed to raise 50 grams of water 1°C, so that in this experiment 150 calories flow from the metal to the water. Since all the heat entering the water must have come from the metal, none coming from the surroundings, the heat flow from the metal must have been 150 calories. Q, the heat flow for the process, must be −150 calories, since that amount of heat had to flow *from the system*, the metal, to the water.

III. System = water. In the previous case we found that 150 calories flowed into the water from the metal. Defining the system to be only the **water** fixes the value of Q to be +150 calories. Both the metal and the container would be part of the surroundings of the system.

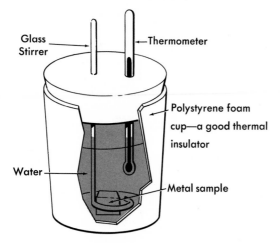

Figure 4.1   A simple heat flow experiment.

Labels: Glass Stirrer, Thermometer, Polystyrene foam cup—a good thermal insulator, Water, Metal sample

For a chemical reaction such as that which occurs between magnesium and oxygen, we can measure accurately and relatively easily the amount of heat that would have to be removed from the hot products in order to cool them to the temperature and pressure of the magnesium and oxygen before the reaction started. (The system in this reaction might be visualized as consisting initially of the magnesium and oxygen confined inside a flexible container; the magnesium is ignited, the reaction to form magnesium oxide occurs, and heat flow is measured as calories that cross the wall of the container as the oxide is cooled.) It is found that for the combustion in oxygen of one gram atomic weight, or one mole, of magnesium, 143.8 kcal of heat must be removed from the magnesium oxide in order to return it to the original temperature of the magnesium and oxygen. That is, for the reaction

We frequently say that an exothermic reaction is one which "evolves heat." Do you see any ambiguity in this statement?

$$\text{Mg(s, 25°C, 1 atm)} + \tfrac{1}{2}\,\text{O}_2\text{(g, 25°C, 1 atm)} \longrightarrow \text{MgO(s, 25°C, 1 atm)}$$

143.8 kcal must be removed from the system. Therefore Q=−143.8 kcal. The reaction is said to be **exothermic**, since heat must move **out** of the products to bring them back to 25°C, the temperature of the reactants.

In a chemical reaction, if heat must flow **into** the system to return the products to the initial temperature, as would be the case if the products were colder than the reactants, the reaction is said to be **endothermic**. Since in most cases a spontaneous chemical reaction results in products that are at a higher temperature than the reactants were before the reaction began, we can say that most spontaneous reactions are exothermic and have associated Q values that are negative.

As with the reaction between magnesium and oxygen just cited, heat flow is usually measured under conditions where the temperature and pressure of the products are made equal to those of the reactants. Most data on heat flow are given at 25°C and 1 atm pressure, and unless otherwise indicated, those conditions may be assumed to apply to both reactants and products of all reactions considered in this chapter.

Laws of Thermochemistry

From studies of the heat flow in many chemical reactions carried out under conditions of constant temperature and pressure, several laws have emerged which apply to all such reactions. These laws form the basis of that area of energy studies known as thermochemistry.

1. The heat flow Q in a chemical reaction is directly proportional to the amount of substance that reacts or is produced.

EXAMPLE 4.1. Consider the reaction (at 25°C and 1 atm)

$$2 H_2(g) + O_2(g) \longrightarrow 2 H_2O(l), \quad Q = -136.6 \text{ kcal}$$

How much heat evolves to the surroundings when 100 g H_2 is burned in oxygen?

SOLUTION. The information given states that when two moles of H_2 are burned, 136.6 kcal are given off to the surroundings. Therefore,

$$2 \text{ moles } H_2 \doteqdot -136.6 \text{ kcal and } 1 \text{ mole } H_2 = 2.016 \text{ g } H_2$$

Converting 100 g H_2 to its equivalent in heat flow, we obtain

$$100 \text{ g } H_2 \times \frac{1 \text{ mole } H_2}{2.016 \text{ g } H_2} \times \frac{-136.6 \text{ kcal}}{2 \text{ moles } H_2} \doteqdot -3390 \text{ kcal}$$

The combustion of 100 g of hydrogen would therefore liberate 3390 kcal of heat to the surroundings.

2. The heat flow Q in a chemical reaction will change its sign but not its magnitude if the reaction is carried out in a direction opposite to that originally studied.

Consider the reaction

$$3 H_2(g) + N_2(g) \longrightarrow 2 NH_3(g), \quad Q = -22.0 \text{ kcal}$$

If the reaction were carried out in the opposite direction,

$$2 NH_3(g) \longrightarrow 3 H_2(g) + N_2(g), \quad Q = +22.0 \text{ kcal}$$

The first reaction is exothermic, and 22.0 kcal must be removed from the system for every two moles of NH_3 formed. Occurring in the opposite direction, the reaction would be endothermic and 22 kcal would have to be furnished to the system for every two moles of NH_3 consumed or for every mole of N_2 produced.

3. If a chemical reaction can be written as the sum of two or more other reactions, the heat flow in the overall reaction is equal to the sum of the heat flows associated with the other reactions.

59

Consider the following chemical reactions:

$$Sn(s) + Cl_2(g) \longrightarrow SnCl_2(s), \quad Q_1 = -83.6 \text{ kcal} \tag{4.1}$$

$$SnCl_2(s) + Cl_2(g) \longrightarrow SnCl_4(l), \quad Q_2 = -46.7 \text{ kcal} \tag{4.2}$$

If we add the equations for Reactions 4.1 and 4.2, we obtain an equation for the reaction by which $SnCl_4(l)$ is formed directly from the elementary substances:

Problem 4.4 at the end of the chapter is a somewhat more sophisticated example of this same principle.

$$Sn(s) + 2 \, Cl_2(g) \longrightarrow SnCl_4(l), \quad Q = Q_1 + Q_2 = -130.3 \text{ kcal} \tag{4.3}$$

The heat flow Q in Reaction 4.3 will be equal to the sum of the heat flows in Reactions 4.1 and 4.2.

This relationship between heat flows in related reactions is called **Hess's Law** and is very useful for determining heat flows in chemical reactions where direct measurements are difficult if not impossible.

4.2 HEATS OF FORMATION OF PURE SUBSTANCES

A large body of information about heat flow has been obtained from thermochemical studies of many chemical reactions. These data are available in the original literature, but by the use of the thermochemical laws in a way that is perhaps not at once obvious, much of this information can be summarized in a simple but practical way. It then becomes possible to use a relatively small amount of summarized data to calculate heat flow in a large number of chemical reactions.

Let us assume that we wish to calculate the heat flow that would be anticipated in the following chemical reaction:

$$4 \, KClO_3(s) \longrightarrow 3 \, KClO_4(s) + KCl(s), \quad Q = ? \tag{4.4}$$

It might be possible to measure Q for Reaction 4.4 directly, but in any event it would be difficult and time consuming. The approach we shall use is to apply Hess's Law to a group of simple reactions, of which Reaction 4.4 is the sum. We are particularly interested in the reactions by which the reactants and the products in Reaction 4.4 were originally obtained from their elements. For Reaction 4.4 the related reactions and their heat flows are:

$$K(s) + \tfrac{1}{2} Cl_2(g) + \tfrac{3}{2} O_2(g) \longrightarrow KClO_3(s), \quad Q_1 = -93.50 \text{ kcal} \tag{4.5}$$

$$K(s) + \tfrac{1}{2} Cl_2(g) + 2 \, O_2(g) \longrightarrow KClO_4(s), \quad Q_2 = -103.6 \text{ kcal} \tag{4.6}$$

$$K(s) + \tfrac{1}{2} Cl_2(g) \longrightarrow KCl(s), \quad Q_3 = -104.4 \text{ kcal} \tag{4.7}$$

Reaction 4.4 can be expressed as a sum involving Reactions 4.5, 4.6, and 4.7, taken in the proper direction with appropriate coefficients for each reaction. Exactly how this is accomplished will be clearer if we consider Reaction 4.4 as occurring in two stages. In the first stage (A) the reactant is dissociated to elementary substances, and in the second (B) the elementary substances are recombined to form new products (Figure 4.2).

According to Hess's Law, the heat flow Q for the overall reaction must equal Q_A plus Q_B, the heat flows associated with the reactions in the first and second stages. Q_A, however, is simply equal to $-4 \, Q_1$, since the reaction in

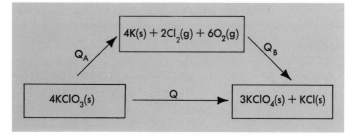

Figure 4.2 Use of an intermediate state in a thermochemical change.

stage A is the reverse of Reaction 4.5, but with 4 moles of $KClO_3$ reacting instead of 1. Similarly, Q_B is equal to 3 Q_2 plus Q_3, since the reaction in stage B is equal to 3 times Reaction 4.6 plus Reaction 4.7. Therefore,

$$Q_A = -4(-93.50 \text{ kcal}) = +374.0 \text{ kcal} \tag{4.8}$$

$$Q_B = 3(-103.6 \text{ kcal}) + (-104.4 \text{ kcal}) = -415.2 \text{ kcal} \tag{4.9}$$

$$Q = Q_A + Q_B = -41.2 \text{ kcal} \tag{4.10}$$

Clearly, by an approach completely analogous to the one we have used for Reaction 4.4, we could calculate the heat flow for *any* reaction for which data such as those in 4.5, 4.6, and 4.7 were available. Heat flows such as Q_1, Q_2, and Q_3, by which pure substances are formed from their elements, take on special significance, since they can be used so conveniently in problems such as the one just illustrated. Such heat flows are called molar **heats of formation** and are given the symbol ΔH_f, for a reason which will become apparent later in this chapter. The molar heat of formation of $KClO_3(s)$, $\Delta H_f \ KClO_3(s)$, is equal to -93.50 kcal, and is defined as the heat flow in the chemical reaction by which one mole of $KClO_3(s)$ is formed from metallic potassium, gaseous chlorine, and gaseous oxygen. The molar heat of formation of $KCl(s)$ is -104.4 kcal; this means that when a mole of $KCl(s)$ is formed from elementary substances, 104.4 kcal are evolved from the system to the surroundings when the $KCl(s)$ is returned to the temperature and pressure of the reactants.

In terms of the notation for molar heat of formation, Equation 4.10 takes the form,

$$Q = -4 \ \Delta H_f \ KClO_3(s) + 3 \ \Delta H_f \ KClO_4(s) + \Delta H_f \ KCl(s) \tag{4.11}$$

We can now generalize Equation 4.11 to include any chemical reaction:

$$Q = \sum \Delta H_f \text{ products} - \sum \Delta H_f \text{ reactants} \tag{4.12}$$

where the sums, Σ, are taken over molar heats of formation of reactants and of products respectively, giving proper attention to the numbers of moles involved. A list of molar heats of formation of some of the more common pure substances at 25°C and 1 atm is given in Table 4.1. From the definition of ΔH_f it is clear that the heat of formation of all elements is necessarily zero.

If the necessary data are available, the heat flow Q in any chemical reaction occurring at 25°C and 1 atm pressure can readily be calculated by means of Equation 4.12. Although heat flow for a chemical reaction is in principle dependent on temperature, the change in Q over rather wide temperature

61

TABLE 4.1 HEATS OF FORMATION (KCAL/MOLE) AT 25°C AND 1 ATM

AgBr(s)	−23.8	$C_2H_2(g)$	+54.2	$H_2O_2(l)$	−44.8	$NH_3(g)$	−11.0
AgCl(s)	−30.4	$C_2H_4(g)$	+12.5	$H_2S(g)$	−4.8	$NH_4Cl(s)$	−75.4
AgI(s)	−14.9	$C_2H_6(g)$	−20.2	$H_2SO_4(l)$	−193.9	$NH_4NO_3(s)$	−87.3
$Ag_2O(s)$	−7.3	$C_3H_8(g)$	−24.8	HgO(s)	−21.7	NO(g)	+21.6
$Ag_2S(s)$	−7.6	n-$C_4H_{10}(g)$	−29.8	HgS(s)	−13.9	$NO_2(g)$	+8.1
$Al_2O_3(s)$	−399.1	n-$C_5H_{12}(l)$	−41.4	KBr(s)	−93.7	NiO(s)	−58.4
$BaCl_2(s)$	−205.6	$C_2H_5OH(l)$	−66.4	KCl(s)	−104.2	$PbBr_2(s)$	−66.3
$BaCO_3(s)$	−291.3	CoO(s)	−57.2	$KClO_3(s)$	−93.5	$PbCl_2(s)$	−85.9
BaO(s)	−133.4	$Cr_2O_3(s)$	−269.7	KF(s)	−134.5	PbO(s)	−52.1
$Ba(OH)_2(s)$	−226.2	CuO(s)	−37.1	KOH(s)	−101.8	$PbO_2(s)$	−66.1
$BaSO_4(s)$	−350.2	$Cu_2O(s)$	−39.8	$MgCl_2(s)$	−153.4	$Pb_3O_4(s)$	−175.6
$CaCl_2(s)$	−190.0	CuS(s)	−11.6	$MgCO_3(s)$	−266	$PCl_3(g)$	−73.2
$CaCO_3(s)$	−288.5	$CuSO_4(s)$	−184.0	MgO(s)	−143.8	$PCl_5(g)$	−95.4
CaO(s)	−151.9	$Fe_2O_3(s)$	−196.5	$Mg(OH)_2(s)$	−221.0	$SiO_2(s)$	−205.4
$Ca(OH)_2(s)$	−235.8	$Fe_3O_4(s)$	−267.0	$MgSO_4(s)$	−305.5	$SnCl_2(s)$	−83.6
$CaSO_4(s)$	−342.4	HBr(g)	−8.7	MnO(s)	−92.0	$SnCl_4(l)$	−130.3
$CCl_4(l)$	−33.3	HCl(g)	−22.1	$MnO_2(s)$	−124.5	SnO(s)	−68.4
$CH_4(g)$	−17.9	HF(g)	−64.2	NaBr(s)	−86.0	$SnO_2(s)$	−138.8
$CHCl_3(l)$	−31.5	HI(g)	+6.2	NaCl(s)	−98.2	$SO_2(g)$	−71.0
$CH_3OH(l)$	−57.0	$HNO_3(l)$	−41.4	NaF(s)	−136.0	$SO_3(g)$	−94.5
CO(g)	−26.4	$H_2O(g)$	−57.8	NaI(s)	−68.8	ZnO(s)	−83.2
$CO_2(g)$	−94.1	$H_2O(l)$	−68.3	NaOH(s)	−102.0	ZnS(s)	−48.5

ranges is ordinarily small, so that heat flow values for chemical reactions up to about 1000°K can also be calculated from the data in Table 4.1 with only a small error. Some illustrations are given in Example 4.2.

EXAMPLE 4.2. Using Table 4.1, calculate the heat changes for the following reactions.

a. $2 MgCl_2(s) + O_2(g) \longrightarrow 2 MgO(s) + 2 Cl_2(g)$

b. $C_3H_8(g) + 5 O_2(g) \longrightarrow 3 CO_2(g) + 4 H_2O(l)$

SOLUTION

a. The amount of heat flow, Q, according to Equation 4.12, must be the difference between the heat of formation of two moles of MgO and the heat of formation of two moles of $MgCl_2$.

$$Q = 2 \Delta H_f MgO − 2 \Delta H_f MgCl_2$$

$$= 2(−143.8 \text{ kcal}) − 2(−153.4 \text{ kcal})$$

$$= −287.6 \text{ kcal} + 306.8 \text{ kcal}$$

$$= +19.2 \text{ kcal}$$

or:

$$2 MgCl_2(s) + O_2(g) \longrightarrow 2 MgO(s) + 2 Cl_2(g); \quad Q = +19.2 \text{ kcal}$$

b. In the same manner:

$$Q = 3\,\Delta H_f\,CO_2(g) + 4\Delta H_f\,H_2O(l) - \Delta H_f\,C_3H_8(g)$$

$$= 3(-94.1\ \text{kcal}) + 4(-68.3\ \text{kcal}) - (-24.8\ \text{kcal})$$

$$= -531\ \text{kcal}$$

It follows that 531 kcal of heat are evolved for the reaction as written, i.e.,

$$C_3H_8(g) + 5\ O_2(g) \longrightarrow 3\ CO_2(g) + 4\ H_2O(l);\quad Q = -531\ \text{kcal}$$

Here, as in a, elementary substances (O_2, Cl_2) are not included. Why?

4.3 THE EXPERIMENTAL MEASUREMENT OF HEAT FLOW

The amount of heat flow in a chemical reaction can be determined by means of an apparatus known as a calorimeter. An instrument of this type, used extensively in studying combustion reactions, is illustrated schematically in Figure 4.3. The device consists of a heavy-walled steel bomb, into which is put a small weighed sample of the material to be burned. The bomb cover is put on and oxygen gas is pumped into the bomb to a pressure of about 25 atm. The bomb is then immersed in a weighed quantity of water, in a container which is itself inside a thermally insulating container, and the whole apparatus is then closed. When the temperature of the water becomes steady, it is measured precisely. The reaction is then initiated by some suitable means; usually a piece of fine iron wire which passes through the sample is ignited by being heated electrically for a moment. The sample burns almost instantly, with the products attaining a very high temperature. The product gases cool rapidly as they hit the bomb wall, which gets warmer and itself exchanges heat with the surrounding water. The highest temperature reached by the water is then measured.

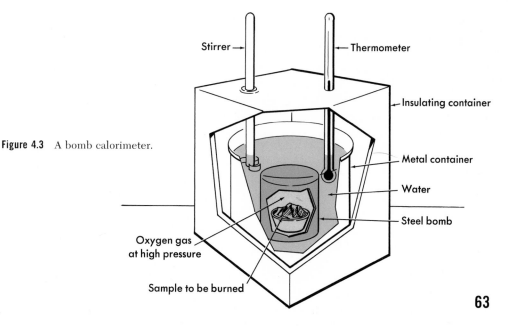

Figure 4.3 A bomb calorimeter.

63

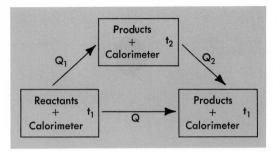

Figure 4.4 Experimental measurement of heat flow in a chemical reaction.

For purposes of analysis, the system in this experiment is most conveniently taken to include the bomb and its contents plus the water with its container. All these components are thermally insulated from the laboratory by the wall of the outside container, and, except for the contents of the bomb, will be referred to as the calorimeter. In Figure 4.4 we again use an intermediate state to assist our understanding of the thermodynamic argument. The change in state which actually occurs in the calorimeter is the one from the lower left to the intermediate state. As a result of combustion, the calorimeter and the reactants at an initial temperature t_1 undergo a change in which the calorimeter and the reaction products end up at a higher temperature t_2. The value of Q we are seeking is that for the reaction at the point where reactants and products are at the same temperature:

$$\text{Reactants } (t_1) \longrightarrow \text{Products } (t_1), \text{ Heat flow} = Q \qquad (4.13)$$

This, however, is the net reaction occurring between the two lower states in Figure 4.4, since the calorimeter is unchanged by that particular change in state.

Under the conditions in the calorimetric experiment, the heat flow Q is given by the relation

$$Q = Q_1 + Q_2 \qquad (4.14)$$

The heat flow Q_2 is simply that for the change by which the calorimeter and the reaction products are cooled from t_2 to t_1. We can determine this value by heating the calorimeter plus products through a small temperature interval with an electrical heater that produces a standard heat input, or by carrying out in the calorimeter a combustion having a known Q value. By either method we can determine a value for the heat capacity C of the calorimeter, which is that amount of heat needed to raise the temperature of the calorimeter plus contents by 1°C. Given the measured value of C,

$$Q_2 = C \times (t_1 - t_2) \qquad (4.15)$$

The value of Q_1 is the heat flow in the reaction occurring in the calorimeter, but because the calorimeter system is inside a thermally insulating container, the value of Q_1 is amazingly easy to calculate. In fact,

$$Q_1 = 0 \qquad (4.16)$$

Any student who recognized that $Q_1 = 0$ for the change in state occurring in the calorimeter before he read this far has an aptitude for thermochemistry that is rare. But indeed, Q_1 must be zero, since by definition Q is the heat which flows into a system, and in the thermally insulated system, Q must equal zero. The calorimeter is designed to prevent heat flow from or into the system during the reaction. A calorimeter is simply a device in which heat flow is known or controlled, and the simplest way one can know the amount of heat flow is to make it zero.

Therefore, for the reaction measured by the bomb calorimeter,

$$Q = Q_1 + Q_2 = 0 + Q_2 = C(t_1 - t_2) \qquad \textbf{(4.17)}$$

Of course the heat flow Q, determined by Equation 4.17, will be proportional to the amount of substance burned. The molar heat flow can be easily calculated from that measured value.

Example 4.3 shows how heat flow is determined by calorimetric means.

Clearly, if the final temperature, t_2, is greater than t_1, the reaction is exothermic and Q is a negative quantity.

EXAMPLE 4.3. A 1.00 g sample of naphthalene, $C_{10}H_8$, was burned in a bomb calorimeter. The temperature of the calorimeter rose from 25.00 to 28.90°C as a result of the combustion. In a separate experiment it was found that it required 2450 calories to raise the temperature of the calorimeter (bomb plus contents plus water plus metal container) by 1.00°C.

  a. Calculate the heat flow for the combustion of the 1.00 g sample of naphthalene.
  b. Calculate the molar heat of combustion of naphthalene.

SOLUTION

  a. By the argument presented in the previous discussion the Q for the reaction is that given by Equation 4.17, and is essentially the amount of heat that would have to flow from the calorimeter in order to cool it from the final temperature 28.90°C to the initial temperature of 25.00°C.

$$Q = C(t_1 - t_2) = 2450 \text{ cal} \times (25.00 - 28.90)$$

$$= -9550 \text{ cal} = -9.55 \text{ kcal}$$

  b. For a mole of naphthalene (128.2 g) the heat flow Q would be

$$-9.55 \frac{\text{kcal}}{\text{g}} \times 128.2 \frac{\text{g}}{\text{mole}} = -1220 \frac{\text{kcal}}{\text{mole}}$$

You may have noted that in the bomb calorimeter the changes all occur at constant volume. That is, the volume of the reactants is necessarily equal to that of the products of the reaction. The measured heat flow is therefore that at constant volume rather than that at constant pressure, which is the usual condition on the change. For most chemical reactions the difference between the two values of heat flow is small compared to the heat flow itself, and in many combustion reactions the difference is as small as the experimental error. In the next section we will consider some fundamentals of thermodynamics and will show how these two kinds of heat flow are related and how their difference can be calculated.

4.4 THE FIRST LAW OF THERMODYNAMICS

The relations we have developed and used in connection with heat flow in chemical reactions will now be derived by a more abstract approach. This approach involves the area in science known as thermodynamics. Thermodynamics deals in a very basic way with the interrelations between heat flow, work effects, and the energies of systems. In this section we will use thermodynamic arguments to deal with heat flow, in particular relating heat flow to energy changes in chemical reactions. In Chapter 13 the laws of thermodynamics will be applied to the problem of spontaneity and equilibrium in chemical reactions.

In thermodynamics we define a sample of matter to be the system in the manner used when we introduced the idea of heat flow. The system is separated from its surroundings by a boundary, across which may flow heat or work. The boundary of the system may be flexible, but its location must be known if meaningful thermodynamic statements are to be made.

We have discussed how heat flow can be measured in a chemical system and will now turn our attention to work effects. In principle, work is done by a system on its surroundings if it changes the surroundings in a way that is equivalent to the lifting of a weight in the surroundings. Work done by or upon a system may be mechanical or electrical, or, as will be the case in this section, may be limited to work of expansion by the system against the pressure exerted by the atmosphere surrounding it.

As an example of the kind of work we will be discussing, let us consider the following reaction at 100°C and one atmosphere:

Note the importance of specifying the physical states of reactants and products.

$$H_2O(l) \longrightarrow H_2O(g), Q = +9.7 \text{ kcal} \tag{4.18}$$

This reaction might be carried out in a container fitted with a tight but frictionless piston, such as that shown in Figure 4.5. If we define the system to be the contents of the cylinder, it should be clear that as the reaction proceeds the system increases in volume from that of a mole of liquid water, about 18 ml, to that of a mole of water vapor at 100°C and 1 atm, about 30 lit. During that change the system does work upon the surrounding atmosphere by pushing it back; we find that the work done under these conditions, where pressure is constant and has the same value both inside and outside the system, is equal to the product of the pressure P and the change in volume of the system:

$$\text{Work} = P(V_{final} - V_{initial})$$

For this change, Work = 1 atm (30 lit − 0.018 lit), or just about 30 lit atm. Since 1 lit atm = 24.2 cal

$$\text{Work} = 30 \text{ lit atm} \times \frac{24.2 \text{ cal}}{1 \text{ lit atm}} = 730 \text{ calories}$$

This is the amount of work W done by the system in Reaction 4.18 as the system undergoes expansion.

Work, like heat flow, has both magnitude and direction. Work done **by** the system on the surroundings is considered to be positive; in the example, W = +730 cal. Work done **upon** a system is negative; if Reaction 4.18 were reversed, work would be done by the atmosphere on the water as condensation proceeded, in the amount of 730 cal. For that change, W would be −730 cal.

A fundamental thermodynamic relation, called the First Law of Thermodynamics, deals directly with heat flow and work effects such as those we have been discussing. The First Law tells us that, associated with any sample

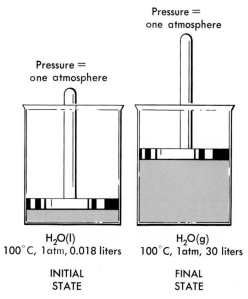

Pressure =
one atmosphere

Pressure =
one atmosphere

Figure 4.5 Work of expansion against a constant pressure.

$H_2O(l)$
100°C, 1 atm, 0.018 liters

$H_2O(g)$
100°C, 1 atm, 30 liters

INITIAL
STATE

FINAL
STATE

of matter, there is a fixed, invariable amount of energy, called **internal energy** E, which remains constant as long as the sample does not change its state. A mole of water, $H_2O(l)$, at 100°C and 1 atm is in a defined state, since the properties which have been specified for it are adequate to fix all of its other properties, such as density, volume, or refractive index. According to the First Law, at 100°C and 1 atm one mole of $H_2O(l)$ will therefore have a definite fixed amount of internal energy E. The internal energy of the water can only be changed by a change in state of the water, for example, a change in temperature, pressure, or state of aggregation. Similarly, by the First Law, a mole of water vapor $H_2O(g)$ at 100°C and 1 atm will also have a fixed amount of internal energy. Since the $H_2O(g)$ is not in the same state at 1 atm and 100°C as the $H_2O(l)$, being of different volume and state of aggregation, we would not expect the energy of water in the two states to be the same.

The essence of the First Law of Thermodynamics is that it allows us to determine unequivocally the difference between the energies of a system in two different states. The system may consist, as here, of a pure substance in two states of aggregation or a pure substance at two different temperatures and pressures; or it may consist of a group of pure substances which undergo a chemical reaction to form another group of substances. All that is necessary to determine the difference in energies is to carry out the change in state of the system under consideration, measuring carefully the values of heat flow Q and work effect W that are observed during the change. The First Law states that the difference ΔE between the internal energy of the system in the final state, E_2, and the internal energy of the system in the initial state, E_1, is given by the equation

$$\Delta E = Q - W = E_2 - E_1 \qquad (4.19)$$

Since the internal energy E is fixed in both the final and initial states, the difference ΔE is also fixed and cannot be affected by, for instance, a change in the manner or path by which the reaction is carried out. For a given change in a system there may be several ways in which the change could occur, with

The minus sign in the equation $\Delta E = Q - W$ arises because Q is + when heat flows *into* the system, and W is + when work flows *out* of the system.

67

In other words, the difference between Q and W is independent of path.
different values of Q and W for each change; the First Law states simply that, for every way, **Q − W will have the same value**.

For the change in state which occurs in Reaction 4.18, Q = 9.7 kcal, W = 0.73 kcal, and $\Delta E = Q - W = 9.0$ kcal (two significant figures), where

$$\Delta E = E_2 - E_1 = E_{H_2O(g)} - E_{H_2O(l)} = 9.0 \text{ kcal} \tag{4.20}$$

By Equation 4.20, we can see that at 100°C and 1 atm,

$$E_{H_2O(g)} = E_{H_2O(l)} + 9.0 \text{ kcal} \tag{4.21}$$

The power of the First Law is that it allows us to relate internal energies of systems in different states in the manner of Equations 4.20 and 4.21. You will note that we do not determine either $E_{H_2O(l)}$ or $E_{H_2O(g)}$, but only their difference; we cannot discover the absolute value of the internal energy of any system by thermodynamics, because the First Law equation is really a difference equation.

One of the primary uses of the First Law of Thermodynamics is to relate heat flow in chemical reactions to other properties of the species involved. In the case of combustion reactions in the bomb calorimeter, the situation is particularly straightforward. In the bomb calorimeter experiment the volume of the system does not change; this means that the system does not experience a work effect, and W = 0. Therefore, by the First Law, in any chemical reaction studied in a bomb calorimeter,

$$\Delta E = Q - W = Q - 0 = Q \tag{4.22}$$

That is, the heat flow which is observed is equal to the difference between the internal energies of the products and reactants in the chemical reaction studied.

Although many heat flow measurements are made in bomb calorimeters, it is relatively easy to make calorimetric measurements at constant pressure, as in the coffee cup calorimeter considered early in this chapter. Particularly where heat flow measurements are made for reactions involving only solutions or condensed phases, it is convenient to use calorimeters which are open to the atmosphere and so operate at constant pressure. It is also true that chemical reactions in general are carried out in open containers, so that it is important for us to be able to deal confidently with heat flows under constant pressure. This is most conveniently done by using another energy function, called the heat energy or, more commonly, the enthalpy, which is given the symbol H. We define the enthalpy H of a system in terms of its internal energy E by the equation

$$H = E + PV \tag{4.23}$$

where P and V are the pressure and volume of the system respectively. From Equation 4.23 it is clear that since both P and V must be known if the state of a system is defined, the change in enthalpy, ΔH, like the change in internal energy, ΔE, which occurs when a system changes its state, is fixed once the initial and final states of the system are fixed. The relation between ΔH and ΔE follows immediately from Equation 4.23:

$$
\begin{aligned}
\text{Final state:} \quad & H_2 = E_2 + P_2V_2 \\
\text{Initial state:} \quad & H_1 = E_1 + P_1V_1 \\
\text{Difference:} \quad & \Delta H = \Delta E + P_2V_2 - P_1V_1 = \Delta E + \Delta(PV)
\end{aligned}
\tag{4.24}
$$

That is, for any change in state, the difference between the change in enthalpy ΔH and the change in internal energy ΔE is equal to $\Delta(PV)$, the change in the PV product as the system goes from the initial to the final state.

For reactions which occur under conditions of constant pressure, as indicated previously, the work of expansion W is given by the relation

$$W = P\Delta V$$

Since, under these conditions, $P_2 = P_1$, it is also true that $\Delta(PV) = P\Delta V$. Substituting these values plus the First Law equation into Equation 4.24, we obtain

$$\Delta H = \Delta E + P\Delta V = Q - W + W = Q \qquad (4.25)$$

For chemical reactions at constant pressure, the heat flow Q is equal to the difference ΔH between the heat energies, or enthalpies, of the reactants and products. That is,

$$\Delta H = H_{products} - H_{reactants} = Q \qquad (4.26)$$

or

$$H_{products} = H_{reactants} + Q$$

Remember that Q in this equation refers to the heat flow at constant pressure. Q in Equation 4.22 represents the heat flow in a constant volume process.

By this analysis we can deduce that for exothermic reactions, where Q is negative, the enthalpy of the products will be **lower** than the enthalpy of the reactants by the magnitude of the heat flow Q. In endothermic reactions, where Q is positive, the products will have an enthalpy which is Q calories **higher** than that of the reactants.

Let us now consider the following reaction in the light of Equation 4.26.

$$\text{Elementary substances at 25°C, 1 atm} \longrightarrow \text{One mole of compound at 25°C, 1 atm}$$

For this reaction,

$$\Delta H = H_{compound} - H_{elementary\ substances} = \Delta H_{f\ compound} = Q \qquad (4.27)$$

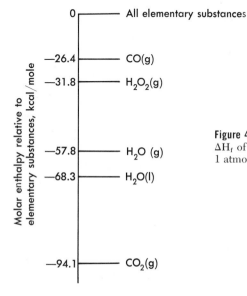

Figure 4.6 Molar enthalpies of formation ΔH_f of some pure substances at 25°C and 1 atmosphere pressure.

It is because heats of formation are really enthalpy changes that the symbol ΔH_f was used in the earlier section on thermochemistry. It is clear from Equation 4.27 that if ΔH_f is negative, the enthalpy of the compound will be lower, by the magnitude of ΔH_f, than the enthalpies of the elements from which it can be made. Similarly, a positive ΔH_f value means the enthalpy of the compound is higher than that of the elements by ΔH_f.

We can indicate graphically the heats of formation of pure substances in a way that may be helpful. In Figure 4.6 we see that the enthalpy of $H_2O(l)$ lies 68.3 kcal/mole below that of the elementary substances, while the enthalpy of $H_2O(g)$ is only 57.8 kcal/mole below the enthalpy of the elements. To carry out the reaction at 25°C

$$H_2O(l) \longrightarrow H_2O(g)$$

requires that the enthalpy of the system increase by $68.3 - 57.8$ kcal, or 10.5 kcal; this amount of heat would have to flow into the system to bring about the change. For the reaction

$$CO(g) + H_2O(g) \longrightarrow H_2(g) + CO_2(g)$$

ΔH would be equal to the increase in enthalpy when a mole of $CO(g)$ and a mole of $H_2O(g)$ are dissociated to the elements minus the decrease when a mole of $CO_2(g)$ is formed from them. Referring to Figure 4.6, we see that:

$$\Delta H = 26.4 + 57.8 - 94.1 = -9.9 \text{ kcal} = Q$$
$$= -\Delta H_f \, CO(g) - \Delta H_f \, H_2O(g) + \Delta H_f \, CO_2(g)$$

In any chemical reaction for which heat of formation data are available, the enthalpy change and heat flow can be calculated by the relation

$$\Delta H = -\sum \Delta H_f \text{ reactants} + \sum \Delta H_f \text{ products} = \text{Heat flow} = Q \qquad \text{(4.28)}$$

This is essentially Equation 4.12, derived here from the First Law instead of Hess's Law and taking into account the fact that the enthalpy change in the reaction is equal to the heat flow.

By Equation 4.24, the difference between the heat flow at constant pressure, which we have shown to be equal to ΔH, and that at constant volume, which is equal to ΔE (recall Equation 4.22), is simply equal to $P\Delta V$, which is just equal to the **work** done by the system during the change in state at constant pressure. For any chemical reaction involving only condensed phases such as solids, liquids, or solutions, the change in volume is very small, so that the work effect is also very small. This means that for all practical purposes, the difference between the heat flow at constant V and at constant P is negligibly small, of the order of a few calories at most.

In combustion reactions in which gases are present, the changes in PV product may be significantly larger. Consider the combustion reaction of hexane, $C_6H_{14}(l)$:

$$C_6H_{14}(l) + \frac{19}{2} O_2(g) \longrightarrow 6 \, CO_2(g) + 7 \, H_2O(l)$$

When this reaction is carried out in a bomb calorimeter at 25°C the observed heat flow Q is -987.7 kcal; therefore, $\Delta E = -987.7$ kcal. If the reaction were carried out at one atmosphere the number of moles of gas present in the system would drop from $9\frac{1}{2}$ to 6. For every mole of gas present in a system at 25°C, there will be an associated value of the PV product of about 24.4 lit atm, or about **600 calories**. For this reaction $\Delta(PV) = 6(600 \text{ cal}) - \frac{19}{2}(600 \text{ cal}) = -2.1$ kcal). Therefore, since

To show that 24.4 lit atm = 600 cal, use the conversion factor given in Chapter 1.

$$\Delta H = \Delta E + \Delta(PV)$$
$$\Delta H = (-987.7 - 2.1) \text{ kcal} = -989.8 \text{ kcal}$$

In this combustion the heat flow at constant volume and that at constant pressure differ by about 2 kcal; this amounts to only 0.2 per cent of the value of Q itself, and is, for all but the best thermochemical experiments, about equal to the expected experimental error. Since in most chemical reactions the heat flow is just about the same whether the reaction is conducted at constant volume or at constant pressure, henceforth we shall usually ignore the distinction between them and use ΔH values to indicate heat flow.

4.5 A PHYSICAL INTERPRETATION OF THE ORIGIN OF HEAT FLOW

By the First Law of Thermodynamics we have related the heat flow in a chemical reaction to the difference between the enthalpies of reactants and products. We might well raise the question as to why such differences exist between substances or to what properties of the substances such differences could be attributed.

If we restrict our attention to gases the answers to these questions follow readily, once the idea is established that any pure substance in a given state has a fixed enthalpy. A molecule of a substance such as hydrogen chloride, HCl, consists of two atoms held together by chemical bonds of substantial stability. If an HCl molecule is dissociated into atoms, enthalpy will have to be furnished to the molecule in an amount sufficient to break the H—Cl bond, and the free atoms will differ in enthalpy from the molecule by the amount of dissociation energy required (Figure 4.7).

Similarly, if a molecule of water, H_2O, is formed from a free O atom and two H atoms, the enthalpy of the molecule, under constant temperature of course, will be lower than that of the atoms by the amount of heat liberated

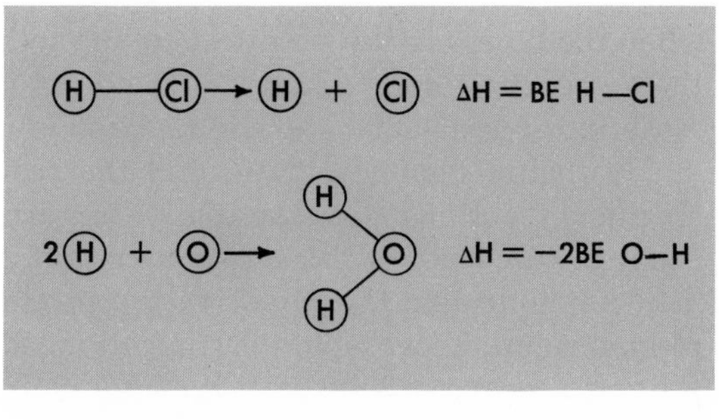

Figure 4.7 Breaking and making chemical bonds.

to the surroundings when the two O—H bonds were formed. In light of this idea, let us consider the reaction by which hydrogen chloride burns in oxygen at 25°C:

$$4 \ HCl(g) + O_2(g) \longrightarrow 2 \ Cl_2(g) + 2 \ H_2O(g)$$

As we have done previously in this chapter, we shall assume again that the reaction occurs through an intermediate state. In the intermediate state in this case, we choose to have the free atoms of the elements (Figure 4.8).

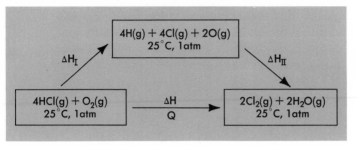

Figure 4.8 A physical interpretation of the origin of enthalpy change.

We may consider that the reaction occurs in two steps, I and II. The overall enthalpy change ΔH for the reaction will be equal to the value of ΔH_I plus the value of H_{II}. (Why?)

The enthalpy change in Step I is essentially equal to the energy that must be supplied, in the form of heat or otherwise, to dissociate four moles of HCl and one mole O_2 to atoms. Step I is highly endothermic.

$$\Delta H_I = 4 \ BE \ H—Cl + BE \ O—O$$

where BE H—Cl is the energy required to break a mole of H—Cl bonds and BE O—O is the energy needed to dissociate a mole of O—O bonds.

In Step II two moles of Cl_2 and 2 moles H_2O are formed from the free atoms. In this step heat would be liberated from the products in an amount equal to the energy of two moles of Cl—Cl bonds and 4 moles of O—H bonds. This step is highly exothermic.

$$\Delta H_{II} = -2 \ BE \ Cl—Cl - 4 \ BE \ O—H$$

$$\Delta H = \Delta H_I + \Delta H_{II} = Q$$

In calculating ΔH from heats of formation, elementary substances (O_2, Cl_2) were omitted. Why not omit them here?

The overall reaction will be exothermic if the heat liberated in Step II is larger than the heat energy required in Step I, or, more loosely speaking, if the bonds in the products are stronger than the bonds in the reactants. If more energy is required to break the bonds in the reactants than is released when the bonds in the products are formed, the reaction will be endothermic. In the reaction considered the bonds in the products are more stable than those in the reactants, and the reaction is indeed exothermic.

Enthalpy changes, then, and the resultant heat flows encountered in chemical reactions, are essentially the result of the bond-breaking and bond-making which go on during the reaction. If the value of ΔH is negative, we may conclude that the bonds in the products are stronger than those in the reactants and, as we noted earlier, in most cases we may also say that the reaction to form products will be a spontaneous one. If ΔH is positive, more energy must be furnished to break bonds than is available from bonds that are made, and the reaction will usually not occur spontaneously. When enthalpy changes are small and the pressure is low, or when the temperature of the system is high, other factors may become important in determining spontaneity of reaction. These factors will be considered in Chapter 13.

4.6 ENERGY CHANGES OTHER THAN HEAT FLOW

Although the change in energy that accompanies a chemical reaction ordinarily manifests itself in the form of heat, as we have seen, it may also be observed as mechanical energy, electrical energy, or light.

Mechanical Energy

When one or more of the products of a spontaneous chemical reaction is a gas, the reaction can, if carried out in a confined space, do mechanical work on the surroundings. The reaction of gasoline vapor with air in an automobile engine produces a miniature explosion in the cylinders; the very high pressure of the hot gaseous products on the pistons causes the engine to move the automobile. When the trigger of a rifle is pulled, a charge of explosive is set off which produces sufficient energy to propel a bullet from the barrel at a speed of perhaps 1200 feet per second. The reaction of alcohol or gasoline with liquid oxygen in a rocket engine can serve as a source of energy adequate to project a satellite into orbit or send a missile to a target 4000 miles away.

The chemical reactions used to do mechanical work also give off considerable heat to the surroundings of the mechanical system. Indeed, the combustion of gasoline in a gasoline stove (rather than in an automobile engine) essentially produces only heat flow. The products of the combustion are CO_2 and H_2O vapor, independent of whether the reaction occurs in an automobile engine, a rocket, or a stove, and at the same temperature these products would have the same enthalpy irrespective of how they were formed. Mechanical devices known as heat engines are designed to harness as much of the mechanical energy available from the combustion as possible, liberating as little heat as possible. The First Law tells us that the sum of the work produced by the system and the heat evolved by the system will be constant for a given chemical change, but it does not put a limitation on the fraction of available energy that can be recovered as useful work. However, a limitation does exist for the conversion of chemical energy into useful work. This limitation is a consequence of the Second Law of Thermodynamics (see Chapter 13).

Electrical Energy

Many spontaneous chemical reactions can be caused to liberate electrical energy. Perhaps the most familiar example of a device that produces electrical energy by chemical reaction is the lead storage battery. The reaction is

Electrochemical reactions will be discussed at greater length in Chapters 20 and 21.

$$Pb(s) + PbO_2(s) + H_2SO_4(l) \longrightarrow 2\ PbSO_4(s) + 2\ H_2O(l)$$

This reaction provides sufficient energy to operate the starting motor of an automobile. A different reaction taking place in an ordinary dry cell may serve to operate a flashlight, a hearing aid, or a child's electric toys. Conversely, many nonspontaneous reactions of considerable industrial importance can be made to occur by the **introduction** of electrical work. For example, the decomposition of sodium chloride to sodium metal and chlorine gas can be accomplished by the passage of an electric current through the molten salt in an electrolysis cell:

$$NaCl(l) \longrightarrow Na(l) + \tfrac{1}{2} Cl_2(g), \Delta H = 98.2 \text{ kcal}$$

Commercial manufacture of such common materials as aluminum, magnesium, and sodium hydroxide is done by electrolysis. In each case electrical work is put into a system to bring about a nonspontaneous reaction.

Light

Occasionally, part of the energy evolved in a chemical reaction is given off as light. We have already mentioned the oxidation of magnesium, which produces a very bright light during the course of the reaction. The light given off carries considerable energy, which is typically dissipated as heat when the light is absorbed by matter that lies in its path.

The energy carried by a beam of light appears to exist in discrete units, called photons, whose energy is given by Planck's equation:

$$E = h\nu \text{ or } E = \frac{hc}{\lambda}$$

where E is the energy of the photon in ergs, ν is the frequency of the light in vibrations per second, h is Planck's constant, 6.62×10^{-27} erg sec, c is the velocity of light, 3.0×10^{10} cm/sec, and λ is the wavelength of light in centimeters. In Figure 4.9 we have indicated how the energy of a light photon

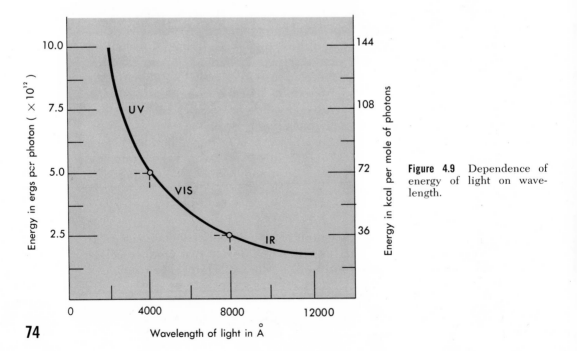

Figure 4.9 Dependence of energy of light on wavelength.

varies with its wavelength. As the wavelength decreases, the energy of the photon increases, approaching infinity at very short wavelengths. Although the energy per photon is small by ordinary standards, the nature of a photon is such that it can transfer all of its energy to an atom or a molecule which absorbs it. (Indeed, it is really because of this property that light is sometimes thought to consist of photons, with particle properties, rather than simply waves.) Light of wavelength longer than about 8000 Å is said to be in the infrared range and has photons of relatively low energy. In the visible region, 4000 to 8000 Å, the photons of light have sufficient energy so that, on striking some molecules, they may be able to break a chemical bond in those molecules. The fragments produced in such a process are typically very reactive, and if the proper substances are present, the fragments may initiate many other reactions, causing an explosion. If a mixture of $H_2(g)$ and $Cl_2(g)$ is exposed to light of a wavelength of 5000 Å or less, Cl_2 molecules will dissociate according to the reaction

$$Cl_2(g) + h\nu \longrightarrow 2\ Cl(g)$$

The chlorine atoms serve to trigger a very rapid, thermodynamically spontaneous reaction in which $HCl(g)$ is the final product. (See Chapter 14 for a further discussion of this reaction.) Light in the ultraviolet, with a wavelength less than 4000 Å, is still more energetic, so that at a wavelength of 2000 Å the photons are able to break all but the most stable of chemical bonds.

Light may also serve as the energy source in nonspontaneous reactions. By far the most important of such reactions is photosynthesis, by which plants are able to convert carbon dioxide and water to carbohydrates. This reaction, which occurs by virtue of the ability of the green plant pigment, chlorophyll, to make use of the energy in the photons in visible light, may be represented by the equation

$$6\ CO_2(g) + 6\ H_2O(l) \longrightarrow C_6H_{12}O_6(s) + 6\ O_2(g), \Delta H = +673\ kcal$$

This reaction is a great deal more complicated than the equation implies; it passes through a great many intermediate steps.

Glucose, $C_6H_{12}O_6$, is but one example of the many substances that are produced by photosynthesis. This reaction provides us with a basic source of food and fuel. When carbohydrates are metabolized in the body, the reverse of the photosynthesis reaction occurs, and the chemical energy stored in the carbohydrates is thereby converted to the thermal and mechanical energy needed to sustain life. Fuels such as wood, coal, and petroleum contain organic matter whose ultimate source was the photosynthesis reaction.

PROBLEMS

4.1 A piece of metal weighing 100 g and at 150°C is dropped into 100 ml of water in a thermally insulated container at 25°C. The temperature of the water after it has attained thermal equilibrium with the metal is 35°C. If there is no exchange of heat with the air surrounding the container, what is the heat flow for the change in state if
a. The system is defined to consist of the water?
b. The system is defined to consist of the metal?
c. The system consists of the water plus the metal plus the container?
(You may assume that the container does not exchange appreciable amounts of heat with the rest of the system or with the surroundings)

4.2 When one gram of aluminum powder is burned in excess oxygen in an open container, it is necessary to remove 7.4 kcal of heat from the products to return them to the initial temperature of the aluminum and oxygen. What is the heat effect Q for
a. $2 \text{ Al}(s) + \frac{3}{2} O_2(g) \longrightarrow Al_2O_3(s)$
b. $Al_2O_3(s) \longrightarrow 2 \text{ Al}(s) + \frac{3}{2} O_2(g)$

4.3 In the combustion of methyl alcohol, $CH_3OH(l)$, to form carbon dioxide, $CO_2(g)$, and water, $H_2O(l)$, 171 kcal is evolved for each mole of alcohol burned at T, P.
a. Calculate the heat flow per gram of methyl alcohol burned.
b. Calculate the heat flow per gram of CO_2 formed.
c. Calculate the heat evolved per mole of H_2O formed.

4.4 For the reactions at 25°C and constant pressure:

$$C_2H_2(g) + \frac{5}{2} O_2(g) \longrightarrow 2 CO_2(g) + H_2O(g); \quad Q_1 = -312 \text{ kcal}$$

$$CH_4(g) + 2 O_2(g) \longrightarrow CO_2(g) + 2 H_2O(g); \quad Q_2 = -211 \text{ kcal}$$

Find Q for the reaction at 25°C and constant pressure:

$$3 CH_4(g) + \frac{7}{2} O_2(g) \longrightarrow C_2H_2(g) + CO_2(g) + 5 H_2O(g)$$

4.5 When 2.17 g of naphthalene, $C_{10}H_8(s)$ is burned in excess oxygen at 25°C and 1 atm, the heat flow in the reaction is −20.8 kcal.
a. What is the molar heat of combustion of naphthalene?
b. What is the molar heat of formation, ΔH_f, of naphthalene?
(You may use data in Table 4.1 if necessary.)

4.6 Using the data given in Table 4.1, calculate the heat flow, Q, at 25°C and constant pressure for the following reactions.
a. $H_2S(g) + 2 O_2(g) \longrightarrow H_2SO_4(l)$
b. $C_2H_6(g) + \frac{7}{2} O_2(g) \longrightarrow 2 CO_2(g) + 3 H_2O(l)$
c. $2 MgO(s) + C(s) \longrightarrow CO_2(g) + 2 Mg(s)$
d. $Al_2O_3(s) + 3 H_2(g) \longrightarrow 2 Al(s) + 3 H_2O(l)$

4.7 Using the data in Table 4.1, calculate the amount of heat evolved when
a. One gram of propane, $C_3H_8(g)$, is burned in excess oxygen to form $CO_2(g)$ and $H_2O(l)$.
b. 4.50 g of $H_2S(g)$ is burned in excess oxygen to form $SO_2(g)$ and $H_2O(l)$.

4.8 A sample of 1.32 g of sucrose, table sugar, $C_{12}H_{22}O_{11}$, is burned in a bomb calorimeter in excess oxygen to form $CO_2(g)$ and $H_2O(l)$. The temperature of the calorimeter rises from 24.62°C to 26.88°C. In a separate experiment, it is found that 2350 cal are required to raise the temperature of the whole calorimeter by 1.00°C.
a. Is this heat flow a ΔE_{comb} or a ΔH_{comb}? Why?
b. What is the heat flow into the calorimeter in this experiment?
c. What is the heat of combustion of sucrose, in kcal/g, under these conditions?
d. What is the heat of formation of sucrose, ΔH_f, in kcal/mole?

4.9 A sample of KNO_3 weighing 3.5 g is dissolved in 50 ml of water in a styrofoam cup calorimeter. The reaction that occurs involves the heat of solution of KNO_3.

$$KNO_3(s) \longrightarrow K^+ + NO_3^-$$

As a result of the solution reaction, the temperature of the water drops from 27.0°C to 25.3°C. The specific heat of the $KNO_3(s)$ is about 0.2 cal/g°C. The specific heat of the calorimeter may be taken to be negligibly small. What is the molar heat of solution of $KNO_3(s)$ under these conditions?

4.10 For the reaction at 25°C,

$$4 NH_3(g) + 3 O_2(g) \longrightarrow 2 N_2(g) + 6 H_2O(l)$$

a. Find ΔH for the reaction, using Table 4.1.
b. Evaluate $\Delta(PV)$ for the reaction.

    c. Calculate ΔE for the reaction.

    d. Find the heat flow, Q, for the reaction when one mole of NH_3 is burned in an open flame at 25°C.

    e. Find the heat flow, Q, when one mole of NH_3 is burned in a bomb calorimeter.

    f. Explain why the two values of Q calculated in d and e are different.

4.11 Find the value of ΔE for each of the reactions in Problem 4.6.

4.12 A system is taken from one state to another by a series of steps.
Step 1. 500 calories of heat are added to the system.
Step 2. The system does 450 calories of work in the surroundings and absorbs 150 calories of heat.
Step 3. 200 calories of work are done upon the system.
What is the change in energy, ΔE, for the overall change in state?

4.13 When one gram of magnesium is burned in an open container, 5.92 kcal of heat is evolved. What is the heat effect, Q, for the reactions
    a. $Mg(s) + \frac{1}{2} O_2(g) \longrightarrow MgO(s)$
    b. $2\ Mg(s) + O_2(g) \longrightarrow 2\ MgO(s)$

4.14 One of the major constituents of gasoline, C_8H_{18}, has a heat of combustion of −1300 kcal/mole.
    a. Calculate the heat flow, Q, when 12.5 g of C_8H_{18} are burned at constant T and P.
    b. If CH_3OH and C_8H_{18} were selling at the same price per pound, which would probably be the more economical fuel to use in an automobile? (Cf Problem 4.3.)

4.15 Given the reactions

$$H_2(g) + Cl_2(g) \longrightarrow 2\ HCl(g);\ Q_1 = -44.2\ kcal$$

$$2\ H_2(g) + O_2(g) \longrightarrow 2\ H_2O(l);\ Q_2 = -136.6\ kcal$$

find the heat flow in the reaction

$$4\ HCl(g) + O_2(g) \longrightarrow 2\ Cl_2(g) + 2\ H_2O(l)$$

4.16 Given the reactions

$$S(s) + O_2(g) \longrightarrow SO_2(g);\ Q_1 = -71.0\ kcal$$

$$2\ SO_2(g) + O_2(g) \longrightarrow 2\ SO_3(g);\ Q_2 = -47.0\ kcal$$

find the heat flow in the reaction

$$S(s) + \frac{3}{2} O_2(g) \longrightarrow SO_3(g)$$

What is the name given to the quantity just calculated?

4.17 Make a diagram on which you indicate the relative molar enthalpies, H, in kcal/mole at 25°C of $S(s)$, $O_2(g)$, $H_2(g)$, $H_2S(g)$, $H_2O(g)$, $H_2O(l)$, $SO_2(g)$, $SO_3(g)$, and $H_2SO_4(l)$. Take the molar enthalpies of the elementary substances to be zero on the diagram.

4.18 Using the diagram from Problem 4.17, find ΔH at 25°C for the following reactions.
    a. $H_2S(g) + \frac{1}{2} O_2(g) \longrightarrow H_2O(g) + S(s)$
    b. $H_2O(l) + SO_3(g) \longrightarrow H_2SO_4(l)$
    c. $2\ SO_2(g) + O_2(g) \longrightarrow 2\ SO_3(g)$

4.19 When one gram of benzoic acid is burned in a calorimeter of the type shown in Figure 4.3, containing 2.30 kg of water, the temperature rises from 24.00°C to 26.50°C. The ΔE of combustion of benzoic acid is −6.315 kcal/g. Calculate
    a. The heat capacity of the calorimeter, in cal/°C.
    b. The ΔE of combustion, in kcal/g, of a certain hydrocarbon, given that 1.23 g of this compound, when burned in the same calorimeter with the same amount of water, brings about a temperature increase of 3.67°C.

77

4.20 One gram of NaI(s) is dissolved in 20 ml of water in a styrofoam cup calorimeter. In the solution process, the temperature rises from 25.00°C to 25.63°C. The specific heat of NaI is about 0.3 cal/g°C; that of water is 1.00 cal/g°C. What is the molar heat of solution of NaI(s) under these conditions?

4.21 Consider the reaction at 25°C

$$KClO_3(s) \longrightarrow KCl(s) + \tfrac{3}{2} O_2(g)$$

a. Determine ΔH, $\Delta(PV)$, and ΔE for the reaction.
b. What is Q when one mole of $KClO_3$ decomposes in an open container?
c. What is Q when one mole of $KClO_3$ decomposes in a bomb calorimeter?

4.22 Find ΔE at 25°C for the reactions in Problem 4.18.

°4.23 A system is carried through a cycle in which it ends at the same state at which it began. The steps in the cycle are:
Step 1. Q = 75 cal, W = 45 cal.
Step 2. 125 cal of heat flow into the surroundings, 200 cal of work is done on the surroundings.
Step 3. Q = −150 cal, W = ?
a. Find the value of ΔE in Step 2.
b. Find the value of ΔE for the cycle.
c. Find the value of W in Step 3. Is work done upon or by the system in this step?

°4.24 A student burns 100 g of acetylene in an oxyacetylene torch. If no heat or gas enters or leaves the laboratory as a result of the reaction, and if all reactants were available within the laboratory, what change occurs in the energy of the laboratory as a result of the reaction? Consider the laboratory to consist of its walls plus its contents. Explain your answer.

°4.25 Given the following bond energies in kcal/mole of bonds,

H—H : 104   N—N : 226   (in N_2)
H—O : 111   O—O : 118   (in O_2)
H—Cl : 103   N—H : 93
Cl—Cl : 58

calculate the value of ΔH for the following reactions.
a. $H_2(g) + Cl_2(g) \longrightarrow 2\ HCl(g)$
b. $N_2(g) + 3\ H_2(g) \longrightarrow 2\ NH_3(g)$
c. $2\ H_2(g) + O_2(g) \longrightarrow 2\ H_2O(g)$

The Physical Behavior
of Gases

Depending on temperature and pressure, pure substances will exist as solids, liquids, or gases. At low temperatures all substances become solids. In some intermediate temperature range they behave as liquids, and at high temperatures they are gaseous. What is "low" or "high" temperature for one substance is not necessarily "low" or "high" temperature for another, so that, while oxygen, ammonia, and carbon dioxide are gases at room temperature, iron, silicon carbide, and tungsten become gases only at temperatures in excess of 2000°C.

It is an experimental fact that in the gaseous state all pure substances exhibit remarkably similar physical behavior. This fact was used to good advantage when Avogadro's Law was applied to the problem of finding atomic weights. Indeed, much of the real foundation of chemistry was laid when scientists realized the far-reaching general laws that could be stated for gases. In this chapter we shall study some of the laws governing the physical behavior of gases.

5.1 SOME GENERAL PROPERTIES OF GASES

When a liquid is heated sufficiently it begins to boil and evaporate. In this process the substance is said to make a transition from the liquid to the gaseous state. During the change in state the particles in the liquid become free from one another and pass into space as molecules of gas. In general this process is accompanied by a great change in volume. If half a cupful of water is evaporated, the resulting water vapor (water in the gas phase) at one atmosphere pressure and 100°C occupies a volume equal to that of a 50-gallon oil drum. Since molecules in the gas phase are the same size as they are in the liquid state, it follows that the distances between them in the gas are much greater than they are in the liquid.

At 1 atm and 100°C, water molecules in the vapor are about 10 diameters apart.

In view of the rather large distances between gas molecules, we might expect that compressing a gas would be fairly readily accomplished. Experimentally we find that this is so, and that in general the volume of a gas varies inversely with the applied pressure; that is, doubling the pressure reduces the volume to about half its previous value. Similarly, if we double the amount (mass) of gas in a container we find that the pressure approximately doubles. Increasing the temperature of a gas in a closed container will increase the pressure of the gas.

Gases can be expanded indefinitely and will always tend to occupy their containers completely and uniformly. If one milliliter of hydrogen gas at one atmosphere pressure is let into an evacuated 10,000-liter container, the hydrogen almost instantly diffuses to give a constant density and pressure throughout the container. The pressure of the gas would be very low under such conditions, about 10^{-7} atm, but could be measured easily. You might wonder whether the situation would be the same if there had already been another gas at one atmosphere in the container. Under such circumstances the amount of hydrogen in any part of the container would ultimately be the same as if the container were initially evacuated, but the time required to attain a homogeneous mixture would be of the order of hours rather than a fraction of a second; the rate of diffusion of the hydrogen molecules would be very much hindered by collisions with the other gas molecules.

All gases mix readily with one another to form completely homogeneous solutions. Ordinary air is such a solution. No one has ever been able to prepare a mixture of gases which tended to settle out into two or more regions of different composition. This situation is very different from that with mixtures of liquids and solids, in which solubilities of one substance in another are usually limited.

5.2 ATMOSPHERIC PRESSURE AND THE BAROMETER

The most common gas we encounter, and the only one known until about 1750, is the air about us. This gas lies over the earth in a blanket about 50 miles thick. Like all earthly matter, the air is subject to the gravitational pull of the earth. The air near the earth is compressed by the weight of the air above it. The pressure of the air at the earth's surface is by no means negligible, amounting to about 14.7 pounds per square inch; this means that our bodies are at all times subject to a rather gigantic force. The pressure of the atmosphere is not constant but varies with the height above sea level and weather conditions. In Denver, Colorado (altitude, 5000 ft above sea level) the atmospheric pressure is about 13.5 pounds per square inch. During a hurricane the pressure may become that low at sea level. Above 10,000 feet, breathing becomes uncomfortable for people not accustomed to such altitudes, and for that reason modern aircraft have pressurized cabins. Military aircraft, which often operate at much higher altitudes, cannot be sufficiently pressurized (why?), and in such aircraft the personnel must wear pressurized suits and oxygen masks. At a height of 10 miles the atmospheric pressure is only about 10 per cent of that at sea level. Uncomfortable is not the word to describe the sensations of an unpressurized human being at such an altitude.

Although the facts of atmospheric pressure are really very simple, they were not clearly understood until about 1650. Men had learned earlier in a very practical way that they could not lift water more than about 33 feet with

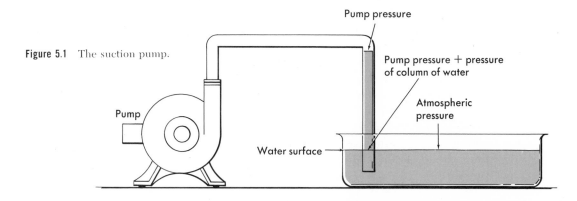

Figure 5.1   The suction pump.

a suction pump; the reason for this was unknown, and the explanation given by philosophers was that "nature abhors a vacuum." Torricelli, an Italian scientist, showed that the limit on pumping heights exists because at the limiting height the pressure exerted by the atmosphere at the water surface is just balanced by the pressure exerted by the raised water column at the water surface. It is a natural law that in any liquid equal pressures must exist at equal heights. In the diagram in Figure 5.1 it is clear that at the water surface,

  pressure outside tube = pressure inside tube
pressure of atmosphere = pressure at pump + pressure of column of water

   Torricelli recognized that the maximum height of the liquid column was essentially a measure of the atmospheric pressure, and he made a simple device for making the measurement conveniently. His device, called a **barometer**, is shown in Figure 5.2; it consists of a closed glass tube filled

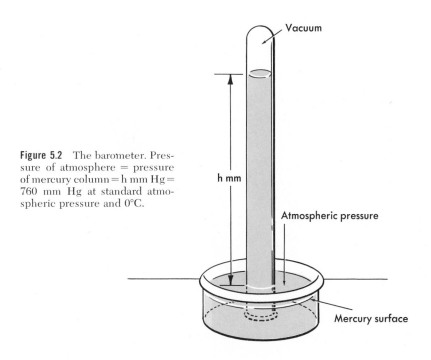

Figure 5.2   The barometer. Pressure of atmosphere = pressure of mercury column = h mm Hg = 760 mm Hg at standard atmospheric pressure and 0°C.

with mercury and inverted over a pool of mercury. Provided that the tube is sufficiently long, liquid mercury flows into the reservoir when the tube is first inverted, leaving essentially a vacuum above the liquid; the height of the liquid column remaining in the tube is then directly proportional to the atmospheric pressure. Since the density of mercury is high, the height of the column is much less than with water and is, at one standard atmosphere pressure, 760 mm if the mercury is at 0°C. It was several years before Torricelli's contemporaries accepted his reasoning and recognized the barometer as the scientific instrument it is. Today we use his barometer, essentially unmodified, as the standard instrument for the measurement of atmospheric pressure.

Using the relation $h_{Hg} \times d_{Hg} = h_{H_2O} \times d_{H_2O}$, it is easily shown that 33 ft of water is equivalent to 760 mm Hg ($d_{Hg} = 13.6$ g/ml).

The Manometer and the Measurement of Gas Pressure

The measurement of the pressure exerted by a gas in a closed container is ordinarily accomplished by a device called a manometer, which is similar but not identical to a barometer (Figure 5.3).

A manometer consists of a glass U-tube partially filled with mercury. One side of the U-tube is connected to the container in which the pressure is to be measured, and the other side is connected to a region of known pressure. The gas in the container will exert a force on the mercury column and will tend to make it go down. This force will be opposed by that created by the gas over the other surface. The difference, as measured by a rule, between the mercury levels at equilibrium is directly proportional to the difference between the gas pressure and the known pressure.

Since pressures in a fluid are equal at equal heights, in Figure 5.3, we can say that at level h_1

$$\text{Pressure}_{known} = \text{Pressure}_{system} + \text{Pressure due to } \Delta h \text{ mm Hg}$$

Ordinarily the known pressure will be expressed in terms of the height of a column of mercury to which it is equivalent; usually the known pressure is

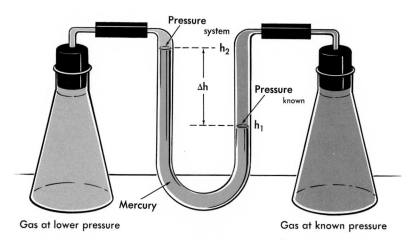

Figure 5.3 The manometer and the measurement of pressure. Pressures at h_1 in right and left sides of manometer are equal. All pressures are in millimeters of mercury.

either that of the air in the laboratory or that of a vacuum. In the former case, the known pressure would be barometric pressure and would be conveniently reported in mm Hg; in the latter case it would be zero. Solving the previous equation for the pressure to be measured, we obtain

$$\text{Pressure}_{\text{system}} = \text{Pressure}_{\text{known}} - \Delta h$$

Since both terms on the right hand side of the equation are in mm Hg, the pressure of the gas in the system will also be in mm Hg. Because of the way in which the measurement of gas pressure is made, gas pressures are usually reported in mm Hg. Given the properties of mercury at the temperature of the manometer, it is possible to find the pressure in absolute units, dynes/cm², but this is rarely done.

5.3 BOYLE'S LAW

One of the first laws to be discovered which related to the behavior of matter was stated in 1660 by Robert Boyle, an English natural philosopher. Boyle performed an experiment with air that is really very simple by modern standards. Using a glass tube and some liquid mercury arranged as in Figure 5.4, he found that the volume of entrapped air varied inversely with the pressure applied to it. Boyle did the experiment in a room in which the temperature was approximately constant, and he needed, in addition to his apparatus, only a barometer and a measuring rule. In order to interpret his data properly Boyle had to understand the role of atmospheric pressure in the experiment. Any schoolboy could repeat the experiment in the laboratory in a few hours, but even in these days it is just possible he might have difficulty in treating the data. Boyle of necessity worked only with ordinary air. When other gases were discovered, however, it was found that they too showed an inverse rela-

Try Problem 5.2 at the end of this chapter.

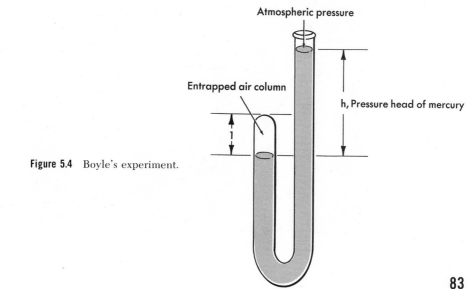

Figure 5.4 Boyle's experiment.

tionship between volume and pressure at constant temperature. This relationship is called Boyle's Law.

Boyle's Law states that: **The volume of a given mass of any gas at constant temperature varies inversely with the pressure.** Alternatively we may state that **the product of the volume times the pressure of a given mass of any gas is a constant at constant temperature**.

Mathematically we would say

pressure × volume = a constant   (for a given mass of gas at constant T)

$$PV = k \quad \text{or} \quad V = k/P \tag{5.1}$$

In Equation 5.1, k is a constant at constant temperature and amount of gas; it may therefore be a function of both temperature and amount of gas.

Boyle's Law can be readily described by a graph. In Figure 5.5 we have plotted the volume of a given amount of gas as a function of pressure when the temperature is held constant. From the graph it is clear that as the gas pressure increases, the volume decreases, in such a way that the product of P × V remains constant. If the pressure is increased fourfold, the volume decreases to one quarter its initial value. If the volume is increased by a factor of ten, the pressure decreases to one tenth its initial value. The actual magnitude of the PV product depends on the units used for the pressure and volume, the amount of gas, and the temperature. The **constancy** of the PV product requires the **constancy** of the amount of gas and the temperature.

Robert Boyle was a remarkable person who, in addition to his discovery of the law which bears his name, made many contributions to scientific thought. One of the most important of these was a book entitled *The Sceptical Chemist* in which he challenged the then prevailing

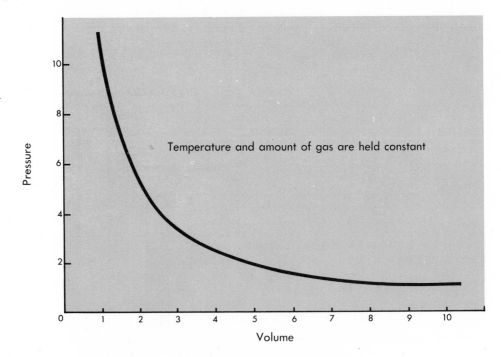

Figure 5.5   Boyle's law.

notion that salt, sulfur, and mercury were the true principles of nature. Although the science of the time was too primitive to allow him to do proper experiments in the area, he advanced the idea that matter was ultimately composed of particles of various sorts which could arrange themselves into groups, and that groups of one kind constituted a chemical substance. Boyle thus used concepts of atomic and molecular theory similar to those we have today and, by some, has been called one of the fathers of modern chemistry. We would prefer to reserve such a title for men like Dalton, Lavoisier, and Priestley, who performed chemical experiments to support their views.

5.4 CHARLES' AND GAY-LUSSAC'S LAW

Boyle's Law, as we have mentioned, was discovered in 1660. It was not until about 100 years later that another quantitative relation for gas behavior was found. This relation involved the way in which the volume of a gas held at constant pressure varied with the temperature. One might wonder why Boyle or his contemporaries did not proceed to investigate the effect of temperature on the properties of gases. Boyle did know that heating a gas would tend to increase its volume and checked to make sure that small temperature changes were unimportant in the experiments that led to his law. At that time, however, the concepts of heat and temperature were not well understood. Any quantitative law involving temperature had to follow the development of the notion of temperature, the creation of a temperature scale, and the construction of a thermometer to measure the temperature. Science did not reach this level until about 1750, when several investigators made these important contributions. Men at that time began to be aware of the existence of gases other than air. Progress in science is made in steps; these steps seem to have a natural order which is only rarely by-passed, and then only by men of extreme insight.

The development of the first balloons large enough to support a man stimulated the search for gases which were lighter than air.

Between about 1780 and 1800 two French scientists, Charles and Gay-Lussac (who, incidentally, were both balloonists at one time or another), independently arrived at the law which bears their names. They found that **when a gas is heated at constant pressure from one given temperature to another, the fractional change in the volume is a constant independent of the gas being studied**. If the two temperatures involved are 0°C and 100°C, the increase in volume for any gas is always just about 37 per cent of the volume at 0°C.

If the change in temperature is 1°C rather than 100°C one finds that the increase in volume is very nearly 0.37 per cent of the volume at 0°C. A change of 2°C increases the volume by about 2×0.37 per cent or 0.74 per cent. The thoughtful student can see that these observations allow us to express the volume of a gas at constant pressure in terms of its centigrade temperature by the following equation:

$$V = V_0 + 0.0037 \, V_0 t = V_0(1 + 0.0037t) \qquad (5.2)$$

in which V is the volume at t°C and V_0 is the volume at 0°C.

This equation can be modified in an interesting manner to yield a very significant result regarding the nature of that property we call temperature. We can say that at t_1°C,

$$V_1 = V_0(1 + 0.0037 \, t_1)$$

and at t_2°C and the same pressure on the same sample of gas,

$$V_2 = V_0(1 + 0.0037 \, t_2)$$

85

Dividing the second equation by the first, we obtain

$$\frac{V_2}{V_1} = \frac{1 + 0.0037\ t_2}{1 + 0.0037\ t_1}$$

Note that
$1/0.0037 = 273$.

Now, if we divide both numerator and denominator of the right side of this equation by 0.0037, we see that

$$\frac{V_2}{V_1} = \frac{273 + t_2}{273 + t_1}$$

The terms $273 + t_1$ and $273 + t_2$ can be very simply interpreted. They would be the *temperatures* as read on a new *temperature scale* on which *all centigrade temperatures were increased by* 273°. Such a temperature scale has some real advantages over the ordinary centigrade scale. For gases the dependence of volume on temperature, in terms of the new scale, is readily expressed. If we let T on the new scale be $273 + t$, the equation becomes

$$\frac{V_2}{V_1} = \frac{T_2}{T_1} \tag{5.3}$$

It has been found that many natural laws are, like the foregoing relation, most simply written in terms of the temperature scale given by

$$T = 273 + t \tag{5.4}$$

and this scale has therefore been adopted for scientific use. T is called the **absolute** temperature, and can always be calculated by Equation 5.4 from the centigrade temperature t. Degrees on the absolute temperature scale are called degrees absolute, °A, or degrees Kelvin, °K, in honor of Lord Kelvin, an English physicist, who arrived at the same scale of temperature on theoretical grounds.

The setting up of the absolute temperature scale completes our study of the temperature concept, but there are a few points which we should clear up before leaving the subject. What has really been done is to *define* the ratio of any two temperatures on the absolute scale to be equal to the ratio of the volumes that a given amount of gas at a fixed, low pressure would have at the two temperatures. If we then keep the same temperature interval as on the centigrade scale, namely 100°, between the freezing and boiling points of water, the absolute scale is completely determined. On this scale the precise value of the freezing point of water is 273.16°A. Absolute temperatures can, at least in principle, be determined with a gas volume thermometer and the foregoing conventions. The centigrade temperature scale is, in the last analysis, defined *after* the absolute scale has been established, the defining relation being

$$t = T - 273.16$$

In practice the scales on the very best centigrade thermometers are made by calibration against a standard gas volume thermometer. Most thermometers, however, are made and calibrated by the procedure described in Chapter 1. This results, rather fortuitously, in fairly accurate thermometers, since the centigrade scale on the ordinary mercury-in-glass thermometer just happens to correspond fairly well to the centigrade scale as determined by the gas volume thermometer.

After studying this discussion of the dependence of gas volume on temperature and the development of the absolute temperature scale, the student might well ask what the law of Charles and Gay-Lussac actually is. Expressed in words, the law states that: **At constant pressure the volume of a given amount of gas is directly proportional to its absolute temperature.** In equation form,

volume = a constant × temperature   (mass of gas and pressure held constant)

$$V = KT \qquad \text{or} \qquad V/T = K \tag{5.5}$$

The actual value of the constant K will depend on the amount of gas in the sample, the pressure, and the units used.

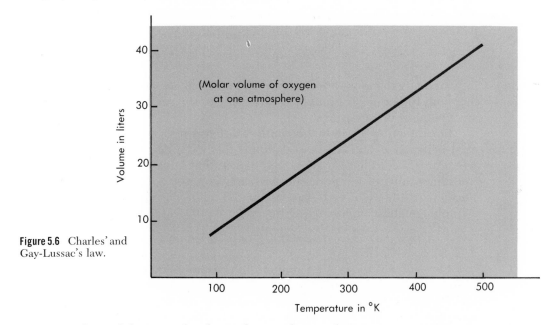

(Molar volume of oxygen at one atmosphere)

Figure 5.6 Charles' and Gay-Lussac's law.

As with Boyle's Law, the dependence of gas volume on temperature as given by Charles' Law can be readily illustrated with a graph. In Figure 5.6 we have plotted the volume of one mole of oxygen at one atmosphere pressure as a function of absolute temperature. Doubling the absolute temperature doubles the volume. Halving the volume at constant pressure would require that the absolute temperature also be halved.

In connection with Charles' Law, the question might arise about the volume the gas would have at very low temperatures. According to that law, as the temperature approaches 0°K the volume of the gas would also approach zero. Since this would be impossible without giving the gas an infinite density at 0°K, we might also wonder about the possibility of reaching the absolute zero of temperature. Actually, it is found that as we approach 0°K, or −273°C, all gases condense to liquids or solids having finite volumes at finite temperatures. It is also true, however, that a very tight theoretical argument can be presented which predicts the impossibility of attaining the absolute zero of temperature. Modern experimental results support this prediction; the lowest temperatures reached under the best conditions are about 0.001°K.

5.5 THE IDEAL GAS EQUATION

The laws of Boyle and Charles relate in two equations the pressure, volume, and temperature of a given amount of gas. We shall now show how these equations can be combined to obtain a more general law for the behavior of gases.

According to Boyle's Law (with temperature and amount of gas constant),

$$V = k/P \qquad (5.1)$$

According to Charles' Law (with pressure and amount of gas constant),

$$V = KT \qquad (5.5)$$

If we double the amount of gas in a container, holding pressure and temperature constant, the volume of the gas will be doubled. In general, the volume of a gas under given conditions will be proportional to the amount (mass) of gas present. This requires that both k and K in Equations 5.1 and 5.5 be themselves proportional to the amount of gas. That is,

$$V = k'a/P = K'aT \qquad (k'a = k;\ \kappa'a = K) \qquad (5.6)$$

in which a is the amount of gas. In Equation 5.6 account is taken of the fact that both Equations 5.1 and 5.5 must apply to the same sample of gas under any given set of conditions.

For Equation 5.6 to be generally valid, it is necessary that a relationship exist between k' and K'; the relation is that k' equals cT and K' equals c/P, where c is a constant for any given gas and has a magnitude determined by the dimensions of temperature, pressure, volume, and amount of gas used. Recognizing the relation between k' and K' allows us to write Equation 5.6 as a single equation, summarizing the general physical behavior of gases,

The volume of a gas (V) is *directly* proportional to the amount (a) and absolute temperature (T); it is *inversely* proportional to pressure (P).

$$V = acT/P \qquad \text{or} \qquad PV = acT \qquad (5.6a)$$

For each gas there is a value of the constant c which can be determined from a single measurement of P, V, T, and a for a sample of the gas under a given set of conditions. Once c is found for a gas, Equation 5.6a becomes an unequivocal relation between the volume, amount, temperature, and pressure for any sample of that gas and allows us to calculate any one of the four properties of the sample, given the other three.

We find that if we agree to express the amount of gas in terms of *moles*, rather than in some other mass unit such as grams, Equation 5.6a takes on a remarkably simple form, in that under those conditions the constant c has the *same value for all gases*. This allows us to write a single equation applicable to all gases and offers such a distinct advantage that the amount of gas in Equation 5.6a is invariably expressed as the number of moles, n. The constant c, which then becomes the same for all gases, is given the symbol R. In terms of the new symbolism, Equation 5.6a takes the form

$$PV = nRT \qquad (5.7)$$

and is called the **Ideal Gas Law**.

The reason that the constant R in the Ideal Gas Law equation is not dependent on the kind of gas being studied follows from a fundamental postulate first stated in 1813 by Amadeo Avogadro, an Italian physicist. He recognized that Gay-Lussac's observations on the relative volumes of gases which react chemically (Section 5.6) could be most simply explained if **equal volumes of different gases under the same conditions of temperature and pressure**

TABLE 5.1 VALUES OF THE GAS CONSTANT R IN VARIOUS UNITS

FOR PRESSURE IN	AND VOLUME IN	R HAS THE VALUE
Atmospheres	liters	0.0821 lit atm*/mole °K
Atmospheres	milliliters	82.1 ml atm/mole °K
Millimeters of mercury	liters	62.3 lit mm Hg/mole °K
Millimeters of mercury	milliliters	6.23×10^4 ml mm Hg/mole °K

* Liter atmosphere

contain equal numbers of molecules. This postulate is now accepted as a basic principle of chemistry and is called Avogadro's Law.

Returning now to Equation 5.6a,

$$V = acT/P$$

we can see that if we take equal *volumes* of any two different gases at the same *temperature* and *pressure*, then P, V, and T for the two gases will obviously have equal values. By Avogadro's Law the number of molecules in the two samples will be equal; since a mole of gas always contains 6.023×10^{23} molecules, the number of moles of gas in the two samples must also be equal, and hence the amount, a, in moles, is the same in the two samples. Therefore, to give the right side of Equation 5.6a the proper magnitude for the two gas samples, it is necessary that c be the same for the two gases and hence for all gases. Equation 5.7, with the proportionality constant expressed as R and the amount of gas expressed in moles, immediately follows.

In the Ideal Gas Law, R is called the **gas constant** and has a numerical value which depends on the units used for pressure and volume. In Table 5.1 we have listed the values of R which are obtained for several common sets of pressure-volume units.

The Ideal Gas Law is one of the more important relations in chemistry. Its importance lies in the fact that it relates in a definite way the pressure, volume, temperature, and amount of *any* gas, as long as the pressure of the gas is not too high. There are many possible applications of the Ideal Gas Law, making it a very powerful tool in the hands of one who understands it.

The following examples illustrate some of the situations in which the Ideal Gas Law can be employed. In all cases we will take the value of R to be 0.0821 lit atm/mole °K and will, where necessary, convert the units of pressure and volume accordingly.

One mole of gas occupies 22.4 lit at STP (0°C, 1 atm).

EXAMPLE 5.1. What volume will a sample of 240 ml of argon occupy if its pressure is increased at 20°C from 0.10 atm to 6.0 atm?

SOLUTION. For the argon in both states, PV = nRT.

Initially: $P_1 = 0.10$ atm; $V_1 = 240$ ml; $n_1 = n$; $T_1 = T$ $P_1V_1 = nRT$

Finally: $P_2 = 6.0$ atm; $V_2 = ?$; $n_2 = n$; $T_2 = T$ $P_2V_2 = nRT$

Since both the temperature and amount of gas are held constant in this problem, the right sides of the two equations are equal, and so

$$P_1V_1 = P_2V_2$$

89

Substituting: 0.10 atm × 240 ml = 6.0 atm × V$_2$

$$V_2 = \frac{0.10 \text{ atm} \times 240 \text{ ml}}{6.0 \text{ atm}} = 4.0 \text{ ml}$$

This problem is typical of many encountered with gases. The state of the gas is changed, with one or more of its variables remaining fixed. By writing the gas law for the two states, one can usually recognize quickly the terms in the two equations which are equal and use that information and the given data to calculate the unknown quantity.

EXAMPLE 5.2. A sample of hydrogen gas, collected at 100°C and 1.00 atm pressure, has a volume of 350 ml. It is subsequently transferred to a 4.00-lit flask and cooled to 25°C. What pressure, in millimeters of mercury, will it exert in the flask?

SOLUTION. For the hydrogen in the two states, PV = nRT.

Initially: P$_1$ = 1.00 atm; V$_1$ = 350 ml × $\frac{1 \text{ lit}}{1000 \text{ ml}}$ = 0.350 lit;

n$_1$ = n; T$_1$ = 273 + 100 = 373°K

$$P_1V_1 = nRT_1$$

Finally: P$_2$ = ?; V$_2$ = 4.00 lit; n$_2$ = n; T$_2$ = 273 + 25 = 298°K

$$P_2V_2 = nRT_2$$

In this problem nR is the same in the two states, and so,

$$\frac{P_1V_1}{T_1} = \frac{P_2V_2}{T_2}$$

Therefore,

$$P_2 = \frac{V_1}{V_2} \times \frac{T_2}{T_1} \times P_1$$

$$P_2 = \frac{0.350 \text{ lit}}{4.00 \text{ lit}} \times \frac{298°K}{373°K} \times 1.00 \text{ atm} = 0.0700 \text{ atm} \times \frac{760 \text{ mm Hg}}{1 \text{ atm}} = 53.2 \text{ mm Hg}$$

This problem is again one that involves a change in state of the gas. Here, however, the pressure, volume, and temperature all change, and the units given are not the same in the two states. By carrying all units in the calculations, we see that we must convert both volumes to the same units, if the volumes are to cancel properly. We could have, of course, converted P$_1$ to millimeters of mercury before making any calculations of P$_2$. *One must not, in any case, ignore the units of the quantities in the calculation,* since they are as important as the numbers themselves, and must be expressed properly if the result is to have the proper units.

Previous examples were concerned with changes in state of a given amount of gas. In such problems, n, the number of moles of gas, is constant, and the value of R need not be known, since R cancels from the calculation. The following problems are somewhat more complex and require full use of the Ideal Gas Law, including the value of R. They are illustrative of the large amount of information that the law allows one to obtain about gases.

EXAMPLE 5.3. 2.50 g of nitrogen gas are introduced into an evacuated 3.00-lit container at −80°C. Find the pressure in atmospheres in the container.

SOLUTION. In this problem only one state is involved, and for it, PV = nRT.

$$P = ?; \quad V = 3.00 \text{ lit}; \quad n = 2.50 \text{ g } N_2 \times \frac{1 \text{ mole}}{28.0 \text{ g } N_2} = 0.0893 \text{ mole}$$

$$R = 0.0821 \text{ lit atm/mole °K}; \quad T = 273 - 80 = 193°K$$

Substituting: $P = \dfrac{nRT}{V} = \dfrac{0.0893 \text{ mole}}{3.00 \text{ lit}} \times 0.0821 \dfrac{\text{lit atm}}{\text{mole °K}} \times 193°K = 0.472 \text{ atm}$

Here all the elements in the Ideal Gas Law enter the calculation directly; if we use 0.0821 lit atm/mole °K for R, the units of all the terms are of necessity those that appear in R, and any quantities which do not have those units must be converted before substituting in the Gas Law.

EXAMPLE 5.4. A lighter-than-air balloon is designed to rise to a height of 25 miles, at which point it will be fully inflated. At that altitude the atmospheric pressure is 2.40 mm Hg and the temperature is −3°C. If the full volume of the balloon is one hundred thousand liters, how many pounds of helium will be needed to inflate the balloon?

SOLUTION. To calculate the number of pounds of helium required, we shall first use the Ideal Gas Law to obtain the number of moles, n, and then convert from moles to grams and finally to pounds.

Since balloons have flexible walls, the volume of the balloon will vary with the applied pressure, the pressure in the balloon being substantially equal to the outside atmospheric pressure. Hence, for the helium at a height of 25 miles,

$$P = 2.40 \text{ mm Hg} \times \frac{1 \text{ atm}}{760 \text{ mm Hg}} = 0.00316 \text{ atm}$$

$$V = 1.000 \times 10^5 \text{ lit}; \quad n = ?; \quad R = 0.0821 \text{ lit atm/mole °K}; \quad T = 273 - 3 = 270°K$$

$$n = \frac{PV}{RT} = \frac{0.00316 \text{ atm} \times 1.000 \times 10^5 \text{ lit}}{\dfrac{0.0821 \text{ lit atm}}{\text{mole °K}} \times 270°K} = 14.3 \text{ moles}$$

Since 1 mole of helium weighs 4.00 grams, the mass of the helium is

$$14.3 \text{ moles} \times \frac{4.00 \text{ g}}{1 \text{ mole}} = 57.2 \text{ g} \times \frac{1 \text{ lb}}{453.6 \text{ g}} = 0.126 \text{ lb}$$

Why three significant figures?

5.6 OTHER APPLICATIONS OF THE IDEAL GAS LAW

There are some problems in which the use of the Ideal Gas Law is not quite so direct as in the previous examples. Two important cases involve the determination of the density of a gas under given conditions and the evaluation of its molecular weight from experimental data. These problems are readily treated if one recognizes the relationships between density or molecular weight and the variables in the Gas Law. These relations allow us to state the law in terms of the density or molecular weight and other measured quantities.

Recalling that

$$\text{density} = \frac{\text{mass}}{\text{volume}}, \qquad \left(d = \frac{g}{V} \right)$$

and that

$$\text{number of moles} = \frac{\text{mass}}{\text{gram molecular weight}}, \qquad \left(n = \frac{g}{M} \right)$$

we can write the Ideal Gas Law as

$$PV = nRT = \frac{g}{M} RT \qquad \text{or} \qquad P = \frac{n}{V} RT = \frac{gRT}{VM} = d \frac{RT}{M} \qquad (5.8)$$

These equations are merely alternate forms of the Gas Law and involve no new concepts. Their use is illustrated in the following examples.

EXAMPLE 5.5. What is the density in grams per liter of sulfur dioxide at 25°C and 300 mm Hg pressure?

SOLUTION. Solving Equation 5.8 for density, $d = \frac{PM}{RT}$

For the SO_2 sample, $\quad P = 300$ mm Hg $\times$ 1 atm/760 mm Hg $= 0.395$ atm

$$M_{SO_2} = (32.07 + 2 \times 16.00) \text{ g/mole} = 64.07 \text{ g/mole}$$

$$R = 0.0821 \text{ lit atm/mole °K}; \quad T = 273 + 25 = 298°K$$

On substitution, $\quad d = \dfrac{0.395 \text{ atm} \times 64.07 \text{ g/mole}}{\dfrac{0.0821 \text{ lit atm}}{\text{mole °K}} \times 298°K} = 1.04 \text{ g/lit}$

EXAMPLE 5.6. A sample of chloroform weighing 0.5280 g is collected as a vapor (gas) in a flask having a volume of 127 ml. At 75°C the pressure of the vapor in the flask is 754 mm Hg. Calculate the molecular weight of chloroform.

SOLUTION. By the alternate form of the Gas Law we see that

$$M = g \frac{RT}{PV}$$

We have merely to express the variables in the equation in the proper units and solve for the gram molecular weight by substitution.

$$g = 0.5280 \text{ g}; \quad R = 0.0821 \text{ lit atm/mole °K}; \quad T = 273 + 75 = 348°K$$

$$P = 754 \text{ mm Hg} \times \frac{1 \text{ atm}}{760 \text{ mm Hg}} = 0.992 \text{ atm}; \quad V = 127 \text{ ml} \times \frac{1 \text{ lit}}{1000 \text{ ml}} = 0.127 \text{ lit}$$

$$M = \frac{0.5280 \text{ g} \times \dfrac{0.0821 \text{ lit atm}}{\text{mole } {}^\circ\text{K}} \times 348{}^\circ\text{K}}{0.992 \text{ atm} \times 0.127 \text{ lit}} = 120 \text{ g/mole}$$

The calculated molecular weight of chloroform is therefore 120.

The data in Example 5.6 are typical of those obtained in one of the simplest experimental methods for the determination of molecular weights. In the experiment a few milliliters of a volatile liquid are placed in a flask fitted with a stopper in which there is a fine orifice (Figure 5.7). The flask is then heated in a water bath to a temperature somewhat above the boiling point of the liquid. The liquid evaporates, and its vapor replaces the air in the flask. After all the liquid has evaporated and the flask is filled with vapor, the flask is removed from the bath and allowed to cool. The vapor condenses and air re-enters the flask. The mass of the vapor is taken to be the difference between the mass of the flask containing the condensed vapor and its mass when dry. The method is an approximate one for several reasons, but when properly done it gives results accurate to within a few per cent.

Another application of the gas laws arises in connection with our interpretation of chemical reactions. We have seen that, given any chemical equation, it is possible to calculate the relative masses of reactants and products as well as to state immediately the relative numbers of moles. Where gases are present it is possible to extend the interpretation of chemical equations to include the volumes of the gases that would be involved under given conditions. The procedure is shown in the following example.

Would this procedure work for sugar? For chlorobenzene (BP = 132°C)?

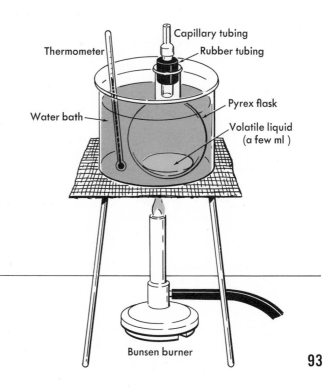

Capillary tubing
Thermometer
Rubber tubing
Pyrex flask
Water bath
Volatile liquid (a few ml)
Bunsen burner

Figure 5.7 Determining molecular weight by vapor density.

EXAMPLE 5.7. One method for the commercial production of chlorine uses the electrolysis of molten sodium chloride. The chemical reaction that occurs is

$$2 \text{ NaCl(l)} \longrightarrow 2 \text{ Na(l)} + \text{Cl}_2\text{(g)}$$

How many liters of chlorine, measured at 25°C and 1 atm, can be produced from one kg of sodium chloride?

SOLUTION. The chemical equation tells us that

$$2 \text{ moles NaCl} \longrightarrow 1 \text{ mole Cl}_2$$

$$117 \text{ g NaCl} \longrightarrow 1 \text{ mole Cl}_2$$

Therefore, 117 g NaCl ≏ 1 mole Cl$_2$ and from 1000 g NaCl we would obtain

$$1000 \text{ g NaCl} \times \frac{1 \text{ mole Cl}_2}{117 \text{ g NaCl}} = 8.55 \text{ moles Cl}_2$$

The chlorine will obey the Gas Law, PV = nRT, in which P = 1 atm; V = ?; n = 8.55 moles; R = 0.0821 lit atm/mole °K; T = 298°K. Hence,

$$V = \frac{nRT}{P} = \frac{8.55 \text{ moles} \times \dfrac{0.0821 \text{ lit atm}}{\text{mole °K}} \times 298\text{°K}}{1 \text{ atm}} = 209 \text{ lit}$$

In problems of this kind it is best to calculate the number of moles of gas produced or used up in the reaction and then to use the Gas Law to find the volume.

A rather interesting interpretation of chemical reactions can be made when the substances involved are all gases whose volumes are measured under the same conditions of temperature and pressure. Consider the following chemical reaction:

$$4 \text{ NH}_3\text{(g)} + 5 \text{ O}_2\text{(g)} \longrightarrow 4 \text{ NO(g)} + 6 \text{ H}_2\text{O(g)}$$

According to the usual interpretation, we would say

$$4 \text{ moles NH}_3 + 5 \text{ moles O}_2 \longrightarrow 4 \text{ moles NO} + 6 \text{ moles H}_2\text{O}$$

If all these gases are measured at the same temperature and pressure, their molar volumes will all be equal, say to some volume V_m liters. Under such conditions it would be true that

$$4 \text{ } V_m \text{ lit NH}_3 + 5 \text{ } V_m \text{ lit O}_2 \longrightarrow 4 \text{ } V_m \text{ lit NO} + 6 \text{ } V_m \text{ lit H}_2\text{O}$$

or, dividing through by V_m, simply

$$4 \text{ lit NH}_3 + 5 \text{ lit O}_2 \longrightarrow 4 \text{ lit NO} + 6 \text{ lit H}_2\text{O}$$

if all the gases are measured at the same temperature and pressure.

Volumes of gases, then, when measured under the same conditions, have the same simple numerical relationships that exist between moles of substances in chemical reactions. The remarkable fact is that, whereas the relation between moles is deduced from theory, that between volumes can be found experimentally. Indeed, in 1808 Gay-Lussac discovered the relationship and stated his Law of Combining Volumes: **In any chemical reaction in-**

volving gaseous substances the volumes of the various gases reacting or produced are in the ratios of small whole numbers. (The gases are measured at the same temperature and pressure.)

In 1811, Avogadro, who saw the implications of Gay-Lussac's Law, stated his famous hypothesis, as we have noted in Chapter 2, can lead to correct and unequivocal values of atomic weights. Unfortunately, the chemists of the time were skeptical of the work of Gay-Lussac and Avogadro, and, by plausible but incorrect arguments, succeeded in discrediting it. Paraphrasing, in modern terms, Dalton's faulty line of reasoning: "It is proposed experimentally that, in the reaction between nitrogen and oxygen to form nitric oxide, one volume of nitrogen combines with one volume of oxygen to form two volumes of nitric oxide. If this is so, then x atoms of nitrogen combines with x atoms of oxygen to yield 2x atoms of nitric oxide. Or, $\frac{1}{2}$ atom of nitrogen combines with $\frac{1}{2}$ atom of oxygen to produce 1 atom of nitric oxide. *But thou canst not split an atom.* Therefore, either the atomic theory is incorrect, or Gay-Lussac's experimental procedure is inaccurate."

Dalton chose to believe the latter, and, in spite of the fact that Gay-Lussac's Law was later proved to be valid, the matter of atomic weights remained confused for 50 years. The situation was not straightened out until 1860, when Cannizzaro, recognizing clearly for the first time the distinction between atoms and molecules, properly interpreted Avogadro's Hypothesis and laid the foundation for the determination of atomic weights (Problem 2.28).

5.7 MIXTURES OF GASES: DALTON'S LAW OF PARTIAL PRESSURES

So far we have considered the physical properties of gaseous systems in which only one component is present. If several substances are present in a gaseous solution, we can still use the gas laws, but must take proper account of the presence of the different substances. Let us assume we have a gaseous mixture of A, B, and C, confined in a container of volume V. The Ideal Gas Law will apply to the mixture, provided that we let n equal the total number of moles present. That is,

$$PV = nRT \quad \text{or} \quad P = \frac{nRT}{V}$$

in which P is the total pressure in the container and n equals $n_A + n_B + n_C$.

The three substances A, B, and C each contribute to the total pressure in the system. Their individual contributions, P_A, P_B, and P_C may be obtained by the Gas Law. We can say that

$$P_A = \frac{n_A RT}{V}; \quad P_B = \frac{n_B RT}{V}; \quad P_C = \frac{n_C RT}{V}$$

P_A, P_B, and P_C are called the **partial pressures** of A, B, and C in the container. The partial pressure of a gas is the pressure the gas would exert if it alone were present in the container at the same temperature as the mixture. It is clear that since

$$P = \frac{nRT}{V} = (n_A + n_B + n_C)\frac{RT}{V} = \frac{n_A RT}{V} + \frac{n_B RT}{V} + \frac{n_C RT}{V}$$

it follows that

$$P = P_A + P_B + P_C \tag{5.9}$$

This equation is a mathematical statement of Dalton's Law of Partial Pressures, first proposed by John Dalton in 1807. In words the law is: **The**

How could you
test Dalton's law
experimentally?

total pressure in a container is equal to the sum of the partial pressures of the component gases.

Dalton's Law makes it possible to handle easily problems in which gaseous solutions are involved. The following examples illustrate typical applications.

EXAMPLE 5.8. A ten-liter flask at 25°C contains a gaseous solution of carbon monoxide and carbon dioxide at a total pressure of 2.00 atm. If 0.20 mole of carbon monoxide is present, find its partial pressure and also that of the carbon dioxide.

SOLUTION. By Dalton's Law $P = P_{CO} + P_{CO_2} = 2.00$ atm

$$P_{CO} = n_{CO} \frac{RT}{V} = \frac{0.20 \text{ mole} \times \dfrac{0.0821 \text{ lit atm}}{\text{mole °K}} \times 298°K}{10 \text{ lit}} = 0.49 \text{ atm}$$

$$P_{CO_2} = P - P_{CO} = (2.00 - 0.49) \text{ atm} = 1.51 \text{ atm}$$

EXAMPLE 5.9. Dry air contains about 78 per cent nitrogen, 21 per cent oxygen, and 1 per cent argon by volume. Find the partial pressures of nitrogen, oxygen, and argon in dry air at 25°C and 1.00 atm pressure.

SOLUTION. Composition of gaseous solutions is frequently given in percentages by volume. Stating that dry air contains 78 per cent nitrogen by volume means that, if air were fractionated into its components and the nitrogen obtained were brought to the same temperature and *total* pressure as the original sample of air, the volume of the nitrogen would be 78 per cent of that of the original sample.

For the whole sample of air

$$P = \frac{nRT}{V} \qquad \text{and} \qquad n = n_{N_2} + n_{O_2} + n_{Ar}$$

For the nitrogen in the sample

$$P_{N_2} = \frac{n_{N_2}RT}{V}$$

Dividing the second equation by the first,

$$\frac{P_{N_2}}{P} = \frac{n_{N_2}}{n}$$

But we can also say that

$$V = \frac{nRT}{P} \qquad \text{and} \qquad V_{N_2} = \frac{n_{N_2}RT}{P} = 0.78 \text{ V}$$

Therefore, it is true that

In general, for
gases, "per cent
by volume" =
"mole per cent."

$$\frac{V_{N_2}}{V} = \frac{n_{N_2}}{n} = \frac{P_{N_2}}{P} = 0.78$$

and so

$$P_{N_2} = 0.78 \text{ P} = 0.78 \times 1 \text{ atm} = 0.78 \text{ atm}$$

By similar reasoning,

$$P_{O_2} = 0.21 \text{ atm} \qquad \text{and} \qquad P_{Ar} = 0.01 \text{ atm}$$

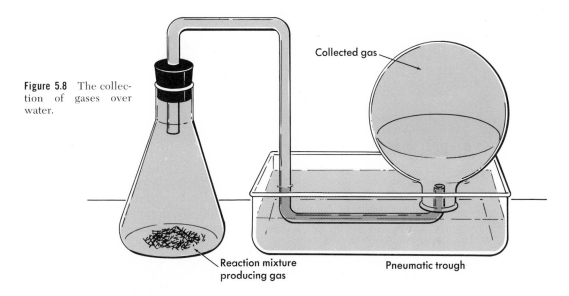

Figure 5.8 The collection of gases over water.

Collected gas

Reaction mixture producing gas

Pneumatic trough

Dalton's Law often finds practical use in experiments involving gases. In order to measure the amount of a gas produced in a chemical reaction we must collect the gas under known conditions. Probably the easiest way of doing this is to let the gas displace water in a system such as that shown in Figure 5.8. In this way we can measure the volume of gas at atmospheric pressure and known temperature. If the gas were pure, we could immediately use the Gas Law to calculate the number of moles produced by the reaction. However, under the conditions of the experiment the gas collected contains water vapor in addition to the gas of interest. The true pressure of the gas produced is, therefore, by Dalton's Law, equal to the total pressure minus the partial pressure of the water vapor. It is found experimentally that the pressure of water vapor in the presence of liquid water is a constant at a given temperature; its value can be obtained from a table and used in the calculation. The following example is illustrative.

The pressure of vapor in equilibrium with its liquid is called the vapor pressure of the liquid and is discussed in detail in Chapter 10.

EXAMPLE 5.10. In a laboratory experiment, concentrated hydrochloric acid was reacted with aluminum. Hydrogen gas evolved and was collected over water at 25°C; it had a volume of 355 ml at a total pressure of 750 mm Hg. How many moles of hydrogen were collected? At 25°C the vapor pressure of water is known to be about 24 mm Hg.

SOLUTION. By Dalton's Law $P = P_{H_2} + P_{H_2O}$

$P_{H_2O} = 24$ mm Hg, the vapor pressure of H_2O at 25°C

Therefore

$$P_{H_2} = P - P_{H_2O} = (750 - 24) \text{ mm Hg} = 726 \text{ mm Hg}$$

$$= 726 \text{ mm Hg} \times \frac{1 \text{ atm}}{760 \text{ mm Hg}} = 0.955 \text{ atm}$$

97

By the Gas Law

$$P_{H_2} = \frac{n_{H_2}RT}{V} \qquad n_{H_2} = \frac{P_{H_2}V}{RT} = \frac{0.955 \text{ atm} \times 0.355 \text{ lit}}{\dfrac{0.0821 \text{ lit atm}}{\text{mole } ^\circ K} \times 298^\circ K} = 0.0139 \text{ mole}$$

$$n_{H_2} = 0.0139 \text{ mole}$$

5.8 REAL GASES

In this chapter we have applied the various gas laws to all gases, tacitly assuming that the laws were obeyed exactly. Under ordinary conditions (and in all problems involving gases in this text) the assumption is a very good one. Actually, however, all gases deviate to some extent from the ideal laws, by amounts that depend on the gas, its temperature, and its pressure. For gases in the vicinity of room temperature and 1 atm pressure the deviation is small, at most a few per cent. Gases like oxygen and hydrogen, which boil far below 25°C, have molar volumes that are within 0.1 per cent of the value calculated by the Ideal Gas Law. Sulfur dioxide and chlorine, which boil at −10° and −35°C respectively, are, at 25°C and 1 atm, not so nearly ideal and have molar volumes that are 2.4 and 1.6 per cent lower than the ideal value.

We can illustrate the behavior of real gases graphically by plotting PV/RT for a mole of gas as a function of pressure. Figure 5.9 is such a graph for several gases at 0°C. For a mole of ideal gas PV/RT would be a constant equal to 1.00 at all pressures. Actually, at relatively low pressures (less than 100 atm) PV/RT is less than 1 in the vicinity of 25°C for all gases except hydrogen and helium, and approaches 1 as the pressure approaches zero. The pressure of most gases at a given volume and temperature is, then, somewhat *less* than would be predicted by the Gas Law. This behavior is caused, we

At high P or low T, the Ideal Gas Law loses much of its usefulness.

believe, by the small attractions (see Chapter 10) that exist between molecules even in the gas phase. As the temperature of the gas is lowered, the energy of motion of the molecules is lowered, and the effect of the attractive forces becomes more important, until, at the boiling point of the substance, the forces are large enough to cause the molecules to condense to a liquid. These forces in all cases tend to reduce the value of PV/RT and, in view of the foregoing line of reasoning, are important at temperatures near the boiling point of the substance. This qualitatively explains why SO_2 is less nearly ideal in its behavior than oxygen at 25°C. As the temperature of a gas is raised, the attractive forces become less and less significant; at very high temperatures the PV/RT behavior of all gases approaches that of hydrogen at 25°C, an effectively high temperature as far as the physical behavior of hydrogen is concerned.

The PV/RT behavior of gases at very high pressures (500 atm and above) is shown at the right of Figure 5.9. For all gases, as the pressure is increased, PV/RT ultimately begins to increase, finally become much greater than the ideal value. Such an effect is best understood by a consideration of the volumes of gas molecules. As we noted, in general the gas volume is much larger than the actual volume of the constituent gas molecules. However, the volume of the molecules is not zero (being roughly the volume of the liquid made by condensing the molecules), and as the pressure increases, the volume of

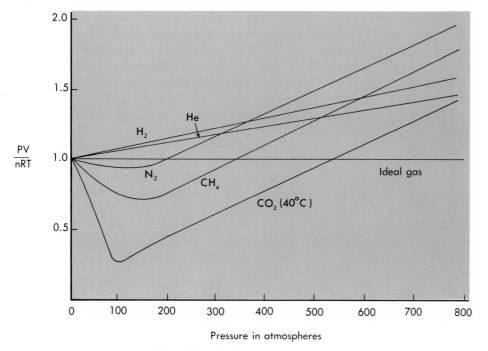

Figure 5.9 The behavior of real gases at 0°C.

the molecules becomes a larger and larger fraction of the gas volume. The volume in which the gas molecules can actually move about is thus decreased by the volume taken up by the molecules, so that the pressure of the gas at a given volume and temperature tends to be somewhat larger than that predicted by the Gas Law. This effect becomes greater at high pressures and accounts for the high values of PV/RT at very high pressures. At intermediate pressures the attractive forces and finite molecular volume effects tend to counteract each other.

It is possible to write equations of state for gases which take account of both intermolecular attractions and finite molecular volumes. Perhaps the best known of these relations is the Van der Waals equation, which for one mole of gas takes the form

$$P = RT/(V - b) - a/V^2 \qquad (5.10)$$

in which b and a are constants selected for each gas to give the best possible agreement between the equation and actual experimental behavior. In spite of its relative simplicity, this equation gives good qualitative prediction of real gas behavior; the constant b can be related to molecular volumes and the constant a to molecular attractions. The Van der Waals equation is used in treating problems in which the nonideality of the gas is important and an analytical relation between P, V, and T is required.

The term a/V^2 corrects for attractive forces, which tend to decrease P; the constant b corrects for molecular volume, which tends to increase P.

5.9 THE KINETIC THEORY OF GASES

The fact that the Ideal Gas Law can be used to summarize reasonably accurately the physical behavior of all gases, whatever their degree of molecular complexity, is a clear indication that the gaseous state of matter is a relatively simple one to attempt to treat from a theoretical point of view. There must be certain properties common to all gases which cause them to follow the same natural law and exhibit so many generally similar characteristics.

With the experimental work on gases by Charles, Gay-Lussac, Graham, and others, enough information on gas behavior became available to allow the development of a theory of gases. Between 1850 and 1880, Maxwell, Boltzmann, Clausius, and others, using the notion that the properties of gases are the result of molecular motions, developed the kinetic theory of gases. The theory was shown to be consistent with the known laws of gas behavior and implied much about gases that was then unknown. Since the time of its creators, who, incidentally, rank among the very best theoreticians the world has known, the theory has had to be modified to only a small extent to make it consistent with quantum theoretical principles. In its present form it is one of the most successful of scientific theories, ranking in stature with the Copernican theory for the motion of the planets and the atomic theory of the nature of matter.

Postulates of the Kinetic Theory of Gases

Here, the word "molecule" includes monatomic species such as He.

1. *Gases consist of molecules in continuous, random motion.* The molecules undergo collisions with one another and with the container walls. The pressure of a gas arises from the forces associated with wall collisions.

2. *Molecular collisions are elastic.* During collisions, there are no frictional losses which result in loss of energy of motion. The temperature of a gas insulated from its surroundings does not change.

3. *The average energy of translational motion of a gas molecule is proportional only to the absolute temperature.* The energy associated with the motion of a molecule from one place to another is dependent on temperature but not on pressure or on the nature of the molecule. Mathematically, this postulate takes the form,

$$\epsilon = (\tfrac{1}{2})mu^2 = cT \tag{5.11}$$

in which ϵ is the average molecular energy and c is a proportionality constant that has the same value for all molecules, whatever their nature.

In addition to these postulates it is often assumed that the volumes of molecules are negligible as compared to container volume and that molecules do not exert forces on each other except by collisions.

The postulates of the kinetic theory of gases are easily stated. Their implications are, however, by no means obvious, and can lead to extremely complicated mathematical equations. In this discussion we will be able to explore only a few of the simpler applications of the theory.

The Ideal Gas Law

Gas pressure, according to the kinetic theory, is the direct result of the forces associated with the collisions of molecules with the container walls. The molecules in a gas move about in a completely random manner, at speeds that are constantly changing because of molecule-molecule and molecule-wall collisions. To treat the forces resulting from molecular motion of this

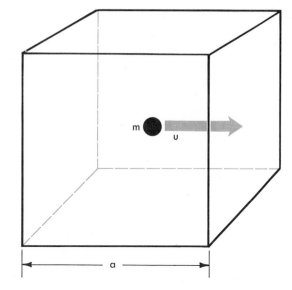

Figure 5.10 A molecule moving in a cubic box. The molecule of mass m travels the distance of the dimension of the container, a, at a constant speed, u.

sort in a rigorous way requires mathematics of a high level of sophistication; of necessity we will limit ourselves to a simplified model that is illustrative of the approach used and yields the same results as the more complete treatment.

Let us first consider a system in which a single molecule is present in a cubic container, as in Figure 5.10. For simplicity we shall assume that the molecule moves at a constant speed u and in a direction perpendicular to the end walls. If the mass of the molecule is m, its momentum, which is the product of mass times velocity in a given direction, will equal mu as the molecule moves in the right-hand direction. After collision with the right-hand wall, the molecule will move to the left with a velocity of $-u$ and hence with a momentum of $-mu$. The change in momentum which occurs at the end walls as a result of a collision will be 2mu for every collision. According to Newton's Laws of Motion,

Momentum is a vector quantity, which has sign as well as magnitude.

$$\text{Force} = \text{Rate of change of momentum}$$

At the right-hand wall the molecule's rate of change of momentum will be equal to 2mu, the change per collision, times the number of collisions occurring at that wall in one second. If the distance between the walls of the container is a, the molecule must travel a distance of 2a between collisions with the right hand wall. As the molecule bounces back and forth across the container at constant speed u, it will hit the wall u/2a times per second. Therefore, by Newton's Laws of Motion, the force f, exerted on the wall by the molecule is given by the relation,

$$f = (2mu)(u/2a) = mu^2/a \qquad (5.12)$$

The molecule we have been considering would only exert force on the end walls of the container, since it would not hit the other walls. It is well known that the properties of a gas are the same in all directions, so that if our simple model is to have the observed properties of a gas, for every molecule moving in the direction we considered, there must be two other molecules

101

present in the container, one moving in a vertical direction and the other moving normal to the plane of the paper and to the plane of the front and rear container walls. For a gas sample containing x molecules, all moving at the same speed in the directions indicated, only x/3 molecules would move in the direction perpendicular to the right-hand wall. The force on that wall will simply be the sum of the forces exerted by the individual molecules that strike it,

$$F = (x/3)f = (x/3)(mu^2/a) = xmu^2/3a \qquad (5.13)$$

Pressure must be the same in all directions; to calculate it, we consider what happens at one wall.

The pressure on the wall is equal to the force divided by the wall area, which for a cubic box of dimension a is a^2. Therefore,

$$P = F/a^2 = xmu^2/3a^3 = xmu^2/3V \qquad (5.14)$$

in which V is the container volume. Equation 5.14 is also obtained in the complete kinetic theory treatment, with u being an average molecular speed. The equation is one of the main results of the theory, and relates the pressure and volume of a gas to the average speed of its molecules.

The Ideal Gas Law follows from Equation 5.14 and the postulate of the kinetic theory that the average energy of translational motion of a gas molecule is dependent only on the absolute temperature and is proportional to it.

$$\epsilon = (\tfrac{1}{2})mu^2 = cT \qquad (5.11)$$

Eliminating mu^2 from Equation 5.14 by using Equation 5.11, we obtain an equation of state for the gas:

$$PV = xmu^2/3 = 2x\epsilon/3 = (\tfrac{2}{3})xcT \qquad (5.15)$$

Recognizing that the number of molecules x equals the number of moles, n, multiplied by Avogadro's number N,

$$PV = (\tfrac{2}{3})nNcT \qquad (5.16)$$

which is the Ideal Gas Law if the proportionality constant c is defined in magnitude to make R, the gas constant, equal to 2Nc/3. This development shows that the Ideal Gas Law is consistent with the kinetic theory, as indeed it should be.

Graham's Law

Characteristic of a successful theory is the fact that it can predict many of the experimental properties of the system to which it is applied. In the previous development the average molecular speed canceled from the equations so that no real evidence regarding molecular speeds was obtained. However, it is possible to test that aspect of the theory in several ways. Probably the simplest of these involves the phenomena of diffusion and effusion.

If two different gases, in separate containers at the same temperature and pressure, are separated by stopcocks from a third gas confined at the same temperature and pressure, they will, when the stopcocks are opened, tend to diffuse into the third gas (see Figure 5.11). The rate of diffusion of each gas will be proportional to the average speed of its molecules. The less massive molecules will, according to Equation 5.11, tend to move more rapidly than the heavier ones, and hence will diffuse more quickly. Hydrogen gas diffuses

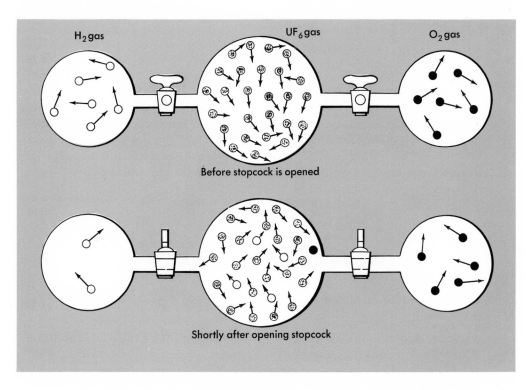

Figure 5.11 Diffusion of gases.

more quickly than does oxygen, as would be expected in light of the foregoing line of reasoning. Uranium hexafluoride, UF_6, with its massive molecules, would diffuse much more slowly than would hydrogen or oxygen.

We can make a quantitative test of the theory if we measure the rate at which a gas will flow into a vacuum through a small opening in its container. This phenomenon is called **effusion,** and here too we would predict that the rate of effusion of a gas would be proportional to the average speed of its molecules. In this case, however, we could measure the rate by noting how rapidly the pressure in the container drops and relating this to the number of molecules leaving the container in a given time interval. If the rates of effusion of two gases, A and B, were measured in the same container under similar conditions, we would predict that

$$\frac{\text{rate of effusion of A}}{\text{rate of effusion of B}} = \frac{\text{average speed of molecule A}}{\text{average speed of molecule B}} = \frac{u_A}{u_B} \quad (5.17)$$

By Equation 5.11, if gases A and B are studied at the same temperature,

$$m_A u_A^2 = 2cT = m_B u_B^2$$

or

$$\frac{u_A^2}{u_B^2} = \frac{m_B}{m_A} = \frac{m_B N}{m_A N} = \frac{M_B}{M_A} \quad (5.18)$$

in which M_B and M_A are the molecular weights of B and A. If Equation 5.18 is solved for the ratio of average molecular speeds and this is substituted in

103

Equation 5.17, it is easily seen that

$$\frac{\text{rate of effusion of A}}{\text{rate of effusion of B}} = \left(\frac{M_B}{M_A}\right)^{\frac{1}{2}} \tag{5.19}$$

This relation, that the rate of effusion of a gas varies inversely as the square root of its molecular weight, was observed experimentally by Thomas Graham in 1828. At that time molecular weights were not thoroughly understood, and Graham stated the relation in terms of the densities of the gases he studied. According to Equation 5.8, the density of a gas and its molecular weight are related, so that if the equation is applied to A and B,

$$\begin{matrix} M_B = d_B RT/P \\ M_A = d_A RT/P \end{matrix} \quad \text{yielding} \quad \frac{M_B}{M_A} = \frac{d_B}{d_A} \tag{5.20}$$

if the densities of gases B and A are obtained under the same conditions of temperature and pressure. If the density ratio in Equation 5.20 is substituted in Equation 5.19, the law reported by Graham is obtained:

$$\frac{\text{rate of effusion of A}}{\text{rate of effusion of B}} = \left(\frac{d_B}{d_A}\right)^{\frac{1}{2}} \tag{5.21}$$

when the densities of gases A and B are measured under the same conditions of temperature and pressure.

Graham's Law gives us an alternate method for measuring the molecular weights of gases, and one of its main uses is in this area. One needs merely to measure the rate of effusion of the unknown gas or vapor against the rate of a known reference gas. The procedure is straightforward and has found some practical application in industry in cases in which the purity of a gas is clearly reflected by its molecular weight. The following example is illustrative.

EXAMPLE 5.11. In an effusion experiment it required 45 seconds for a certain number of moles of an unknown gas to pass through a small orifice into a vacuum. Under the same conditions it required 18 seconds for the same number of moles of oxygen to effuse. Find the molecular weight of the unknown gas.

SOLUTION. The rates of effusion are inversely proportional to the times needed for the flow. Hence, by Equation 5.19,

Note that the time of effusion is directly proportional to $M^{1/2}$.

$$\frac{\text{rate}_X}{\text{rate}_{O_2}} = \frac{t_{O_2}}{t_X} = \left(\frac{M_{O_2}}{M_X}\right)^{\frac{1}{2}} \qquad \frac{18}{45} = \left(\frac{32}{M_X}\right)^{\frac{1}{2}} = 0.40$$

Squaring both sides, $\dfrac{32}{M_X} = 0.16$, and therefore $M_X = \dfrac{32}{0.16} = 200$

During World War II Graham's Law had a rather unexpected application in connection with a very complicated chemical problem. It had been found that the isotope of uranium having a mass of 235, ^{235}U, has a nucleus unstable to collisions with neutrons. Such collisions result in a splitting of the uranium nucleus into lighter fragments (fission) and the liberation of large amounts of energy in the form of heat and γ-rays (see Chapter 23). It became necessary to separate ^{235}U from its much more plentiful (but not fissile) isotope, ^{238}U. Because of the great chemical similarity of the isotopes of an element, the chemical resolution of uranium into its isotopes was not feasible, and some physical method was sought. Since the rate of diffusion of a gas varies with its molecular weight, the composition of a gas mixture coming through an orifice will not be quite the same as that in the original sample, and hence the resolution of a gas mixture by successive diffusions is possible, at least in principle. Preliminary diffusion experiments with uranium hexafluoride, UF_6, a volatile uranium compound, indicated that $^{235}UF_6$ could indeed be separated from $^{238}UF_6$ by diffusion, and an enormous plant was built for the purpose in Oak Ridge, Tennessee. In the process, UF_6 diffuses many thousands of times through porous barriers, with the lighter fractions moving on to the next stage and the heavier fractions being recycled through earlier stages.

Molecular Speeds

Of greater theoretical interest than the relative molecular speeds treated by Graham's Law are the actual average speeds of molecules. These can be calculated from Equation 5.11 after substitution of the value of c required to make the kinetic theory consistent with the Ideal Gas Law. Solving for u^2 in Equation 5.11 and letting c equal 3R/2N, (p. 102) we obtain

$$u^2 = 2cT/m = 2(3R/2N)(T/m) = 3RT/mN = 3RT/M$$

or

$$u = (3RT/M)^{\frac{1}{2}} \tag{5.22}$$

in which M is the gram molecular weight of the gas and R is the gas constant. From Equation 5.22 it is clear that molecular speeds increase with increasing temperature, decrease with increasing molecular weight, and are not dependent on gas pressure. Simple substitution into the equation allows one to determine the average speed of any gas molecule at any given temperature.

EXAMPLE 5.12. Find the average speed of an oxygen molecule in air at room temperature (25°C).

SOLUTION. In order to give u the proper dimensions, cm/sec in the metric system, the value of R used in Equation 5.22 must be in absolute units. R will always have the dimensions of energy per mole per degree Kelvin, and in absolute units has a value of 8.31×10^7 ergs per mole per °K. Using this value of R,

One erg equals 1 g cm²/sec².

$$u = (3RT/M)^{\frac{1}{2}} = \left[\frac{3 \times \left(8.31 \times 10^7 \frac{ergs}{mole\,°K}\right) \times 298°K}{32\ g\ mole} \right]^{\frac{1}{2}}$$

$$= 4.82 \times 10^4 \text{ cm/sec} = 482 \text{ m/sec or about 1000 miles per hour!}$$

According to the kinetic theory, molecular speeds are, on the average, very high by ordinary standards. It has been possible to test this prediction by several direct experiments, and the quantitative data obtained agree very well with the theoretical values. Qualitatively, the very high average speed of molecules appears reasonable when one considers the speed of sound in air. Since sound is propagated by molecular motion, one would expect that the speed of sound and that of the molecules in the gas through which it passes would be roughly equal. The speed of sound is about 800 miles an hour, which indeed is very close to the average speed of a nitrogen or oxygen molecule at 25°C.

Although it is useful to be able to calculate an average speed for a molecule, one must remember that not all the molecules will be moving at the identical speed. The motion of molecules in a gas is utterly chaotic. In the course of a second a molecule will typically undergo millions of collisions with other molecules, with each collision resulting in a change of the molecule's speed and direction of motion. In view of this situation you might well wonder whether one could hope to be able to determine anything more than an average speed; in fact, the calculation of the average speed itself seems a rather remarkable achievement.

Surprisingly, we do know considerably more about molecular motion in a gas than would first appear. In 1860 James Clerk Maxwell showed that different possible speeds were distributed among molecules in a definite way, and by a careful analysis of the gaseous system he derived a mathematical expression for this distribution. In Figure 5.12 the relation he obtained is illustrated graphically for oxygen gas at 25° and 1000°C. On the graph the relative number of oxygen molecules having the speed u is plotted as a function of the speed. We see that there are very few molecules having speeds near 1 m/sec, and that the number having a given speed increases rapidly with

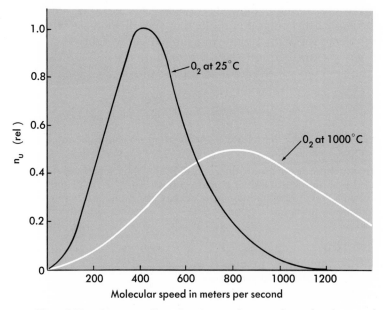

Figure 5.12   The Maxwellian distribution function for molecular speeds.

speed up to a maximum at about 400 m/sec at 25°C. This is the most probable speed of an oxygen molecule at 25°C, with the majority of the molecules having speeds between 200 and 600 m/sec. Above about 400 m/sec, the number of molecules moving at any particular speed decreases, so that the likelihood of finding a molecule moving at about 800 m/sec is only about one-tenth as large as the likelihood of finding a molecule moving at about 400 m/sec. For speeds in excess of about 1200 m/sec the number of molecules with such speeds drops off rapidly. As the temperature of the gas increases, the speeds of the molecules increase, and the distribution curve for molecular speeds is displaced to the right and becomes broader. From the curves at 25° and 1000°C it is clear that the probability of a molecule having a given high speed is enormously larger at high temperatures than it is at low. This property of gases has a great effect on the rate at which they undergo chemical reactions and will be discussed further in Chapter 14.

Maxwell was a British scientist who spent most of his life as a teacher, first in Scotland and later in England at King's College and Cambridge University. His scientific abilities were apparent from his early youth; by the time he was 19 he had published two papers, one in mathematics and one in physics.

Maxwell's analysis of the molecular speed problem was but one of his many contributions. In his work in thermodynamics he developed several fundamental equations which bear his name. His greatest successes were in connection with the theory of light and electricity, where he discovered and formulated the general equations of the electromagnetic field, on which much of the present theory of electricity, magnetism, and light is based.

The number of truly outstanding theoreticians the world has known is very small, certainly amounting to less than one hundred. James Clerk Maxwell belongs among the elite of this group. He was truly an intellectual giant, to be ranked with Newton and Einstein.

PROBLEMS

5.1 Nitrogen gas confined in a 450 ml container at 25°C has a pressure of 0.86 atm. What would be the pressure exerted by the gas at 25°C if it were expanded to a volume of 1200 ml?

5.2 In Boyle's original experiment he measured the length (directly proportional to the volume) of a gas column as a function of the pressure head of mercury (Figure 5.4). The following table of data was obtained by his procedure:

Length of gas column (l) in cm	Pressure head (h) in cm Hg
50	0.0
45	8.6
40	18.7
35	32.3
30	50.6
25	75.8

Make calculations from these data which demonstrate the validity of Boyle's Law (barometric pressure = 75.0 cm Hg).

5.3 A gas in a cylinder is heated at a constant pressure of 0.75 atm from 25°C to 175°C. If the volume of the gas was initially 2.70 lit, what is its final volume?

5.4 A 25.0 lit sample of steam at 100°C and 1.00 atm pressure is cooled to 25°C and expanded until the pressure is 20 mm Hg. If no water condenses, calculate the final volume of the water vapor.

5.5 Calculate the volume in liters of a mole of ideal gas at 0°C and 1.00 atm. (This volume is sometimes called the molar volume of a gas at standard temperature and pressure, STP, and in some texts is used as the basis of calculations in problems involving gases. The Ideal Gas Law is a much more powerful tool for treating such problems, and is always used in more advanced studies of gas behavior.)

5.6 How many grams will the steam (H_2O) filling a 50 gallon oil drum at 100°C and 1.00 atm weigh? (1 gal = 3.78 lit)

5.7 How many grams will the liquid water filling a 50 gal oil drum at 25°C and 1.00 atm weigh?

5.8 Calculate the density in grams per liter of carbon dioxide gas, CO_2, at 25°C and 1.00 atm pressure.

5.9 Six ml of a volatile organic liquid were vaporized at 97°C in a 250 ml flask with a fine-holed stopper. After all the liquid was vaporized, the flask was cooled and weighed with the condensed liquid. The liquid was found to weigh 0.890 g. The barometric pressure in the laboratory was 748 mm Hg. Calculate the molecular weight of the liquid from these data.

5.10 The molecular weight of the liquid calculated in Problem 5.9 will be inaccurate for several reasons. Name as many reasons as you can, indicating the one that you think would be most important.

5.11 A sample of wet hydrogen, saturated with water vapor at 23°C, exerts a total pressure of 754 mm Hg.
a. What is the partial pressure of the hydrogen at 23°C? (vp water = 21 mm Hg)
b. If the sample occupies 250 ml, calculate the total mass, in grams, of H_2.

5.12 In the reaction $2 SO_2(g) + O_2(g) = 2 SO_3(g)$, how many liters of SO_3 will be produced from 24 lit of SO_2 if each of the gases is measured at 300°C and 0.5 atm pressure?

5.13 When ammonium nitrite in aqueous solution is heated, the following reaction occurs.

$$NH_4^+ + NO_2^- = N_2(g) + 2 H_2O(l)$$

If the nitrogen produced by the reaction is collected over water at 23°C and a total pressure of 745 mm Hg, how much ammonium nitrite would have to react to evolve 2.50 lit of wet gas?

5.14 How many grams of $O_2(g)$ can be obtained from 5.00×10^6 lit of air at 27°C and 740 mm Hg if air is 21 per cent oxygen and 79 per cent nitrogen by volume?

5.15 It took 74 sec for a given amount of a certain gas to effuse through a porous plug at 25°C and 108 sec for the same number of moles of nitrogen gas to effuse under the same conditions. Estimate the molecular weight of the gas.

5.16 For a mole of ideal gas, plot graphs of

a. P vs. V at 127°C
b. P vs. T at V = 20 lit
c. PV vs. P at T = 27°C
d. PV vs. T

5.17 Compare the mean speed of an ammonia, NH_3, molecule with that of a benzene, C_6H_6, molecule when they are both present as gases in the same container. Compare the mean kinetic energies of the two molecules under these conditions.

5.18 On the average, how fast are gaseous water molecules moving at 100°C?

5.19 A sample of dry air collected in a McLeod gauge of 225 ml volume is compressed at constant temperature to a volume of 0.075 ml, where its pressure is 1.0 cm Hg. What was the pressure in mm Hg of the gas when collected?

5.20   A certain gas exerts a pressure of 1.20 atm when it is confined in a 10.0 lit container at 27°C. If the container is compressed to 5.40 lit and heated to 127°C, what pressure will the gas exert?

5.21   Calculate the mass of air (MW = 29) in lbs at 25°C and 1.00 atm in a room whose dimensions are 4.0 m by 6.0 m by 3.0 m.

5.22   What is the density of ammonia gas, NH_3, at 50°C and 2.50 atm?

5.23   A student carries out the experiment described in Problem 5.9. He finds that the vapor filling a 240 ml flask at 100°C and 750 mm Hg weighs 1.240 g. Calculate the molecular weight of the liquid.

5.24   A gaseous mixture of hydrogen and oxygen at 57°C and a total pressure of 2.5 atm contains 60 mole per cent hydrogen. What are the partial pressures of the two gases in the mixture?

5.25   A 0.236 g sample of $XH_2(s)$ reacts with water according to the following equation:

$$XH_2(s) + 2\ H_2O(l) = X(OH)_2(s) + 2\ H_2(g)$$

The hydrogen which is evolved is collected over water at 23°C and occupies a volume of 286 ml at 746 mm Hg total pressure. Find the number of moles of H_2 produced and the atomic weight of X. At 23°C the vapor pressure of water is 21 mm Hg.

5.26   A 0.100 g sample of an Al-Zn alloy when reacted with HCl evolves 100 ml of hydrogen, measured at 25°C and one atm. Calculate the per cent of Al in the alloy (1 mole Al $\rightarrow \frac{3}{2}$ mole H_2; 1 mole Zn $\rightarrow$ 1 mole H_2).

5.27   Given the composition of air as in Problem 5.14, calculate an average "molecular weight" for air.

5.28   Calculate the ratio of the rates of diffusion of UF_6 derived from the uranium-235 and uranium-238 isotopes respectively.

5.29   What properties of a gas determine

a. The average speed of its molecules?
b. The average energy of its molecules?
c. The pressure exerted by the gas?

5.30   How hot would a sample of carbon dioxide gas have to be if its molecules were to have the same average speed as gaseous water molecules at 27°C? What would the temperature be if they were to have the same average kinetic energy?

°5.31   In a classic experiment the density of carbon dioxide was carefully measured at 0°C at several low pressures, with the following results:

Pressure (atm)	Density (g/lit)
1.00000	1.97676
0.66667	1.31485
0.33333	0.65596

Use the Ideal Gas Law to calculate the apparent molecular weight of carbon dioxide at each of these pressures. Since the accuracy of the data exceeds that possible using a slide rule, use either a desk calculator or five place logarithms for the calculation. (R = 0.082056 lit atm/mole°K and 0°C = 273.16°K.) The differences between the values of MW obtained result from deviations of the gas from the ideal. Make a graph of $MW_{apparent}$ vs. pressure, and extrapolate the (straight) line to zero pressure. The value of $MW_{apparent}$ at zero pressure is the best value for the molecular weight of CO_2, since the gas must behave ideally at that pressure. Assuming the molecular formula of carbon dioxide and the atomic weight of oxygen, calculate the atomic weight of carbon from the molecular weight of CO_2 you obtained.

109

The approach employed in this problem is called the limiting density method for molecular weight determination. It was used extensively in the early part of this century to find molecular and atomic weights.

°5.32 At 27°C a certain gas has a density of 4.69 g/lit at 2.00 atm pressure and a density of 9.58 g/lit at 4.00 atm pressure. Use the approach of Problem 5.31 to find the best value obtainable from these data for the molecular weight of the gas.

°5.33 Estimate the relative magnitudes of the Van der Waals constant b for the following gases: O_2, SO_2, H_2, CCl_4. Do the same for the constant a for these gases. You may use any physical properties of these gases as found in a handbook as a basis for your ranking.

°5.34 The molar heat capacity of a gas is equal to the number of calories required to raise the temperature of the gas by one degree centigrade. At constant volume the molar heat capacity of all monatomic gases is essentially independent of temperature and equal to 3 cal/°C. Can you suggest why this is so? Hint: At constant volume, all the heat absorbed by a gas is used to increase the energy of its molecules.

°5.35 Absolute temperature scales can be set up on bases other than the one described in this chapter. For example, we can readily define an absolute scale that is based on the Fahrenheit rather than the centigrade temperature scale. Find the absolute temperature of the normal boiling point of water on such a scale.

°5.36 Absolute temperatures can be measured experimentally on a gas-volume thermometer. With a given thermometer the volume of gas at 0.00°C at low pressure was 150.0 ml. At the same pressure in a system having a temperature to be measured, the gas volume was found to be 185.6 ml. What is the value of the temperature of the system in °C? in °K?

°5.37 A mole of oxygen gas is confined in a 5 liter container at 27°C. State the effect, qualitatively and then quantitatively, of each of the following changes on

a. The average molecular speed.
b. The average molecular energy.
c. The average momentum associated with an average molecule-wall collision.
d. The pressure of the gas.

The changes to be considered are

(1) The gas is compressed at 27°C to a volume of 1 liter.
(2) The temperature of the gas is raised to 327°C.
(3) Another mole of oxygen is added to the container.
(4) A mole of hydrogen is substituted for the oxygen.

The Periodic Classification
of the Elements

Thus far in our study of chemistry we have spent most of our efforts on the laws that govern chemistry. These laws allow us to predict the temperature-volume-pressure relationships in gases and the weight relations in known chemical reactions, and to find atomic weights and chemical formulas from experimental data. They do not, however, tell us much about the physical and chemical properties of specific chemical substances. Thus we find ourselves in the rather anomalous position of being able to calculate how much phosphine, PH_3, could be produced by the reaction of phosphorus and sodium hydroxide, given the equation for the reaction, without having any idea about the chemical properties of phosphine or even knowing whether it is a solid, a liquid, or a gas. Clearly, a knowledge of chemistry must include familiarity with the actual properties of chemical substances as well as an understanding of theoretical principles. One would hope that some broad generalizations of chemical behavior could be found, similar in character to the generalizations which were possible in the treatment of the physical properties of gases. This chapter illustrates how such generalizations can be shown to exist and how they can be used to classify and predict the chemical and physical properties of pure substances.

6.1 THE PERIODICITY OF PROPERTIES
OF THE ELEMENTS

A search for possible generalizations regarding chemical behavior would probably begin with a study of the elementary substances, simply because there are only about a hundred elements and millions of pure substances. To expect at the outset that an examination of the properties of the elements would have fruitful results might well be considered a case of extreme optimism. So far our chemical theory has taken the elements as separate, independent entities. We have said very little about the nature of atoms or the possible relationships between different kinds of atoms.

We begin our investigation by listing the first 25 elements in order of their atomic weights. With the notion that there may be some obvious, if unexpected, relations between the properties of the elementary substances, we also list several of their readily measured physical properties, including the density of the solid form of the element, its melting point, the normal boiling point, and the ionization energy, which is essentially the energy required to remove one electron from the atom of the element. These quantities are tabulated in Table 6.1 and are presented in graphic form in Figures 6.1 and 6.2.

Even a casual inspection of the graphs reveals that there is a certain amount of regularity in the properties of these substances. The properties do not vary randomly, but rather show a rough tendency to vary in some sort of cycle, being successively relatively high, then relatively low. The same

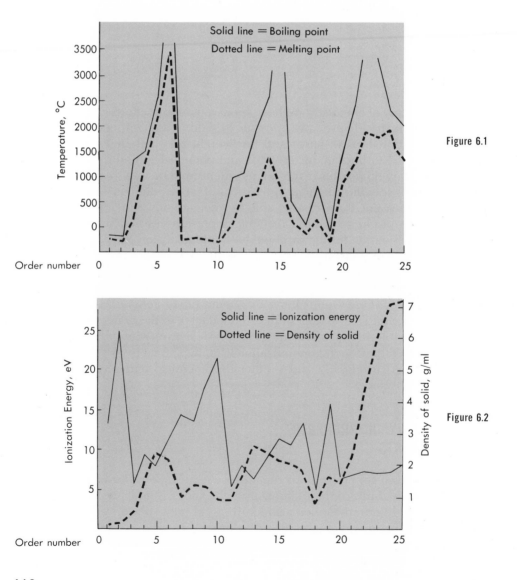

Figure 6.1

Figure 6.2

Some physical properties of the lighter elements.

TABLE 6.1   SOME PHYSICAL PROPERTIES OF THE LIGHTER ELEMENTS

ELEMENT	ORDER NO.	ATOMIC WEIGHT	DENSITY OF SOLID (g/ml)	MELTING POINT (°C)	BOILING POINT (°C)	IONIZATION ENERGY (eV)
H	1	1.008	0.08	−259	−253	13.6
He	2	4.003	0.13	−272	−269	24.6
Li	3	6.939	0.53	186	1340	5.4
Be	4	9.012	1.8	1350	1500	9.3
B	5	10.81	2.5	2300	2550	8.3
C	6	12.011	2.26	3500	?	11.3
N	7	14.007	1.03	−210	−196	14.5
O	8	16.00	1.43	−218	−183	13.6
F	9	19.00	1.3	−223	−187	17.4
Ne	10	20.183	1.0	−249	−246	21.6
Na	11	22.990	0.97	98	880	5.1
Mg	12	24.312	1.74	651	1110	7.6
Al	13	26.98	2.71	660	1800	6.0
Si	14	28.09	2.33	1420	2600	8.1
P	15	30.974	2.2	590	?	10.9
S	16	32.064	2.1	119	445	10.4
Cl	17	35.453	1.9	−102	−35	13.0
K	18 (19)	39.102	0.86	62	760	4.3
Ar	19 (18)	39.948	1.65	−189	−186	15.8
Ca	20	40.08	1.55	810	1215	6.1
Sc	21	44.96	~2.5	1200	~2400	6.6
Ti	22	47.90	4.5	1800	>3000	6.8
V	23	50.94	6.0	1710	~3000	6.7
Cr	24	52.00	7.1	1890	2500	6.8
Mn	25	54.94	7.2	1260	1900	7.4

length of cycle is clearly present in the melting and boiling points, with minima at elements having order numbers of about 2, 10, and 19 and maxima at about numbers 6, 14, and 22. The density minima occur at order numbers 1, 11, and 18 and the maxima at 5, 13, and 25.

The ionization energy graph shows the most striking periodic character. The minima occur at order numbers 3, 11, and 18, and the maxima at numbers 2, 10, and 19. The steps in the curve from number 3 to number 10 are repeated between numbers 11 and 19, *except* that the ionization energy of element 18, potassium, is much too low. In fact, if elements number 18 and 19 are interchanged, the general appearance of the curve in the two cycles is very much the same. In the table the elements are ordered by their atomic weights; K, no. 18, = 39.102 and Ar, no. 19, = 39.948. These two atomic weights are correct to at least four places, so experimental error has not given the wrong order. However, potassium is also out of order in the graphs of the other properties, which would all be much more regular if elements no. 18 and 19 were interchanged. It seems possible that there is a natural order of the elements which is almost, but not quite, that of their atomic weights, and that, in

113

the natural order, potassium is no. 19. With the notion that potassium, for some as yet unknown reason, is out of place in our list of elements, we shall arbitrarily interchange elements no. 18 and 19 in the list.

When this is done the regularity of the cycles in all the physical properties improves markedly.

Physical Property	Maxima at Nos.	Minima at Nos.
Ionization energy	2, 10, 18	3, 11, 19

	Minima at Nos.	Maxima at Nos.
Melting point	2, 10, 18	6, 14, 22
Boiling point	2, 10, 18	6, 14, 22
Density	1, 11, 19	5, 13, 25

The cycle for ionization energy is now perfect, being eight elements long between maxima and between minima, with the smaller steps in the curve also spaced by eight elements. The other properties repeat in cycles varying in length from seven to twelve elements, with eight again being a reasonable average. The cycles for the four properties are also either completely in phase or out of phase.

This kind of variation is observed with many other physical properties of the elements. There does seem to be a periodic, or cyclic, character to the progression of such properties. In all cases potassium and argon fit into the cycles best when arranged in positions 19 and 18 respectively.

So far our inquiry has been limited to the physical properties of the elementary substances. Our main concern, however, is with an organization, or generalization, of the chemical behavior of both elementary and compound substances. We now look into the question of whether the chemical properties of the elements show any regularities of the sort seen in the physical properties. Let us first examine the formulas of their chlorides.

In Figure 6.3 the data on the chlorides are presented graphically. Several elements form more than one chloride; the graph has been drawn both to go through the points for those elements which form only one or no compounds and also to be as symmetrical as possible. The cycles for composition of the chlorides are seen to be perfectly regular and eight elements in length. Similar graphs can be drawn for the oxides of these elements and for their other binary compounds. In all cases, the cycles are regular and the period is eight elements long.

An extension of this kind of study to include all the elements reveals the existence of a similar, though somewhat less simple, periodicity of their chemical and physical properties. The experimental evidence supporting these regularities of properties is now indeed overwhelming, and can be summarized in the following statement, usually called the Periodic Law: **The elements, if arranged in an order which closely approximates that of their atomic weights, exhibit an evident periodicity of properties.**

6.2 HISTORY OF THE PERIODIC LAW

Several early chemists suspected the existence of the general relations we have been discussing, but their supporting evidence at the time was so

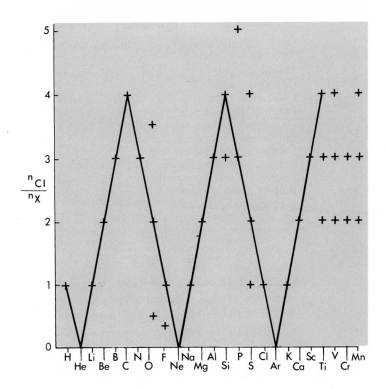

Figure 6.3 Atom ratios in the chlorides of the lighter elements.

sketchy and the idea that such regularities were possible seemed so implausible that scientists paid them very little attention.

In 1817 Döbereiner reported that when certain groups of similar elements were arranged in groups of three, the atomic weight and other properties of the middle element were very close to the mean of the properties of the other two. The "triads" of elements then known were very limited in number, with that of calcium, strontium, and barium being the best example, but in later years other "triads," such as lithium, sodium, and potassium, were discovered. About 1860 Newlands, an English chemist, proposed that the elements occurred in sets of seven, but his "octaves," as he called them, had such a resemblance to the musical scale that the chemists of the time were so busy laughing that they could not see that Newlands was right.

The noble gases (group 8A) were not known at the time of Newlands or Mendeleev.

Between 1869 and 1871 Mendeleev, a Russian chemist, wrote a classic series of papers on the periodic properties of the elements. He included all the elements that had been discovered up to that time, about 70 in number, and used an approach very similar to the one presented in this chapter. He was bold enough to assert that gaps in his arrangement corresponded to elements yet to be discovered and that experimental errors in atomic weights were responsible for the interchanges of positions of some of the elements. By noting the positions of the gaps, Mendeleev was able to predict many of the properties of the missing elements. Gallium, scandium, and germanium were among the elements predicted by Mendeleev, and these were indeed discovered in later years. In Table 6.2 we have listed the properties of ger-

115

TABLE 6.2 THE PREDICTED AND OBSERVED PROPERTIES OF GERMANIUM

PROPERTY	PREDICTED BY MENDELEEV	OBSERVED
Atomic weight	72	72.60
Density of metal	5.5	5.36
Color of metal	dark gray	gray
Formula of oxide	GeO_2	GeO_2
Density of oxide	4.7	4.703
Formula of chloride	$GeCl_4$	$GeCl_4$
Density of chloride	1.9	1.887
Boiling point of chloride	below 100°C	86°C
Formula of ethyl compound	$Ge(C_2H_5)_4$	$Ge(C_2H_5)_4$
Boiling point of ethyl compound	160°C	160°C
Density of ethyl compound	0.96	Slightly less than 1.0

manium (then called ekasilicon) as predicted by Mendeleev, and the actual properties of that element as observed on its discovery in 1886. The agreement is truly remarkable. The success of predictions such as these removed any doubt of the validity of the concept of the periodicity of the properties of the elements and their compounds; it also explains why Mendeleev is given credit for the Periodic Law.

6.3   THE PERIODIC TABLE

In order to emphasize the periodicity of their properties, Mendeleev arranged the elements in horizontal rows of such length that elements of similar chemical properties fell directly beneath each other. This arrangement is still used and is called the **periodic table**. In the periodic table the elements in a given cycle, or period, are arranged in a horizontal row. The cycles for the lighter elements are most clearly shown in Figure 6.3. There seems to be a sequence of only two elements followed by two very clear cycles of eight elements beginning with lithium, element no. 3. The second cycle of eight elements in Figure 6.3 begins with sodium, element no. 11. Sodium is similar in its properties to lithium: both are soft metals of low density and melting point, both have relatively low ionization energies, both form chlorides with the formula XCl and oxides with the formula X_2O. To indicate the similarity between sodium and lithium, sodium is placed immediately below lithium in the periodic table. If this is done, magnesium, element number 12, falls below its counterpart, beryllium, aluminum falls below boron, and the rest of the elements in the second cycle of eight fall below their counterparts in the first cycle.

Elements such as boron, carbon, and neon, which belong to the same cycle, are said to be in the same **period**; elements that have similar properties, such as lithium and sodium, oxygen and sulfur, or neon and argon, are said to belong to the same **family** or **group**. To distinguish the families of elements, they are given numbers: lithium and sodium belong to group 1A, beryllium and magnesium to group 2A, and so on up to neon and argon, which belong to group 8A. Since hydrogen and helium are considered to be the only elements in the first cycle, they are said to make up the **first period**. Boron and fluorine are then in the **second period**, and magnesium, aluminum, and phosphorus are in the **third period**.

In Table 6.3 we have summarized these conventions by setting up the periodic table for the first 20 elements, indicating their groups and their periods. In this table we have included the first two elements, hydrogen and helium. Helium is clearly in the same family as neon and argon; these elements are all inert, or noble, gases. Hydrogen is put in group 7A, although, as we shall note later, it might equally well be put in group 1A. In the table we have put potassium and calcium, elements 19 and 20, in their proper places; on the basis of their chemical properties they are clearly the first elements in the fourth period, with potassium in group 1A and calcium in group 2A.

At this point one might well be led to believe that the periodic table would be simply a rectangular array of the elements, eight elements across, with as many periods as needed to give a place to every element. This does not turn out to be the case. Following calcium there is a series of 10 elements, all metals, which do not seem to belong to any of the eight groups in the first three periods. If the fourth period were the same length as the third, element 26, iron, would be a noble gas like argon, neon, and helium, but it certainly has none of their properties. Figure 6.2 shows that the ionization energies of these 10 elements do not follow the two previous cycles.

After these 10 elements we again encounter a series of elements which complete the fourth period begun by potassium and calcium. The period ends with elements 35 and 36, bromine and krypton. Bromine has chemical properties similar to those of chlorine and fluorine and belongs to group 7A. Krypton is a typical noble gas and hence is a member of group 8A.

The elements following krypton are rubidium, with properties that make it a member of group 1A, and strontium, which fits well into group 2A. These two elements, then, begin the fifth period. Again in this period there is a series of ten elements that are not members of any of the A groups, followed by six elements that fit into groups 3A through 8A and complete the period. The last element in the fifth period is no. 54, xenon, a noble gas. The next element is cesium, which has all the properties of a 1A metal and begins the sixth period.

The overall structure of the periodic table for the first 54 elements can now be seen. The form is not as simple as we might have first thought, but it nonetheless has a high degree of symmetry. The first period contains two elements; the second and third, eight elements each; the fourth and fifth, 18 elements each. The two series of 10 metals, in the fourth and fifth periods, fall into ten new groups, called the **transition metals**. The transition metals, taken collectively, have more similar properties than do the A elements taken col-

TABLE 6.3 THE LIGHTER ELEMENTS IN THE PERIODIC TABLE

GROUP	1A	2A	3A	4A	5A	6A	7A	8A
1st period							H	He
2nd period	Li	Be	B	C	N	O	F	Ne
3rd period	Na	Mg	Al	Si	P	S	Cl	Ar
4th period	K	Ca						

PERIODIC CHART OF THE ELEMENTS

1A	2A	3B	4B	5B	6B	7B	8	8	8	1B	2B	3A	4A	5A	6A	7A	8A
1 **H** 1.00797 ±0.00001																	2 **He** 4.0026 ±0.0005
3 **Li** 6.939 ±0.0005	4 **Be** 9.0122 ±0.00005											5 **B** 10.811 ±0.003	6 **C** 12.01115 ±0.00005	7 **N** 14.0067 ±0.00005	8 **O** 15.9994 ±0.0001	9 **F** 18.9984 ±0.00005	10 **Ne** 20.183 ±0.0005
11 **Na** 22.9898 ±0.00005	12 **Mg** 24.312 ±0.0005											13 **Al** 26.9815 ±0.00005	14 **Si** 28.086 ±0.001	15 **P** 30.9738 ±0.00005	16 **S** 32.064 ±0.003	17 **Cl** 35.453 ±0.001	18 **Ar** 39.948 ±0.005
19 **K** 39.102 ±0.0005	20 **Ca** 40.08 ±0.005	21 **Sc** 44.956 ±0.0005	22 **Ti** 47.90 ±0.005	23 **V** 50.942 ±0.0005	24 **Cr** 51.996 ±0.001	25 **Mn** 54.9380 ±0.00005	26 **Fe** 55.847 ±0.003	27 **Co** 58.9332 ±0.00005	28 **Ni** 58.71 ±0.005	29 **Cu** 63.54 ±0.005	30 **Zn** 65.37 ±0.005	31 **Ga** 69.72 ±0.005	32 **Ge** 72.59 ±0.005	33 **As** 74.9216 ±0.00005	34 **Se** 78.96 ±0.005	35 **Br** 79.909 ±0.002	36 **Kr** 83.80 ±0.005
37 **Rb** 85.47 ±0.005	38 **Sr** 87.62 ±0.005	39 **Y** 88.905 ±0.0005	40 **Zr** 91.22 ±0.005	41 **Nb** 92.906 ±0.005	42 **Mo** 95.94 ±0.005	43 **Tc** (99)	44 **Ru** 101.07 ±0.005	45 **Rh** 102.905 ±0.0005	46 **Pd** 106.4 ±0.05	47 **Ag** 107.870 ±0.003	48 **Cd** 112.40 ±0.005	49 **In** 114.82 ±0.005	50 **Sn** 118.69 ±0.005	51 **Sb** 121.75 ±0.005	52 **Te** 127.60 ±0.005	53 **I** 126.9044 ±0.0005	54 **Xe** 131.30 ±0.005
55 **Cs** 132.905 ±0.0005	56 **Ba** 137.34 ±0.005	57 **La** 138.91 ±0.005	72 **Hf** 178.49 ±0.005	73 **Ta** 180.948 ±0.0005	74 **W** 183.85 ±0.005	75 **Re** 186.2 ±0.005	76 **Os** 190.2 ±0.005	77 **Ir** 192.2 ±0.05	78 **Pt** 195.09 ±0.005	79 **Au** 196.987 ±0.0005	80 **Hg** 200.59 ±0.005	81 **Tl** 204.37 ±0.005	82 **Pb** 207.19 ±0.005	83 **Bi** 208.980 ±0.0005	84 **Po** (210)	85 **At** (210)	86 **Rn** (222)
87 **Fr** (223)	88 **Ra** (226)	89 †**Ac** (227)	104 (257)														

Lanthanum Series

58 **Ce** 140.12 ±0.005	59 **Pr** 140.907 ±0.0005	60 **Nd** 144.24 ±0.005	61 **Pm** (147)	62 **Sm** 150.35 ±0.005	63 **Eu** 151.96 ±0.005	64 **Gd** 157.25 ±0.005	65 **Tb** 158.924 ±0.0005	66 **Dy** 162.50 ±0.005	67 **Ho** 164.930 ±0.0005	68 **Er** 167.26 ±0.005	69 **Tm** 168.934 ±0.0005	70 **Yb** 173.04 ±0.005	71 **Lu** 174.97 ±0.005

Actinium Series

90 **Th** 232.038 ±0.0005	91 **Pa** (231)	92 **U** 238.03 ±0.005	93 **Np** (237)	94 **Pu** (242)	95 **Am** (243)	96 **Cm** (247)	97 **Bk** (247)	98 **Cf** (249)	99 **Es** (254)	100 **Fm** (253)	101 **Md** (256)	102 **No** (253)	103 **Lw** (257)

Atomic Weights are based on C^{12}—12.0000 and Conform to the 1961 Values lectively. However, it is possible to note the existence of a cycle of properties in each of the two series of transition metals in the fourth and fifth periods, which justifies their classification into 10 families, each with one member in the fourth and one in the fifth periods.

In the sixth period, there is a total of 32 elements. Following barium, element no. 56, in group 2A, there is a series of 14 elements, called the **rare earths**, or lanthanides, then a third series of 10 transition metals, and then six group A elements to complete the period. The situation in the sixth period appears to be repeated in the seventh, but that period is incomplete and contains several elements that do not exist in nature and whose properties are not well known.

In Table 6.4 and inside the front cover of this book we have constructed the periodic table in accordance with our discussion. The transition metals are inserted between groups 2A and 3A. The rare earths, elements 58 through 71, would rightly appear between the first and second transition metals in the sixth period, but to keep the length of the table manageable, they are listed as a separate sequence, along with the equivalent series in the seventh period, at the bottom of the table.

Many other forms of the periodic table have been proposed. Different numbers are sometimes used for the various groups.

6.4   SOME OBSERVATIONS ON THE PERIODIC TABLE

For purposes of discussion the periodic table can be broken up into several regions or classes of elements. The first and largest class includes those elements in groups 1A through 8A; these elements are often called the representative elements (sometimes the noble gases in group 8A are classified separately). The second class includes the transition elements, which we have designated as being in groups 1B through 7B, and the three families collectively classified as group 8. A third class consists of the elements called the **lanthanides** and the related series in the seventh period called the **actinides**; this group of elements is sometimes called the inner transition elements.

The representative elements include some metals and all of the nonmetals and metalloids. The metals in this class, all of which have characteristic luster and good electrical conductivity, include all of groups 1A and 2A and some of the elements in groups 3A through 5A. The most active metals in this class are the alkali metals, group 1A, all of which are relatively soft, low-melting, ductile substances. These elements are highly electropositive, which means that their atoms readily donate electrons to electron-accepting, or electronegative, atoms. The metals in group 2A, the alkaline earths, are somewhat similar in their properties to the alkali metals, but are somewhat less active chemically, being less electropositive. These metals are harder, stronger, and higher-melting than those in group 1A. Binary compounds of the 1A and 2A metals are primarily ionic in nature.

On the far right of the table the 8A elements, consisting of the noble gases, are all gaseous at room temperature and are nonmetals. They all have high chemical stability and do not participate in chemical reactions at all readily; until recent years these elements were thought to occur only uncombined chemically and were often called the inert gases. In Chapter 8 we will discuss some of the compounds which some of these elements are now known to form. The elements in group 7A, called the halogens, have widely varying

119

physical properties. Fluorine and chlorine are gases at room temperature, bromine is a liquid, and iodine and astatine are solids. Except for astatine, which is very rare and radioactive and which has not been well characterized, all the halogens are nonmetals. These elements are highly electronegative and readily form ionic compounds with elements in groups 1A and 2A.

Between the active metals in groups 1A and 2A and the nonmetals in 7A and 8A are four families of elements with properties ranging from those of classic nonmetals to those of typical metals. The first members of groups 3A, 4A, 5A, and 6A are not metals, but the higher members in each group exhibit increasing metallic properties. As the group number increases, the metallic characteristics become apparent in later and later periods. Except for boron, all of group 3A consists of metals; tin and lead in group 4A and bismuth in group 5A would also be classified as metals. Boron (3A), silicon and germanium (4A), arsenic and antimony (5A), and tellurium (6A), have properties intermediate between those of metals and nonmetals. These substances are all solids, and all have at least one form with some metallic luster, and some electrical conductivity, which increases with temperature, rather than decreases as it does with metals. These elements, which lie on a diagonal in the periodic table (see Table 6.4), are called **metalloids**. They often have different crystal modifications with widely different electrical properties. Carbon can exist as graphite, a semiconductor, or as diamond, nonconducting and a nonmetal. Arsenic, antimony, and tin exist in both metallic and nonmetallic forms.

These elements are used in semiconductor devices such as transistors.

The true nonmetals among the elements are relatively few in number and are clustered in the upper right hand corner of the periodic table. Nitrogen, oxygen, hydrogen, the first two halogens, and the noble gases are the only elements that are gases under ordinary conditions. Phosphorus, sulfur, bromine, and iodine are the only other familiar elements which do not exhibit at least some metallic character.

The large middle portion of the periodic table is taken up by the transition elements, which are all metals. In general, these metals have higher densities and higher melting points than the metals of the representative elements. Almost without exception, these metals exhibit several different equivalent weights in their possible compounds; put another way, their combining ratios with elements such as oxygen and the halogens can have several values. For example, oxides of manganese are known with the following formulas: MnO, Mn_2O_3, MnO_2, MnO_3, and Mn_2O_7. MnO is an ionic solid melting at about 1600°C; Mn_2O_7 is a dark red molecular substance melting below -20°C and exploding at about 70°C. As the O:Mn atom ratio increases, the molecular character of the oxide increases. This trend is common to compounds of all the transition metals; for compounds having the general formula M_AX_B, where X is oxygen or a halogen, as the ratio B/A becomes larger the compound becomes more molecular and has lower melting and boiling points.

Many of the compounds formed by the transition metals are highly colored, in contrast to the primarily colorless compounds formed by the metals among the representative elements. One of the characteristic properties of the transition metals is their ability to form complex ions, of which $Cu(NH_3)_4^{2+}$ is a classic example; Chapter 19 will be devoted to the properties of these species.

The inner transition elements, which include the lanthanide and the actinide series, are also all metals. The rare earths, or lanthanides, with atomic numbers from 58 to 71, all have similar chemical and physical properties, and until the advent of ion-exchange columns they could not be efficiently separated. These elements are all relatively uncommon, not because they are especially scarce, but because of the difficulty in resolving them from one another. The rare earths are used in industry mainly in the form of "misch metal," which is a mixture of these metals with a small amount of iron. In this material, cerium predominates, with smaller amounts of lanthanum and neodymium also present in moderate concentrations. Misch metal is useful in small amounts (about 1 per cent) as an alloying agent with magnesium, where it serves to improve high temperature properties. It is used in an alloy with iron in the "flints" for cigarette lighters and other devices which produce sparks by friction.

Among the actinides, all of which are radioactive, only uranium and thorium exist in any appreciable amounts in nature. The other elements, from neptunium through lawrencium, were all first observed in the products of controlled nuclear reactions, and in many cases have been produced in only very small amounts. Uranium and thorium are used in the fuel elements in nuclear reactors and nuclear weapons, and these applications have been responsible at least in part for the interest in these, the heaviest elements in the periodic table.

The lanthanides are all highly electropositive elements, whose atoms readily donate electrons to electronegative species. The compounds formed by these elements all appear to be ionic. (The use of misch metal in flints reflects the high heat of formation of the oxides of these metals.) The very high chemical similarity of these elements can be attributed to the fact that the ions formed from their atoms all tend to have the same charge (+3) in solutions of their compounds and also have very nearly the same size. The metals are so electropositive (only slightly less so than the 1A and 2A metals) that obtaining the metals from their compounds is difficult and can only be accomplished by electrolysis of their molten salts or by reaction of the salt with an active metal such as calcium or lithium.

Cerium (no. 58) is more abundant than boron, cobalt or tin.

6.5 PREDICTIONS ON THE BASIS OF THE PERIODIC TABLE

One of the factors that established the validity of the periodic classification of the elements was Mendeleev's ability to predict the properties of some undiscovered elements. His principle was simply that the properties within given groups vary in a regular fashion, and that the unknown element would therefore have properties that are a properly chosen average of the properties of the adjacent elements in the same family group. This principle is a useful one, and we shall now show how Mendeleev might have predicted the properties of germanium on the basis of the properties of the related elements. Table 6.5 gives the portion of the periodic table in the vicinity of germanium.

TABLE 6.5 GERMANIUM IN THE PERIODIC TABLE

3A	4A	5A
Al	Si	P
Ga	Ge	As
In	Sn	Sb

EXAMPLE 6.1. Predict the formula of germanium chloride.

SOLUTION. Germanium is in group 4A, in the same family as silicon and tin, and, by our general principle, will tend to form compounds having the same formulas as those of silicon and tin. The chlorides of those two elements are known to have the following formulas:

$$SiCl_4, \quad Si_2Cl_6 \quad \text{and} \quad SnCl_2, \quad SnCl_4$$

Since $SiCl_4$ and $SnCl_4$ both exist, a likely formula for the germanium compound would be $GeCl_4$; the possibility that Ge_2Cl_6 and $GeCl_2$ also exist is suggested by the other formulas. Actually the ordinary chloride observed is $GeCl_4$. $GeCl_2$ and Ge_2Cl_6 have been reported but are less stable.

EXAMPLE 6.2. Estimate the atomic weight of germanium.

SOLUTION. The atomic weights of elements surrounding germanium are

Al	26.98	Si	28.09	P	30.97
Ga	69.72	Ge	–	As	74.92
In	114.82	Sn	118.69	Sb	121.75

If our approach is correct, gallium and arsenic should have atomic weights that are the averages of the atomic weights of aluminum and indium, and of phosphorus and antimony, respectively.

$$\text{predicted AW of Ga} = \frac{26.98 + 114.82}{2} = 70.90 \quad (\text{observed AW} = 69.72)$$

$$\text{predicted AW of As} = \frac{30.97 + 121.75}{2} = 76.36 \quad (\text{observed AW} = 74.92)$$

The predicted values are close to those observed, but are both high, the first by 1.18 units and the second by 1.44 units. We would expect that the similarly predicted value for germanium would also be high, by about 1.3 units. For Ge,

$$\text{simple predicted value} = \frac{28.09 + 118.69}{2} = 73.39$$

Correcting this value by the anticipated error of 1.3 units, we would make a "best estimate" of 72.1 for the atomic weight of germanium. The observed value is 72.59.

The procedures used in these examples can be employed to predict reasonably well the properties of germanium and its compounds, or for that matter, the properties of any other element and its compounds. In all cases the prediction is based on the properties of the related substances. (It must be admitted that germanium turns out to be an ideal element to use as an illustration of the power of the approach. Mendeleev was "lucky" in the sense that great scientists with good ideas recognize a promising situation when it arises.)

Correlation of Chemical Formulas

The possibilities for prediction that arise from the existence of the periodic table are particularly evident when one considers the formulas of chemical substances. The approach used here is essentially that of Mendeleev, but it has results that are more quantitatively reliable than are the predictions of physical properties.

Let us first examine the formulas of the group 1A chlorides and bromides:

LiCl	NaCl	KCl	RbCl	CsCl	FrCl
LiBr	NaBr	KBr	RbBr	CsBr	FrBr

These formulas and indeed all the formulas of all alkali halides, the binary compounds between the 1A and 7A elements, can be represented by the general formula MX, where M is any 1A element and X is any 7A element. Thus, in order to be able to remember the formula of any alkali halide, one needs only to know the formula for one of them and to realize the condition that the periodic table imposes.

A condition similar to that which fixes the formulas of the 1A–7A binary compounds is obeyed by the formulas of many inorganic compounds. This condition allows the correlation of the chemical formulas of substances in a very powerful way, and in its most general form is embodied in the following rule: **Given the formula of any inorganic compound, the replacement of an element in the compound by another from the same group will often give the correct chemical formula for another compound.**

Given the formula for magnesium chloride, $MgCl_2$, one would correctly predict, on the basis of this rule, that the other 2A chlorides would have the formulas $CaCl_2$, $SrCl_2$, and $BaCl_2$, and that the 7A binary magnesium compounds would be MgF_2, $MgBr_2$, and MgI_2. The general formula for 2A–7A binary compounds would be predicted to be MX_2, as is observed.

The general relation is really far broader than this. It would imply, for instance, that since calcium molybdate has the formula $CaMoO_4$, that of barium molybdate would be $BaMoO_4$, that calcium chromate would have the formula $CaCrO_4$, and that strontium tungstate would have the formula $SrWO_4$. These formulas are all correct and follow directly on the substitutions of elements in the same family in the basic formula. It should be clear that the relation allows one to generalize chemical formulas and to make a great many formula predictions on the basis of a relatively small amount of information.

Like many general rules, the relation we have been using has many exceptions. The difficulty is that in some cases the relation predicts the existence of substances that do not occur or are extremely unstable. A simple example of such a failure is observed if carbon monoxide, CO, is used as a reference substance. The rule would predict the existence of a sulfide of carbon with the formula CS; this compound, if it exists at all, appears to be highly unstable. The fact that we would also predict the existence of substances with the formulas PbS and SnO, both of which are well known, may

123

For example, none of the other 5A elements form oxides analogous to NO or N₂O.

increase our confidence in the relation but does not remove the fact that an exception occurs. It is found by experience that exceptions such as that mentioned occur most frequently when one or more of the elements involved is the first number of a group of nonmetals (groups 4A to 8A) or is the first number of a group in a transition series. With practice and due allowance for its limitations, the student will find the general relation for systematizing chemical formulas a useful one.

6.6 ANOMALIES AND LIMITATIONS OF THE PERIODIC CLASSIFICATION

If one is to make the most effective use of the periodic table, he must be aware of and take account of the various anomalies and limitations inherent in it. We have already considered one of these in the previous section on prediction of chemical formulas.

Another of the anomalies of the table arises in the positions of some of the elements. In order to preserve the cyclic character of the properties of the elements, we found it necessary to interchange the positions of argon and potassium. There are three other instances in which similar interchanges are required if the elements are arranged in order of increasing atomic weights: the elements involved are iodine and tellurium, cobalt and nickel, and thorium and protoactinium. This anomaly disappears when the reasons for the existence of the periodic table become known, and the atomic number rather than atomic weight is used in ordering the elements. This important matter is discussed in the next chapter.

Another anomaly in the table is the position of hydrogen. We have placed hydrogen in group 7A, along with fluorine, bromine, etc., but in some tables it is listed as the first element in group 1A. It has some of the properties of the halogens but also behaves in some respects like the alkali metals. It forms compounds with the alkali metals, called hydrides, which are analogous to the alkali halides, and contains the H^- ion, but its ion in aqueous solution is positively charged, as are the alkali metal ions. The properties of hydrogen result from its being the first, and probably the simplest, of the elements.

A similar anomaly is observed for the first member of each family of elements belonging to an A group. It is found that the chemical properties of these light elements differ from those of the family to which they belong to a much greater extent than would be anticipated on the basis of typical family

TABLE 6.6 SOME PROPERTIES OF THE HALIDES OF SILVER AND HYDROGEN

SUBSTANCE	SOLUBILITY IN WATER (moles/lit 25°C)	SUBSTANCE	NORMAL BOILING POINT (°C)	EXTENT OF DISSOCIATION IN WATER
AgF	14.6	HF	19	slight
AgCl	1.3×10^{-5}	HCl	−84	high
AgBr	5.8×10^{-7}	HBr	−67	high
AgI	9.3×10^{-9}	HI	−35	high

behavior. Fluorine, the first well-established member of group 7A, forms compounds with other elements which have markedly different properties from the compounds of the other halogens with those elements. In Table 6.6 the properties of the silver and hydrogen compounds with the halogens are listed. Clearly, the listed properties of the fluorides are not consistent with those of the other halides.

Similarly, nitrogen, the first member of group 5A, differs markedly from phosphorus, arsenic, and antimony in the same group. It is much less reactive chemically than any of the latter elements, and, fortunately for man, will not readily form compounds even with oxygen, which combines readily with the other members of group 5A. One can easily discover that oxygen, carbon, beryllium, and lithium, all first members of their families, differ in properties from their families in unpredictable ways.

PROBLEMS

6.1 What is meant by the term periodicity? What evidence can you cite for the existence of a periodicity of properties of the elements?

6.2 Was the periodic table the result of theoretical developments? How did it originate?

6.3 What sorts of similarities would you anticipate among the elements of group 2A? group 7A? What sorts of differences would be likely to be found within these groups?

6.4 The suggestion that the elements exhibit a periodicity of properties must have met with a great deal of skepticism at the time it was made. What evidence supported this suggestion? Why was it convincing?

6.5 Of what use is the periodic table to chemists?

6.6 What are some of the anomalies in the periodic table?

6.7 Look up the formulas of the hydrogen compounds of elements 3 through 17. For these compounds, plot the number of atoms of hydrogen per atom of the element vs. the order number of the element. Compare your graph to Figure 6.3.

6.8 How would you use the periodic table and any other sources of information to predict the properties of arsine, AsH_3? For example, how would you go about predicting its boiling point?

6.9 A certain metal is known to have an atomic weight in the range 85 to 90. It forms a chloride of empirical formula MCl_3 and an oxide of empirical formula M_2O_3. Identify the metal.

6.10 Consider the element yttrium. Predict, from the properties of any appropriate substances, the following properties of yttrium and its compounds.
a. Density of the metal.   d. Formula of yttrium oxide.
b. Atomic radius of the metal.   e. Density of yttrium oxide.
c. Melting point of the metal.
Check your predictions by looking up in an appropriate reference source the properties listed.

6.11 Give the atomic number of an element which
a. Is chemically similar to sulfur but has an atomic weight approximately four times that of sulfur.
b. Resembles silicon quite closely in metallic character.
c. Forms a chloride in which the per cent of chlorine by weight is very close to that in NaCl.

6.12 Given the chemical formulas H_2O, HCl, $CaCO_3$, Na_3PO_4, CO_2, Fe_2O_3, As_2O_3, and $KMnO_4$, predict likely formulas for the following substances.
  a. Hydrogen sulfide
  b. Barium silicate
  c. A sulfide of carbon
  d. Hydrogen fluoride
  e. Potassium phosphate
  f. Antimony sulfide
  g. Lithium arsenate
  h. Potassium thioantimonate
  i. Rubidium thiophosphate
  j. Sodium perrhenate
  (When sulfur is substituted for oxygen in a compound, the prefix *thio* is often used.)

6.13 Given the following known substances, which of the possible predicted substances is more likely to exist. Give your reasoning.

Known Substance	Possible Substances
SF_6	OF_6 or TeF_6
PCl_5	NCl_5 or $AsCl_5$
WCl_6	WBr_6 or $CrCl_6$
XeF_4	RnF_4 or HeF_4
Na_2SiF_6	Na_2SiCl_6 or Na_2GeF_6

6.14 State precisely what is meant by a periodic function. Which of the following are periodic functions of x?
  a. $\sin x$
  b. x^2
  c. $\log x$
  d. Average daily temperature, where x = month of the year.
  e. Expenditures by political parties, where x = year (1966, 1967, . . .).
  f. Intelligence quotient (I.Q.), where x = age of individual.

6.15 Plot the melting points, boiling points, and ionization energies of the 7A elements fluorine through iodine versus their atomic weights. Draw smooth curves through your data for each property. Use the curves to predict the melting point, boiling point, and ionization energy of astatine, the last member of this group.

6.16 Consider the element strontium in light of the periodic table. From the properties of any appropriate substances, predict the following properties of strontium and its compounds:
  a. Atomic weight.
  b. Density of the solid metal.
  c. Melting point of the metal.
  d. Formulas of the chloride and oxide.
  e. Density of the oxide.
  f. Melting point of the chloride.

6.17 Given the formulas SeF_6, BrF_5, $SbCl_3$, Sc_2O_3, $NaCl$, $CaCl_2$, Na_2O_2 (sodium peroxide), $NaMnO_4$ (sodium permanganate), Na_2CrO_4 (sodium chromate), predict the formula of:
  a. lanthanum sulfide
  b. bismuth iodide
  c. krypton fluoride
  d. potassium molybdate
  e. barium peroxide
  f. calcium pertechnate

6.18 What properties of elements can you name which are *not* periodic functions of their atomic numbers?

°6.19 The oxide of a certain element contains 31% by weight of oxygen. The oxide does not dissolve in nitric acid or sulfuric acid, but does dissolve in concentrated hydrochloric acid when heated to form a volatile chloride. The element is a solid (mpt = 960°C) and is a poor conductor of electricity. Identify the element.

The information you will need to solve some of these problems can be found in:

Handbook of Chemistry and Physics, Chemical Rubber Co.
Handbook of Chemistry, N. A. Lange, McGraw-Hill Book Co.
International Critical Tables, McGraw-Hill Book Co.
Chemical Periodicity, R. T. Sanderson, Reinhold Publishing Corp.

7

The Structure of Atoms

The periodic classification of the elements was the result of the realization that experimentally obtained chemical knowledge could be correlated and organized in a useful manner. The periodic table is based on experimental fact, and at the time of its origin it had no theoretical basis. It was not until the end of the nineteenth century that we began to have a definite theory for the periodic properties of the elements. This theory arose from new concepts of the nature and structure of atoms.

7.1 ELECTRONS AND NUCLEI— THE COMPONENTS OF ATOMS

In Chapter 2 we briefly described the structure of atoms and the particles of which they are composed. We now wish to begin a study of chemical bonding, and to do this we will need to discuss our present ideas about the nature of atoms in considerable detail and consider the experiments and theories on which these ideas are based.

The first convincing demonstration of the existence of subatomic particles came from experiments involving the conduction of electricity through gases at low pressures. When an apparatus of the type shown in Figure 7.1 is partially evacuated and connected to a source of high voltage such as a spark coil, an electric current flows through the tube. Associated with the flow are colored streaks or rays of light, which appear to originate at the negative electrode (cathode). The properties of these so-called **cathode rays** were studied extensively during the last two decades of the nineteenth century. In particular, it was discovered that the rays could be bent by both electric and magnetic fields. From a careful study of the extent of this deflection, J. J. Thomson was able to demonstrate in 1897 that the rays consist of a stream of negatively charged particles, which he called **electrons**. Thomson was able to measure the charge-to-mass ratio of both these particles and the positive ions which are formed simultaneously with the electrons in the discharge tube. From his data he was able to conclude that the electron has a mass several orders of magnitude less than that of the lightest atom and must therefore be a subatomic particle.

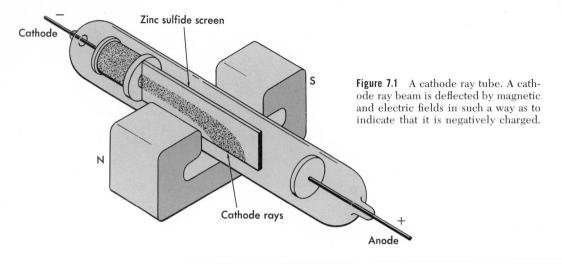

Cathode

Zinc sulfide screen

S

Figure 7.1 A cathode ray tube. A cathode ray beam is deflected by magnetic and electric fields in such a way as to indicate that it is negatively charged.

N

Cathode rays

Anode

The existence of electrons has been confirmed in many experiments. Electrons are among the particles given off by atoms undergoing radioactive decay; they are emitted by active metal surfaces at high temperatures or upon irradiation by light of short wavelength; their addition to or removal from atoms is responsible for the phenomenon known as electrolysis. Electrons are among the most ubiquitous of particles and are a component of all known atoms.

In 1909 Millikan, working at the University of Chicago, experimentally measured the charge on the electron. He found that tiny oil drops in ionized air would tend to pick up an electric charge, and could then be kept from falling under the action of gravity by application of an electric field of appropriate direction and intensity. He found that the charge acquired by the drops was always a multiple of a smallest charge, which he was able to determine by measuring the rates of rise and fall of a given drop as, under a given field, it gained and lost charge. The smallest charge, which he concluded must be that on a single electron, in ordinary units is equal to 1.60×10^{-19} coulombs.

As we mentioned, electrons have a mass much less than that of any atom. From the e/m values of Thomson and the charge as measured by the method of Millikan, one can calculate the mass of an electron to be about $\frac{1}{1837}$ of that of an H atom, or about 9.108×10^{-28} grams. The diameter of an electron is about $\frac{1}{10000}$ that of an atom, or about 1×10^{-12} cm. Although we can state reasonably accurately many of the properties of electrons, to say that we know all that we would like to know about them is not true. As we shall see in this and later chapters, some aspects of the story of the electron remain to be written.

How would you calculate the mass of an H atom? an electron?

The Atomic Nucleus

In 1911 Ernest Rutherford and his students performed a series of experiments that profoundly influenced our ideas regarding the nature of atoms. Using a radioactive source, they let a beam of high-energy α-particles (He^{2+} ions) fall on a piece of thin gold foil. Then, with a fluorescent screen, they observed the degree to which the beam was scattered by the foil. Most of the α-particles went through the foil almost undeflected. A few, however, were found to be reflected back from the foil at acute angles. The scientists measured the relative numbers of α-particles reflected at different angles by counting on the screen the scintillations caused by individual α-particles.

By a beautiful mathematical analysis of the rather large electrostatic forces required to explain the observed back scattering, Rutherford was able to conclude that:

1. The scattering was caused by a relatively massive particle with a diameter not exceeding 10^{-12} cm.

2. The charge on the particle was positive and had a magnitude equal to about half the atomic weight of the element of which the foil was made.

The experiment was repeated with foils of different elements, from carbon to gold, and the results confirmed perfectly those obtained with gold. These were the experiments which established that an atom contains a tiny, positively charged massive center, called the atomic **nucleus**.

X-ray Spectra and Atomic Number

Rutherford's work established that the mass of an atom and the mass of its nucleus are essentially equal, since electrons were known to have only a tiny mass. His experiments, however, did not allow him to assign a precise positive charge to the nucleus. The magnitude of this charge was found in experiments with x-ray spectra.

If a beam of high-energy electrons, accelerated by about 10,000 volts of potential, falls on a metal target, the atoms of the metal will be excited and will emit radiant energy. This radiation differs from visible light in that it has a much shorter wavelength. Visible light has a wavelength of between 4000 and 8000 Å, according to its color. The radiation emitted by the metal atoms has a wavelength of only about 1 Å.

Radiant energy generated by this method has many of the properties of ordinary light, but its very short wavelength enables it to penetrate matter to a much greater degree. Wilhelm Roentgen, who in 1895 did the first experiments with these rays, did not realize the nature of the radiation he was observing and so called it "x-rays."

In 1912 it was proved that x-rays are indeed electromagnetic waves, just as light waves are, and in 1913 Bragg developed a method for measuring their wavelengths. That year a young English physicist, H. Moseley, investigated the x-rays emitted by different elements. He found, as had Bragg, that only at certain wavelengths did they have high intensity (we would say that the x-ray spectrum contained only a few lines) and that the positions of the strongest lines seemed to have some relation to the atomic weight of the element being studied.

In attempting to find a relationship between the wavelengths of the x-rays emitted by an atom and its other properties, Moseley plotted (of all things) the square root of the reciprocal of the shortest observed wavelength, $(1/\lambda)^{\frac{1}{2}}$, for each element against its atomic weight (Figure 7.2). The line he obtained was nearly straight, but the points for several elements fell off the curve. He then plotted the reciprocal of the square root of the wavelength

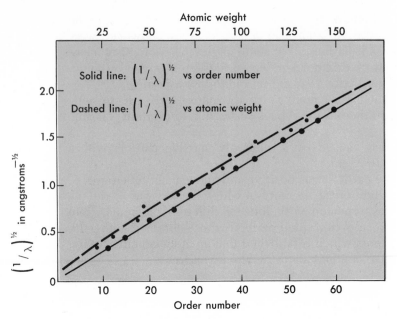

Figure 7.2 Moseley's data showing the relationship between x-ray wavelengths and order numbers of the elements.

against the **order number** of the element in the periodic table and found again that the line obtained was nearly straight, but that this time the points for **all the elements lay exactly on the line**!

Moseley's discovery was indeed significant. It established that the order number of an element in the periodic table could be found experimentally, and showed that the order number, based on properties, was actually correct. It also gave an unequivocal method for determining whether all the elements in a given region in the periodic table had been discovered, since breaks would occur on the graph at any missing elements.

See Problem 7.33.

The equation of the line relating x-ray wavelength to the order number of the element is similar to an equation obtained by Niels Bohr in his treatment of the hydrogen atom. (This will be discussed in Section 7.4.) Though we cannot go into the theory of x-ray spectra here, it was soon clear that Moseley's relation was most reasonably interpreted if the **order number of an element is numerically equal to the charge on its atomic nucleus**. This conclusion is completely consistent with Rutherford's work and is now an accepted part of atomic structural theory.

The realization that the order number of an element and its nuclear charge are the same simplifies our concept of the atom considerably. Since atoms are electrically neutral, the number of extranuclear electrons is equal to the nuclear charge, and hence also equal to the order number of the element. Since the order number of an element is of such significance it is given a special name, the **atomic number** of the element. Hydrogen has an atomic number equal to 1. Sodium, eleventh in the list of elements, has atomic number 11; chlorine has atomic number 17. Moseley's work thus allows us to state that:

All sodium atoms have 11 electrons, and all sodium nuclei have a charge of +11.

All chlorine atoms have 17 electrons, and all chlorine nuclei have a charge of +17.

All hydrogen atoms have 1 electron and a nucleus having a charge of +1.

The **element** to which **an atom belongs** is, therefore, determined by its **number of electrons**, or its **nuclear charge**, and **not by its mass**. Stated in another completely equivalent way, the **chemical properties** of an element are determined by the **number of electrons** possessed by its **atoms** and **not by the masses of those atoms**.

The Structure of Atomic Nuclei

In our discussion of methods for measuring atomic weights in Chapter 2, we noted that the masses of the isotopes of a given element were very nearly integral if one takes the mass of the ^{12}C isotope to be 12.0000. The rule is a general one and is illustrated in Table 7.1 for the isotopes of oxygen and magnesium.

TABLE 7.1   RELATIVE MASSES OF THE ISOTOPES OF OXYGEN AND MAGNESIUM

| | OXYGEN | | | MAGNESIUM | |
Isotope	Mass	Per Cent Abundance	Isotope	Mass	Per Cent Abundance
^{16}O	15.9949	99.76	^{24}Mg	23.9847	78.6
^{17}O	16.9990	.04	^{25}Mg	24.9858	10.1
^{18}O	17.999	.20	^{26}Mg	25.9814	11.3

In view of the fact that isotopic masses are so nearly integral, it seems reasonable to propose that the nuclei themselves are made up from an integral number of more basic particles. Two such particles, sometimes called fundamental particles, which appear to be reasonable components of atomic nuclei are the **proton** and the **neutron**. Their properties are given in Table 7.2. The proton is the nucleus of the ordinary hydrogen atom. The neutron is an uncharged particle having about the same mass as a proton. The neutron was discovered by Chadwick in 1932 and is produced in many reactions between nuclei.

How could one detect a beam of neutrons?

TABLE 7.2   PROPERTIES OF TWO FUNDAMENTAL PARTICLES

PARTICLE	CHARGE	MASS NUMBER	ATOMIC WEIGHT ($^{12}C = 12.0000$)
Proton	+1	1	1.00728
Neutron	0	1	1.00867

An interpretation of atomic nuclei on the basis of neutrons and protons is very straightforward. Each nucleus has a positive charge Z, equal to the atomic number of the element to which it belongs. It has a **mass number** A, equal to the integer nearest its atomic weight. Since the proton has a charge of +1 and is the only charged particle in the nucleus, there must be Z protons in all nuclei of atoms of atomic number Z. These contribute Z units of mass to the nucleus, since each proton has a mass number of unity. The remaining mass is made up by neutrons. If the total mass of the nucleus is A, $A - Z$ units will be due to neutrons. Since each neutron also has a mass number of unity, there will be $A - Z$ neutrons in nuclei of mass number A and atomic number Z.

Let us now describe the nuclei of the elements oxygen and magnesium. All oxygen atoms have atomic number 8; the three isotopic atoms have mass numbers 16, 17, and 18. Therefore, all oxygen nuclei contain 8 protons. The ^{16}O nucleus contains $A - Z$, $16 - 8$, or 8 neutrons. The ^{17}O nucleus contains $17 - 8$, or 9 neutrons, and the ^{18}O nucleus will contain $18 - 8$, or 10 neutrons.

Magnesium has atomic number 12 and isotopes of mass numbers 24, 25, and 26. Hence ^{24}Mg nuclei contain 12 protons and 12 neutrons, ^{25}Mg nuclei contain 12 protons and 13 neutrons, and ^{26}Mg nuclei contain 12 protons and 14 neutrons. We have summarized these results in Table 7.3.

TABLE 7.3 COMPOSITION OF NUCLEI OF THE ISOTOPES OF OXYGEN AND MAGNESIUM

ISOTOPE	ATOMIC NUMBER = Z	MASS NUMBER = A	NUMBER OF PROTONS = Z	NUMBER OF NEUTRONS = A − Z
^{16}O	8	16	8	8
^{17}O	8	17	8	9
^{18}O	8	18	8	10
^{24}Mg	12	24	12	12
^{25}Mg	12	25	12	13
^{26}Mg	12	26	12	14

Although the nuclear model based on the proton and neutron as fundamental particles is satisfactory in many ways, it cannot really be said that nuclei contain protons and neutrons in the same way that water molecules contain hydrogen and oxygen atoms. Within the nucleus there is an interplay of enormous forces that prevent our identifying any specific particles, and although we could form, in principle, the ^{16}O nucleus from 8 protons and 8 neutrons, the protons and neutrons would lose their identity in the act of combination. In spite of a great deal of time and effort spent on its investigation, the structure of the atomic nucleus is still not thoroughly understood. In our discussions involving nuclei we shall assume the validity of our simple model and shall find it adequate for our needs.

7.2 THE ELECTRONS IN ATOMS

The properties of electrons in atoms and molecules have been the subject of extensive research, both experimental and theoretical, during all of this century, but it must be admitted that at present our knowledge of the detailed electronic structure of all but the simplest atoms is still incomplete. Much progress has been made on this problem, but much more work remains to be accomplished. In this book we shall present some of the common current models for electron arrangements in atoms and molecules, and shall find them very useful in interpreting the chemical and physical properties of pure substances.

The main obstacle to our understanding of the properties of chemical substances in terms of the electrons and nuclei of which they are composed is that small particles, such as atoms, molecules, nuclei, and particularly electrons, appear to obey different laws regarding energy and motion than do larger objects, such as billiard balls and rotating bicycle wheels. Systems with which we are ordinarily concerned, with masses many, many times those of atoms and molecules, follow exactly the laws of motion first formulated by Isaac Newton. These constitute that part of physics called **classical mechanics**. Small particles obey the laws of a somewhat different kind of mechanics, called **quantum mechanics**. (Actually it turns out that classical mechanics is in a very real sense a special case of quantum mechanics and is valid for all but those particles which have exceedingly small masses.)

One cannot specify both the momentum and the position of an electron (Uncertainty Principle).

Quantum mechanics is part of a general theory, called the **quantum theory**, which had its beginnings early in this century. Like the atomic theory, the quantum theory has evolved considerably during its development; some of its original postulates have been retained, while others have been modified or discarded. It is at present the fundamental theory used to explain the behavior of electrons and other small particles. Some experiments which led to the quantum theory will be considered in the next section, after we have stated and discussed two of the underlying principles of the theory that are of chemical interest.

The Quantum Theory

1. Atoms and molecules can only exist in certain states, characterized by definite amounts of energy. When an atom or molecule changes its state, it must absorb or emit an amount of energy just sufficient to bring it to another state.

Atoms and molecules can, as we have seen, possess various kinds of energy. One form of energy of particular importance when considering atomic structure arises from the motion of electrons about the atomic nucleus and from the charge interactions among the electrons and between the electrons and the nucleus. This kind of energy is called **electronic energy**. Only certain values of electronic energy are allowed to an atom. When an atom goes from one allowed electronic state to another, it must absorb or emit just enough energy to bring its own energy to that of the final state.

Analogous considerations apply to the other forms of energy possessed by atoms, molecules, and other small particles. Translational energy of motion, rotational and vibrational energy, in addition to electronic energy, are subject to the limitations of the quantum theory.

The energy of systems that can exist only in discrete states is said to be **quantized**. A change in the energy level of such a system involves the absorption or emission of a **quantum** of energy.

2. The allowed energy states of atoms and molecules can be described by sets of numbers called quantum numbers.

The mathematical solutions to problems regarding the energies of atoms and molecules as obtained by quantum mechanics usually result in sets of integral numbers, which serve to denote the allowed states and allow calculation of their energies. These numbers are called **quantum numbers**.

Associated with each electronic state of an atom is a group of quantum

Discussed in detail in Section 7.6.

numbers that identify the state. In the usual model, the quantum numbers are associated with the individual electrons in the atom. Each electron is assigned a set of quantum numbers according to a set of rules. A statement of all the quantum numbers of all the electrons in an atom is used to designate the energy level, or quantum state, of the atom.

Similarly, the various translational, rotational, vibrational, and electronic energy states of molecules or other small particles can be described by the appropriate sets of quantum numbers. Ordinarily only one kind of energy and one set of quantum numbers are considered at a time, but to specify completely the state of a water molecule would require a statement of the quantum numbers associated with its translational energy, its rotational energy, its vibrational energy, and its electronic energy.

7.3 EXPERIMENTAL BASIS OF THE QUANTUM THEORY

Now that we have stated some of the principles of the quantum theory it should be helpful to your understanding of the theory to examine some of the experiments and theoretical relations on which it is based. There are actually a great many experiments which are best explained by the theory; among them should be mentioned black body radiation, the photoelectric effect, atomic spectra, and the several kinds of molecular spectra. Though these are all important phenomena, we shall restrict our attention here mainly to atomic spectra, postponing a discussion of the other experiments to more advanced courses of study in physics and chemistry. Atomic spectra bear directly on the problem of atomic structure and nicely illustrate an area of experiment in relation to the general quantum theory.

Atomic Spectra

As we saw in Section 7.1 atoms emit x-rays when they are hit by high-energy electrons. The emission of radiant energy by excited atoms is a general phenomenon. The nature of the radiation emitted depends on the excitation which is used.

If we should heat any metal in a furnace or in a flame, it will, depending on its temperature, give off visible light; at 1000°C it will look red, at 1500°C it will appear essentially white. If the emitted light is examined with a spectroscope, a device which breaks up light into its component colors, the light is found to contain essentially all colors. More precisely, we would say that, over the region in which the metal radiates, its **spectrum** is **continuous**, containing light at all *wavelengths*.

Not all emitters of light radiate at all wavelengths. If we observe in a spectroscope the light emitted by sodium chloride when it is placed in a flame, say from a Bunsen burner, we see only a few bright lines, which indicate the few wavelengths at which sodium atoms, excited by the flame, are emitting light. In this case we are seeing the **atomic spectrum** of sodium, which, since it contains light at only a few wavelengths, is said to be **discrete**.

Atomic spectra are emitted when atoms are mildly excited; this can be accomplished by a flame, by a spark, or by electrons which have been accelerated by falling through a few volts of potential. Our common fluorescent

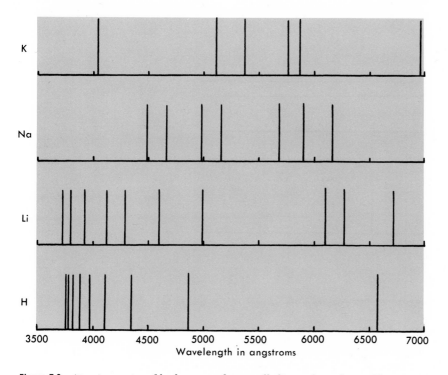

Figure 7.3 Atomic spectra of hydrogen and some alkali metals in the visible region.

lights and mercury vapor highway lights give off light of this sort, with well-resolved components (or lines) which can be seen through a band spectroscope (Figure 7.3).

The existence and character of atomic spectra are readily treated by the quantum theory. According to the theory, an atom, say the sodium atom, can exist only in certain states, which occur at levels of energy characteristic of the sodium atom. By absorbing or emitting energy in the proper amount, the atom can make transitions from one energy level to another. When a gas containing sodium atoms is heated to a high temperature or is exposed to fast-moving electrons, some of the sodium atoms will, by collisions, tend to absorb energy and become excited, moving in the process to higher allowed electronic energy levels. Once excited, the atoms will be unstable and will tend to return to lower electronic energy levels, ultimately reaching the lowest allowed level, which is usually called the ground state of the atom. Under excitation, the sodium gas will quickly reach a state of dynamic equilibrium, in which some atoms are releasing energy at the same rate that it is being absorbed by others.

One of the ways in which electronically excited sodium atoms can lose energy is by radiating it as light. In a process in which an excited atom makes a transition to a lower energy level, a photon of light may be emitted. The photon will have some of the characteristics of a particle, and will have an energy given by an equation you may recall from Chapter 4,

$$E = hc/\lambda$$

Atomic spectra can be used to identify known elements and discover new ones; helium was first found in the solar spectrum in 1868.

135

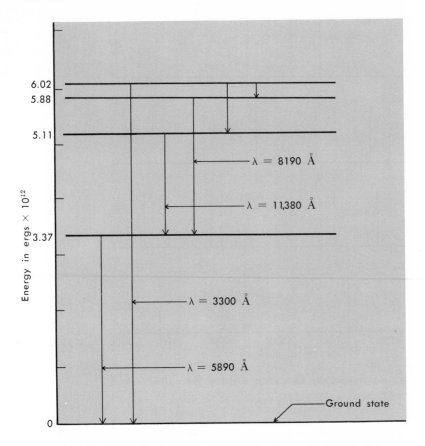

Figure 7.4 Some transitions and spectral lines for a few low-lying energy levels of the sodium atom.

in which E is the photon energy, h is Planck's constant, c is the velocity of light, and λ is the wavelength of the photon. This equation was first proposed by Einstein in 1905, in the very early days of the quantum theory; it was based on his assumption that interchanges of energy between radiation and matter take place in discrete units called light quanta, with energies given by the foregoing equation. The relation is a basic postulate of the quantum theory and is consistent with many different kinds of experiments.

Atomic spectra are believed to consist of the radiation resulting from transitions made by electronically excited atoms from higher to lower allowed energy levels. In Figure 7.4, we have arranged, in somewhat simplified form, some of the lower electronic energy levels of the sodium atom, along with the transitions that are observed between those levels. Each transition gives rise to a line in the atomic spectrum of sodium, with the wavelength we have indicated. One of the lines in the sodium spectrum in Figure 7.3 is the result of a transition shown in Figure 7.4. The actual energies of the levels were determined by the observation that in general several spectral lines are associated with each allowed energy level; this means that, if the theory is correct, there must be mathematical relations between the energies of photons emitted by sodium atoms. By using these relations it is possible to use the atomic spectrum of an element to determine an unequivocal set of energy levels to be associated with its atoms. In spite of the fact that some atomic

Einstein's equation can be used to calculate the wavelengths of the lines in Figure 7.4 ($h = 6.63 \times 10^{-27}$ erg sec, $c = 3.00 \times 10^{10}$ cm/sec).

spectra are very complex, containing thousands of lines, the spectra of essentially all the atoms and many of the ions of the elements have been unraveled and explained in terms of a relatively small number of energy levels.

Regularities in Atomic Spectra

Beginning about 1880 it was recognized that there is a certain amount of order in atomic spectra. It was found possible, in some cases, to sort out the lines in a spectrum into series; several lines could then be assigned to each series on the basis of wavelength, intensity, and breadth. In the atomic spectrum of sodium one discusses the so-called principal (intense), sharp, and diffuse series. It was also found that the spectra of elements in the same group in the periodic table were qualitatively similar.

In 1885 Balmer found a very remarkable relationship between the wavelengths of the lines in the atomic spectrum of hydrogen (Figure 7.3). He showed that one could mathematically express the wavelengths of the nine then known lines in the atomic spectrum of hydrogen by the equation

$$\lambda = b[n^2/(n^2 - 4)]$$

in which λ is the wavelength, b is a constant, and n is an integer and has the values 3, 4, 5, and so on, for the first, second, third, . . ., lines in the hydrogen spectrum.

TABLE 7.4 CALCULATED AND OBSERVED WAVELENGTHS (IN Å) IN THE BALMER SERIES FOR HYDROGEN

$$\lambda_{calc} = 3646.00 \ n^2/(n^2 - 4)$$

LINE	n	λ_{obs}	λ_{calc}	LINE	n	λ_{obs}	λ_{calc}
1	3	6562.79	6562.80	6	8	3889.06	3889.07
2	4	4861.33	4861.33	7	9	3835.40	3835.40
3	5	4340.47	4340.48	8	10	3797.91	3797.92
4	6	4101.74	4101.75	9	11	3770.06	3770.65
5	7	3970.07	3970.09				

Balmer's formula for the series of lines which now bears his name predicts the wavelengths exceedingly well. In Table 7.4 we have tabulated the calculated and observed wavelengths of the first nine lines in the series. Balmer's equation undoubtedly gives the most nearly exact prediction of experimental quantities in all physical science, and its discovery gave others great incentive to look for other such relations. To this day none have been found that approach the Balmer formula in exactness.

7.4 THEORIES OF THE ELECTRONIC STRUCTURE OF ATOMS

By the beginning of this century knowledge of atomic structure had advanced to the point at which scientists could begin to speculate on the way in which positive and negative charges were arranged in atoms. Part of the problem was solved when Rutherford demonstrated the existence of atomic nuclei. It was only two years later, in 1913, that Niels Bohr presented a theory for the structure of the hydrogen atom that added greatly to our ideas regarding the behavior of the electrons in atoms.

Bohr based his approach on Rutherford's nuclear atom and on Planck's suggestion that atoms and other small particles can only possess certain definite amounts of energy. By making some rather drastic assumptions about the motion of the hydrogen electron about its nucleus, Bohr was able to derive a relation for the possible energies of the hydrogen atom. By Bohr's theory, the energy levels of the hydrogen atom are given by the equation

$$E = -R/n^2$$

in which R is a constant obtainable from the theory and n is a quantum number equal to 1, 2, 3, 4. . . . In Figure 7.5 we have shown the energy levels of hydrogen as given by Bohr's equation. According to the theory, the atomic spectrum of hydrogen arises from transitions made by hydrogen atoms from higher to lower energy levels. When the lower state is that for which $n = 2$, these transitions can be shown to predict the Balmer series. When the lower state is that associated with $n = 1$, or 3, or 4, other series are predicted and are verified by experiment.

Bohr assumed that a hydrogen atom consists of an effectively stationary central positive nucleus, the proton, about which an electron moves in a circular orbit. By calculating the conditions under which the electrostatic attractive force between proton and electron was balanced by the centrifugal force due to the motion of the electron, he was able to find a relation for the total energy of the atom. To make the predicted spectrum of hydrogen agree with that which is observed, he had to postulate a condition on the motion of the electron.

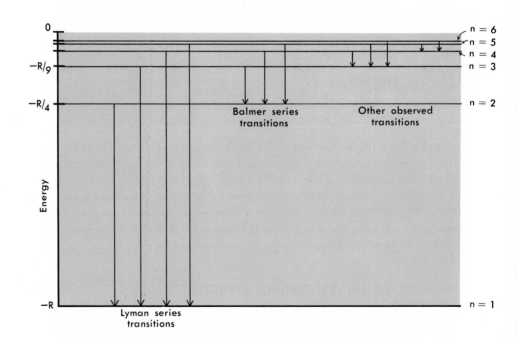

Figure 7.5   Some energy levels and transitions of the hydrogen atom.

Mathematically, we can summarize Bohr's theory for hydrogen as follows:
The force of electrostatic attraction between proton and electron is, by Coulomb's Law,

$$e^2/r^2 \qquad (7.1)$$

in which e is the charge on the electron (Figure 7.6). The centrifugal force due to orbital motion of the electron is, by Newton's Law,

$$mv^2/r \qquad (7.2)$$

At equilibrium, the two forces are equal:

$$e^2/r^2 = mv^2/r, \quad \text{or} \quad mv^2r = e^2 \qquad (7.3)$$

The kinetic energy of motion of the electron is

$$\tfrac{1}{2}mv^2 \qquad (7.4)$$

The potential energy due to electrostatic forces between proton and electron is

$$-e^2/r \qquad (7.5)$$

By convention, a negative potential energy corresponds to an attractive force.

The total energy of the atom is the sum of the kinetic and potential energies:

$$E = \tfrac{1}{2}mv^2 + (-e^2/r) \qquad (7.6)$$

By Equation 7.3,

$$\tfrac{1}{2}mv^2 = \tfrac{1}{2}e^2/r,$$

so, for the total energy,

$$E = -e^2/2r \qquad (7.7)$$

This was as far as Bohr could go without making any new assumptions. To solve for the specific energies he assumed arbitrarily that, for the electron,

$$mvr = nh/2\pi \qquad (7.8)$$

in which n is an integer and may take on any integral value, and h is Planck's constant, known at that time to have a value of about 6.6×10^{-27} erg sec.
To solve for allowed values of r, we square Equation 7.8, obtaining

$$m^2v^2r^2 = n^2h^2/4\pi^2 \qquad (7.9)$$

Dividing Equation 7.9 by Equation 7.3, we obtain

$$r = n^2h^2/4\pi^2e^2m \qquad (7.10)$$

Now substituting for r in Equation 7.7:

$$E = -e^2/2(n^2h^2/4\pi^2e^2m) = -2\pi^2e^4m/n^2h^2 = -R/n^2 \qquad (7.11)$$

Equation 7.11 is Bohr's classic equation for the energy of a hydrogen atom. It predicts that the atom can exist only at energies given by integral values of the quantum number n.
By the quantum theory, atomic spectra arise from changes in the energy of an atom, according to the equation

$$\Delta E = E_2 - E_1 = hc/\lambda \qquad (7.12)$$

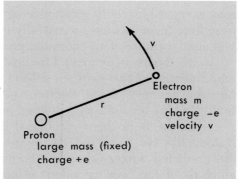

Figure 7.6 Bohr's model of the hydrogen atom.

in which c is the velocity of light, λ is the wavelength of the emitted light, and the subscripts 2 and 1 refer to higher and lower energy states respectively. Substituting Equation 7.11 in Equation 7.12,

$$E_2 - E_1 = -R((1/n_2{}^2) - (1/n_1{}^2)) = hc/\lambda = R(n_2{}^2 - n_1{}^2)/n_1{}^2 n_2{}^2 \tag{7.13}$$

Solving Equation 7.13 for the wavelength,

$$\lambda = \frac{hc}{R}\left[\frac{n_1{}^2 n_2{}^2}{n_2{}^2 - n_1{}^2}\right] = \frac{h^3 c}{2\pi^2 m e^4}\left[\frac{n_1{}^2 n_2{}^2}{n_2{}^2 - n_1{}^2}\right] \tag{7.14}$$

Letting $n_1 = 2$, and substituting known modern values of h, c, m, and e into Equation 7.14

$$\lambda = \frac{(6.6251 \times 10^{-27})^3 \times 2.9979 \times 10^{10} \times 10^8}{2 \times (3.1416)^2 \times 9.1085 \times 10^{-28} \times (4.8028 \times 10^{-10})^4}\left[\frac{4n^2}{n^2 - 4}\right]$$

$$\lambda = 3645\left[\frac{n^2}{n^2 - 4}\right] \quad (\text{in Å}) \tag{7.15}$$

The agreement between Equation 7.15 and Balmer's formula is remarkable, considering the complexity of Bohr's equation. There can be little doubt that, irrespective of the theory from which it follows, Equation 7.11 describes the energies that can be associated with the hydrogen atom very well indeed.

In addition to allowing a prediction of the energy levels of the hydrogen atom, the theory is explicit in its prediction of the radii of the electron orbits. By Equation 7.10

$$r = \frac{n^2 h^2}{4\pi^2 e^2 m}$$

Substituting values for the physical constants, we obtain

$$r = \frac{n^2 \times (6.6251 \times 10^{-27})^2}{4 \times (3.1416)^2 \times (4.8028 \times 10^{-10})^2 \times 9.1085 \times 10^{-28}} = 0.5292 \; n^2 \times 10^{-8} \; \text{cm}$$

$$= 0.5292 \; n^2 \; \text{Å}$$

The calculated value of about 0.53 Å for the radius of the hydrogen atom in its ground state agrees favorably with estimates of the size of the atom on the basis of the kinetic theory of gases, and is of the same order of magnitude as interatomic distances in molecules and solids. According to the theory, the size of the atom would increase as the value of n^2, making the size of the radius for n equals 2 and 3, four and nine times as large, respectively, as the ground state radius.

7.5 MODERN QUANTUM THEORY

Bohr's theory for the structure of the hydrogen atom (and all other one-electron species like He^+ and Li^{2+}) was highly successful, and scientists of the day must have thought that they were on the verge of being able to predict all of the allowed energy levels of all of the atoms. However, it soon became obvious that the extension of Bohr's ideas to atoms with two or more electrons gave, at best, only qualitative agreement with experiment. We have at present no theory which successfully predicts the allowed energies of atoms in general, in spite of an enormous amount of research on the problem.

In his theory Bohr treated the hydrogen atom much like a classical system, consisting of two particles subject to classical forces. His postulate about the value for mvr of the electron was surely not consistent with classical theory, but the idea of a well-defined path for the electron was not discarded.

Since the Bohr theory produced inaccurate predictions of the energies of atoms like He, Li, or indeed any species with more than one electron, and also predicted wrong values for the energies of some molecular systems to which the theory was extended, an extensive search was begun about 1920 for other approaches to the problem of energies of electrons in atoms and molecules. In 1924 Louis de Broglie, a young French physicist, made a suggestion that was to have far-reaching results.

You may recall that in Chapter 4 we noted that in some experiments light rays behave as though light has some of the properties of a stream of particles. In particular, atomic spectra, the photoelectric effect, and the initiation of chemical reactions by light of short wavelength seem to indicate that light energy exists in quanta, with energy equal to hc/λ, and that energy in this amount can be transferred to atoms or molecules as though the energy were confined in a particle-like packet.

Louis de Broglie, reasoning as physicists do, suggested in 1924 that if light rays have particle properties, then perhaps particles may, under some circumstances, exhibit wave properties. By an argument which is beyond the level of this text, de Broglie predicted the wavelength associated with a particle of mass m moving at velocity v to be given by the relation

$$\lambda = h/mv \qquad (7.16)$$

where h equals Planck's constant, 6.625×10^{-27} erg sec.

Within a few years, Davisson and Germer, working at the Bell Telephone Laboratories, tested de Broglie's prediction by diffracting electrons of known energy from crystals. They established that a beam of electrons did indeed have wave properties, and that the wavelength to be associated with electrons of known mv value was exactly that predicted by the de Broglie equation.

Some Implications of the Wave Properties of Particles

The de Broglie relation is a key to the wave properties of small particles, much as the Einstein equation is a key to the particle properties of light. Although it cannot be used directly to solve problems involving atoms and molecules as generally as the wave equation to be presented in the next section, it can very easily be applied to a few simple problems, and yields solutions completely consistent with those obtained by the wave equation. We shall now consider one problem involving de Broglie-type waves, so as to show you the kinds of waves that appear to be associated with small particles, how quantum numbers arise in the wave treatment, and how the allowed energy levels of a system can be found. We shall then apply the solution obtained to several small systems to illustrate how the observed properties of small particles are correlated with their masses and locations.

The system we shall work with is very similar to that treated in the section on kinetic theory of gases in Chapter 5, namely, a particle confined to move in one dimension between two impenetrable walls. We assume, as in the previous development, that the particle has only kinetic energy and moves across the space allowed to it at a constant speed v. By the classical relations of mechanics, the energy ϵ and momentum are given by the relations

$$\epsilon = \tfrac{1}{2}mv^2 \qquad \text{and} \qquad \text{momentum} = mv$$

The de Broglie relation states that there must be a wave associated with the particle. To one first encountering this notion, it is certainly not clear how the wave would arise or how it might be represented. Our theory also tells us that the wave must just fit into the space allowed to the particle; in this prob-

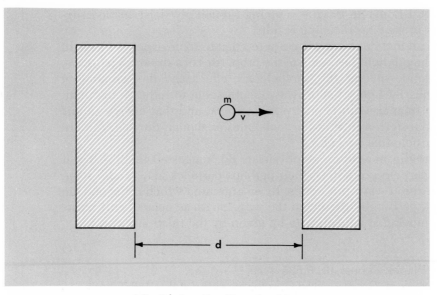

Figure 7.7

A Particle in a One-Dimensional Box

lem the wave can have finite amplitudes only inside the system, and the wave amplitude at the walls must be zero. Examples of some waves which meet these conditions are shown in Figure 7.8.

Analytic functions called **wave functions** can be written which give the amplitude of each wave as a function of position; the waves of Figure 7.8 would result from plotting the amplitude in the wave function against position

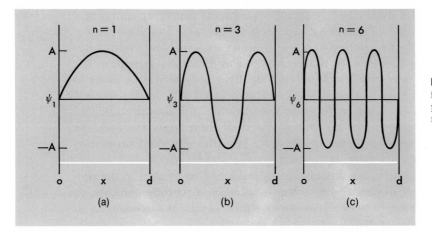

Figure 7.8 Some wave functions allowed to a particle in a one-dimensional box.

in the container. Wave functions are often given the symbol ψ, and for the wave in Figure 7.8a, ψ would obey the following equation:

What is the value of ψ when $x = 0$? $x = d/2$? $x = d$?

$$\psi = A \sin \frac{\pi x}{d} \qquad (7.17)$$

where x is the distance from the left hand wall and d is the distance between walls.

The physical significance of wave functions turns out to be rather surprising. According to the current theory, the square of the value of ψ at any point is equal to the relative probability of finding the particle at that point. If we could examine experimentally a system which had a wave function like that in Figure 7.8a, we would be most likely to find the particle in the middle of the box, and would essentially never be able to detect it at the wall! A strange situation indeed, but one which, insofar as it can be tested, is consistent with the behavior of such systems.

When we treated this problem by the kinetic theory of gases, we assumed that we could, if we wished, predict the path a single molecule would follow, given constant speed and perfectly reflecting walls. When we use the wave approach, all we can hope to do in this regard is to determine the waves to be associated with a particle, and so predict the likelihood of finding it at any given point in space. We cannot predict its path as Bohr attempted to do in his approach to the H atom problem. This difference between classical and quantum systems appears to be fundamental, and makes our "understanding" of quantum systems more difficult.

Although the paths of particles are not obtainable by the wave approach, their energies are. From the properties of the waves in Figure 7.8, we can see that for this system there are many waves that satisfy the condition that the amplitude be zero at the walls. The wavelengths λ of allowed waves all can be seen to satisfy the relation

$$\lambda = \frac{2d}{n} \tag{7.18}$$

where d is the distance between the walls and n is an integer equal to 1, 2, 3, . . . The number n is the **quantum number** for this system and arises here naturally, as a result of the condition that the waves must fit properly in the space allowed to the particle.

By combining de Broglie's Equation 7.16 with the values of λ in Equation 7.18, we obtain,

$$\lambda = \frac{h}{mv} = \frac{2d}{n} \quad \text{or} \quad v = \frac{nh}{2md} \tag{7.19}$$

Substituting now into the equation for the energy of the particle,

$$\epsilon = \tfrac{1}{2}mv^2 = \tfrac{1}{2}m\,(n^2h^2/4\ m^2d^2) = n^2h^2/8\ md^2 \tag{7.20}$$

Equation 7.20 tells us that a particle confined to move in one direction between two walls will have allowed energies which are limited to certain definite values and depend on the mass of the particle, the interval d through which it can move, and the quantum number n. The energy levels available to such a particle are shown in Figure 7.9. For each allowed energy there will be an associated value of the quantum number n and a wave function, ψ_n, with n loops, analogous to ψ_3 in Figure 7.8.

The energy for any given system will vary inversely both with the mass of the moving particle and the square of the size of its container. Particles of low mass confined in small regions will, by the nature of Equation 7.20, tend to have much larger minimum energies (n = 1) than heavier particles in larger systems.

143

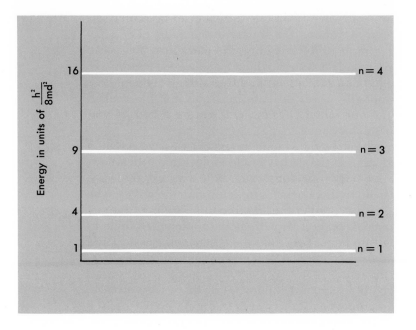

Figure 7.9 Lower-lying energy levels of a particle in a one-dimensional box.

Some Systems for which the One-dimensional Particle in a Box can Serve as a Model

The system we have been considering is a relatively simple one, but it can nevertheless serve as a qualitative model in several problems of chemical interest. We shall apply the model first to a gas molecule, then to the electron in a hydrogen atom, and finally to electrons in nuclei.

System I. An Average Gas Molecule. We could find the energy levels of an ordinary gas molecule moving in one dimension from Equation 7.20 simply by substituting appropriate values for the mass of the molecule and the size of the container in which the molecule moves. For an oxygen, O_2, molecule, moving between walls 100 cm apart, we would obtain,

$$\epsilon_n = \frac{n^2 h^2}{8\ md^2} = \frac{(6.63 \times 10^{-27})^2\ n^2}{8\ (32.0/6.02\ \times 10^{23})\ (100)^2} = 1.03 \times 10^{-35}\ n^2\ \text{ergs}$$

The minimum energy allowed to the molecule is therefore about 1×10^{-35} ergs, which is a very small energy by any standards.

The average energy in one direction of an O_2 molecule as it moves about in a gas is about $\frac{1}{2} RT/N_0$, so that at 25°C,

$$\epsilon_{ave} = \frac{1\ (8.3 \times 10^7\ \text{ergs/mole °K}) \times 298°\text{K}}{2\ (6.02 \times 10^{23}\ \text{molecules/mole})} = 2.06 \times 10^{-14}\ \text{ergs/molecule}$$

A molecule of O_2 in the $n = 1$ state would, by collision with average molecules, tend to change its energy and n value, until it too became an average molecule. Under such conditions its quantum number could be obtained from the relation

$$\epsilon_{ave} = 2.06 \times 10^{-14} \text{ ergs} = \epsilon_n = 1.03 \times 10^{-35} \text{ n}^2 \text{ ergs}$$

$$n^2_{ave} = 2 \times 10^{21} \quad \text{or} \quad n_{ave} = 4.5 \times 10^{10}$$

We would expect, by this analysis, that O_2 molecules in the air would ordinarily be in states with very high n values. In the section on the kinetic theory we were able to treat molecules as classical particles. This means that in their motion in a container gas molecules actually behave classically. We can therefore deduce that if a particle has a high quantum number n, it will behave as a classical particle. This indeed turns out to be the case, on both theoretical and experimental grounds. Gas molecules, tennis balls, and people, as they move about in space, typically are high quantum-number particles and obey the laws of classical mechanics as well as quantum mechanics. Since the mathematics necessary to treat a classical system is usually simpler than that needed to deal with the system quantum mechanically, we use the classical approach whenever it is applicable.

Can you see why the O_2 molecule will behave classically if n is large?

System II. An Electron in a Hydrogen Atom. In the previous system the gas molecule was confined by its container wall. We could also confine a particle with some sort of potential wall, such as would arise if a negatively charged particle were in the electric field of a positively charged particle. In a hydrogen atom the electron is confined by the positive charge of the proton to move in a sphere about 1 Å in diameter. Although the potential energy barrier to the electron is not of the same form that an impenetrable wall would create, and the electron can actually move in three directions rather than one, a crude model of the H atom would be that it contains an electron moving in a one-dimensional box about 1 Å in length.

Substituting into Equation 7.20 the mass of the electron and the size of the H atom, we obtain for the kinetic energy of the electron in the atom,

$$\underset{\substack{\epsilon_{electron} \\ \text{(in H atom)}}}{} = \frac{n^2 h^2}{8 \text{ md}^2} = \frac{(6.63 \times 10^{-27})^2 \text{ n}^2}{8 \times 9.1 \times 10^{-28} \times (1 \times 10^{-8})^2} = 6.0 \times 10^{-11} \text{ n}^2 \text{ ergs}$$

The minimum energy (n = 1) of an electron in an H atom by this model is about 6×10^{-11} ergs, or about 6×10^{24} times as large as that obtained for the O_2 molecule. The tremendous difference arises because of the very low mass of the electron and the very small space in which it is confined to move.

The energy required to raise an H atom to the next level (n = 2) is so great that all atoms will ordinarily be in the ground state.

An idea of the adequacy of the model we have used is obtained by comparing the energy for ionization of the H atom with the minimum kinetic energy of the electron we have just obtained. The ionization energy of the atom is about 13.5 electron volts, or about 2.1×10^{-11} ergs, which is roughly equal to the kinetic energy of the electron as found from the model, as it must be if the electron is to be held near the nucleus. This simple approach to the energy of an atom is, then, at least qualitatively correct. The results correctly imply that electrons in atoms and in molecules will generally be in their lowest allowed energy states and will not obey classical laws. The quantum properties of chemical systems are observed because electrons in atoms and molecules have very high minimum and excited state energies and so exist in energy states with low quantum numbers.

System III. An Electron in an Atomic Nucleus. Our nuclear model involves protons and neutrons in atomic nuclei but excludes electrons. When the authors of this text were the age of the typical student presently studying freshman chemistry, atomic nuclei were considered to consist of electrons and protons, which could also formally satisfy the mass and charge properties of nuclei.

The simple model for an electron in a nucleus would be very similar to that for an electron in an H atom, except that instead of being restricted to a path of 1 Å, or 1×10^{-8} cm, d_H, it would be confined to move in a much smaller space, namely the diameter d_{nuc} of the nucleus, or about 1×10^{-12} cm. Since energy varies as the inverse of the box dimension squared, the minimum energy allowed to an electron in a nucleus is about 1×10^8 times as large as the energy of the electron in the H atom. That is

$$\epsilon_{\substack{\text{electron} \\ \text{(in nucleus)}}} = \epsilon_{\substack{\text{electron} \\ \text{(in H atom)}}} \times \left(\frac{d_H{}^2}{d_{nuc}{}^2}\right) = 6 \times 10^{-11} \text{ ergs} \left(\frac{1 \times 10^{-8}}{1 \times 10^{-12}}\right)^2 = 6 \times 10^{-3} \text{ ergs}$$

The stabilities of atomic nuclei are well known. To completely disintegrate a typical nucleus, say that of a carbon atom, would require about 12×10^{-5} ergs or about 1×10^{-5} ergs per particle produced. The minimum energy of an electron in a nucleus would have to be about 1000 times the energy that binds a particle in the nucleus; hence we must conclude that the binding energy of nuclei is simply not adequate to hold electrons within those nuclei. As Dalton might have put it, "Thou canst not confine an electron to an atomic nucleus." It was on this kind of argument that the model for nuclei was changed to exclude electrons.

A proton, because of its larger mass, can be confined to the nucleus (Problem 7.22).

The Schrödinger Wave Equation

Even before the experimental proof of de Broglie's theory was established, Erwin Schrödinger had applied de Broglie's ideas to systems of small particles and in 1926 developed an equation, now called the Schrödinger wave equation, which allowed him to determine the energy levels and wave properties of the hydrogen atom and a few other relatively simple systems. The solutions obtained by Schrödinger agree exceedingly well with experiment. Unlike the Bohr theory, the wave equation can also be applied to other atoms and to molecules. Unfortunately the extension of the wave equation to these other systems has proved to be more difficult in fact that in principle, since the form of the equation for such systems is in general so complicated mathematically as to be insoluble. However, in the relatively few cases in which a satisfactory solution has been possible, there is excellent agreement with experiment. This has led scientists to believe that the wave equation approach to atomic and molecular properties is a correct albeit complex one.

Although we shall not carry out any mathematical problems using the wave equation, we shall write it down so that you may have some idea of its nature. For a one-particle system, the equation takes the form

$$\frac{\partial^2 \psi}{\partial x^2} + \frac{\partial^2 \psi}{\partial y^2} + \frac{\partial^2 \psi}{\partial z^2} + \frac{8\pi^2 m}{h^2}(E - V)\psi = 0 \qquad (7.21)$$

where m is the mass of the particle, E and V are its total and potential energies respectively, h is Planck's constant, and the first three terms are partial derivatives with respect to the coordinates (x, y, z) of the particle of ψ, the wave function to be associated with the particle.

The wave equation is a formidable mathematical relation, as is undoubtedly obvious even to the novice. The quantum mechanical problem is to solve the equation for the wave functions ψ allowed to the system. In general for a given problem there are a great many ψ functions that satisfy Equation 7.21. In Figure 7.8 we sketched some of these functions for the one-dimensional particle in a box problem. Those functions were obtained by the rigorous solution of the wave equation for that problem. Similar wave functions are found in the exact solution to the H atom problem and to other problems which have been treated exactly.

Figure 7.10 Electron cloud surrounding the nucleus of a typical atom.

As we noted, the interpretation of the wave function in a given state is that its square at any point in space is proportional to the probability of finding the particle at that point. In problems involving electrons, we can interpret the value of ψ^2 as being proportional to the electric charge density at the point. "Electron cloud" diagrams showing charge density in atoms and molecules are frequently drawn on this basis. In Figure 7.10 we have drawn schematically the "electron cloud" surrounding the nucleus of a typical atom.

7.6 ELECTRON ARRANGEMENTS IN ATOMS

Quantum Numbers of Electrons

The Schrödinger wave equation can readily be set up for the problem of the hydrogen atom. In that problem we consider the hydrogen nucleus to be fixed, and we examine the energy E of the electron in its allowed energy states. The potential energy V of the electron is obtained from the laws of electrostatics and is simply $-e^2/r$, where e is the electronic charge and r is the distance from the nucleus to the electron. Knowing this, we can immediately write the wave equation

$$\frac{\partial^2\psi}{\partial x^2} + \frac{\partial^2\psi}{\partial y^2} + \frac{\partial^2\psi}{\partial z^2} + \frac{8\pi^2 m}{h^2}\left(E + \frac{e^2}{r}\right)\psi = 0 \qquad (7.22)$$

Much of the modern theory of chemistry arises from solutions of this equation.

Schrödinger solved this equation for the energy levels of the H atom, obtaining the identical relation obtained by Bohr with his theory 15 years earlier. Schrödinger also obtained the wave functions and quantum numbers associated with each of the allowed states of the atom.

147

Schrödinger found that the electron in the H atom could be described by three quantum numbers, which are now called **n, l,** and **m**ₗ. There are three such numbers because the electron requires three coordinates to determine its motion. Given the values of these quantum numbers, the state of the electron, its energy, and its wave function are determined. Quantum numbers are also used in describing other atoms, where they furnish similar but less quantitative information about the electrons. We shall now discuss the quantum numbers of electrons as they are used with atoms in general.

The first or **principal** quantum number, given the symbol **n**, is of primary importance in determining the energy of the electron. For the H atom, the energy is completely fixed by the value of **n**:

$$E = -\frac{R}{n^2}$$

In other atoms, in which there are more electrons, the energy of each electron is dependent mainly (but not completely) on the value of the principal quantum number of the electron. As the quantum number increases, the energy of the electron increases, and the average distance of the electron cloud from the nucleus also increases. The principal quantum number is always integral and can take on the values

$$n = 1, 2, 3, 4, 5, \ldots, \qquad \text{but not } 0$$

These levels are often referred to as K, L, M, and so forth.

In an atom, electrons having the same value of **n** move about in roughly the same region, and are said to be in the same **level** or **shell**.

Each level of electrons in an atom includes one or more **sublevels** or **subshells**. The sublevels are denoted by the second quantum number, **l**, called the **angular momentum** quantum number. (Angular momentum is equal to the moment of the momentum, and in the Bohr model, for example, is equal to mvr for the electron.) The general geometric contour of the electron cloud for the electron is determined by **l**. The quantum number **l** for an electron is related to the quantum number **n** for the electron in that state; the relation between the quantum numbers is:

$$l = 0, 1, 2, \ldots (n - 1)$$

How many different l values are there when n = 1? n = 2? n = 3?

An electron with a principal quantum number **n** can, by the preceding equation, have **n** different **l** values, and so can exist in **n** sublevels.

The third quantum number **m**ₗ is often called the **magnetic orbital** quantum number. This quantum number is associated with the orientation of the electron cloud with respect to a given direction, usually one which is imposed on the atom by a strong magnetic field. The quantum number **m**ₗ has very little effect on the energy of an electron in any atom. Its importance derives from its influence on the orientation of the electron cloud and from its relation to the angular momentum quantum number **l**. For a given value of **l**, **m**ₗ can have any integral value between 1 and −1. That is:

$$m_l = l, l - 1, l - 2, \ldots, 0, -1, -2, \ldots, -l$$

How many different m₁ values are there for l = 0? l = 1? l = 2?

For a given value of **l**, **m**ₗ can assume $2l + 1$ different values. Electrons in an atom having the same values of **n, l,** and **m**ₗ are said to be in the same **orbital**. Within a given sublevel of quantum number **l** there will be $2l + 1$ orbitals.

In addition to its three quantum numbers **n, l,** and **m**ₗ, each electron appears to have a fourth quantum number m_s called its magnetic **spin** quantum number, or simply the spin quantum number. The spin quantum number can be loosely associated with the spin of the electron about its own axis. It was introduced into the theory to make the properties of atoms consistent with experiment. The quantum number m_s is not related to the values of **n, l,** or m_l for an electron, and can have one of two possible values, $+\frac{1}{2}$ or $-\frac{1}{2}$, depending on the direction of rotation of the electron about its axis. Two electrons in the same orbital having m_s values of $+\frac{1}{2}$ and $-\frac{1}{2}$ are said to be **paired**.

The final property of electrons in atoms which we must consider relates to the possible sets of quantum numbers which the electrons in an atom can possess. As we have said, each electron in an atom will have four quantum numbers to describe it; an electron in an atom will be in a state determined by its quantum numbers **n, l,** m_l, and m_s. For instance, according to the rules we have cited, we might have an electron in a sodium atom with the quantum numbers 2, 0, 0, $+\frac{1}{2}$. The question we might ask is, How many electrons in the atom can be assigned this particular set of quantum numbers? The **Pauli exclusion principle** answers this question by telling us: **No two electrons in any atom can have the same set of four quantum numbers.** This condition was first stated by Pauli in 1925, again to make the theory of atomic structure consistent with experimental observations.

The Pauli exclusion principle, plus the conditions on the four quantum numbers for electrons, allows us to say a great deal about the manner in which electrons can fill the levels, sublevels, and orbitals available to them in atoms. You will recall that,

> in each **level**, containing electrons with given quantum number **n**, there are **n sublevels**, each with different quantum numbers **l**
> in each **sublevel**, with a given quantum number **l**, there are $2l + 1$ **orbitals**, each with a different quantum number m_l.

By the Pauli principle, within a given orbital there can be only **two** electrons, one with a spin quantum number of $+\frac{1}{2}$, and the other with a spin quantum number of $-\frac{1}{2}$. If there were more than two electrons in an orbital, they could not all have a different set of quantum numbers, and the Pauli principle would be violated. Since the population of each electronic orbital is thus limited to two electrons, the population of each sublevel, and of each level, is also limited.

In Table 7.5 we have summarized the relations between the quantum numbers for some of the lower electronic levels, sublevels, and orbitals in atoms. From the Table we can see that the following combinations of quantum numbers **n, l,** m_l, and m_s are among those which are allowed,

$$1, 0, 0, +\tfrac{1}{2} \qquad 2, 1, -1, +\tfrac{1}{2} \qquad 3, 2, 1, +\tfrac{1}{2}$$

but we see that the following combinations are **not** allowed:

$$1, 1, 0, +\tfrac{1}{2} \qquad 2, -1, 1, 0 \qquad 3, 4, 2, +\tfrac{1}{2}$$

Table 7.5 includes all allowed quantum number combinations up to **n** = 3. Each sublevel contains one or more orbitals, each of which has a capacity of two electrons. An **l** = 0 sublevel includes only one orbital (m_l = 0) and hence can contain two electrons. An **l** = 1 sublevel has three associated orbitals (m_l = 1, 0, −1), and so can hold six electrons. An **l** = 2 sublevel can con-

149

TABLE 7.5 ALLOWED SETS OF QUANTUM NUMBERS FOR ELECTRONS IN ATOMS

Level n	1	2				3								
Sublevel l	0	0	1			0	1			2				
Orbital m_l	0	0	1	0	−1	0	1	0	−1	2	1	0	−1	−2
Spin m_s $\uparrow = +\frac{1}{2}, \downarrow = -\frac{1}{2}$	⇅	⇅	⇅	⇅	⇅	⇅	⇅	⇅	⇅	⇅	⇅	⇅	⇅	⇅

tain ten electrons. Similarly each level has one or more associated sublevels, and will have a capacity equal to the sum of those of its sublevels. The $n = 1$ level, containing only an $l = 0$ sublevel and one orbital, has a capacity of two electrons. The $n = 3$ level, with three sublevels which include a total of nine orbitals, can hold up to 18 electrons. The capacities of the electronic sublevels and levels up to $n = 5$ are summarized in Table 7.6. It is interesting to note that the total capacity of each electron level is equal to $2n^2$, where n is the principal quantum number for the level.

TABLE 7.6 CAPACITIES OF ELECTRONIC LEVELS AND SUBLEVELS IN ATOMS

LEVEL n	TOTAL NUMBER OF ELECTRONS IN LEVEL	NUMBER OF ELECTRONS IN SUBLEVELS					
		$l =$ 0	1	2	3	4	5
1	2	2	—	—	—	—	—
2	8	2	6	—	—	—	—
3	18	2	6	10	—	—	—
4	32	2	6	10	14	—	—
5	50	2	6	10	14	18	—

Electron Configurations of Atoms

In order to describe the electronic state of an atom we need to specify as completely as we can the sets of quantum numbers of all of its electrons. This specification is called the **electron configuration** of the atom and is of considerable use in connection with its chemical bonding properties. We have considered the rules governing the four quantum numbers of electrons in atoms in order that we might be able to predict the electron configuration of any atom.

Once the allowed sets of quantum numbers for electrons have been determined, as in Table 7.5, all that is required to establish the electron configuration of an atom is to assign its electrons to orbitals in the proper order. Electrons will enter the possible orbitals in order of their increasing energy, filling all the lower lying orbital positions before entering any of higher energy.

You will recall that the energy of an electron is dependent almost completely on the values of its n and l quantum numbers, increasing sharply with increasing n value and only slightly as the l value goes up. The m_l and m_s quantum numbers of an electron have only a very small effect on its energy.

It is then the electronic **sublevels**, characterized by different **n, l** values, which have appreciably different energies in atoms. Although the energies of the different sublevels in atoms cannot be calculated exactly, their relative magnitudes are known and are qualitatively quite similar in different atoms. In Figure 7.11 we have shown the relative energies of some of the lower lying sublevels for electrons in atoms.

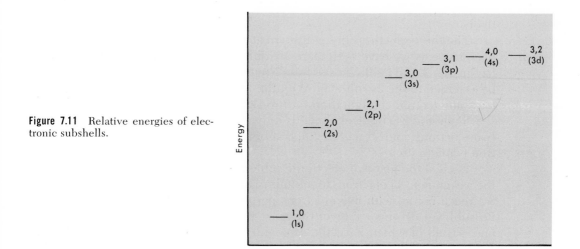

Figure 7.11 Relative energies of electronic subshells.

On this basis we can say that the two following sets of electron quantum numbers are associated with the lowest lying energy level in an atom:

$$1, 0, 0, +\tfrac{1}{2} \qquad\qquad 1, 0, 0, -\tfrac{1}{2}$$

These are the only two sets which have **n** = 1, and both necessarily have an **l** value of 0.

A hydrogen atom with its one electron will ordinarily be in one or the other of the two states described by the above sets of quantum numbers. The complete electron configuration of the H atom would then be simply

$$1, 0, 0, +\tfrac{1}{2} \qquad \text{or} \qquad 1, 0, 0, -\tfrac{1}{2}$$

Since only the values of **n** and **l** effect electron energies appreciably, the above description is often abbreviated to omit reference to the values of the **m$_l$** and **m$_s$** quantum numbers. This leaves the configuration specification of hydrogen as

$$1, 0$$

For some reason, probably traditional, we do not express electron configurations in quite this way. The notation for the principal quantum number, **n**, is as we have written it, but a new symbol is used for the second quantum number, **l**. Electrons for which **l** = 0 are called **s** electrons; if **l** = 1 they are called **p** electrons; for **l** = 2 they are called **d** electrons; if their **l** value is 3, they are called **f** electrons. The symbols s, p, d, and f come from the sharp, principal, diffuse, and fundamental series observed in atomic spectra early in this century. Spectroscopists were among the first to describe atoms in this

151

way, and their notation has persisted. On this basis the electron in the ordinary H atom is a 1s electron, and the electron configuration of the H atom is simply 1s. In the helium atom, with two electrons, we have an electron in each state in the lowest lying level with quantum number descriptions

$$1, 0, 0, +\tfrac{1}{2} : 1, 0, 0, -\tfrac{1}{2}$$

Since the two electrons are in the same sublevel, the electron configuration is written $1s^2$, meaning there are two electrons with **n, l** values of 1, 0 respectively.

The three electrons in the lithium atom cannot all have 1s quantum numbers, since as we have seen there are only two sets of four quantum numbers having **n** = 1 and **l** = 0. We see from Figure 7.11 that the third electron would be assigned to the 2s sublevel. With the first two electrons in the 1s level, we have an electron configuration for the Li atom of $1s^2 2s$.

Since the 2s sublevel contains only one orbital ($\mathbf{m_l}$ = 0), it can hold only two electrons, and so will be filled at beryllium, atomic number 4, with electron configuration $1s^2 2s^2$.

Each of the first two electronic sublevels contain only one orbital, so that the assignment of electrons to orbitals in atoms up to Be is unequivocal. When we reach boron, with five electrons per atom, we begin to fill orbitals in the 2p sublevel. In this sublevel there are three orbitals, with $\mathbf{m_l}$ values of 1, 0, and −1, all of which have the same energy. The manner in which these orbitals are populated by electrons has an effect on the magnetic and chemical properties of atoms and must be considered.

A 2p electron may be described by any one of the six sets of quantum numbers listed below:

$2, 1, 1, +\tfrac{1}{2}$	$2, 1, 0, +\tfrac{1}{2}$	$2, 1, -1, +\tfrac{1}{2}$
$2, 1, 1, -\tfrac{1}{2}$	$2, 1, 0, -\tfrac{1}{2}$	$2, 1, -1, -\tfrac{1}{2}$
$\mathbf{m_l} = 1$	$\mathbf{m_l} = 0$	$\mathbf{m_l} = -1$

In each of the three columns the two sets of quantum numbers would belong to a pair of electrons in an orbital in the 2p energy sublevel.

The first four electrons in the B atom enter and fill the 1s and 2s orbitals. The fifth electron will assume one of the six sets of quantum numbers listed above. Each set, being of equal energy, will be equally probable, and since we ordinarily cannot distinguish between the sets, we would say simply that the electron is in a 2p orbital, and write the electron configuration of the B atom as $1s^2 2s^2 2p$.

When we come to the carbon atom, with its six electrons, the situation is somewhat different. The two 2p electrons in the C atom could both go into the same orbital, with paired spins (spin quantum numbers of $+\tfrac{1}{2}$ and $-\tfrac{1}{2}$), or they could go into different orbitals with either unpaired parallel spins (both spin quantum numbers having the same sign) or with paired opposed spins (spin quantum numbers having opposite signs). The experimental properties of the carbon atom indicate that there is an energy difference between the different possible arrangements and that the most stable configuration is the one in which the *two electrons are unpaired, with parallel spins, and so are in different orbitals.*

The condition we observe on the electron arrangement in the carbon atom can be generalized to include other atoms for which similar situations exist. There is a general governing principle for all such cases, **Hund's rule**, which states that

In an atom in which orbitals of equal energy are to be filled by electrons, the order of filling is such that as many electrons remain unpaired as possible.

In order to keep the electron arrangement notation determined by Hund's rule as simple as possible, a system which avoids listing all four quantum numbers of the electrons is commonly employed. One simply lists the orbitals pertinent to the problem in some manner and indicates the electron populations and spins present in each orbital. In the case of the carbon atom, there would be five orbitals to be considered, one 1s, one 2s, and three 2p. These might be denoted by boxes, as in the following diagram. The populations and electron spins in each orbital are indicated by arrows: ↑ means an electron is present with spin quantum number equal to $+\frac{1}{2}$; ↓ means there is an electron in the orbital with spin quantum number equal to $-\frac{1}{2}$. By this system, the electron arrangement in the carbon atom would be

1s   2s   2p

The interpretation of the above representation, which we will call an **orbital diagram**, is that the 1s orbital contains two electrons which are paired (opposed spins), the 2s orbital also contains two electrons with opposed spins, and the two 2p electrons are in different orbitals and are unpaired (parallel spins). Since the other possible arrangements in the 2p orbitals which are consistent with Hund's rule are equivalent to the one given, they are not stated. (These arrangements can be distinguished and confirmed experimentally in studies of atomic spectra obtained in strong magnetic fields.)

An orbital diagram is intermediate in detail between the abbreviated form, in which only sublevel populations are listed, and the limiting form, in which all four quantum numbers for every electron are listed. For many purposes it is not necessary to consider the populations and spin arrangement in equivalent orbitals, and under such conditions the abbreviated electron configuration is used. For carbon, then, the electron configuration would be stated as $1s^2 2s^2 2p^2$. If a more detailed description is helpful, we will use the orbital diagram notation, listing orbitals, electron populations and spins.

The orbital diagrams for atoms beyond carbon which contain 2p electrons in the outermost sublevel follow immediately by the approach we used for carbon. In Figure 7.12 we have shown the orbital diagrams for all the atoms from boron to neon. From the Figure you can see that the neon atom has no unpaired electrons, boron and fluorine both have one, carbon and oxygen two, and in the nitrogen atom the three 2p electrons can all remain unpaired. As we shall see in succeeding chapters, the orbital diagram notation for electronic arrangements in atoms is useful in discussions of chemical bonding in many chemical substances. Hund's rule applies to electronic structures in any atoms, ions, or molecules in which there are several orbitals with equal, or nearly equal, energies. Those arrangements will be energetically preferred where the maximum number of electrons are unpaired.

Using the building up, or Aufbau, principle, adding electrons one by one as atomic number increases (taking due regard of the Pauli principle and Hund's rule), one can readily determine the electron configurations and or-

How many unpaired electrons are there in an Al atom? Si? P?

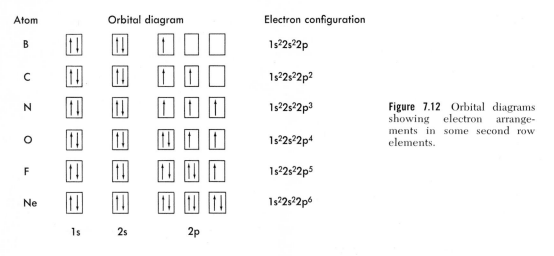

Figure 7.12 Orbital diagrams showing electron arrangements in some second row elements.

bital diagrams for the rest of the elements in the periodic table. The electrons are assigned to the sublevels in order of increasing energy, in general filling each sublevel to capacity before beginning the next. The order of sublevels is that in Figure 7.12; a complete list in proper order is given as follows:

1s, 2s, 2p, 3s, 3p, 4s, 3d, 4p, 5s, 4d, 5p, 6s, 4f, 5d, 6p, 7s, 5f, 6d.

You will note several cases where there is an "overlap" between principal energy levels. For example, the 4s sublevel ($n = 4$, $l = 0$) appears to be slightly lower in energy than the 3d sublevel ($n = 3$, $l = 2$). This means that in potassium and calcium, electrons enter the 4s level in preference to the 3d. The electronic configuration of potassium is $1s^2 2s^2 2p^6 3s^2 3p^6 4s^1$ rather than $1s^2 2s^2 2p^6 3s^2 3p^6 3d^1$.

Let us determine the electron configuration and the orbital diagram for the iron atom, atomic number 26. There are two 1s electrons, two 2s, six 2p, two 3s, six 3p, and two 4s electrons, filling the first six sublevels and requiring 20 electrons; these electrons will of necessity be paired in the orbitals in which they are found (why?). The remaining six electrons will go into the 3d sublevel, with its five equivalent orbitals, according to Hund's rule. The electron configuration of iron, Fe, atoms is, therefore, $1s^2 2s^2 2p^6 3s^2 3p^6 4s^2 3d^6$. The complete orbital diagram for the Fe atom is

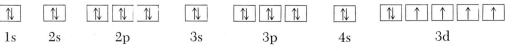

In applying Hund's rule we placed the 3d electrons one by one in the orbitals available, keeping the electron spins all parallel (spin quantum number the same, $+\frac{1}{2}$ chosen arbitrarily). Having filled the orbitals with the first five such electrons, we put the sixth electron, with opposite spin, in the first orbital. In the atom we would anticipate that there would be four unpaired electrons.

The electron configurations of ions are analogous to those of atoms and are determined in the same general way. For the most part, electron configurations and orbital diagrams depend only on the number of electrons in the species and not on its charge or on the charge of its nucleus. The electron configurations of ions of the lighter elements are the same as those of the

How many unpaired electrons are there in Co? Ni?

154

atoms with the same number of electrons. The electron configuration of the Na^+ ion (10 electrons) is the same as that of the neon atom, $1s^2 2s^2 2p^6$; this is also the configuration of the Mg^{2+}, F^-, and O^{2-} ions, all of which have 10 electrons.

An exception to this general rule occurs with many of the transition metals. When the iron atom loses two electrons to form Fe^{2+}, the two 4s electrons are removed, not two of the 3d electrons. In the formation of the ions of all of the transition metals it appears that it is the outermost s electrons which are first removed.

In Table 7.7 we have listed the electron configurations of the elements of atomic number 1 through 101. The configurations generally follow the rules we have given. The major exceptions occur for those elements which have configurations with near-full or near half-full sublevels. By Hund's rule we might surmise that half-full sublevels would be relatively stable; it is also true that a full sublevel, be it s, or p, or d, confers a measure of stability on the electronic structure of an atom. Where 3d sublevels are involved, with energies near those of the 4s sublevel, the effect is large enough in chromium (atomic number 24) to allow promotion of a 4s electron to the 3d sublevel, giving the Cr atom five 3d electrons, all with parallel spins. In the copper atom, with 29 electrons, a 4s electron is promoted to the 3d sublevel, thereby filling that sublevel. In the fifth and sixth periods these effects are somewhat more pronounced.

Geometric Representations of Electron Charge Clouds

Each of the states in which an electron can exist is characterized by a specific wave function as well as a set of quantum numbers and a definite energy. The wave functions of the electron in the H atom were found by Schrödinger and, like the quantum numbers of the electron, have been used in dealing qualitatively with the properties of multielectronic species, including atoms, ions, and molecules.

You will recall that the value at any point of the square of the wave function ψ associated with an electron is proportional to the probability of finding the electron at that point. The value of ψ^2 reflects the relative amount of electric charge at the point, or the charge density at that point. The electron cloud sketched in Figure 7.10 indicates that the charge density in the atom is largest near the nucleus and drops off rapidly as the distance from the nucleus is increased.

The electron charge cloud associated with a single electron is dependent, as you might expect, on the values of **n**, **l**, and **m_l** for the state under consideration. The dependence on **n** can be expressed in a function that includes only r, the distance of the electron from the nucleus, and so is always spherically symmetric. As **n** increases, the charge cloud moves on the average farther from the nucleus and takes on a very roughly shell-like structure, with maxima and minima that depend in their position and magnitude on the value of **n**. The charge clouds for an electron with different **n** values and a value of $l = 0$ (an s electron) are shown in Figure 7.13. The charge cloud for any s electron is spherically symmetric.

For states in which l is not equal to zero the associated electron clouds

TABLE 7.7 THE ELECTRON CONFIGURATIONS OF THE ATOMS OF THE ELEMENTS

Element	Atomic Number	1s	2s	2p	3s	3p	3d	4s	4p	4d	4f	5s
H	1	1										
He	2	2										
Li	3	2	1									
Be	4	2	2									
B	5	2	2	1								
C	6	2	2	2								
N	7	2	2	3								
O	8	2	2	4								
F	9	2	2	5								
Ne	10	2	2	6								
Na	11	Neon core			1							
Mg	12	Neon core			2							
Al	13	Neon core			2	1						
Si	14	Neon core			2	2						
P	15	Neon core			2	3						
S	16	Neon core			2	4						
Cl	17	Neon core			2	5						
Ar	18	2	2	6	2	6						
K	19	Argon core						1				
Ca	20	Argon core						2				
Sc	21	Argon core					1	2				
Ti	22	Argon core					2	2				
V	23	Argon core					3	2				
Cr	24	Argon core					5	1				
Mn	25	Argon core					5	2				
Fe	26	Argon core					6	2				
Co	27	Argon core					7	2				
Ni	28	Argon core					8	2				
Cu	29	Argon core					10	1				
Zn	30	Argon core					10	2				
Ga	31	Argon core					10	2	1			
Ge	32	Argon core					10	2	2			
As	33	Argon core					10	2	3			
Se	34	Argon core					10	2	4			
Br	35	Argon core					10	2	5			
Kr	36	2	2	6	2	6	10	2	6			
Rb	37	Krypton core										1
Sr	38	Krypton core										2
Y	39	Krypton core								1		2
Zr	40	Krypton core								2		2
Nb	41	Krypton core								4		1
Mo	42	Krypton core								5		1
Tc	43	Krypton core								6		1
Ru	44	Krypton core								7		1
Rh	45	Krypton core								8		1
Pd	46	Krypton core								10		
Ag	47	Krypton core								10		1
Cd	48	Krypton core								10		2

TABLE 7.7 *Continued.*

Element	Atomic Number	4d	4f	5s	5p	5d	5f	6s	6p	6d	7s
In	49	10		2	1						
Sn	50	10		2	2						
Sb	51	10		2	3						
Te	52	10		2	4						
I	53	10		2	5						
Xe	54	10		2	6						
Cs	55	10		2	6			1			
Ba	56	10		2	6			2			
La	57	10		2	6	1		2			
Ce	58	10	2	2	6			2			
Pr	59	10	3	2	6			2			
Nd	60	10	4	2	6			2			
Pm	61	10	5	2	6			2			
Sm	62	10	6	2	6			2			
Eu	63	10	7	2	6			2			
Gd	64	10	7	2	6	1		2			
Tb	65	10	9	2	6			2			
Dy	66	10	10	2	6			2			
Ho	67	10	11	2	6			2			
Er	68	10	12	2	6			2			
Tm	69	10	13	2	6			2			
Yb	70	10	14	2	6			2			
Lu	71	10	14	2	6	1		2			
Hf	72	10	14	2	6	2		2			
Ta	73	10	14	2	6	3		2			
W	74	10	14	2	6	4		2			
Re	75	10	14	2	6	5		2			
Os	76	10	14	2	6	6		2			
Ir	77	10	14	2	6	9					
Pt	78	10	14	2	6	9		1			
Au	79	10	14	2	6	10		1			
Hg	80	10	14	2	6	10		2			
Tl	81	10	14	2	6	10		2	1		
Pb	82	10	14	2	6	10		2	2		
Bi	83	10	14	2	6	10		2	3		
Po	84	10	14	2	6	10		2	4		
At	85	10	14	2	6	10		2	5		
Rn	86	10	14	2	6	10		2	6		
Fr	87	10	14	2	6	10		2	6		1
Ra	88	10	14	2	6	10		2	6		2
Ac	89	10	14	2	6	10		2	6	1	2
Th	90	10	14	2	6	10		2	6	2	2
Pa	91	10	14	2	6	10	2	2	6	1	2
U	92	10	14	2	6	10	3	2	6	1	2
Np	93	10	14	2	6	10	5	2	6		2
Pu	94	10	14	2	6	10	6	2	6		2
Am	95	10	14	2	6	10	7	2	6		2
Cm	96	10	14	2	6	10	7	2	6	1	2
Bk	97	10	14	2	6	10	9	2	6		2
Cf	98	10	14	2	6	10	10	2	6		2
Es	99	10	14	2	6	10	11	2	6		2
Fm	100	10	14	2	6	10	12	2	6		2
Md	101	10	14	2	6	10	13	2	6		2

(The 1s, 2s, 2p, 3s, 3p, 3d, 4s, 4p subshells are filled as the *Krypton core*.)

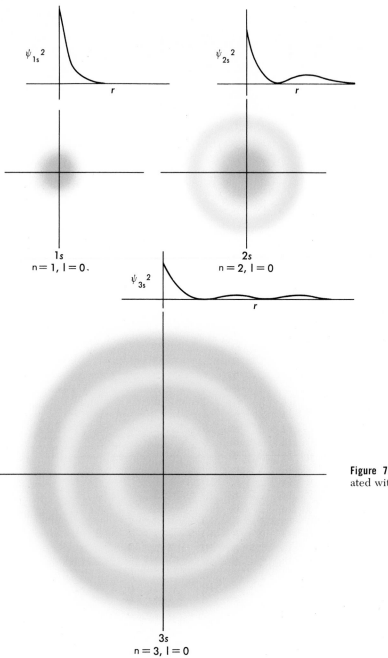

ψ_{1s}^2

ψ_{2s}^2

r

r

1s
n = 1, l = 0.

2s
n = 2, l = 0

ψ_{3s}^2

r

3s
n = 3, l = 0

Figure 7.13 Charge clouds associated with s orbitals.

About 90 per cent of the electron density is concentrated within the region shown by the charge cloud.

are more complicated in their geometric structure. If we are working with p electrons, where l = 1, the value of $\mathbf{m_l}$ is no longer restricted to zero, but may take on the values +1, 0, and −1. We would anticipate that there would be three different electron clouds possible for l = 1 states, one for each possible value of $\mathbf{m_l}$. This is indeed the case. The three clouds are identical in their overall structure but differ in their orientation with respect to a given set of axes, whose position might be established, for instance, by a strong magnetic field. The three clouds may be considered to be concentrated along the x, y,

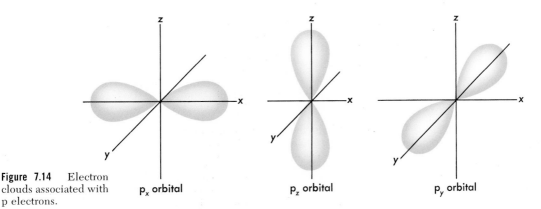

Figure 7.14 Electron clouds associated with p electrons.

p_x orbital p_z orbital p_y orbital

and z axes in the manner indicated in Figure 7.14. As is apparent from their structure, the energies associated with the three clouds would tend to be very nearly identical.

An electron with any one of the charge clouds in Figure 7.14 is said to be in a p orbital, since it would have an **l** value of 1. There are three possible p orbitals for an electron having the quantum numbers **n**, 1, **m**; they are often called p_x, p_y, and p_z orbitals. (The actual structure of the charge cloud for a p orbital will depend on the value of **n** as well as that of m_l, and in Figure 7.14 **n** has been taken to be 2.)

It is possible to draw charge clouds for orbitals associated with a d electron. In this case there are five different orbitals, corresponding to the five possible values for m_l. We shall consider d orbitals in some detail in Chapter 19.

The charge clouds we have drawn so far have been for single electrons. Qualitatively, one can assume that the charge cloud for an atom containing several electrons would be equal to the superposition of charge clouds associated with the electrons in each occupied orbital in the atom. In succeeding chapters we will make extensive use of electron configurations, orbital diagrams, and geometric properties of atomic orbitals in our discussions of chemical bonding.

Some Comments on the Electron Configuration Theory. The theory of electron configurations associates the electrons in an atom with quantum states appropriate to their energies and space properties. The electron configuration and orbital diagram of an atom describes, as best we can at present, the properties of an atom in terms of the properties of its electrons. Associated with each electron quantum state is its set of quantum numbers and a cloud of negative charge, whose density at any point is proportional to the probability of finding the electron at that point in space. An atom, according to the model, consists of a positive nucleus surrounded by a cloud of negative charge resulting from the superposition of the charge clouds associated with each of its electrons.

Since the charge clouds for different sets of quantum numbers, or different quantum states, may differ in their properties, particularly in the distance from the atomic nucleus at which they have appreciable densities, our notion of the electronic structure of the atom is not quite so nebulous as it might first appear. Theory tells us that electrons with high principal quantum number **n** have high energy and, on the average, are much more likely to be found in regions relatively distant from the nucleus than are electrons with low **n** values. On this basis we can assign electrons to regions, which are usually called "shells" or "levels," one shell or level for each value of **n**, with shells associated with low **n** values near the nucleus and those with higher **n** values more distant. To discriminate between electrons in the same level (same **n** value) but different values of the quantum number **1**, we use the term "same sublevel" or "same subshell" to denote those electrons with the same values of both **n** and **1**. Since the regions over which the charge clouds for electrons have appreciable density are not at all sharply defined, "shell" terminology is somewhat unfortunate, since to the novice it implies more than it should. We use the shell notation occasionally because it is convenient, and we must simply remember the meaning of the term in light of this discussion. In this and subsequent chapters we frequently use the terms "level" and "sublevel" in place of "shell" and "subshell"; this terminology is perhaps better in principle, in view of the fact that electron configurations describe energy levels, but it has the disadvantage that when used in this context "level" must be taken to mean that group of states having a given **n** value. Similarly, "sublevel" will mean that group of states with the same values of both **n** and **l**.

There is one difficulty with the electron configuration theory that should be mentioned. The theory is based on the assumption that the electrons in an atom can be described by assigning them quantum numbers. This means that in the theory we consider the electrons to have properties as individuals. (Indeed the quantum numbers we use and the interpretations we give them are based on the wave mechanical solution of the one-electron problem, and hence involve only the electron-nucleus interactions.) In such an atomic model the electronic properties of the atom will be simply equal to the sum of the properties of all the electrons in the atom, with individual electron properties determined by solutions to one-electron problems. Such a model is clearly incorrect, since in an atom there are important electron-electron charge interactions and electron-electron spin interactions, as well as the electron-nucleus charge interaction covered by the model. So far, the theory of atomic structure has been unable to properly treat electron-electron interactions in atoms. Such interactions are usually considered to be "averaged in" when one interprets electron configurations. The qualitative structure of atoms, involving the existence of shells and their populations, does not seem to be appreciably altered by such an averaging; this explains the usefulness of electron configurations. Clearly, however, the necessity for such an averaging renders impossible the use of electron configurations in quantitative calculations of atomic energy levels and atomic dimensions.

7.7 EXPERIMENTAL SUPPORT FOR ELECTRON CONFIGURATIONS

The electronic configurations of atoms are based on theoretical evidence, particularly the wave mechanical solution to the hydrogen atom problem, and on experimental evidence, mainly atomic spectra. The configurations are also closely related, as we shall see in the next section, to the chemical properties of the elements.

It would take us too far afield to show the detailed relations between atomic spectra and electron configurations. However, similar relations are obtained in investigations of the ionization energies of atoms and ions. The ionization energy of a particle is a direct measure of the energy required to remove an electron from that particle. A given element will have several ionization energies, one for each electron in its atom. The first ionization energy of sodium is equal to the energy required to remove the outermost, 3s, electron from the atom; the second ionization energy measures the energy required to remove a 2p electron from Na^+; the third, the energy needed to remove a second 2p electron from Na^{+2}, and so on. As the charge on the ion increases, the ionization energy increases. In Figure 7.15 we have plotted the square roots of the ionization energies of sodium as a function of the electron being removed.

According to the theory, the electron configuration of sodium, atomic number 11, would be $1s^2 2s^2 2p^6 3s$. Electrons would be present in three levels, with **n** values of 1, 2, and 3, with higher-energy electrons having higher **n** values. The first electron to be removed, being a 3s electron, would be expected to come off relatively easily. Then we would predict that there would be a group of eight electrons, in the 2p and the 2s sublevels, each of which would be somewhat less easily removed than the one before it, but which

With Na, there is a "jump" in ionization energy between the first and second electrons. Where would this jump occur with Mg? Al?

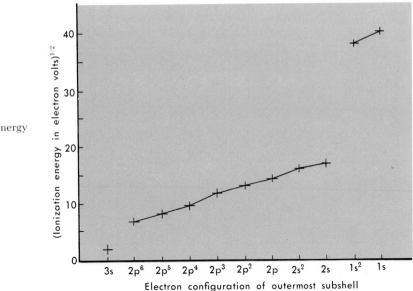

Figure 7.15 Ionization energy of sodium.

would not be grossly different in character from one another as far as energy of removal is concerned. The last two electrons, in the $\mathbf{n} = 1$ level, would be very tightly bound and could be removed only with difficulty.

Figure 7.15 clearly supports the theory. The ionization energies are grouped as expected, indicating the three levels in which electrons are found; with careful examination one can discern a difference between the 2p and the 2s electrons.

Using a somewhat different approach it is possible to show the resemblances between species having what our theory would predict to be the same electronic configuration. Let us consider, for example, the series of species

$$\text{Li} \quad \text{Be}^+ \quad \text{B}^{2+} \quad \text{C}^{3+} \quad \text{N}^{4+} \quad \text{O}^{5+} \quad \text{F}^{6+}$$

These seven species all belong to different elements, but all have three electrons, and therefore should all have the same electron configuration, $1s^2 2s$. The electronic structures of these species should all be similar to one another, differing only in the charge on the central nucleus, which would vary from $+3$ for Li to $+9$ for F^{6+}. This charge increase should make it harder to remove an electron from F^{6+} than from Li, but the change in ionization energy for that electron should, if the theory is valid, be gradual and smooth as one proceeds from Li to F^{6+}.

In Figure 7.16 we have plotted the square roots of the ionization energies versus nuclear charge for each of the species in the series mentioned. As you can see, the line obtained is essentially straight, clearly indicating that those species are closely related as far as electronic structure is concerned. Species such as these, with the same number of electrons, are said to form an **isoelectronic series**, and, as here, typically show resemblances that can be attributed to their having the same electron configurations.

The properties of the ionization energies of atoms and ions are consistent

161

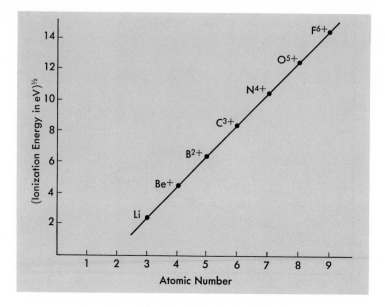

Figure 7.16 Ionization energies of an isoelectronic series.

with the theory of electron configurations and support the validity of the classifications we have made. The quantum theory does not at present allow us to predict quantitatively the ionization energies of isoelectronic series, but it does afford an excellent qualitative explanation of the experimental observations.

The Electronic Structure of Atoms and the Periodic Table

In the previous section we associated the electronic configurations of atoms with one of their physical properties, their ionization energies. Probably the most striking support of such configurations, however, comes from a study of the chemical properties of the elements.

When substances undergo an exothermic chemical reaction, a rearrangement of the electrons in the reacting atoms or molecules takes place, and the result is a system which is somewhat more stable than the reactants. The difference in stability is given by the energy, or heat, which must be removed from the reaction products to lower their temperature to that of the reactants. In chemical changes the energy liberated is of the order of 100 kcal per mole of reactant. The formation of a mole of liquid water from gaseous hydrogen and oxygen, a vigorous reaction, evolves about 68 kcal. The combustion of a mole of methane evolves 210 kcal. In terms of electron volts per molecule these energies are, respectively, about 3 and about 9 eV. These energies are about of the same order of magnitude as the ionization potentials of the electrons in the outermost shell, or level, of an atom, but are considerably smaller than the ionization potentials of the inner electrons.

From these observations one can conclude with some confidence that when atoms take part in chemical reactions it is only the outer electrons that are appreciably affected. The stability of inner shells is simply too high for them to be appreciably influenced by only a few electron volts of energy.

This reasoning leads us to believe that, since only the outermost shells of electrons participate in chemical reactions, the chemical properties of the atoms are to be associated primarily with their electron configurations in the outermost shell.

As soon as one examines the periodic table in the light of the outer electron configurations of atoms it becomes obvious that the relationship between chemical properties and configuration of electrons is close indeed. Consider, for instance, the electron configurations of the alkali metals (Table 7.8). For

TABLE 7.8 ELECTRON CONFIGURATIONS OF THE ALKALI METALS

ELEMENT	ATOMIC NUMBER	CONFIGURATION
Li	3	$1s^22s$
Na	11	$1s^22s^22p^63s$
K	19	$1s^22s^22p^63s^23p^64s$
Rb	37	$1s^22s^22p^63s^23p^63d^{10}4s^24p^65s$
Cs	55	$1s^22s^22p^63s^23p^63d^{10}4s^24p^64d^{10}5s^25p^66s$

each of the alkali metal atoms the electron configuration in the outermost shell is ns. These are the only elements with such a configuration and, by their chemical properties, clearly belong in the same family.

As with the alkali metal family, each family of atoms has its own particular electron configuration in its outermost shell. The difference in electronic structure between members of a given family is only in the number of inner shells. Indeed, if the elements are arranged in a table according to the outer electron configurations of their atoms, with atoms having the same outer configuration placed in the same vertical column and atoms with the same number of shells in the same row, the table of all the elements has a form identical to that of the periodic table.

What family has the configuration ns^2? ns^2np^6?

The basis on which we constructed the periodic table was primarily chemical. The foundations on which we built the electron configurations of the atoms are partly theoretical and partly physical. The fact that the periodic table can be derived by grouping atoms by outer electron configuration is strong evidence in support of the following very important generalization: **The chemical properties of an atom are determined primarily by the electron configuration in its outermost shell. The periodicity of properties of the elements is the result of the periodicity of the electron configurations in their outermost shell.**

7.8 THE CORRELATION OF ATOMIC AND PHYSICAL PROPERTIES WITH ELECTRON CONFIGURATIONS

A knowledge of electronic configurations in atoms is extremely useful in all areas of chemistry. It allows us to correlate and predict many of the physical and chemical properties of atoms and molecules in a truly remarkable way. In the next few chapters we shall discuss the applicability of the theory to problems of chemical bonding. Here we shall show how electron configurations can be used to aid our understanding of atomic sizes.

According to our model, the electrons in an atom are arranged in shells and subshells around the nucleus, at distances which increase as their principal quantum numbers increase. The electrons and the nucleus obey, within quantum-mechanical limits, the laws of electrostatics. Electrons, being of like charge, repel each other; electrons and nuclei, because of their unlike charges, are attracted to each other. Mathematically, these forces are:

For electron-electron repulsion,

$$F_R = e^2/r^2$$

and for electron-nucleus attraction,

$$F_A = Ze^2/r^2$$

in which $-e$ is the electronic charge, Ze is the nuclear charge, and r is the interparticle distance. In addition, the force between a charge and a group of charged concentric spheres is proportional to the total net charge of the spheres (Figure 7.17). In this atomic model (admittedly very simplified), Figure 7.17 can be considered to resemble an atom, with its nuclear charge (q_1) shielded from the outermost electrons ($-Q$) by the charge on the inner electron shells ($-q_2 - q_3$).

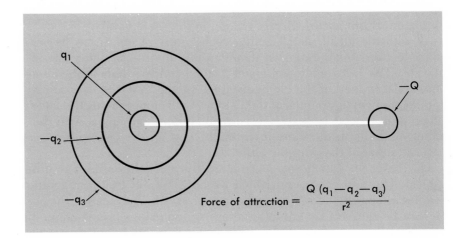

Figure 7.17 Shielding of charge of nucleus by inner electronic subshells.

On the basis of our model let us now consider sizes of atoms. In any given group in the periodic table we can predict that atomic diameters will increase as the number of electron shells increases. The reason for the increase is clearly shown by the examination of the electronic structures of the group 1A elements (Table 7.9). Each of these elements has a single s electron outside a closed shell or a closed p subshell. These subshells are relatively closer to the nucleus than is the outermost electron, and so shield the electron from much of the nuclear charge. If the shielding were perfect the outermost s electron would move in a field essentially of a +1 charge and the atom would behave as would a hydrogen atom, its electron having the same quantum number as the outermost s electron. Since the average distance of the elec-

TABLE 7.9 RADII OF THE ALKALI METALS

	Li	Na	K	Rb	Cs
Configuration in outermost shell	2s	3s	4s	5s	6s
Atomic radius in Å	1.52	1.86	2.31	2.44	2.62

tron from the hydrogen nucleus increases rapidly with increasing values of **n**, the size of the alkali metal atoms should increase with increasing principal quantum number of the outermost electron, and hence should increase with increasing period. The shielding is not perfect, since the atomic shells and subshells are not sharply defined, but the general line of reasoning is valid enough to properly predict the increase in atomic size with period.

A similar argument can be applied to atoms of elements in the same period, that is, with the same number of shells but differing configurations in the outermost shell. A good example would be the third period shown in Table 7.10.

TABLE 7.10 RADII OF ELEMENTS IN THE THIRD PERIOD

	Na	Mg	Al	Si	P	S	Cl	Ar
Outer configuration	3s	3s²	3s²3p	3s²3p²	3s²3p³	3s²3p⁴	3s²3p⁵	3s²3p⁶
Atomic radius in Å	1.86	1.60	1.43	1.17	1.10	1.04	0.99	(1.54)

In each of these atoms the electron configuration in the inner shells is the same, $1s^2 2s^2 2p^6$. Since electrons in the same shell do not shield the nucleus from each other, the shielding effect on the outermost, often called **valence**, electrons is also of the same magnitude in all of these atoms. This means that since the nuclear charge of the atoms increases from left to right, the valence electrons of these atoms are attracted by an increasing positive charge and hence move closer and closer to the nucleus. The ever-decreasing size of the atoms across the row supports this reasoning. Argon appears to be an exception to the rule; the reason for the exception is that the radii of the other elements are those observed for bonded atoms, whereas the argon radius was found from the density of solid argon. (See Chapter 8 for a discussion of the methods used in determining atomic dimensions.)

Other facts about atomic radii are made reasonable by analysis in terms of electron configurations. In parts of some rows of the periodic table the electrons go into inner shells rather than the outermost one. As one proceeds across such a period from one element to the next, both the nuclear charge and the population of the inner shell increase by the same amount, making for only a small change in the electric field attracting the outermost electrons. One would expect that for such atoms the radii would change gradually and only by a relatively small amount. The transition metals, where inner d electrons are being added, are examples of series of elements which show this behavior. In Figure 7.18 we have shown the dimensions of the atoms of the transition metals in the fourth, fifth, and sixth periods. In each period the

The size of the positive charge would be Z − 10 if the shielding were perfect.

165

Period

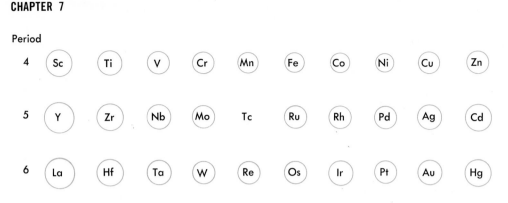

Figure 7.18 Atomic radii of transition metals.

atomic radii of these elements decrease slowly to a minimum at about the middle of the period and then gradually increase toward the end.

The last two series of transition metals are especially interesting, in that the radii of atoms in the same family, for example, Mo and W are virtually the same, even though the atoms in the sixth period contain 32 more electrons than those in the fifth. The explanation lies in the fact that between lanthanum and hafnium in the sixth period lie the 14 lanthanide elements. In this long series of elements the electrons being added gradually fill the deep-lying 4f subshell. In this sequence (see Table 7.11) there is a gradual

TABLE 7.11   ATOMIC RADII (Å) OF THE LANTHANIDE ELEMENTS

Ce	Pr	Nd	Pm	Sm	Eu	Gd	Tb	Dy	Ho	Er	Tm	Yb	Lu
1.82	1.82	1.82	–	1.83	2.04	1.79	1.77	1.77	1.76	1.75	1.74	1.93	1.74

The anomalous sizes of Eu and Yb reflect the different crystal structures of these elements.

slow decrease in atomic size, reflecting the failure of the shielding effect of the added 4f electrons to completely compensate for the increasing nuclear charge. The total effect of the decrease in atomic size through the lanthanide series is quite substantial and is called the *lanthanide contraction*. Fortuitously, it is of just about the right magnitude to make the transition metal atoms in the sixth period almost exactly equal in size to those in the same family in the fifth. The identity of atomic (and ionic) size make for a great chemical similarity in each family of elements in these series, and accounts, for example, for the great difficulty in separating such elements as zirconium and hafnium from each other.

PROBLEMS

7.1   What experimental evidence can you cite for the existence of

    a. Extranuclear electrons?   b. Atomic nuclei?   c. Isotopes?

7.2   What likelihood is there that a new element will be discovered with an atomic weight between those of two known elements somewhere in the middle of the periodic table, say between silver and cadmium? State your reasoning.

7.3   How might one produce an atomic spectrum for potassium? Of what practical use is such a spectrum? Of what theoretical use? How might atomic spectra be used to determine the composition of an alloy containing several different metals?

7.4   How do the laws governing the motion and energy of an electron in a sodium atom differ from those governing the motion and energy of a planet in the solar system?

7.5   What does an "electron cloud" diagram, such as that in Figure 7.10, represent? How does the electron cloud picture of the hydrogen atom differ from Bohr's model?

7.6   An electron in an atom might have the quantum numbers: 3, 2, 1, $+\frac{1}{2}$. What symbol is given to each of the quantum numbers in this set? What properties of the electron are implied by its quantum numbers? In the usual notation, what kind of electron would this be (i.e., 6p, 1s, 4f)?

7.7   What is meant by the statement, "The electron configuration of the nitrogen atom is $1s^2 2s^2 2p^3$?"

7.8   What experimental observations demonstrate that not all of the electrons in a typical atom can be in the 1s sublevel? That is, how do we know that the electron configuration of an atom of atomic number x is not $1s^x$?

7.9   What is an orbital? Describe the geometry of a p orbital. How many electrons can there be in a given p orbital? In what ways, if any, would these electrons differ?

7.10   What are isoelectronic species? Name a species isoelectronic with the K atom; with the Al^{3+} ion. How do the properties of isoelectronic species support the theory of electronic configurations?

7.11   In which of the following would you have more confidence concerning long-term scientific usefulness?

    a. The periodic table or electron configurations.
    b. Atomic energy levels or electron clouds.
    c. p orbitals or atomic number.

7.12   How does the electron configuration theory rationalize the fact that the rare earths are so nearly alike in their properties?

7.13   Characterize each of the following as to charge, mass, and size.

    a. Electron   b. Proton   c. Alpha particle   d. Sodium atom
    e. ^{24}Na nucleus

7.14   Calculate the lowest energy possible for a proton confined to move in one dimension in a 0.1 Å box. Compare this energy to kT at 25°C. This calculated energy would be approximately that of a hydrogen atom confined to move along the H—Cl bond in an HCl molecule, and so would be about that of the lowest vibrational energy of HCl. On this basis, would you expect the vibration of HCl at room temperature to follow classical or quantum mechanical laws?

7.15   What are the electron configurations of the following?

    a. C   b. Ne   c. Cl⁻

What would be the electronic configurations of each of these species in its first excited state?

7.16 Give the electronic configurations and orbital diagrams for the following species.

a. Cr   b. Cr^{3+} c. Zn^{2+}

How many unpaired electrons would there be in each of these species?

7.17 Which two of the following orbital diagrams for the carbon atom are equivalent to each other?

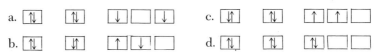

7.18 Give a set of four quantum numbers for each of the four electrons in the Be atom (ground state). How would you modify these sets if the atom were in its first excited state?

7.19 There are 18 electrons which have the quantum number $n = 3$. List the values of the other three quantum numbers for each of these electrons.

7.20 Identify the element(s) which

a. Falls in the first transition series and has four unpaired electrons per atom.
b. Has no electron with a quantum number $l = 1$.
c. Has an atomic number less than 20, and has no unpaired electrons.

7.21 Characterize each of the following as to nuclear charge, number of neutrons and protons in the nucleus, and number of extranuclear electrons.

a. $^{30}_{14}Si$ b. $^{13}_{6}C$ c. Al^{3+} d. $^{52}Cr^{3+}$

7.22 Show, using Equation 7.20, that it would be reasonable to have protons in the nucleus.

7.23 What are the electronic configurations of the following?

a. S   b. Na^+ c. Zn^{2+}

What would be the electronic configuration of each of these species in its first excited state?

7.24 Give the electron configuration and orbital diagram of each of the following.

a. Mg^{2+} b. Fe^{3+} c. Zr^{4+}

7.25 Explain why each of the following orbital diagrams for the ground state of the carbon atom is incorrect.

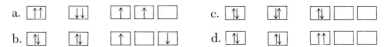

7.26 Give a possible set of four quantum numbers for each of the electrons in the F atom; the F^- ion.

7.27 Consider the 14 electrons in the 4f sublevel. Give the four quantum numbers of each of these electrons.

7.28 Identify the element(s) which

a. Has three unpaired electrons, for each of which the quantum number $n = 2$.
b. Cannot have an electron with a quantum number m_l other than 0.
c. Has one unpaired electron, shows metallic properties, but is not a transition element.

°7.29 Given that the ionization energy of the Li atom is 5.36 eV and that of the C^{3+} ion is 64.22 eV, calculate the ionization energy of the F^{6+} ion and that of He^- ion. The latter quantity is called the electron affinity of the helium atom and is a measure of the tendency for that atom to acquire an electron.

°7.30 Using Bohr's equation (7.15), calculate the wavelengths of the first three lines in the Balmer series. Compare your results with those in Table 7.4. Modify Equation 7.15 appropriately, and calculate the wavelength of the first line in the Lyman series, $n = 2$ to $n = 1$.

°7.31 Using quantum theory principles, calculate the wavelengths of the lines associated with the two transitions in the sodium atom shown with unspecified wavelengths in Figure 7.4.

°7.32 Iron atoms have a characteristic x-ray wavelength at 1.932 Å; the analogous line in the x-ray spectrum of silver occurs at 0.558 Å. An unknown metal has the line at 0.708 Å. Find the atomic number of the metal.

°7.33 X-ray spectra are thought to be caused by the removal of inner electrons from an atom by a beam of very energetic electrons. When, for example, an $n = 1$ electron is removed from the atom, electrons in higher shells tend to make transitions to the $n = 1$ level, thereby lowering the energy of the atom and emitting x-ray photons. The situation is similar to that occurring in hydrogen atoms when the Lyman series is excited. One of the strong lines in the x-ray spectrum of an atom is the result of an $n = 2$ to $n = 1$ transition, the same transition responsible for the first line in the Lyman series. In the x-ray transition, the electron which makes the quantum jump is not appreciably influenced in its energy by any atomic electrons except the remaining 1s electron, which shields the atomic nucleus and decreases its effective charge by one unit. (Shielding here is essentially perfect.)

Develop a relation, analogous to Equation 7.14, for a one-electron system in which the nuclear charge is Z rather than 1. Using this relation, predict the wavelength in the copper x-ray spectrum which is associated with the $n = 2$ to $n = 1$ transition.

°7.34 Suppose that the spin quantum number could have the values $+\frac{1}{2}$, 0, and $-\frac{1}{2}$. Assuming that the rules governing the possible values of the other quantum numbers and the order in which the various sublevels are filled remain unchanged,

a. What would be the capacity for electrons of an s sublevel? A p sublevel?
b. What would be the capacity of an $n = 3$ level?
c. What would be the electron configuration of the element with atomic number 7? 14?
d. What would be the atomic number of the element which would most closely resemble in chemical and physical properties the element having atomic number 7?
e. Using electron configurations, construct a "periodic table" for the first 40 elements.

8

Chemical Bonding

In Chapter 2 we referred briefly to the forces which hold atoms together in elementary and compound substances. Now that we have discussed the electronic structure of atoms, we are in a better position to understand the nature of these forces, called chemical bonds. In this chapter we shall consider two different types of bonds, both of which arise from interactions between electrons of atoms.

1. **Ionic** bonds are the electrostatic forces that hold together oppositely charged ions formed by a transfer of electrons from a metal to a nonmetal atom.

2. **Covalent** bonds arise when two atoms share an electron pair.

8.1 IONIC BONDING. FORMATION OF SODIUM FLUORIDE FROM THE ELEMENTS

When sodium metal is exposed to fluorine gas, a vigorous, exothermic reaction occurs:

$$Na(s) + \tfrac{1}{2} F_2(g) \longrightarrow NaF(s), \Delta H = -136 \text{ kcal} \qquad (8.1)$$

The product of this reaction is the compound sodium fluoride, which has a relatively high melting point (1000°C) and is a good conductor of electricity in the molten state or in water solution. These properties are characteristic of ionic compounds, which consist of oppositely charged ions held together by strong electrostatic forces. In sodium fluoride there are an equal number of Na^+ and F^- ions. The manner in which these ions are arranged in the sodium fluoride crystal is shown in Figure 8.1.

From Equation 8.1 we see that the formation of one mole of sodium fluoride from the elements evolves 136 kcal of heat. To understand where this energy comes from, let us imagine a plausible, three-step sequence for this reaction:

$$
\begin{array}{lll}
(1) & Na(s) \;+ \tfrac{1}{2} F_2(g) \longrightarrow Na(g) \;+ F(g) & ; \Delta H = \;+45 \text{ kcal} \\
(2) & Na(g) \;+ F(g) \longrightarrow Na^+(g) + F^-(g) & ; \Delta H = \;+38 \text{ kcal} \\
(3) & \underline{Na^+(g) + F^-(g) \longrightarrow NaF(s)} & ; \underline{\Delta H = -219 \text{ kcal}} \\
& Na(s) \;+ \tfrac{1}{2} F_2(g) \longrightarrow NaF(s) & ; \Delta H = -136 \text{ kcal}
\end{array}
$$

(8.1)

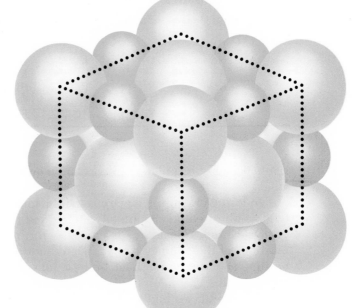

Figure 8.1 Ionic arrangement in sodium fluoride crystal: in the crystal each sodium ion (black sphere) is surrounded by six fluoride ions (light sphere) and each fluoride ion by six sodium ions.

In step 1 the bonds holding sodium atoms together in sodium metal and those holding fluorine atoms to one another in the F_2 molecule are broken, forming gaseous sodium and fluorine atoms. This process, as one would expect, is endothermic. Step 2 involves an electron transfer from sodium to fluorine atoms to form Na^+ and F^- ions. Interestingly enough, this process is also endothermic. The energy which must be absorbed to ionize a mole of sodium atoms ($\Delta H = +118$ kcal) is somewhat greater than that evolved when the electrons released in the formation of Na^+ ions are acquired by fluorine atoms to form F^- ions ($\Delta H = -80$ kcal). On the other hand, step 3, in which gaseous Na^+ and F^- ions come together to form the ionic solid sodium fluoride, is strongly exothermic.

From this point of view, we can attribute the evolution of energy in the overall reaction to the exothermic nature of the final step, the formation of the ionic crystal lattice from the gaseous ions. This enthalpy change, often referred to as the **lattice energy**, reflects the strong electrostatic forces that hold oppositely charged ions to one another in the crystal. The fact that the lattice energy of a stable ionic compound is a large negative number explains why its formation from the elements represents an exothermic, spontaneous reaction.

How would the lattice energy of CaF_2 compare with that of NaF?

8.2 CHARGES OF IONS FOUND IN IONIC COMPOUNDS

Monatomic Ions with Noble Gas Electronic Structures

The formation of an ionic compound involves the transfer of electrons. When reaction occurs, there is a tendency for ions that have particularly stable electronic structures to be formed. One structure which is particularly

TABLE 8.1   IONS WITH NOBLE GAS CONFIGURATIONS

NOBLE GAS	−2 ION	−1 ION	+1 ION	+2 ION	+3 ION
He		H⁻	Li⁺	Be²⁺	
Ne	O²⁻	F⁻	Na⁺	Mg²⁺	Al³⁺
Ar	S²⁻	Cl⁻	K⁺	Ca²⁺	Sc³⁺
Kr	Se²⁻	Br⁻	Rb⁺	Sr²⁺	Y³⁺
Xe	Te²⁻	I⁻	Cs⁺	Ba²⁺	La³⁺
Rn	Po²⁻	At⁻	Fr⁺	Ra²⁺	Ac³⁺

stable is that of completely filled s and p orbitals, as exemplified by the noble gases.

In losing or gaining electrons to form ions, atoms of the A-group elements tend to fill or empty completely their s and p orbitals so as to acquire noble gas structures. Consider, for example, the monatomic ions formed by the reactions of the nonmetals oxygen and fluorine with the metals sodium, magnesium, and aluminum. Each of these ions has the electronic structure of the nearest noble gas, neon (2nd row, Table 8.1).

Boron, the first member of group 3A, does not form +3 ions.

In general, we find that when metals with 1, 2, or 3 more electrons than the preceding noble gas react with nonmetals, they tend to form <u>positive ions</u> (**cations**) with charges of +1, +2, and +3 respectively. The nonmetals that form simple ionic compounds are ordinarily those in groups 6A or 7A of the periodic table. Elements in these groups form <u>negative ions</u> (**anions**) with charges of −2 and −1 respectively, having electronic structures identical with that of the following noble gas.

Table 8.1 can be used to predict the simplest formula of a variety of ionic compounds (Example 8.1).

EXAMPLE 8.1.   Predict the simplest formulas of
    a. rubidium bromide b. potassium sulfide c. aluminum oxide

SOLUTION.   In each case we obtain the simplest formula by taking account of the fact that an ionic compound must be electrically neutral.
    a. Equal numbers of Rb^+ and Br^- ions are required. Simplest formula = RbBr.
    b. Two K^+ ions are required to balance one S^{2-} ion. Simplest formula = K_2S.
    c. Two Al^{3+} ions balance three O^{2-} ions. Simplest formula = Al_2O_3.

Monatomic Cations with Non-Noble Gas Structures

The metals located to the right of group 3B in the periodic table would have to lose four or more electrons to acquire a noble gas structure. The ionization energy required to form a cation with a charge greater than +3 is prohibitively large. One can calculate, for example, that the formation of a mole of Ti^{4+} ions from titanium metal would require the absorption of more than

2000 kcal of energy. Yet, in many of the compounds of these metals, there is evidence for the existence of monatomic positive ions. For example, silver fluoride, AgF, is a high-melting solid (mp = 435°C) with a high electrical conductivity in the liquid state. We interpret this evidence to mean that in this compound there are Ag^+ ions combined with an equal number of F^- ions. The same sort of evidence leads us to believe that the compounds CuCl and $MnCl_2$ are made up of ions (Cu^+ and Cl^-; Mn^{2+} and Cl^-).

The electronic structures of some of the more important ions of the transition and post-transition metals are listed in Table 8.2. Note that when the transition metals form cations, the outer s electrons are lost first. For example, the Ni^{2+} ion is formed from a nickel atom by the loss of two 4s electrons; the Ag^+ ion is derived from the silver atom by the removal of a 5s electron.

TABLE 8.2 MONATOMIC POSITIVE IONS WITH NON–NOBLE GAS STRUCTURES

POPULATIONS OF PRINCIPAL LEVELS

Ion	Total e⁻	1st	2nd	3rd	4th	5th	6th
Mn^{2+}	23	2	8	$3s^2\ 3p^6\ 3d^5$			
Fe^{3+}	23	2	8	$3s^2\ 3p^6\ 3d^5$			
Co^{2+}	25	2	8	$3s^2\ 3p^6\ 3d^7$			
Ni^{2+}	26	2	8	$3s^2\ 3p^6\ 3d^8$			
Cu^{2+}	27	2	8	$3s^2\ 3p^6\ 3d^9$			
Cu^+	28	2	8	$3s^2\ 3p^6\ 3d^{10}$			
Zn^{2+}	28	2	8	$3s^2\ 3p^6\ 3d^{10}$			
Ag^+	46	2	8	18	$4s^2\ 4p^6\ 4d^{10}$		
Cd^{2+}	46	2	8	18	$4s^2\ 4p^6\ 4d^{10}$		
Sn^{2+}	48	2	8	18	$4s^2\ 4p^6\ 4d^{10}$	$5s^2$	
Tl^+	80	2	8	18	32	$5s^2\ 5p^6\ 5d^{10}$	$6s^2$
Pb^{2+}	80	2	8	18	32	$5s^2\ 5p^6\ 5d^{10}$	$6s^2$

It may be noted from Table 8.2 that the electronic structures of many (e.g., Cu^+, Zn^{2+}, Ag^+, Cd^{2+}, Sn^{2+}, Tl^+, Pb^{2+}) but by no means all (e.g., Mn^{2+}, Fe^{2+}, Co^{2+}, Ni^{2+}, Cu^{2+}) of these ions are characterized by a completed sublevel. Just as such ions as Na^+, Mg^{2+}, and Al^{3+} have the completed p-level characteristic of the noble gas structure, so Cu^+, Zn^{2+}, Ag^+, and Cd^{2+} have a completed d-level; Sn^{2+}, Pb^{2+}, and Tl^+ have a completed s-level. Two of the ions listed in Table 8.2, Mn^{2+} and Fe^{3+}, have a half-filled sublevel ($3d^5$), a structure that is particularly stable, as pointed out in Chapter 7.

It should be emphasized that although many of the transition and post-transition metals are capable of forming positive ions such as those listed above, the bonding in many of their compounds is not primarily ionic. We find that many of the binary compounds of these metals have melting points and electrical conductivities in the liquid state which are inconsistent with a simple ionic structure. An example of such a compound is $SnCl_4$; though we might be tempted to postulate the existence of discrete Sn^{4+} ions in this compound, the experimental evidence clearly shows it to be molecular rather than ionic in nature.

Polyatomic Ions

Many of the ions most frequently encountered in general chemistry contain more than one atom. The structures of certain ions of this type will be discussed later in this chapter. The names and charges of a few of the more common polyatomic ions are given in Table 8.3.

TABLE 8.3 POLYATOMIC IONS

Ammonium, NH_4^+	Hydrogen sulfate, HSO_4^-	Dichromate, $Cr_2O_7^{2-}$	Phosphate, PO_4^{3-}
Mercurous, Hg_2^{2+}	Hydrogen carbonate, HCO_3^-	Chromate, CrO_4^{2-}	
	Hydroxide, OH^-	Sulfate, SO_4^{2-}	
	Nitrate, NO_3^-	Sulfite, SO_3^{2-}	
	Chlorate, ClO_3^-	Carbonate, CO_3^{2-}	
	Perchlorate, ClO_4^-	Peroxide, O_2^{2-}	
	Acetate, $C_2H_3O_2^-$		
	Cyanide, CN^-		
	Permanganate, MnO_4^-		

8.3 SIZES OF MONATOMIC IONS

The fact that it is impossible to obtain a sample of matter made up of only one kind of ion poses a fundamental difficulty in the experimental determination of ionic sizes. By x-ray diffraction (Chapter 10) one can determine the distance between the centers of touching ions in a crystal, but only that. In the case of an ionic compound such as RbI, where Rb^+ and I^- ions are in contact with each other (Figure 8.2), the measured distance is that between the centers of oppositely charged ions. In other words,

$$d = 3.64 \text{ Å} = r_{Rb^+} + r_{I^-}$$

Without further information, it is impossible to decide how the internuclear distance of 3.64 Å is to be divided between the two ions.

One way out of this dilemma is to carry out measurements on crystals in which the anion is much larger than the cation. In certain crystals of this type, of which lithium iodide is one example, there is anion-anion rather than anion-cation contact (Figure 8.2). In this case the internuclear distance is directly related to the radius of the anion:

$$d = 4.32 \text{ Å} = 2\, r_{I^-}$$

The radius of the iodide ion can be immediately calculated; once it is known, the radius of the rubidium ion can be deduced and the problem of establishing ionic radii begins to unravel.

As you might expect, the size of an ion depends to some extent upon its environment.

Several other methods have been suggested for the determination of the radii of individual ions. They lead to values which are internally self-consistent but differ slightly according to the approach followed in obtaining them. One such list of ionic radii (taken from L. Pauling, *The Nature of the Chemical Bond*, Cornell University Press, Ithaca, N.Y., 3rd edition, 1960) is given in Table 8.4. A more extensive table is given in Appendix 3.

TABLE 8.4 ATOMIC AND IONIC RADII

ATOMS	PROTONS	ELECTRONS	ATOMIC RADIUS (Å)	ION	PROTONS	ELECTRONS	IONIC RADIUS (Å)
H	1	1	0.37	H^-	1	2	2.08
He	2	2	0.93				
Li	3	3	1.52	Li^+	3	2	0.60
Be	4	4	1.11	Be^{2+}	4	2	0.31
O	8	8	0.66	O^{2-}	8	10	1.40
F	9	9	0.64	F^-	9	10	1.36
Ne	10	10	1.12				
Na	11	11	1.86	Na^+	11	10	0.95
Mg	12	12	1.60	Mg^{2+}	12	10	0.65
Al	13	13	1.43	Al^{3+}	13	10	0.50
S	16	16	1.04	S^{2-}	16	18	1.84
Cl	17	17	0.99	Cl^-	17	18	1.81
Ar	18	18	1.54				
K	19	19	2.31	K^+	19	18	1.33
Ca	20	20	1.97	Ca^{2+}	20	18	0.99
Sc	21	21	1.60	Sc^{3+}	21	18	0.81
Se	34	34	1.17	Se^{2-}	34	36	1.98
Br	35	35	1.14	Br^-	35	36	1.95
Kr	36	36	1.69				
Rb	37	37	2.44	Rb^+	37	36	1.48
Sr	38	38	2.15	Sr^{2+}	38	36	1.13
Y	39	39	1.80	Y^{3+}	39	36	0.93
Te	52	52	1.37	Te^{2-}	52	54	2.21
I	53	53	1.33	I^-	53	54	2.16
Xe	54	54	1.90				
Cs	55	55	2.62	Cs^+	55	54	1.69
Ba	56	56	2.17	Ba^{2+}	56	54	1.35
La	57	57	1.87	La^{3+}	57	54	1.15

Table 8.4 suggests several generalizations concerning the sizes of ions:

1. A series of particles such as O^{2-}, F^-, Ne, Na^+, Mg^{2+} and Al^{3+}, all of which have the same electronic structure, show a steady decrease in radius (1.40, 1.36, 1.12, 0.95, 0.65, 0.50 Å) with increasing nuclear charge (8, 9, 10, 11, 12, 13). The increased nuclear charge tends to pull the outer electrons closer to the nucleus.

2. The radii of negative ions are larger than those of the corresponding nonmetal atoms.

3. Positive ions are considerably smaller than the atoms from which they are derived. The conversion of a metal atom to a positive ion gives it a surplus of nuclear charge over electron cloud charge, and pulls in the remaining electrons.

4. Within a given vertical group of the periodic table, the ionic radius ordinarily increases with increasing atomic number. This same effect is, of course, observed with atomic radii.

Why should the Cl^- ion be larger than the Cl atom?

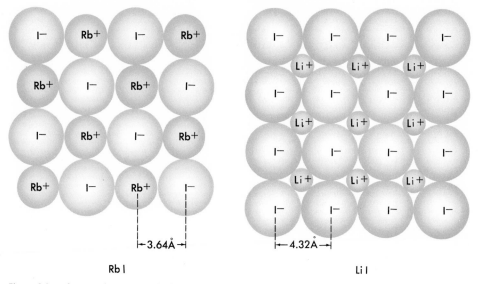

Figure 8.2 Planes of ions in rubidium iodide (cation-anion contact) and lithium iodide (anion-anion contact).

8.4 THE COVALENT BOND. FORMATION OF THE H_2 MOLECULE

Most chemical bonds cannot be adequately described in terms of a simple transfer of electrons from one atom to another. It is unreasonable to suppose that the bonds holding identical atoms to each other in the H_2 or the F_2 molecule could be formed in this way. An electron transfer seems somewhat more plausible when atoms of the elements hydrogen and fluorine come together to form the compound substance hydrogen fluoride, HF. Yet the physical properties of hydrogen fluoride indicate that it, like the elementary substances hydrogen and fluorine, has a molecular rather than an ionic structure. For example, hydrogen fluoride, like hydrogen and fluorine but unlike sodium fluoride, is a gas at room temperature and atmospheric pressure. Again, like hydrogen and fluorine but unlike sodium fluoride, liquid HF is a nonconductor of electricity. For these reasons we believe that hydrogen fluoride consists of HF molecules rather than H^+ and F^- ions.

The forces which hold the atoms together in H_2, F_2, and HF are referred to as **covalent** or **electron-pair** bonds. In the H_2 molecule this bond is often represented as

$$H : H \quad \text{or} \quad H—H$$

with the understanding that the two dots or the straight line drawn between the two hydrogen atoms represents the covalent bond that holds the molecule together. This simple way of representing the electronic structure of the H_2 molecule can be misleading if it is taken to imply that the two electrons are located at a fixed position between the two nuclei. Quantum mechanical calculations suggest that a more accurate map of the electron density in the H_2 molecule would resemble that shown in Figure 8.3. At any instant the elec-

176

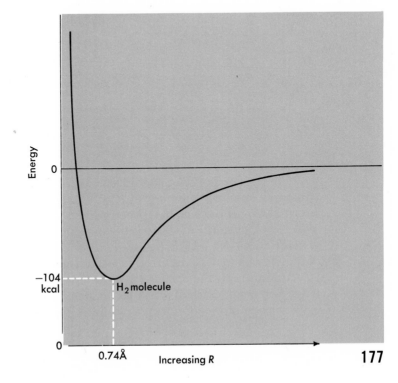

Figure 8.3 Electron density in an H_2 molecule.

trons may be located at any of various points between or around the two nuclei. On the average, however, there is a much greater probability of finding the electrons between the two nuclei than at the far ends of the molecule.

The question arises as to why the sharing of electrons between two nuclei should result in increased stability. Why, for example, should the H_2 molecule be more stable than two isolated hydrogen atoms? The first plausible explanation for the strength of the covalent bond was put forth by two physicists, W. A. Heitler and F. London, in 1927. By combining the wave functions of two hydrogen atoms, using the principles of quantum mechanics, they were able to calculate the interaction energy as a function of the distance of separation. A Heitler-London energy plot for the interaction between two hydrogen atoms is shown in Figure 8.4.

At large distances of separation (far right, Figure 8.4), the system consists of two isolated hydrogen atoms which do not interact with each other. As the

Figure 8.4 Energy of a system of two hydrogen atoms as a function of internuclear distance. The minimum of the curve represents the equilibrium distance of separation.

Energy

0

−104 kcal

H_2 molecule

0

0.74Å

Increasing R

atoms come closer together (moving to the left in Figure 8.4), they experience an attraction which leads gradually to an energy minimum. At the point corresponding to an internuclear distance of 0.74 Å and an attractive energy of about 104 kcal/mole H_2, the system is in its most stable state. If we attempt to bring the atoms closer together, forces of repulsion become increasingly important and the energy curve rises steeply.

The existence of the energy minimum shown in Figure 8.4 is directly responsible for the stability of the hydrogen molecule. According to the Heitler-London model, the attractive forces which bring about this minimum result from two factors:

1. At distances in the vicinity of 0.74 Å, the electrostatic attraction between the electron of one atom and the nucleus of the other (electron 1 → nucleus 2, electron 2 → nucleus 1) exceeds the electrostatic repulsion between particles of like charge (electron 1 + electron 2, nucleus 1 + nucleus 2). At distances appreciably less than 0.74 Å, the forces of repulsion between particles of like charge predominate.

Chemical bonds are believed to result from the overlap of orbitals.

2. When two hydrogen atoms come together to form a molecule, the electron density is spread over the entire volume of the molecule instead of being confined to a particular atom. Quantum mechanical calculations tell us that increasing the volume available to an electron decreases its kinetic energy and in this way imparts stability to the system. This principle is ordinarily described in terms of an **overlap** of the charge clouds of the two electrons. We say that in the H_2 molecule, the 1s orbitals of the individual atoms overlap in such a way that an electron originally confined to a single orbital can now spread itself over both orbitals. Calculations suggest that this factor is the principal source of stability in the H_2 molecule.

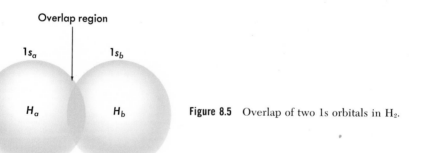

Figure 8.5   Overlap of two 1s orbitals in H_2.

The Heitler-London model has been extended to describe the electronic structures of a large number of both elementary and compound substances. Important contributions have been made by L. Pauling at California Institute of Technology and J. C. Slater at Massachusetts Institute of Technology. The general approach, which emphasizes the overlap between electron orbitals of individual atoms, is referred to as the **atomic orbital** or **valence bond** method.

According to the valence bond model, covalent bond formation occurs when two atoms, **each of which has an unpaired electron**, come together. Thus, covalent bond formation can occur between two hydrogen atoms:

$$\underset{_1H}{\overset{1s}{\boxed{\uparrow}}} + \underset{_1H}{\overset{1s}{\boxed{\uparrow}}} \longrightarrow H_2$$

two fluorine atoms:

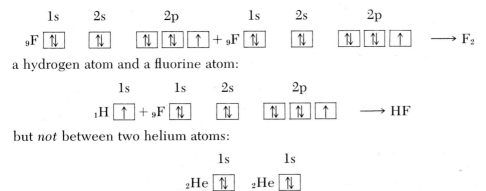

a hydrogen atom and a fluorine atom:

but *not* between two helium atoms:

neither of which has a half-filled orbital available for bond formation.

Throughout most of this chapter, we shall use the atomic orbital approach, in part because of its simplicity, but more important because it can account satisfactorily for the properties of most of the covalently bonded substances with which we deal in general chemistry. It does, however, suffer from certain rather serious deficiencies. In particular, it tends to underemphasize the extent to which the energy levels of electrons in atoms are modified by covalent bond formation.

Within the past decade an alternative approach to covalent bonding, developed originally by F. Hund and R. S. Mulliken around 1930, has become increasingly popular. This approach is known as the **molecular orbital** method. In the derivation of the electronic structure of a molecule by this method, all the electrons are considered to belong to the molecule as a whole. These electrons are then distributed among a set of energy levels, called molecular orbitals, which are analogous to the atomic orbitals used to describe the energies of electrons in isolated atoms. We shall present an introduction to molecular orbital theory in Section 8.10.

What would a Heitler-London plot (Fig. 8.4) look like for two He atoms?

8.5 POLAR AND NONPOLAR COVALENT BONDS. ELECTRONEGATIVITY VALUES

For the F_2 molecule, containing 18 electrons and an equal number of protons, it is impossible to obtain from quantum mechanics an exact picture of the electron distribution. Nevertheless, one can arrive at a reasonably accurate picture of the bonding electrons. A "density map" for these two 2p electrons, one contributed by each atom, closely resembles that shown in Figure 8.3 for the H_2 molecule. There is a high probability of finding the bonding electrons in the region between the two nuclei. Moreover, they are as likely to be found in the vicinity of one nucleus as the other. The remaining electrons in the molecule appear to be relatively unaffected by the formation of the covalent bond. Half of them still "belong" to one fluorine atom, and half to the other.

The electron distribution in the HF molecule differs in one important respect from that in molecules of elementary hydrogen and fluorine. Experimental evidence indicates that the density of the electron cloud formed by the overlap of the bonding electrons is concentrated in the vicinity of the fluorine nucleus. In other words, the two electrons making up the covalent bond are closer, on the average, to the fluorine than to the hydrogen nucleus. As a result, the fluorine atom carries a partial negative charge, while the hydrogen atom has a partial positive charge of equal magnitude. The molecule as a whole is electrically neutral.

Covalent bonds such as that in HF in which the shared electrons are unsymmetrically distributed are referred to as **polar**. Symmetrical bonds such as those in H_2 and F_2 are called **nonpolar**. The polarity of the covalent bond in HF is a consequence of the fact that fluorine has a greater attraction for electrons than does hydrogen. Since atoms of two different elements always differ in their affinity for electrons, covalent bonds between unlike atoms are always polar. Only when the atoms joined are identical is the bond nonpolar.

Figure 8.6 Electron density in a polar molecule (only valence electrons are shown).

The extent of polarity of a covalent bond can be interpreted in terms of the relative **electronegativity** (attraction for electrons) of the two atoms involved. Where the atoms are identical, as in the F_2 molecule, the difference in electronegativity is zero and the bond is nonpolar. In the ClF molecule, the bonding electrons are only slightly displaced toward the fluorine, which has a slightly greater attraction for electrons than does chlorine. When the two atoms differ widely in electronegativity, as in the HF molecule, the polarity of the bond is much more pronounced.

An ionic bond may be regarded as an extreme case of a polar covalent bond in which the bonding electrons are so much closer to the more electronegative atom that, for all practical purposes, an electron transfer has taken place. The term **partial ionic character** is often used to describe the extent of polarity of a covalent bond. Thus, we may say that the bond joining two different atoms is 50 per cent ionic to indicate that it is halfway between a pure ionic and a nonpolar covalent bond. In another case, in which the atoms involved differ more widely in electronegativity, the bond may be described as 90 per cent ionic, implying that the transfer of electrons to the more electronegative element is almost complete.

The difference in electronegativity between atoms of two different elements can be estimated in various ways. One method, based upon calculations of bond energy (Section 8.6), leads to the scale of electronegativity

TABLE 8.5 ELECTRONEGATIVITY VALUES

H						
2.1						
Li	Be	B	C	N	O	F
1.0	1.5	2.0	2.5	3.0	3.5	4.0
Na	Mg	Al	Si	P	S	Cl
0.9	1.2	1.5	1.8	2.1	2.5	3.0
K	Ca	Sc	Ge	As	Se	Br
0.8	1.0	1.3	1.8	2.0	2.4	2.8
Rb	Sr	Y	Sn	Sb	Te	I
0.8	1.0	1.2	1.8	1.9	2.1	2.5
Cs	Ba	La–Lu	Pb	Bi	Po	At
0.7	0.9	1.0–1.2	1.9	1.9	2.0	2.2

values listed in Table 8.5. On this scale, each element is assigned a number ranging from 4.0 for the most strongly electronegative element, fluorine, to 0.7 for the element having the least attraction for electrons, cesium. Among the A-group elements, electronegativity values increase from left to right in the periodic table and ordinarily decrease as one moves down in a given group.

The greater the difference in electronegativity between two elements, the more ionic will be the bond between them. The relationship between these two variables is shown graphically in Figure 8.7. Note that a difference of 1.7 electronegativity units corresponds to a bond with approximately 50 per cent ionic character. Such a bond might be described as being halfway between a pure covalent and a pure ionic bond.

It is important to point out that the electronegativity of an element may vary from one of its compounds to another, depending upon its valence state. For example, the electronegativity of

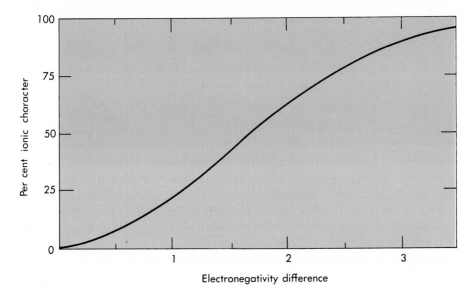

Figure 8.7 Relation between ionic character and electronegativity difference.

phosphorus in PCl_3 is quite different from that of the same element in PCl_5. The numbers given in Table 8.5 should not be regarded as an exact measure of an element's attraction for electrons, but rather as a rough guide that gives us an estimate of the degree of polarity of a covalent bond.

It is clearly an oversimplification to refer to a bond between two elements as being ionic or covalent. Consider, for example, the bonding in compounds formed by the reaction of a 1A or 2A metal with a nonmetal in group 6A or 7A. The difference in electronegativity values ranges from a minimum of 0.6 for the beryllium-tellurium pair to a maximum of 3.3 for cesium and fluorine. The percentage of ionic character in the bonds formed between these pairs of elements shows a corresponding variation, ranging from about 10 per cent for BeTe to 95 per cent for CsF.

The difference in electronegativity between oxygen or fluorine on the one hand and a 1A or 2A metal on the other in all cases exceeds 1.7 units, the value corresponding to 50 per cent ionic character. In this sense the bonding in the oxides and fluorides of these metals is predominantly ionic. The same statement applies to the oxide and fluoride of aluminum in group 3A, where electronegativity differences of 2.0 and 2.5 units correspond to about 65 per cent and 80 per cent ionic character in Al_2O_3 and AlF_3 respectively. The situation is quite different with the chloride, bromide, and iodide of aluminum; in each of these three compounds, electronegativity differences less than 1.7 units imply that the bonding is predominantly covalent.

8.6   COVALENT BOND DISTANCES AND BOND ENERGIES

Two of the most important characteristics of a covalent bond joining two atoms are the bond distance and the bond energy. The bond distance is defined as the **equilibrium distance between the nuclei of the two atoms joined by the bond**. Thus, when we find that the two fluorine nuclei in the F_2 molecule are 1.28 Å apart, we may say that the F—F bond distance is 1.28 Å. The bond energy is defined as **the amount of energy absorbed per mole** (more precisely, the change in enthalpy, **ΔH**, per mole) **when a particular kind of bond is broken in the gas state**. The fact that ΔH for the dissociation of F_2

$$F_2(g) \longrightarrow 2\ F(g)$$

is +37 kcal means that the F—F bond energy is 37 kcal. When we learn that the heat of dissociation of the Cl_2 molecule is 58 kcal,

$$Cl_2(g) \longrightarrow 2\ Cl(g), \Delta H = +58\ \text{kcal}$$

we deduce that the Cl—Cl bond is some 21 kcal/mole stronger than the F—F bond.

Interatomic distances in covalently bonded solids can be determined by x-ray diffraction measurements; electron diffraction and spectroscopy can be used to obtain bond distances in gaseous molecules. The application of these techniques allows us to calculate the **covalent radius** of the element, which is defined as one-half the bond distance between identical atoms. For example, having determined that the bond distance in the F_2 molecule is 1.28 Å, it follows that

$$\text{covalent radius F} = 1.28\ \text{Å}/2 = 0.64\ \text{Å}$$

The atomic radius (Chapter 7) of a nonmetal is taken to be its covalent radius. This should be kept in mind in interpreting trends in atomic radii, which reflect not only the "sizes" of atoms (a nebulous concept at best) but also the strength of the bond between them. The fact that the atomic radius of neon is 1.12 Å as compared to 0.64 Å for fluorine does not necessarily mean that the neon atom is "larger" than the fluorine atom, but only that the forces between neon atoms in the solid element are weaker than those between fluorine atoms in the covalently bonded F_2 molecule.

Bond energies, like bond distances, can be calculated from spectroscopic data. Alternatively, they can be obtained from thermochemical measurements. For example, to find the C—H bond energy, we might obtain data for the thermal dissociation of methane, CH_4, into gaseous atoms:

$$\begin{matrix} & H & \\ & | & \\ H- & C & -H \ (g) \longrightarrow C(g) + 4 \ H(g); \ \Delta H = +396 \ \text{kcal} \\ & | & \\ & H & \end{matrix}$$

C—H bond energy = 396 kcal/4 = 99 kcal

The C—C bond energy could then be calculated from similar data for ethane, C_2H_6:

$$\begin{matrix} & H & H & \\ & | & | & \\ H- & C & C & -H \ (g) \longrightarrow 2 \ C(g) + 6 \ H(g); \ \Delta H = +677 \ \text{kcal} \\ & | & | & \\ & H & H & \end{matrix}$$

677 kcal = C—C bond energy + 6(C—H bond energy)

C—C bond energy = 677 kcal − 6(99 kcal) = 83 kcal

How could the O—H bond energy be determined?

Ordinarily it is found that the length of a particular kind of bond and the energy required to break it are approximately the same in different molecules. Consider, for example, the first three compounds listed in Table 8.6 (ethane, propane, and butane). The measured C—C bond distance and the calculated C—C bond energies in these three compounds are virtually identical. Whenever we find a large change in bond energy or bond distance, we suspect a change in bond type. This is the case with the last two compounds listed in Table 8.6 (ethylene, C_2H_4, and acetylene, C_2H_2). The distance between the carbon atoms in ethylene (1.33 Å) is significantly shorter than the C—C bond distance (1.54 Å). At the same time, the bond holding the carbon atoms together in this compound is significantly stronger than the C—C bond. These effects become even more pronounced in acetylene.

TABLE 8.6 Distances and Bond Energies between Carbon Atoms

	Bond Distance (Å)	Bond Energy (kcal/mole)
C_2H_6	1.54	83
C_3H_8	1.54	83
C_4H_{10}	1.54	83
C_2H_4	1.33	143
C_2H_2	1.20	196

This evidence is interpreted to mean that there are **multiple** bonds between the carbon atoms in ethylene and acetylene. Specifically, we say that there is a **double bond** in ethylene, consisting of two pairs of electrons joining the two carbon atoms,

and a **triple bond** in acetylene (three pairs of electrons)

$$H—C\equiv C—H$$

Experimentally, it is always found that double or triple bonds are stronger than single bonds between the same two atoms. This leads to a larger bond energy and a shorter bond distance. Thus we find that the triple bond holding the nitrogen atoms together in the N_2 molecule:

$$: N\equiv N :$$

has a bond energy of 225 kcal/mole and a bond distance of 1.10 Å, as compared to values of 38 kcal/mole and 1.47 Å for the N—N single bond.

Just as one can detect a double bond by an increase in bond energy and a decrease in bond distance, it is possible to estimate the polarity of a bond by measuring these two quantities. Consider, for example, the H—F bond in the hydrogen fluoride molecule. If the bond were nonpolar, we might expect the bond distance and bond energy to be the average of those for the H—H bond (0.74 Å and 104 kcal/mole) and the F—F bond (1.28 Å and 37 kcal/mole).

$$\text{average bond distance} = (0.74 \text{ Å} + 1.28 \text{ Å})/2 = 1.01 \text{ Å}$$

$$\text{average bond energy} = (104 \text{ kcal/mole} + 37 \text{ kcal/mole})/2 = 71 \text{ kcal/mole}$$

It is found experimentally that the H—F bond is shorter (bond distance = 0.92 Å) and stronger (bond energy = 135 kcal/mole) than that calculated for a nonpolar bond. This implies that the partial ionic character brought about by the polarity of the HF molecule strengthens the H—F bond.

TABLE 8.7 CALCULATED AND OBSERVED BOND ENERGIES OF HX MOLECULES (KCAL/MOLE)

	X = F	X = Cl	X = Br	X = I
H—H	104	104	104	104
X—X	37	58	46	36
H—X (calc)	71	81	75	70
H—X (obs)	135	103	88	71
Δ	64	22	13	1

The bond energy of a polar covalent bond is particularly sensitive to the extent of polarity. Consider, for example, the bonds in the four hydrogen halides (Table 8.7).

The "extra bond energy" Δ associated with bond polarity is seen to decrease as the bond becomes less polar; it has its maximum value in the strongly polar HF molecule and virtually disappears in HI, where the two atoms have nearly equal electronegativities. Observations of this sort led

TABLE 8.8 BOND ENERGIES (KCAL/MOLE)

H—H	104	I—I	36	H—Si	70
C—C	83	H—F	135	Br—Cl	52
C=C	143	H—Cl	103	I—Cl	50
C≡C	196	H—Br	88	O—Cl	49
N—N	38	H—I	71	S—Cl	60
N≡N	225	H—O	111	N—Cl	48
O—O	33	H—Se	66	P—Cl	79
S—S	51	H—S	81	C—Cl	79
F—F	37	H—N	93	Si—Cl	86
Cl—Cl	58	H—P	76	C—O	84
Br—Br	46	H—C	99	C=O	173

Pauling to suggest a relationship between Δ and the difference in electro-negativity, ΔX, between atoms of two different elements.

$$\Delta = 23 \ (\Delta X)^2 \tag{8.2}$$

The electronegativity values listed in Table 8.5 were calculated with the aid of this equation. For example, substituting 22 for Δ in Equation 8.2, we obtain for the difference in electronegativity between H and Cl:

$$\Delta X = \left(\frac{22}{23}\right)^{\frac{1}{2}} = (0.96)^{\frac{1}{2}} = 0.98$$

8.7 LEWIS STRUCTURES. THE OCTET RULE

The concept of the covalent or electron-pair bond was first introduced by the American physical chemist G. N. Lewis in 1916. To rationalize the stability of this bond, he pointed out that atoms, by sharing electrons, can acquire a stable, noble-gas configuration. For example, when two hydrogen atoms, each with one electron, combine to form an H_2 molecule,

$$H \cdot + H \cdot \longrightarrow H : H$$

each hydrogen atom acquires a share in the two electrons and, in that sense, attains the electronic configuration of the noble-gas helium (atomic number = 2). Similarly, when two fluorine atoms, each with seven electrons in its outermost principal energy level (n = 2), combine to form the F_2 molecule,

$$: \ddot{F} \cdot + \cdot \ddot{F} : \longrightarrow : \ddot{F} : \ddot{F} :$$

they acquire the neon structure with eight electrons in the outermost level.

The structures written above are referred to as **Lewis structures** (or, in the vernacular, "flyspeck formulas"). In writing a Lewis structure for an atom or molecule, one includes only the electrons in the outermost principal energy level, the so-called valence electrons. For the A group elements, the number

185

of valence electrons is given by the group number. The Lewis structures of the atoms of the elements in the third period of the periodic table are:

1A	2A	3A	4A	5A	6A	7A	8A
Na·	·Mg·	·Àl·	·S̈i·	·P̈·	·S̈·	:C̈l·	:Àr:

In the Lewis structure for a molecule, a covalent bond may be represented as a pair of "electron dots" written between the bonded atoms or as a straight line connecting the two atoms. Lewis structures for the simple molecules formed by hydrogen with the nonmetals of the third period are:

SiH₄	PH₃	H₂S	HCl

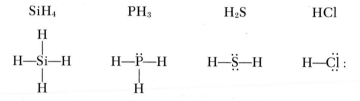

Many of the polyatomic ions listed in Table 8.3 can be assigned simple Lewis structures. For example, the OH⁻ and NH₄⁺ ions can be represented as

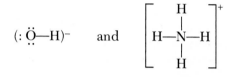

In both of these ions, hydrogen atoms are joined by covalent bonds to non-metal atoms (O, N). In both ions there are eight valence electrons. In the case of the hydroxide ion, this is one more than the number associated with the neutral atoms $(6 + 1 = 7)$, in agreement with the -1 charge of the ion. The $+1$ charge of the NH₄⁺ ion is accounted for when one realizes that four hydrogen atoms and one nitrogen atom would supply 9 valence electrons $(4 + 5 = 9)$, one more than the number present in the ion.

The NH₄⁺ ion is an interesting example of a species containing a **coordinate covalent** bond, in which one atom has donated both of the bonding electrons. The NH₄⁺ ion can be formed by the reaction of ammonia with a proton:

$$
\text{H—N̈:} + \text{H}^+ \longrightarrow \left[\text{H—N—H} \right]^+ \tag{8.3}
$$

From Equation 8.3 it is clear that the fourth N—H bond is formed by the nitrogen atom donating its unshared pair of electrons to the H⁺ ion. Experimentally, the four bonds in the NH₄⁺ ion are identical. We shall have more to say about coordinate covalent bonding in Chapter 19 in connection with a discussion of complex ions.

The rule that molecules or ions in which each atom has a noble-gas configuration are particularly stable is often referred to as the **octet rule**. Non-metals, with the exception of hydrogen, achieve a noble-gas structure by acquiring an "octet" of electrons. As we shall see later, many stable, covalently bonded species "violate" the octet rule. Nevertheless, we shall find it quite useful in suggesting plausible electronic structures for molecules and polyatomic ions.

It must be emphasized that Lewis structures are deficient in many respects. In particular, they tell us nothing about the geometry of a molecule or ion; neither do they tell us what orbitals are occupied by the shared and unshared electrons. However, we shall find that the ability to write Lewis structures is a necessary prerequisite for understanding the principles of molecular geometry and orbital electronic configurations, to be presented in Sections 8.8 through 8.10.

Rules for Writing Lewis Structures

Lewis structures for many molecules and ions are readily written by inspection. However, for more complex species, it is helpful to follow the general procedure outlined as follows:

1. **Draw a skeleton structure for the molecule or ion, joining atoms by single bonds.** In some cases, only one arrangement of atoms is possible; in others, chemical or physical evidence must be used to decide between two or more alternative structures.

2. **Count the number of valence electrons.** For a molecule, this can be done by simply adding the valence electrons contributed by the various atoms making up the molecule. For a polyatomic ion, this total must be adjusted to take into account the charge of the ion. A -1 ion, for example, would have one valence electron in addition to those of the neutral atoms.

3. **Deduct two valence electrons for each single bond written in step 1. Distribute the remaining valence electrons so as to give each atom a noble-gas structure, if possible.** If you find that you have "too few electrons to go around," form multiple bonds so as to make more effective use of the ones you have. The formation of a double bond compensates for a deficiency of two electrons; a triple bond, a deficiency of four electrons.

The application of these rules is illustrated in Example 8.2.

EXAMPLE 8.2. Draw Lewis structures for the following species:
     a. ClO^- b. $SO_4{}^{2-}$ c. N_2F_2

SOLUTION.

    a. There is only one skeleton structure possible for the ClO^- ion:

$$(Cl\text{—}O)^-$$

The total number of valence electrons is obtained by adding one (the charge of the ion) to the number contributed by the chlorine and oxygen atoms.

$$\text{no. of valence electrons} = 7 + 6 + 1 = 14$$

Deducting two electrons for the covalent bond written in the skeleton structure leaves us 12 to work with. Putting 6 electrons around each atom, we satisfy the octet rule and arrive at a reasonable electronic structure for the ClO^- ion:

$$(:\overset{..}{\underset{..}{Cl}}\text{—}\overset{..}{\underset{..}{O}}:)^-$$

    b. Various skeleton structures could be written for the $SO_4{}^{2-}$ ion. However, in oxyanions such as $SO_4{}^{2-}$, $NO_3{}^-$, and $CO_3{}^{2-}$, we ordinarily find that the oxygen atoms are symmetrically arranged around the central, nonmetal atom. Following this

Another way to derive Lewis structures is by analogy; the ClO^- ion has the same number of valence e^- (14) and the same structure as Cl_2.

187

general rule, we write the skeleton structure:

The number of valence electrons is found by adding the charge of the ion to the total contributed by the sulfur and oxygen atoms:

$$\text{no. of valence electrons} = 6 + 4(6) + 2 = 32$$

Deducting 8 electrons for the four covalent bonds in the skeleton structure leaves 24. Putting 6 electrons around each oxygen atom gives us a plausible Lewis structure for the sulfate ion.

c. One could imagine various structures for the N_2F_2 molecule. Chemical evidence indicates that there is a bond between the two nitrogen atoms. Moreover, it is generally observed that the fluorine atom never forms more than one covalent bond. This reduces our choice of skeleton structures to two:

This rules out skeletons such as F—N—F—N.

F—N—N—F   or   N—N—F
                                        |
                                        F

structure 1   structure 2

The total number of valence electrons is found by adding those contributed by the two nitrogen atoms to those of the two fluorines:

$$\text{no. of valence electrons} = 2(5) + 2(7) = 24$$

Deducting 6 valence electrons for the three bonds written in the skeleton structures leaves a total of 18. Distributing these in a symmetrical manner, we obtain:

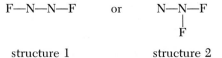

structure 1   structure 2

Other distributions could be written, but they do not affect our conclusion that there are two fewer electrons than the number required to give each atom a noble-gas structure. To remedy this deficiency, we form a double bond between the two nitrogen atoms.

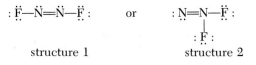

structure 1   structure 2

Both of these structures appear plausible in terms of the octet rule. Physical evidence indicates that the more symmetrical structure, the first, is the correct one.

It should be clear from these examples that writing Lewis structures for ions or molecules is not as simple as students sometimes suppose. It requires a certain amount of skill in manipulating electrons; it may also require a considerable amount of knowledge of the physical and chemical properties of the species. Nevertheless, a beginning student, with a little practice, can come up with plausible Lewis structures for a great many molecules and ions. An extensive research program may be required to decide whether the structure he has deduced is indeed correct.

Resonance

It was discovered many years ago that the Lewis structures of certain molecules do not adequately explain certain of their properties. For example, consider the sulfur dioxide molecule. Following the rules outlined above, we can readily arrive at the following Lewis structure for SO_2:

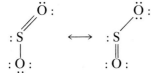

This structure implies that there are two different kinds of bonds in the SO_2 molecule. We would expect that the oxygen attached by a double bond would be closer to the sulfur atom than the single-bonded oxygen. Experimentally, it is found that the two bond distances are exactly the same.

The sulfur to oxygen distance in SO_2, 1.43 Å, is intermediate between that calculated for a single and a double bond. This suggests that each of the bonds in the SO_2 molecule is a "hybrid," intermediate between a single and a double bond. There is no simple way to indicate this in writing the electronic structure of the molecule. What is commonly done is to write two structures:

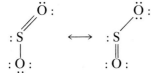

with the understanding that the actual electronic structure is intermediate between these. Structures such as these are referred to in valence bond terminology as **resonance** forms. The word resonance is used to describe the phenomenon in which a single electronic structure does not suffice to describe the properties of a substance.

A similar situation arises with SO_3; experimental evidence indicates that all three bonds in the molecule are equivalent. The sulfur to oxygen distance is the same in all three cases. This evidence is interpreted to mean that the three resonance forms

make equal contributions to the actual structure of the molecule. We may say that each bond in SO_3 has "$\frac{1}{3}$ double bond" and "$\frac{2}{3}$ single bond" character.

In a slightly different sense, we may call HF a "resonance hybrid" of the two structures H—F and H+F−.

189

Another species which can be described as a "resonance hybrid" is the nitrate ion, for which three equivalent structures, analogous to those for SO_3, can be written.

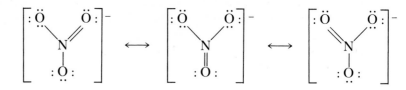

How many resonance forms would you predict for NO_2^-?

We shall encounter other examples of molecules and ions which can be represented in terms of the valence-bond model as hybrids of two or more resonance forms. It is important to point out that:

1. Resonance forms do not imply different kinds of molecules. Sulfur dioxide contains only one type of molecule with a structure intermediate between those written here.

2. Resonance can be anticipated when it is possible to write two or more Lewis structures which appear to be about equally plausible. In the case of the nitrate ion, the three structures we have written are equivalent. One could, in principle, write many other structures, including:

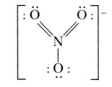

However, since this structure involves putting 10 electrons around the nitrogen atom, it seems unlikely that it would make a major contribution to the true structure of the nitrate ion.

3. In writing resonance forms, it is not permissible to change the relative positions of atoms. For example, a structure such as

$$(: \ddot{\text{O}}\!-\!\ddot{\text{O}}\!-\!\ddot{\text{N}}\!=\!\ddot{\text{O}} :)^-$$

could not be a resonance form of the nitrate ion. We know from experiment that in this ion the nitrogen atom is bonded to three oxygen atoms. While the structure just written is entirely plausible, it implies the existence of a distinct species with properties entirely different from those of the nitrate ion.

4. Molecules or ions which can be represented as resonance hybrids are found to be more stable than one would otherwise predict. For example, the SO_2 molecule is a more stable species than one would predict for a molecule having one double and one single bond. Putting it another way, the observed sulfur-oxygen bond energy is greater than one would predict by taking the average for a $S\!=\!O$ and a $S\!-\!O$ bond. This extra energy is sometimes referred to as the "resonance energy."

Failure of the Octet Rule

For one class of molecular substances, it is impossible to write Lewis structures in which each atom obeys the octet rule. These are substances whose molecules contain an odd number of valence electrons. Perhaps the simplest example is nitric oxide, molecular formula NO, in which there are a total of 11 valence electrons. In valence bond terminology, the NO molecule is regarded as a resonance hybrid with two contributing structures:

$$\cdot \ddot{N} = \ddot{O} : \quad \longleftrightarrow \quad : \ddot{N} = \ddot{O} \cdot$$

One would expect the structure at the left to be the more important since it assigns the electron deficiency (7 valence electrons rather than 8) to the less electronegative atom, nitrogen. Another **odd-electron molecule** is nitrogen dioxide, NO₂. Of the several resonance structures which can be written for NO₂, those of the type:

Very few odd-electron species are stable.

are probably most important. Such a structure is suggested by the experimental observation that nitrogen dioxide can be formed by heating dinitrogen tetroxide, N₂O₄, which has the structure:

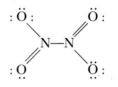

Fission of the N—N bond in the N₂O₄ molecule would give the structure written for NO₂.

Species such as NO and NO₂, in which there is an unpaired electron, are **paramagnetic**; that is, they show a weak attraction toward a magnetic field. Elementary oxygen is also paramagnetic, which suggests that the conventional Lewis structure

$$: \ddot{O} = \ddot{O} :$$

is incorrect, since it requires that all the electrons be paired. The paramagnetism of oxygen could be explained by the structure

$$: \ddot{O} - \ddot{O} :$$

in which there are two unpaired electrons. However, this structure, like the one written previously, is unsatisfactory. In the first place, it does not conform to the octet rule; much more important, it does not agree with experimental evidence. The distance between the two oxygen atoms in O₂ (1.21 Å) is considerably smaller than that ordinarily observed with an O—O single bond (1.48 Å). These properties of oxygen are difficult to explain in terms of valence-bond theory. As we shall see in Section 8.10, the molecular orbital approach leads to a more satisfactory picture of the electron distribution in the O₂ molecule.

There are many other species for which Lewis structures written to conform to the octet rule are unsatisfactory. Examples include the fluorides of beryllium and boron. Although one could write multiple bonded structures

191

Can you write
such a structure
for BeF₂?

for these molecules in accordance with the octet rule, experimental evidence strongly supports the structures

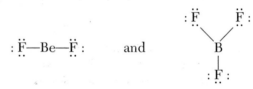

in which the central atom is surrounded by 4 and 6 valence electrons respectively, rather than 8.

At the opposite extreme, certain of the halides of the 5A and 6A elements have structures in which the central atom is surrounded by more than 8 valence electrons. In PF_5 and PCl_5, the phosphorus atom is joined by single bonds to each of five halogen atoms and consequently must be surrounded by 10 bonding electrons. An analogous structure for SF_6 requires that a sulfur atom have 12 valence electrons around it. We shall have more to say in Section 8.9 about the electronic structures of molecules such as BeF_2, BF_3, and SF_6, which do not "conform" to the octet rule.

8.8   MOLECULAR GEOMETRY

In Section 8.6 we considered two important characteristics of the covalent bond — bond distance and bond energy. When we are dealing with molecules in which there are two or more covalent bonds, we must be concerned with another factor, the angle between the bonds. In the case of the H_2O molecule, for example, it is important to know whether the three atoms are in a straight line, as would be the case if the angle between the bonds were 180°, or arranged in a triangular pattern, corresponding to a bond angle of less than 180°.

The ability to predict molecular geometries has long been regarded as an essential criterion of any successful theory of covalent bonding. We shall see in Section 8.9 how the valence bond model, by allowing for the hybridization of atomic orbitals, is able to account for the geometries of many important molecules and polyatomic ions. We shall present here an alternative approach, which requires only a knowledge of the Lewis structure of the species. It was first suggested by Sidgwick and Powell in 1940 and has since been modified by R. J. Gillespie of McMaster University.

Electron Pair Repulsion

The major features of the geometry of molecules and polyatomic ions can be predicted by using a very simple principle.

For a more detailed treatment,
read the Journal of
Chemical Education,
Vol. 40, p.
295, 1963.

The electron pairs surrounding an atom, because of electrostatic repulsion, are oriented so as to be as far from each other as possible.

To illustrate the application of this principle, let us consider cases in which a central atom is surrounded by 2, 3, 4 or 6 electron pairs, which are the most common situations. (For simplicity, we shall show only the electrons surrounding the central atom.)

Two Pairs of Electrons (Example = BeF_2).   Obviously, two pairs of electrons are as far apart as possible when they are joined to the central atom in a straight line. Consequently, we would predict that the beryllium fluoride molecule would be **linear**, as indeed it is:

$$F—Be—F \text{ (bond angle} = 180°)$$

Three Pairs of Electrons (Example = BF_3). Three electron pairs surrounding an atom will be as far apart as possible when they are directed toward the corners of an **equilateral triangle**. This results in a planar structure for molecules such as BF_3, with bond angles of 120°,

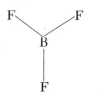

Four Pairs of Electrons (Examples = CH_4, NH_3, H_2O). The geometrical structure which puts four electron pairs as far apart as possible is a regular **tetrahedron** (bond angle = 109°). In the methane molecule, the four hydrogen atoms will be located at the corners of a tetrahedron, equidistant from the central carbon atom (Figure 8.8).

In ammonia, as in methane, the four electron pairs will be directed toward the corners of a regular tetrahedron. Since there are only three hydrogen atoms, one of the corners will appear to be unoccupied (actually it is occupied by a nonbonding electron pair). The three hydrogen atoms are located at the vertices of an equilateral triangle, with the nitrogen atom directly above the center of the triangle. (The geometrical figure formed by the nitrogen atom and the three hydrogen atoms is referred to as a triangular pyramid.)

The geometry of the water molecule can be derived from that of ammonia by removing another hydrogen atom. We deduce that the H_2O molecule is bent, with a bond angle of 109°.

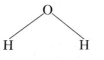

Geometries of more complicated molecules, such as ethane (C_2H_6), are readily derived. The ethane molecule (Figure 8.8) can be regarded as two tetrahedra joined at a corner.

Six Pairs of Electrons (Example = SF_6). It can be shown that 6 electron pairs will be a maximum distance apart when they are directed toward the corners of a regular octahedron (Figure 8.8). We shall find in Chapter 19 that octahedral geometry is extremely common in species referred to as complex ions.

It is found experimentally that the bond angles in molecules in which the central atom has an octet structure and in which there is at least one unshared pair tend to be somewhat less than the predicted value of 109°. For example, the measured bond angle in NH_3 is 107°; in H_2O, it is 105°. This effect has been attributed to the influence of the unshared pair of electrons in these molecules. The electron cloud formed by the unshared pair in the NH_3 molecule may be expected to spread out over a greater volume than that of the three pairs involved in covalent bond formation. Consequently, it should tend to repel the clouds formed by the bonding electrons more strongly than they repel each other, thereby drawing the three shared pairs slightly closer together and reducing the angles between them. This effect should be more pronounced in the H_2O molecule, in which there are two unshared pairs of electrons around the oxygen atom as compared to the lone unshared pair surrounding the nitrogen atom in NH_3.

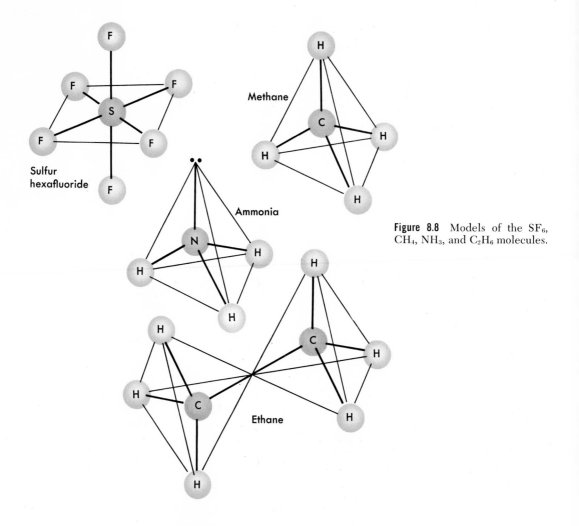

Figure 8.8 Models of the SF_6, CH_4, NH_3, and C_2H_6 molecules.

One might expect the distortion of the tetrahedral angle by unshared electron pairs to be particularly important when bonding electrons are relatively far removed from each other. Under these conditions, repulsive forces between the bonding electrons are relatively weak, and it should be somewhat easier to reduce the bond angle. Experimentally, this prediction is confirmed. Compare, for example, the bond angle of 107° in NH_3 to that of 103° in NF_3. In the latter molecule, the bonding electrons, concentrated around the highly electronegative fluorine atoms located at the far corners of the molecule, are much farther apart than in NH_3. Again, consider the PH_3 molecule, in which both the increased size and reduced electronegativity of the central atom (P vs. N) tends to drive the three pairs of bonding electrons apart and thereby weaken the repulsive forces between them. The H—P—H bond angle in PH_3 is only 93°, some 14° less than the H—N—H angle in NH_3. A similar effect may explain why the bond angles in H_2S, H_2Se, and H_2Te (92°, 91°, 89°) are so much smaller than that in H_2O (105°).

Molecules Containing Multiple Bonds

The rules discussed in the previous section can be applied to molecules in which double or triple bonds are present if we assume that **so far as molecular geometry is concerned, a multiple bond behaves as if it were a single electron pair**. Thus, we find that the CO_2 molecule, like BeF_2, is linear,

$$O=C=O$$

while the SO_3 molecule and the NO_3^- ion, like the BF_3 molecule, are triangular:

This principle can be applied to deduce the geometries of the acetylene and ethylene molecules. In acetylene, where there is a triple bond between the carbon atoms, each of the bond angles must be 180°, leading to a linear structure:

$$H—C≡C—H$$

The ethylene molecule, with a double bond between the two carbon atoms, has the geometry to be expected if each carbon atom had only three pairs of electrons around it.

The six atoms are located in a plane with bond angles of 120°.

Polarity of Molecules

We have previously considered the polarity of covalent bonds; now that we have some idea of the spatial distribution of atoms in molecules, we are in a position to consider the polarity of molecules themselves. A polar molecule is one in which there is a separation of positive and negative charge, that is, + and − poles. A simple example is the HF molecule: the fact that the bonding electrons are somewhat closer to the fluorine atom gives it a partial negative charge, while the hydrogen atom acts as a positive pole. In general, any diatomic molecule in which the two atoms differ from each other will be polar. The more electronegative atom will act as a negative pole, while the less electronegative atom serves as the positive pole. Examples include HCl and HF; the convention of writing the symbol of the more electronegative element last tells us that the atom written last in the formula will act as the negative pole.

When a molecule contains more than two atoms, we must know the bond angles before we can decide whether the molecule is polar or nonpolar. Consider, for example, the two triatomic molecules BeF_2 and H_2O. The linear BeF_2 molecule is nonpolar; the two polar Be—F bonds cancel each other.

$$(−) \ (+) \ (−)$$

$$F—Be—F$$

From a slightly different viewpoint, we can deduce that this molecule is nonpolar because the centers of positive and negative charge coincide. In contrast, the nonlinear water molecule is polar, since the two polar covalent bonds do not cancel each other.

195

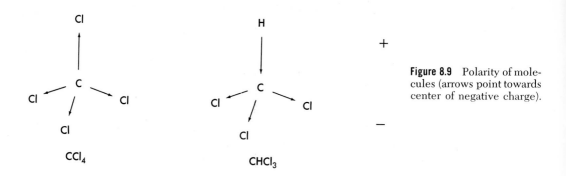

Figure 8.9 Polarity of molecules (arrows point towards center of negative charge).

The negative pole is located at the oxygen atom. The positive pole is midway between the two hydrogen atoms.

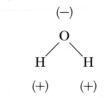

Another example of a molecule which is nonpolar despite the presence of polar bonds is CCl_4. The four C—Cl bonds are polar, with the bonding electrons displaced slightly toward the chlorine atoms. However, since the four chlorines are arranged about the central carbon atom in a symmetrical, tetrahedral pattern, the polar bonds cancel each other. If one of the chlorine atoms in CCl_4 is replaced by hydrogen, the symmetry of the molecule is destroyed. The chloroform molecule, $CHCl_3$, is polar.

One can distinguish polar from nonpolar molecules by studying their behavior in an electrical field. When a substance made up of polar molecules is placed in a field such as that between the plates of a condenser (Figure 8.10), the molecules act as **dipoles**, tending to line up with their positive poles oriented toward the negative plate and their negative poles toward the posi-

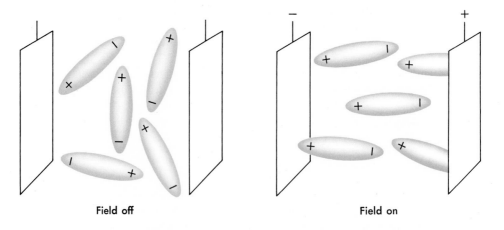

Field off   Field on

Figure 8.10   Orientation of polar molecules in an electrical field.

TABLE 8.9 Dipole Moments (Debye Units)°

H_2	0	HF	2.00	H_2O	1.84	CH_3Cl	1.86
Cl_2	0	HCl	1.03	H_2S	0.92	CH_2Cl_2	1.59
CO_2	0	HBr	0.79	NH_3	1.46	$CHCl_3$	1.15
CH_4	0	HI	0.38	PH_3	0.55	CCl_4	0

° A molecule in which unit + and − charges are separated by 0.21 Å has a dipole moment of one Debye unit.

tive plate. This behavior may be described by saying that the molecule has a **dipole moment** which causes it to turn into a preferred position in an electrical field. Nonpolar molecules, having no permanent centers of positive and negative charge, show little tendency to orient themselves in an electrical field and are said to have a dipole moment of zero.

8.9 HYBRID ATOMIC ORBITALS

Simple valence-bond theory requires the presence of unpaired electrons in an atom if it is to participate in covalent bond formation. It would seem reasonable to suppose that the number of bonds that a given atom can form would be governed by the number of unpaired electrons present in its valence shell. Let us consider the elements Be, B, and C. The orbital diagrams of these elements, following Hund's rule (Chapter 7), must be:

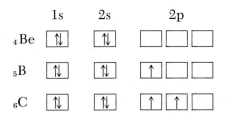

Clearly, these three atoms have 0, 1, and 2 unpaired electrons respectively. Yet we have seen that atoms of Be, B, and C form 2, 3, and 4 covalent bonds respectively (BeF_2, BF_3, CH_4). How do we explain this discrepancy between the number of unpaired electrons and the number of bonds? In order to answer questions such as this within the framework of the valence bond model, it is necessary to invoke a new kind of atomic orbital, the so-called **hybrid bond orbital**.

sp Hybrid Bonds

The fact that a beryllium atom forms two covalent bonds could be explained by assuming that prior to reaction, one of the 2s electrons is promoted to the 2p level, giving the structure:

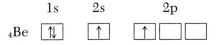

Now that the beryllium atom has two unpaired electrons, it can form two covalent bonds by sharing two electrons contributed by two fluorine atoms.

197

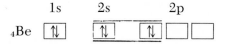

(Lines drawn above and below indicate the orbitals involved in bond formation.)

One might object to the promotion of a 2s electron to a 2p orbital as being improbable from an energy standpoint. However, the energy evolved when the "excited" beryllium atom forms two stable bonds with an element such as fluorine should greatly exceed the relatively small amount of energy absorbed in promoting the electron.

There is, however, a more fundamental objection to the model we have just described. It implies that two different bonds are formed; one of these might be called an "s bond" and the other a "p bond." Experimentally, it is found that the two bonds are identical; the Be-F distance, for example, is the same for both bonds. This leads us to believe that the two orbitals used for bond formation by beryllium must be equivalent. In valence-bond terminology, we describe this situation by saying that an s and a p orbital are "hybridized" to give two new bonding orbitals, described as **sp hybrids**.

It should be clearly understood that the hybrid sp orbitals have their own unique properties, quite different from those of the individual s and p orbitals from which they are formed. It can be shown, by arguments based on wave mechanics, that even though an s orbital is spherical in nature and p orbitals are located at right angles to each other, two sp orbitals are directed at an angle of 180° to each other.

sp² Hybrid Bonds

The formation of three covalent bonds by boron can be explained in a manner analogous to that described for beryllium. We can imagine the formation of the three bonds as occurring in three steps.

1. Promotion of an electron from the 2s to the 2p level.

2. Formation of three **sp² hybrid orbitals** from the three individual atomic orbitals.

3. The filling of these three orbitals with three electrons contributed by an element such as fluorine to give

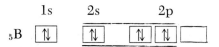

The sp² hybrid orbitals have their own characteristic geometry. Quantum mechanical calculations suggest that they should be oriented in a plane, at angles of 120° to each other. In this manner, atomic orbital theory explains the experimental observation that the BF_3 molecule is planar and contains three equivalent covalent bonds directed toward the corners of an equilateral triangle.

sp³ Hybrid Bonds

Extending the ideas just presented to carbon, we are able to explain the formation of four equivalent covalent bonds by carbon.

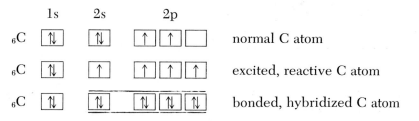

The four sp³ orbitals differ only in direction. Each orbital contains contributions from an s and three p orbitals.

The four equivalent orbitals used by carbon in bond formation are described as **sp³ hybrids**. One can show by calculations that these orbitals should be directed toward the corners of a tetrahedron. This agrees with the observed fact that carbon tends to form four equivalent bonds oriented at tetrahedral angles to each other.

There is a good deal of evidence to suggest that sp³ hybridization occurs in many atoms which have more than four valence electrons. The fact that the bond angles in water and ammonia are very nearly tetrahedral suggests that whenever an atom is surrounded by an octet of electrons, the four electron pairs occupy sp³ hybrid orbitals. If the bonds formed by oxygen in water or nitrogen in ammonia were pure "p bonds," one would expect them to be oriented at right angles to each other.

sp³d² Hybrid Bonds

We have seen that certain nonmetals, notably phosphorus and sulfur, are capable of forming molecules in which the central atom is bonded to more than four atoms. Specifically, sulfur forms the compound SF_6; in this molecule, the sulfur atom is bonded to six fluorines. This requires that six orbitals of the sulfur be occupied by bonding electrons. The evidence indicates that the orbitals involved are **sp³d² hybrids**, formed by the combination of one 3s orbital, three 3p orbitals, and two 3d orbitals.

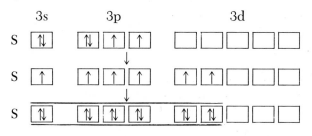

The six sp³d² orbitals are directed toward the corners of a regular octahedron. We shall find in Chapter 19 that this structure is an extremely common one. Indeed, if we were to tabulate the structures of all inorganic ions and molecules, octahedral and tetrahedral configurations taken together would far outnumber all others.

In Table 8.10 are listed the formulas of various nonmetal fluorides in which the central atom is surrounded by more than eight valence electrons.

199

TABLE 8.10 Binary Fluorides Involving d Orbitals

One d Orbital Used			Two d Orbitals Used		
5A	6A	7A	5A	6A	7A
PF_5	SF_4	ClF_3	—	SF_6	—
AsF_5	SeF_4	BrF_3	—	SeF_6	BrF_5
SbF_5	TeF_4	—	—	TeF_6	IF_5

Examining Table 8.10, one might speculate whether it could be extended to include the noble gases, which lie directly to the right of the halogens in the periodic table. As heretical as this idea might have seemed to chemists a generation ago, it has recently been found that fluorides of certain of the noble gases can indeed be prepared and have the formulas that one would predict by extrapolation from Table 8.10. Xenon difluoride, XeF_2, analogous in electronic structure to SbF_5, TeF_4, and the yet unknown compound IF_3, can be prepared from the elements by a photochemical reaction. It is a stable, molecular solid which sublimes without decomposition at 25°C. The compounds KrF_4 and XeF_4, which fall in the series $SeF_6 \rightarrow BrF_5 \rightarrow KrF_4$ and $TeF_6 \rightarrow IF_5 \rightarrow XeF_4$, have also been prepared.

Bonding Orbitals in Molecules Containing Multiple Bonds

We have seen in Section 8.8 that the geometry of the C_2H_4 molecule is similar in many respects to that of the BF_3 molecule. Both molecules are planar; in both cases, the bonds around the carbon or boron atoms are oriented at angles of 120° to each other. This evidence suggests that the bonding orbitals utilized in the two molecules must be similar. If we recall that in the BF_3 molecule the central boron atom is joined to the three fluorine atoms by sp^2 hybrid bonds, it seems reasonable to suppose that a carbon atom in C_2H_4 also forms three such bonds (two to hydrogen, one to the other carbon atom). In other words, it appears that the geometry of the C_2H_4 molecule is fixed by a framework of sp^2 bonds, three of which extend from each carbon atom.

The question remains as to the spatial distribution of the two remaining electrons that the Lewis structure of C_2H_4 requires to be located between the two carbon atoms. The two electrons, one from each carbon atom, must occupy the unhybridized p orbitals remaining after the formation of sp^2 hybrid bonds.

The valence-bond model suggests that the overlap of these two p orbitals results in a fourth bond with an electron density concentrated in two sausage-shaped regions, one above the plane of the molecule and the other below. This concept is shown in Figure 8.11. The shaded areas indicate the region where the electron density is high.

According to this picture, the "double bond" joining the carbon atoms in ethylene is composed of two distinct parts having quite different electron distributions. In the sp^2 hybrid bond, the electron cloud is completely symmetrical about the axis joining the two atoms. Such bonds are often described as **sigma (σ) bonds**. All the hydrid bonds we have discussed (sp, sp^2, sp, sp^3d^2) are of this type. In the other bond joining the two carbon atoms, the electron density is not symmetrical about the C—C axis. Instead, it is concentrated in two particular regions above and below this axis. Bonds in which the electron

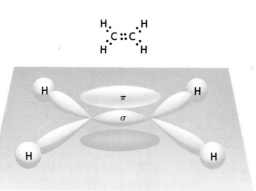

Figure 8.11 Valence bond model
of the ethylene molecule.

distribution is not symmetrical along a line joining the bonded atoms are
referred to as **pi (π) bonds**.

Recalling that the acetylene molecule, C_2H_2, like the BeF_2 molecule, is
linear, we conclude that the framework in C_2H_2 consists of sp hybrid bonds,
two per carbon atom. Each carbon atom is joined to the other carbon and to a
hydrogen atom by an sp sigma bond. The two remaining bonds joining the
carbon atoms are a pair of pi bonds, with the electron density "sausages"
oriented at 90° angles to each other.

Figure 8.12 Valence bond model
of the acetylene molecule.

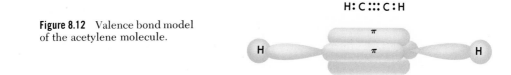

This picture of multiple bonding explains the empirical rule discussed
in Section 8.8 concerning the geometry of species containing double or triple
bonds. The bond angles in such species are determined by the framework of
sigma bonds. The pi bonds have relatively little influence on bond angles and
hence on molecular geometries.

8.10 MOLECULAR ORBITALS

The valence-bond model of the covalent bond presented in this chapter
explains a great many of the structural features of molecular substances. Most
important, it accounts at least qualitatively for the stability of the covalent
bond in terms of an overlap of atomic orbitals. By introducing the concept of
hybridization, valence-bond theory can explain the number of bonds formed
by nonmetal atoms and the angles between these bonds. If we modify
valence-bond theory to include the concept of resonance, it is even possible
to explain the properties of molecules such as SO_2 for which a conventional
Lewis structure is inadequate.

A major weakness of the valence-bond approach has been its inability to
explain the magnetic properties of many simple molecules. One example
cited previously is molecular oxygen, which simple valence-bond theory

predicts to have the Lewis structure

$$: \ddot{O}{=}\ddot{O} :$$

Experimentally, oxygen is found to be paramagnetic, with two unpaired electrons. Another molecule for which valence-bond theory predicts **diamagnetism** (no unpaired electrons) is B_2:

$$: B{-}B :$$

Magnetic studies show this molecule to contain two unpaired electrons.

The deficiencies of the valence-bond theory are related to an oversimplification inherent in its approach. It assumes that the electrons in a molecule can be located in the atomic orbitals of the individual atoms. For example, in the CH_4 molecule a pair of bonding electrons is considered to spend part of its time in the 1s orbital of a hydrogen atom and the remainder in one of the sp^3 orbitals of the carbon atom. Clearly, this is an approximation; it would be more satisfactory to locate these electrons in a **molecular orbital** characteristic of the molecule as a whole.

The molecular orbital approach to deducing the electronic structures of molecules, referred to briefly in Section 8.4, involves three fundamental operations.

1. The atomic nuclei are first located at known positions relative to one another. For the H_2 molecule, this would mean placing two protons 0.74 Å apart (the H—H bond distance).

2. An attempt is then made to calculate from quantum mechanics the energies of the various molecular orbitals available for occupancy by the electrons in the molecule. In practice, it is impossible to make precise calculations of molecular orbital energies for any but the simplest of molecules. Ordinarily, all that can be done is to deduce the relative energies of different molecular orbitals. For simple diatomic molecules of elementary substances (H_2, N_2, O_2, F_2), we can do this by combining the atomic orbitals of the individual atoms according to certain rules. In doing this, we always find that the **number of molecular orbitals formed is exactly equal to the total number of atomic orbitals combined** (e.g., two s orbitals, one from each atom, yield two molecular orbitals; six p orbitals, three from each atom, give six different molecular orbitals).

3. Once the relative energies of the available molecular orbitals have been established, valence electrons are distributed among these orbitals in much the same way that electrons in atoms are distributed among atomic orbitals. We find, for example, that:

    a. Each molecular orbital can hold a maximum of two valence electrons.

    b. Electrons go into the lowest available molecular orbitals. High-energy orbitals start to fill only when lower energy orbitals have their complement of two electrons.

    c. When two molecular orbitals of equal energy are available, one electron goes into each to give two half-filled orbitals before electron pairing occurs.

Another example of Hund's rule.

We shall now consider the distribution of valence electrons among molecular orbitals in the diatomic molecules of the elementary substances in the first two periods of the periodic table.

First Period Elements (H, He). Combination of 1s Atomic Orbitals

Molecular orbital (MO) calculations show that the combination of two 1s atomic orbitals leads to the formation of two molecular orbitals, one of which has an energy lower than that of the atomic orbitals from which it is formed (Figure 8.13); placing electrons in this orbital gives a species which is more stable than the combination of two isolated atoms. For this reason, the lower molecular orbital in Figure 8.13 is referred to as a **bonding orbital**. The other molecular orbital has a higher energy than the corresponding atomic orbitals; electrons entering it find themselves in an unstable, higher-energy state. It is referred to as an **antibonding orbital**. The "center of gravity," or average MO energy, lies exactly at the level of the atomic orbitals from which the molecular orbitals were formed.

The electron density corresponding to these two molecular orbitals is shown at the bottom of Figure 8.13. It will be observed that the bonding orbital has a high electron density in the region between the nuclei, which accounts for its stability. In the antibonding orbital, the probability of finding an electron between the nuclei is very small; the electron density is concentrated at the far ends of the "molecule." Since the nuclei are less shielded from each other than they are in the isolated atoms, the antibonding orbital is unstable with respect to the individual atomic orbitals. The electron density in both the bonding and antibonding orbitals is symmetrical with respect to the internuclear axis; both of these orbitals are sigma orbitals. In MO notation, these molecular orbitals are designated as σ_{1s} and σ_{1s}^{*} respectively; the asterisk is used to represent the antibonding orbital.

The order in which electrons enter these molecular orbitals is indicated in Figure 8.14. Four species, H_2^+, H_2, He_2^+, and He_2, containing 1, 2, 3, and 4 valence electrons respectively, are considered. It will be noted that the number of unpaired electrons in the first three species (1, 0, 1, respectively)

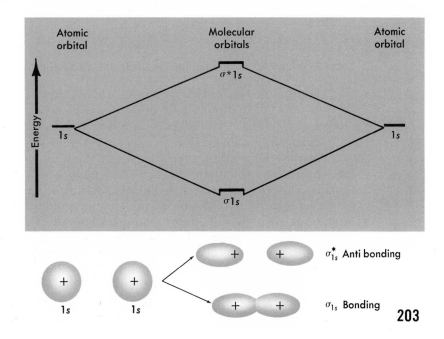

Figure 8.13 Molecular orbitals formed by combining two 1s orbitals.

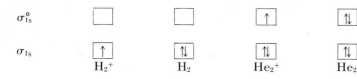

σ_{1s}^{*}

σ_{1s}

H$_2$⁺   H$_2$ He$_2$⁺   He$_2$

Figure 8.14 Electron distribution in H$_2$⁺, H$_2$, He$_2$⁺, and He$_2$.

is in agreement with experimental observation (Table 8.11). Moreover, the distribution of electrons among bonding and antibonding orbitals can be correlated with the observed bond energies. In the H$_2$⁺ ion, there is a single electron in the bonding orbital; this corresponds to one-half of an electron-pair bond. In H$_2$ there are two electrons in the bonding orbital, yielding one electron pair bond with a large bond energy. With the He$_2$⁺ ion we start to fill the antibonding orbital. Putting two electrons in a bonding orbital and one in an antibonding orbital is equivalent to:

$$\frac{2-1}{2} = \frac{1}{2} \text{ electron pair bond}$$

As expected, the bond energy is less than that in H$_2$. Finally, in the hypothetical He$_2$ molecule, both orbitals are filled. Hence the bonding and antibonding orbitals cancel each other and the net number of electron pair bonds is zero. Experimentally, the He$_2$ molecule has never been detected.

TABLE 8.11 OBSERVED PROPERTIES OF THE SPECIES H$_2$⁺, H$_2$, He$_2$⁺

	NO. UNPAIRED ELECTRONS	BOND ENERGY (KCAL/MOLE)
H$_2$⁺	1	61
H$_2$	0	103
He$_2$⁺	1	60

Second Period Elements. Combination of 2s and 2p Orbitals

The combination of two 2s atomic orbitals, one from each atom, leads to the formation of two molecular orbitals completely analogous to those discussed above. These orbitals are designated as σ_{2s} (sigma, bonding, 2s) and σ_{2s}^{*} (sigma, antibonding, 2s) respectively (Figure 8.15).

The rules for combining p orbitals are somewhat more complicated. Combining three 2p orbitals from each of two atoms gives a total of six molecular orbitals. Three of these are described as bonding orbitals since their energies are lower than those of the individual atomic orbitals. Of this set of bonding orbitals, one is higher in energy than the other two. This orbital, in which the electron density is symmetrical about the line joining the nuclei (Figure 8.15), is described as a σ_{2p} orbital. The other two bonding orbitals, which have equal energies, are referred to as π_{2p} orbitals. You will note from Figure 8.15 that the electron density in these orbitals is not symmetrical with respect to the line joining the nuclei; hence their designation as π orbitals.

The three antibonding orbitals formed by the combination of 2p atomic orbitals have energies which are higher than those of the isolated atomic orbitals from which they are formed. In terms of a physical picture, the electron density in these orbitals is concentrated away from the internuclear region. Putting electrons in these orbitals reduces the stability of a species. Of the three antibonding orbitals, one, designated as σ_{2p}^*, is higher in energy than the other two π_{2p}^* orbitals. The "center of gravity" principle suggests such a symmetric arrangement of bonding and antibonding orbitals. The electron density in the σ_{2p}^* and one of the π_{2p}^* orbitals is shown in Figure 8.15.

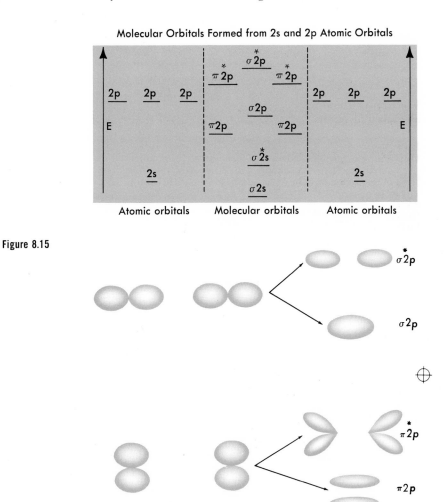

Molecular Orbitals Formed from 2s and 2p Atomic Orbitals

Figure 8.15

Let us now consider the order in which these molecular orbitals are filled in the diatomic molecules of the elementary substances in the second period. In Li_2 there are two valence (2s) electrons available. They go into the σ_{2s} bonding orbital to give a stable molecule with one electron pair bond and no unpaired electrons (cf. Table 8.12). (The inner 1s electrons of lithium are not considered in this treatment; they may be considered to occupy orbitals almost identical in nature to the 1s atomic orbitals of the individual atoms.)

In the hypothetical molecule Be_2, both the σ_{2s} and σ_{2s}^* orbitals are filled.

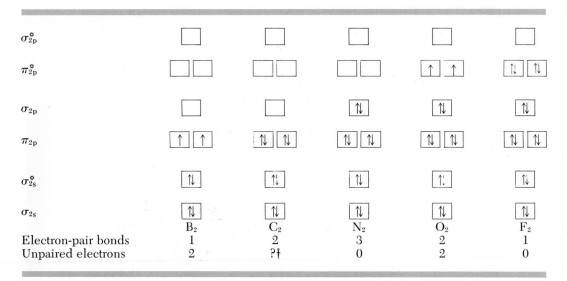

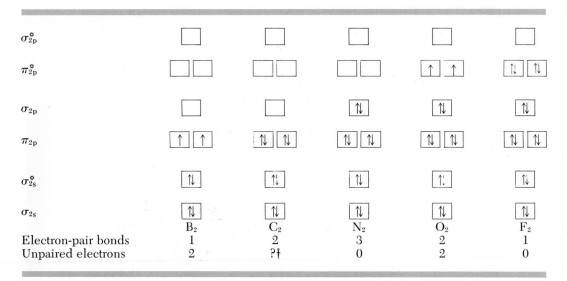

	B_2	C_2	N_2	O_2	F_2
Electron-pair bonds	1	2	3	2	1
Unpaired electrons	2	?†	0	2	0

† The ground state shown for C_2 should have no unpaired electrons. There is evidence, however, that the C_2 molecule is paramagnetic, with 2 unpaired electrons. The paramagnetism may be explained on the basis of an excited state in which a $\pi 2p$ electron has been promoted to the $\sigma 2p$ level, which is only slightly higher in energy.

Figure 8.16 Molecular orbitals occupied in diatomic molecules of the non-metals in the second period.

This leads to an unstable species with no net bonding. The electron distributions in the other diatomic molecules of elements in this period (B_2, C_2, N_2, O_2, and F_2) are shown in Figure 8.16 along with the calculated number of electron pair bonds and unpaired electrons. Note that in B_2 and again in O_2, where two molecular orbitals of equivalent energy are available to accommodate two electrons, the electrons split up, one to each orbital (Hund's rule).

The experimentally observed properties of the diatomic molecules listed in Table 8.12 are in reasonably good agreement with the MO structures given in Figure 8.16. Note particularly:

1. The correlation between the calculated number of electron pair bonds (Figure 8.16) and the corresponding bond energies (Table 8.12).

TABLE 8.12 PROPERTIES OF THE DIATOMIC MOLECULES $Li_2 \rightarrow F_2$

MOLECULE	BOND ENERGY (KCAL/MOLE)	NO. UNPAIRED ELECTRONS
Li_2	25	0
Be_2	—	—
B_2	69	2
C_2	150	?
N_2	225	0
O_2	118	2
F_2	37	0

2. The fact that MO theory predicts correctly that there should be two unpaired electrons in the B_2 and O_2 molecules. You will recall that valence bond theory was unable to explain satisfactorily the paramagnetism of oxygen.

The number of unpaired electrons in C_2 has not been established definitely.

Other Molecules

The simplified MO theory can readily be applied to polar diatomic molecules formed by two nonmetals in the second period of the Periodic Table. Consider, for example, the CO molecule (10 valence electrons) and the NO molecule (11 valence electrons).

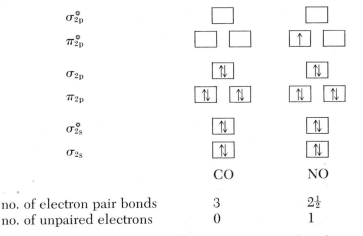

	CO	NO
no. of electron pair bonds	3	$2\frac{1}{2}$
no. of unpaired electrons	0	1

The theory can also be used to deduce the electron distribution in the simple diatomic molecules of elements in successive periods of the periodic table. The electronic structure of the Cl_2 molecule, for example, would be essentially identical to that of F_2, except that the molecular orbitals would be formed by the combination of 3s and 3p rather than 2s and 2p orbitals.

The MO theory as applied to diatomic molecules can be extended to predict the electron distribution in molecules such as HF, where the molecular orbitals are formed by combining atomic orbitals in different valence shells (1 and 2). We shall not attempt to do this, nor shall we consider the application of MO theory to molecules containing more than two atoms. The principal factor which has tended to discourage chemists from using molecular orbital theory to discuss electronic structure has been the difficulty in applying it to polyatomic molecules. Within the past decade, several approximation methods have been developed to extend MO theory to molecules containing more than two atoms, and the popularity of this approach has increased accordingly.

PROBLEMS

8.1 Predict the simplest formula of the ionic compound formed by

    a. Magnesium and iodine   c. Lithium and oxygen
    b. Scandium and fluorine   d. Calcium and sulfur

8.2 An element of atomic number 56 exists in its compounds as a +2 ion. Without referring to the periodic table, give the atomic number of

a. Four other elements which form ions of the same electronic structure.
b. An element which forms a +2 ion with a smaller radius.
c. An element which forms a +2 ion with a larger radius.
d. An element which forms a −1 ion with a larger radius.

8.3 Complete and balance the following equations.

a. $K(s) + Te(s) \longrightarrow$
b. $Mg(s) + H_2(g) \longrightarrow$
c. $Sc(s) + O_2(g) \longrightarrow$
d. $Al(s) + F_2(g) \longrightarrow$

8.4 Give the simplest formulas of

a. Ammonium sulfide   c. Sodium chromate   e. Sodium hydrogen carbonate
b. Potassium sulfate   d. Silver phosphate   f. Barium peroxide

8.5 The heat of formation of NaI is −69 kcal/mole. Given

$$Na(s) \longrightarrow Na(g) \qquad\qquad \Delta H = +26 \text{ kcal}$$
$$\tfrac{1}{2} I_2(s) \longrightarrow I(g) \qquad\qquad \Delta H = +25 \text{ kcal}$$
$$Na(g) \longrightarrow Na^+(g) + e^- \qquad\qquad \Delta H = +118 \text{ kcal}$$
$$Na^+(g) + I^-(g) \longrightarrow NaI(s) \qquad\qquad \Delta H = -161 \text{ kcal}$$

calculate ΔH for the reaction $I(g) + e^- \longrightarrow I^-(g)$.

8.6 Compare the Na^+ and Na^{2+} ions in regard to

a. Size.
b. Energy absorbed in formation from Na atoms.
c. Lattice energy for the formation of an ionic compound with F^- ions.
d. Stability in ionic compounds.

8.7 Given that the F—F bond energy is 37 kcal/mole and that the bond distance is 1.28 Å, draw a plot similar to Figure 8.4 for F_2.

8.8 From the graph given in Figure 8.7, predict the approximate percentage of ionic character in each of the following bonds.

a. H—Cl   b. P—Cl   c. B—N

8.9 Using Table 8.5, arrange the following bonds in order of decreasing polarity.

F—F, F—N, Cl—O, P—S

8.10 For the reaction $CH_3Cl(g) \rightarrow C(g) + 3 H(g) + Cl(g)$, ΔH is +376 kcal. Taking the C—H bond energy to be 99 kcal/mole, calculate the C—Cl bond energy.

8.11 Using the data in Table 8.8, find ΔH for the reaction

$$N_2(g) + 3 H_2(g) \longrightarrow 2 NH_3(g)$$

8.12 Draw Lewis structures for

a. H_2S b. ClO_3^- c. CO_3^{2-} d. C_2HCl

8.13 Draw Lewis structures for

a. PO_4^{3-} b. SO_3^{2-} c. N_2F_4 d. N_2O_3

8.14 According to the valence-bond model, which of the following species would you expect to be paramagnetic?

a. CO_2 b. ClO_2 c. NO^+ d. BN

8.15 Of the species listed in Problems 8.12 and 8.13, which would you expect to show resonance? Draw appropriate resonance structures.

8.16 Describe, either in words or by means of a figure, the geometry of the following species.

a. NO_2^- b. N_2F_2 c. SO_3 d. CO_3^{2-}

8.17 Classify each of the following molecules as polar or nonpolar.

a. H_2S b. O_2 c. CO_2 d. CH_2Cl_2 e. C_2H_2

8.18 List the hybrid orbitals used by the element whose symbol appears first in each of the following species.

a. NCl_3 b. BCl_3 c. $BeCl_2$ d. PF_5 e. NO_3^-

8.19 What hybrid orbitals are used by carbon in

a. C_2H_6 b. C_2H_2 c. CO_2 d. CO_3^{2-}

8.20 Indicate, by means of a diagram similar to Figure 8.14 or 8.16, the occupancy of molecular orbitals in

a. Li_2^+ b. O_2^+ c. O_2^- d. BN   e. BeF

Calculate the number of electron-pair bonds and unpaired electrons in each species.

8.21 Predict the formulas of the ionic compounds formed by

a. Magnesium and hydrogen   d. Barium and oxygen
b. Scandium and sulfur   e. Lithium and tellurium
c. Potassium and astatine   f. Lanthanum and oxygen

8.22 A certain element of atomic number 37 exists in all its compounds as a +1 ion. Without referring to the periodic table, give the atomic numbers of

a. Two other elements which form cations with this same electronic structure.
b. Two other elements which form anions with this same electronic structure.
c. An element whose atoms have this same electronic structure.
d. Three other elements which form +1 ions of smaller radius than that formed by element 37.
e. Two other elements which form +1 ions of larger radius than that formed by element 37.

8.23 Complete and balance the following equations.

a. $Na(s) + S(s) \longrightarrow$
b. $K(s) + I_2(s) \longrightarrow$
c. $Al(s) + O_2(g) \longrightarrow$
d. $Ba(s) + F_2(g) \longrightarrow$
e. $Sr(s) + O_2(g) \longrightarrow$

8.24 Give the simplest formula of each of the following compounds.

a. Ammonium bromide   d. Scandium acetate   g. Calcium hydroxide
b. Ammonium sulfate   e. Aluminum nitrate   h. Nickel chloride
c. Magnesium carbonate   f. Lithium phosphate   i. Zinc sulfate

8.25 From the graph given in Figure 8.7, predict the approximate percentage of ionic character in each of the following bonds.

a. H—F   b. H—C   c. H—N

8.26 Using the electronegativity table given in the text, arrange the following bonds in order of increasing polarity.

Mg—O   S—O   C—H   As—F   N—Cl

8.27 Compare the bond energies and bond distances of

a. N—N vs. N≡N
b. O—O vs. O=O
c. H—Cl vs. the average for H—H and Cl—Cl

8.28 Assuming that each atom in the following molecules has a noble-gas configuration, draw electron dot structures for

    a. $AsCl_3$ b. SeO_2 c. CH_3Cl d. HCN   e. Al_2I_6 f. O_3

8.29 Assuming that each atom in the following polyatomic ions has a noble-gas configuration, draw electron dot structures for

    a. NO_3^- b. PO_4^{3-} c. BrO_3^- d. O_2^{2-} e. SO_3^{2-}

8.30 Of the molecules and ions included in Problems 8.28 and 8.29, which would you expect to exhibit resonance? Explain your reasoning.

8.31 Give an example of a particle, other than one specifically mentioned in this chapter, which

    a. Contains a double bond.   d. Exhibits resonance.
    b. Contains a triple bond.   e. Uses d orbitals in bonding.
    c. Contains a coordinate covalent bond.

8.32 Which of the following species would you *not* expect to obey the octet rule?

    a. SiO_4^{4-} b. I_3^- c. S_2^{2-} d. SF_5^- e. $COCl_2$ f. BeF_3^-

8.33 Sketch geometric representations of the following molecules.

    a. CCl_4 b. NCl_3 c. C_2H_2 d. H_2S e. BF_3 f. SeF_6

8.34 Which of the following molecules would you expect to be polar?

    a. CF_4 b. P_4 c. CH_2Cl_2 d. CH_3Cl e. CO_2 f. $BeBr_2$

8.35 List the hybrid orbitals used by the element whose symbol is written first in each of the following species.

    a. C_2H_6 b. NH_4^+ c. SiF_6^{2-} d. BeI_2 e. BF_4^- f. BBr_3

8.36 Indicate the orbitals used in bonding by the central atom in

    a. KrF_4 b. ClF_3 c. IF_5 d. SiF_6^{2-} e. SiF_4

8.37 Indicate the electronic structure in MO notation of

    a. F_2^+ b. N_2^{2-} c. NO^+ d. H_2^-

For each species, state the number of electron pair bonds and the number of unpaired electrons.

°8.38 If the radius ratio of cation to anion was less than about 0.41, one would have anion-anion rather than cation-anion contact in a crystal of the type shown in Figure 8.1. Derive, by simple geometry, the precise value of the radius ratio for which there will be both cation-anion and anion-anion contact.

°8.39 A certain compound analyzes as follows: 54.5% C, 13.6% H, 31.8% N. The density of its vapor at 227°C and 1 atm pressure is about 2.1 g/lit. Suggest at least two possible structural formulas for this compound. How could you determine experimentally which of these formulas is correct?

°8.40 There are three known compounds of molecular formula $C_2H_2Cl_2$, but only one of formula C_2H_3Cl. Explain this in terms of molecular structure (cf. Chapter 24).

°8.41 A compound called diborane has the molecular formula B_2H_6. Suggest a reasonable electronic structure for this compound (cf. L. Pauling, *The Nature of the Chemical Bond*, Cornell University Press, Ithaca, N.Y., 3rd edition, 1960).

°8.42 In the PCl_5 molecule, there are five chlorine atoms bonded to phosphorus. Suggest a reasonable geometry. Predict the electronic structure around the phosphorus atom in terms of the valence-bond model.

°8.43 One of the less commonly known interhalogen compounds is iodine pentafluoride, IF_5. How many pairs of electrons are there around the central iodine atom in a molecule of IF_5? What symmetry would you expect these electrons to assume? What would be the resulting structure to be predicted for the IF$_5$ molecule?

°8.44   Iodine heptafluoride, IF$_7$, is one of the very few substances in which seven atoms are bound to a central atom. By extension of the reasoning used in rationalization of the structures of substances like PCl$_5$ and SF$_6$, suggest a likely structure for IF$_7$.

°8.45   A molecule whose structure is still in doubt is XeF$_6$. This molecule does not appear to have octahedral symmetry. Considering your solution to Problem 8.44, suggest a likely alternative structure for XeF$_6$.

Physical Properties in Relation to Structure

The physical properties of a compound or elementary substance are determined largely by two factors:

1. The nature of the structural units (ions, molecules, or atoms) of which the substance is composed.

2. The strength of the forces (interionic, intermolecular, or interatomic) between these particles.

In this chapter, we shall consider how such physical properties as melting point, boiling point, hardness, and electrical conductivity are related to structure. The substances which we shall consider will be subdivided into four categories:

1. Ionic substances (such as NaCl, MgF_2, BaO, KNO_3).
2. Molecular substances (such as Cl_2, CH_4, H_2O).
3. Macromolecular substances (such as C, SiO_2, BN).
4. Metallic substances (such as Li, Be, Fe).

9.1 IONIC SUBSTANCES

We pointed out in Chapter 8 that an ionic crystal is held together by strong electrostatic forces between oppositely charged ions. The characteristic physical properties of ionic substances such as sodium chloride or barium oxide can be explained in terms of these forces.

Melting Point. Thermal Decomposition of Polyatomic Ions

Ionic compounds are invariably solids at room temperature; their melting points vary from several hundred °C to over 2000° C. These high melting points reflect strong interionic forces; high temperatures are required to separate oppositely charged ions to give the freedom of motion characteristic of the liquid state.

According to Coulomb's Law, the attractive force between oppositely charged particles is directly proportional to their charges and inversely proportional to the square of the distance between them. Applied to ions in a crystal lattice, the relation becomes:

$$f = \text{constant} \times \frac{q_1 \times q_2}{(r_+ + r_-)^2} \qquad \begin{array}{l} (q_1, q_2 = \text{charges of} + \text{and} - \text{ions}) \\ (r_+, r_- = \text{radii of} + \text{and} - \text{ions}) \end{array} \qquad (9.1)$$

According to Equation 9.1, interionic forces should be greatest for small ions of high charge. The melting points of ionic compounds might be expected to show the same effect. If we compare the compounds NaCl and BaO, for which the sums of the ionic radii are nearly the same but the charge products differ by a factor of four, we find that barium oxide has a considerably higher melting point (Table 9.1). The size effect is illustrated by a comparison of MgO and BaO; the smaller radius of the Mg^{2+} ion as compared to the Ba^{2+} ion results in a stronger attraction for the O^{2-} ion and, consequently, a higher melting point.

Equation 9.1 cannot be used for compounds where there is an appreciable amount of covalent bonding.

TABLE 9.1 EFFECT OF IONIC SIZE AND CHARGE ON MELTING POINT

COMPOUND	CHARGES OF IONS	$r_+ + r_-$ (Å)	MELTING POINT (°C)
NaCl	+1, −1	0.95 + 1.81 = 2.76	800
BaO	+2, −2	1.35 + 1.40 = 2.75	1920
MgO	+2, −2	0.65 + 1.40 = 2.05	2800

Compounds in which a polyatomic ion is present frequently decompose before they melt. Consider, for example, the behavior upon heating of calcium carbonate, $CaCO_3$, which consists of Ca^{2+} and CO_3^{2-} ions. At temperatures in the vicinity of 800° C, the following reaction becomes spontaneous:

$$CaCO_3(s) \longrightarrow CaO(s) + CO_2(g) \qquad (9.2)$$

From a structural standpoint, we can interpret this reaction in terms of a decomposition of the CO_3^{2-} ion:

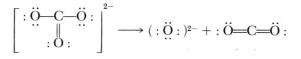

Calcium hydroxide, $Ca(OH)_2$, undergoes an analogous reaction at about 600°C:

$$Ca(OH)_2(s) \longrightarrow CaO(s) + H_2O(g) \qquad (9.3)$$

Here, it is the polyatomic OH^- ion which decomposes:

$$2 \, (: \overset{..}{\underset{..}{O}} : H)^- \longrightarrow (: \overset{..}{\underset{..}{O}} :)^{2-} + H - \overset{..}{\underset{..}{O}} - H$$

Another type of ionic compound which decomposes upon heating is one that contains water of hydration. An example is $BaCl_2 \cdot 2H_2O$, which loses water at about 100° C to form anhydrous barium chloride:

$$BaCl_2 \cdot 2H_2O(s) \longrightarrow BaCl_2(s) + 2H_2O(g) \qquad (9.4)$$

213

Efflorescence may
not occur in moist
air where the
vapor pressure of
water is high
enough to reverse
reactions such as
9.4.

Certain hydrated salts lose water upon exposure to dry air at room temperature; this process is referred to as **efflorescence**. A familiar example is copper sulfate pentahydrate, $CuSO_4 \cdot 5H_2O$. Blue crystals of this salt lose water to form first the lower hydrates, $CuSO_4 \cdot 3H_2O$ (blue) and $CuSO_4 \cdot H_2O$ (white), and finally, upon heating, anhydrous copper sulfate.

Electrical Conductivity

Perhaps the best evidence for the existence of ions in a compound such as sodium chloride is the ability of the substance to conduct an electric current when melted or dissolved in water. In either the molten state or water solution, the current is carried by the ions (Na^+ and Cl^-), which are free to move under the influence of an electrical field.

We should emphasize that the mechanism of conduction in molten salts is entirely different from that in metals. In the latter, it is electrons rather than ions which carry the current. The fact that most metals have conductivities at least a thousand times greater than that of molten sodium chloride implies that ions experience more resistance to flow than do electrons.

Metallic and ionic conductors differ in still another important respect. The passage of a direct electric current through fused or dissolved salts results in a chemical reaction at the electrodes. With molten sodium chloride, the reaction products are elementary sodium and chlorine. A somewhat more complicated reaction takes place when a direct current passes through a water solution of sodium chloride (cf. Chapter 20). In contrast, no chemical change whatsoever occurs when electrons flow through a metallic conductor.

Ionic compounds are ordinarily nonconductors in the solid state at room temperature. However, as the temperature is raised, there is usually a significant increase in conductivity. Solid sodium chloride at the melting point (800°C) has a conductivity about 10^{10} times that at 25°C. The increase in conductivity with increase in temperature can be explained in terms of the introduction of defects into the crystal lattice (Chapter 10).

Water Solubility

Ionic compounds as a class tend to be relatively soluble in water. There are, however, a great many ionic compounds (e.g., MgF_2, $BaSO_4$) which are almost quantitatively insoluble. We shall postpone until Chapter 16 a general discussion of the factors which determine the water solubility of ionic solids.

As one might expect, ionic substances are generally insoluble in nonpolar solvents such as benzene or carbon tetrachloride. The fact that such compounds as aluminum chloride, $AlCl_3$, and mercury(II) chloride, $HgCl_2$, have appreciable solubilities in organic solvents leads us to believe that the bonding in these compounds is not ionic.

9.2 MOLECULAR SUBSTANCES

Molecular substances differ generally from ionic compounds in two respects:

1. They are poor conductors of electricity. This is, of course, a conse-

quence of the fact that neutral molecules cannot carry an electrical current. It is true that water solutions of certain polar molecular compounds are electrical conductors. In all such cases, however, the current is carried by ions formed by a chemical reaction with water.

2. Molecular substances as a class are relatively volatile. Many of them are gases at room temperature and atmospheric pressure; others are liquids or solids of appreciable vapor pressure. In contrast, ionic compounds have extremely low vapor pressures at room temperature.

The volatility or, more specifically, the melting points and boiling points of molecular substances are directly related to the forces between molecules.

HCl, for example, reacts with water to give H_3O^+ and Cl^- ions (Chap. 17).

Interatomic vs. Intermolecular Forces

The generally low melting and boiling points of molecular substances are a direct consequence of the weak forces between molecules. They imply nothing about the forces within a molecule, which may be very strong. Consider, for example, elementary hydrogen, which is made up of diatomic molecules. To melt or boil hydrogen, it is necessary only to overcome the attractive forces holding the molecules together, rigidly in the solid, loosely in the liquid state. The low melting point ($-259°C$) and boiling point ($-253°C$) of hydrogen reflect the weakness of these forces. They reveal nothing about the strength of the covalent bond between hydrogen atoms in the H_2 molecule, since this bond remains intact when hydrogen melts or boils. The greater strength of interatomic as opposed to intermolecular forces is implied by the fact that at a temperature of 2400°C, only about one per cent of the H_2 molecules are dissociated into atoms. A comparison of the energy of dissociation, 104 kcal/mole, with the energy of sublimation, 0.122 kcal/mole, gives another indication of the relative magnitudes of these forces.

Trends in Melting and Boiling Points

Among molecular substances we find that the melting point, boiling point, heat of fusion, and heat of vaporization all tend to increase with molecular weight. Consider, for example, the halogens (Table 9.2). A further illustration of this effect appears in a comparison of the volatilities of different hydrocarbons. Those of low molecular weight, such as CH_4, C_2H_6, C_3H_8, and C_4H_{10} are gases at room temperature and atmospheric pressure. Hydrocarbons containing from 5 to 18 carbon atoms are liquids of steadily decreasing volatil-

TABLE 9.2 PHYSICAL PROPERTIES OF THE HALOGENS

ELEMENT	MOLECULAR WEIGHT	MELTING POINT (°C)	BOILING POINT (°C)	HEAT OF FUSION (kcal/mole)	HEAT OF VAPORIZATION (kcal/mole)
F_2	38	-223	-187	0.38	2.80
Cl_2	71	-102	-35	1.63	4.79
Br_2	160	-7	59	2.59	7.17
I_2	254	113	184	4.01	10.50

ity (bp $C_5H_{12} = 36°C$, $C_{18}H_{38} = 308°C$). Paraffin wax is a mixture of solid hydrocarbons of even higher molecular weight.

Another factor affecting the volatility of molecular substances is their polarity. Compounds built up of polar molecules melt and boil at slightly higher temperatures than nonpolar substances of comparable molecular weight.

The effect of polarity on melting or boiling point is ordinarily small enough to be obscured by differences in molecular weight. For example, in the series $HCl \rightarrow HBr \rightarrow HI$, boiling point increases steadily with molecular weight despite decreasing polarity. In molecular compounds in which hydrogen is bonded to a small, highly electronegative atom (N, O, F), polarity has a more pronounced effect on volatility. Hydrogen fluoride, despite its low molecular weight, has the highest boiling point of all the hydrogen halides. Water and ammonia also have abnormally high boiling points compared to those of the other hydrides of the 6A and 5A elements. In these three cases, the effect of polarity reverses the normal trend to be expected on the basis of molecular weight alone.

TABLE 9.3 BOILING POINTS OF POLAR VS. NONPOLAR SUBSTANCES

NONPOLAR			POLAR		
Formula	Molecular Weight	Boiling Point (°C)	Formula	Molecular Weight	Boiling Point (°C)
N_2	28	−196	CO	28	−192
SiH_4	32	−112	PH_3	34	−85
GeH_4	77	−90	AsH_3	78	−55
Br_2	160	59	ICl	162	97

As one might expect, the substances that are most difficult to condense to liquids or solids are those in which the basic structural unit is both light in weight and nonpolar. Such substances include the nonmetallic elements of low molecular weight (H_2, bp $= -253°C$; N_2, bp $= -196°C$; O_2, bp $= -183°C$; F_2, bp $= -187°C$) and the lower members of the noble gas group (He, bp $= -269°C$; Ne, bp $= -246°C$; Ar, bp $= -186°C$).

An explanation of these trends in volatility lies in the nature of the forces holding molecular substances together. These forces, for reasons which we shall now consider, increase in magnitude with the size and polarity of the molecule.

Types of Intermolecular Forces

Dipole Forces. The effect of polarity on the physical properties of molecular substances is readily explained in terms of the dipole forces existing between polar molecules. It has been pointed out that such molecules tend to line up in an electrical field. A similar orientation exists in a crystal composed of polar molecules. In solid iodine chloride, for example, the ICl molecules are aligned in such a way that the iodine atom (+ pole) of one molecule is adjacent to the chlorine atom (− pole) of the next molecule.

The electrostatic attraction holding neighboring molecules together in an iodine chloride crystal is similar in origin to that between adjacent ions in solid sodium chloride. However, the dipole forces between polar molecules in ICl are an order of magnitude weaker than the ionic bonds in NaCl. In the former case, the unequal electronegativities of iodine and chlorine produce only partial + and − charges within the molecule. In NaCl, on the other hand, a complete transfer of electrons leads to ions with full + and − charges.

When iodine chloride is heated to 27°C the comparatively weak dipole forces are no longer able to hold the molecules in rigid alignment and the solid melts. Dipole forces remain significant in the liquid state, in which the polar molecules are still relatively close to each other. Only in the gas, in which the molecules are very far apart, do the electrical forces become negligible. Consequently, the boiling points as well as the melting points of polar compounds such as ICl are higher than those of nonpolar substances of comparable molecular weight (Table 9.3).

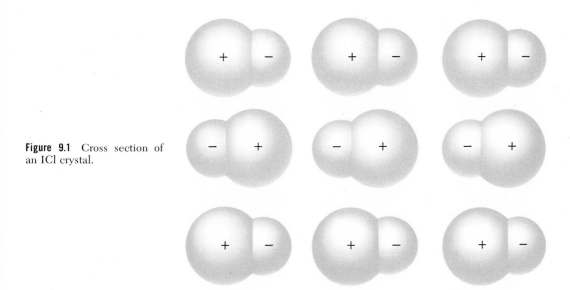

Figure 9.1 Cross section of an ICl crystal.

Hydrogen Bonds. The abnormal properties of hydrogen fluoride, water, and ammonia result from the presence in these substances of an unusually strong type of intermolecular force. This attractive force, exerted between the hydrogen atom of one molecule and the fluorine, oxygen, or nitrogen atom of another, is sufficiently unique to be given a special name, the hydrogen bond. There are two reasons why the hydrogen bond is stronger than ordinary dipole forces:

1. The difference in electronegativity between hydrogen (2.1) and fluorine (4.0), oxygen (3.5), or nitrogen (3.0) is great enough to cause the bonding electrons in HF, H_2O, and NH_3 to be markedly displaced from the hydrogen. Consequently, the hydrogen atoms in these molecules, insofar as their interaction with adjacent molecules is concerned, behave almost like bare protons. The hydrogen bond is strongest in HF, in which the difference in electronegativity is greatest, and weakest in NH_3, in which the difference in electronegativity is smallest.

217

2. The small size of hydrogen allows the fluorine, oxygen, or nitrogen atom of one molecule to approach the hydrogen atom of another molecule very closely. It is significant that hydrogen bonding appears to be limited primarily to compounds containing these three elements, all of which have comparatively small atomic radii. The larger chlorine and sulfur atoms, with electronegativities (3.0, 2.8) similar to that of nitrogen, show little or no tendency to form hydrogen bonds in such compounds as HCl and H_2S.

Many of the abnormal properties of water are a consequence of hydrogen bonding. Its high surface tension and viscosity are directly related to the strength of these intermolecular forces. The comparatively high heat capacity of water can be explained in terms of the energy required to break hydrogen bonds, which become less numerous as the temperature is raised. When water boils, the remaining hydrogen bonds are broken. This requires the absorption of 10.5 kcal/mole, more than twice the heat of vaporization of hydrogen sulfide.

When water freezes to ice, an open, hexagonal pattern of molecules results. Each oxygen atom in the ice crystal is bonded to four hydrogens, two by ordinary covalent bonds at a distance of 0.99 Å and two by hydrogen bonds 1.77 Å in length. The large proportion of "empty space" in the ice structure explains the curious fact that ice is less dense than liquid water. Indeed, water starts to decrease in density if cooled below 4°C, which indicates that the transition from a closely packed to an open structure occurs gradually over a temperature range rather than taking place abruptly at the freezing point. It is believed that even in water at room temperature, some of the molecules are oriented in an open, icelike pattern. More and more molecules assume this pattern as the temperature is lowered. Below 4°C the transition to the open structure predominates over the normal contraction on cooling, and liquid water expands as its temperature is lowered toward 0°C.

Hydrogen bonding occurs in a great many molecular substances containing hydrogen and fluorine, oxygen, or nitrogen. Hydrogen cyanide, HCN, is an example of a ternary compound in which hydrogen bonding between adjacent molecules has a profound effect on physical properties. The boiling point of hydrogen cyanide is 26°C, some 222°C higher than that of elementary nitrogen, a nonpolar substance of comparable molecular weight.

Dispersion (London) Forces.* The two types of intermolecular forces already discussed can exist only between polar molecules. A different type of attractive force must be postulated to explain the existence in the liquid and solid states of such nonpolar substances as bromine and iodine. Since the melting and boiling points of nonpolar substances tend to increase with molecular weight, it may be inferred that the magnitude of this force increases with the mass or size of the molecule. The fact that the volatility of polar as well as nonpolar substances decreases with increasing molecular weight suggests that this third type of intermolecular force must be common to all molecular substances.

Without hydrogen bonding, water would boil at about −200°C.

Many molecules (C_3H_8, $CHCl_3$) form cage-like hydrates, similar in structure to ice, and often stable above 0°C.

---

* These forces are sometimes referred to as Van der Waals forces. Properly speaking, Van der Waals forces include all the intermolecular forces referred to in this section, as well as one other, the force between a dipole in one molecule and an induced dipole in an adjacent molecule. It is the sum of all these forces which determines the magnitude of the deviation of real gases from ideal behavior (cf. Van der Waals equation, Chapter 5).

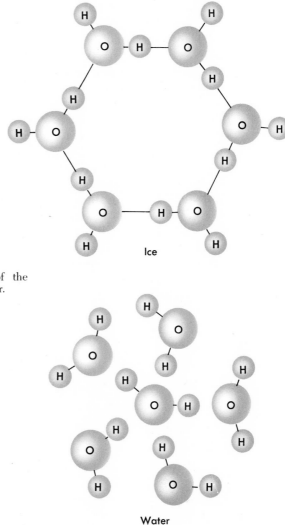

Figure 9.2 Plane projection of the geometry of ice and liquid water.

The intermolecular force whose characteristics have just been deduced is known as a dispersion force. Its origin is more difficult to visualize than that of the dipole force or hydrogen bond. Like them, it is basically electrical in nature. However, while hydrogen bonds and dipole forces arise from an attraction between permanent dipoles, dispersion forces are due to what might be called temporary or instantaneous charge separations.

It has been pointed out that over a period of time, the two bonding electrons in a nonpolar molecule such as H_2 are as close to one nucleus as to the other; the molecule has no permanent dipole moment. However, at any given moment, the electron cloud may be concentrated at one end of the molecule (position 1A in Figure 9.3). A fraction of a second later, it may be located at the opposite end of the molecule (position 1B). The situation is analogous to that of a person watching a tennis match from a position directly in line with the net. At one instant, his eyes are focused on the player to his left; a moment later, they shift to the player on his right. Over a period of time, he looks to

219

one side as often as the other; the "average" position of focus of his eyes is straight ahead.

The instantaneous concentration of the electron cloud on one side or the other of the center sets up a temporary dipole in the H_2 molecule. This in turn induces a similar dipole in an adjacent molecule. When the electron cloud in the first molecule is at position 1A, the electrons in the second molecule are attracted to 2A. As the first electron cloud shifts to 1B, the electrons of the second molecule are pulled back to 2B. These temporary dipoles, both oriented in the same direction, lead to an attractive force between the molecules, the dispersion force.

The strength of dispersion forces depends upon the ease with which the electronic distribution in a molecule can be distorted or "polarized" by a temporary dipole set up in an adjacent molecule. As one might expect, the ease of polarization depends primarily upon the size of the molecule. Large molecules in which the electrons are far removed from the nuclei are more readily polarized than small, compact molecules in which the nuclei exert greater control over the position of the electrons. In general, molecular size and molecular weight parallel each other; hence, the experimental observation that dispersion forces increase in magnitude with molecular weight.

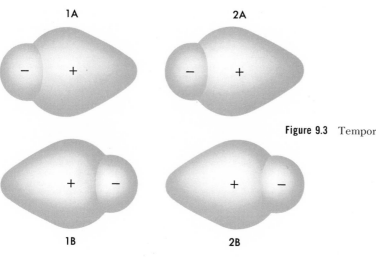

Figure 9.3 Temporary dipoles in H_2 molecules.

Of the three types of intermolecular forces, the only one operative between nonpolar molecules is, of course, the dispersion force. Even with polar molecules, this force is often the most important. For example, it is estimated that in HBr about 95 per cent of the attractive force between molecules is of the dispersion type; the other 5 per cent is accounted for by dipole forces. Even in HCl, dipole forces contribute only about 15 per cent to the total intermolecular attractive forces. Only where hydrogen bonding is involved do dispersion forces play a minor role. In water, about 80 per cent of the intermolecular attraction can be attributed to hydrogen bonding and only 20 per cent to dispersion forces.

9.3 MACROMOLECULAR SUBSTANCES

Covalent bond formation need not, and often does not, lead to the formation of small discrete molecules. Instead, it can result in macromolecular substances, in which all the atoms are held together by a network of covalent bonds. Consider, for example, the element carbon, each atom of which has four valence electrons. One could readily imagine the formation of a structure in which each carbon atom is held to four others by covalent bonds.

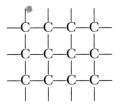

Such a structure would contain no small discrete molecules; instead the entire crystal might be described as one huge **macromolecule**, C_n, where n is of the order of Avogadro's number.

In a similar manner, oxygen and silicon atoms, which tend to form two and four covalent bonds respectively, could form a huge macromolecule, a tiny portion of which might resemble:

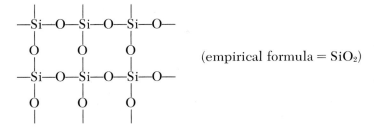

(empirical formula $= SiO_2$)

Experimentally, it is found that both elementary carbon and silicon dioxide exist in nature as macromolecular solids. The crystal structures of diamond and graphite (C) and of quartz (SiO_2)* are considerably more complex than those shown above, which would require an unlikely bond angle of 90°. The three-dimensional network structures of these substances are shown in Figures 9.4 and 9.9.

Macromolecular substances, like ionic compounds, are high-melting

TABLE 9.4 PROPERTIES AND BONDING IN LiF, BeO, BN, C

	MELTING POINT	PER CENT IONIC CHARACTER
LiF	870° C	89
BeO	2600° C	63
BN	3000° C	22
C (diamond)	~3500° C	0

---

* Solid carbon exists in only two crystalline phases, graphite and diamond, but silicon dioxide exists in some 23 different solid modifications, of which quartz is one. All the others are based on the same network of alternate silicon and oxygen atoms, but differ in the regularity and packing of the networks.

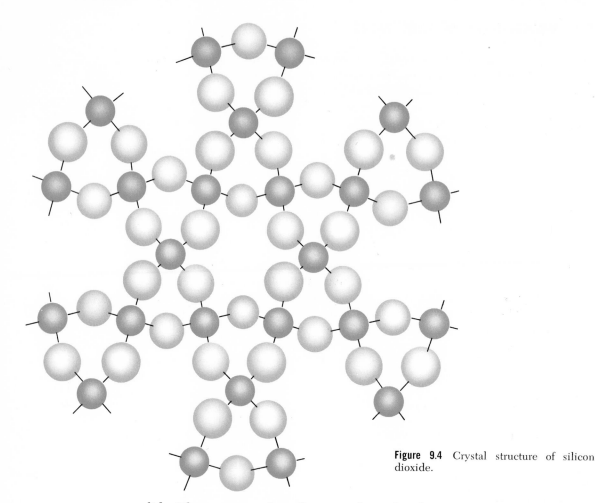

Figure 9.4 Crystal structure of silicon dioxide.

solids. Elementary carbon, for example, melts above 3500° C, quartz at about 1470° C. The high melting points of these substances reflect the strength of the covalent bonds holding atoms together in the crystal (C—C bond energy = 83 kcal/mole, Si—O = 107 kcal/mole). They are comparable to, or higher than, those of ionic compounds, in which adjacent particles are held together by electrostatic bonds. In this respect, it is interesting to compare the properties of the compounds in the series LiF, BeO, BN, and elementary carbon, in which there is a gradual transition from ionic to covalent bonding.

Macromolecular substances differ from ionic compounds in two other respects:

1. They are invariably insoluble in water, or, for that matter, any other solvent with which they do not react.

2. In the liquid state they are poor electrical conductors, since there are no charged particles present to carry the current. The semiconductivity of certain of these substances in the solid state will be discussed in Chapter 10.

9.4 METALS

The metals, which comprise about three fourths of the known elements, show a unique type of bonding different from any considered up to this point.

Before discussing metallic bonding, it may be helpful to summarize the more important properties of metals. From these properties, it is possible to deduce a great deal about the arrangement of particles in metals and the forces between these particles.

General Properties

It is difficult if not impossible to draw a sharp dividing line between the metallic elements on the one hand and nonmetals on the other. Nevertheless, one can list several general properties which are possessed to a greater or lesser degree by all metals. These include:

1. **High Electrical Conductivity.** The metallic elements as a group have electrical conductivities greater by several orders of magnitude than those of typical nonmetals. Lead, one of the poorest metallic conductors, has a conductivity nearly 4000 times that of the metalloids silicon and germanium.

The high conductivity of metals implies the presence of a large number of electrons which are relatively free to move under the influence of an electric field. This in turn indicates a type of bonding quite different from that in the nonmetals, in which the outermost electrons are tied up in covalent bonds.

2. **High Thermal Conductivity.** Of all solids, metals are by far the best conductors of heat. Frying pans and kettles are made of metals such as copper, aluminum, or iron; the insulating handles of these utensils are constructed from such nonmetallic materials as wood, glass, or plastic.

TABLE 9.5 RELATIVE ELECTRICAL CONDUCTIVITIES OF SOLID ELEMENTS (Pb = 1)

METALS		NONMETALS	
Ag	13.6	graphite	0.016
Cu	13.0	Si	0.00026
Al	8.4	Ge	0.00025
Zn	3.7	I	1.7×10^{-14}
Fe	2.2	diamond	4.4×10^{-20}
Pb	1.0	S	1.1×10^{-22}

Comparison of Tables 9.5 and 9.6 reveals that the various metals fall in the same order whether arranged according to decreasing thermal or electrical conductivity. This suggests that the mechanism of conduction of heat in metals is similar to that of electrical conduction.

TABLE 9.6 RELATIVE THERMAL CONDUCTIVITIES OF SOLIDS (Pb = 1)

METALS		OTHER SOLIDS	
Ag	12.2	graphite	0.087
Cu	11.1	glass	0.022
Al	6.1	wood	0.013
Zn	3.2	sand	0.012
Fe	2.0	S	0.009
Pb	1.0	paper	0.004

223

3. Luster. Polished metal surfaces are excellent reflectors of light and hence present a shiny appearance. Most metals appear to have a silvery white color when illuminated by ordinary light, indicating that light of all wavelengths is being reflected. A few metals, notably gold and copper, absorb some light in the blue region of the spectrum and hence have a yellow or red color.

4. Ductility, Malleability. Most metals are ductile (capable of being drawn out into wire) and malleable (capable of being hammered into thin sheets). Nonmetallic solids tend to shatter if drawn out or hammered. The ductility and malleability of metals indicates that in these elements it is relatively easy for layers of atoms to slide over one another.

5. Emission of Electrons. All metals, if heated to sufficiently high temperatures, eject electrons. This property, referred to as **thermionic emission**, is used in the ordinary vacuum tube in which electrons emitted by a hot metal filament travel to a positively charged plate.

Many metals emit electrons at room temperature when exposed to light of the proper frequency. The 1A metals in particular are sensitive to light. This so-called **photoelectric effect** is used in an "electric eye" photocell to convert light into electrical energy. A beam of light striking a negatively charged cesium surface causes electrons to flow to a positively charged collecting plate. When the light is cut off temporarily, the flow of current stops, a switch is thrown, and a traffic light turns green or a supermarket door opens.

The photoelectric and thermionic effects offer additional evidence for a metallic structure in which the electrons are relatively mobile.

6. Formation of Positive Ions in Reactions. Metals, in reacting with nonmetallic elements, lose electrons to form positive ions. This property, exhibited to varying degrees by all metals, offers chemical evidence in support of the idea that electrons are readily separated from the metal lattice.

The Metallic Bond

The types of chemical bonding previously considered arise from interactions between valence electrons of atoms. These are the electrons which are bound least tightly to the nucleus. Accordingly, it is reasonable to suppose that it is the valence electrons of the metallic elements which participate in bond formation between metal atoms. On this basis, an atom of a 1A metal would have one electron available for bonding, an atom of a 2A metal two, and so on.

It is by no means obvious how interactions between valence electrons can lead to stable chemical bonds between metal atoms. The possibility of ionic bonding, which would require the transfer of electrons from one atom to another, can be discarded immediately. Since all the atoms in a metal are alike, it is unreasonable to suppose that one atom would give up electrons to another.

A more likely possibility would be for adjacent metal atoms to share electrons, thereby forming a covalent bond. This is, of course, precisely what happens with nonmetallic elements. However, there is a fundamental objection to the formation of conventional electron-pair bonds in metals; there simply are not enough valence electrons to go around. Consider, for example, the metal cesium, each atom of which has one valence electron available for bond formation. In a crystal of cesium, each cesium atom is surrounded by eight others, all of which have equally valid claims to the single available electron. In order for a cesium atom to form ordinary covalent bonds with its eight neighbors, it would need eight valence electrons rather than one.

There remains one alternative for a cesium atom; it can share its valence electron equally with all eight neighboring atoms. In this way, a cesium atom forms what might be described as one-eighth of an electron-pair bond with each of these atoms. If we introduce the concept of resonance, we can picture the valence electron of cesium as being shared equally by each of the eight surrounding atoms. In barium, in which two valence electrons are available to share with eight nearest neighbors, one can expect to find two-eighths or one-fourth of an ordinary electron-pair bond between adjacent atoms.

Electron-Sea Model. The picture of metallic bonding that we have just presented implies that the bonding electrons are not tied down to a particular pair of atoms, as is the case in ordinary covalent bonding, but are spread out over a relatively wide region. This concept has led to what is known as the electron-sea model of metallic bonding. From this point of view, the metallic

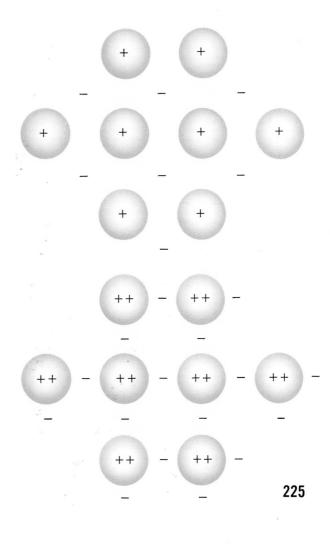

Figure 9.5 Schematic diagram of electronic structure of metals (cesium and barium).

lattice can be visualized as consisting of a regular array of positive ions (i.e., metal atoms minus their valence electrons) anchored in position like bell buoys in a mobile "sea" of electrons. The valence electrons can wander throughout the lattice in much the same way that gas molecules are able to move freely throughout their container. The metal ions are held together by a sort of "electron glue" composed of the valence electrons.

Many of the characteristic properties of metals are explained quite adequately by this admittedly oversimplified picture of metallic bonding. In particular, the high mobility of the valence electrons explains the high electrical conductivity of metals. Under the influence of an electrical field, it is possible for these electrons to move through the lattice, thereby conducting a current. The high thermal conductivity of metals is another consequence of the freedom of motion of the valence electrons. Heat energy is transferred through the metal by collisions between electrons. Since the valence electrons are responsible for the conduction of both thermal and electrical energy, it is hardly surprising that the metals that are the best electrical conductors are also the best conductors of heat (Tables 9.5 and 9.6).

The property of metallic luster is readily explained in terms of the electron-sea model. Since the electrons are not tied to a particular bond with a characteristic energy, they are able to absorb and re-emit light of all wavelengths. A metal thereby becomes a virtually perfect reflector; the silvery color that we associate with metals can be attributed to their high reflectance.

The picture of a metallic lattice as consisting of a regular array of positive ions embedded in a sea of mobile electrons offers a simple explanation of the malleability and ductility of metals. Consider, for example, a metal crystal in which the atoms in adjacent layers have the positions indicated in 1A, Figure 9.6. Under an applied stress, the ions in the upper layer move through a distance equivalent to one atomic diameter, taking the position shown in 1B. Since the valence electrons are mobile, they are free to move along with the metal ions to give a structure entirely equivalent to the original one. For this reason, most metals offer comparatively little resistance to distortions produced by the slippage of crystal planes. Contrast this situation to that which exists in ionic crystals. If one attempts to move one layer of ions across another, ions of like charge come in contact with each other (position 2B, Figure 9.6). The strong electrostatic repulsions created by this situation shatter the crystal.

Would you expect SiO₂ to be ductile?

Given sufficient energy, it is possible for valence electrons to escape completely from the metallic lattice. This energy can be supplied as light (photoelectric effect) or heat (thermionic effect). Alternatively, a nonmetal atom approaching close to the surface of a metal may be able to detach an electron. In doing so, the nonmetal atom is converted to a negative ion, while a positive ion is left behind in the lattice. The formation of an ionic compound can be visualized quite simply as being due to a flow of electrons from the metal lattice to nonmetal atoms.

Finally, the electron-sea model provides a reasonable explanation for the trends in the physical properties of the metals that are observed when we move across the Periodic Table from left to right. Calcium, with two valence electrons per atom, is harder, stronger, and higher-melting than potassium, which has one valence electron per atom. There is a further increase in hardness, strength and melting point when we move from calcium to scandium,

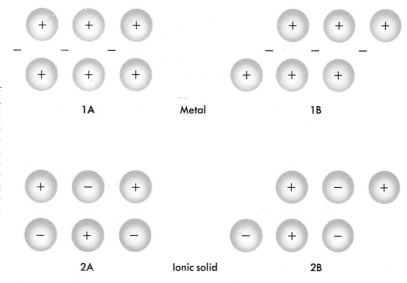

Figure 9.6 Movement of crystal planes. In an ionic solid, movement through one ionic diameter brings ions of like charge in contact with each other. In metals, in which the negative charges are free to move, the movement of a plane of atoms produces little change in the crystal structure.

suggesting that scandium metal is made up of Sc^{3+} ions embedded in a sea of electrons, three per ion. The toughest, strongest, and highest-melting elements are the transition metals located in groups 4B through 8B. The properties of these metals may be rationalized in terms of the large number of valence electrons per atom, which produces a lattice consisting of ions of relatively high charge (possibly as high as +6).

Even though the electron-sea model qualitatively explains many of the characteristic properties of metals, it is inadequate in certain respects. In particular, it is difficult to explain, in terms of this model, why the molar heat capacity of metals is not considerably higher than that of nonmetals. One would suppose that if the electrons in a metal were all free to move throughout the crystal, they should make a significant contribution to the heat capacity. In the case of calcium, for example, one can calculate that the heat capacity on the basis of a free-electron model should be nearly twice the observed value of 6.2 cal/mole.

Band Theory

The electron-sea model of metallic bonding is an extension of the valence-bond approach discussed in Chapter 8. The concept of resonance structures, in which an electron pair is shared, in effect, between several neighboring atoms, leads naturally to a model of the metallic state in which the valence electrons are not tied down to a particular metal atom. An alternative explanation of the properties of metals can be developed by using the molecular orbital approach to derive what is known as the band theory of the metallic bond.

In Section 8.10 we discussed the electronic structure of the gaseous Li_2 molecule in terms of the molecular orbitals built up by combining two 2s atomic orbitals, one from each lithium atom. To extend this approach to lithium metal, we must realize that we are dealing here not with two lithium atoms and two atomic orbitals, but rather with a huge number of lithium atoms and a correspondingly large number of atomic orbitals. The question which we must answer is the following: If we bring a large number, N, of lithium atoms together, each with a 2s atomic orbital, how many molecular orbitals do we form and what are their energies relative to one another?

Quantum mechanical calculations allow us to answer this question. We find that when n atomic orbitals are combined, an equal number, n, of molecular orbitals are formed. This situation is illustrated in Figure 9.7, where n = 2, 4, 8, and 16 respectively.

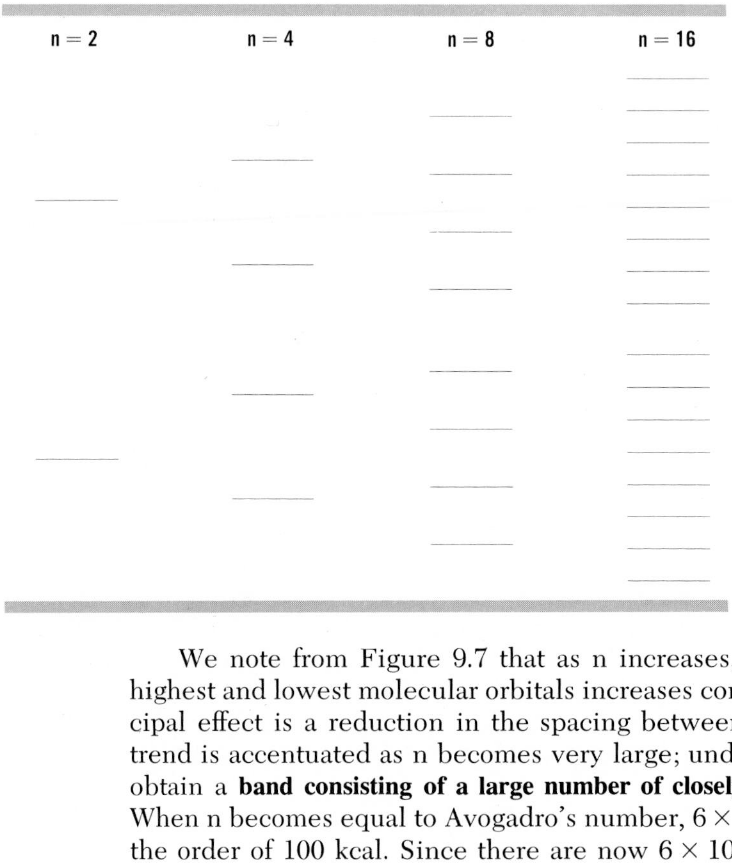

n = 2   n = 4   n = 8   n = 16

Figure 9.7 Molecular orbitals formed by combining **n** atomic orbitals.

We note from Figure 9.7 that as n increases, the spacing between the highest and lowest molecular orbitals increases comparatively little; the principal effect is a reduction in the spacing between individual orbitals. This trend is accentuated as n becomes very large; under these circumstances we obtain a **band consisting of a large number of closely spaced molecular orbitals**. When n becomes equal to Avogadro's number, 6×10^{23}, the total spacing is of the order of 100 kcal. Since there are now 6×10^{23} orbitals in all, it follows that the spacing between individual levels is

$$\frac{100 \text{ kcal}}{6 \times 10^{23}} = 1.7 \times 10^{-22} \text{ kcal}$$

Let us now consider what will happen when we bring together Avogadro's number of lithium atoms in a sample of lithium metal. The 2s atomic orbitals, one per atom, will combine to give a band of 6×10^{23} closely spaced energy levels (molecular orbitals). This band of levels is available for occupancy by the 6×10^{23} valence electrons associated with the lithium atoms. We recall from Chapter 8 that these electrons will tend to fill the lowest energy levels available, two per level. This means that only the lower half of the levels in the band are filled. This will, of course, be true regardless of the number of atoms present; i.e., in any sample of lithium metal, half the energy levels in the valence band are unfilled.

When a crystal of lithium with the ground state electronic structure just described is placed in an electrical field, a few electrons acquire enough energy to move up into higher, unoccupied levels within the 2s band. These high-energy electrons are the ones that will carry an electrical current. The high conductivity of lithium can be attributed to the presence of a large number of energy levels differing only slightly in energy from the occupied levels. Comparatively little energy is required to promote the few electrons needed to conduct an electric current or transfer thermal energy through the metal.

While it is relatively easy to promote a few electrons in lithium to higher, unoccupied levels, most of the electrons are located in levels buried so deeply within the 2s band that they cannot readily be promoted. This explains why the electrons in a metal under ordinary conditions make only a small contribution to the heat capacity. When the temperature of the metal is raised, the vast majority of electrons remain in their original energy levels and so absorb no heat.

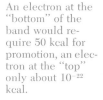

An electron at the "bottom" of the band would require 50 kcal for promotion, an electron at the "top" only about 10^{-22} kcal.

The electronic distribution in a sample of beryllium, in which each atom has two valence electrons, is somewhat more complex than that of lithium. One might expect to find all the energy levels in the 2s band filled with two electrons each. It turns out, however, that the band of molecular orbitals formed by combining the 2p atomic orbitals of beryllium overlaps the 2s band. The lower levels of the 2p band have energies somewhat less than those of the higher levels of the 2s band. This overlap is a direct consequence of the spreading of energies that results when a large number of atomic orbitals are combined to give molecular orbitals. The 2s band covers a range of energy great enough to cause it to overlap with the 2p band, even though the latter has a higher average energy.

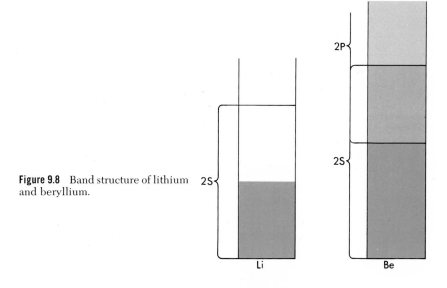

Figure 9.8 Band structure of lithium and beryllium.

Because of this overlap, the 2s band of beryllium is not completely filled with electrons. Beryllium, like lithium, has a large number of empty levels differing only slightly in energy from the filled ones. The amount of energy required to promote electrons is relatively small, so beryllium, like lithium, is a good electrical conductor.

229

9.5 ⌐ALLOTROPY⌐

Many elementary substances are capable of existing in two different forms in the same physical state. This general phenomenon is referred to as **allotropy**; the two different forms of the element are referred to as allotropic modifications or, more simply, **allotropes**.

The allotropes of an element may differ in either molecular composition or crystal structure. A familiar example of the first type of allotropy is that shown by elementary oxygen, which can exist as either diatomic or triatomic molecules, O_2 (ordinary oxygen) or O_3 (ozone). On the other hand, the allotropic forms of solid crystalline sulfur, rhombic and monoclinic sulfur, differ only in crystal form; they both consist of S_8 molecules. The so-called "white" and "gray" forms of tin, like rhombic and monoclinic sulfur, differ in crystal form. In this case, however, the difference in crystal structure appears to reflect a more fundamental difference in bonding. White tin, the form we are most familiar with, has typical metallic properties. Gray tin, which forms spontaneously at temperatures below 13°C, has a macromolecular structure similar to that of diamond and the other nonmetals in group 4A.

Gray tin is very brittle; tin objects often disintegrate to a powder at low temperatures.

To a greater or lesser extent, the allotropes of an element differ from each other in their physical properties. In certain cases they may even show quite different chemical properties. In discussing the allotropes of oxygen, sulfur, carbon, and phosphorus, we shall be particularly interested in how their properties reflect differences in structure and interatomic forces.

Oxygen

When ordinary oxygen O_2 is subjected to a silent electrical discharge, it is partially converted to ozone O_3:

$$3 \ O_2(g) \longrightarrow 2 \ O_3(g); \qquad \Delta H = +69.0 \text{ kcal} \qquad (9.5)$$

The same reaction occurs, also to a limited extent, in the air around a spark coil or high voltage transformer. Ultraviolet light can also serve as a source of energy to bring about this reaction.

A layer of ozone formed by the interaction of oxygen with the ultraviolet portion of sunlight is known to exist in the upper atmosphere, about 14 miles above the surface of the earth. This ozone layer forms only a tiny fraction of the total atmosphere; it has been estimated that if it were concentrated in a pure band at sea level, it would be less than a tenth of an inch thick. Nevertheless, it exerts a profound influence on life as we know it. Ozone strongly absorbs ultraviolet radiation, which is harmful to body cells. At the other end of the spectrum, it absorbs infrared radiation. The greater part of the energy radiated from the earth's surface is in the infrared range; the presence of the ozone layer in the atmosphere helps to prevent this energy from escaping into outer space. Water vapor and carbon dioxide behave similarly; were it not for the presence of these three gases in the atmosphere, our climate would be a great deal colder.

The physical properties of ozone and ordinary oxygen differ considerably (Table 9.7). Ozone is also much more reactive chemically than oxygen, as might be inferred from the direction of the energy change in Reaction 9.5. It

TABLE 9.7 PHYSICAL PROPERTIES OF O_2 AND O_3

	MELTING POINT (°C)	BOILING POINT (°C)	DENSITY AT STP (G/LIT)	SOLUBILITY IN WATER AT STP (MOLES/LIT)
O_2	−218	−183	1.429	0.00218
O_3	−251	−112	2.145	0.00022

is used for the controlled oxidation of hydrocarbons to make organic chemicals and on a small scale to purify drinking water by oxidation, which destroys harmful bacteria and other organic material.

Chlorine is cheaper than ozone but has an objectionable taste and odor.

The ozone molecule can be described in valence-bond terminology as a resonance hybrid of the two contributing forms

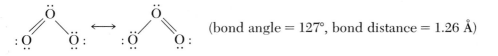

(bond angle = 127°, bond distance = 1.26 Å)

The two oxygen atoms at the ends of the molecule are equidistant from the central atom. The bond distance is intermediate between that calculated for a single and a double bond.

Sulfur

The rhombic and monoclinic allotropes of crystalline sulfur are both composed of puckered octagonal ring molecules of formula S_8:

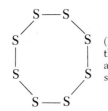

(In this diagram and those that follow, the two unshared electron pairs associated with each sulfur atom are not shown.)

The two forms differ only in the pattern in which the molecules pack in the crystal.

Rhombic and monoclinic sulfur are in equilibrium with each other at a temperature of 96°C. Above this temperature, the monoclinic form is stable; below 96°C, the transformation to rhombic sulfur is spontaneous. Monoclinic crystals, separated from molten sulfur at its freezing point (119°C), may be preserved for some time if the melt is quickly cooled to room temperature, where the rate of conversion to the rhombic form is quite slow.

The free-flowing, pale yellow liquid formed when sulfur melts is composed of the same S_8 molecules that are present in the solid. However, upon heating to about 160°C, sulfur undergoes an unusual transformation. The liquid becomes so viscous that it cannot be poured; at the same time its color changes to a deep reddish-brown. These changes result from the formation of another form of liquid sulfur in which large numbers of atoms are linked together to form a long-chain **polymer**. Two steps are involved in this transition. First, the S_8 ring molecules break up to form eight-membered chains:

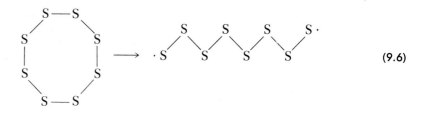

(9.6)

in which the terminal sulfur atoms have unpaired electrons. Two such chains adjacent to each other can then link together to form a double chain of 16 atoms:

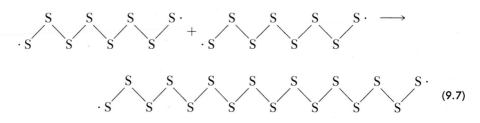

(9.7)

Since this process leaves unpaired electrons on the terminal sulfur atoms, the chains can continue to grow to 24, 32, 40, . . . sulfur atoms. Liquid sulfur at temperatures between 160°C and 250°C contains a high proportion of such polymeric chains, varying in length from eight to several thousand atoms. Entanglements between neighboring chains prevent free flow, producing a highly viscous liquid. The deep color is attributed to the absorption of light by the unpaired electrons at the ends of the chains.

The high viscosity of long-chain hydrocarbons used in lubricating oils is due to a similar effect.

Above 250°C the viscosity of liquid sulfur decreases; at the boiling point, 445°C, it again becomes free-flowing. Over this temperature range, the long chains break into smaller fragments:

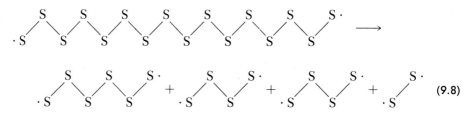

(9.8)

The shorter chains show less tendency to overlap and interlock with each other, thereby reducing the viscosity. The color steadily deepens as the boiling point is approached, reflecting the increased number of unpaired electrons produced when the chains break.

If liquid sulfur at a temperature between 160° and 250°C is quickly cooled by being poured into water, a rubbery mass commonly referred to as "plastic sulfur" separates. This material consists of long-chain molecules which did not have time to rearrange to the ring molecules characteristic of solid sulfur at room temperature. Over a period of a day or two, the plastic sulfur loses its elasticity as it is transformed to rhombic crystals.

Carbon

The arrangement of atoms in the two crystalline forms of carbon, diamond and graphite, is shown in Figure 9.9. In diamond, each carbon is covalently bonded to four other atoms, arranged tetrahedrally. The nonvolatility of diamond (sublimes at ~3500°C) reflects the difficulty of breaking these bonds. Diamond is extremely hard; it will scratch all other minerals. Tips of cutting tools and dies are often set with tiny diamonds; these tools can be used repeatedly without losing their cutting edge.

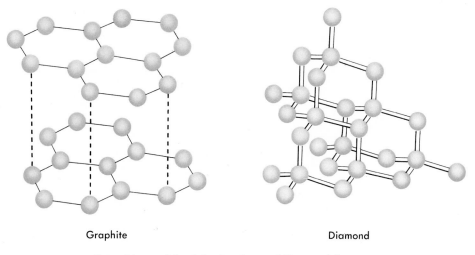

Graphite   Diamond

Figure 9.9   Models of the Graphite and Diamond Structures.

The layer structure shown by graphite reflects the fact that the carbon atoms are themselves bonded in layers. Each atom is joined to three neighboring atoms in the same plane by strong covalent bonds; one of these is a double bond which resonates among the three possible positions. There are no bonds between layers; adjacent planes of carbon atoms are held together by weak dispersion forces. The ability of these layers to slip readily over one another makes graphite soft. When one writes with a "lead" pencil, which is really made of graphite, thin layers of graphite rub off on the paper. In spite of its softness, graphite has an extremely high melting point (~3500°C). In order to melt graphite, the strong bonds within the layers must be broken.

Diamond has a considerably higher density (3.51 g/ml) than graphite (2.26 g/ml). The lower density of graphite results from the wide spacing between adjacent layers (3.40 Å). Within a given layer, the atoms in graphite are slightly closer together than in diamond (1.42 Å vs. 1.54 Å).

At room temperature and atmospheric pressure, graphite is the stable form of carbon. Under these conditions, diamond should, in principle, be transformed to graphite. Fortunately for the owners of diamond rings, this transition occurs at an infinitesimal rate under ordinary conditions. At high temperatures, the rate of conversion increases markedly; a sample of diamond maintained at 1800°C changes rapidly into graphite. For understandable reasons, no one has ever become excited over the commerical possibilities of this process. The more difficult task of converting graphite to diamond has aroused more enthusiasm.

Since diamond has a greater density than graphite, its formation is favored by high pressures. One can calculate that at 25°C, a pressure of 8000 atm should be sufficient to accomplish this transition. Higher temperatures are required to bring about the conversion of graphite to diamond in a finite amount of time. Unfortunately, at high temperatures the pressure required for conversion increases; at 2800°C, a pressure of 100,000 atm is needed. In 1954 scientists at the General Electric laboratories were able to achieve sufficiently high temperatures and pressures to bring about the conversion of graphitic material to diamond. The synthetic diamonds produced by this process are extremely small, but are suitable for industrial purposes.

Several amorphous forms of carbon can be prepared; lampblack and charcoal are two of the more familiar. These materials show no ordered crystal pattern and contain varying amounts of impurities. However, x-ray studies have shown that the carbon atoms in the so-called amorphous forms are arranged in an irregular hexagonal pattern similar in many respects to that of graphite.

Phosphorus

The element phosphorus can exist in three different allotropic forms known as white, red, and black phosphorus. Of these, the white and red modifications are by far the more familiar. Black phosphorus is formed only at very high pressures. "Yellow" phosphorus, believed at one time to be a distinct allotropic form of the element, is now known to be a mixture resulting from a slow transition, at room temperature, of white phosphorus to the more stable red form.

White phosphorus is a soft, waxy substance with a low melting point (44°C) and boiling point (280°C). It is readily soluble in such nonpolar organic solvents as carbon tetrachloride and carbon disulfide. The chemical reactivity of white phosphorus is so great that it is commonly stored under water to protect it from the oxygen of the atmosphere. A piece of white phosphorus exposed to air in a dark room glows as a result of the evolution of light accompanying its exothermic oxidation.

Many "phosphorescent" materials emit light by a different mechanism involving electronic transitions.

The high volatility, solubility in nonpolar solvents, and chemical reactivity of white phosphorus show it to be a molecular substance. Molecular weight determinations reveal that there are four atoms per molecule. X-ray diffraction measurements indicate a tetrahedral structure for the P_4 molecule with each phosphorus atom bonded to three others (Figure 9.10).

White phosphorus is extremely toxic; as little as 0.1 g taken internally can be fatal. Direct contact with the skin produces painful burns which are slow to heal. Inhaling the vapor leads to decay of the bone structure, particularly around the nose and jaw. To make matters worse, phosphorus vapor is a cumulative poison; persons exposed to low concentrations may show no symptoms of phosphorus poisoning for several years. Despite its toxicity, white phosphorus was at one time widely used in the manufacture of matches. Laws were eventually passed prohibiting its use for this purpose, and a nontoxic sulfide of phosphorus, P_4S_3, is now used instead.

The properties of red phosphorus contrast strikingly with those of the white form. Red phosphorus is higher melting (mp = 590°C at 43 atm), insoluble in organic solvents, and stable to oxidation at temperatures below

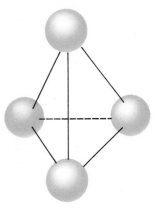

Figure 9.10   Model of a P_4 molecule.

about 250°C. The low volatility of red phosphorus makes it much less toxic than the white form.

As its physical and chemical properties imply, red phosphorus is a polymeric solid in which large numbers of phosphorus atoms are joined by a network of covalent bonds. Each phosphorus atom is joined to three others, as in white phosphorus, but bonding extends throughout the entire crystal instead of being confined to a discrete molecule. To melt or dissolve red phosphorus, it is necessary to break covalent bonds between phosphorus atoms. These bonds must also be broken before reaction with oxygen can occur.

PROBLEMS

9.1   Predict which member of the following pairs will have the higher melting point.

a. NaCl or NaBr   b. CaO or KCl   c. CaO or CaS

9.2   Write balanced equations for the reactions to be expected when the following compounds are heated strongly.

a. $Mg(OH)_2$ b. $NiCO_3$ c. $Na_2SO_4 \cdot 7H_2O$

9.3   Predict which compound in each of the following pairs will have the higher boiling point. In each case, explain your reasoning.

a. NaCl or ICl   c. BN or HF   e. C_6H_{14} or C_8H_{18}
b. SiH_4 or CH_4 d. HF or HBr   f. Carbon or sulfur

9.4   What kind of attractive force or chemical bond must be broken to

a. Melt lithium fluoride?   d. Melt sodium?
b. Boil water?   e. Dissolve Ar in water?
c. Boil carbon tetrachloride?   f. Dissolve LiF in water?

9.5   Explain, in terms of interatomic or intermolecular forces, why

a. Sodium is more malleable than sodium chloride.
b. Ammonia has a higher boiling point than phosphine.
c. Carbon dioxide sublimes readily.
d. Polar molecules tend to boil at higher temperatures than nonpolar molecules of similar molecular weight.

9.6   Explain, in terms of electronic structure and bonding, why

a. Diamond is more dense than graphite.

235

b. White phosphorus is more volatile than red phosphorus.

c. Metals are good conductors of electricity.

d. Graphite and lead are both malleable, while diamond is not.

9.7 a. It was at one time believed that ozone had a ring structure

What evidence can you think of which would tend to support such a structure?

What experimental observations appears to refute this structure?

b. If someone were to propose that ozone had the following structure,

how would you go about testing this possibility?

9.8 The molecular weight of sulfur at 25°C has been found to be 256, within experimental error. In each of the following temperature regions, would you expect the molecular weight of sulfur to be approximately 256, significantly less than 256, or significantly greater than 256?

a. 25–96°C. b. 96–119°C. c. 119–250°C. d. 250–445°C. e. 445–1000°C.

9.9 Predict which member of each of the following pairs will have the higher melting point.

a. BaO or CaO   b. KI or NaBr   c. LiF or NaI

9.10 Give the formula of an ionic compound, other than one listed in the text, which

a. Decomposes on heating to evolve CO_2.

b. Decomposes on heating to evolve H_2O.

c. Has a higher melting point than BaO.

9.11 Predict which compound in each of the following pairs will have the higher boiling point. In each case, explain your reasoning.

a. LiCl or CCl_4 c. CO_2 or SiO_2 e. CCl_4 or $SiCl_4$
b. PH_3 or AsH_3 d. HCl or F_2 f. HCN or HCl

9.12 What kind of attractive force of chemical bond must be broken to

a. Melt calcium chloride?   d. Boil hydrogen fluoride?
b. Melt carbon dioxide?   e. Dissolve I_2 in water?
c. Melt silicon dioxide?   f. Dissolve I_2 in CCl_4?

9.13 Explain, in terms of intermolecular forces, why

a. Most molecular compounds are volatile.

b. Hydrogen fluoride has a higher boiling point than hydrogen bromide.

c. Symmetrical molecules usually boil at lower temperatures than unsymmetrical molecules of comparable mass.

d. An ice cube floats in a glass of water.

9.14 Give examples of elements which have allotropic forms differing in

a. Crystal structure in the solid state.

b. Molecular weight in the gaseous state.

c. Molecular weight in the liquid state.

d. Type of bonding in the solid state.

9.15 Explain, in terms of electronic structure and bonding, why

a. The carbon atoms within a layer of graphite are closer together than are those in diamond.

  b. The coordination number of the carbon atoms in graphite is lower than that in diamond.

  c. Graphite is a very soft substance even though it has a high melting point.

  d. Graphite can be converted to diamond at high pressures.

9.16 Explain, first in terms of the electron-sea model and then in terms of the band theory, why

  a. Metals are good conductors of heat.

  b. Metals have shiny surfaces.

  c. Metals tend to form positive ions in reacting with nonmetals.

9.17 Explain in terms of electronic structure and bonding why

  a. White phosphorus is more reactive chemically than red phosphorus.

  b. White phosphorus is more soluble in organic solvents than is red phosphorus.

  c. Neither form of phosphorus is appreciably soluble in water.

°9.18 The density of gaseous hydrogen fluoride at 28°C and 1 atm is 2.30 g/lit. What information does this yield concerning intermolecular forces in hydrogen fluoride?

°9.19 The electrical conductivity of lead is 10^{22} times that of sulfur; the thermal conductivity of lead is only about 100 times that of sulfur. Can you suggest an explanation for the great difference in these two ratios?

°9.20 Look up in a handbook or other reference source the melting points of the metals in the fourth period, starting with potassium and going through zinc. Assuming that there is a correlation between melting point and number of valence electrons, what can you say concerning the variation in this number as one moves along from potassium to zinc?

°9.21 Assuming that the ozone in the atmosphere would occupy a layer 0.1 in thick if it were concentrated at the surface of the earth (radius = 4000 miles), calculate an approximate value for the total mass of the ozone in the atmosphere.

10

Liquids and Solids: Changes in State

In Chapter 5, we discussed the laws governing the physical behavior of gases and the interpretation of these laws in terms of the kinetic theory. In succeeding chapters we have had frequent occasion to refer to substances in the liquid and solid states. We are now ready to discuss the properties of these two condensed states of matter. In doing so, we shall be particularly interested in:

1. The particle structure of liquids and solids and its influence upon their physical properties.

2. The equilibria between the gaseous, liquid, and solid phases of a pure substance.

10.1 NATURE OF THE LIQUID STATE

The structure of liquids is less well established than that of gases or solids. Despite a great deal of research in this area, we still do not have a clear picture of the way in which molecules are arranged in even the most common liquid, water. We do, however, have a reasonably detailed knowledge of the average distances between atoms or molecules in a liquid. Moreover, we can estimate with considerable accuracy the magnitude of the forces between particles in a liquid. It is these two aspects of liquid structure which we shall now consider.

At ordinary temperatures and pressures, the molecules in a liquid are much closer together than in a gas (Figure 10.1). The closer approach of molecules in the liquid as opposed to the gaseous state provides a simple interpretation of many of the observed differences in behavior of these two states of matter.

1. The density of liquids is ordinarily much greater than that of gases. Compare, for example, the density of liquid water at 100°C and 1 atm, 0.958 g/ml, to that of water vapor at the same temperature and pressure, as cal-

culated from the Ideal Gas Law:

$$d = \frac{PM}{RT} = \frac{(1\ atm)(18.0\ g/mole)}{\left(82.1\ \dfrac{ml\ atm}{mole\ °K}\right)(373°K)} = 0.000588\ g/ml$$

What is the ratio of the density of $H_2O(l)$ to $H_2O(g)$?

2. Liquids diffuse into each other much more slowly than do gases. If one pours concentrated sulfuric acid carefully down the inside of a beaker containing water, two layers can be maintained almost indefinitely despite the fact that the two liquids are soluble in each other. The small amount of free space between molecules in the liquid tends to prevent one liquid from diffusing into another. In contrast, diffusion in the gaseous state, in which the molecules are widely separated, occurs much more rapidly.

3. Liquids are much less compressible than gases. At 25°C, an increase in pressure from 1 to 2 atm decreases the volume of a sample of liquid water by only 0.0045 per cent. The same change in pressure decreases the volume of an ideal gas by 50 per cent. In the gas, in which most of the space is unoccupied to begin with, an increase in pressure drastically reduces the volume; liquids, in which the molecules are much closer together, are virtually incompressible.

4. Liquids expand less when heated than do gases. The density of liquid water at 100°C is only about 4 per cent less than that at 0°C. An ideal gas, in going through this same temperature change at constant pressure, expands by $\frac{100}{273}$ or about 37 per cent.

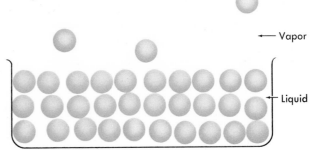

← Vapor

← Liquid

Figure 10.1 Spacing of particles in liquid and vapor at ordinary pressures.

Since the attractive forces between molecules are inversely related to the distances separating them, we would expect intermolecular forces to be much stronger in the liquid than in the gaseous state. This is, of course, the case; we find that energy must be supplied to overcome these forces and thereby convert a liquid to a vapor. One can calculate that the attractive energy due to Van der Waals forces between carbon tetrachloride molecules in the vapor at 25°C and 1 atm is of the order of a few *calories* per mole. In the liquid, on the other hand, it amounts to several *kilocalories* per mole.

The amount of heat required to vaporize a mole of liquid against a constant external pressure (e.g., in an open container) is referred to as the **heat of vaporization**. The heat of vaporization is a measure of the strength of the intermolecular forces holding molecules in the liquid state. Compare, for example, the heat of vaporization of water, 10.5 kcal/mole,* to that of methane, 2.0 kcal/mole. The difference between these two quantities reflects the strong

---

* The heat of vaporization of a liquid varies to some extent with temperature. In the case of water, it is 10.7 kcal/mole at 0°C, 10.5 kcal/mole at 25°C, and 9.7 kcal/mole at 100°C.

hydrogen bonding in water superimposed upon the dispersion forces which account for all the intermolecular attraction in methane.

The attractive force acting on a molecule at the surface of a liquid is considerably smaller than that in the interior, where the number of surrounding molecules is approximately twice as great. A molecule at the surface experiences an unbalanced attractive force tending to pull it into the body of the liquid. This explains why liquids tend to achieve as small a surface as possible. Small drops of water falling through air or through a liquid with which water is immiscible take on the shape of a sphere, thereby achieving the smallest possible surface-to-volume ratio.

The work required to expand the surface of a liquid by unit area is referred to as its surface tension (γ). This quantity is ordinarily expressed in ergs/cm² ($\equiv$ dynes/cm). The surface tension, like the heat of vaporization, is a measure of the strength of intermolecular forces. Water, in which these forces are comparatively strong, has a high surface tension, one that is considerably greater than that of most organic liquids. This explains, at least in part, why organic liquids tend to "wet" or spread out over solid surfaces much more readily than does water. Mercury, with a surface tension even higher than that of water, is a notoriously poor wetting agent; it forms spherical drops even on clean glass surfaces.

TABLE 10.1 HEATS OF VAPORIZATION AND SURFACE TENSIONS OF LIQUIDS

LIQUID	ΔH_{vap} (KCAL/MOLE)*	γ(ERGS/CM²) AT 20°C.
Mercury	14.2	475
Water	9.7	72.8
Pyridine	8.5	38.0
Benzene	7.3	29.0
Ether	6.2	17.0

* The heat of vaporization is given at the normal boiling point (cf. Table 10.2).

10.2 LIQUID-VAPOR EQUILIBRIUM

All of us are familiar with the process of evaporation, in which a volatile liquid, placed in an open container, spontaneously passes into the vapor phase. Let us now consider this physical change from a molecular viewpoint.

According to the kinetic theory, the molecules in a liquid will have a certain average kinetic energy at a given temperature. Some of the molecules, however, have energies considerably above the average. A few of these fast-moving, high-energy molecules will manage to escape through the surface of the liquid, passing into the vapor state. In an open container these molecules tend to diffuse into the surrounding atmosphere rather than to return to the liquid. The escape of high-energy molecules from the liquid lowers the temperature and hence the average kinetic energy of its molecules—evaporation is a cooling process. However, under ordinary conditions, heat flows to the liquid from the surroundings, restoring the temperature and average kinetic energy to their original values. More high-energy molecules escape from the liquid into the atmosphere; the process of evaporation continues until all the liquid is gone.

Vapor Pressure

Let us now consider what happens when a volatile liquid is placed in a closed, evacuated container maintained at a constant temperature (Figure 10.2). As before, some of the molecules in the liquid spontaneously pass through the surface into the space above it. This time, however, molecules in the vapor are unable to escape from the container; eventually some of them re-enter the liquid. At first the movement of molecules is primarily in one direction, from liquid to vapor. Gradually, however, as the concentration of molecules in the vapor builds up, the rates of vaporization and condensation approach each other. Within a few minutes, we reach a state of **equilibrium**, at which these two rates become equal. From this point on, the concentration of molecules in the gas phase has a certain fixed value which does not change with time.

The process which we have described cannot, of course, be followed by direct visual observation. We cannot "see" molecules escape from the liquid, nor can we see their concentration in the gas phase build up to its equilibrium value. It is, however, possible to follow the process by observing what happens to the mercury levels in the manometer shown in Figure 10.2. At the beginning of the experiment, there are no molecules in the space above the liquid and the mercury levels are equal. As vaporization occurs, molecules leave the liquid and exert a pressure in the space above. This pressure can be read on the manometer. This pressure steadily increases until equilibrium is reached; at that point it becomes constant, and the mercury levels no longer change in position.

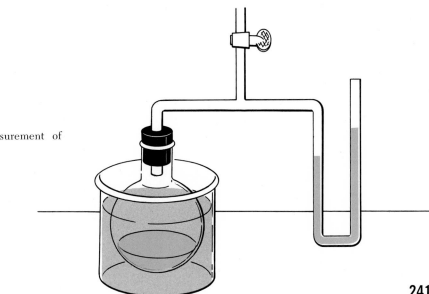

Figure 10.2   Measurement of vapor pressure.

The pressure of vapor in equilibrium with a pure liquid at a given temperature is referred to as its **vapor pressure**. The apparatus shown in Figure 10.2 can be used to measure vapor pressure. It is found, for example, that if one places liquid benzene in the flask and allows it to come to equilibrium with its vapor at 50°C, the pressure read on the manometer is 272 mm Hg. It follows that the vapor pressure of liquid benzene at 50°C is 272 mm Hg. If one uses liquid water at 50°C, the observed vapor pressure is somewhat lower, 92.5 mm Hg.

Vapor pressures of liquids are often measured in the presence of air. One can calculate that the vapor pressure of water in equilibrium with air at 50°C is about 0.018 mm Hg greater than that of water in contact with pure water vapor. Since this effect is so small, it is ordinarily neglected. Indeed, it is a good approximation to say that the vapor pressure of a liquid is unaffected by the presence of an "inert" gas, i.e., one which is not appreciably soluble in the liquid. Only at very high pressures would such a gas be likely to change the vapor pressure by a detectable amount.

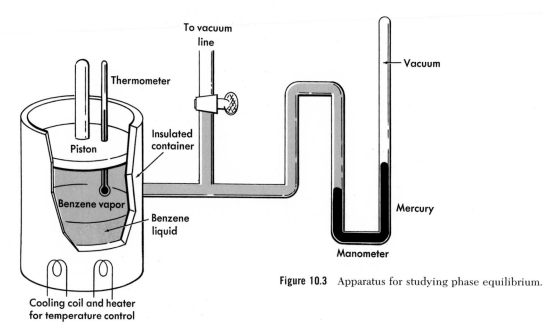

Figure 10.3 Apparatus for studying phase equilibrium.

Independence of Volume. As the above discussion implies, a liquid at a given temperature has a certain fixed vapor pressure that is characteristic of the substance. It is important to point out that **the pressure exerted by a vapor in equilibrium with a liquid is independent of the volume occupied by the vapor**. Students sometimes fail to grasp all of the implications of this deceptively simple statement. To understand what it means, consider the experiment shown schematically in Figure 10.3. The apparatus is similar to that in Figure 10.2, except that the container is equipped with a piston which can be raised or lowered to change the volume of the system. If we allow liquid benzene to come to equilibrium with its vapor at 50°C, with the piston in the position shown, we read a pressure of 272 mm Hg on the manometer. If we now slowly lift the piston, the pressure will remain constant as long as there is any liquid present. Molecules will escape from the liquid to maintain a constant concentration and therefore a constant pressure in the gas phase. The liquid level

The total number of molecules in the vapor increases to keep their concentration constant.

in the container will drop steadily as vaporization occurs to maintain equilibrium. Eventually, all of the liquid will "disappear." If we continue to raise the piston beyond this point, equilibrium with the liquid can no longer be maintained. Then and only then will the vapor behave as an ordinary gas; its pressure will decrease as the volume is increased.

This process can, of course, be reversed. If we start with the entire container filled with benzene vapor at a low pressure, let us say 100 mm Hg, and gradually lower the piston, the pressure will increase steadily. When the pressure reaches 272 mm Hg, liquid will suddenly appear in the system. No further change in pressure will occur as the volume of the system is reduced by lowering the piston. Condensation of vapor molecules will proceed, maintaining a constant concentration and therefore a constant pressure in the gas phase.

EXAMPLE 10.1. Suppose that in the experiment just described, we start with one mole of benzene at 50°C in a 5.0 liter container.

    a. To what value must the volume be increased in order for the liquid to just disappear? (vp = 272 mm Hg)

    b. What will be the pressure read on the manometer when the volume is 5.0 lit? 20.0 lit? 100 lit?

    c. From the results of the calculations in a and b, plot pressure in this system as a function of volume.

SOLUTION.

    a. The liquid will "disappear" when the one mole of benzene originally present is completely vaporized. The question we must answer is: What volume will be occupied by one mole of benzene vapor at 50°C and 272 mm Hg? This is readily calculated from the Ideal Gas Law:

$$V = nRT/P$$
$$= \frac{(1 \text{ mole})(0.0821 \text{ lit atm/mole °K})(323°K)}{(272/760) \text{ atm}} = 73.9 \text{ lit}$$

    b. At a volume of 5.0 lit we have liquid-vapor equilibrium; p = 272 mm Hg

At a volume of 20.0 lit we have liquid-vapor equilibrium; p = 272 mm Hg

At a volume of 100 lit we have **only** vapor. The pressure of benzene vapor can now be calculated from the Ideal Gas Law:

$$P = \frac{nRT}{V} = \frac{(1 \text{ mole})(0.0821 \text{ lit atm/mole °K})(323°K)}{100 \text{ lit}}$$
$$= 0.265 \text{ atm} = 201 \text{ mm Hg}$$

The gas laws cannot be used to calculate the pressure of a vapor in equilibrium with a liquid.

    c.

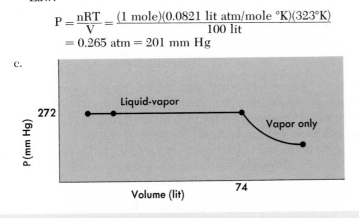

243

Dependence on Temperature. The vapor pressure of a liquid always increases with temperature. When the temperature is raised, the fraction of molecules having kinetic energies great enough to escape from the liquid rises accordingly. Consequently, the concentration of molecules in the vapor, or the equilibrium vapor pressure, increases. The effect is shown in Figure 10.4 for a particular liquid, water. Note that in the plot of vapor pressure against temperature, the slope increases rapidly as one goes to higher temperatures. In the case of water, for example, the vapor pressure increases by only about 88 mm Hg in going from 0 to 50°C; in going from 50 to 100°C, it increases by 668 mm Hg.

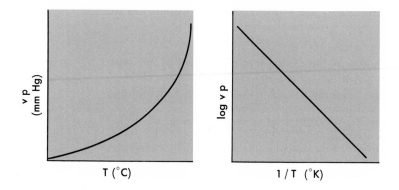

Figure 10.4 Effect of temperature on vapor pressure of water.

It is found experimentally that if one plots the logarithm of the vapor pressure of a liquid ($\log_{10}P$) vs. the reciprocal of the absolute temperature ($1/T$), a straight line results (Figure 10.4). Mathematically,

This is a linear relation of the form $y = Ax + B$, where $y = \log_{10}P$ and $x = 1/T$.

$$\log_{10}P = \frac{A}{T} + B \qquad (10.1)$$

From thermodynamic considerations it can be shown that the constant A in this equation, which represents the slope of the linear plot, is related to the heat of vaporization of the liquid. Specifically,

$$A = -\Delta H_{vap}/2.3\ R \qquad (10.2)$$

where ΔH_{vap} is the heat of vaporization in calories per mole and R is the gas law constant in cal/mole °K (1.99 cal/mole °K). Substituting in Equation 10.1 we obtain

$$\log_{10}P = \frac{-\Delta H_{vap}}{(2.3)(1.99)T} + B \qquad (10.3)$$

Heats of vaporization are ordinarily determined experimentally by taking advantage of the relationship given in Equation 10.3. If we measure the vapor pressure of a liquid at a series of temperatures and plot $\log_{10}P$ vs. $1/T$, ΔH_{vap} can be calculated directly from the slope.

Equation 10.3 can also be used to calculate the vapor pressure of a liquid at one temperature from that at another temperature, assuming that ΔH_{vap} is known. For this purpose, it is convenient to write

$$\log_{10}P_2 = \frac{-\Delta H_{vap}}{(2.3)(1.99)T_2} + B$$

$$\log_{10}P_1 = \frac{-\Delta H_{vap}}{(2.3)(1.99)T_1} + B$$

where the subscripts 2 and 1 represent two different temperatures. Subtracting, we obtain

$$\log_{10}P_2 - \log_{10}P_1 = \frac{-\Delta H_{vap}}{(2.3)(1.99)} \left(\frac{1}{T_2} - \frac{1}{T_1}\right)$$

or, in an equivalent but more convenient form,

$$\log_{10} \frac{P_2}{P_1} = \frac{\Delta H_{vap}}{(2.3)(1.99)} \left(\frac{T_2 - T_1}{T_2 T_1}\right) \qquad \textbf{(10.4)}$$

The use of Equation 10.4, known as the Clausius-Clapeyron equation, is illustrated in Example 10.2.

Under what conditions might we use this equation rather than 10.3 to determine ΔH?

EXAMPLE 10.2. The vapor pressure of benzene at 50°C is 272 mm Hg; its heat of vaporization is 7300 cal/mole. Calculate the vapor pressure at 60°C.

SOLUTION. Substituting in Equation 10.4, we have

$$\log_{10} \frac{P_2}{272} = \frac{7300}{(2.3)(1.99)} \frac{(333 - 323)}{(333)(323)} = +0.148$$

$$\log_{10}P_2 - 2.435 = +0.148; \quad \log_{10}P_2 = 2.583; \quad P_2 = 383 \text{ mm Hg}$$

Critical Temperature

We have just considered the effect of temperature upon a liquid-vapor system at equilibrium, pointing out that the vapor pressure increases with temperature. One might wonder whether there is any upper limit to the existence of the liquid state. In other words, is it possible to reach a temperature so high that the liquid can no longer exist because of the extreme agitation of the molecules?

To answer this question, we might conduct the following experiment. Let us seal a sample of benzene in a pressure-tight metal container fitted with a window through which observations can be made (Figure 10.5). The apparatus is immersed in an oil bath whose temperature can be raised at a controlled rate; the pressure within the apparatus is read on a gauge. As the temperature is increased, the pressure steadily rises. The level of the liquid drops somewhat as vaporization occurs. Nothing spectacular happens until we reach a temperature of 289°C, at which point the vapor pressure, as read on the gauge, is about 47.9 atm. Suddenly, as we pass this temperature, the liquid phase completely disappears. We are no longer able to see the liquid-vapor surface; the entire container is filled with benzene vapor. This phenomenon is completely reversible; if we cool the container to 289°C, the liquid appears as suddenly as it disappeared on heating.

The behavior described for benzene is typical of all liquids. For every substance, there is a temperature called the **critical temperature**, above which

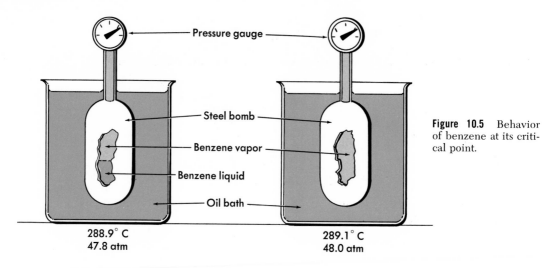

Figure 10.5 Behavior of benzene at its critical point.

Pressure gauge

Steel bomb

Benzene vapor

Benzene liquid

Oil bath

288.9° C
47.8 atm

289.1° C
48.0 atm

the liquid phase cannot exist. The pressure at that temperature, which is the maximum vapor pressure that the liquid can exert, is referred to as the critical pressure. Critical temperatures and pressures vary widely from one substance to another. For water, the critical temperature is 374°C; the vapor pressure of water at that temperature (its critical pressure) is 218 atm. On the other hand, oxygen has a critical temperature of −119°C; that of helium is −268°C, only five degrees above absolute zero. The very low critical temperatures of substances like oxygen and helium make their liquefaction difficult, since they must be cooled below their critical temperature before they can be made to condense, regardless of the applied pressure.

At the critical temperature, the density of the vapor becomes equal to that of the liquid.

The existence of the critical temperature can be rationalized in terms of a kinetic-molecular model of liquid-vapor equilibrium. One might expect that if the kinetic energy of the molecules were increased sufficiently, they would no longer be able to stay in the condensed, liquid state. By increasing the temperature, we reach a critical kinetic energy, the point at which the attractive forces between molecules are no longer able to hold them together in the liquid.

Boiling Point

If a liquid is heated in an open container rather than in a closed system of the type we have been discussing, it will still tend to vaporize so as to establish its vapor pressure above the surface. If the temperature is raised sufficiently, bubbles of vapor will form in the liquid and rise to the surface. When this happens, we say that the liquid is boiling.

We find experimentally that the temperature at which a liquid boils depends upon the pressure above it. To understand why this is the case, let us refer to Figure 10.6. For a vapor bubble to form within the liquid, the pressure within the bubble, P_1, must be at least equal to the pressure above the liquid, P_2.*

---

* Actually, for the bubble to be stable, the pressure within it must be slightly greater than the total pressure at the point where the bubble is formed, which in turn will be slightly greater than the pressure at the surface.

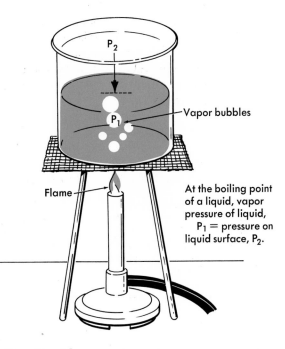

Figure 10.6   A boiling liquid.

At the boiling point of a liquid, vapor pressure of liquid, P_1 = pressure on liquid surface, P_2.

$$P_1 = P_2 \quad \text{for a boiling liquid}$$

But P_1 is simply the vapor pressure of the liquid. Consequently, we deduce that **a liquid boils at a temperature at which its vapor pressure becomes equal to the pressure above its surface**. If the pressure at the surface is constant, the boiling point of the liquid must also be constant. If the pressure is equal to one standard atmosphere (760 mm Hg), the liquid boils at a temperature referred to as its **normal boiling point**. In the case of water, the vapor pressure reaches 760 mm Hg at a temperature of exactly 100°C. The normal boiling point of benzene, 80.1°C, is somewhat lower; that of oxygen, −183°C, is a great deal lower.

Since a liquid will boil whenever its vapor pressure is equal to that at its surface, the boiling point can be changed by a change in the applied pressure. Both water and benzene will boil at room temperature in a container which is being evacuated with a vacuum pump. Chemists often purify compounds which are unstable at their normal boiling points by boiling them at low pressures (and, hence, low temperatures) and condensing their vapors. Of course it is possible to raise the boiling point of a liquid by increasing the applied pressure. At 2 atm pressure water boils at about 120°C. The sensible housewife often uses a pressure cooker, a device in which the pressure on water can be raised to over 1 atm. Under the conditions in the cooker, the water rises to a temperature above 100°C and the cooking reaction takes place more rapidly. It should now be obvious why, when the boiling point of a liquid is recorded, the pressure on the liquid must also be recorded.

The normal boiling point of a liquid is, of course, always lower than its critical temperature. However, if we examine a series of liquids, we find that these two temperatures are directly related to each other. For most liquids, we find that

$$\text{normal boiling point (°K)} \approx \tfrac{2}{3} \times \text{critical temperature (°K)} \qquad (10.5)$$

Could you boil a liquid by heating it in a sealed container?

Does the water in a pressure cooker actually boil?

We can also relate, at least qualitatively, the normal boiling point of a liquid to its vapor pressure at room temperature. Liquids such as ether, which have a high vapor pressure at 20°C, will reach a vapor pressure of 760 mm Hg at a comparatively low temperature. Relatively nonvolatile liquids such as mercury must be heated to a high temperature to reach this same vapor pressure.

TABLE 10.2 PHYSICAL PROPERTIES OF LIQUIDS

	VAPOR PRESSURE AT 20°C		CRITICAL TEMPERATURE	NORMAL BOILING POINT	HEAT OF VAPORIZATION	$\dfrac{\Delta H_{vap}}{T_b}$
Mercury	0.0012	mm Hg	1750°K	630°K	14,200 cal/mole	22.0
Water	17.5	mm Hg	647°K	373°K	9,700 cal/mole	26.0
Benzene	75	mm Hg	562°K	353°K	7,300 cal/mole	20.7
Bromine	75	mm Hg	561°K	332°K	7,160 cal/mole	21.6
Ether	442	mm Hg	467°K	308°K	6,200 cal/mole	20.1
Ethane	27000	mm Hg	305°K	184°K	3,700 cal/mole	20.1
Oxygen	————		154°K	90°K	1,630 cal/mole	18.2

As Table 10.2 indicates, there is also a correlation between the heat of vaporization of a liquid and its normal boiling point. For many liquids, we find that the heat of vaporization (cal/mole) is approximately 21 times the normal boiling point.

$$\frac{\Delta H_{vap}}{T_b} \approx 21 \text{ cal/mole } °K \tag{10.6}$$

This generalization, known as **Trouton's rule**, holds to within 5 to 10 per cent for most liquids that are not hydrogen-bonded. For water and other liquids in which hydrogen bonding makes a major contribution to the intermolecular forces, the heat of vaporization is considerably higher than that calculated by Trouton's rule.

10.3 NATURE OF THE SOLID STATE

The density of a substance in the solid state is ordinarily somewhat greater than that of the liquid (Table 10.3). This implies that the particles in the solid are slightly closer together than they are in the liquid. The difference is small; the increase in density on solidification seldom exceeds 5 to 10 per cent. A few substances, of which water is the most notable example (cf. Chapter 9, Section 9.2), expand on freezing because of the formation of a relatively "open" solid structure.

The fact that heat must be absorbed to melt a solid implies that the attractive forces between particles are stronger in the solid than in the liquid state. It should be noted, however, that the heat absorbed on melting, referred to as the **heat of fusion**, is ordinarily only a small fraction of the heat of vaporization (Table 10.4). This indicates that the difference in attractive forces in the solid and liquid states is relatively small; the major portion of these forces is

248

TABLE 10.3   DENSITIES OF SOLIDS VS. LIQUIDS[*]

	SOLID (G/ML)	LIQUID (G/ML)
Mercury	14.2	13.6
Benzene	1.005	0.894
Sodium chloride	2.0	1.55
Water	0.917	1.000
Bromine	3.9	3.2

[*] Densities given at the melting point.

not overcome until the molecules pass into the vapor, in which they are widely separated from one another.

The most obvious structural difference between solids on the one hand and liquids on the other is the rigidity characteristic of the solid state. While the particles in a liquid have sufficient translational energy to allow them to move past one another, motion of this type cannot ordinarily occur in a solid. In a perfect crystal, the particles are restricted to vibrating about a fixed point. The amplitude of this vibration increases with temperature, but never becomes great enough for one particle to slip past another.

How would the rate of diffusion of a solid compare to that of a liquid?

TABLE 10.4   SOME TYPICAL HEATS OF FUSION AND VAPORIZATION[*]

	HEAT OF FUSION (KCAL/MOLE)	HEAT OF VAPORIZATION (KCAL/MOLE)	$\Delta H_{fus}/\Delta H_{vap}$
Sodium chloride	6.9	43	0.16
Bromine	2.6	9.0	0.29
Benzene	2.55	8.27	0.31
Water	1.44	10.7	0.13
Mercury	0.56	13.5	0.04

[*] Values given at the melting point.

Crystal Structure

Solids tend to crystallize in definite geometric forms which can frequently be seen by the naked eye. In ordinary table salt, we can distinguish small, cubic crystals of sodium chloride; large, beautifully formed crystals of such minerals as quartz (SiO_2) and fluorite (CaF_2) are found in nature. Even in finely powdered solids, it is ordinarily possible to observe distinct crystal forms under a microscope.

The existence of crystals with distinct geometric forms implies that the particles making up the crystal are arranged in a definite, three-dimensional pattern. One can sometimes deduce what this pattern must be from the crystal form. A great deal more information concerning the packing of particles can be obtained by x-ray diffraction, which is one of the most powerful structural tools available to the chemist.

X-rays are diffracted by a crystal in somewhat the same way that ordinary light is spread out by a diffraction grating. Those of you who have had a course

249

in physics will recall that visible light striking a glass plate ruled with a large number of closely spaced parallel lines is broken up by diffraction into a series of beams, provided the spacing between the lines is of the same order of magnitude as the wavelength of the light. In 1912 Von Laue suggested that crystals, which are made up of layers of particles a few angstroms apart, should act as a diffraction grating for x-rays, which have wavelengths of 0.1 to 10 Å. This reasoning was soon confirmed experimentally. Figure 10.7 shows a simple apparatus used in x-ray diffraction measurements.

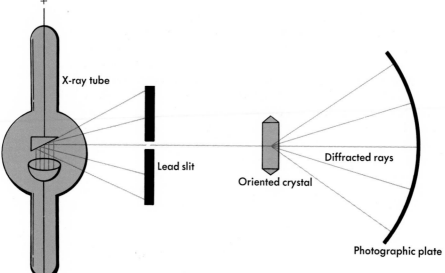

Figure 10.7 X-ray diffraction from a single crystal.

In 1913 W. L. Bragg adapted the technique of x-ray diffraction to the determination of crystal structures. The principle he used is embodied in the so-called Bragg equation which relates the angle of diffraction to the distance between successive layers of atoms or ions in the crystal:

$$\sin \theta = \frac{n\lambda}{2d} \tag{10.7}$$

In this equation, θ is the angle between the beam and the layer of atoms, λ is the wavelength of the x-rays, d is the distance between successive layers, and n is an integer (1, 2, 3, . . .) known as the order number of the particular diffracted beam.

By measuring the angles at which x-rays of known wavelength are diffracted by a crystal, we can calculate from Equation 10.7 the distances between planes of atoms or ions. In this way, we can often obtain values for atomic and ionic radii. It is more difficult to deduce the geometric pattern in which the particles are arranged. The basic problem is that the x-ray beam "sees" not one but many different series of layers of particles oriented at various angles to each other. Consequently, an x-ray beam, in passing through a crystal, is broken up into a large number of diffracted beams. Nevertheless, by studying the relative intensities of the various beams, x-ray crystallographers have been able to unravel the particle structure of a wide variety of crystals.

Unit Cells

The basic information which comes out of x-ray diffraction studies concerns the dimensions and geometric form of what is known as the **unit cell**, the smallest unit which, repeated over and over again in three dimensions, generates the crystal lattice. Perhaps the simplest unit cell to visualize is the **simple cubic cell**, which consists of eight atoms (or molecules, or ions) located at the corners of a cube. Two other types of cubic cells are shown in Figure 10.8. One of these, in which there is an atom at the center of the cube, is referred to as a **body-centered cubic cell**. The third cell, in which there are atoms at the center of each face of the cube, is called a **face-centered cubic cell**.

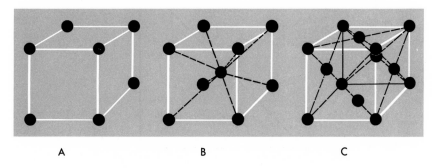

Figure 10.8 Cubic packing of various metals. *A*, simple; *B*, body-centered; *C*, face-centered.

In attempting to visualize how a crystal lattice is built up from a unit cell, two limitations of the simple diagrams shown in Figure 10.8 must be kept in mind.

1. The atoms in a unit cell are not separated to the extent that Figure 10.8 might imply. In a simple cubic cell, identical atoms touch along the edge of the cube. In a face-centered cubic cell, the atoms are touching along a face diagonal; that is, two atoms at opposite corners of a face touch the atom in the center of the face. In a body-centered cubic cell, there is contact along a diagonal passing through the center of the cube; that is, two atoms at opposite corners of the cube touch the atom at the center.

2. Certain of the atoms shown in the unit cell do not belong exclusively to that cell. When the crystal lattice is built up, these atoms are shared by adjacent cells. For example, an atom at the corner of a cubic cell will form a part of eight different cubic cells that touch at that point. In this sense, only $\frac{1}{8}$ of such an atom should be assigned to a particular unit cell. An atom in the face of a cubic cell is shared by a second cube adjacent to the first; only $\frac{1}{2}$ of it should be assigned to a particular cell, and so on.

The number of atoms to be assigned to each of the unit cells shown in Figure 10.8 is as follows:

An atom located along the edge of a cube would be shared by how many cubes?

Simple cubic: 8 atoms $\times \frac{1}{8} = 1$ atom per unit cell.
Face-centered cubic: 8 atoms $\times \frac{1}{8} + 6$ atoms $\times \frac{1}{2} = 4$ atoms per unit cell.
Body-centered cubic: 8 atoms $\times \frac{1}{8} + 1$ atom $= 2$ atoms per unit cell.

EXAMPLE 10.3. Copper crystallizes in a structure with a face-centered cubic unit cell 3.63 Å on an edge.

        a. Calculate the atomic radius of copper.

        b. Calculate the density of copper.

SOLUTION.

        a. It will be helpful here to draw a face of the unit cell.

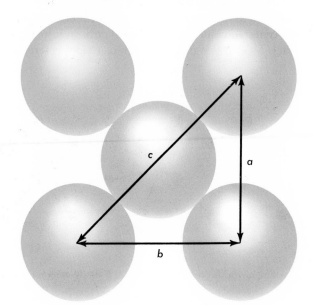

The distances a and b are each 3.63 Å. The distance c by the Pythagorean theorem is

$$c = \sqrt{2}a = (1.41)(3.63 \text{ Å}) = 5.12 \text{ Å}$$

But, from the figure, c = 4r, where r is the radius of the metal atom.

$$4r = 5.12 \text{ Å}; \; r = 1.28 \text{ Å}$$

        b. In order to calculate the density of copper, we note that there are four atoms per unit cell in a face-centered cube.

$$\text{mass 4 copper atoms} = 4 \text{ atoms} \times \frac{63.5 \text{ g}}{6.02 \times 10^{23} \text{ atoms}} = 42.2 \times 10^{-23} \text{ g}$$

$$\text{volume of unit cell} = (3.63 \text{ Å})^3 = 47.8(\text{Å})^3 = 47.8 \times 10^{-24} \text{ cc}$$

$$\text{density} = \frac{42.2 \times 10^{-23} \text{ g}}{47.8 \times 10^{-24} \text{ cc}} = 8.83 \text{ g/cc}$$

(The observed density of copper is 8.92 g/cc at 20°C.)

Types of Packing

When the atoms that form a crystal lattice are all of the same size, as is the case with elementary substances, there is a tendency for them to pack as closely as possible. Consequently, we seldom find metals or other elements crystallizing in a simple cubic structure, which leaves a great deal of empty space between the atoms. A body-centered cubic structure, in which each atom has eight nearest neighbors, represents a more closely packed structure with less "wasted space" between the atoms. Approximately 20 different metals crystallize in a pattern with a body-centered cubic unit cell. A still more efficient method of packing identical atoms is represented by the face-centered cubic structure, in which each atom touches 12 other atoms. Face-centered cubic unit cells are characteristic of about 24 metals.

There are actually two different ways in which one can pack atoms of equal radius most efficiently, that is, leaving as little empty space as possible. One of these "closest-packed" arrangements is equivalent to the face-centered cubic structure discussed above. The other, which is equally efficient, is known as an **hexagonal close-packed** structure. It is characteristic of 30 different metals.

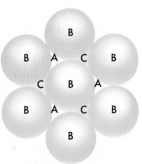

Figure 10.9 Packing pattern of atoms in closest-packed structures.

Closest packing: hexagonal = ABA
cubic   = ABC

To understand why there are two closest-packed arrangements, consider Figure 10.9. Here a portion of a single layer of spheres, packed as tightly as possible, is indicated by the circles marked B. A second closest-packed layer can be superimposed on the first if we place spheres so that their centers lie directly above the points labeled A. There are now two different ways in which we can add a third layer of spheres, below the plane of the paper:

1. The spheres can be placed so that they line up exactly with the top layer; i.e., their centers fall directly below the points labeled A. In this structure, the pattern repeats itself every two layers; one might represent this as ABABAB. . . . This arrangement is called **hexagonal closest-packing**.

2. The spheres in the third layer can be placed so that their centers lie directly beneath the points marked C. In this structure the pattern repeats itself every three layers; the arrangement can be represented as ABCABC. . . . It can be shown that this structure gives a face-centered cubic unit cell; it is sometimes referred to as **cubic closest-packing**.

In Figure 10.10 an attempt has been made to indicate the three-dimensional appearance of crystals in which there is cubic closest-packing. Note particularly the face-centered cubes that are outlined in the cubic close-packed structure. The closest-packed layers in this arrangement are made visible by cutting away a corner of the structure.

The geometry of ionic crystals, in which there are two different kinds of ions of different radius, is necessarily more complicated than that of metallic

253

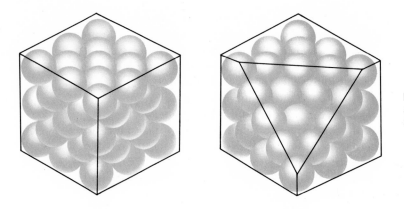

Figure 10.10 Arrangement of atoms in cubic closest packing.

crystals. However, it is possible to visualize the packing in certain ionic crystals in terms of the simple types of unit cells discussed above. A particularly simple case is the lithium chloride crystal, where the large Cl^- ions (r = 1.81 Å) form a face-centered cubic lattice with the small Li^+ ions (r = 0.60 Å) fitting snugly into holes in the lattice, one Li^+ ion per Cl^- ion (Figure 10.11).

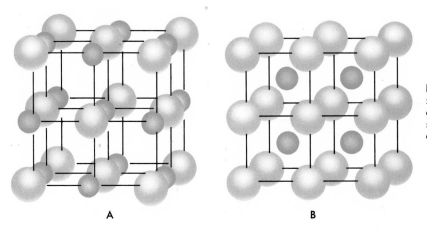

Figure 10.11 Arrangement of ions in lithium chloride (A) and cesium chloride (B) crystals.

A   B

The structure of a crystal of sodium chloride is similar to that of LiCl, except that the Na^+ ions (0.95 Å) are slightly too large to fit into a close-packed lattice composed of Cl^- ions touching each other along the faces of a cube. Consequently the face-centered cubic array of Cl^- ions found in LiCl is slightly expanded to accommodate the Na^+ ions. Even though the Cl^- ions are no longer touching each other, the symmetry of the face-centered cubic structure is maintained.

The anions cannot touch when the radius ratio of cation to anion is greater than 0.41 (cf. Problem 8.38).

In a crystal of cesium chloride, the cations are too large (r of Cs^+ = 1.69 Å) to fit into a face-centered cubic array of Cl^- ions. Instead, cesium chloride crystallizes in the body-centered cubic structure shown at the right of Figure 10.11. We can visualize this structure as an arrangement of Cl^- ions in a simple cubic array with Cs^+ ions located at the center of each cube.

EXAMPLE 10.4. Calculate the distance between adjacent Cl^- ions in:
a. LiCl   b. CsCl

SOLUTION.

a. In LiCl the Cl^- ions are touching. The distance between the centers of these ions should be twice the ionic radius of Cl^-.

$$d = 2 \, r \, Cl^- = 3.62 \text{ Å}$$

b. In CsCl the Cl^- ions do not touch each other. Instead, there is cation-anion contact along the diagonal of the cube. The length of the diagonal (l) is obtained by multiplying the radius of the central Cs^+ ion by two and adding the radii of the two Cl^- ions at the corners of the cube.

$$l = 2rCs^+ + rCl^- + rCl^-$$
$$= 2(1.69 \text{ Å}) + 2(1.81 \text{ Å}) = 7.00 \text{ Å}$$

Using the relation between the side of a cube s and the length of its diagonal l, we find that

$$l = \sqrt{3} \, s; \quad s = l/\sqrt{3} = \frac{7.00 \text{ Å}}{1.73} = 4.05 \text{ Å}$$

This distance, 4.05 Å, must represent the separation between the centers of two Cl^- ions located at the corners of the cube. Note that this distance is somewhat larger than in LiCl (3.62 Å), where the anions are touching each other.

Ionic crystals in which there are unequal numbers of positive and negative ions introduce a further complication so far as their geometry is concerned. A common structure for ionic crystals in which there are two ions of one type for every ion of opposite charge is that shown in Figure 2.2 (Chapter 2) for CaF_2. This structure is similar to that of CsCl, except that there are only half as many cations. Only half the cubes have Ca^{2+} ions at their center; the others are "empty."

Crystal Defects

A perfect crystal, in which all of the particles are exactly where they are supposed to be and every lattice site is occupied by the proper particle, is an extremely useful model for discussing the structure of solids. However, a perfect crystal, like an ideal gas, is an abstraction; the crystals we work with in the laboratory contain imperfections, no matter how carefully they are prepared. These defects, even if relatively few in number, can profoundly affect the physical and chemical properties of a solid substance.

The three most important types of crystal defects are

1. **Vacant sites** in the crystal lattice.

2. **Interstitial** atoms or ions, i.e., particles occupying positions other than those to be found in a perfect lattice.

3. **Foreign atoms or ions** incorporated into the crystal lattice.

Ionic Crystals. Two types of defects commonly found in ionic crystals are illustrated in Figure 10.12. In one of these, known as a Frenkel defect, an ion is displaced from its normal site to an interstitial position. Ordinarily it is

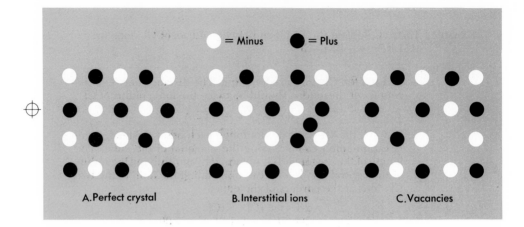

Figure 10.12 A two-dimensional representation of some common defects in ionic lattices.

the cation which moves, since cations are usually smaller than anions. The migration of a cation and an anion from the interior to the surface of the crystal, leaving two vacancies in the lattice, leads to what is known as a Schottky defect.

From the standpoint of chemistry, the most important effect of Frenkel and Schottky defects is their tendency to promote diffusion of atoms or ions through the crystal lattice. Reactions involving solids usually proceed by a diffusion-controlled mechanism whose rate is increased tremendously by the presence of lattice vacancies. One such reaction which has been studied extensively is that between silver and molten sulfur. It has been shown that the silver sulfide formed in this reaction is "permeable" to silver atoms. That is, silver atoms are able to diffuse through the silver sulfide so as to come in contact with sulfur atoms and react with them. Were it not for diffusion, the reaction would stop as soon as a thin layer of silver sulfide formed at the silver-sulfur interface.

Another type of defect that is found in many of the sulfides and oxides of the transition metals is illustrated at the left of Figure 10.13. In nickel oxide that has been prepared by heating in an excess of oxygen at high temperatures, it is found that there are occasional lattice sites from which Ni^{2+} ions are missing. In removing a Ni^{2+} ion from the lattice one must, of course, maintain electrical neutrality. One way to do this is to convert two Ni^{2+} ions in the vicinity of a vacancy to Ni^{3+} ions. When this happens, the crystal becomes non-stoichiometric in the sense that it no longer contains equal numbers of moles of nickel and oxygen. We might, for example, have 97 Ni^{2+} and 2 Ni^{3+} ions for every 100 O^{2-} ions, in which case the "simplest formula" would be $Ni_{99}O_{100}$, which is slightly different from the predicted stoichiometric formula, NiO. Very careful analysis of nickel oxide crystals of this type shows that the atom ratio of nickel to oxygen is indeed slightly smaller than 1:1.

Crystals such as that described above have electrical conductivities that are considerably higher than those associated with perfect ionic lattices. Electrical conductivity in non-stoichiometric nickel oxide is believed to be caused primarily by the ease with which an electron can be transferred from a Ni^{2+}

to a Ni^{3+} ion. In practice, however, crystals of this type are of limited value as semiconductors because it is extremely difficult to produce large deviations from stoichiometry. Nonstoichiometric crystals are black because of ready absorption of light by their rather free electrons. Manganese dioxide, MnO_2, a major constituent of dry cell batteries, is another such non-stoichiometric compound and is exceedingly black. Many other oxides and sulfides of the transition metals behave in this way.

A more effective way of increasing the electrical conductivity of an ionic crystal is to introduce certain foreign ions into it. At the Philips Laboratories in Holland, Verwey has produced crystals with a conductivity 10 billion times that of the pure material by introducing Li^+ ions into nickel oxide. Since the radii of Li^+ and Ni^{2+} are so similar (0.70 Å vs. 0.74 Å), it is possible to substitute large quantities of Li^+ ions for Ni^{2+} ions (up to 10 per cent or more). Every time a Li^+ ion is introduced into the crystal, a Ni^{2+} ion is converted to Ni^{3+} so as to maintain electroneutrality (Figure 10.13). Here, as in non-stoichiometric nickel oxide, exchange of electrons between Ni^{2+} and Ni^{3+} ions is believed to be responsible for the enhanced conductivity.

If 10 out of 100 Ni^{2+} ions were replaced by Li^+, what would be the ratio of Ni^{3+} to Ni^{2+}?

Figure 10.13 Crystal defects in nickel oxide.

Ni^{2+}	O^{2-}	Ni^{3+}	O^{2-}	Ni^{2+}	O^{2-}	Ni^{2+}	O^{2-}
O^{2-}		O^{2-}	Ni^{2+}	O^{2-}	Li^+	O^{2-}	Ni^{2+}
Ni^{3+}	O^{2-}	Ni^{2+}	O^{2-}	Ni^{3+}	O^{2-}	Ni^{2+}	O^{2-}
O^{2-}	Ni^{2+}	O^{2-}	Ni^{2+}	O^{2-}	Ni^{2+}	O^{2-}	Ni^{2+}

Macromolecular Crystals

Extremely pure samples of germanium and silicon are intrinsic semiconductors, but the degree of conductivity is very low because electrons must be displaced from covalent bonds. This situation changes drastically when small numbers of foreign atoms (as few as 1 in 10^6) are introduced into the crystal. In order to understand how this comes about, let us examine how the crystal structure of a 4A element is disturbed by the introduction of atoms of a 5A or 3A element.

An atom of arsenic or antimony with its five valence electrons can fit into the tetrahedral lattice of germanium or silicon because it has a similar size. If it gives up its extra electron, this electron is relatively free to move throughout the crystal under the influence of an electrical field, giving rise to an **n-type semiconductor** (current carried by the flow of negative charge). If an atom of boron or other element with three valence electrons is introduced into the lattice, a quite different situation arises. An electron deficiency is created at the site occupied by the foreign atom; it is surrounded by seven valence electrons rather than eight. Physicists describe the introduction of such a defect in terms of the formation of a "positive hole" in the lattice. In an electrical field, an electron moves from a neighboring atom to fill this hole. In so doing, it creates an electron deficiency around the atom which it leaves. Conduction in this type of defect crystal, a **p-type semiconductor**, can be thought of as a movement of positive holes through the lattice.

257

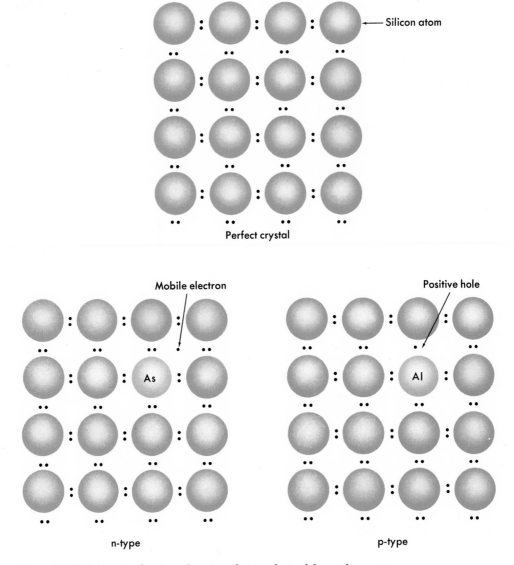

Figure 10.14   Schematic drawing of semiconductors derived from silicon.

A semiconductor in which there is a junction between an electron-rich and an electron-deficient region acts as a rectifier, capable of converting alternating to direct current. Such an "n-p" junction can be formed by starting with a pure silicon disc and introducing a trace of boron on one side of the disc and a trace of arsenic on the other. These impurities are allowed to diffuse into the silicon at high temperatures until the region of p-type semiconductor meets the region of n-type semiconductor. Electrons flow readily from the electron-rich region containing arsenic to the electron-deficient area created by the presence of boron atoms. Potentials as high as 1000 volts are incapable of bringing about electron flow in the opposite direction.

Perhaps the best known of all semiconductor devices is the transistor, in which n-p-n or p-n-p junctions are created by forming a sandwich of alternate electron-rich and electron-poor regions. A transistor amplifies an electric current in much the same way as a vacuum tube. The small size of transistors makes them ideal for use in hearing aids, miniature radios, and electronic components of missiles or satellites. Their increasing use in television and radio receivers, in which they compete with vacuum tubes, reflects the greater durability and lower energy consumption of transistors.

The electrical conductance or, conversely, the resistance of all types of semiconductors is extremely sensitive to temperature. The resistance of a typical semiconductor decreases by 4 per cent per °C at room temperature; that of copper wire increases by only one tenth of that amount under the same conditions. Semiconductor devices known as thermistors have been developed to take advantage of this property. Accurate resistance readings taken in a circuit including thermistors make it possible to estimate temperature changes to 0.0001°C or better. The small size of thermistors makes them suitable for temperature measurements in restricted spaces. A thermistor inserted into the tip of a glass probe can be used to measure temperatures in various parts of the human body. Thermistors can be taped to the skin or soldered to metal surfaces to measure surface temperatures.

Quite recently it has been found that a semiconductor whose surface contains an n-p junction can convert radiant energy to electrical energy. This principle is used in the solar battery, which promises to be an important power source in the interplanetary space vehicles of the future. Light striking a crystal of germanium or silicon impregnated with an electron-rich impurity such as arsenic ejects some of the loosely held electrons. These electrons are collected at the surface, which is coated with a thin, transparent layer of a p-type semiconductor. The electrons then pass through an electrical circuit where their energy is used to do useful work.

10.4 SOLID-VAPOR EQUILIBRIUM

We ordinarily think of the transition from the solid to the vapor state as occurring in two steps, the melting of the solid to a liquid followed by vaporization of the liquid. However, it is entirely possible for a solid to pass directly to the vapor state; this type of phase change is referred to as **sublimation**. All of us have seen this process occur with solid carbon dioxide (Dry Ice), which seems to disappear when exposed at room temperature. Iodine is another substance which sublimes readily; if crystals of iodine are placed in a stoppered bottle at room temperature, one can observe the purple color of iodine vapor as it forms above the crystals. It is perhaps less generally realized that the substance water can and often does sublime. On a winter day when the temperature is well below freezing, snow passes directly to water vapor without going through the liquid as an intermediate. High quality dehydrated foods are now being prepared by the process known as freeze-drying, in which water vapor is removed from the frozen material at very low temperatures and pressures.

Vapor Pressure of Solids

A solid placed in a closed container at a constant temperature will establish equilibrium with its vapor. When equilibrium is reached, we find that the vapor exerts a constant, characteristic pressure, referred to as the vapor pressure (or sometimes the sublimation pressure) of the solid. The vapor pressure of a solid, like that of a liquid, increases with temperature. To study the variation of vapor pressure with temperature, we can use the same apparatus employed with liquids (Figure 10.2). If we introduce solid benzene at a temperature of, let us say, 0°C, we find that upon reaching equilibrium, it establishes a vapor pressure of 24 mm Hg. Upon raising the temperature to 5.5°C, the vapor pressure of the solid rises to 36 mm Hg; at temperatures below 0°C, the vapor pressure decreases slowly, reaching 10 mm Hg at −12°C. In Figure 10.15 we have shown the vapor pressure curves for both solid and liquid benzene. It will be noted that the two curves have the same general form. The vapor pressure curve of the solid is, of course, displaced toward lower temperatures and pressures.

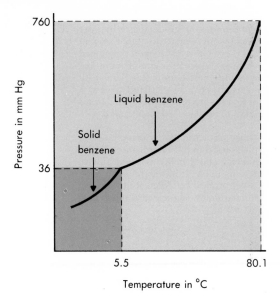

Figure 10.15 Vapor pressures of solid and liquid benzene.

Heat of Sublimation

The amount of heat absorbed when a solid passes directly to the vapor state at a constant pressure is referred to as its heat of sublimation. We find experimentally that for a given amount of a substance at a particular temperature, the heat of sublimation (ΔH_{subl}) is equal to the sum of the heat of fusion (ΔH_{fus}) plus the heat of vaporization (ΔH_{vap})

$$\Delta H_{subl} = \Delta H_{fus} + \Delta H_{vap} \tag{10.8}$$

For 1 mole of benzene at 5.5°C, these quantities are respectively 10.82 kcal, 2.55 kcal, and 8.27 kcal.

Equation 10.8 illustrates the principle referred to earlier that the heat change (more exactly, the enthalpy change) for a process depends only upon the final and initial states and not upon the path. Thus, the heat required to convert solid benzene to benzene vapor at 5.5°C and 36 mm Hg is independent of whether this phase change occurs directly (sublimation) or passes through the liquid as an intermediate (fusion and evaporation).

10.5   SOLID-LIQUID EQUILIBRIUM

Referring to Figure 10.15, we note that the vapor pressure curves of solid and liquid benzene intersect at a temperature of 5.5°C and a pressure of 36 mm Hg. This is referred to as the **triple point** of benzene because it is the only temperature and pressure at which solid, liquid, and vapor can exist in equilibrium with each other in a system containing only pure benzene.

Each pure substance has a characteristic triple point. That of water is 0.01°C and 4.58 mm Hg, the point at which ice, liquid water, and water vapor are in equilibrium with each other. The triple point of carbon dioxide occurs at −56.6°C and 5.1 atm. This explains why a piece of Dry Ice, exposed to the atmosphere in a warm room, sublimes instead of melting; its sublimation pressure reaches 1 atm at a temperature of −78°C, far below the triple point. In general, substances which have high triple point pressures, such as carbon dioxide, iodine, and naphthalene, are readily sublimed.

Although the solid, liquid, and gaseous phases of a pure substance can be in equilibrium with each other at only one temperature, given by the triple point, equilibrium between the solid and liquid phases of a pure substance at other temperatures can be attained by increasing the external pressure. Consider, for example, a sample of benzene, originally at the triple point temperature and pressure, 5.5°C and 36 mm Hg. If we increase the pressure on the benzene we observe that (1) the vapor phase disappears, and (2) in order to maintain equilibrium between the solid and liquid phases, the temperature must be raised slightly. If the pressure is increased to one atmosphere, we find that the temperature has to be increased by about 0.026°C to maintain equilibrium between solid and liquid benzene. At an external pressure of 50 atm the equilibrium temperature is 1.3°C higher than the triple point, or 6.8°C.

The temperature at which the liquid and solid phases of a pure substance are in equilibrium with each other at a given pressure is referred to as the **melting point** of the solid, or, alternatively, the **freezing point** of the liquid. Melting points are ordinarily measured in an open container at atmospheric pressure. Consequently, the melting points which we find recorded in the literature will ordinarily represent the temperature at which solid and liquid are in equilibrium at an external pressure of 1 atm. For most substances, the melting point under these conditions is virtually identical with the triple point. The temperature at which ice and water are at equilibrium with each other in an open container in contact with air is defined as exactly 0°C; it will be recalled that the triple point of water is +0.01°C.

Even though the effect of pressure upon the melting point of a substance is very small, we are often interested in the direction of this effect. We would like to be able to predict, without resort to experiment, whether the melting point of a substance will be increased or decreased by increasing the external pressure. This can be done very simply by applying the principle that an **increase in pressure will favor the formation of the more dense phase**. In the case of benzene, in which the solid is more dense than the liquid (Table 10.3), an increase in pressure favors the solid. This means that at higher pressures, the solid will become stable at temperatures above the normal melting point; i.e., the melting point will be raised by an increase in pressure. The behavior of benzene is typical of that of most substances, since the solid phase is usually more dense than the liquid.

We recall that water is unusual in that the liquid phase is more dense than the solid; an ice cube floats in a glass of water. We would then predict that the melting point of ice should decrease with pressure; i.e., at high pressures the liquid should be stable at temperatures below the normal melting point. This is confirmed experimentally; the melting point of ice decreases by about 1°C for every 134 atm of applied pressure. This effect can be demonstrated strikingly by suspending two heavy weights from a wire stretched across a block of ice. In time, the wire passes completely through the block of ice, which appears to be unchanged. What happens is that the pressure

Recall the effect of pressure on the graphite-diamond transition (Chapter 9).

CHAPTER 10

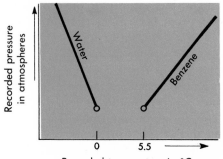

Figure 10.16 Effect of pressure on the melting points of water and benzene.

exerted by the weights melts a thin layer of ice around the wire. As the wire falls, the pressure above it drops and the ice re-forms.

10.6 PHASE DIAGRAMS

We have now completed our discussion of the various possible equilibria that are observed in a system containing one pure substance. It is possible to summarize all that has been said in a graphic way, by means of what is known as a phase diagram. In Figure 10.17 the phase diagram for benzene (a typical pure substance), is illustrated. One can regard this diagram as being constructed by putting together the various curves denoting solid, liquid, and vapor relationships for pure benzene.

The coordinates in a phase diagram are the temperature and the pressure (the scales of both coordinates have been adjusted to bring all the important points on the diagram). The purpose of the diagram is to describe the state in which benzene will exist under any given conditions of temperature and pressure. Properly interpreted, the diagram does this very satisfactorily. For example, let us determine the state in which benzene would be found at 200°C and 2 atm pressure. We locate the point on the diagram corresponding to these conditions (A) and note that it is in the region labeled Vapor; this means that benzene under such conditions will exist only as a vapor. Any set of conditions of temperature and pressure falling in the vapor region will find benzene occurring in only one phase and that will be vapor. Similarly, benzene in any state lying in the regions labeled Liquid or Solid (B or C, for example) will exist solely as a liquid or a solid respectively. For sets of conditions lying on the lines in the phase diagram, benzene can exist in two phases in equilibrium. Along the line separating the liquid and vapor regions, systems containing benzene in both the liquid and the vapor phases can exist; this line is indeed the vapor pressure curve for liquid benzene and can be used to find the vapor pressure at any temperature. The normal boiling point of a liquid will always be on that line at the temperature corresponding to a vapor pressure of one atm. For benzene this point is at 80.1°C (D). The vapor pressure curve ends at the critical temperature of the liquid, since above that temperature the liquid does not exist, and of course has no vapor pressure. On the diagram this is E (289°C and 47.90 atm).

Along the solid-vapor line two-phase equilibria involving solid and gaseous benzene can occur. This line is the sublimation pressure curve for ben-

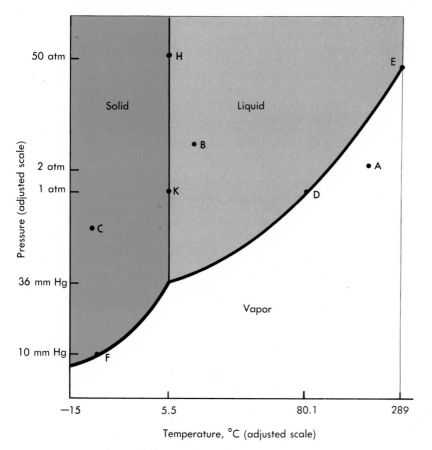

Figure 10.17   The phase diagram for benzene.

zene and from it we can say that the sublimation pressure of benzene at −12°C is about 10 mm Hg (F). The solid-vapor and liquid-vapor equilibrium lines meet at the triple point of benzene, 5.5°C and 36 mm Hg, where solid, liquid, and vapor coexist in equilibrium. Clearly this is the only point at which this can occur. Rising nearly vertically from the triple point is the solid-liquid line, essentially the melting-point curve for benzene as a function of total pressure. The line is nearly straight and inclined slightly to the right. We see that the melting point of benzene at 1 atm (K) is about 5.5°C and at 50 atm (H) is about 6.8°C.

Within an area, only one phase can exist; at the boundary between two areas, two phases are in equilibrium.

EXAMPLE 10.5.   Using Figure 10.17, describe what happens when:
    a. A sample of benzene, originally at 0°C, is heated at a constant pressure of 1 atmosphere.
    b. A sample of benzene, originally at 0°C, is heated at a constant pressure of 50 atmospheres.
    c. A sample of benzene, originally at a pressure of 10 mm Hg, is compressed at a constant temperature of 20°C.

SOLUTION.
    a. The point corresponding to 0°C and 1 atm is located in the solid region of the phase diagram. Heating this system at

263

constant pressure corresponds to moving horizontally across the diagram. Between 0°C and 5.5°C, we are simply supplying heat to raise the temperature of the solid. If we supply the heat of fusion at 5.5°C (point K), the sample is melted. The temperature of the liquid is then increased until we arrive at point D, corresponding to 80.1°C, whereupon the liquid vaporizes. Further heating simply raises the temperature of the vapor.

b. The description here is similar to that in a. The sample melts at a slightly higher temperature, about 6.8°C. The liquid boils at a considerably higher temperature, 289°C, the critical temperature of benzene.

c. The point corresponding to 10 mm Hg and 20°C is located in the vapor region; the sample starts off as benzene vapor. Compressing the system at constant temperature means that we move upward along a line parallel to the vertical axis. As we do so, the volume of the gas steadily decreases. When we reach the vapor-liquid boundary at a pressure of about 75 mm Hg, the liquid suddenly appears. Condensation occurs at this constant pressure until all of the vapor is gone. At pressures above 75 mm Hg we are dealing with a system that is entirely liquid; increasing the pressure simply compresses the liquid slightly.

The phase diagrams for other pure substances are similar in appearance to that for benzene and are interpreted in the same way. Depending on whether the solid is more or less dense than the liquid, the slope of the solid-liquid equilibrium line will be positive, as it is for benzene, or negative, as it is for water.

10.7 NONEQUILIBRIUM PHASE BEHAVIOR

The phase changes we have considered in this chapter have been assumed to take place under equilibrium conditions. In practice, the phase changes that we observe in the laboratory or in the world around us occur at temperatures and pressures at least slightly removed from equilibrium. When experiments of the type we have described are done, we frequently find that the systems do not behave in quite the manner predicted by the phase diagram. A liquid being vaporized may not boil smoothly, particularly at low pressures. Instead, it may superheat to temperatures above the calculated boiling point and then "bump" and boil furiously. A very pure liquid on cooling does not start to freeze exactly at its freezing point; it has to be supercooled below that temperature before crystallization occurs. Cooling a vapor to a temperature at which its pressure slightly exceeds the vapor pressure may not bring about condensation.

Nonequilibrium behavior is thought to be caused by the system being unable to respond fast enough to maintain equilibrium. For a cooled liquid to crystallize, some centers, or nuclei, on which crystallization can occur are needed. These centers may be dust particles, other solid impurities, or small crystals of the substance itself. In a very pure liquid, such centers will be missing and the liquid will supercool. As an extreme example, very pure water can be supercooled to as low as −40°C without freezing.

For a liquid to boil, there must be centers at which bubbles can form. These centers may be microscopic bubbles of dissolved gas, dust particles, or sharp crystal edges or corners. When a liquid is to be distilled, a small quantity of an inert solid such as marble or porcelain is usually added. These solids offer a large surface on which vapor bubbles can form. In the absence of such a surface, pronounced superheating can occur. Water that has been thoroughly purged of dust particles and gas bubbles has been heated in an open capillary tube to a reported temperature of 270°C without boiling.

From a practical standpoint, one of the most frustrating examples of non-equilibrium behavior is the failure of water in the atmosphere to condense and fall as rain at equilibrium temperatures and pressures. A cloud in the upper atmosphere consists of many billions of tiny water droplets, each approximately 0.1 mm in diameter. These droplets will not spontaneously coalesce to form drops large enough to fall through the atmosphere. Precipitation is ordinarily initiated by the formation of tiny ice crystals, which grow at the expense of the droplets. This means that for precipitation to occur, the temperature within the cloud must be reduced to at least 0°C. In practice, because of supercooling, ice crystals do not form unless the temperature drops to at least −15°C. This means that the water droplets in clouds, particularly on warm days or in hot, arid regions, frequently evaporate without condensing into large drops. To prevent this, clouds are sometimes seeded with condensation nuclei.

One popular method of cloud seeding involves the use of Dry Ice. The sublimation of solid carbon dioxide absorbs enough heat from the cloud to lower the temperature below that required for ice crystal formation. Another substance that is frequently used is finely divided silver iodide, which has a crystal structure similar to that of ice. The presence of silver iodide presumably tends to prevent supercooling and hence allows ice crystals to form at temperatures close to 0°C.

Urea, which has a large, positive heat of solution, has also been used.

PROBLEMS

10.1 Explain, in terms of the kinetic-molecular model, why

    a. Liquids are much less compressible than gases.
    b. Liquids have a surface tension.
    c. The vapor pressure of a liquid is independent of volume.
    d. The vapor pressure of a liquid increases with temperature.
    e. Heat is absorbed when a liquid vaporizes.

10.2 The density of mercury at its normal boiling point, 357°C, is 12.76 g/ml.

    a. What is the volume of one mole of liquid mercury at 357°C?
    b. Using the Ideal Gas Law, calculate the volume of one mole of mercury vapor at 357°C and 1 atm.
    c. Assuming that mercury atoms can be treated as spheres with a radius of 1.55 Å, find the volume of a mole of mercury atoms.
    d. From your answers to a, b, and c, calculate the percentages of the total volume of liquid and gaseous mercury at 357°C and 1 atm that is occupied by the atoms.

265

10.3   A sample of water vapor in equilibrium with liquid water at 20°C is expanded from a volume of 1.0 liter to a volume of 2.0 liters. Assuming equilibrium is maintained at 20°C, what change, if any, occurs in

a. The number of molecules in the vapor?
b. The concentration of molecules in the vapor?
c. The pressure exerted by the vapor?
d. The average kinetic energy of the molecules in the vapor?

10.4   A sample of benzene vapor at 80°C and 200 mm Hg pressure is cooled at constant volume.

a. What will be its pressure at 50°C (vp benzene at 50°C = 272 mm Hg)
b. What will be its pressure at 20°C (vp benzene at 20°C = 75 mm Hg)

10.5   A sample of 0.100 mole of benzene is placed in a 1.00 liter container at 50°C (vp benzene = 272 mm Hg). The volume of the container is slowly increased, while constant temperature is maintained.

a. At what volume does the sample become entirely vapor?
b. What will be the pressure of the vapor when the volume is 2.00 lit? 10.0 lit?

10.6   The vapor pressure of water at 25°C is 23.8 mm Hg. What will be its vapor pressure at 50°C? The average heat of vaporization of water in this temperature range is about 10.4 kcal/mole.

10.7   Given the following data for heptane,

T (°C)	20	30	40	50
vp (mm Hg)	35.5	58.4	92.1	141

calculate, by a graphical method, the heat of vaporization of heptane.

10.8   Two liquids A and B have vapor pressures at 20°C of 12 mm Hg and 242 mm Hg respectively. Which of these liquids would you expect to have

a. The higher normal boiling point?   c. The higher surface tension at 20°C?
b. The higher critical temperature?   d. The larger heat of vaporization?

10.9   If 4.00 g of steam at 100°C is added to 20.0 g of ice at 0°C, calculate the final temperature of the system, given that the heat of vaporization of water at 100°C is 540 cal/g, the heat of fusion of ice is 80 cal/g at 0°C, and the specific heat of liquid water is 1.00 cal/g °C.

10.10   Explain why

a. The normal melting point of a solid is almost identical with the triple point temperature.
b. External pressure has a much greater effect on the boiling point than on the melting point.
c. A substance whose triple point pressure is relatively high often sublimes when it is heated rather than melting.
d. It is possible for a liquid to freeze and boil at the same time.

10.11   A sample of pure benzene is at a pressure of 36 mm Hg and a temperature of 5.5°C. Referring to the phase diagram of benzene (Figure 10.17), state precisely what will happen if

a. Heat is removed from the system at constant pressure.
b. Heat is added to the system at constant pressure.
c. The pressure is increased at constant temperature.
d. The temperature is increased at constant pressure.

10.12   A beam of x-rays obtained by bombarding platinum with electrons has a wavelength of 0.182 Å. Calculate the angle of diffraction for n = 1, 2, and 3 when this x-ray beam strikes a crystal in which the layers of atoms are 1.60 Å apart.

10.13 Chromium crystallizes in a structure with a body-centered cubic unit cell 2.89 Å on an edge.

 a. Calculate the atomic radius of chromium.
 b. Calculate the density of chromium.

10.14 Show how it would be possible to determine Avogadro's number from x-ray diffraction measurements on a crystal. State precisely what experimental quantities would have to be known (cf. Problem 10.13).

10.15 Determine the distance between F^- ions in

 a. A crystal of NaF, which has a structure similar to that of NaCl, in which Na^+ ions and F^- ions touch each other along the face diagonal of a cube.
 b. A crystal of CaF_2 in which a Ca^{2+} ion at the center of a cube touches F^- ions at the corners.

10.16 Suppose a small number of atoms of each of the following elements were incorporated into a silicon crystal. Would you expect this to result in the formation of a semiconductor and, if so, would it be of the n- or p-type?

 a. Sn   b. B   c. P   d. Se   e. Be

10.17 The density of liquid xenon at its normal boiling point, −109°C, is 3.06 g/ml.

 a. What is the volume of one mole of liquid xenon at −109°C and 1 atm?
 b. Using the Ideal Gas Law, calculate the volume of a mole of xenon vapor at −109°C and 1 atm.
 c. Assuming that xenon atoms can be treated as spheres with a radius of 1.90 Å, find the volume of a mole of xenon atoms.
 d. From your answers to a, b, and c, determine the percentages of the total volume in liquid and gaseous xenon at −109°C and 1 atm that is occupied by the atoms.

10.18 If in Problem 10.3 the temperature were increased instead of the volume, what would be your answers to a, b, c, and d?

10.19 A tin can, originally full of air at 20°C and 750 mm Hg, is half-filled with water and immediately stoppered.

 a. What will be the total pressure inside the can when equilibrium is established at 20°C?
 b. If the stoppered can is heated to 50°C, what will be the total pressure inside it?

10.20 A certain liquid has a vapor pressure of 240 mm Hg at 20°C. A sample of 0.100 mole of this liquid is placed in an evacuated, one-liter container at 20°C.

 a. Using the Ideal Gas Law, calculate the number of moles of vapor which will be present at equilibrium under these conditions (neglect the volume of the liquid).
 b. If the volume of the container is slowly increased, maintaining a constant temperature, at what point will all the liquid be vaporized?
 c. What will be the pressure inside the container when the volume is 2 liters?
 d. What will be the pressure inside the container when the volume is 10 liters?

10.21 The heat of vaporization of benzene is 7300 cal/mole. Its normal boiling point is 80°C. Estimate its vapor pressure at 50°C.

10.22 Calculate the total amount of heat that must be absorbed to convert a mole of ice at 0°C to steam at 100°C.

 a. Using the data given in Problem 10.9.
 b. Using the data given in Table 10.4 and taking the specific heat of water vapor to be 0.45 cal/g °C. Follow the path: water (s, 0°C) → water (l, 0°C) → water (g, 0°C) → water (g, 100°C).

267

10.23 Construct a phase diagram for water. Using your diagram, predict what will happen when

a. A sample originally at $-15°C$ and 2.0 mm Hg is heated at constant pressure.
b. A sample originally at $-5°C$ and 760 mm Hg is heated at constant pressure.
c. A sample originally at $25°C$ and 10 mm Hg is compressed at constant temperature.

10.24 Silver crystallizes in a face-centered cubic structure. The atomic radius of silver is 1.44 Å. What are the dimensions of the unit cell and the density of silver metal?

10.25 Determine the simplest formula of an ionic compound in which

a. The unit cell consists of a cube in which there are cations (C) at each corner and an anion (A) at the center of the cube.
b. The unit cell consists of a cube in which there are cations (C) at each corner and anions (A) at the center of each face.

10.26 Metallic manganese at a temperature of $1130°C$ and a pressure of 1 atm undergoes a change in crystal structure from face-centered cubic to body-centered cubic. Would you expect this transition to be accompanied by an increase or decrease in density? Would you expect the transition temperature to be raised or lowered by an increase in pressure?

10.27 If you wanted to prepare a semiconductor from CdO, what metal cation would you suggest as a likely substitute for Cd^{2+}? Consider both size and charge.

°10.28 Calculate the fraction of empty space or "free volume" in a crystal made up of simple cubic unit cells; body-centered cubic unit cells; face-centered cubic unit cells. Consider the volume of the atoms assigned to each cell in comparison to the volume of the cell itself.

°10.29 Potassium has a body-centered cubic structure; its density is 0.86 g/cc. If it were to crystallize in a simple cubic structure, what would its density be?

°10.30 The energy of vaporization, ΔE, is actually a more direct measure of the intermolecular forces in a liquid than is the heat of vaporization, ΔH. Why? Using the equation $\Delta H = \Delta E + P\Delta V$ and the Ideal Gas Law, show that for the vaporization of 1 mole of liquid,

$$\Delta H_{vap} = \Delta E_{vap} + RT$$

Calculate ΔE_{vap} for water at $20°C$.

°10.31 Using the data in Table 10.1, calculate the amount of work in ergs which must be done to create one sq cm of surface in liquid water. Convert this amount of energy to kilocalories. Assuming that approximately half as much energy must be supplied to bring a molecule to the surface as to vaporize it, determine the amount of energy in kilocalories required to bring one molecule of water to the surface. On this basis, what would be the average area occupied by a water molecule at the surface? From what you know about the atomic radii of hydrogen and oxygen, does the number you have just calculated seem physically reasonable?

°10.32 A certain liquid has a normal boiling point of $50°C$. Estimate

a. Its critical temperature.
b. Its heat of vaporization.
c. Its vapor pressure at $25°C$.

°10.33 The density of gold is 19.2 g/cc; its atomic radius is 1.44 Å. Use this data to decide whether gold crystallizes in a body-centered cubic or face-centered cubic structure.

°10.34 In a certain solid, the unit cell is a cube with (A) atoms at the corners, (B) atoms at the center of each edge and (C) atoms in the centers of each cube. What is the simplest formula of the solid?

Solutions

Up to this point we have been concerned primarily with the structures and properties of pure substances. From now on we shall be dealing increasingly with chemical changes taking place in solution. In this chapter we shall develop the background for a discussion of solution chemistry by studying the structure and physical properties of solutions.

A solution was defined in Chapter 1 as a homogeneous mixture of two or more substances. In this connotation, the adjective "homogeneous" ordinarily means uniform to visual observation by eye or microscope. From a structural point of view, homogeneity implies that:

1. The particles of the various components are of molecular size, of the order of 50 Å or less in diameter.

2. These submicroscopic particles, which may be atoms, ions, or molecules, are distributed in a more or less random pattern showing little if any long-range order.

Both of these criteria must be met by a true solution. In a solution formed by adding benzene (C_6H_6) to toluene (C_7H_8), the particles are individual molecules 6 to 8 Å in diameter. These molecules are distributed in a random pattern throughout the liquid phase; there is no tendency for large clusters of benzene or toluene molecules to aggregate in one portion of the solution. In a suspension, either or both of these criteria are not met. To cite a familiar example, fog is formed by the clustering of thousands upon thousands of water molecules into droplets large enough to be visible to the eye, so fog is not a solution of water vapor in air.

11.1 SOLUTION PHASES

Solutions may exist in any of the three states of matter: gas, solid, or liquid. The physical and chemical properties of gaseous solutions have been treated previously (Chapter 5) and will not be discussed further here.

Solid solutions are more common than is generally realized; many familiar alloys fall into this category. The "nickel" coin is actually a solid solution containing 25 per cent by weight of nickel dissolved in copper. Two structural types of solid solutions can be distinguished:

1. **Substitutional solid solutions**, in which one type of particle is substituted, more or less randomly, for another. Metals whose atoms are of about the same size tend to form solutions of this type. If we cool a melt containing nickel (at. rad. = 1.24 Å) and copper (at. rad. = 1.28 Å), a completely homogeneous solid phase is produced regardless of the proportions of the two metals. On the other hand, lead (at. rad. = 1.75 Å) is completely insoluble in solid copper. Microscopic examination of an alloy prepared by heating these two metals together shows it to be a heterogeneous mixture of crystals of the two elements.

2. **Interstitial solid solutions**, in which small, foreign atoms fit into empty spaces in a regular crystal lattice. A commercially important example of an interstitial solid solution is that formed by carbon (at. rad. = 0.77 Å) with iron (at. rad. = 1.26 Å). Below 912°C iron crystallizes in a body-centered cubic structure (α-iron or ferrite) in which the interstices are too small to accommodate carbon atoms. At 912°C the iron undergoes a crystal modification to a face-centered cubic structure (γ-iron or austentite) with larger interstices capable of dissolving up to 2 per cent by weight of carbon. If an alloy of this composition is cooled rapidly by quenching, the carbon remains in solution, giving a hard, brittle steel.

The chemist most frequently deals with solutions in the liquid rather than the gaseous or solid state. Frequently, liquid solutions are subdivided according to the physical states of the pure components. We may have a solution in which the components are liquids (gasoline, alcohol-water). One component may be a gas (soda water) or a solid (salt water). It is even possible to have a liquid solution in which neither component, at the temperature and pressure of the solution, is a liquid. A familiar example is the solution formed when calcium chloride or sodium chloride is used to "melt" ice below 0°C.

The remainder of this chapter will be devoted to a discussion of the prop-

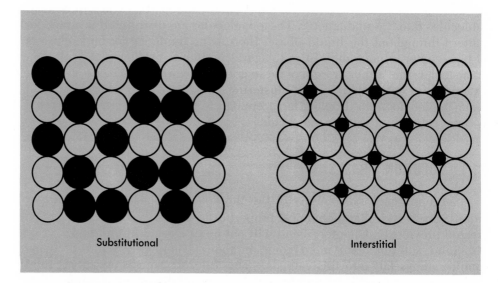

Substitutional   Interstitial

Figure 11.1   Two-dimensional representations of two kinds of solid solution.

erties and structures of liquid solutions, particularly those in which water is a major constituent. Before making such a study, it will be helpful to consider some of the terms used to describe the composition of solutions.

11.2 SOLUTION TERMINOLOGY

Solvent and Solute

While no rigid rules dictate which component of a solution shall be called the solute and which the solvent, certain conventions are ordinarily followed. In a gas-liquid or solid-liquid solution, the liquid is referred to as the solvent. This choice reflects the way in which we visualize the solution process; it is natural to think of carbon dioxide or sodium chloride as "dissolving" in water rather than the reverse.

If both components of a solution are themselves liquids, the designations "solute" and "solvent" are more ambiguous and are seldom used without qualification. Frequently, the component present in the greater amount is referred to as the solvent. In a solution containing 1 g of ethyl alcohol in 100 g of water, we ordinarily speak of the water as the solvent and the alcohol as the solute. If the amounts of alcohol and water are more nearly equal or if one wishes to consider a series of alcohol-water solutions covering the entire composition range, the choice becomes less clear-cut and must be specified.

Can you suggest why most reactions are carried out in solution?

Dilute and Concentrated Solutions

Although the adjectives dilute and concentrated are frequently used to refer to the composition of solutions, they are seldom given an exact meaning. Neither, for that matter, can they be restricted to a particular concentration range; about all that one can say is that a dilute solution of A in B contains a lower concentration of A than does a concentrated solution of A in the same solvent.

In a few cases the terms dilute and concentrated have taken on a quantitative meaning. Dilute and concentrated solutions of certain acids and bases, labeled as such in the general chemistry laboratory, have the concentrations specified in Table 11.1.

TABLE 11.1 CONCENTRATIONS OF LABORATORY ACID AND BASE SOLUTIONS

		SOLUTE	MOLES SOLUTE PER LITER	PER CENT SOLUTE (BY WEIGHT)	DENSITY (G/ML)
Hydrochloric acid	conc.	HCl	12	36	1.18
	dilute		6	20	1.10
Nitric acid	conc.	HNO_3	16	72	1.42
	dilute		6	32	1.19
Sulfuric acid	conc.	H_2SO_4	18	96	1.84
	dilute		3	25	1.18
Ammonia	conc.	NH_3	15	28	0.90
	dilute		6	11	0.96

Saturated, Unsaturated, and Supersaturated Solutions

A saturated solution of a solute A is one which is in equilibrium with undissolved A. A simple way to prepare a saturated solution at 25°C of, let us say, sodium chloride in water, is to bring an excess of the solid into contact with water at that temperature and stir until no more NaCl goes into solution. The resulting solution, which contains 36.2 g of sodium chloride per 100 g of water, is said to be saturated with sodium chloride. Addition of further solid sodium chloride fails to change its concentration in solution.

If you attempt to carry out an experiment of the type just described with alcohol and water at 25°C, you will find that equilibrium cannot be established. Regardless of the relative amounts of alcohol and water added, only one layer forms. The two liquids are soluble in each other in all proportions; there is no such thing as a saturated solution of alcohol in water or water in alcohol. As we shall see later, this behavior is not uncommon with liquid-liquid solutions; one might look upon it as the rule rather than the exception.

An **unsaturated** solution contains a lower concentration of solute than a saturated solution. A solution at 25°C containing less than 36.2 g of sodium chloride per 100 g of water is said to be unsaturated with respect to sodium chloride. It is not in a state of true equilibrium; if an unsaturated solution is brought into contact with solute its concentration increases to approach saturation.

A **supersaturated** solution represents a metastable situation in which the solution actually contains more than the equilibrium concentration of solute. Supersaturated solutions most commonly arise when a solid is dissolved in a hot liquid and the solution is cooled. As an illustration of one method of preparing a supersaturated solution, consider a specific example, that of sodium acetate, $NaC_2H_3O_2$, dissolved in water. At 20°C a saturated solution contains 46.5 g of sodium acetate per 100 g of water; at higher temperatures the solubility of sodium acetate is considerably greater. If we heat 80 g of this salt with 100 g of water until it is completely dissolved (a temperature of about 50°C is required) and then cool carefully, without shaking or stirring, to 20°C, the excess solute remains in solution. This supersaturated solution can be maintained indefinitely so long as there are no nuclei upon which crystallization can start (cf. the phenomenon of supercooling, described in Chapter 10). If a small seed crystal of sodium acetate is then added, crystallization takes place until equilibrium is attained by the formation of a saturated solution.

11.3 CONCENTRATION UNITS

The physical and chemical properties of solutions depend to a large extent upon the relative amounts of solute and solvent present. For this reason, in any quantitative work involving solutions it is important to specify concentrations. This can be done by stating either the relative amounts of solute and solvent or, alternatively, the amount of one component relative to the total amount (mass or volume) of solution.

In this chapter, we shall have occasion to use three different concentration units in each of which the amount of solute is expressed in moles. These are: the mole fraction, molality, and molarity.

Mole Fraction

The mole fraction of component A in a solution, designated X_A, is given by the expression

$$X_A = \frac{\text{no. of moles of A}}{\text{total no. moles of all components}} \qquad (11.1)$$

Example 11.1 illustrates how this defining equation can be used to calculate the mole fractions of substances in solution or to give directions for the preparation of a solution containing components at specified mole fractions.

EXAMPLE 11.1

    a. Calculate the mole fractions of ethyl alcohol, C_2H_5OH, and water, H_2O, in a solution prepared by adding 50.0 g of ethyl alcohol to 50.0 g of water.

    b. How should one prepare an ethyl alcohol–water solution in which the mole fraction of ethyl alcohol is 0.300?

SOLUTION

    a. Since 1 mole $C_2H_5OH = 46.0$ g and 1 mole $H_2O = 18.0$ g

$$\text{no. of moles } C_2H_5OH = 50.0 \text{ g} \times \frac{1 \text{ mole}}{46.0 \text{ g}} = 1.09$$

$$\text{no. of moles } H_2O = 50.0 \text{ g} \times \frac{1 \text{ mole}}{18.0 \text{ g}} = 2.78$$

Using Equation 11.1,

$$X\ C_2H_5OH = \frac{1.09}{1.09 + 2.78} = 0.282$$

$$X\ H_2O = \frac{2.78}{1.09 + 2.78} = 0.718$$

The sum of the mole fractions of all components must equal one.

    b. A mole fraction of C_2H_5OH of 0.300 requires that there be 0.300 mole of C_2H_5OH in a total of 1.000 mole of solution; this in turn means that for every 0.300 mole of C_2H_5OH there must be $(1.000 - 0.300) = 0.700$ mole of H_2O.

$$.300 \text{ mole } C_2H_5OH = .300 \times 46.0 \text{ g } C_2H_5OH = 13.8 \text{ g } C_2H_5OH$$

$$.700 \text{ mole } H_2O = .700 \times 18.0 \text{ g } H_2O = 12.6 \text{ g } H_2O$$

One could prepare such a solution by adding alcohol to water in a mass ratio of $13.8 : 12.6$. Depending upon the amount of solution required, one might add 13.8 g of C_2H_5OH to 12.6 g of H_2O, 27.6 g of C_2H_5OH to 25.2 g of water, and so on.

If the molecules of a solute are not dissociated or changed in any other way, the mole fraction becomes identical with the molecule fraction. Thus, in an alcohol-water solution in which the mole fraction of alcohol is 0.3, three out of every 10 molecules in the solution are alcohol molecules; the other seven are water molecules (Figure 11.2).

273

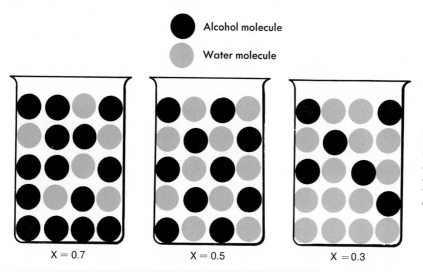

Figure 11.2 Schematic drawing showing relative numbers of alcohol and water molecules in solutions of various mole fractions of alcohol.

X = 0.7   X = 0.5   X = 0.3

Molality

Molality (m) is defined as the number of moles of solute per kilogram of solvent. Mathematically:

$$m = \frac{\text{no. of moles solute}}{\text{no. of kg solvent}} \qquad (11.2)$$

EXAMPLE 11.2. Calculate the molality of a solution prepared by dissolving 20.4 g of NaCl in 192 g of water.

SOLUTION

$$1 \text{ mole NaCl} = 58.45 \text{ g NaCl}$$

$$\text{no. of moles NaCl} = 20.4 \text{ g NaCl} \times \frac{1 \text{ mole NaCl}}{58.45 \text{ g NaCl}} = 0.349$$

$$\text{no. of kg H}_2\text{O} = 0.192$$

From Equation 11.2,

$$\text{molality NaCl} = \frac{0.349}{0.192} = 1.82$$

As Example 11.2 implies, we can prepare a solution of a given molality by dissolving a known weight of solute in a predetermined weight of solvent. The precision with which the concentration is known is limited only by that of the balance used to make the weighings. One of the advantages of molality as a concentration unit is that it is independent of temperature; a one molal solution prepared at 20°C will retain the same molality at 100°C, provided there is no loss of solute or solvent on heating.

Molarity

Concentrations of reagents in the general chemistry laboratory are most often specified in terms of molarity (M), which is defined as the number of moles of solute per liter of solution, or

$$M = \frac{\text{no. of moles solute}}{\text{no. of liters solution}} \qquad (11.3)$$

EXAMPLE 11.3

    a. What is the molarity of a solution prepared by dissolving 16.0 g of $BaCl_2$ in sufficient water to give 450 ml of solution?

    b. How should one prepare 20.0 lit of 6.0 M NaOH solution, starting with solid NaOH?

    c. How should one prepare 1.0 lit of 0.50 M NaOH solution, starting with the solution in b?

SOLUTION

    a. 1 mole $BaCl_2 = 137.3$ g $+ 70.9$ g $= 208.2$ g

$$\text{no. of moles } BaCl_2 = 16.0 \text{ g } BaCl_2 \times \frac{1 \text{ mole } BaCl_2}{208.2 \text{ g } BaCl_2} = 0.0769$$

no. of lit solution $= 0.450$

By Equation 11.3

$$M = \frac{0.0769}{0.450} = 0.171$$

    b. From Equation 11.3 we have:

no. of moles NaOH $=$ (M NaOH) (volume of solution in liters)

$$= 6.0 \frac{\text{moles}}{\text{lit}} \times 20.0 \text{ lit} = 120 \text{ moles NaOH}$$

It follows that one should weigh out 120 moles (4800 g) of NaOH and dissolve in sufficient water to give 20.0 lit of solution.

    c. Clearly, the more concentrated solution (6.0 M) must be diluted with water to give a 0.50 M solution. The question is: What volume of 6.0 M solution should one start with to prepare one liter of 0.50 M solution? To answer this question, we note that the number of moles of solute is not changed by dilution. In the final solution, we have:

$$0.50 \frac{\text{mole}}{\text{lit}} \times 1.0 \text{ lit} = 0.50 \text{ mole NaOH}$$

All that remains is to calculate what volume of 6.0 M NaOH must be taken to give 0.50 mole of NaOH:

$$\text{volume (in liters)} = \frac{\text{no. moles solute}}{\text{molarity}} = \frac{0.50}{6.0} = 0.083 \text{ lit}$$

Hence, one should take 0.083 lit (83 ml) of the 6.0 M solution and dilute with water to give a final volume of 1.0 lit of the 0.50 M solution.

This technique is frequently used to prepare dilute reagents.

Since the volume of a liquid is more easily measured than its mass, laboratory reagents are usually made up to a specified molarity rather than a given molality. Furthermore, as we shall see later (Chapter 18), molarity is a very convenient unit for calculations involving the relative quantities of two solutions that react with each other. Consequently, we find that many of the standard solutions used in analytical laboratories have their concentrations expressed in molarity, often to three or more significant figures. Since the volume of a solution ordinarily increases with temperature, it follows that, for very exact work, the temperature at which a solution of a given molarity was prepared should be specified. The molarity of a solution can readily be converted to molality if its density is known (Example 11.4).

EXAMPLE 11.4. The density of a 0.0600 M solution of potassium iodide in water is 1.006 g/ml at 20°C. Calculate the molality of the solution.

SOLUTION. In order to obtain the molality, we must know the number of moles of KI per kilogram of water. Let us base our calculations on one liter of solution.

$$\text{no. of moles KI} = 0.0600$$

To obtain the mass of water in one liter of solution, we first note that since the solution has a density of 1.006 g/ml, the total mass of one liter of solution must be 1006 g. To find the mass of water, we must subtract from 1006 g the mass of KI. Noting that one mole of KI weighs 166 g, we have:

$$\text{mass of KI} = 0.0600 \text{ mole} \times \frac{166 \text{ g}}{\text{mole}} = 10.0 \text{ g}$$

Hence,

$$\text{mass of water} = 1006 \text{ g} - 10 \text{ g} = 996 \text{ g}$$

To calculate the molality:

$$m = \frac{\text{no. of moles KI}}{\text{no. of kg water}} = \frac{0.0600}{0.996} = 0.0602$$

As this example implies, the molarity of a solute in a dilute water solution is virtually the same as its molality. For more concentrated solutions or with a solvent other than water, concentrations expressed in these two units may differ widely from each other.

A fourth concentration unit, which happens to be particularly useful in quantitative analysis, is normality, defined as the number of gram equivalent weights of solute per liter of solution:

$$N = \frac{\text{no. of GEW solute}}{\text{no. of liters solution}}$$

We shall postpone further discussion of normalities of solutions to Chapter 18, Section 18.2.

11.4 PRINCIPLES OF SOLUBILITY

The prediction of the extent to which one substance will dissolve in another is one of the most fascinating and, at the same time, one of the most frustrating problems in chemical theory. The following brief discussion of this topic deals with solubility in the liquid state and is subdivided according to whether the solute is a liquid, a gas, or a solid.

Liquid-Liquid

In discussing the solubility of two liquids in each other, it is sometimes stated that "like dissolves like." A more meaningful way of expressing this same idea is to say that substances that have similar molecular structures and, consequently, intermolecular forces of about the same magnitude will be soluble in each other in all proportions. An illustration of this rule is furnished by the liquid aliphatic hydrocarbons, of general formula C_nH_{2n+2} (C_5H_{12}, C_6H_{14} ... $C_{18}H_{38}$), all of which are completely miscible with each other. Molecules of these nonpolar substances are held together by dispersion forces which increase only slightly with molecular size. The forces between C_5H_{12} molecules in pure liquid pentane are very nearly the same as those between C_5H_{12} and C_8H_{18} molecules in a solution of pentane in octane. A pentane molecule readily passes into solution in octane because it undergoes virtually no change in environment in the solution process.

Moderate differences in polarity between solute and solvent seem to have little effect on solubility. Aliphatic hydrocarbons of the type just discussed are as soluble in chloroform ($CHCl_3$, dipole moment = 1.15 Debye units) as they are in each other or in carbon tetrachloride (CCl_4, dipole moment = 0). This reflects the weakness of dipole forces, which make only a minor contribution to the total attractive force between $CHCl_3$ molecules. The fact that the heats of vaporization and surface tensions of chloroform and carbon tetrachloride are almost identical implies that the intermolecular forces in $CHCl_3$ and CCl_4 are nearly the same. It is about as easy to "break into" the liquid structure of $CHCl_3$ as it is with CCl_4, and consequently the two liquids show similar solvent properties despite a considerable difference in polarity.

If one attempts to dissolve a hydrocarbon in water, the situation is quite different. In order to dissolve appreciable quantities of, let us say, pentane in water, one would have to break the strong intermolecular hydrogen bonds holding water molecules together. The large amount of energy required to do this would not be restored in any step of the solution process; the attractive forces in solution between C_5H_{12} and H_2O molecules are probably weaker than those between C_5H_{12} molecules in liquid pentane. Consequently, pentane and other hydrocarbons are only very slightly soluble in water; at 25°C the mole fraction of pentane in a saturated water solution is 0.00003. It is commonly supposed that the few molecules of pentane that dissolve do so by fitting, with minimum distortion, into "holes" in the hydrogen-bonded water structure. This is suggested by the fact that the heats of solution per mole of most liquid hydrocarbons in water are very nearly zero.

Of the relatively few organic liquids which dissolve readily in water, the majority are oxygen-containing compounds of low molecular weight. Two familiar examples are the alcohols containing one and two carbon atoms,

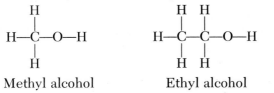

Methyl alcohol   Ethyl alcohol

277

both of which are soluble in water in all proportions. Methyl and ethyl alcohol each contain an $-OH$ group, as does water. Even more important, both these compounds are known to be hydrogen-bonded in the liquid state. Consequently, it is hardly surprising that they dissolve readily in water. One would expect the intermolecular forces between alcohols and water in solution to be roughly comparable to those in the pure liquids.

As the number of carbon atoms in the alcohol molecules increases, we find that the solubility in water decreases. The compound n-butyl alcohol, C_4H_9OH, has a limited solubility in water; its mole fraction in a saturated water solution at 20°C is only about 0.02. Octyl alcohol, $C_8H_{17}OH$, is extremely insoluble (X = 0.0008 in a saturated water solution at 20°C). The same trend is observed with many other types of organic compounds; there is a general tendency for solubility to decrease with an increase in chain length. This effect can be explained most simply in terms of the large number of hydrogen bonds that must be broken if a long-chain molecule is to be inserted into the water structure. A considerable amount of energy must be absorbed to break these bonds, making the solution process less spontaneous and decreasing the solubility.

Solid-Liquid

Unlike liquid-liquid pairs, in which we frequently find that the two components are soluble in each other in all proportions, it is always possible to prepare a saturated solution of a solid in a liquid. In other words, there is always a finite limit to the solubility of a solid in a liquid.

The limited solubility of solids in liquids is readily explained if one imagines the process by which a solid A goes into solution in a liquid solvent B as taking place in two steps:

From a thermodynamic standpoint, it does not matter whether the solution process actually follows this path; only final and initial states are important.

1. Solid A melts to form pure liquid A:

$$A(s) \longrightarrow A(l)$$

2. Liquid A dissolves in B to form a liquid solution:

$$A(l) + B(l) \longrightarrow \text{liquid solution of A in B}$$

Step 1 is clearly nonspontaneous at any temperature at which solid A is the stable phase. It follows that the combination of Step 1 and Step 2, which represents the solution process for a solid, is less spontaneous than Step 2 by itself, which is the only step involved when a liquid dissolves.

Following the reasoning we have just outlined, we can deduce that the relative solubilities of different solids in a given solvent at room temperature should be inversely related to their melting points. The greater the difference between the melting point of a solid and the temperature at which its solubility is measured, the less spontaneous Step 1 will be. Consequently, high-melting solids should be less soluble than solids whose melting points are closer to room temperature. The validity of this reasoning is confirmed by data such as those in Table 11.2, which compare the solubilities of a series of solid hydrocarbons in a nonpolar hydrocarbon solvent, benzene.

In comparing the relative solubilities of a given solid in a series of liquid solvents, we must take into account the factors discussed previously in con-

TABLE 11.2 Solubilities of Solid Hydrocarbons in Benzene at 25°C*

Solute	Melting Point (°C)	X Solute
Anthracene	218	0.008
Phenanthrene	100	0.21
Naphthalene	80	0.26
Biphenyl	69	0.39

* Mole fraction of solid in saturated solution.

nection with liquid-liquid systems. Nonpolar solids are most soluble in solvents of low polarity; hydrogen-bonded liquids are poor solvents for solid hydrocarbons. Compare, for example, the solubility of naphthalene in benzene at 20°C (X naphthalene = 0.24) to that in methyl alcohol (X naphthalene = 0.018) and in water (X naphthalene = 4.8×10^{-6}). Ionic solids are considerably more soluble in water than they are in organic solvents. We shall postpone until Chapter 16 a general discussion of the solubilities of ionic compounds in water.

Gas-Liquid

Following the reasoning outlined for solid-liquid and liquid-liquid solutions, we arrive at the following conclusions regarding the solubilities of gases in liquids:

1. The higher the boiling point of the gas (i.e., the closer it is to the liquid state) the more soluble it will be in a given solvent.

2. The best solvent for a given gas will be the one whose intermolecular forces are most similar to those of the gaseous solute.

Which would be the more soluble in water, H_2 or O_2? He or N_2?

TABLE 11.3 Solubility of the Noble Gases in Benzene and Water at 25°C and 1 atm*

Gas	Boiling Point (°C)	Solvent	
		Benzene	Water
He	−269	0.76×10^{-4}	0.069×10^{-4}
Ne	−246	1.14×10^{-4}	0.082×10^{-4}
Ar	−186	8.9×10^{-4}	0.25×10^{-4}
Kr	−152	27.3×10^{-4}	0.45×10^{-4}
Xe	−109	110×10^{-4}	0.86×10^{-4}
Rn	−62	310×10^{-4}	1.63×10^{-4}

* Mole fraction of gas in saturated solution.

Both of these principles are illustrated by the solubility data for the noble gases given in Table 11.3. Note the steady increase in solubility with boiling point (He < Ne < Ar < Kr < Xe < Rn) for both solvents. The reduced solubility of all these gases in water as compared to benzene reflects the strong intermolecular hydrogen bonding in water; in the nonpolar solvent benzene, as in the noble gases themselves, the attractive forces between particles are of the dispersion type.

11.5 EFFECT OF TEMPERATURE AND PRESSURE ON SOLUBILITY

The mutual solubilities of two substances A and B depend not only upon their chemical and physical properties, but also upon the external conditions of temperature and pressure. The effects of these two variables upon solubility are readily deduced if we regard the solution process as an equilibrium:

$$A + B \rightleftharpoons \text{solution}$$

and apply the principles discussed in Chapter 10 concerning the influence of temperature and pressure upon physical equilibria.

Temperature

In any equilibrium, an increase in temperature favors the endothermic process. Applied to the solution process, this means that if heat is absorbed when A dissolves in B,

$$A + B + \text{heat} \rightleftharpoons \text{solution}$$

an increase in temperature will increase the solubility of A in B. Conversely, if dissolving A in B liberates heat,

$$A + B \rightleftharpoons \text{solution} + \text{heat}$$

an increase in temperature will favor the reverse process; i.e., it will reduce the solubility. Unfortunately, there are no infallible rules for predicting whether the formation of a solution from the pure components will absorb or evolve heat.

The solubility of most solids in liquids increases with temperature. To explain this, it is convenient, once again, to imagine the solution process as occurring in two steps:

1. $A(s) \longrightarrow A(l)$
2. $A(l) + B(l) \longrightarrow$ liquid solution of A in B

Step 1, the melting of the solid, is invariably endothermic. If the heat change in Step 2 is endothermic, as will ordinarily be the case, the net solution process will absorb heat, and solubility will increase with temperature. Exceptions are known, particularly when the components interact strongly with each other.

It is difficult to predict in advance whether the solubility of a gas in a liquid will increase or decrease with temperature. Many examples of both

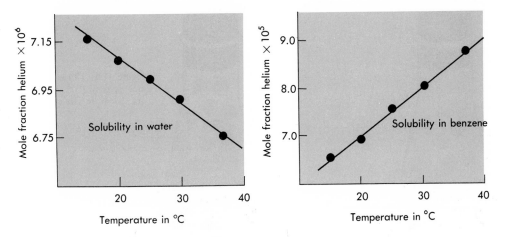

Figure 11.3 Temperature dependence of solubility of helium in water and benzene.

types of behavior are known (Figure 11.3). To appreciate the problems involved, imagine, as before, a two-step process:

1. $A(g) \longrightarrow A(l)$
2. $A(l) + B(l) \longrightarrow$ liquid solution of A in B

Step 1, the condensation of a gas to a liquid, is exothermic. Step 2, as already pointed out, is usually endothermic. Under these conditions, the direction of the net heat change depends upon the relative amounts of heat evolved in Step 1 and absorbed in Step 2. In the case of helium dissolving in benzene, the heat absorbed in forming a cavity in the solvent large enough to accommodate an He atom (Step 2) is greater than the heat evolved when the gas condenses: the overall process is endothermic and solubility increases with temperature. In the helium-water system, the heat change in Step 2 virtually vanishes, presumably because the few gas molecules which dissolve do so by fitting into cavities already present in the water structure. Consequently, the overall process is exothermic and solubility decreases with increasing temperature.

Pressure

Pressure has little effect on solubility equilibria involving only condensed phases, i.e., liquids and solids. The volume change accompanying the solution process in liquid-liquid or solid-liquid systems is so small that very large pressure changes are required to produce a detectable change in solubility.

Experimentally, it is found that at moderate pressures the solubility of a gas in a liquid is directly proportional to its partial pressure in the gas phase over the solution.

$$C_a = k\, P_a \qquad (11.4)$$

in which P_a = partial pressure of gas, C_a = concentration of gas in solution, and k is a constant. A simple kinetic explanation of Equation 11.4 (Henry's Law) is shown in Figure 11.4. Doubling the partial pressure of a gas over a

281

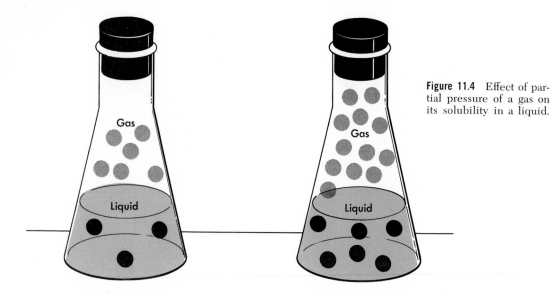

Figure 11.4 Effect of partial pressure of a gas on its solubility in a liquid.

solution amounts to doubling its concentration in the gas phase. In order to maintain equilibrium, the concentration of gas in the liquid phase must also be doubled.

EXAMPLE 11.5. The solubility of oxygen in water at 20°C and one atmosphere pressure is 1.38×10^{-3} moles/lit. Calculate the number of grams of oxygen present in ten liters of water saturated with air (mole fraction $O_2 = 0.210$) at a total pressure of 740 mm Hg and a temperature of 20°C.

SOLUTION. In order to use Henry's Law, we must first calculate the partial pressure of oxygen above the solution. From Dalton's Law:

$$P_{O_2} = (X_{O_2}) \times P_{tot} = 0.210 \times \frac{740}{760} \text{ atm} = 0.204 \text{ atm}$$

Applying Henry's Law at the two pressures (0.204 atm and 1 atm),

$$\frac{\text{concentration } O_2 \text{ at } 0.204 \text{ atm}}{\text{concentration } O_2 \text{ at } 1 \text{ atm}} = \frac{0.204}{1.000}$$

$$\text{concentration } O_2 \text{ at } 0.204 \text{ atm} = 0.204 \times 1.38 \times 10^{-3} \text{ mole/lit}$$

$$= 2.81 \times 10^{-4} \text{ mole/lit}$$

Having calculated the solubility of oxygen under these conditions, we can readily obtain the number of grams of O_2 in 10.0 liters:

no. of moles $O_2 = 2.81 \times 10^{-4}$ mole/lit $\times$ 10.0 lit $= 2.81 \times 10^{-3}$ mole

no. of grams $O_2 = 2.81 \times 10^{-3}$ mole $\times$ 32.0 g/mole $= 0.0899$ g O_2

The influence of partial pressure upon gas solubility is utilized in the commercial production of carbonated beverages. Beer, certain wines, and many soft drinks are bottled under a carbon dioxide pressure slightly greater than 1 atm. When the container is open, exposing the liquid to air, in which the partial pressure of CO_2 is very small (0 to .001 atm), the carbon dioxide bubbles out of solution, giving a froth or "head." Pressurized containers for whipped cream work on a similar principle. Opening a valve causes dissolved gas (N_2O) to come out of solution, carrying the liquid with it as a foam.

The excruciatingly painful and sometimes fatal affliction known as the "bends" is another consequence of the effect of pressure on gas solubility. Compressed air breathed by divers working far below the surface of water dissolves in the body fluids and tissues. If the diver ascends rapidly to the surface, the excess air comes out of solution in the form of tiny bubbles which impair circulation and affect nerve impulses. One way of minimizing this effect is to substitute a mixture of helium and oxygen for the compressed air supplied to the diver. Helium, which has a much lower boiling point than nitrogen, is only about one-fifth as soluble in body fluids as nitrogen. Consequently, much less gas comes out of solution on decompression.

11.6 ELECTRICAL CONDUCTIVITIES OF WATER SOLUTIONS

Solutes are often classified as to the conductivity of their water solutions into three categories: nonelectrolytes, strong electrolytes, and weak electrolytes.

Nonelectrolytes. Substances which dissolve as molecules and hence give water solutions which are nonconducting are classed as nonelectrolytes. Most but not all covalent solutes are of this type. Examples include methyl alcohol, CH_3OH; sugar, $C_{12}H_{22}O_{11}$; and urea, $CO(NH_2)_2$.

Strong Electrolytes. Substances which exist in water solution almost exclusively in the form of ions are referred to as strong electrolytes. Their electrical conductivities, even at concentrations as low as 0.1 M, are at least 100,000 times that of pure water. Conduction is the result of movement of ions through the solution.

Any compound which exists as ions in the solid state ordinarily acts as a strong electrolyte in water solution. Compounds in this category include NaCl (Na^+, Cl^-), BaI_2 (Ba^{2+}, I^-), and $Cu(NO_3)_2$ (Cu^{2+}, NO_3^-). Certain covalently bonded substances react with water to form high concentrations of ions. Strong electrolytes of this type include hydrogen chloride (HCl molecules in the gas state, H^+ and Cl^- ions in solution) and nitric acid (HNO_3 molecules in the pure liquid, H^+ and NO_3^- ions in solution). We shall have more to say about this type of substance when we discuss acids and bases in Chapter 17.

Weak Electrolytes. Substances which exist in water solution as an equilibrium mixture of ions and molecules are called weak electrolytes. An example is hydrogen fluoride: a water solution of this compound contains a relatively small number of H^+ and F^- ions in equilibrium with HF molecules (cf. Chapter 17).

11.7 COLLIGATIVE PROPERTIES OF DILUTE SOLUTIONS*

Certain properties of solutions are found to depend primarily upon the concentrations of solute particles rather than upon the nature of these par-

---

* As pointed out earlier in this chapter, the adjective dilute does not have a precise quantitative meaning. The laws developed in this section are best regarded as limiting laws, which are approached more and more closely as the solution becomes more dilute. In practice, Equations 11.5 to 11.9 are generally applicable to within at most a few per cent at concentrations as high as 1 molal.

ticles. These are called **colligative** properties. In discussing such properties, it is convenient to distinguish between solutions of nonelectrolytes on the one hand and electrolytes on the other.

Nonelectrolytes

Vapor Pressure of Solvent. It is invariably found that the vapor pressure of a solvent above a solution of a nonelectrolyte is lower than that of the pure solvent at the same temperature. Moreover, careful studies carried out with dilute solutions of nonvolatile solutes show that the amount by which the vapor pressure of the solvent is lowered is:

1. *Directly proportional* to the *concentration* of solute. For example, the vapor pressure of water above a 0.10 M solution of sugar at 25°C is 0.043 mm Hg less than that of pure water; the vapor pressure lowering in a 0.20 M sugar solution at the same temperature is 0.086 mm Hg.

2. *Independent* of the *nature* of the solute. A 0.10 M solution of urea at 25°C, like a 0.10 M solution of sugar, has a vapor pressure 0.043 mm Hg less than that of pure water.

These two observations, which are generally valid for all dilute solutions of nonelectrolytes, show vapor pressure lowering to be a colligative property.

The relationship between solvent vapor pressure and concentration can be expressed mathematically as

$$P_1 = X_1 P_1^\circ \qquad (11.5)$$

Equation 11.6 can be derived from 11.5 by substituting $X_1 = 1 - X_2$ and rearranging.

or $$P_1^\circ - P_1 = \text{vapor pressure lowering} = X_2 P_1^\circ \qquad (11.6)$$

in which P_1 = vapor pressure of solvent in solution; P_1° = vapor pressure of pure solvent; X_1 and X_2 = mole fractions of solvent and solute respectively. Equation 11.5, known as Raoult's Law, lends itself more readily to a simple kinetic explanation, while Equation 11.6 is somewhat more useful in calculations of the type illustrated by Example 11.6.

EXAMPLE 11.6. Calculate the vapor pressure lowering and the vapor pressure of a solution containing 100 g of sugar, $C_{12}H_{22}O_{11}$, in 500 g of water at 25°C.

SOLUTION. In order to use Equation 11.6 we need to know the vapor pressure of pure water (23.76 mm Hg at 25°C) and the mole fraction of sugar. To obtain the mole fraction of sugar, we note that since the molecular weight of sugar is 342,

$$\text{no. of moles sugar} = 100 \text{ g} \times \frac{1 \text{ mole}}{342 \text{ g}} = 0.292$$

$$\text{no. of moles H}_2\text{O} = 500 \text{ g} \times \frac{1 \text{ mole}}{18.0 \text{ g}} = 27.8$$

$$\text{X sugar} = \frac{0.292}{0.292 + 27.8} = 0.0104$$

Therefore, vapor pressure lowering $= 0.0104 \times 23.76$ mm Hg $= .247$ mm Hg

vapor pressure of solution $= (23.76 - 0.25)$ mm Hg $= 23.51$ mm Hg

Equation 11.5 (Raoult's Law) may be interpreted in a manner entirely analogous to that used earlier with Henry's Law of gas solubilities. Considering the equilibrium between solvent molecules in solution and in the gas phase (Figure 11.5), it seems entirely reasonable that the concentration in the gas phase, as measured by the solvent vapor pressure, should be directly proportional to the concentration in solution, as measured by the mole fraction of solvent. In this case we are able to evaluate the proportionality constant as being the vapor pressure of the pure solvent, i.e., the vapor pressure when the mole fraction of solvent is one.

Raoult's Law can in principle be applied to solutions of nonelectrolytes in which both components are volatile. If both solute and solvent obey Raoult's Law,

$$P_1 = X_1 P_1^{\circ} \quad \text{and} \quad P_2 = X_2 P_2^{\circ}$$

The total vapor pressure over such a solution is obtained by adding the two partial pressures:

$$P_1 + P_2 = P\text{ tot} = X_1 P_1^{\circ} + X_2 P_2^{\circ}$$

Solutions which obey this equation are said to be **ideal**.

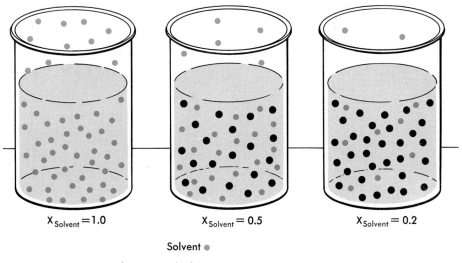

$X_{Solvent} = 1.0 \qquad X_{Solvent} = 0.5 \qquad X_{Solvent} = 0.2$

Solvent •

Solute (non volatile) ●

Figure 11.5 Molecular interpretation of Raoult's Law.

In practice, very few solutions behave ideally. Although the solvent ordinarily obeys Raoult's Law in dilute solution, the solute seldom does so. Instead, the solute vapor pressure follows a more general expression equivalent to Henry's Law

$$P_2 = k X_2$$

in which k is a constant that may be greater or less than P_2°, the vapor pressure of pure solute, but is seldom equal to it. It is not difficult to see why this should be the case. When a small amount of solute is dissolved in a large amount of solvent, the solute molecules find themselves in an environment quite different from that of the pure solute. The **escaping tendency** of the solute, or its vapor pressure, is determined by the attractive forces between the solute molecule and the solvent particles which surround it. These forces may be greater or less than those in the pure solute, but they will rarely be exactly equal to the solute-solute attractive forces.

If the solute is volatile (methyl alcohol), it will contribute to the vapor pressure and will ordinarily lower the boiling point.

Boiling Point. A water solution of a nonvolatile solute invariably boils at a higher temperature than pure water at the same external pressure. Moreover, the boiling point elevation, in dilute solution, is found to be directly proportional to solute concentration. The general relation between these two quantities for water solutions of nonelectrolytes is usually expressed in terms of the molality of solute

$$\Delta T_b = 0.52°C \text{ (m)} \tag{11.7}$$

where ΔT_b is the boiling point elevation in °C and m is the molality. An analogous relationship is found to apply to all dilute solutions of nonelectrolytes, regardless of the solvent. The constant appearing in Equation 11.7 is a function of the solvent; it is generally larger than 0.52°C for organic solvents (Table 11.4).

TABLE 11.4 MOLAL FREEZING POINT AND BOILING POINT CONSTANTS

SOLVENT	FREEZING POINT (°C)	FREEZING POINT CONSTANT (°C)	BOILING POINT (°C)	BOILING POINT CONSTANT (°C)
Water	0	1.86	100	0.52
Acetic acid	17	3.90	118	2.93
Benzene	5.50	5.10	80	2.53
Cyclohexane	6.5	20.2	81	2.79
Camphor	178	40.0	208	5.95

The elevation of the boiling point associated with solutions of nonvolatile solutes is readily explained in terms of a lowering of the vapor pressure. Since the solution, at any given temperature, has a vapor pressure lower than that of the pure solvent, a higher temperature must be reached before the solution boils, that is, before its vapor pressure becomes equal to the external pressure. Moreover, since the extent of vapor pressure lowering is directly proportional to solute concentration, one would expect to find the same type of linear relationship between boiling point elevation and solute concentration. Figure 11.6 illustrates this reasoning graphically.

Freezing Point. Dilute water solutions freeze at temperatures below 0°C. The lowering of the freezing point in a dilute solution, like the lowering of the vapor pressure and elevation of the boiling point, is directly proportional to solute concentration. The equation for water solutions of nonelectrolytes is

All colligative properties derive from the lowering of solvent vapor pressure.

$$\Delta T_f = 1.86°C \text{ (m)} \tag{11.8}$$

The constant in this equation, like that in Equation 11.7, depends on the solvent (Table 11.4).

The freezing point depression, like the boiling point elevation, is a direct consequence of the lowering of the solvent vapor pressure by a solute. The freezing point of a solution from which pure solvent crystallizes out* is that temperature at which the solvent has the same vapor pressure in the two phases, liquid solution and solid solvent. Since the solvent vapor pressure in solution is depressed, its vapor pressure will become equal to that of the solid solvent at a lower temperature. Again, since vapor pressure lowering is directly proportional to solute concentration, one would expect to find a similar relationship between freezing point lowering and concentration (Figure 11.6).

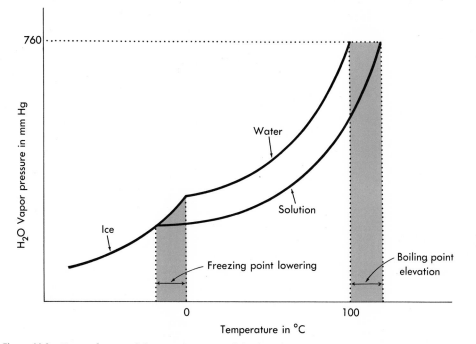

Figure 11.6 Dependence of freezing point and boiling point of aqueous solutions on vapor pressure lowering.

The use of Equations 11.7 and 11.8 in calculations involving freezing and boiling points of solutions of nonelectrolytes is illustrated by Example 11.7.

EXAMPLE 11.7. Calculate the freezing point and boiling point, at 760 mm Hg, of a solution of 2.60 g of urea, $CO(NH_2)_2$ in 50.0 g of water.

SOLUTION. To use equations 11.7 and 11.8, we must first calculate the molality. Noting that the molecular weight of urea is 60.0, we have

$$\text{no. moles urea} = 2.60/60.0 = 0.0433$$

$$m = \frac{\text{no. moles urea}}{\text{no. kg water}} = \frac{0.0433}{0.050} = 0.866$$

Hence:

$$\Delta T_b = 0.52°C \times 0.866 = 0.45°C; \quad bp = 100.00°C + 0.45°C = 100.45°C$$
$$\Delta T_f = 1.86°C \times 0.866 = 1.61°C; \quad fp = 0.00°C - 1.61°C = -1.61°C$$

---

* If a solid solution separates on freezing, this argument will not be valid. In practice, it is almost always pure ice that separates when dilute aqueous solutions freeze.

Osmotic Pressure.   Imagine an experiment in which two beakers, one containing pure water, the other a sugar solution, are placed under a bell jar (Figure 11.7). As time passes, it is found that the liquid level in the beaker containing the solution rises, while the level of pure water in the other beaker falls. Eventually, by evaporation and condensation, all this water is transferred to the solution; at the end of the experiment, the beaker that contained pure water is empty.

The driving force behind the process just described is the difference in vapor pressure of water in the two beakers. Water moves spontaneously from a region in which its vapor pressure is high (pure water) to a region in which its vapor pressure is low (sugar solution). The air in the bell jar is permeable only to water molecules; the nonvolatile solute is unable to move from one beaker to the other.

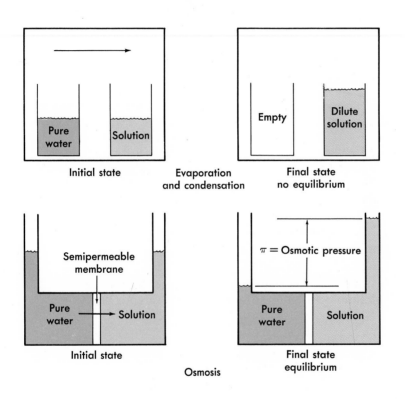

Initial state

Evaporation and condensation

Final state no equilibrium

Semipermeable membrane

Osmosis

Initial state

$\pi =$ Osmotic pressure

Final state equilibrium

Figure 11.7   Movement of water toward region of lower vapor pressure via evaporation or osmosis.

The apparatus shown at the bottom of Figure 11.7 can be used to achieve a result similar to that found in the bell jar experiment. Here, the sugar solution is separated from the pure water by a semipermeable membrane, which may be an animal bladder, a slice of vegetable tissue, or a piece of parchment. This membrane, by a mechanism which is poorly understood, allows solvent molecules to pass through it preferentially. As before, water moves from a region in which its vapor pressure or mole fraction is high to a region in which its vapor pressure or mole fraction is low. This process, taking place through a semipermeable membrane, is referred to as **osmosis**. In general, the term osmosis is used to describe any process in which one component of a solution moves preferentially through a barrier or membrane which is permeable only to it. The component moves from a region in which its mole fraction is high toward a region in which its mole fraction is low.

The passage of water molecules through a membrane into a solution may be prevented by applying pressure to the solution. The external pressure which is just sufficient to prevent osmosis is referred to as the **osmotic pressure** of a solution. Osmotic pressure may be measured in an apparatus such as that shown in Figure 11.8. The inner, porous tube A contains within it a strong, semipermeable membrane consisting of a film of copper(II) ferrocyanide, $Cu_2Fe(CN)_6$. This insoluble film is formed by allowing solutions containing Cu^{2+} and $Fe(CN)_6^{4-}$ ions to diffuse into each other through the walls of the tube. The tube is filled with pure water; the compartment B surrounding the tube is filled with the solution whose osmotic pressure is to be measured. Pressure is applied to the solution at C so as to maintain a constant level D in the capillary attached to the tube containing pure water. This pressure is, by definition, the osmotic pressure.

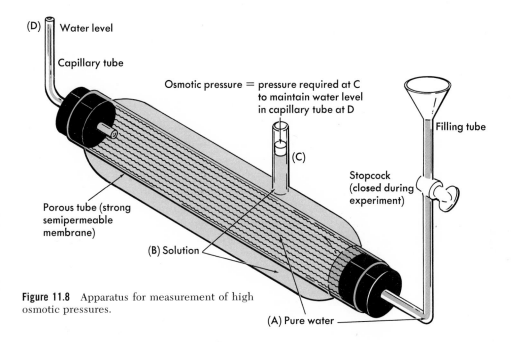

Figure 11.8 Apparatus for measurement of high osmotic pressures.

It is found experimentally that the osmotic pressure π of a dilute non-electrolyte solution, like vapor pressure lowering, boiling point elevation, and freezing point lowering, is a colligative property. The osmotic pressure is directly proportional to the concentration of solute and independent of its nature; the equation relating osmotic pressure of a solution to its concentration may be written in the form

Since molarity (M) is equal to the number of moles per liter (n/V), $\pi V = nRT$.

$$\pi = MRT \qquad (11.9)$$

in which π = osmotic pressure, M = molarity, R = gas constant (0.0821 lit atm/mole °K), and T = temperature in °K.

Substitution into Equation 11.9 shows that the osmotic pressure, even in dilute solution, is comparatively great. In a 0.10 M solution at 25°C, for example,

$$\pi = 0.10 \, \frac{\text{moles}}{\text{lit}} \times 0.0821 \, \frac{\text{lit atm}}{\text{mole °K}} \times 298°K = 2.45 \text{ atm}$$

A pressure of 2.45 atm is equivalent to a column of water 83 feet high. This may give some indication of the driving force behind osmosis and the difficulties involved in measuring osmotic pressure accurately with ordinary membranes.

Osmosis plays a vital role in many biological processes. Nutrient and waste materials are transported by osmosis through the cell walls of animal tissues, which show varying degrees of permeability to different solutes. A striking example of a natural osmotic process is afforded if we observe under a microscope the behavior of blood cells placed in pure water. Water passes through the walls to dilute the solution inside the cell, which swells and eventually bursts, releasing its red pigment. If the blood cells are placed in a concentrated sugar solution, the reverse process occurs; the cells shrink and shrivel up as water moves out into the sugar solution. To avoid effects such as these, solutions used in intravenous feeding must be carefully adjusted in concentration so that they have the same osmotic pressure as the solution inside the cells.

Plant as well as animal cell walls can act as semipermeable membranes. Flowers immersed in sugar or salt solution wilt as they are dehydrated by osmosis; if transferred to pure water, they appear to regain their freshness as water moves back into the cells. It has been suggested that the flow of sap up a tree results from osmosis; the movement of water through the roots into the concentrated nutrient solution inside could easily develop an osmotic pressure sufficient to push fluid to the top of the highest tree.

Determination of Molecular Weights from Colligative Properties

In previous chapters we discussed the determination of molecular weights from gas density data. While this method works quite well for gases or volatile liquids, it cannot be applied to solids such as sugar or urea which decompose on heating. An alternative approach, applicable to a wide variety of nonelectrolytes, involves the measurement of the colligative properties of their solutions. For example, from the measurement of the freezing point lowering of a solution of known composition, we can calculate the molecular weight of the solute, making use of Equation 11.8. Example 11.8 illustrates the calculations involved.

What will happen to the freezing point of this solution as more and more ice separates?

EXAMPLE 11.8. A solution of 1.250 g of a certain nonelectrolyte in 20.0 g of water freezes at −1.06°C. Calculate the molecular weight of the solute.

SOLUTION. From the information given, we can readily calculate the molality, using Equation 11.8. Knowing the molality, we can then calculate the number of moles of solute present. Since the number of grams of solute is given, it should then be possible to determine the number of grams per mole, i.e., the gram molecular weight of the solute.

Step 1. $\Delta T_f = 1.86°C \times m = 1.06°C$

$$m = \frac{1.06}{1.86} = 0.570 = \frac{\text{no. of moles solute}}{\text{no. of kg solvent}}$$

Step 2. no. of moles solute = m × no. of kg solvent

$$= (0.570)(0.0200) = 0.0114$$

Step 3.   0.0114 moles = 1.250 g

$$\text{no. of grams in 1 mole} = 1 \text{ mole} \times \frac{1.250 \text{ g}}{.0114 \text{ mole}} = 110 \text{ g}$$

MW = 110

It is, of course, possible to determine molecular weights from measurements of the vapor pressure lowering, boiling point elevation, or the osmotic pressure of a solution, using the appropriate equation (11.6 to 11.9) relating the particular colligative property to the concentration of solute. Freezing point lowerings are perhaps most commonly used because the effect is comparatively large (compare, for example, the constants for the freezing point lowering, 1.86°C, and the boiling point elevation, 0.52°C, for water solutions) and readily measured. For nonelectrolytes that are insoluble in water, it is usually possible to choose a suitable organic solvent. Camphor, which has a particularly large freezing point depression constant, 40°C, is often used.

Osmotic pressure measurements are frequently used to determine the molecular weights of polymeric materials, in which the molar concentration of solute is ordinarily extremely low. Consider, for example, a solution containing 10 g of a polymer of molecular weight 10,000 dissolved in 1 kg of water. We can calculate that such a solution would have a molality of 0.0010; its osmotic pressure would be approximately 0.025 atm (19 mm Hg) as compared to a freezing point lowering of only 0.0019°C. The principal difficulty associated with osmotic pressure measurements has always been that of finding a membrane which is both semipermeable and strong enough to withstand pressure. Recently, instruments have come on the market which are capable of measuring osmotic pressures of solutions of polymers covering a wide range of molecular weights.

Electrolytes

We have seen that in dilute solutions of nonelectrolytes, the vapor pressure lowering, boiling point elevation, freezing point depression, or osmotic pressure are directly proportional to the concentration of solute particles. If we extend this relationship to electrolyte solutions, it would seem that at a given molality a salt such as sodium chloride should have a greater effect on the colligative properties of water than a nonelectrolyte such as sugar. If we dissolve a mole of sugar in water, we obtain *one* mole of solute molecules; one mole of NaCl, on the other hand, produces *two* moles of ions. Again, we would predict that a 1 molal solution of $CaCl_2$ (*three* moles of ions per mole of solute) should have a lower vapor pressure, a higher boiling point, a lower freezing point, and a greater osmotic pressure than a 1 molal solution of NaCl.

Qualitatively, the predictions we have just made are confirmed experimentally. A 1 molal solution of sodium chloride has a lower vapor pressure than a 1 molal solution of sugar; the vapor pressure of a 1 m solution of calcium chloride is still lower. Many electrolytes form saturated solutions whose vapor pressures are so low that the solids pick up water when exposed to moist air. This phenomenon, known as **deliquescence**, is common with very

A beaker ⅔ full of conc. H₂SO₄ frequently overflows if allowed to stand overnight. Why?

soluble electrolytes of high charge. Calcium chloride, for example, forms a saturated solution with a vapor pressure only 20 per cent that of pure water; if the dry salt is exposed to air in which the relative humidity is greater than 20 per cent it deliquesces. The saturated solution continues to take on water until its vapor pressure is equal to that of the water in the air, that is, until equilibrium is reached.

The freezing point of an electrolyte solution, like the vapor pressure, is normally lower than that of a nonelectrolyte at the same molality. Sodium chloride or calcium chloride is commonly used to remove ice from the highway after a snowstorm.

Quantitatively, one might expect that the equations (11.6 to 11.9) which relate the colligative properties of nonelectrolyte solutions to their concentrations could be adapted to electrolytes by introducing a multiplier, n, equal to the number of moles of ions formed from 1 mole of electrolyte. Thus, for the freezing point lowering and boiling point elevation, we would have

$$\Delta T_f = n(1.86°C)m \tag{11.10}$$

$$\Delta T_b = n(0.52°C)m \tag{11.11}$$

in which n = 2 for NaCl, KCl, MgSO₄, n = 3 for CaCl₂, H₂SO₄, and so on. Experimentally, it is found that the observed freezing point lowerings, even in quite dilute solution, are somewhat smaller than these equations would predict (Table 11.5).

TABLE 11.5 FREEZING POINTS OF SOLUTIONS OF KCl AND MgSO₄

m	ΔT_f OBSERVED		n CALCULATED (EQUATION 11.10)	
	KCl	MgSO₄	KCl	MgSO₄
0.005	0.0182	0.0158	1.96	1.70
0.01	0.0361	0.0301	1.94	1.62
0.02	0.0714	0.0573	1.92	1.54
0.05	0.175	0.132	1.88	1.42
0.10	0.346	0.246	1.86	1.32
0.20	0.682	0.454	1.83	1.22
0.50	1.67	1.00	1.80	1.08

It is evident from the data in Table 11.5 that for both KCl and MgSO₄, n, as calculated from the freezing point lowering, is less than the expected value of 2, approaching this as a limit in very dilute solution. Stated another way, at finite concentrations, the observed freezing point lowering is less than that calculated from Equation 11.10 with n = 2. It may also be observed that the deviations from ideal behavior are considerably greater for MgSO₄, in which we are dealing with +2 and −2 ions, than with KCl (+1, −1 ions). Conductivity data for solutions of these two salts show precisely the same trends.

The deviations of colligative properties of electrolyte solutions from the simple relationships predicted on the basis of completely independent solute particles have been explained in various ways. In 1887 Arrhenius suggested that strong electrolytes were incompletely dissociated in solution. According to the theory of Arrhenius a solution of potassium chloride should consist of an equilibrium mixture of K^+ ions, Cl^- ions, and KCl molecules:

$$KCl \rightleftharpoons K^+ + Cl^-$$

At high concentrations, enough KCl molecules would be present to make the conductivity or freezing point lowering appreciably less than that calculated on the basis of complete dissociation.

In the first two decades of the twentieth century, x-ray studies conducted on solid salts such as potassium chloride showed them to be made up of individual ions (K^+, Cl^-) rather than molecules. This evidence tended to discredit the Arrhenius picture of electrolyte solutions. If K^+ and Cl^- ions exist as separate entities in the solid state, it is difficult to understand why they should combine to form molecules in solution. By 1920 it was generally conceded that strong electrolytes were completely dissociated in solution and that there must be some other explanation for the observed anomalies in conductivities and colligative properties.

A quantitative explanation of the properties of dilute solutions of strong electrolytes was put forth by Debye and Hückel in 1923. Their theory was based on the idea that, as a result of electrical attraction between positive and negative ions, there are in solution around a given ion more ions of opposite than of like charge. An ion in solution will surround itself with an **ionic atmosphere** containing an excess of oppositely charged ions. The existence of such an atmosphere lowers the conductivity of the solution by reducing the mobility of the ions; it has a similar though less obvious effect on such colligative properties as freezing point lowering and boiling point elevation. One would expect, in agreement with experiment, that ionic atmosphere effects would be most pronounced in concentrated solutions of electrolytes of high charge.

By taking into account the effect of the ionic atmosphere, Debye and Hückel were able to derive an equation for the variation with concentration of such properties as conductivity and freezing point lowering. The equations they proposed worked quite well in very dilute solutions, below about 0.01 m. Various modifications of the Debye-Hückel equations, notably that of Onsager for the conductivity of electrolyte solutions, extended their range to as high as 0.1 m. In more concentrated solutions, however, all attempts to extend the Debye-Hückel treatment have proved unsatisfactory.

Recently, in attempting to explain the properties of concentrated solutions of electrolytes, chemists have gone back to an earlier concept proposed by Bjerrum in 1909. He suggested that in concentrated solutions, a considerable fraction of the electrolyte could be tied up in the form of "ion-pairs." Thus when solid $MgSO_4$ dissolves to give a concentrated solution, it may dissolve primarily as $Mg^{2+}SO_4^{2-}$ ion pairs rather than as individual Mg^{2+} and SO_4^{2-} ions. The strength of the electrostatic forces holding the ion pair together should be directly related to the charge of the ions. Ions of high charge such as Mg^{2+} would be expected to form ion pairs more readily than ions of low charge such as K^+. The ratio of ion pairs to free ions increases with concentration, in agreement with the observed trends in conductivity and colligative properties.

It is now generally realized that a comprehensive structural model of electrolyte solutions must include not only ion-ion interactions (the ionic atmosphere of Debye or the ion pair of Bjerrum) but also interactions between ions and polar solvent molecules. In this connection, there is considerable evidence to indicate that ions may distort the peculiar hydrogen-bonded structure of water. Until we know more about the molecular structure of liquid water, it seems unlikely that we will be able to give a completely satisfactory explanation for the properties of electrolyte solutions.

Despite the observed deviations of electrolyte solutions from Equation 11.10, it is possible to use freezing point measurements to decide upon the mode of ionization of a salt in water solution. To illustrate the principles involved, consider the compound $NaHCO_3$ (Example 11.9).

EXAMPLE 11.9. The freezing point of a 0.010 m solution of $NaHCO_3$ is $-0.038°C$. On the basis of this information, decide whether this compound ionizes in water as

$$NaHCO_3(s) \longrightarrow Na^+ + HCO_3^- \qquad (n = 2)$$

or $\qquad NaHCO_3(s) \longrightarrow Na^+ + H^+ + CO_3^{2-} \qquad (n = 3)$

SOLUTION. Let us calculate the value of n in Equation 11.10.

$$0.038°C = n(1.86°C)(0.010)$$

$$n = \frac{0.038}{0.0186} = 2.0$$

Clearly, the first mode of ionization is indicated. This is confirmed by conductivity measurements which show that the conductivity of 0.010 M $NaHCO_3$ is very nearly equal to that of 0.01 M NaCl but considerably lower than that of 0.01 M Na_2CO_3.

PROBLEMS

11.1 Explain briefly what is meant by each of the following terms.

a. Supersaturated solution
b. Raoult's Law
c. Henry's Law
d. Interstitial solid solution
e. Colligative property

f. Ionic atmosphere
g. Osmotic pressure
h. Substitutional solid solution
i. Ion pair
j. Weak electrolyte

11.2 The solubility of sodium acetate in water at 20°C is 46 g per 100 g of water. A solution containing 100 g of sodium acetate in 150 g of water is prepared at a high temperature and cooled to 20°C. When a crystal of sodium acetate is added to this supersaturated solution, the excess solute comes out of solution. How many grams of sodium acetate crystallize?

11.3 Describe how you would prepare

a. 10.0 liters of 6.0 M NaOH.
b. A 2.0 m solution containing 60.0 g of sugar, $C_{12}H_{22}O_{11}$, in water.
c. A water solution of CH_3OH in which the mole fraction of CH_3OH is 0.250.

11.4 Calculate the mole fraction of each substance and the molality of KNO_3 in a solution containing 16.0 g of KNO_3 in 50.0 g of water.

11.5   Find

a. The number of grams of NaBr required to prepare 12.0 lit of 2.00 M solution.
b. The volume of 3.50 M KNO$_3$ that can be prepared from 150 g of KNO$_3$.
c. The molarity of a solution prepared by dissolving 20.0 g of CuSO$_4$·5H$_2$O in a total of 125 ml of solution.

11.6   You wish to prepare 100 ml of 1.00 M KOH solution. Describe how this could be done, starting with

a. Solid KOH.
b. A 2.50 M KOH solution.
c. A 0.50 M KOH solution.

11.7   A solution is prepared by dissolving 85.1 g of NaBr in enough water to form one liter of solution. The density of the solution is 1.063 g/ml. Calculate the molarity, molality, and mole fraction of NaBr.

11.8   For each of the following pairs of substances, predict which will be the more soluble, first in benzene and then in water.

a. CH$_4$(bp = −161°C) or H$_2$(bp = −253°C)
b. NaCl or CCl$_4$
c. Naphthalene (mp = 80°C) or 2-methyl naphthalene (mp = 35°C)
d. H$_2$O$_2$ or C$_6$H$_{14}$

11.9   The solubility of CO$_2$ in water at 10°C is approximately 0.053 mole/lit at one atmosphere pressure.

a. Calculate the number of grams of CO$_2$ in a bottle containing 200 ml of a soft drink bottled under a CO$_2$ pressure of 1.20 atm at 10°C.
b. If this bottle is opened at 10°C in air, where the partial pressure of CO$_2$ is 1.0 mm Hg, how many grams of CO$_2$ comes out of solution? What volume does this CO$_2$ occupy at one atmosphere pressure?

11.10   What is the vapor pressure of water above a solution at 25°C (vp pure water = 23.76 mm Hg) containing 12.0 g of urea, CO(NH$_2$)$_2$, in 100 g of water?

11.11   The vapor pressure of pure benzene at 30°C is 118 mm Hg; that of toluene is 37 mm Hg. Assuming that benzene and toluene form an ideal solution, what is the total pressure over a solution containing 50.0 g of both liquids (C$_6$H$_6$ and C$_7$H$_8$)?

11.12   Determine the boiling point at 760 mm Hg and the freezing point of the following solutions.

a. 12.0 g of sugar, C$_{12}$H$_{22}$O$_{11}$, in 80.0 g of water.
b. 6.00 g of naphthalene, C$_{10}$H$_8$, in 120 g of benzene.

11.13   How many quarts of CH$_3$OH (d = 0.792 g/ml) should be added per gallon of water (d = 1.00 g/ml) to make an antifreeze that will protect an automobile radiator down to 0°F?

11.14   When 1.000 g of a certain nonelectrolyte is dissolved in 10.0 g of water, the freezing point of the solution is found to be −0.420°C. Calculate the molecular weight of the nonelectrolyte.

11.15   The freezing point of a 0.010 M solution of an electrolyte of empirical formula USO$_6$ is −0.036°C. The freezing point of a 0.010 M solution of a similar salt of empirical formula UBr$_2$O$_2$ is −0.054°C. Deduce what ions are present in these solutions.

11.16   Calculate the osmotic pressure of a solution prepared by dissolving 5.00 g of sugar, C$_{12}$H$_{22}$O$_{11}$, in enough water to give 200 ml of solution at 25°C.

295

11.17 A solution prepared by dissolving 1.00 g of a certain polymer to give 100 ml of solution has an osmotic pressure of 38.0 mm Hg at 20°C. What is the molecular weight of the polymer?

11.18 Referring to Table 11.1,

a. Suggest a reason for the fact that the molarity of dilute H_2SO_4 is set at one-half the molarity of dilute HCl and dilute HNO_3.
b. Confirm by calculation that a solution of hydrochloric acid containing 36 per cent by weight of HCl and having a density of 1.18 g/ml is 12 M in HCl.
c. Describe how one would prepare 12 lit of dilute NH_3 from concentrated NH_3.

11.19 The solubility of potassium nitrate in water at 20°C is 2.77 moles/lit. A solution of this salt is prepared by dissolving 155 g at 80°C to form 375 ml of solution. Upon cooling to 20°C, no solid separates until a tiny crystal of KNO_3 is added, whereupon all of the excess solute comes out of solution. How many grams of solute separate?

11.20 Describe how you would prepare each of the following solutions.

a. A 1.20 molar solution of NaCl in water.
b. A 1.20 molal solution of NaCl in water.
c. A 0.20 M solution of $NiSO_4$ in water (the stockroom does not carry anhydrous $NiSO_4$; it does, however, have $NiSO_4 \cdot 6\ H_2O$).
d. A solution of KOH in CH_3OH in which the mole fraction of KOH is 0.20.

11.21 Calculate the mole fractions of all substances and the molality of the underlined substance in each of the following solutions.

a. 1.00 mole of $\underline{C_2H_5OH}$ and 6.00 moles of water.
b. 1.80 moles of $\underline{NaBr}$ and 512 g of water.
c. 16.0 g of CCl_4, 12.9 g of $CHCl_3$, and 19.1 g of C_6H_{14}.

11.22 Complete the following table. All the data refer to water solutions.

Solute	Grams Solute	Moles Solute	Volume Solution	Molarity
$NaNO_3$	25	—	—	1.2
$NaNO_3$	—	—	16 lit	0.023
KBr	91	—	450 ml	—
$_\wedge$Br	—	0.420	—	1.8

11.23 If you were asked to prepare 50 ml of 0.30 M KCl solution and were given a bottle of solid KCl, a supply of distilled water, a 0.20 M KCl solution, and a 0.40 M solution of KCl, describe four different methods which you could use. Which of these methods would be the fastest?

11.24 A certain solution is prepared by dissolving 15.2 g of NaCl in 197 g of water. The density of the resulting solution is 1.012 g/ml. Calculate the mole fraction of NaCl, the molality of NaCl, and the molarity of NaCl.

11.25 Predict which member of each of the following pairs of solutes will be more soluble in carbon tetrachloride at 20°C and 1 atm.

a. Benzene or hydrogen peroxide.
b. Anthracene or biphenyl (cf. Table 11.2).
c. Helium or argon (cf. Table 11.3).
d. Methane or propane.
e. Sodium fluoride or anthracene.
Repeat your predictions for water as a solvent.

11.26 Water saturated with air (20% O_2, 80% N_2) at 20°C contains 8.9×10^{-3} g/lit of dissolved oxygen. Estimate the solubility of pure oxygen in water at a pressure of 25 atm and a temperature of 20°C.

11.27 Calculate the vapor pressure of benzene above a solution containing 10.0 g of naphthalene, $C_{10}H_8$, in 100 g of benzene, C_6H_6, at 25°C. The vapor pressure of pure benzene at 25°C is 97.0 mm Hg.

11.28 Two organic liquids A and B have vapor pressures at 25°C of 150 mm Hg and 250 mm Hg respectively. Assuming Raoult's Law applies to both components, draw a graph of

  a. The partial pressure of A over the solution vs. the mole fraction of A from $X_A = 0$ to $X_A = 1$.
  b. The partial pressure of B vs. mole fraction of A.
  c. The total pressure over the solution vs. the mole fraction of A.

11.29 A beaker containing 20 g of sugar in 100 g of water and another containing 10 g of sugar in 100 g of water are placed under a bell jar and allowed to stand until equilibrium is reached. How much water will be transferred from one beaker to another?

11.30 Calculate the boiling points at 760 mm Hg and the freezing points of the following solutions.

  a. 50.0 g of sugar, $C_{12}H_{22}O_{11}$, in 50.0 g of water.
  b. 0.32 g of glucose, $C_6H_{12}O_6$, in 20.0 g of water.
  c. 16.0 g of NaCl in 185 g of water, assuming ideal behavior.

11.31 How many quarts of ethylene glycol, $C_2H_6O_2$ (density = 1.12 g/ml), should be added to five gallons of water (density = 1.0 g/ml) to make up an antifreeze which will protect an automobile radiator down to $-20°C$? At what temperature will this solution boil?

11.32 A student determines the molecular weight of a certain nonelectrolyte by dissolving 5.23 g of it in 168 g of water and measuring the freezing point of the solution to be $-0.510°C$. What is the molecular weight of the nonelectrolyte?

11.33 When 5.30 g of an organic solute are dissolved in 200 g of benzene, the freezing point of the solution is 4.20°C.

  a. What is the molecular weight of the solute?
  b. If the compound contains 9.4% H and 90.6% C, what is its molecular formula?

11.34 When placed in water, ammonia ionizes as follows: $NH_3 + H_2O \rightarrow NH_4^+ + OH^-$. Calculate the per cent ionization of NH_3 in a 0.010 molal solution from the fact that it freezes at $-0.0193°C$.

11.35 A storage battery contains a solution of H_2SO_4 (38 g of H_2SO_4 per 100 g of water). At this concentration, the apparent value of n is 2.50. At what temperature will the battery contents freeze?

11.36 Arrange the following water solutions in order of decreasing freezing point, assuming ideal behavior.

0.01 m sugar   0.01 m Li_2SO_4 0.02 m urea   0.01 m AlF_3

11.37 What is the osmotic pressure of a solution containing 5.00 g of sugar, $C_{12}H_{22}O_{11}$, in one liter of solution at 20°C?

*11.38 An inorganic chemist prepares a compound which he believes to be $[Co(NH_3)_5F]F_2 \cdot H_2O$. The compound could, however, be $[Co(NH_3)_5H_2O]F_3$. Describe two different physical methods he might use to distinguish between these two possibilities.

*11.39 Assuming that osmosis is responsible for sap rising in a tree, determine the approximate height to which the sap can rise if it is 0.10 M in sugar and the water outside the tree contains dissolved solids equivalent to a 0.02 M solution of sugar.

*11.40 A cell containing 1.68 g of sugar, $C_{12}H_{22}O_{11}$, in 20.0 g of water and another containing 2.45 g of a nonvolatile nonelectrolyte in 24.0 g of water are placed in an evacuated container and allowed to come to equilibrium. It is found that the total mass of the sugar solution at equilibrium is 24.9 g. What is the molecular weight of the nonelectrolyte?

*11.41 The Debye-Hückel theory, applied to the calculation of the freezing point lowering ΔT_f of a solution of KCl, predicts the relation

$$\Delta T_f = (\Delta T_f)_{ideal} (1 - 0.39 \sqrt{m})$$

in which $(\Delta T_f)_{ideal}$ is the freezing point lowering predicted by Equation 11.10 with $n = 2$, and m is the molality. Calculate ΔT_f from this equation at m = 0.005, 0.01, and 0.05 and compare to the values given in Table 11.5.

Equilibrium in
Chemical Systems

In Chapter 10 we discussed the physical equilibria that are possible in one-component systems. We pointed out that under certain conditions a pure substance can exist in more than one state of aggregation in the same container. If solid iodine is introduced into a closed vessel at 100°C, a portion of it passes directly into the gas state by the process called sublimation. At first, the violet color of the iodine vapor intensifies as molecules move predominantly from the solid to the gas phase. After a short time, the intensity of the violet color stabilizes, implying that the rates of sublimation and condensation have become equal. We are now at a position of **physical equilibrium**: experiment shows that at this temperature the partial pressure of iodine vapor in equilibrium with solid iodine is 45.6 mm Hg (0.0600 atm). Alternatively, we could say that the equilibrium concentration of iodine vapor, as calculated from the Ideal Gas Law, is

$$\frac{n}{V} = \frac{p}{RT} = \frac{(0.0600 \text{ atm})}{\left(0.0821 \frac{\text{lit atm}}{\text{mole °K}}\right)(373 \text{ °K})} = 0.00196 \text{ mole/lit}$$

This equilibrium concentration in the gas phase is a characteristic property of the pure substance iodine at 100°C. From a slightly different point of view, we can say that the position of the equilibrium between solid and gaseous iodine at 100°C is such that the concentration of iodine vapor is 0.00196 moles/lit.

When a system contains substances that react chemically, a situation completely analogous to that described previously can arise. Consider, for example, what happens when a sample of calcium carbonate is introduced into a closed container at a temperature of 800°C. The calcium carbonate partially decomposes by way of the reaction

$$CaCO_3(s) \longrightarrow CaO(s) + CO_2(g)$$

As the concentration of carbon dioxide builds up in the gas phase, some of the

CO_2 molecules react with calcium oxide:

$$CaO(s) + CO_2(g) \longrightarrow CaCO_3(s)$$

Eventually, the rates of these two competing reactions become equal; we arrive at a position of **chemical equilibrium**. Experiment shows that at 800°C the position of this equilibrium is such that the concentration of carbon dioxide is 0.0036 moles/lit. The concentration of carbon dioxide is a characteristic property of the $CaCO_3$-CaO-CO_2 equilibrium at 800°C.

It should be obvious that there is a close resemblance between the chemical equilibrium expressed by the equation

$$CaCO_3(s) \rightleftharpoons CaO(s) + CO_2(g)$$

and the physical equilibrium

$$I_2(s) \rightleftharpoons I_2(g)$$

Unfortunately, the great majority of systems at chemical equilibrium cannot be described as simply as the $CaCO_3$-CaO-CO_2 system. To illustrate the complications involved, consider the system

$$2 \ HI(g) \rightleftharpoons H_2(g) + I_2(g) \tag{12.1}$$

Equation 12.1 implies that if HI is put in a closed container, some but not all of it will dissociate.

Here, the equilibrium concentration of iodine in the gas phase will not be a fixed value at a particular temperature. Instead, it will depend upon the concentrations of the other two gases, hydrogen and hydrogen iodide, involved in the equilibrium system.

In general, we always find that when a system at chemical equilibrium involves two or more species in solution, in either the gas or the liquid phase, the concentrations of these species are interrelated. Throughout the remainder of this chapter, we shall be concerned with the nature of this relationship. To be specific, we shall be interested in arriving at expressions which will enable us to predict the equilibrium concentration of one species relative to that of others. In this chapter we shall deal with equilibria in gaseous systems. In later chapters the principles governing chemical equilibria will be extended to liquid systems, particularly to aqueous solutions.

12.1 AN EXAMPLE OF CHEMICAL EQUILIBRIUM. THE HI-H₂-I₂ SYSTEM

At a temperature of 520°C the three substances hydrogen iodide, hydrogen, and iodine, are gaseous; if placed in a container they will react chemically to a state of equilibrium:

$$2 \ HI(g) \rightleftharpoons H_2(g) + I_2(g) \tag{12.1}$$

To illustrate the principles of chemical equilibrium, let us consider a series of experiments in which HI-H₂-I₂ systems are set up in various ways, always at the same temperature, 520°C, and always in the same container, which will be taken to have a volume of 10.0 liters. In these experiments we shall admit various amounts of HI, H₂, and I₂ and determine the concentrations of these three species after equilibrium has been reached.

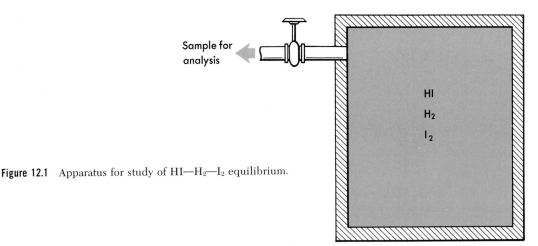

Sample for
analysis

HI

H₂

I₂

Figure 12.1 Apparatus for study of HI—H₂—I₂ equilibrium.

10.0 liter container
maintained at 520°C

System I. Two (2.00) moles of HI are admitted to the evacuated 10.0 liter container. The HI partially decomposes, at first rapidly, then more slowly, to form H_2 and I_2. After a period of time, the rates of the forward and reverse reactions become equal; the concentrations of the three substances reach their equilibrium values and do not change thereafter.

To ascertain the position of the equilibrium under these conditions, we might measure the number of moles of iodine present. If we do this, we find that there are 0.20 moles of I_2 in the 10.0 liter container. Referring to Equation 12.1, we note that for every mole of I_2 formed, one mole of H_2 is formed simultaneously. It follows, then, that 0.20 moles of H_2 must be present at equilibrium in this system. Again, Equation 12.1 tells us that two moles of HI are required to form one mole of I_2; this means that 0.40 moles of HI must have decomposed to account for the 0.20 moles of I_2 found at equilibrium. Consequently, of the 2.00 moles of HI that we started with,

$$(2.00 - 0.40) \text{ moles} = 1.60 \text{ moles}$$

of HI must remain at equilibrium.

From this information, we can readily calculate the concentrations of I_2, H_2, and HI at equilibrium:

$$[I_2] = 0.20 \text{ moles}/10.0 \text{ liters} = 0.020 \text{ mole/liter}$$
$$[H_2] = 0.20 \text{ moles}/10.0 \text{ liters} = 0.020 \text{ mole/liter}$$
$$[HI] = 1.60 \text{ moles}/10.0 \text{ liters} = 0.160 \text{ mole/liter}$$

The square brackets, here and elsewhere throughout the remainder of the text, are used to represent equilibrium concentrations in moles/liter.

System II. Another possible way to achieve equilibrium in the HI-H₂-I₂ system would be to start with the elements H_2 and I_2 and allow them to react together to form HI. Let us suppose that we introduce into the evacuated 10.0 liter container at 520°C, one (1.00) mole of H_2 and one (1.00) mole of I_2. In this case, reaction occurs to form HI; the concentrations of H_2 and I_2 decrease with time. Experimentally we find that equilibrium is attained when 1.60 moles of HI has been formed. According to Equation 12.1, this requires the

301

consumption of 0.80 moles of I_2 (one mole of $I_2 \rightarrow$ two moles of HI) and 0.80 moles of H_2. Consequently, in addition to 1.60 moles of HI, we must have, at equilibrium,

$$1.00 - 0.80 = 0.20 \text{ moles } I_2$$
and
$$1.00 - 0.80 = 0.20 \text{ moles } H_2$$

We deduce that the equilibrium concentrations of the three substances are:

$$[I_2] \ = 0.20 \text{ moles/10.0 liters} = 0.020 \text{ mole/liter}$$
$$[H_2] = 0.20 \text{ moles/10.0 liters} = 0.020 \text{ mole/liter}$$
$$[HI] = 1.60 \text{ moles/10.0 liters} = 0.160 \text{ mole/liter}$$

You will note that the equilibrium concentrations of the three substances in System II are precisely the same as those in System I. This is hardly surprising in view of the fact that the two systems have the same overall composition; in both cases two gram atomic weights of hydrogen and two gram atomic weights of iodine are present. The only difference between the two systems is that in one case all the hydrogen and all the iodine were originally present as HI, while in the other they were originally present as the elementary substances H_2 and I_2. We approached equilibrium from opposite directions in the two systems and, logically enough, reached the same final state. These two systems illustrate the general principle that the final equilibrium state of a system of fixed overall composition is independent of the nature and amounts of the species initially present.

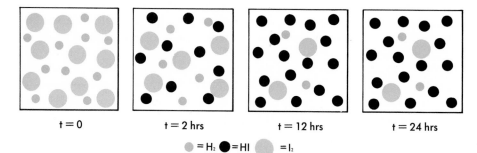

| t = 0 | t = 2 hrs | t = 12 hrs | t = 24 hrs |

● = H₂   ⬤ = HI   ◯ = I₂

Figure 12.2 Attainment of equilibrium in HI—H_2—I_2 (System II). The small circles represent H_2 molecules, the medium-sized circles HI molecules, and the large circles I_2 molecules.

System III. Let us now study the position of the equilibrium attained by a system that has an overall composition quite different from that of Systems I and II. To be specific, let us start with one mole of HI, one mole of H_2, and no I_2. Qualitatively, we might predict that some of the HI would decompose to form some I_2 and, at the same time, an equivalent amount of H_2. That is, at equilibrium we would expect to have somewhat less than one mole of HI, a small but definite amount of I_2, and somewhat more H_2 than the one mole we started with.

Experiment confirms these predictions. At 520°C in a 10.0 liter container, we find that at equilibrium there are present about 0.970 moles of HI, 0.015 moles of I_2, and 1.015 moles of H_2. The equilibrium concentrations are:

$$[HI] = 0.970 \text{ moles/10.0 liters} = 0.0970 \text{ mole/liter}$$
$$[I_2] \ = 0.015 \text{ moles/10.0 liters} = 0.0015 \text{ mole/liter}$$
$$[H_2] = 1.015 \text{ moles/10.0 liters} = 0.1015 \text{ mole/liter}$$

It should be obvious at this stage, if it were not at the beginning of this discussion, that there are an infinite number of systems that we could start with in studying the position of the equilibrium between HI, H_2, and I_2. Life is too short to look at all of these systems, but perhaps we would have patience enough to carry out experiments with the five systems listed in Table 12.1.

TABLE 12.1 EQUILIBRIUM AT 520°C IN THE HI-H_2-I_2 SYSTEMS (V = 10.0 LITERS)

SYSTEM	INITIAL No. OF MOLES			EQUILIBRIUM No. OF MOLES			EQUILIBRIUM CONCENTRATION		
	HI	H_2	I_2	HI	H_2	I_2	[HI]	[H_2]	[I_2]
I	2.00	0	0	1.60	0.20	0.20	0.160	0.020	0.020
II	0	1.00	1.00	1.60	0.20	0.20	0.160	0.020	0.020
III	1.000	1.000	0	0.970	1.015	0.015	0.0970	0.1015	0.0015
IV	1.00	1.00	1.00	2.40	0.30	0.30	0.240	0.030	0.030
V	0	2.000	1.000	1.894	1.053	0.053	0.1894	0.1053	0.0053

Looking at the data in Table 12.1 we might wonder whether there could be any quantitative relationship between the equilibrium concentrations of H_2, I_2, and HI that would be valid for all five systems. Amazingly enough, there is. *If one multiplies the equilibrium concentration of H_2 by that of I_2 and divides by the square of the equilibrium concentration of HI, a number is obtained* (approximately 0.016) *that is the same for all the systems.* That is:

Systems I, II: $\dfrac{[H_2] \times [I_2]}{[HI]^2} = \dfrac{(0.020)(0.020)}{(0.160)^2} = \dfrac{4.0 \times 10^{-4}}{2.56 \times 10^{-2}} = 0.016$

System III: $\dfrac{[H_2] \times [I_2]}{[HI]^2} = \dfrac{(0.1015)(0.0015)}{(0.0970)^2} = \dfrac{1.5 \times 10^{-4}}{0.941 \times 10^{-2}} = 0.016$

System IV: $\dfrac{[H_2] \times [I_2]}{[HI]^2} = \dfrac{(0.030)(0.030)}{(0.240)^2} = \dfrac{9.0 \times 10^{-4}}{5.76 \times 10^{-2}} = 0.016$

System V: $\dfrac{[H_2] \times [I_2]}{[HI]^2} = \dfrac{(0.1053)(0.0053)}{(0.1894)^2} = \dfrac{5.6 \times 10^{-4}}{3.59 \times 10^{-2}} = 0.016$

Experimentally, this simple relationship is found to hold for any equilibrium system containing H_2, I_2, and HI at 520°C. Regardless of the relative amounts of the three gases we start with or the volume of the container, we eventually arrive at an equilibrium whose position is described by the condition that

$$\frac{[H_2] \times [I_2]}{[HI]^2} = 0.016 \text{ (at 520°C)}.$$

Further experiments with many different systems containing HI, H_2, and I_2 at various temperatures leads us to the following conclusion:
At any given temperature, the quantity

$$\frac{[H_2] \times [I_2]}{[HI]^2}$$

is a constant, independent of the amounts of H_2, I_2, and HI that one starts with or of the volume of the container. This constant is referred to as the equilibrium constant K_c for the reaction.

$$2 \text{ HI(g)} \rightleftharpoons \text{H}_2(g) + \text{I}_2(g)$$

At 520°C the numerical value of K_c is 0.016; at 800°C it is a somewhat larger number, 0.024.

12.2 THE GENERAL FORM OF THE EQUILIBRIUM CONSTANT EXPRESSION

We have seen that for the reaction

$$2 \text{ HI(g)} \rightleftharpoons \text{H}_2(g) + \text{I}_2(g)$$

the equilibrium constant expression takes the form

$$K_c = \frac{[\text{H}_2] \times [\text{I}_2]}{[\text{HI}]^2}$$

At this stage, you might well ask how one can predict in advance, for a given reaction, what the mathematical form of the expression for K_c will be. The answer to this question may become a little more obvious when one learns that for the reaction

$$2 \text{ H}_2\text{O(g)} \rightleftharpoons 2 \text{ H}_2(g) + \text{O}_2(g)$$

the ratio

$$\frac{[\text{H}_2]^2 \times [\text{O}_2]}{[\text{H}_2\text{O}]^2}$$

has a constant value at a given temperature; that is,

$$K_c = \frac{[\text{H}_2]^2 \times [\text{O}_2]}{[\text{H}_2\text{O}]^2}$$

and that for the reaction

$$\text{N}_2(g) + 3 \text{ H}_2(g) \rightleftharpoons 2 \text{ NH}_3(g)$$
$$K_c = \frac{[\text{NH}_3]^2}{[\text{N}_2] \times [\text{H}_2]^3}$$

From this example, one can deduce the Law of Chemical Equilibrium, which can be stated as follows:

For any reaction, there is a condition governing the relative concentrations of products and reactants present at equilibrium at a particular temperature. This condition may be expressed by a number, called the equilibrium constant, which takes the form of a quotient. The numerator of this quotient is obtained by multiplying together the equilibrium concentrations of products, each raised to a power equal to its coefficient in the equation used to describe the reaction. The denominator is obtained in the same manner, using the equilibrium concentrations of reactants.

Translating this paragraph into the language of algebra, we can say that for the general gas-phase reaction,

$$a\text{A(g)} + b\text{B(g)} \rightleftharpoons c\text{C(g)} + d\text{D(g)}$$

where A, B, C, and D represent different substances and a, b, c, and d are the coefficients of the balanced equation

$$K_c = \frac{[C]^c \times [D]^d}{[A]^a \times [B]^b}$$ (12.2)

Express K_c for the reaction $SO_3(g) \rightarrow SO_2(g) + \frac{1}{2} O_2(g)$.

The Law of Chemical Equilibrium, sometimes called the Law of Mass Action, was first suggested in 1846 by Guldberg and Waage, Norwegian chemists, in a rather ambiguous form. It was later shown that the Law follows rigorously from thermodynamics. It has been checked thoroughly by experiment, and, properly interpreted, can be applied to any system in chemical or physical equilibrium.

We should point out that the condition on the relative concentrations of products and reactants present at equilibrium in a particular system can be expressed in an infinite number of ways. Consider, for example, the HI-H_2-I_2 system. We have chosen to express the condition on this equilibrium at 520°C by saying that

$$\frac{[H_2] \times [I_2]}{[HI]^2} = 0.016$$

Clearly, it would be equally valid to invert this expression and say that

$$\frac{[HI]^2}{[H_2] \times [I_2]} = \frac{1}{0.016} = 63$$

Alternatively, we could operate on the original equation by taking the square root of both sides, writing

$$\frac{[H_2]^{\frac{1}{2}} \times [I_2]^{\frac{1}{2}}}{[HI]} = (0.016)^{\frac{1}{2}} = 0.13$$

Each of the three equations we have written, and many others that we could write, is a valid way of expressing the condition governing the equilibrium concentrations of H_2, I_2, and HI at 520°C. Clearly, it would be ambiguous at the very least to say that "the equilibrium constant for the HI-H_2-I_2 system is 0.016." The equilibrium constant for a particular system takes on meaning only when it is associated with an equation written to describe that reaction. Thus, we have

$$2\,HI(g) \rightleftharpoons H_2(g) + I_2(g), \quad K_c = \frac{[H_2] \times [I_2]}{[HI]^2} = 0.016$$

or

$$H_2(g) + I_2(g) \rightleftharpoons 2\,HI(g), \quad K_c' = \frac{[HI]^2}{[H_2] \times [I_2]} = 63$$

or

$$HI(g) \rightleftharpoons \tfrac{1}{2} H_2(g) + \tfrac{1}{2} I_2(g), \quad K_c'' = \frac{[H_2]^{\frac{1}{2}} \times [I_2]^{\frac{1}{2}}}{[HI]} = 0.13$$

12.3 APPLICATIONS OF THE EQUILIBRIUM CONSTANT EXPRESSION

A knowledge of the equilibrium constant for a particular reaction is valuable to the chemist who wishes to carry out that reaction in the laboratory or, on a large scale, in the chemical plant. The types of problems to which it can be applied are conveniently classified into three categories. The equilibrium constant can be used to predict:

1. The direction in which a chemical system will move in order to reach equilibrium.
2. The extent to which a chemical reaction will occur.
3. The effect that a change in conditions will have upon a chemical system at equilibrium.

We shall now consider in turn each of these applications of the equilibrium constant expression. (The third application will be discussed in Sections 12.4 and 12.5.)

Prediction of the Direction of Reaction

Consider the general gas-phase reaction

$$aA(g) + bB(g) \rightleftharpoons cC(g) + dD$$

for which

$$K_c = \frac{[C]^c[D]^d}{[A]^a[B]^b}$$

The fact that it is possible to write an equilibrium constant, which has a definite numerical value, for this system means that if we start with a mixture containing *only* reactants, A and B, a reaction must occur to form products C and D. In order to achieve equilibrium, the concentrations of C and D must increase by a finite amount from their original value of zero; the concentrations of A and B must decrease by a corresponding amount. By the same token, if we start with a mixture containing only C and D, the reverse reaction must occur until the concentration ratio satisfies the equilibrium constant expression for the equilibrium constant.

Let us suppose now that one starts with a mixture containing all but one of the species A, B, C, and D, involved in this system. Specifically, consider what happens when we start with a mixture of A, B, and D, containing no C. One can readily see that since the equilibrium constant K_c is a finite number other than zero, the reaction must proceed from left to right to form at least a small amount of C. In doing so, the concentrations of A and B will necessarily decrease and that of D will increase. In other words, the equilibrium concentrations of C and D will be greater than their original concentrations, while those of A and B will be reduced.

If one starts with a mixture containing all four species, A, B, C, and D, it is not quite so obvious whether reaction will proceed in the forward or reverse direction. This decision can be made, however, if the original concentration quotient is compared to that which must be established at equilibrium. Specifically, we can distinguish three possibilities:

Note the impor-
tance of distin-
guishing between
equilibrium con-
centrations and
original concen-
trations.

1. If $\dfrac{(\text{orig. conc. C})^c\,(\text{orig. conc. D})^d}{(\text{orig. conc. A})^a\,(\text{orig. conc. B})^b} < K_c$

The reaction must proceed to the right, i.e., $aA + bB \rightarrow cC + dD$, in order that equilibrium be established. In other words, the concentrations of C and D increase at the expense of A and B until the quotient becomes equal to K_c.

2. If $\dfrac{\text{(orig. conc. C)}^c \text{ (orig. conc. D)}^d}{\text{(orig. conc. A)}^a \text{ (orig. conc. B)}^b} > K_c$

the reaction must proceed to the left, i.e., $cC + dD \rightarrow aA + bB$, in order that equilibrium be established. In this case the concentrations of C and D decrease and those of A and B increase until equilibrium is achieved.

3. If $\dfrac{\text{(orig. conc. C)}^c \text{ (orig. conc. D)}^d}{\text{(orig. conc. A)}^a \text{ (orig. conc. B)}^b} = K_c$

the system is already at equilibrium and no reaction occurs in either direction.

These general principles are illustrated in Example 12.1.

EXAMPLE 12.1. Among the products that come out of the exhaust system of an automobile are carbon dioxide and the extremely toxic gas, carbon monoxide. In the presence of oxygen, the following equilibrium is established:

$$CO_2(g) \rightleftharpoons CO(g) + \tfrac{1}{2} O_2(g)$$

At 100°C, K_c for this reaction is about 10^{-36}. Assume that, in the gas coming out of the exhaust, the concentration of CO is 1% that of CO_2. Predict the direction in which reaction will occur to approach equilibrium

    a. When there is no oxygen present originally.

    b. When the concentration of oxygen is 10^{-2} mole/lit (its normal concentration in air).

SOLUTION

    a. If there is no oxygen present originally, reaction must proceed from left to right (i.e., some CO_2 must decompose) to establish equilibrium. Notice that this is true despite the fact that K_c for this reaction is an extremely small number. Actually, the amount of CO_2 that decomposes is minute; the concentration of CO at equilibrium is only slightly greater than it was to begin with. From a practical standpoint, our most important conclusion is that, in the absence of oxygen, lethal amounts of CO may be present.

    b. Here, we need to compare the original concentration quotient

$$\frac{\text{(orig. conc. CO) (orig. conc. } O_2)^{\frac{1}{2}}}{\text{(orig. conc. } CO_2)}$$

to that required for equilibrium ($K_c = 10^{-36}$). We note from the statement of the problem that

$$\text{(orig. conc. CO)} = (0.01) \text{ (orig. conc. } CO_2)$$
$$\text{(orig. conc. } O_2) \ = 10^{-2}$$

Consequently, the original concentration quotient must be

$$(0.01) (10^{-2})^{\frac{1}{2}} = (0.01) (0.10) = 10^{-3}$$

Since this number, 10^{-3}, is greater than K_c (10^{-36}), reaction must take place from right to left, with carbon monoxide being converted to carbon dioxide.

At what conc. O_2 would no reaction occur in either direction?

Prediction of the Extent of Reaction

From the magnitude of the equilibrium constant, we can predict not only the direction in which a reaction will take place but also the extent to which it will occur. When we find, for example, that the equilibrium constant for the reaction

$$2 \ O_3(g) \rightleftharpoons 3 \ O_2(g)$$

is a very large number at 25°C,

$$K_c = \frac{[O_2]^3}{[O_3]^2} = 10^{55}$$

we deduce that the reaction goes virtually to completion at room temperature. Stated another way, ozone (O_3) is a very unstable substance which decomposes spontaneously to ordinary oxygen (O_2).

On the other hand, when the equilibrium constant for a reaction is very small, as is the case in the thermal decomposition of hydrogen fluoride at 2000°C

$$2 \ HF(g) \rightleftharpoons H_2(g) + F_2(g)$$

$$K_c = \frac{[H_2] \times [F_2]}{[HF]^2} = 10^{-13}$$

we conclude that there will be very little tendency for reaction to occur. The amounts of elementary hydrogen and fluorine that are in equilibrium with hydrogen fluoride at 2000°C are extremely small. Indeed, hydrogen fluoride is one of the most stable compounds known, as far as thermal decomposition is concerned. This explains, at least in part, the continuing interest in rocket propellants that react to produce hydrogen fluoride.

When the equilibrium constant is neither an extremely large nor an extremely small number, as is the case in the reactions

$$N_2(g) + O_2(g) \rightleftharpoons 2 \ NO(g), \quad K_c = 0.10 \text{ at } 2000°C$$
$$N_2(g) + 3 \ H_2(g) \rightleftharpoons 2 \ NH_3(g), \quad K_c = 10 \text{ at } 300°C$$

we can expect to obtain an equilibrium mixture containing appreciable amounts of both reactants and products. It is for reactions of this type that we are particularly interested in calculating the extent of reaction. We would like to know, for example, whether either or both of the two preceding reactions represent feasible processes for converting atmospheric nitrogen to useful compounds of that element. Example 12.2 illustrates how an industrial chemist might carry out calculations of this type.

EXAMPLE 12.2

  a. If we start with one mole of N_2 and one mole of O_2 in a 10-liter container at 2000°C, how much NO will be present at equilibrium?

  b. If one mole of N_2 and three moles of H_2 are allowed to come to equilibrium in a two-liter container at 300°C, how much ammonia will be present?

SOLUTION

a. Some of the N_2 and O_2 will combine to form NO until the concentrations of the three species satisfy the expression for the equilibrium constant. Let us represent by x the number of moles of N_2 that react. The balanced equation tells us that an equal number of moles, x, of oxygen must be consumed in the reaction. Furthermore, since two moles of NO are formed for every mole of N_2 that reacts, we must form 2x moles of NO. It may be helpful to summarize our reasoning in the form of a table.

No. of Moles	Original	Change	Equilibrium
N_2	1	$-x$	$1 - x$
O_2	1	$-x$	$1 - x$
NO	0	$+2x$	$2x$

Expressions for the equilibrium concentrations are readily set up, since we know that the volume of the container is ten liters.

$$[NO] = \frac{2x}{10}; \quad [N_2] = \frac{1 - x}{10}; \quad [O_2] = \frac{1 - x}{10}$$

Substituting into the expression for the equilibrium constant,

$$K_c = 0.10 = \frac{[NO]^2}{[N_2] \times [O_2]} = \frac{(2x/10)^2}{\left(\dfrac{1 - x}{10}\right)\left(\dfrac{1 - x}{10}\right)}$$

In this case (but not in general), volume cancels (see Section 12.4).

Simplifying, $\qquad 0.10 = \dfrac{(2x)^2}{(1 - x)^2}$

All that remains is to solve this expression for x. This is easily done by taking the square root of both sides of the equation.

$$(0.10)^{\frac{1}{2}} = 0.31 = \frac{2x}{1 - x}$$

Solving, we find that $\quad 0.31 - 0.31x = 2x$
$$x = 0.13$$

We were asked to calculate the number of moles of NO at equilibrium. We recall from our table that this quantity is 2x. Consequently,

no. moles of NO at equilibrium $= 0.26$

b. Here we proceed in a manner completely analogous to that followed in part a. If we choose x to be the number of moles of N_2 that reacts to form NH_3, it follows that the number of moles of H_2 consumed must be 3x (*one* N_2 + *three* H_2). The number of moles of NH_3 formed must be 2x (*one* $N_2 \rightarrow two$ NH_3). Our table takes the form:

No. of Moles	Original	Change	Equilibrium
N_2	1	$-x$	$1 - x$
H_2	3	$-3x$	$3 - 3x$
NH_3	0	$+2x$	$2x$

Recalling that the volume of the container is two liters, we have

$$[NH_3] = \frac{2x}{2}; \quad [N_2] = \frac{1 - x}{2}; \quad [H_2] = \frac{3 - 3x}{2}$$

309

Substituting into the equilibrium constant expression,

$$K_c = 10 = \frac{[NH_3]^2}{[N_2] \times [H_2]^3} = \frac{(2x/2)^2}{\left(\frac{1-x}{2}\right)\left(\frac{3-3x}{2}\right)^3}$$

Simplifying, $\quad 10 = \dfrac{(2x)^2\,(2)^2}{(1-x)\,(3-3x)^3}$

$$10 = \frac{(2x)^2\,(2)^2}{(1-x)\,(1-x)^3\,(3)^3}$$

or $\quad\quad \dfrac{(2x)^2}{(1-x)^4} = \dfrac{3^3 \times 10}{2^2} = 68$

At first glance, this seems to be a rather formidable expression to solve for x. However, if we take the square root of both sides, we obtain an expression that is easier to work with.

$$\frac{2x}{(1-x)^2} = (68)^{\frac{1}{2}} = 8.2$$

We now have a quadratic equation which we can solve by rearranging to the form

$$ax^2 + bx + c = 0$$

and applying the quadratic formula,

$$x = \frac{-b \pm \sqrt{b^2 - 4ac}}{2a}$$

Following this procedure, we obtain

$2x = 8.2 - 16.4x + 8.2\,x^2$

$8.2\,x^2 - 18.4x + 8.2 = 0$; i.e., a = 8.2, b = −18.4, c = 8.2

$$x = \frac{18.4 \pm \sqrt{(-18.4)^2 - 4\,(8.2)(8.2)}}{16.4}$$

$$= \frac{18.4 \pm \sqrt{339 - 269}}{16.4} = \frac{18.4 \pm 8.4}{16.4}$$

$$= 0.61 \text{ or } 1.63$$

Of the two answers, only 0.61 is plausible. Remember, we had only one mole of N_2 to start with; we could hardly have used up 1.63 moles of N_2! Since x must be 0.61, it follows that the amount of NH_3 at equilibrium, the quantity we were asked to solve for, must be 2x, or

$$2\ (0.61 \text{ mole}) = 1.22 \text{ moles}$$

It will always turn out that one of the answers is physically impossible.

Let us summarize what we have learned from this problem:

1. Some equilibrium calculations are more difficult than others. (If you feel that part b was rather difficult, see what happens if you start with one mole of N_2 and *one* mole of H_2).

2. Both of the reactions considered here would appear to be feasible methods of "fixing" atmospheric nitrogen, at least at the temperature specified. In the first case, 0.26 mole of NO was produced from one mole of nitrogen; in the second reaction, 1.2 moles of NH_3 were formed from the same quantity of nitrogen. In both cases, appreciable amounts of reactants were left at equilibrium, but these could be recycled if necessary to produce more NO or more NH_3. It turns out, for reasons which we shall examine more closely in Chapter 15, that the formation of ammonia (the Haber process) is a more practical way to form useful nitrogen compounds from atmospheric nitrogen than is the synthesis of nitric oxide.

12.4 EFFECT OF CHANGES IN CONCENTRATION UPON THE POSITION OF AN EQUILIBRIUM

Once a system has attained chemical equilibrium, it is possible to change the position of that equilibrium in various ways. One of the more obvious ways of changing the position of the equilibrium is to change the concentration of one or more of the substances present in the equilibrium mixture. This may be done by increasing or decreasing the number of moles of one of these substances. Alternatively, one might compress or expand the vessel in which the equilibrium mixture is contained. In either case, we frequently observe that the position of the equilibrium changes; that is, reaction occurs in either the forward or reverse direction.

Changes in the Number of Moles of Reactants or Products

Using the equilibrium constant expression, we can readily deduce the effect of adding or removing a substance on the position of an equilibrium in a gaseous system. Consider the general equilibrium system

$$aA(g) + bB(g) \rightleftharpoons cC(g) + dD(g)$$

$$K_c = \frac{[C]^c \times [D]^d}{[A]^a \times [B]^b}$$

Let us suppose that after equilibrium has been established in this system, we add a fixed amount of substance C, thereby increasing the number of moles and consequently the concentration of C. The system is no longer at equilibrium; by increasing the concentration of C, we have increased the concentration quotient above that required to satisfy the equilibrium constant expression:

$$\frac{(\text{conc. C})^c (\text{conc. D})^d}{(\text{conc. A})^a (\text{conc. B})^b} > K_c$$

By comparing the conc. product to K_c, we can always predict which way the reaction will go.

In order to restore equilibrium, part of the C which has been added reacts:

$$cC(g) + dD(g) \longrightarrow aA(g) + bB(g)$$

In this way, the concentrations of C and D decrease from the values they had immediately after the equilibrium was disturbed; the concentrations of A and B increase by corresponding amounts. The concentration ratio decreases until it becomes equal in magnitude to K_c; at this point, equilibrium is re-established, and no further reaction occurs.

Following the same line of reasoning, we can deduce that the addition of substance A to an equilibrium mixture of A, B, C, and D would cause the concentration ratio to drop below the value required for equilibrium. In order to re-establish equilibrium, part of the A added would react:

$$aA(g) + bB(g) \longrightarrow cC(g) + dD(g)$$

In general, we can say that:

The addition of a reaction product to a system at equilibrium disturbs that system by temporarily increasing the concentration quotient so that it exceeds K_c; the reverse reaction occurs to a sufficient extent to restore equilibrium. Addition

311

of a reactant to a system at equilibrium temporarily reduces the concentration quotient to a value below K_c; equilibrium is restored by reaction taking place in the forward direction.

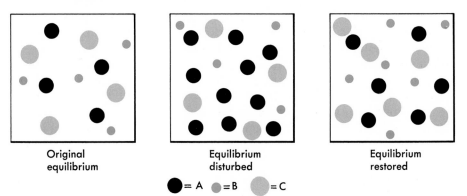

Original equilibrium Equilibrium disturbed Equilibrium restored

● = A ● = B ○ = C

Figure 12.3 Effect of adding reactant to the equilibrium system $2 A \rightleftharpoons B + C$. The medium-sized circles represent A molecules, the small circles B molecules, and the large circles C molecules.

EXAMPLE 12.3. The system $2 ICl(g) \rightleftharpoons I_2(g) + Cl_2(g)$ is at equilibrium at 227°C when 0.050 moles of I_2, 0.050 moles of Cl_2, and 1.60 moles of ICl are present in a 2.0 liter container.

    a. Determine K_c.

    b. Determine the number of moles of each substance present when equilibrium is restored following the addition of one more mole of ICl.

SOLUTION

a. $$K_c = \frac{[I_2] \times [Cl_2]}{[ICl]^2} = \frac{(0.050/2.0)(0.050/2.0)}{(1.60/2.0)^2} = 1.0 \times 10^{-3}$$

b. At the instant the ICl is added, we have 0.050 moles of I_2, 0.050 moles of Cl_2, and 2.60 moles of ICl. To restore equilibrium, some of the ICl reacts: $2 ICl(g) \rightarrow I_2(g) + Cl_2(g)$. If we represent by x the number of moles of I_2 formed by this reaction, we have:

No. of Moles	Original	Change	Final
ICl	2.60	−2x	2.60 − 2x
I_2	0.050	+x	0.050 + x
Cl_2	0.050	+x	0.050 + x

In a 2.0 liter container,

$$1.0 \times 10^{-3} = \frac{\left(\dfrac{0.050 + x}{2.0}\right)^2}{\left(\dfrac{2.60 - 2x}{2.0}\right)^2} = \frac{(0.050 + x)^2}{(2.60 - 2x)^2}$$

To solve, we extract the square root of both sides:

$$\frac{0.050 + x}{2.60 - 2x} = (1.0 \times 10^{-3})^{\frac{1}{2}} = (10 \times 10^{-4})^{\frac{1}{2}} = 3.1 \times 10^{-2}$$

Solving, x = 0.029; No. of moles I_2 = 0.079, Cl_2 = 0.079, ICl = 2.54.

Changes in Volume

It is also possible to deduce from the equilibrium constant expression what will happen to a system at equilibrium when the volume is changed. The reasoning is a little more difficult to follow here, because a change in volume produces an instantaneous change in the concentrations of all the species present at equilibrium. To illustrate the principle involved, it is helpful to consider a specific equilibrium, that involving two gases, NO_2 and N_2O_4, produced when concentrated nitric acid is heated with certain metals, including copper and silver.

$$2\ NO_2(g) \rightleftharpoons N_2O_4(g)$$

$$K_c = \frac{[N_2O_4]}{[NO_2]^2}$$

Since we are interested in the effect of a change in volume upon the position of this equilibrium, let us introduce the volume explicitly into the expression for K_c. We can do this by realizing that the concentration of any species, in moles per liter, is equal to the number of moles of that species, n, divided by the volume, V:

$$[N_2O_4] = \frac{n\ N_2O_4}{V}; \quad [NO_2] = \frac{n\ NO_2}{V}$$

$$K_c = \frac{(n\ N_2O_4/V)}{(n\ NO_2/V)^2}$$

Combining terms, we obtain

$$K_c = \frac{(n\ N_2O_4)}{(n\ NO_2)^2} \times V$$

Now suppose that in an equilibrium mixture of NO_2 and N_2O_4, we *increase* the volume, perhaps from one liter to two liters. Since K_c remain constant, it follows that the ratio

$$\frac{n\ N_2O_4}{(n\ NO_2)^2}$$

must *decrease* to restore equilibrium. If the volume were doubled, the ratio would have to decrease to one-half its original value. The only way that this can happen is for the reaction

$$N_2O_4(g) \longrightarrow 2\ NO_2(g)$$

to take place, decreasing the number of moles of N_2O_4 and increasing the number of moles of NO_2.

Following the same line of reasoning, if we were to *decrease* the volume available to the system at equilibrium, the mole ratio

$$\frac{n\ N_2O_4}{(n\ NO_2)^2}$$

would have to *increase* to compensate for this change. To accomplish this, more N_2O_4 would be formed via the reaction

$$2\ NO_2(g) \longrightarrow N_2O_4(g)$$

313

The analysis we have carried out for the NO_2-N_2O_4 system can be generalized to any equilibrium system.

When the position of an equilibrium is disturbed by an increase in the volume, reaction takes place in a direction so as to increase the number of moles of gas (e.g., $N_2O_4 \rightarrow 2\,NO_2$); **when the volume is decreased, the reaction which decreases the number of moles of gas** (e.g., $2\,NO_2 \rightarrow N_2O_4$) **takes place.**

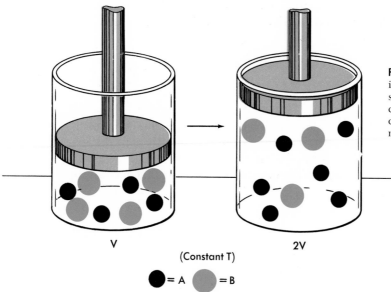

V   (Constant T)   2V

● = A   ● = B

Figure 12.4 Effect of volume increase on the equilibrium system $2\,A \rightleftharpoons B$. The small circles represent A molecules, the large circles B molecules.

The application of this principle to various reactions involving gases is illustrated in Table 12.2. Two of the entries in this table are particularly deserving of comment:

1. For reactions such as

$$2\,HI(g) \rightleftharpoons H_2(g) + I_2(g)$$

where there are the same number of moles on both sides of the equation, a change in volume has no effect upon the position of the equilibrium. This is a consequence of the fact that when one writes the equilibrium expression for such a system, the volume does not appear in the relation between K_c and the number of moles of H_2, I_2, and HI.

$$K_c = \frac{[H_2] \times [I_2]}{[HI]^2} = \frac{(n\;H_2/V)(n\;I_2/V)}{(n\;HI/V)^2} = \frac{(n\;H_2)(n\;I_2)}{(n\;HI)^2}$$

TABLE 12.2 EFFECT OF CHANGE IN VOLUME ON THE POSITION OF A CHEMICAL EQUILIBRIUM

SYSTEM	V INCREASES	V DECREASES
$2\,NO_2(g) \rightleftharpoons N_2O_4(g)$	⟵	⟶
$N_2(g) + 3\,H_2(g) \rightleftharpoons 2\,NH_3(g)$	⟵	⟶
$2\,NO_2(g) \rightleftharpoons 2\,NO(g) + O_2(g)$	⟶	⟵
$2\,HI(g) \rightleftharpoons H_2(g) + I_2(g)$	0	0
$C(s) + H_2O(g) \rightleftharpoons CO(g) + H_2(g)$	⟶	⟵

2. It is the change in the number of moles of **gas** which is important in determining the direction in which the equilibrium shifts to compensate for a volume change; the number of moles of solids or liquids involved in the equilibrium equation are omitted. This is a consequence of the fact that the concentrations of solids or liquids do not appear in the expression for K_c (see Section 12.6).

EXAMPLE 12.4. In a mixture of NO_2 and N_2O_4,

$$2 \ NO_2(g) \rightleftharpoons N_2O_4(g)$$

1.76 moles of N_2O_4 and 0.40 moles of NO_2 are present when equilibrium is attained at 25°C in a 20.0 liter container.
  a. Calculate K_c for this reaction.

  b. Calculate the number of moles of NO_2 and N_2O_4 in the equilibrium mixture obtained when the volume is reduced to 10.0 lit.

SOLUTION

a. $$K_c = \frac{[N_2O_4]}{[NO_2]^2} = \frac{(1.76/20.0)}{(0.40/20.0)^2} = \frac{0.0880}{(0.020)^2} = 220$$

b. We deduce that a decrease in volume will favor the formation of N_2O_4, since the reaction $2 \ NO_2(g) \rightarrow N_2O_4(g)$ results in a decrease in the number of moles. If we represent by x the number of moles of N_2O_4 formed as the result of this shift in the position of the equilibrium, we have

No. of Moles	Original	Change	Final
N_2O_4	1.76	+x	1.76 + x
NO_2	0.40	−2x	0.40 − 2x

In a 10.0 liter container, the final equilibrium concentrations will be

$$[N_2O_4] = (1.76 + x)/10.0; \quad [NO_2] = (0.40 − 2x)/10.0$$

$$K_c = 220 = \frac{(1.76 + x)/10.0}{\left(\frac{0.40 − 2x}{10.0}\right)^2}$$

$$22 = \frac{(1.76 + x)}{(0.40 − 2x)^2}$$

Solving this quadratic equation (recall Example 12.2, part b), we obtain x = 0.055; therefore,

no. of moles NO_2 = 0.40 − 0.11 = 0.29
no. of moles N_2O_4 = 1.76 + 0.055 = 1.82

Changes in volume of gaseous systems at equilibrium at a constant temperature inevitably result in changes in pressure. For example, when we change the volume of the NO_2-N_2O_4 system discussed in Example 12.4 from 20.0 lit to 10.0 lit we are momentarily increasing the pressure by a factor of two. Instead of saying that the shift in the position of the equilibrium comes about because of a change in volume, we could ascribe the shift to the pres-

315

sure change that accompanies the decrease in volume. Thus, we could say that an *increase* in pressure shifts the position of this equilibrium in such a way as to *decrease* the number of moles of gas (2 $NO_2 \rightarrow N_2O_4$).

In discussing the effect of pressure on the position of an equilibrium, we must be very careful to specify how the pressure change is achieved. We might increase the total pressure of the NO_2-N_2O_4 system by changing the total number of moles of gas at constant volume. If we did this by adding N_2O_4 to the system, the equilibrium would shift in the direction $N_2O_4 \rightarrow 2\,NO_2$. If the number of moles of gas were increased by adding a substance such as helium or hydrogen to the equilibrium mixture, the position of the equilibrium would remain unchanged; there would be no tendency for the forward or reverse reaction to occur. *Only if the pressure change results from a change in volume can we state with confidence that an increase in pressure will favor the reaction that decreases the number of moles, or conversely, that a decrease in pressure will favor the reaction that increases the number of moles.*

Another way to increase the pressure would be to increase the temperature, in which case K_c would change.

Le Chatelier's Principle

We have discussed changes in equilibrium systems, both qualitatively and quantitatively, in terms of the expression for the equilibrium constant. Alternatively, one can deduce qualitatively how the position of an equilibrium will shift with a change in conditions by applying Le Chatelier's Principle, which states that:

If a system in chemical equilibrium is altered in any way, the system will shift so as to minimize the effect of the change.

To illustrate the use of Le Chatelier's Principle, consider what happens when ICl is added to the equilibrium system

$$2\,ICl(g) \rightleftharpoons I_2(g) + Cl_2(g)$$

Le Chatelier's Principle tells us that the equilibrium will shift in such a way as to consume part of the ICl which is added. This is accomplished by the reaction

$$2\,ICl(g) \longrightarrow I_2(g) + Cl_2(g)$$

It will be recalled that we came to the same conclusion by considering the equilibrium expression, and that, working with K_c, we were able to calculate precisely how much of the added ICl was removed by this reaction (cf. Example 12.3).

As a further example, consider the effect of compressing the equilibrium system

$$2\,NO_2(g) \rightleftharpoons N_2O_4(g)$$

On the basis of Le Chatelier's Principle, we would predict that this equilibrium would shift to the right upon compression. In this way, the increase in total pressure is minimized by decreasing the number of moles of gas. By the same token, if the volume of the NO_2-N_2O_4 system were increased, the resulting drop in pressure could be partially erased by an increase in the number of moles. Le Chatelier's Principle implies that the reaction $N_2O_4 \rightarrow 2\,NO_2$ (1 mole $\rightarrow$ 2 moles) would occur under these conditions. Of course, we can come to the same conclusion by considering the expression for the equilibrium constant for the reaction (recall Example 12.4 and the discussion that precedes it).

Le Chatelier's Principle can also be used to predict qualitatively the effect of a change in temperature upon the position of a chemical equilibrium. We shall consider such changes in Section 12.5.

12.5 THE EFFECT OF TEMPERATURE ON CHEMICAL EQUILIBRIA

So far our attention has been directed toward equilibrium systems at a constant temperature. Under such conditions, the equilibrium constant K_c is a fixed number, independent of any changes in concentrations of reactants or products. However, if one changes the temperature, the magnitude of the equilibrium constant changes.

The direction in which the equilibrium constant shifts as the temperature is increased is readily predicted by Le Chatelier's Principle. In an equilibrium system where we have two competing reactions going on, one would expect an increase in temperature to favor the endothermic reaction. In this way, heat is absorbed and the increase in temperature is minimized. This means that in an equilibrium system, **if the forward reaction is endothermic, an increase in temperature will favor that reaction and therefore will increase the equilibrium constant**. An example is the system

$$N_2(g) + O_2(g) \rightleftharpoons 2\ NO(g)$$

where the forward reaction is endothermic

$$N_2(g) + O_2(g) \longrightarrow 2\ NO(g); \quad \Delta H = +43.2\ kcal$$

ΔH can, of course, be calculated from heat of formation data (Chap. 4).

The value of K_c for this equilibrium increases from about 10^{-30} at room temperature to 0.1 at 2000°C. Clearly, if one hopes to prepare NO in good yield from the elements, it is desirable to work at as high a temperature as possible.

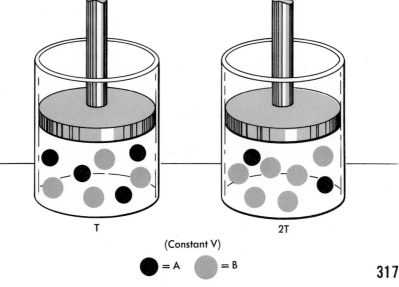

Figure 12.5 Effect of temperature increase on the equilibrium system A + heat ⇌ B. The small circles represent A molecules, the large circles B molecules.

T   2T

(Constant V)

● = A   ⬤ = B

317

By the same reasoning, one can deduce that **if the forward reaction is exothermic, an increase in temperature will favor the reverse reaction and therefore will decrease the equilibrium constant**. Consider, for example, the system

$$N_2(g) + 3\ H_2(g) \rightleftharpoons 2\ NH_3(g)$$

where the forward reaction evolves heat:

$$N_2(g) + 3\ H_2(g) \longrightarrow 2\ NH_3(g), \quad \Delta H = -22.0\ \text{kcal}$$

K_c for this reaction decreases from 5×10^8 at 25°C to somewhat less than 10^3 at 200°C and about 0.5 at 400°C.

We are interested, of course, not only in the direction in which the equilibrium constant changes as the temperature increases, but also the extent to which it changes. The laws of thermodynamics provide us with a relationship which can be used to calculate the equilibrium constant at one temperature from its value at another temperature.

We shall postpone a discussion of this relationship to Chapter 13.

12.6 EQUILIBRIA INVOLVING SOLIDS OR LIQUIDS IN ADDITION TO GASES

Reactions taking place in the gas phase often include solids or liquids as reactants or products. An example considered earlier in this chapter is the decomposition of calcium carbonate:

$$CaCO_3(s) \rightleftharpoons CaO(s) + CO_2(g)$$

In discussing this reaction, we pointed out that the concentration of CO_2 in equilibrium with $CaCO_3$ and CaO is a fixed number at a given temperature. In other words, for this equilibrium at constant temperature,

$$[CO_2] = \text{constant} = K_c$$

This expression for K_c does not include terms for either of the solids, $CaCO_3$ and CaO, that take part in the reaction. It is found experimentally that so long as these two solids are present, regardless of their amounts, the concentration of CO_2 has a fixed value (e.g., 0.0036 moles/lit at 800°C).

In a similar manner, if one studies the equilibrium involved in the formation of tin from its principal ore, cassiterite (SnO_2), using carbon monoxide as a reducing agent,

$$\tfrac{1}{2}\ SnO_2(s) + CO(g) \rightleftharpoons \tfrac{1}{2}\ Sn(s) + CO_2(g)$$

it is found that so long as there are sufficient amounts of the two solids SnO_2 and Sn to establish equilibrium, the ratio of the concentration of CO_2 to that of CO is a fixed number, dependent only on temperature:

$$K_c = \frac{[CO_2]}{[CO]}$$

This ratio is not affected by adding more SnO_2 to the system or withdrawing part of the tin produced in the reaction.

Still another example of this type is the equilibrium between the constituents of "water gas" (a fuel produced by the reaction of steam with hot coal):

$$CO_2(g) + H_2(g) \rightleftharpoons CO(g) + H_2O(l) \qquad (12.3)$$

For temperatures and pressures at which water can exist as a liquid, the equilibrium constant for the system can be written in the form

$$K_c = \frac{[CO]}{[CO_2] \times [H_2]}$$

Here again we see that there is no term in the equilibrium expression involving the concentration of water. Experimentally, it is found that adding or removing water from the system does not change the position of the equilibrium, provided some liquid water remains. In general, we find that **in any system where a product or reactant is present as a solid or liquid, the position of the equilibrium is independent of the amount of that substance. Furthermore, its concentration need not, and ordinarily does not, appear in the expression for K_c.**

To understand why these statements are valid, let us consider how the equilibrium represented by Equation 12.3 might be established. Suppose we admit the two gases CO_2 and the H_2 into a container at a total pressure of one atmosphere and a temperature of 50°C. The following gas-phase reaction occurs:

$$CO_2(g) + H_2(g) \longrightarrow CO(g) + H_2O(g)$$

As this reaction proceeds, the partial pressure of water vapor increases. However, as soon as it reaches 92.5 mm Hg, the vapor pressure of water at 50°C, liquid water starts to condense. From that point on, the partial pressure of water vapor remains at 92.5 mm Hg, regardless of how much water vapor is formed. The concentration of water vapor, like its partial pressure, remains constant; a gas-law calculation indicates that

$$[H_2O] = \frac{n\ H_2O}{V} = \frac{p\ H_2O}{RT} = \frac{(92.5/760)\ atm}{\left(0.0821\ \dfrac{lit\ atm}{mole\ °K}\right)(323°K)} = 0.00460\ \frac{mole}{liter}$$

This will be the concentration of water vapor when equilibrium is reached. At constant temperature, nothing that one can do to the equilibrium system can change the concentration of water vapor, provided some liquid water is present. Adding or removing one of the reactants or products will have no effect; neither will changing the volume of the container. Hence, the general expression for the equilibrium constant

Adding more water would not change its pressure or concentration in the gas phase (Chap. 10).

$$K = \frac{[CO] \times [H_2O]}{[CO_2] \times [H_2]}$$

can be simplified to read

$$K = \frac{[CO] \times 0.00460}{[CO_2] \times [H_2]} \quad \text{or} \quad \frac{K}{0.00460} = \frac{[CO]}{[CO_2] \times [H_2]}$$

Consequently, one can define the equilibrium constant for the system to be the ratio of the two numbers $K/0.00460$. That is,

$$K_c = \frac{K}{0.00460} = \frac{[CO]}{[CO_2] \times [H_2]}$$

This convention is ordinarily followed in dealing with equilibria involving liquids or solids. The gas-phase concentration of a substance present as a liquid or solid, which must be a constant at a particular temperature, is incorporated into the equilibrium constant expression and consequently does not appear explicitly in K_c.

PROBLEMS

In solving certain of these problems, Table 4.1 (p. 62) may prove useful.

12.1   Formulate expressions for the equilibrium constant K_c for each of the following reactions.

    a. $2\ CO(g) + O_2(g) \rightleftharpoons 2\ CO_2(g)$
    b. $2\ HCl(g) \rightleftharpoons H_2(g) + Cl_2(g)$
    c. $4\ NH_3(g) + 7\ O_2(g) \rightleftharpoons 4\ NO_2(g) + 6\ H_2O(g)$
    d. $SnO_2(s) + 2\ H_2(g) \rightleftharpoons Sn(s) + 2\ H_2O(g)$
    e. $2\ H_2(g) + O_2(g) \rightleftharpoons 2\ H_2O(l)$

12.2   Consider the equilibrium $N_2(g) + 3\ H_2(g) \rightleftharpoons 2\ NH_3(g)$. An equilibrium mixture at 300°C in a 5.0 liter container consists of 1.0 mole of NH_3, 0.10 mole of N_2, and 3.0 moles of H_2. Calculate K_c.

12.3   One mole of NO_2 is placed in a 10.0 liter container at 50°C. Part of it reacts to form N_2O_4; when equilibrium is reached 0.36 mole of N_2O_4 is present. Calculate K_c for the reaction

$$2\ NO_2(g) \rightleftharpoons N_2O_4(g)$$

12.4 K_c for the reaction $2\ HI(g) \rightleftharpoons H_2(g) + I_2(g)$ is 0.0156 at 520°C. Predict the direction in which each of the following systems, confined in a 1.0 liter container at 520°C, will move to achieve equilibrium.

    a. 1.00 mole HI + 0.010 moles H_2
    b. 1.00 mole HI, 1.00 mole H_2, + 1.00 mole I_2
    c. 1.00 mole HI, 0.10 mole H_2, + 0.10 mole I_2

12.5   Find the number of moles and the concentration of HI, H_2, and I_2 at equilibrium in each of the systems listed in Problem 12.4.

12.6   Two moles of H_2 and two moles of I_2 are introduced into a 5.0 liter container at 520°C and react to equilibrium. Using the equilibrium constant given in Problem 12.4, determine

    a. The number of moles of HI produced by the reaction.
    b. The final concentration of I_2.
    c. The percentage of the I_2 that is used up in the reaction.

12.7   For the reaction $PCl_5(g) \rightleftharpoons PCl_3(g) + Cl_2(g)$, $K_c = 0.050$ at 150°C. If 1.00 mole of PCl_5 is introduced into a 10.0 liter container at 150°C and allowed to react to equilibrium, calculate the numbers of moles and the concentrations of PCl_5, PCl_3 and Cl_2 at equilibrium.

12.8   For the reaction $2\ NO_2(g) \rightleftharpoons N_2O_4(g)$, $K_c = 220$ at 25°C. If 1.00 mole of NO_2 and 1.00 mole of N_2O_4 are introduced into an 11.0 liter container at 25°C, determine

    a. The equilibrium concentrations of NO_2 and N_2O_4.
    b. The density, in g/lit, of the equilibrium system.
    c. The original pressure and the pressure of the system at equilibrium.

12.9   Consider the equilibrium $N_2(g) + 3\ H_2(g) \rightleftharpoons 2\ NH_3(g)$. A mixture of these three substances is allowed to reach equilibrium at 200°C in an 18.0 liter container. Predict the direction in which the system will move to reestablish equilibrium if

a. One mole of H_2 is added to the container.
b. One mole of NH_3 is added to the container.
c. The volume of the container is changed to 20.0 lit.
d. The pressure is increased by compressing the mixture.
e. The pressure is increased by adding helium to the container.

12.10 An equilibrium mixture of PCl_5, PCl_3, and Cl_2 at a certain temperature in a 10.0 liter container consists of 1.00 mole of PCl_5, 0.30 mole of PCl_3, and 0.80 mole of Cl_2.

a. Calculate K_c for the reaction $PCl_5(g) \rightleftharpoons PCl_3(g) + Cl_2(g)$.
b. An additional mole of PCl_5 is added to this mixture. Determine the number of moles of each substance present when equilibrium is reestablished.
c. The volume of the original equilibrium mixture is decreased to 5.0 liters. Find the number of moles of PCl_3 when equilibrium is reestablished.

12.11 Predict the direction in which the following equilibria will shift

$$4 NH_3(g) + 5 O_2(g) \rightleftharpoons 4 NO(g) + 6 H_2O(g)$$
$$2 HF(g) \rightleftharpoons H_2(g) + F_2(g)$$
$$4 NH_3(g) + 5 O_2(g) \rightleftharpoons 4 NO_2(g) + 6 H_2O(l)$$

a. When the volume of the system is decreased.
b. When the temperature of the system is increased.

12.12 Consider the equilibrium $H_2(g) + I_2(s) \rightleftharpoons 2 HI(g)$. At 25°C, K_c for this system is 0.35. If one mole of H_2 and one mole of $I_2(s)$ are introduced into a 2.00 liter container at 25°C, what will be the final concentration of HI?

12.13 Formulate expressions for the equilibrium constant K_c for each of the following reactions.

a. $2 NO(g) + O_2(g) \rightleftharpoons 2 NO_2(g)$
b. $CH_4(g) + 2 H_2S(g) \rightleftharpoons CS_2(g) + 4 H_2(g)$
c. $CH_4(g) + 2 O_2(g) \rightleftharpoons CO_2(g) + 2 H_2O(l)$
d. $CaC_2(s) + 2 H_2O(l) \rightleftharpoons Ca(OH)_2(s) + C_2H_2(g)$

12.14 Consider the equilibrium $4 NH_3(g) + 5 O_2(g) \rightleftharpoons 4 NO(g) + 6 H_2O(g)$.

a. At a certain temperature, 1.00 mole of NH_3, 2.00 moles of O_2, 1.50 moles of NO, and 2.00 moles of H_2O are present at equilibrium in a 10.0 liter container. Calculate K_c.
b. At a different temperature, 2.00 moles of NH_3 and 3.00 moles of O_2 are introduced into a 10.0 liter container. At equilibrium, the concentration of NO is 0.16 mole/liter. Calculate K_c.

12.15 K_c for the reaction $SO_2(g) + \frac{1}{2} O_2(g) \rightleftharpoons SO_3(g)$ is 25 at 600°C. Predict the direction in which each of the following systems, confined in a 5.0 liter container at 600°C, will move to achieve equilibrium.

a. 2.00 moles SO_2 plus 2.00 moles SO_3.
b. 1.00 mole SO_2, 0.50 mole O_2, and 1.00 mole SO_3.
c. A mixture in which there are 0.010 moles of O_2 and in which the number of moles of SO_3 is twice the number of moles of SO_2.

12.16 Two moles of HI and one mole of H_2 are admitted to an evacuated 10.0 liter container at 520°C. Using the equilibrium constant given in Problem 12.4, find the equilibrium concentrations of HI, H_2, and I_2. How many grams of I_2 are formed in the reaction?

12.17 To the system in Problem 12.16 is added another mole of HI. Predict qualitatively and then quantitatively, what happens to the concentrations of HI, H_2, and I_2 in the container.

12.18 For the reaction $N_2(g) + 3 H_2(g) \rightleftharpoons 2 NH_3(g)$, K_c is 0.50 at 400°C.

   a. If one starts with 1.00 mole of N_2 and 3.00 moles of H_2 in a 20.0 liter container at 400°C, what is the concentration of NH_3 at equilibrium?
   b. If one starts with 1.00 mole of NH_3 in a 5.0 liter container at 400°C, calculate the final concentrations of NH_3, N_2 and H_2.

12.19 Consider the gaseous reaction $CO(g) + H_2O(g) \rightleftharpoons CO_2(g) + H_2(g)$. At 420°C a mole of CO and a mole of H_2O are introduced into an evacuated 5.0 liter container. After reaction to equilibrium, the system is found to contain 0.75 mole of CO_2.

   a. Calculate K_c at 420°C.
   b. The volume of the container is increased to 20 liters. What happens to the value of K_c? to the amount of CO_2 present? to the concentration of CO_2?
   c. Enough CO_2 is added to the system in a to double its equilibrium concentration. How much H_2 is present after equilibrium is reestablished? How much CO_2 was added?

12.20 For the reaction $2 SO_2(g) + O_2(g) \rightleftharpoons 2 SO_3(g)$, $K_c = 5.8 \times 10^3$ at 600°C.

   a. If two moles of SO_2 and one mole of O_2 are introduced into a 12.0 liter container at 600°C, find the equilibrium concentration of SO_3.
   b. If one mole of SO_2 and one mole of SO_3 are introduced into a 10.0 liter container at 600°C, find the number of moles of O_2 at equilibrium.

12.21 When 1.00 mole of N_2O_4 at 150°C is introduced into a 4.0 liter container, 80 per cent of it dissociates to form NO_2.

   a. Determine K_c for the reaction $2 NO_2(g) \rightleftharpoons N_2O_4(g)$.
   b. If the volume of the container is increased to 8.0 liters, how many moles of NO_2 will be present when equilibrium is reestablished?
   c. Determine the density of the equilibrium mixture in b, the partial pressures of both components, the total pressure, and the apparent molecular weight.

12.22 Write an equation for a reaction in which

   a. K_c increases with temperature.
   b. A change in volume has no effect on the position of the equilibrium.
   c. The concentration of a gaseous reactant is a fixed number at a particular temperature.
   d. The reverse reaction is spontaneous when all reactants and products are at a concentration of one mole/liter.

°12.23 Given that K_c for the reaction $N_2(g) + 3 H_2(g) \rightleftharpoons 2 NH_3(g)$ is 9.5 at 300°C, how many grams of ammonia are formed when one mole of nitrogen and one mole of hydrogen are introduced into a 10.0 liter container and allowed to come to equilibrium?

°12.24 For the reaction $ZnO(s) + H_2(g) \rightleftharpoons Zn(s) + H_2O(g)$, $K_c = 0.030$ at 1000°C. One mole of ZnO is added to an evacuated 10.0 liter container at 1000°C. Enough hydrogen is then added to give a pressure of one atmosphere. Find the number of moles of zinc and water formed as the reaction proceeds to equilibrium.

°12.25 For the reaction $SO_2(g) + \frac{1}{2} O_2(g) \rightleftharpoons SO_3(g)$, $K_c = 70$ at 600°C. Sulfur dioxide and oxygen, each at one atmosphere pressure, are introduced into a 20.0 liter container at this temperature. How many moles of SO_3 are formed?

°12.26 At 25°C and one atmosphere pressure the density of a mixture of NO_2 and N_2O_4 is 3.23 g/lit. Calculate K_c for $2 NO_2(g) \rightleftharpoons N_2O_4(g)$.

Thermodynamics and the
Spontaneity and Equilibrium
Properties of Chemical Reactions

In Chapter 4 we discussed energy effects in chemical reactions, mainly in terms of the heat flow that occurs when a reaction is carried out. It was shown that a characteristic molar heat of formation, ΔH_f, can be associated with every pure substance. This can be determined by measuring the heat flow when that substance is formed at constant temperature and pressure from elementary substances. By proper interpretation of the values of the heats of formation of the participants in a chemical reaction, we were able to calculate the heat flow Q in that reaction. The fact that the heat of formation of a pure substance, like its temperature and pressure, is a characteristic of that substance in a given state, was shown to be a direct consequence of the First Law of Thermodynamics.

In Chapter 12 we saw that a chemical reaction generally proceeds to some equilibrium condition that is characterized by an equilibrium constant, K_c, that puts limitations on the relative magnitudes of the equilibrium concentrations of the reactants and products. It would seem reasonable to suggest that the equilibrium conditions in a chemical reaction, like the heat flow in that reaction, might be related somehow to the relative energies of the reactants and products. The equilibrium state would be one of minimal possible energy for the system, under the conditions allowed to it. There is indeed such a relation, and it will be the purpose of this chapter to present the nature of that relation and to show as clearly as possible how it arises. The results we shall obtain will come from thermodynamic arguments, and in particular will be concerned with some of the consequences of the Second Law of Thermodynamics.

13.1 SPONTANEITY OF CHEMICAL REACTIONS: FREE ENERGY

Before attempting to deal with chemical equilibrium, it will be helpful if we consider briefly a broader matter, that of the *direction* in which a chemical reaction will tend to proceed spontaneously.

Some questions which the practicing chemist would like to be able to answer regarding spontaneity of chemical changes might include the following: Can we decompose silver oxide, Ag_2O, to the elements by heating it in an open container? If so, what temperature would be required to bring about the reaction? Will lead and bromine react to form lead bromide at ordinary temperatures? Would it be sensible to try to produce metallic aluminum by treating the oxide, Al_2O_3, with hydrogen? Would the reaction tend to proceed more readily at high or at low temperatures?

A hundred years ago chemists felt that they had the answer to questions such as those we have just posed. The prevailing idea was that all spontaneous chemical reactions were exothermic; that is, one simply had to evaluate ΔH for the reaction. If ΔH was negative the reaction was spontaneous, and if ΔH was positive, it was not.

It turns out that indeed most spontaneous chemical reactions are exothermic, but there are some that are not. At 25°C and 1 atm, we might cite the following examples of exothermic spontaneous reactions:

$$4 \, Ag(s) + O_2(g) \longrightarrow 2 \, Ag_2O(s), \qquad \Delta H = -14.62 \text{ kcal} \qquad (13.1)$$

$$Pb(s) + Br_2(l) \longrightarrow PbBr_2(s), \qquad \Delta H = -66.3 \text{ kcal} \qquad (13.2)$$

However, we should also mention a rather important spontaneous endothermic reaction, which occurs at one atmosphere at any temperature above 0°C:

$$H_2O(s) \longrightarrow H_2O(l), \qquad \Delta H = +1.44 \text{ kcal} \qquad (13.3)$$

Another common spontaneous endothermic reaction is the solution of many salts in water. A typical example would be

$$KNO_3(s) \longrightarrow K^+ + NO_3^-, \qquad \Delta H = +9.35 \text{ kcal}$$

It is also observed that the temperature and pressure (or concentration) of reactants and products have a marked influence on reaction spontaneity. At temperatures below 0°C, the following reaction is certainly spontaneous, as anyone who lives in Minnesota well knows:

$$H_2O(l) \longrightarrow H_2O(s)$$

Above about 190°C silver oxide dissociates spontaneously at 1 atm to elementary substances:

$$2 \, Ag_2O(s) \longrightarrow 4Ag(s) + O_2(g)$$

As we saw in the previous chapter, changes in concentration of reactants or products will cause spontaneous shifts in equilibrium systems. Related effects are caused by pressure changes. At pressures below about 24 mm Hg the following reaction is spontaneous at 25°C,

$$H_2O(l) \longrightarrow H_2O(g)$$

whereas at pressures above 24 mm Hg the reaction is spontaneous in the opposite direction.

Enthalpy changes for chemical reactions are almost independent of both temperature and pressure. Clearly, although it is often true that ΔH is negative for spontaneous reactions, the governing criterion for reaction spontaneity must be more involved than a simple correlation of spontaneity with the sign of the enthalpy change.

J. Willard Gibbs, an American theoretician, showed late in the nineteenth century that the proper criterion for the spontaneity of a chemical reaction was the capacity of that reaction to produce useful work. He proved that **if, at constant temperature and pressure, a chemical reaction can in principle or in practice be used to perform useful work, that reaction is thermodynamically spontaneous**. If work has to be done on the system to make the reaction occur, that reaction is not spontaneous.

The word "useful" excludes expansion work, discussed in Chapter 4.

J. Willard Gibbs was a professor of mathematical physics at Yale University from 1871 until his death in 1903. His contributions to the science of thermodynamics, particularly his clear analysis of the laws governing chemical equilibrium, have earned him his place among the great scientists. He also published extensively in the area of statistical mechanics, and in mathematics he did significant work in the area that is now known as vector analysis.

Gibbs was a quiet person by nature, one who thought out his ideas without consultation with other scientists. His great ability was first recognized by foreign scientists, particularly Ostwald and Le Chatelier, who translated his works into German and French. Gibbs is still relatively unknown in the United States, although in the minds of many his ability would rank him above any other theoretician this country has yet produced.

The meaning of Gibbs' criterion for reaction spontaneity is perhaps best illustrated by a chemical reaction that is carried out under conditions where useful work is clearly produced. A simple way in which this can be done is to let the reaction occur in a voltaic cell producing electrical work. Let us consider the following reaction:

$$Pb(s) + Br_2(l) \longrightarrow PbBr_2(s) \tag{13.2}$$

This reaction can serve as a source of electrical energy if it is carried out in the electrical cell illustrated in Figure 13.1. It is found that under optimal conditions as much as 62.1 kcal of useful electrical work can be obtained when a mole of $PbBr_2(s)$ is formed at 25°C by Reaction 13.2. It follows that this reaction is thermodynamically spontaneous, which is indeed the experimental observation. Ordinarily, when the reaction occurs on mixing lead and bromine in an open container, no useful work is done. It is the fact that it is *possible* to carry the reaction out in such a way as to accomplish useful work that establishes its spontaneity.

In contrast to Reaction 13.2, consider the decomposition of water

$$H_2O(l) \longrightarrow H_2(g) + \tfrac{1}{2} O_2(g) \tag{13.3}$$

which is known to be nonspontaneous at 25°C and 1 atm. One can show that in order for this reaction to occur under these conditions at least 56.7 kcal of work must be expended on the system. This work may be furnished electrically; it is found experimentally that 56.7 kcal of electrical energy is required to decompose one mole of water.

The relation between the spontaneity of a process and its capacity to produce useful work may further be illustrated by certain physical changes. All of us are familiar with the spontaneous tendency of water to run downhill. Work may be obtained from this process if the water is caused to flow through a turbine used to operate an electric generator. If the water is allowed to fall directly, no work is produced, but this does not alter our conclusion that the process is *capable* of producing useful work.

Although this discussion is rather theoretical, it has direct relevance to

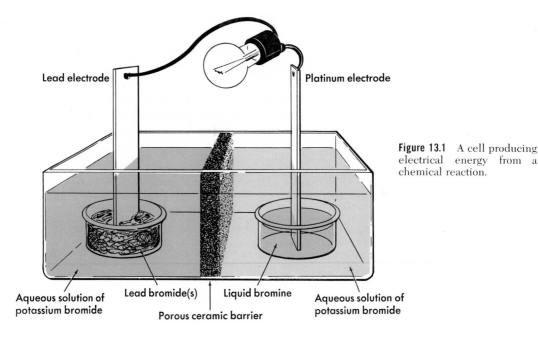

Figure 13.1 A cell producing electrical energy from a chemical reaction.

Lead electrode

Platinum electrode

Aqueous solution of potassium bromide

Lead bromide(s)

Liquid bromine

Aqueous solution of potassium bromide

Porous ceramic barrier

some very important practical problems. A substantial part of the mechanical work done in our industrialized society depends on chemical reactions as the source of energy. The automobile engine, the jet engine, and the steam turbine, all get their energy from the work available from the chemical combustion reactions of petroleum products or coal. Small electrical devices are often powered by cells similar in design to that shown in Figure 13.1. The fundamental laws governing the recovery of useful work from chemical changes have a very direct application to the continuing problem of efficient design of work-producing devices.

Maximum Useful Work

As we have seen, a chemical change may occur spontaneously with or without the simultaneous production of useful work. The amount of work produced by the change may vary from zero up to some maximum amount. When $PbBr_2(s)$ is formed at 25°C and 1 atm from the elements, the reaction may produce any amount of useful work up to but not exceeding 62.1 kcal. If $PbBr_2(s)$ is dissociated to the elements it is found that at least 62.1 kcal of work must be put into the system to carry out the change.

Similarly, when water is dissociated to the elements,

$$H_2O(l) \longrightarrow H_2(g) + \tfrac{1}{2} O_2(g)$$

at least 56.7 kcal of work must be expended to make the change occur. If a mole of water is formed from the elements, up to 56.7 kcal of useful work can be produced.

There is, then, a maximum amount of useful work, which we will call W', which can be produced by a spontaneous chemical reaction occurring at constant temperature and pressure. As you might expect, this maximum amount of work would be observed when the change is carried out slowly, against a resisting force only slightly less than the force driving the reaction. The reaction carried out in the opposite, nonspontaneous, direction would require the expenditure of work in an amount at least as great as W'.

You may recall from our discussion of heat flow Q and work effect W in Chapter 4 that the sign on Q and W is used to indicate direction of flow. Heat flow *into* a system is positive (+). Work produced *by* a system on its surroundings is also positive (+). Heat flow **from** a system, and work done *on* a system, are both negative (−).

Using these conventions and considering the criterion for spontaneity of chemical reactions, we can see that for any reaction, at T and P,

> if W' is positive, the reaction is spontaneous, and
> if W' is negative, the reaction is not spontaneous.

If, by chance, W' = 0, the system is in equilibrium, and will not undergo any change.

If W' < 0, work must be done to make the reaction go.

Free Energy Change, ΔG

We have seen that associated with the heat flow Q in a chemical reaction occurring at constant temperature and pressure and doing no useful work, there is a change, ΔH, in the enthalpy of a system, such that,

$$\Delta H = Q = H_{products} - H_{reactants} \qquad (13.4)$$

where $H_{products}$ and $H_{reactants}$ are properties of the products and reactants respectively, which are fixed by the states of the substances involved.

By a tight thermodynamic argument it can be shown that a similar relation exists between W', the maximum amount of useful work that can be produced by a chemical change at constant temperature and pressure, and a property called the **free energy** G that can be associated with the reactants and products of the reaction. The relation is

$$\Delta G = -W' = G_{products} - G_{reactants} \qquad (13.5)$$

As implied by Equation 13.5, there is associated with every pure substance in a given state a property called the free energy G which relates directly to **the capacity of that pure substance to produce useful work**. The change in free energy ΔG, which occurs when a chemical change takes place at constant temperature and pressure, is equal to −W', the negative of the amount of useful work that can be produced by the change. A decrease in free energy, ΔG < 0, will correspond to a positive value of W' and hence will be associated with a spontaneous reaction. The thermodynamic conditions for spontaneity in terms of free energy changes can therefore be expressed as follows:

If ΔG is −, then W' is + and the reaction is spontaneous.
If ΔG is +, then W' is − and the reaction is not spontaneous; the reverse reaction is spontaneous.
If ΔG is 0, then W' is 0 and the reaction is at equilibrium.

If the free energy of the products is less than that of the reactants, the reaction is spontaneous.

These criteria are valid for chemical or physical changes that occur under conditions of constant temperature and pressure for every reactant and product.

The free energy change for a reaction, like the changes in internal energy

327

and enthalpy, is independent of the path by which the reaction takes place. However, unlike ΔE and ΔH, ΔG can vary considerably with both temperature and pressure. Consider, for example, Reaction 13.1:

$$4 \text{ Ag(s)} + O_2\text{(g)} \longrightarrow 2 \text{ Ag}_2O\text{(s)} \tag{13.1}$$

It is found experimentally that ΔG for this reaction (Table 13.1) increases with rising temperature and decreases with rising pressure. In contrast, ΔH for the reaction remains virtually constant at -14.62 kcal over the whole range of temperatures and pressures in Table 13.1.

TABLE 13.1 VARIATION OF ΔG WITH TEMPERATURE
AND PRESSURE FOR THE REACTION
$4 \text{ Ag(s)} + O_2\text{(g)} \rightarrow 2 \text{ Ag}_2O\text{(s)}$

	ΔG in kcal			
T(°C)	P = 10 atm	P = 1 atm	P = 0.01 atm	P = 0.0001 atm
25	−6.54	−5.18	−2.44	+0.30
100	−4.50	−2.80	+0.62	+4.04
190	−2.12	0.00	+4.24	+8.48
300	+0.92	+3.54	+8.78	+14.02
500	+6.35	+9.88	+16.96	+24.04

From Table 13.1 it is clear that at low temperatures and high pressures Reaction 13.1 tends to be spontaneous ($\Delta G < 0$). However, even at 25°C, at very low pressures Ag_2O(s) will tend to dissociate spontaneously. At high temperatures the tendency to dissociate increases tremendously, so that at 300°C the dissociation is spontaneous even at pressures well over 1 atm. At 190°C and 1 atm, ΔG equals zero, and under these conditions silver oxide will be in thermodynamic equilibrium with oxygen gas and solid metallic silver.

When writing chemical equations we ordinarily consider that all reactants and products are at the same temperature and at *1 atm pressure*. ΔG for a reaction occurring under these conditions is a characteristic of the reaction and is given the symbol $\Delta G°$. You should keep in mind, however, that as with Reaction 13.1, changes in pressure could have a very marked effect on the free energy change for that reaction.

We saw in Chapter 4 how one can evaluate the changes in enthalpy, ΔH, in chemical reactions from the molar heats of formation, ΔH_f, of the reactants and products. It is possible to carry out analogous calculations for free energy changes, $\Delta G°$, in chemical reactions if the free energies of formation, $\Delta G_f°$, of the reactants and products are available. The molar free energy of formation, $\Delta G_f°$, of a pure substance is equal to the free energy change in the reaction at 1 atm by which the substance is formed from the elements. In Table 13.2 are listed the molar free energies of formation $\Delta G_f°$ of some of the common pure substances at 25°C and 1 atm pressure. (Can you suggest how these values were obtained?)

From examination of Table 13.2 you will note that the free energy of formation $\Delta G_f°$ is negative for most pure substances at 25°C. This reflects the fact that most compounds are stable with respect to decomposition into their elements at 25°C and 1 atm pressure. In principle, any pure substance with

TABLE 13.2 Free Energies of Formation (kcal/mole) at 25°C, 1 atm

$AgBr(s)$	−22.9	$CO(g)$	−32.8	$H_2O(g)$	−54.6	$NH_4Cl(s)$	−48.7
$AgCl(s)$	−26.2	$CO_2(g)$	−94.3	$H_2O(l)$	−56.7	$NO(g)$	+20.7
$AgI(s)$	−15.9	$C_2H_2(g)$	+50.0	$H_2S(g)$	−7.9	$NO_2(g)$	+12.4
$Ag_2O(s)$	−2.6	$C_2H_4(g)$	+16.3	$HgO(s)$	−14.0	$NiO(s)$	−51.7
$Ag_2S(s)$	−9.6	$C_2H_6(g)$	−7.9	$HgS(s)$	−11.7	$PbBr_2(s)$	−62.1
$Al_2O_3(s)$	−376.8	$C_3H_8(g)$	−5.6	$KBr(s)$	−90.6	$PbCl_2(s)$	−75.0
$BaCl_2(s)$	−193.8	$CoO(s)$	−51.0	$KCl(s)$	−97.6	$PbO(s)$	−45.1
$BaCO_3(s)$	−272.2	$Cr_2O_3(s)$	−250.2	$KClO_3(s)$	−69.3	$PbO_2(s)$	−52.3
$BaO(s)$	−126.3	$CuO(s)$	−30.4	$KF(s)$	−127.4	$Pb_3O_4(s)$	−147.6
$BaSO_4(s)$	−323.4	$Cu_2O(s)$	−35.0	$MgCl_2(s)$	−141.6	$PCl_3(g)$	−68.4
$CaCl_2(s)$	−179.3	$CuS(s)$	−11.7	$MgCO_3(s)$	−246	$PCl_5(g)$	−77.6
$CaCO_3(s)$	−269.8	$CuSO_4(s)$	−158.2	$MgO(s)$	−136.1	$SiO_2(s)$	−192.4
$CaO(s)$	−144.4	$Fe_2O_3(s)$	−177.1	$Mg(OH)_2(s)$	−199.3	$SnCl_4(l)$	−113.3
$Ca(OH)_2(s)$	−214.3	$Fe_3O_4(s)$	−242.4	$MgSO_4(s)$	−280.5	$SnO(s)$	−61.5
$CaSO_4(s)$	−315.6	$HBr(g)$	−12.7	$MnO(s)$	−86.8	$SnO_2(s)$	−124.2
$CCl_4(l)$	−16.4	$HCl(g)$	−22.8	$MnO_2(s)$	−111.4	$SO_2(g)$	−71.8
$CH_4(g)$	−12.1	$HF(g)$	−64.7	$NaCl(s)$	−91.8	$SO_3(g)$	−88.5
$CHCl_3(l)$	−17.1	$HI(g)$	0.3	$NaF(s)$	−129.3	$ZnO(s)$	−76.1
$CH_3OH(l)$	−38.7	$HNO_3(l)$	−19.1	$NH_3(g)$	−4.0	$ZnS(s)$	−47.4

a negative free energy of formation can be formed directly from the elementary substances. For example, in the reaction

$$\tfrac{1}{2}\,N_2(g) + \tfrac{3}{2}\,H_2(g) \longrightarrow NH_3(g), \quad \Delta G° = \Delta G_f°\,NH_3 = -4.0 \text{ kcal} \qquad \textbf{(13.6)}$$

The free energy change $\Delta G°$ in the reaction, which by definition is equal to the molar free energy of formation of ammonia, $\Delta G_f°\,NH_3(g)$, is equal to −4.0 kcal. By our thermodynamic criteria, the reaction by which ammonia is formed from the elements is spontaneous. (In practice, Reaction 13.6 occurs very slowly, so that at 25°C and 1 atm a mixture of N_2 and H_2 may exist for many years without producing a significant amount of NH_3. Many spontaneous reactions occur very slowly, particularly at low temperatures. We would say that such reactions are in principle spontaneous, but kinetically they are so slow as not to occur.)

How do you interpret the fact that $\Delta G_f°$ for NO_2 is positive?

Free energy changes at 1 atm can be calculated from $\Delta G_f°$ data in a manner completely analogous to that used to find enthalpy changes from ΔH_f data. The relation we use is:

$$\Delta G° = \sum \Delta G_f° \text{ products} - \sum \Delta G_f° \text{ reactants} = -W' \qquad \textbf{(13.7)}$$

EXAMPLE 13.1. Find the free energy change $\Delta G°$ at 25°C for the following reactions:

a. $Al_2O_3(s) + 3\,H_2(g) \longrightarrow 2\,Al(s) + 3\,H_2O(l)$

b. $4\,NH_3(g) + 5\,O_2(g) \longrightarrow 4\,NO(g) + 6\,H_2O(l)$

SOLUTION

a. $\Delta G^{\circ} = \sum \Delta G_f^{\circ}$ products $- \sum \Delta G_f^{\circ}$ reactants

$= 2 \Delta G_f^{\circ} Al(s) + 3 \Delta G_f^{\circ} H_2O(l) - \Delta G_f^{\circ} Al_2O_3(s)$
$- 3 \Delta G_f^{\circ} H_2(g)$

The free energies of formation of the elementary substances are, by the definition of ΔG_f°, zero. Therefore,

$$\Delta G^{\circ} = 3 (-56.7) - (-376.8) = +206.7 \text{ kcal}$$

This reaction does not occur at 25°C and 1 atm.

b. $\Delta G^{\circ} = 4 \Delta G_f^{\circ} NO(g) + 6 \Delta G_f^{\circ} H_2O(l) - 4 \Delta G_f^{\circ} NH_3(g)$
$= 4 (20.7) + 6 (-56.7) - 4 (-4.0) = -241.4 \text{ kcal}$

The reaction is spontaneous at 25°C and 1 atm.

13.2 DEPENDENCE OF ΔG ON TEMPERATURE. THE CONCEPT OF ENTROPY

The free energy change ΔG represents the maximum amount of useful work that can be obtained from a chemical reaction carried out at constant temperature and pressure. On the other hand, ΔH represents the heat flow in that same reaction when it is carried out directly so that no useful work is done. These two kinds of energy changes are related to each other in a rather subtle way which was first clearly understood by Gibbs. We shall now consider the nature of this relationship.

Distinction Between ΔG and ΔH

In order to give some physical interpretation to the difference between ΔG and ΔH, let us return to Reaction 13.2:

$$Pb(s) + Br_2(l) \longrightarrow PbBr_2(s) \tag{13.2}$$

As pointed out previously, this reaction can be carried out in two quite different ways. If it is carried out in an open container, no work is done (W = 0); all the energy is evolved in the form of heat (Q). When the reaction is carried out in a voltaic cell in such a way that a maximum amount of useful work (W') is done, it is accompanied by a certain heat flow (Q'), which differs numerically from Q. But according to the First Law of Thermodynamics, the difference between the amount of heat absorbed and the amount of work done must be the same for the two paths. That is,

In general there is a difference between the amount of useful work that can be obtained from a reaction and the amount of heat flow associated with it.

$$Q - W = Q' - W' \tag{13.8}$$

Setting W = 0, equating −W' to ΔG, and realizing that Q must equal ΔH,

$$\Delta H = Q' + \Delta G$$

or

$$\Delta G = \Delta H - Q' \tag{13.9}$$

330

For Reaction 13.2, ΔG is −62.1 kcal and ΔH is −66.3 kcal. Equation 13.9 tells us that Q' = −4.2 kcal. Experimentally, we find that 4.2 kcal of heat is evolved when this reaction takes place in a voltaic cell to produce a maximum amount of work.

Equations 13.8 and 13.9 are valid for any reaction. The change in the free energy, ΔG, is always equal to the difference between heat flow (ΔH) when the reaction is carried out at constant pressure without production of any useful work and heat flow (Q') when the maximum amount of work is done. The quantity Q', being the difference between ΔH and ΔG, depends only on the states of the reactants and products. From one point of view Q' might be considered as that part of the enthalpy change, ΔH, which is not available for doing useful work; we must subtract it from ΔH to obtain ΔG.

TABLE 13.3 Values of ΔG, ΔH, and Q' for Certain Reactions at 25°C, 1 atm

Reaction	ΔG (kcal)	=	ΔH (kcal)	−	Q' (kcal)
$Pb(s) + Br_2(l) \rightarrow PbBr_2(s)$	−62.1		−66.3		−4.2
$\frac{1}{2} H_2(g) + \frac{1}{2} Cl_2(g) \rightarrow HCl(g)$	−22.8		−22.1		+0.7
$CO(g) + \frac{1}{2} O_2(g) \rightarrow CO_2(g)$	−61.5		−67.7		−6.2
$NO(g) \rightarrow \frac{1}{2} N_2(g) + \frac{1}{2} O_2(g)$	−20.7		−21.6		−0.9
$CaCO_3(s) \rightarrow CaO(s) + CO_2(g)$	+31.1		+42.5		+11.4
$H_2O(l) \rightarrow H_2(g) + \frac{1}{2} O_2(g)$	+56.7		+68.3		+11.6
$Ag_2O(s) \rightarrow 2 Ag(s) + \frac{1}{2} O_2(g)$	+2.59		+7.31		+4.72

Depending on the particular reaction, Q' may be either positive or negative (see Table 13.3). The fact that it is never (except by coincidence) equal to zero, explains why ΔG and ΔH differ in magnitude. The magnitude of Q', unlike that of ΔH, is quite sensitive to changes in temperature and, in many cases, to changes in pressure. This explains why the sign of ΔG and consequently the direction in which a reaction proceeds spontaneously may change with temperature and pressure.

The Origin of Q'. Entropy Changes

At first sight it may well seem strange that there is a difference between the heat flow that can be obtained from a reaction and the amount of useful work that reaction can produce. One might easily conclude that since a chemical reaction involves a specific amount of energy change, that change might be made up of heat flow or work effects in any combination, including those situations in which all the energy change is observed as either heat flow or work produced or absorbed. The First Law of Thermodynamics, indeed, puts no restriction on the form in which energy changes may be observed, and simply states that

$$\Delta E = Q - W$$

The fact that W', the useful work, has a value that is not predictable from the value of ΔE, clearly indicates that there is some other effect which must be taken into account when we deal with the flow of heat and useful work associated with physical and chemical changes, an effect that is embodied in Q'.

If we study different reactions, looking specifically at the sign and magni-

331

tude of Q', we find that some generalizations can be made. For reactions in which gases are formed from condensed phases, Q' is almost invariably positive. If the number of moles of gaseous products exceeds the number of moles of gaseous reactants, Q' is also usually positive. As might be expected, since Q' is determined by the states of reactants and products, if a reaction produces a decrease in the numbers of moles of gas (Δn_g is negative), Q' is negative (see Table 13.3). When reactions are studied at different temperatures, Q' is found to vary and to be very nearly proportional to the absolute temperature of the reaction.

The second law puts a restriction on the conversion of heat into work.

These observations, plus some rather careful reasoning based on the Second Law of Thermodynamics, led Gibbs and others to deduce that

$$Q' = T\Delta S \tag{13.10}$$

where ΔS is the change in a quantity called the **entropy** S of the system. By Equations 13.10 and 13.9 it is clear that entropy, like enthalpy and free energy, is a property of a substance which is determined by the state of the substance.

The entropy of a substance arises from a rather different source than its internal energy or enthalpy. Entropy is associated with what might best be described as the **degree of disorder** in the substance. Some substances, such as those which exist as hard crystals, are highly ordered and have low entropy. Others, which occur as gases, are highly disordered and have high entropy. Factors which contribute to the disorder of a substance and hence increase its entropy are lack of structure, large volume available to all the particles, high particle mobility, and weak interparticle forces.

There is an inherent tendency, common to inanimate substances (and probably to animate objects like students and professors), to become disorganized. This tendency toward disorder, or a state of maximum entropy, actually provides a driving force for many chemical reactions and physical processes. In some cases the change in entropy furnishes essentially the entire driving force for the process. We shall now consider some specific changes in state, in which the entropy change plays an important role.

Let us take two evacuated flasks of equal volume connected by a closed stopcock. Into one of the flasks we admit carbon dioxide to a pressure of 1 atm and to the other we add nitrogen, until its pressure also becomes 1 atm. We bring both gases to 25°C and then open the stopcock connecting the flasks. We find that the two gases tend to mix spontaneously, so that after a period of time the nitrogen and carbon dioxide form a homogeneous gas solution in both flasks (Figure 13.2). During the mixing reaction the temperature and pressure do not change, indicating that ΔH for the reaction is zero. However, ΔG for the reaction cannot be zero, since the reaction is clearly spontaneous. It can be shown by thermodynamic arguments that for this reaction at 25°C,

$$\Delta G = -410 \text{ calories}$$

for each mole of gas in the system. This accounts for the spontaneous mixing that occurs in the experiment.

Since ΔH in the reaction equals zero,

$$\Delta G = -T\Delta S \quad \text{and} \quad \Delta S = \frac{410 \text{ cal}}{298°K} = 1.38 \text{ cal/°K} = 1.38 \text{ eu} \tag{13.11}$$

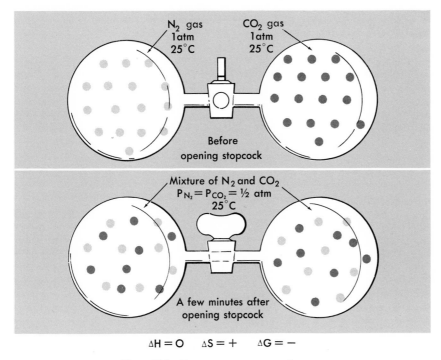

$$\Delta H = 0 \qquad \Delta S = + \qquad \Delta G = -$$

Figure 13.2 Spontaneous mixing of gases.

The entropy change ΔS in the reaction is positive and equal to 1.38 entropy units (calories/°K) per mole of gas present.

During the mixing reaction the molecules of gas become more disorganized; whereas before the reaction we could say that all the CO_2 molecules were in one of the flasks, after the mixing they may be found in either flask. The entropies of both gases increase when the volume allowed to them increases.

Some spontaneous chemical reactions involve essentially no change in enthalpy ($\Delta H = 0$), and are driven to an equilibrium state by the increase in entropy which occurs. One class of such reactions is described by the term "scrambling"; these reactions involve a randomization of chemical bonds. As an example, we might consider a mixture of two gases, $SnCl_4$ and $SnMe_4$, where Me represents a CH_3 group. If these two gases are at a sufficiently high temperature, they will react in such a way as to scramble the Cl and Me groups in the molecules in a random way, forming the species $SnCl_3Me$, $SnCl_2Me_2$, and $SnClMe_3$, until an equilibrium is reached with the reactants. Since the interchange of a Cl atom on one molecule with an Me group on another will occur without any change in enthalpy, we can actually calculate the probability of finding any particular kind of molecule in the equilibrium system. (This calculation is discussed in some detail in Problem 13.26 at the end of the chapter.) In Figure 13.3 we have shown the molecular structures of the various species and their relative amounts at equilibrium. In reactions of this kind, the equilibrium system has a considerably higher entropy than the reactants but the same enthalpy. The increase in disorder is due to the formation of the new species and to the fact that they are less ordered than the reactants. One finds that the entropy increases in this reaction by about 4.9 eu.

Osmosis is another process in which the driving force is an increase in entropy.

333

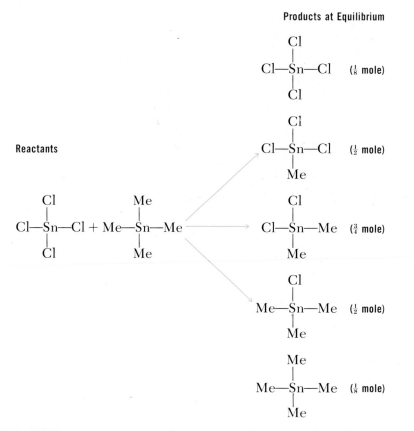

Products at Equilibrium

Reactants

Figure 13.3 A reaction driven by an increase in entropy ($\Delta H = 0$, $\Delta S = 4.9$ cal, $\Delta G = -4.9$ T cal).

Qualitatively it is very easy to predict the effect of physical changes on the entropy of a pure substance, since the entropy will increase whenever the degree of disorder increases. An increase in entropy of a pure substance will always result from (1) an increase in the temperature of a solid, liquid, or gas (which increases molecular motion), (2) the melting of a solid (which in-

TABLE 13.4 ABSOLUTE MOLAR ENTROPIES OF SOME REPRESENTATIVE SUBSTANCES

	$S°$(eu) at 25°C and 1 atm	Comments
Kr(g)	39.2	Simple monatomic gas; disorder due to motion of molecules through space
O_2(g)	49.0	Diatomic gas; disorder of rotational as well as translational motion
SF_6(g)	69.5	Large molecule; disorder of vibrational as well as rotational and translational motion
Hg(l)	18.5	Liquid metal
Br_2(l)	36.4	Diatomic molecules in liquid
CCl_4(l)	51.3	Large molecules in liquid
H_2O(l)	16.7	Small molecules in liquid
Na(s)	12.2	Soft metal
Cr(s)	5.7	Hard metal
C(diamond)	0.6	Very hard macromolecular substance with lowest known molar entropy
NaCl(s)	17.3	Typical ionic solid
I_2(s)	27.9	Diatomic molecules in solid

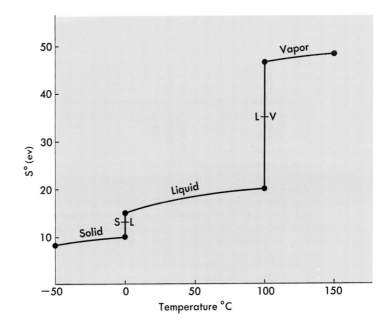

Figure 13.4 The molar entropy of water at 1 atmosphere pressure.

creases freedom of molecules to move about), (3) the vaporization of a liquid (vapor molecules have more volume available to them), or (4) a decrease in the pressure of a gas (which leaves more volume for molecules). It is possible to use thermodynamic theory to find quantitatively the change in entropy for each of these four processes, but our concern here is primarily with the qualitative effects we have noted. Using the Third Law of Thermodynamics (you may be glad to hear that there are only three, so far), one can actually determine absolute entropies for pure substances. In Table 13.4 are listed the absolute entropies of some pure substances. From the values given, it is clear that entropy increases as one goes from solids to liquids to gases, and that it also tends to increase with molecular complexity.

The third law sets the entropy of a perfect crystal at 0°K to be zero.

In Figure 13.4 we have shown graphically how the entropy of a mole of pure water at 1 atm changes with temperature. As expected, the largest changes occur when the water changes from one state of aggregation to another, but there is a gradual slow increase in entropy whenever the temperature is raised.

For chemical reactions it is often difficult to predict the sign of the entropy change. Since gases have large entropies in general, however, one can be quite sure that ΔS will have the same sign as Δn_g, the change in the number of moles of gas resulting from the reaction; if Δn_g is zero, the value of ΔS will usually be small but may be positive or negative. For reactions involving only condensed phases, ΔS will often be positive or negative, depending on whether the total number of moles of substances present increases or decreases, respectively, but exceptions to this rule are more common than when gases are involved in the reaction (Table 13.3).

Gibbs-Helmholtz Equation

Having recognized the existence of and the physical basis for the entropy property, we are ready to consider in a quantitative way the effect of temperature on the spontaneity of chemical reactions. By combining Equations

335

13.9 and 13.10 we obtain

$$\Delta G = \Delta H - T\Delta S \qquad (13.12)$$

This relation, known as the Gibbs-Helmholtz equation, is one of the most important in all of chemical thermodynamics. It tells us that the spontaneity of a chemical or physical change depends on two factors, the enthalpy change for the reaction and the product of the temperature and the entropy change for the reaction. Depending on the relative magnitudes and signs of these factors, ΔG will be negative (spontaneous reaction), positive (reaction will not proceed), or zero (reacting system in equilibrium).

You may recall from Chapter 4 that the enthalpy change ΔH in a chemical reaction is essentially the difference between the energy required to break all the chemical bonds in the reactants and the energy required to break all the bonds in the products. If the bonds in the products are more stable than those of the reactants, ΔH will be negative, the reaction will be exothermic, and ΔG in Equation 13.12 will tend to be negative. Since ΔH has a value that is nearly independent of temperature, its contribution to ΔG will be effectively constant. The magnitude of ΔS, like that of ΔH, will be essentially independent of the temperature at which a reaction occurs. This means that the term $T\Delta S$ will be directly proportional to temperature. Consequently, ΔG, unlike ΔH and ΔS, will be strongly temperature dependent and will indeed be a linear function of temperature.

Changes in pressure do not significantly affect the value of ΔH. However, for reactions in which there is a change in the number of moles of gas, ΔS will vary markedly with pressure. For such reactions, the absolute magnitude of ΔS will become greater as the pressure is lowered. This means that at low pressures the $T\Delta S$ term will make a greater contribution to ΔG. This effect may be large enough to change the sign of ΔG and hence the direction of spontaneity of the reaction (see Table 13.1).

At high T and low P, the $T\Delta S$ term becomes more important.

Under ordinary conditions, that is, in the vicinity of 25°C and 1 atm pressure, the value of $T\Delta S$ will usually be much smaller than the magnitude of ΔH, so that the sign of ΔG, and hence the spontaneity of a reaction, will be determined by the energy effect associated with the making and breaking of chemical bonds in the reaction. This explains the early observation that most spontaneous reactions are exothermic. Under conditions of either high temperature or low pressure, or both, the temperature-randomness change product, $T\Delta S$, will increase and may become large enough to overcome the influence of ΔH. In this way it may control the spontaneity of the reaction.

Many chemical and physical changes can best be understood by a simple qualitative application of Equation 13.12 and the line of reasoning just presented. For example, consider the vaporization of water:

$$H_2O(l) \rightleftharpoons H_2O(g) \qquad (13.13)$$

This reaction is endothermic in the amount of about 10 kcal/mole; a substantial amount of heat is required to vaporize water. By the general rule just cited, this implies that water would not vaporize spontaneously at room temperature when under atmospheric pressure. However, the randomness of water vapor is much higher than that of liquid water, so that the entropy change in the reaction is positive. As the temperature of the reaction is increased, the value of $T\Delta S$ will increase whereas that of ΔH will remain nearly constant. We would predict that at some temperature, $T\Delta S$ at one atm would become equal to ΔH; at that temperature, which happens to be 100°C, water liquid and water vapor can coexist at equilibrium. Above 100°C Reaction 13.13 will be spontaneous at 1 atm.

At pressures less than 1 atm, ΔS would be larger than at the higher pressure (greater randomness of the gas), and thus the effect of $T\Delta S$ would be larger. This means that if the pressure were lowered sufficiently at any temperature, $T\Delta S$ could be made equal to or larger than ΔH. At 25°C, $T\Delta S$ becomes equal to ΔH at a pressure of about 24 mm Hg. This is the water-water vapor equilibrium pressure ($\Delta G = 0$), which you would recognize as the vapor pressure of water. Above this pressure Reaction 13.13 is not spontaneous at 25°C, since then the value of ΔG is positive; below that pressure water at 25°C will tend to vaporize.

Given literature values for ΔH_f and ΔG_f°, it is possible to deal in a quantitative way with the effect of temperature on the spontaneity of chemical reactions at 1 atm. To illustrate the procedure, let us return to Reaction 13.1, which we have discussed previously.

$$4 \text{ Ag(s)} + O_2\text{(g)} \rightleftharpoons 2 \text{ Ag}_2\text{O(s)} \tag{13.1}$$

From Table 4.1 and Table 13.2 we find that, for this reaction at 25°C and 1 atm,

$$\Delta H = -14,620 \text{ cal} \quad \text{and} \quad \Delta G^\circ = -5,200 \text{ cal}$$

Substituting these values in Equation 13.12, we obtain, at 25°C and 1 atm,

$$\Delta G^\circ = \Delta H - T\Delta S^\circ; \quad -5,200 = -14,620 - 298 \, \Delta S^\circ$$

In this equation the superscript $^\circ$ means that the change is evaluated at 1 atm pressure; since ΔH is not a function of pressure, its superscript is often omitted. The change in entropy in the reaction is readily found:

$$\Delta S^\circ = \frac{-14,620 \text{ cal} + 5,200 \text{ cal}}{298°K} = -31.6 \text{ cal/}°K = -31.6 \text{ eu} \tag{13.14}$$

Since the entropy unit (eu) is ordinarily expressed in cal/°K, you must remember to express both ΔG° and ΔH in **calories** when calculating ΔS° by Equation 13.12. As you would expect, the entropy of the products in this reaction is less than that of the reactants ($\Delta n_g = -1$).

Recalling that ΔH and ΔS are *not temperature dependent*, we can say that, for this reaction at 1 atm and a given temperature T,

$$\Delta H = -14,620 \text{ cal} \quad \text{and} \quad \Delta S^\circ = -31.6 \text{ eu}$$

The value of ΔG° at T is given by Equation 13.12.

$$\Delta G^\circ \text{ (cal)} = -14,620 - T(-31.6) = -14,620 + 31.6 \text{ T} \tag{13.15}$$

Equation 13.15 describes quantitatively the spontaneity of Reaction 13.1 at 1 atm as a function of temperature. As the temperature increases, ΔG° becomes less negative and the driving force for the reaction decreases. The temperature at which ΔG° becomes zero is readily calculated:

$$\Delta G^\circ = 0 = -14,620 + 31.6 \text{ T}, \quad T = \frac{14,620}{31.6} = 463°K \quad \text{or} \quad 190°C$$

337

At 190°C silver oxide is in equilibrium with silver and oxygen at 1 atm. Below 190°C Reaction 13.1 is spontaneous; above that temperature silver oxide will tend to decompose spontaneously to the elements at one atm. By Equation 13.15 one can easily find $\Delta G°$ at any reasonable temperature; the values obtained at several temperatures are listed in Table 13.1 (P = 1 atm). In Figure 13.5 we have illustrated graphically the variation of $\Delta G°$ with temperature.

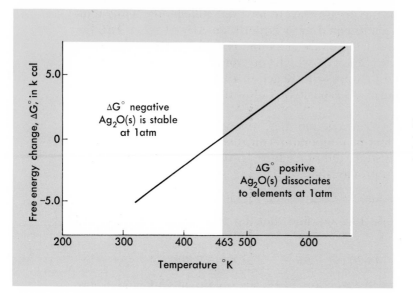

Figure 13.5 Temperature dependence of $\Delta G°$ for the reaction 4 Ag(s) + $O_2(g) \rightarrow$ 2 Ag$_2$O(s).

Equation 13.12 can be applied to any chemical reaction for which thermodynamic data is available. In using this equation there are a few precautions to be noted and a few inherent conditions to be observed. The problem is to find an equation analogous to Equation 13.15 for the reaction being studied. From this equation $\Delta G°$ can be found at any selected temperature. Both $\Delta G°$ and ΔH can be found at 25°C from the tables for $\Delta G_f°$ and ΔH_f for pure substances. ΔH can be assumed to be independent of temperature and pressure, and thus can be used directly in Equation 13.12 at any reasonable temperature (say from −200°C to at least 1000°C). $\Delta S°$ in cal/°K (eu) is calculated at 25°C and 1 atm from the values of $\Delta G°$ and ΔH, both expressed in calories. $\Delta S°$ can be assumed to be temperature independent and thus can be substituted back into Equation 13.12 to give the needed equation for $\Delta G°$ as a function of temperature. The use of Equation 13.12 is limited to furnishing information on reaction spontaneity at 1 atm, since the literature values are at 1 atm and both ΔG and ΔS are pressure dependent. The following example shows how these ideas can be applied in a specific case.

To find $\Delta G°$ at T, first find ΔH and $\Delta G°$ at 25°C. Then use equation 13.12.

EXAMPLE 13.2. For the reaction $MnO_2(s) + 2 H_2(g) \rightarrow Mn(s) + 2 H_2O(g)$ at 1 atm,

a. Find $\Delta G°$ for the reaction at 100°C. Is the reaction spontaneous at 100°C?

b. Find the temperature at which the reaction is in equilibrium at 1 atm.

c. Is the reaction at 1 atm pressure spontaneous at 200°C? 400°C?

SOLUTION

a. We must first find the values of $\Delta G°$ and ΔH at 25°C from Tables 13.2 and 4.1.

$$\Delta G° = 2\ \Delta G_f°\ H_2O(g) - \Delta G_f°\ MnO_2(s)$$
$$= 2(-54.6\ kcal) - (-111.4\ kcal) = +2.2\ kcal$$
$$\Delta H = 2\ \Delta H_f\ H_2O(g) - \Delta H_f\ MnO_2(s)$$
$$= 2(-57.8\ kcal) - (-124.5\ kcal) = +8.9\ kcal$$

By Equation 13.12, $\Delta G° = \Delta H - T\Delta S°$

At 25°C, $2200 = 8900 - 298\ \Delta S°$

$$\Delta S° = \frac{6700\ cal}{298°K} = +22.5\ cal/°K = +22.5\ eu$$

At T, $\qquad\qquad \Delta G° = 8900 - 22.5\ T \qquad$ in cal

At 100°C (373°K), $\Delta G° = 8900 - 22.5(373) = +500\ cal$

Since $\Delta G°$ is positive at 100°C, the reaction at 1 atm is not spontaneous.

b. As the temperature increases, $\Delta G°$ becomes smaller. $\Delta G°$ becomes equal to zero at a temperature which can be found from the previously derived equation,

$$\Delta G° = 8900 - 22.5\ T = 0$$

$$T = \frac{8900}{22.5}\ °K = 395°K,\ or\ 122°C$$

The reacting system will be in equilibrium at 122°C at a pressure of 1 atm.

c. Below 122°C, $\Delta G°$ will be positive and the reaction would not occur if P_{H_2O} and P_{H_2} were 1 atm. Above 122°C, $\Delta G°$ will be negative and the reaction will be thermodynamically spontaneous. Therefore the reaction would be expected to occur spontaneously at both 200°C and 400°C.

13.3 REACTION SPONTANEITY AND EQUILIBRIUM AS A FUNCTION OF PRESSURE: THE RELATION BETWEEN $\Delta G°$ AND THE EQUILIBRIUM CONSTANT FOR A REACTION

In discussing the spontaneity and equilibrium conditions for chemical reactions by Equation 13.12 we were of necessity limited to systems maintained at one atm pressure. You may wonder whether it is possible to analyze chemical reactions thermodynamically at any other pressure.

The fact is, as is perhaps evident to some of you, that we have already dealt with the pressure dependence of reaction spontaneity and equilibrium. In the previous chapter, using the concept of an equilibrium constant for a reaction, we investigated gaseous equilibria rather extensively; in those discussions concentrations of gases, rather than their pressures, were employed in the calculations. However, given gas pressures and the values of K_c and T

339

for a reaction, you could clearly convert pressures to concentrations and then apply the properties of K_c to find whether the reaction would proceed spontaneously under the conditions given and what the equilibrium state of the reacting mixture would be.

The problem, then, is really to be able to find the value of the equilibrium constant for the reaction at any given temperature. Since $\Delta G°$ at any temperature reflects the driving force, or the degree of spontaneity of a reaction when all reactants and products are at 1 atm pressure, you might expect that there would be a relation between $\Delta G°$ and the equilibrium constant for the reaction at that temperature. A development similar to that used in the previous section reveals that a simple relation does indeed exist, and takes the form:

If $\Delta G°$ is negative, K_p will be greater than one. Explain.

$$\Delta G° = -2.30 \ RT \ log_{10}K_p \qquad (13.16)$$

where R is the gas constant, T is the absolute temperature, and 2.30 is the factor used to convert from natural logarithms, present in the original derived equation, to common logarithms.

The equilibrium constant K_p differs in form from the equilibrium constant K_c used in Chapter 12 in that K_p involves partial pressures in atmospheres rather than concentrations in moles/liter. Consider, for example, the reaction

$$PCl_5(g) \rightleftharpoons PCl_3(g) + Cl_2(g) \qquad (13.17)$$

for which K_p has the form

$$K_p = \frac{P_{PCl_3} \times P_{Cl_2}}{P_{PCl_5}} \qquad (13.18)$$

where P_{PCl_3}, P_{Cl_2}, and P_{PCl_5} represent equilibrium partial pressures in atmospheres of PCl_3, Cl_2, and PCl_5 respectively. It will be recalled from Chapter 12 that K_c for this system has the form

$$K_c = \frac{[PCl_3] \times [Cl_2]}{[PCl_5]} \qquad (13.19)$$

where $[PCl_3]$, $[Cl_2]$, and $[PCl_5]$ represent equilibrium concentrations in moles/liter of PCl_3, Cl_2, and PCl_5 respectively.

The two forms of the equilibrium constant are related by an equation which is readily derived by applying the Ideal Gas Law. For the PCl_5-PCl_3-Cl_2 system, we have

$$P_{PCl_3} = \frac{n_{PCl_3} \times RT}{V} = [PCl_3] \times RT$$

Similarly, $\qquad P_{Cl_2} = [Cl_2] \times RT; \quad P_{PCl_5} = [PCl_5] \times RT$

Consequently, $\qquad \dfrac{P_{PCl_3} \times P_{Cl_2}}{P_{PCl_5}} = \dfrac{[PCl_3] \times RT \times [Cl_2] \times RT}{[PCl_5] \times RT}$

$$K_p = K_c \times RT \qquad (13.20)$$

For any reaction, K_p and K_c will be related by an equation similar to 13.20. In the general case

$$a \ A(g) + b \ B(g) \rightleftharpoons c \ C(g) + d \ D(g)$$

$$K_p = \frac{(P_C)^c \times (P_D)^d}{(P_A)^a \times (P_B)^b}; \quad K_c = \frac{[C]^c \times [D]^d}{[A]^a \times [B]^b}$$

But,

$$P_C = [C] \times RT; \quad P_D = [D] \times RT$$
$$P_A = [A] \times RT; \quad P_B = [B] \times RT$$

So

$$\frac{(P_C)^c \times (P_D)^d}{(P_A)^a \times (P_B)^b} = \frac{[C]^c \times [D]^d}{[A]^a \times [B]^b} \frac{(RT)^c \times (RT)^d}{(RT)^a \times (RT)^b}$$

$$K_p = K_c \times (RT)^{c+d-a-b}$$

$$K_p = K_c \times (RT)^{\Delta n_g} \qquad (13.21)$$

where Δn_g is the change in the number of moles of gas in the reaction. (As with K_c, the expression for K_p does not include pure solids or pure liquids). Thus we have

$$PCl_5(g) \rightleftharpoons PCl_3(g) + Cl_2(g); \Delta n_g = 1; \quad K_p = K_c(RT)$$

$$2\ HI(g) \rightleftharpoons H_2(g) + I_2(g); \Delta n_g = 0; \quad K_p = K_c$$

$$4\ NH_3(g) + 5\ O_2(g) \rightleftharpoons 4\ NO(g) + 6\ H_2O(l); \quad \Delta n_g = -5; \quad K_p = K_c(RT)^{-5}$$

Using Equation 13.21, one can readily determine K_p from a known value of K_c. Alternatively, K_c can be calculated if K_p is known (Example 13.3).

EXAMPLE 13.3. For the reaction, $PCl_5(g) \rightleftharpoons PCl_3(g) + Cl_2(g)$, K_p has the value 1.8 at a temperature of 250°C. Find K_c for the reaction at that temperature.

SOLUTION. From Equation 13.21,

$$K_p = K_c\ (RT)^{\Delta n_g} = K_c\ (RT)^1$$
$$K_c = K_p/RT$$

In calculations of this type, where we are converting from pressures in atmospheres to concentrations in moles/liter, R will have the value 0.0821 lit atm/mole °K.

$$K_c = \frac{1.8}{(0.0821)(523)} = 0.042$$

In any equilibrium problem it is in principle possible to work with either K_c or K_p. Since you are more familiar with K_c, you may wish to use it in most cases, but you should be aware of the approach that might be employed if K_p were used in the calculation. In some problems, K_p has a very simple form, and since K_p is most readily found from values of $\Delta G°$, it is often advantageous to use it directly rather than converting it to K_c by Equation 13.21.

EXAMPLE 13.4. PCl_5 is pumped into an empty container at 250°C until its partial pressure at equilibrium is 1.0 atm. Find the partial pressures of PCl_3 and Cl_2 in the container ($K_p = 1.8$).

SOLUTION. Some of the PCl_5 pumped into the container will dissociate:

$$PCl_5(g) \rightleftharpoons PCl_3(g) + Cl_2(g); \quad K_p = \frac{P_{PCl_3} \times P_{Cl_2}}{P_{PCl_5}} = 1.8$$

The partial pressures of the gases at equilibrium must satisfy the equilibrium constant expression for K_p in the same way that concentrations must satisfy the equilibrium constant expression for K_c. We are told that P_{PCl_5} is 1.0 atm at equilibrium. By the stoichiometry of the reaction, we conclude that when PCl_5 dissociates, the number of moles of PCl_3 formed must equal the number of moles of Cl_2 formed. That is,

$$n_{PCl_3} = n_{Cl_2}$$

Since we are working at a fixed temperature in a container of definite volume, it must also be true that

$$P_{PCl_3} = P_{Cl_2}$$

Knowing that P_{PCl_5} is 1.0 atm, we have

$$K_p = \frac{P_{PCl_3} \times P_{Cl_2}}{P_{PCl_5}} = \frac{(P_{PCl_3})^2}{1.0} = 1.8$$

$$P_{PCl_3} = (1.8)^{\frac{1}{2}} = 1.3 \text{ atm} = P_{Cl_2}$$

To have solved this problem by converting K_p to K_c, then changing all gas pressures to concentrations, solving for equilibrium concentrations, and finally converting back to pressures, would have been a cumbersome approach, to say the least!

Having acquired a bit of familiarity with the expression for K_p, its relation to K_c, and its possible application to problems involving chemical equilibria, we can now apply Equation 13.16 to a specific chemical reaction to find the equilibrium constant K_p at some specified temperature. The calculation can be carried out most readily at 25°C, the temperature at which free energy of formation data (Table 13.2) are available.

EXAMPLE 13.5. For the reaction $PCl_5(g) \rightleftharpoons PCl_3(g) + Cl_2(g)$, calculate K_p at 25°C, using the data in Table 13.2.

SOLUTION. Solving Equation 13.16 for $log_{10}K_p$,

$$log_{10}K_p = \frac{-\Delta G^\circ}{2.30 \text{ RT}}$$

$$\Delta G^\circ = \Delta G_f^\circ \text{ PCl}_3 - \Delta G_f^\circ \text{ PCl}_5$$

$$= -68.4 \text{ kcal} - (-77.6 \text{ kcal}) = 9.2 \text{ kcal} = 9200 \text{ cal}$$

$$log_{10}K_p = \frac{-9200 \text{ cal}}{(2.30)(1.99 \text{ cal/°K})(298°K)} = -6.74 = 0.26 - 7.00$$

$$K_p = 1.8 \times 10^{-7}$$

Why do the units of R differ from those used in Example 13.3?

It is possible to combine Equations 13.12 and 13.16 analytically to obtain the overall governing equation for K_p as a function of temperature:

$$\Delta G^\circ = \Delta H - T\Delta S^\circ = -2.30 \text{ RT } log_{10}K_p$$

$$log_{10}K_p = \frac{-\Delta H}{2.30 \text{ RT}} + \frac{\Delta S^\circ}{2.30 \text{ R}} \qquad \text{(13.22)}$$

This equation is used to calculate K_p from tables giving ΔG_f° and ΔH_f.

In this equation, both ΔH and ΔS° can be obtained from data in Tables 4.1 and 13.2. Since they do not vary with temperature, they may be substituted directly into Equation 13.22 to calculate K_p at the desired temperature. In a very real sense, Equation 13.22 is a general equation from which the spontaneity and equilibrium condition of any reaction can be predicted. Among other things, it enables us to calculate the value of the equilibrium constant at a particular temperature (Example 13.6).

EXAMPLE 13.6

a. Find the equation for the vapor pressure of water as a function of temperature.

b. Use this equation to calculate the vapor pressure of water at 100°C.

SOLUTION

a. The vapor pressure of water is the pressure at which liquid water and its vapor are in equilibrium:

$$H_2O(l) \rightleftharpoons H_2O(g); \quad K_p = P_{H_2O}$$

Some equilibrium constant expressions have a very simple form.

For this reaction, the equilibrium constant is simply the vapor pressure of water. To find K_p we need to find the values of ΔH and $\Delta S°$ for the reaction.

At 25°C and 1 atm,

$$\Delta H = \Delta H_f \, H_2O(g) - \Delta H_f \, H_2O(l)$$

$$= -57.8 \text{ kcal} - (-68.3 \text{ kcal}) = 10.5 \text{ kcal}$$

$$\Delta G° = \Delta G_f° \, H_2O(g) - \Delta G_f° \, H_2O(l)$$

$$= -54.6 \text{ kcal} - (-56.7 \text{ kcal}) = 2.1 \text{ kcal}$$

To obtain $\Delta S°$, we apply the relation

$$\Delta G° = \Delta H - T\Delta S°$$

Substituting the values of $\Delta G°$ and ΔH at 25°C,

$$2100 = 10{,}500 - 298 \, \Delta S°$$

$$\Delta S° = \frac{8400 \text{ cal}}{298°K} = 28.2 \text{ cal/°K}$$

Having calculated both ΔH and $\Delta S°$, we are now ready to substitute in Equation 13.22.

$$\log_{10} P_{H_2O} = \log_{10} K_p = \frac{-10{,}500}{(2.30)(1.99) \, T} + \frac{28.2}{(2.30)(1.99)}$$

$$\log_{10} P_{H_2O} = \frac{-2300}{T} + 6.18$$

You may recall that in Chapter 10, we wrote a similar equation to describe the variation of the vapor pressure of a liquid with temperature (Equation 10.3).

b. At 100°C (373°K), we have

$$\log_{10} P_{H_2O} = \frac{-2300}{373} + 6.18 = 6.18 - 6.18 = 0.00$$

$$P_{H_2O} = 1.0 \text{ atm}$$

This is, of course, the vapor pressure of water at its normal boiling point, 100°C.

We have seen from this example that Equation 13.22 can be used to calculate K_p at any given temperature from ΔH_f and ΔG_f° data at 25°C. Frequently, we are faced with a somewhat different problem: given K_p at one temperature and ΔH for the reaction, to calculate K_p at another temperature. Calculations of this sort are most simply carried out with the aid of the following equation, which can be derived from Equation 13.22 in precisely the same way that Equation 10.4, Chapter 10, was obtained from 10.3:

$$\log_{10} \frac{K_p(\text{at } T_2)}{K_p(\text{at } T_1)} = \frac{\Delta H \,(T_2 - T_1)}{2.30 \; RT_2 T_1} \tag{13.23}$$

where T_2 and T_1 represent two different temperatures (one ordinarily considers T_2 to be the higher temperature).

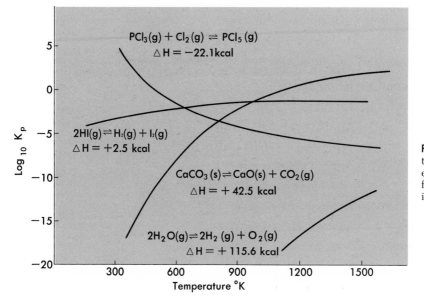

Figure 13.6 Temperature dependence of the equilibrium constant K_p for some typical chemical reactions.

Equation 13.23 tells us that if ΔH is positive (endothermic reaction),

$$\log_{10} \frac{K_p(\text{at } T_2)}{K_p(\text{at } T_1)} > 0; \quad K_p(\text{at } T_2) > K_p(\text{at } T_1)$$

and we deduce that K_p will increase with temperature. Conversely, if ΔH is a negative quantity, we conclude that K_p must decrease with increasing temperature (see Figure 13.6). It will be recalled that we came to this same conclusion by applying Le Chatelier's Principle in Chapter 12. The quantitative application of Equation 13.23 is illustrated by Example 13.7.

EXAMPLE 13.7. Given that K_p for the reaction $PCl_5(g) \rightleftharpoons PCl_3(g) + Cl_2(g)$ is 1.8×10^{-7} at 25°C and that ΔH is 22.2 kcal, calculate K_p at 250°C.

SOLUTION. Substituting in Equation 13.23, we have

$$\log_{10} \frac{K_p(\text{at } 250°C)}{K_p(\text{at } 25°C)} = \frac{22,200 \ (523 - 298)}{(2.30)(1.99)(523)(298)} = 7.0$$

$$\frac{K_p \text{ at } 250°C}{K_p \text{ at } 25°C} = 10^7$$

$$K_p \text{ at } 250°C = 10^7 \ (1.8 \times 10^{-7}) = 1.8$$

With Equation 13.23 we complete the investigation of spontaneity and equilibrium that we set up as our task for this chapter. Using Equations 13.22 and 13.12, we can deal intelligently with the decomposition of silver oxide or the conditions under which lead bromide could, in principle, be formed by the spontaneous reaction between lead and bromine. We can say qualitatively at least why KNO_3 may dissolve in water even though the process is endothermic. The possibility of reducing Al_2O_3 with hydrogen, or with any other substance, can be studied, and the dependence of the spontaneity of the reaction on temperature and pressure can be determined.

The power of chemical thermodynamics to deal with problems of chemical stability, the spontaneity of chemical reactions, and the equilibria in chemical systems, is impressive. The concepts which form the basis of the approach we have used are among the more profound that man has yet developed. Once understood, however, these concepts lead to relatively simple mathematical relations, which can be used with profit even by relative beginners. You will find that any understanding you can attain of the qualitative and quantitative aspects of thermodynamics will prove valuable whenever you are called upon to deal with the factors which influence the equilibrium states in chemical reactions.

PROBLEMS

13.1 For the following reaction at 25°C and 1 atm pressure,

$$2 \ Ag(s) + Cl_2(g) \longrightarrow 2 \ AgCl(s)$$
$$\Delta H = -60.8 \text{ kcal} \qquad \Delta G = -52.4 \text{ kcal}$$

a. Interpret ΔG and ΔH in terms of measurable experimental quantities.
b. Is the reaction spontaneous under the given conditions? Why?

13.2 The molar free energy of formation, ΔG_f°, of $H_2S(g)$ is -7.9 kcal at 25°C, whereas that of $NO_2(g)$ is $+12.4$ kcal at that temperature. On the basis of these values, what can you conclude about the stability of H_2S and NO_2 at 25°C with respect to their decomposition to the elementary substances?

13.3 One can be certain that a reaction that is thermodynamically non-spontaneous will not occur unless work is done upon the reacting system. One cannot, however, be sure that a reaction which is spontaneous under given conditions will actually take place under those conditions. Explain why this is the case.

13.4 Calculate $\Delta G°$ at 25°C for the following reactions, using the data in Table 13.2.

a. $Cu_2O(s) + H_2(g) \longrightarrow 2 \ Cu(s) + H_2O(l)$
b. $C_2H_6(g) + O_2(g) \longrightarrow C_2H_2(g) + 2 \ H_2O(g)$

345

13.5   In view of your answers to Problem 13.4, comment on

   a. The feasibility of reducing Cu_2O with hydrogen at moderate temperatures.
   b. The possibility of making acetylene, C_2H_2, by oxidation of ethane, C_2H_6.

13.6   Predict the sign of the entropy change in the following reactions.

   a. $2 CO(g) + O_2(g) \longrightarrow 2 CO_2(g)$
   b. $Mg(s) + Cl_2(g) \longrightarrow MgCl_2(s)$
   c. $Al(s) \longrightarrow Al(l)$
   d. $I_2(s) \longrightarrow I_2(g)$
   e. $CH_4(g) + 2 O_2(g) \longrightarrow CO_2(g) + 2 H_2O(l)$

13.7   a. Using data in Tables 13.2 and 4.1, find ΔH, $\Delta G°$, and $\Delta S°$ at 25°C for the reaction $Fe_2O_3(s) + 3 H_2(g) \rightarrow 2 Fe(s) + 3 H_2O(g)$.
   b. What would be the values of ΔH, $\Delta G°$, and $\Delta S°$ at 1500°C for this reaction?
   c. Would this reaction proceed more or less readily at high than at low temperatures? Estimate the temperature at which the reaction mixture would be in equilibrium at 1 atm.

13.8   Find K_p at 25°C for the following reactions, using the data in Table 13.2.

   a. $2 SO_2(g) + O_2(g) \rightleftharpoons 2 SO_3(g)$
   b. $CH_4(g) + Cl_2(g) \rightleftharpoons CH_3Cl(g) + HCl(g)$
   Formulate expressions for K_p for each of these reactions.

13.9   Calculate the value of K_c for each of the reactions in Problem 13.8.

13.10   At 250°C, K_p for the reaction $PCl_5(g) \rightleftharpoons PCl_3(g) + Cl_2(g)$ is 1.8.

   a. PCl_5 is pumped into an originally empty container at 250°C. When equilibrium is attained, the partial pressure of Cl_2 is 0.5 atm. What are the equilibrium pressures of PCl_3 and PCl_5?
   b. One-half a mole of PCl_5 is pumped into an originally empty 10 liter container at 250°C. Find the partial pressures of Cl_2, PCl_3 and PCl_5 at equilibrium.

13.11   A suggested industrial process involves the reduction of $SnO(s)$ by hydrogen:

$$SnO(s) + H_2(g) \longrightarrow Sn(s) + H_2O(g).$$

   a. Would the reaction be spontaneous at 25°C and 1 atm?
   b. Calculate K_p for the reaction at 300°C.
   c. Some SnO is treated with H_2 in a closed container at 300°C until an equilibrium is reached. At that time $P_{H_2} = 2.5$ atm. Calculate P_{H_2O} in the container.
   d. Would you expect that H_2 gas at a pressure of 1 atm, passed over SnO at 300°C in a tube open to the atmosphere would reduce the oxide to the metal?

13.12   Calcium carbonate can be decomposed by the endothermic reaction

$$CaCO_3(s) \rightleftharpoons CaO(s) + CO_2(g); \quad \Delta H = +42.5 \text{ kcal}$$

The equilibrium pressure of CO_2 at 25°C is 2×10^{-23} atm. Calculate its equilibrium pressure at 700°C.

13.13   a. Find K_p at 25°C for the scrambling reaction shown in Figure 13.3.
   b. Calculate $\Delta G°$ for the reaction at 300°C.

13.14   From data in the tables, calculate ΔH and $\Delta G°$ at 25°C for the reaction $Pb(s) + PbO_2(s) \rightarrow 2 PbO(s)$. What physical significance can you attach to the two values just calculated? (Cite experiments which would in principle yield those values.)

13.15   Above about 500°C, mercury(II) oxide breaks down spontaneously in an open container:

$$HgO(s) \longrightarrow Hg(g) + \tfrac{1}{2} O_2(g)$$

What, if anything, does this behavior tell us about

a. The enthalpy H of HgO at 500°C as compared to that of the elements to which it decomposes?
b. The free energy G of HgO at 500°C as compared to that of the elements to which it decomposes?

13.16   For the reaction $2 Ag(s) + Br_2(l) \rightarrow 2 AgBr(s)$, (25°C, 1 atm), $\Delta H = -47.6$ kcal and $\Delta G° = -45.8$ kcal.

a. What experimentally measurable quantity would be numerically equal to the difference between ΔH and $\Delta G°$, namely, -1.8 kcal?
b. Calculate the entropy change, $\Delta S°$, in eu, for the reaction.
c. What does this value of $\Delta S°$ tell you about the relative degree of order in the reactants and products?

13.17   Calculate $\Delta G°$ at 25°C for the following reactions, using the data in Table 13.2.

a. $2 NaF(s) + Cl_2(g) \longrightarrow 2 NaCl(s) + F_2(g)$
b. $PbO_2(s) + 2 Zn(s) \longrightarrow Pb(s) + 2 ZnO(s)$

13.18   In view of the answers to Problem 13.17, comment on

a. The likelihood of obtaining $F_2(g)$ if you treat NaF with Cl_2.
b. The use of zinc to reduce PbO_2 to the metal.

13.19   Arrange the following substances in order of increasing entropy.

$$N_2O_4(g), Na(s), NaCl(s), Br_2(l), Br_2(g)$$

13.20   Calculate $\Delta G°$, ΔH, and $\Delta S°$ at 800°K for the reaction $NiO(s) + H_2(g) \rightarrow Ni(s) + H_2O(g)$.

13.21   Find K_p at 25°C for the following reactions, using the data in Table 13.2.

a. $SnO_2(s) + 2 H_2(g) \rightleftharpoons Sn(s) + 2 H_2O(l)$
b. $CuS(s) + O_2(g) \rightleftharpoons Cu(s) + SO_2(g)$
Formulate expressions for K_p for each of these reactions.

13.22   Calculate the value of K_c for each of the reactions listed in Problem 13.21.

13.23   At 25°C, K_p for the reaction $N_2O_4(g) \rightleftharpoons 2 NO_2(g)$ is 0.10.

a. In an equilibrium system at 25°C, the partial pressure of N_2O_4 is 0.50 atm. Calculate the partial pressure of NO_2.
b. If one mole of N_2O_4 is added to a ten liter container at 25°C, find the partial pressure of NO_2 at equilibrium.

13.24   At 25°C the molar free energy of formation of $Br_2(g)$ is 0.75 kcal, whereas its enthalpy of formation is 7.34 kcal.

a. Calculate K_p at 25°C for the reaction $Br_2(l) \rightleftharpoons Br_2(g)$.
b. Find the vapor pressure of Br_2 at 25°C in mm Hg.
c. Find the normal boiling point of $Br_2(l)$.
d. Calculate the vapor pressure of $Br_2(l)$ at 100°C.
e. Would Br_2 under a pressure of 10 atm at 100°C be liquid or gaseous?

13.25   Given that K_p for the reaction

$$SO_2(g) + \tfrac{1}{2} O_2(g) \rightleftharpoons SO_3(g)$$

is 2×10^{12} at 25°C and that ΔH is $-23,500$ cal, calculate K_p at 300°C.

°13.26   In the scrambling reaction which occurs when $SnCl_4$ and $SnMe_4$ are mixed, one can say that at equilibrium the probability of any given bond in a molecule chosen at random being an Sn—Cl bond is $\tfrac{1}{2}$, since there are equal numbers of Sn—Cl and Sn—Me bonds present, and they are randomly distributed in the molecules.

a. What would be the probability of choosing at random a molecule having four Sn—Cl bonds? Four Sn—Me bonds?

347

b. Consider the bare Sn atom with its four bonds, —Sn—. How many different molecules can you draw containing the Sn atom and Cl atoms and Me groups (0 to 4 Cl atoms plus 4 to 0 Me groups), assuming that the bonds on the Sn atom are distinguishable, for instance, by their position on the paper. Draw all the different molecules. How many would have the formula $SnCl_3Me$? The formula $SnCl_2Me_2$? The formula $SnClMe_3$? $SnCl_4$? If each different molecule you drew is equally likely to occur in the mixture, what would be the relative numbers of moles of each possible substance in the mixture? Do your results agree with the data in Figure 13.3?

*13.27 It is often the case that ΔH and ΔS for a reaction will have the same sign. Can you suggest a reason why this might be so?

*13.28 The reduction of aluminum oxide by carbon is not at present thought to be industrially feasible. Consider the reduction reaction thermodynamically, calculating in particular the temperature at which equilibrium would be attained at 1 atm pressure. Why is this reduction reaction not used? Would hydrogen be a better reagent to use for the reduction?

*13.29 Copper forms two common oxides, both of which can be purchased commercially. These are copper(I) oxide, Cu_2O, and copper(II) oxide, CuO.
a. Which oxide of copper is thermodynamically stable in air at 25°C?
b. Which oxide of copper is stable at 1100°C in an open container?
Hint: Consider the reactions $Cu(s) + \frac{1}{2} O_2(g) \rightarrow CuO(s)$ and $2 CuO(s) \rightarrow Cu_2O(s) + \frac{1}{2} O_2(g)$.

Rates of Reaction

In Chapter 13 we considered how one can use equilibrium constants or free energy changes to predict the direction and extent of reaction. Problems of this type fall in the general area of **chemical thermodynamics**. A knowledge of thermodynamic principles is essential to the chemist who wishes to carry out a reaction in the laboratory. They help him to choose reactant concentrations and temperatures at which the reaction is feasible.

It is often found, however, that reactions that are spontaneous in principle proceed so slowly as to be of little practical value. Consider, for example, the reaction

$$2 \; H_2(g) + O_2(g) \longrightarrow 2 \; H_2O(l) \tag{14.1}$$

The fact that ΔG for this reaction at 25°C and one atmosphere pressure is a negative number (−113.4 kcal) tells us that it should occur spontaneously under these conditions. Indeed, K_c for this reaction is so large ($\sim 10^{83}$) that hydrogen and oxygen, even at pressures far less than one atmosphere, should react to form water in nearly 100 per cent yield. However, a mixture of these two elements, held at 25°C and one atmosphere pressure for long periods of time, shows no evidence of reaction. Only if the mixture is sparked or exposed to a heated platinum wire does it react at an appreciable rate.

Many of the most familiar substances in the world around us are unstable from a thermodynamic viewpoint. According to thermodynamic calculations, all of the common fuels, including wood, coal, and petroleum products, should be oxidized to carbon dioxide and water upon exposure to air. The same is true of the organic compounds that make up the living cells of our body. Life persists in spite of reactions that are thermodynamically feasible but proceed at an infinitesimal rate under the conditions of temperature and pressure that prevail in our environment.

It's somewhat sobering to realize that people are thermodynamically unstable.

We conclude that there is no direct relationship between the rate of a reaction and the thermodynamic driving force, as expressed in terms of such quantities as ΔG or K_p. In order to predict how rapidly reactions will occur, we need quite a different set of principles, which fall in the general area of **chemical kinetics**. Taken together, the two sciences of thermodynamics and kinetics are capable of telling us a great deal about the course a reaction will take and the extent to which it will occur in a given time.

In this chapter we shall first consider how such factors as reactant concentrations and temperature affect reaction rates. We shall then discuss one of the theories that has been developed to explain these experimental observations. Finally, we shall show how the principles of chemical kinetics can be employed to deduce the path or **mechanism** through which reactants pass as they are converted to products. Throughout this chapter we shall deal almost exclusively with reactions involving gases; in later chapters we shall have more to say concerning reaction rates in aqueous solutions.

14.1 MEANING OF REACTION RATE

In order to discuss "reaction rates" intelligently, one must understand precisely what is meant by the term. **The rate of reaction is a positive quantity that tells us how the concentration of a reactant or product changes with time.** To see what this statement means, consider a specific reaction, that between carbon monoxide and nitrogen dioxide:

$$CO(g) + NO_2(g) \longrightarrow CO_2(g) + NO(g) \qquad (14.2)$$

The rate of reaction can be taken to be the change in concentration per unit time of one of the products, for example, carbon dioxide:

$$\text{rate} = \frac{\text{change of concentration of } CO_2}{\text{time interval}} = \frac{\Delta \text{conc } CO_2}{\Delta t} \qquad (14.2a)$$

Alternatively, the rate could be expressed in terms of a reactant:

$$\text{rate} = -\frac{\Delta \text{conc } CO}{\Delta t} \; \underset{}{\doteq} \; \frac{\Delta \text{ conc}.CO_2}{\Delta t} \quad \text{from stoicheom} \qquad (14.2b)$$

Here, the minus sign is introduced to make the rate a positive quantity; the concentration of CO decreases with time.

Because of the stoichiometry of Equation 14.2, the two rate expressions, 14.2a and 14.2b, are equivalent. That is, since one mole of CO_2 is formed for every mole of CO consumed, the rate at which the concentration of CO_2 increases, $\Delta \text{conc } CO_2 / \Delta t$, is exactly equal to the rate of decrease of CO concentration, $-\Delta \text{conc } CO / \Delta t$. If the stoichiometry were such that different numbers of moles of reactants or products were taking part in the reaction, the rate would be based on the change in concentration of any one substance chosen arbitrarily.

See problems 14.1 and 14.15.

We note from the rate expressions that the units in which the rate is expressed will be those of concentration divided by time. We shall always express concentrations in moles/liter. Time, on the other hand, may be given in seconds, minutes, hours, days, or years. Consequently, the rate of reaction will have the units:

$$\text{moles lit}^{-1} \text{ time}^{-1} = \text{moles lit}^{-1} \text{ sec}^{-1}, \text{ moles lit}^{-1} \text{ min}^{-1}, \ldots$$

Returning to the reaction between carbon monoxide and nitrogen dioxide, let us consider how we might determine the rate of this reaction as expressed by Equation 14.2b. Clearly, we need to know how the concentration of CO changes with time. If we admit 0.10 mole of CO and 0.10 mole of NO_2 to a one-liter container at 430°C and measure the concentration of CO as a function of time, we accumulate the data plotted in Figure 14.1. Note that the concentration of CO drops rapidly in the initial stages of the reaction. As equilibrium is approached, the concentration of CO decreases more slowly and eventually reaches a constant value. As it happens, K_c for this reaction is so large ($\sim 10^{17}$) that the reaction goes virtually to completion and the concentration of CO drops almost to zero.

If K_c were very small, what would this plot look like?

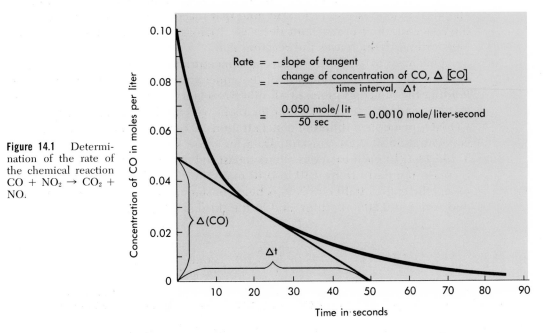

Figure 14.1 Determination of the rate of the chemical reaction $CO + NO_2 \rightarrow CO_2 + NO$.

Looking at Figure 14.1, it is clear that the rate of reaction is itself a function of time. The concentration of CO is decreasing more rapidly at 20 seconds than it is at 40 seconds. To determine the rate at a particular time, we must know the slope of the curve at that point. We can find this by drawing a tangent to the curve at the point of interest and taking the slope of the tangent line. The tangent at t = 20 seconds has a slope of approximately 0.050 moles/lit (distance along the vertical axis) divided by 50 seconds (distance along the horizontal axis). That is, at t = 20 seconds,

$$\text{rate} = -\frac{\Delta \text{conc CO}}{\Delta t} = -(\text{slope of tangent}) = \frac{0.050 \text{ moles/lit}}{50 \text{ sec}} = 0.0010 \frac{\text{moles}}{\text{lit sec}}$$

At t = 40 seconds the tangent would have a smaller slope and the rate would be less than 0.0010 moles/lit sec. When equilibrium is reached, the tangent becomes a straight line parallel to the horizontal axis, and the calculated rate is, as one would expect, zero.

The rate of reaction can be expressed in the language of calculus as the derivative of the concentration with respect to time. For the reaction of carbon monoxide with nitrogen dioxide,

$$\text{rate} = -\frac{d \, (\text{conc CO})}{dt}$$

This expression is less cumbersome to work with than Equation 14.2b. It is particularly useful in discussions of the effect of concentration upon reaction rate. However, for the benefit of those students who are not familiar with calculus, we will use algebraic relations such as Equation 14.2b.

351

14.2 DEPENDENCE OF REACTION RATE UPON CONCENTRATION

In discussing the reaction of carbon monoxide with nitrogen dioxide we pointed out that the rate of reaction decreases with time. From a slightly different point of view, we could say that the rate decreases as the concentration of CO or NO_2 decreases. This observation is generally valid for a variety of chemical reactions. We ordinarily find that reactions proceed more slowly as the concentrations of reactants decrease. Increasing the concentration of reactants ordinarily increases the reaction rate.

One way to study the effect of concentration upon reaction rate is to measure the rate as a function of the concentration of a particular reactant, holding the concentrations of all other reactants constant. We might, for example, conduct a series of experiments in which we measure the rate of the CO-NO_2 reaction at different concentrations of carbon monoxide, holding the concentration of NO_2 constant. Data for three such series are presented in Table 14.1. In each case the rate is measured at concentrations of CO which increase regularly from 0.10 to 0.40 mole/liter. In the first series of experiments the concentration of NO_2 is held constant at 0.10 mole/liter, in the second series at 0.20 mole/liter, and in the third at 0.30 mole/liter.

TABLE 14.1 RATES OF REACTION (MOLES/LIT SEC)
$CO(g) + NO_2 \longrightarrow CO_2(g) + NO(g)$ at 430°C.

	SERIES 1			SERIES 2			SERIES 3	
conc CO	conc NO_2	rate	conc CO	conc NO_2	rate	conc CO	conc NO_2	rate
0.10	0.10	0.012	0.10	0.20	0.024	0.10	0.30	0.036
0.20	0.10	0.024	0.20	0.20	0.048	0.20	0.30	0.072
0.30	0.10	0.036	0.30	0.20	0.072	0.30	0.30	0.108
0.40	0.10	0.048	0.40	0.20	0.096	0.40	0.30	0.144

Examining the vertical columns in Table 14.1, we conclude that the rate of reaction is directly proportional to the concentration of carbon monoxide. If, for example, the concentration of CO is doubled (e.g., from 0.10 to 0.20 moles/lit), the rate also is doubled (from 0.012 to 0.024 moles/lit sec in series 1; from 0.024 to 0.048 moles/lit sec in series 2, and so forth). We can deduce the effect of NO_2 concentration upon rate by examining the horizontal rows of data in Table 14.1. We note, for example, that when the concentration of CO is held constant at 0.10 mole/lit (first horizontal row), the rate increases in direct proportion to the concentration of NO_2. We conclude that the rate of this reaction is directly proportional to the concentrations of both CO and NO_2. Mathematically,

$$\text{rate} = k \text{ (conc CO)(conc } NO_2) \tag{14.3}$$

The constant of proportionality k in Equation 14.4 can be calculated from the observed rate at known concentrations of CO and NO_2. Substituting in Equation 14.3 at the point where the concentrations of CO and NO_2 are both 0.10 mole/lit, we have

$$k = \frac{\text{rate}}{(\text{conc CO})(\text{conc NO}_2)} = \frac{0.012 \text{ mole/lit sec}}{(0.10 \text{ mole/lit})(0.10 \text{ mole/lit})} = 1.2 \text{ (mole/lit)}^{-1} \text{ sec}^{-1}$$

Reaction rates (mole/lit sec) at 430°C can now be calculated from the expression

$$\text{rate} = 1.2 \text{ (conc CO)(conc NO}_2) \tag{14.3a}$$

Rate expressions analogous to 14.3a have been established experimentally for a large number of reactions. In general, for the reaction

$$aA(g) + bB(g) \longrightarrow \text{products}$$

we find, from experimental measurements, that

$$\text{rate} = k \text{ (conc A)}^m \text{ (conc B)}^n \tag{14.4}$$

The powers m and n to which the concentrations of reactants are raised in the rate expression describe the **order** of the reaction. Thus, if m in Equation 14.4 is 1, we say that the reaction is "first order with respect to reactant A." If n were 2, we would say that the reaction is "second order in B." The overall order of the reaction is the sum of the exponents m and n. Thus, a "first order reaction" is one in which $m + n = 1$, a "second order reaction" one in which $m + n = 2$, and so on.

What would it mean if a reaction were zero order in B?

It should be emphasized that the order of a reaction must be determined experimentally by a study of the effect of reactant concentration upon rate. In particular, the reaction order can not be deduced from the stoichiometry of the reaction. Frequently, it is found that the exponents m and n in Equations 14.4 are integers (0, 1, or 2); in many reactions, however, they are simple or complex fractions (Table 14.2).

Write the rate equation for a zero-order reaction.

TABLE 14.2 ORDERS OF SOME TYPICAL REACTIONS INVOLVING GASES

REACTION	m	n	OVERALL ORDER
$N_2O_4(g) \longrightarrow 2 \text{ NO}_2(g)$	1	—	first
$2 \text{ N}_2O_5(g) \longrightarrow 4 \text{ NO}_2(g) + O_2(g)$	1	—	first
$CH_3CHO(g) \longrightarrow CO(g) + CH_4(g)$	2	—	second (at low pressures)
$C_2H_4(g) + H_2(g) \longrightarrow C_2H_6(g)$	1	1	second
$2 \text{ NO}(g) + 2 \text{ H}_2(g) \longrightarrow N_2(g) + 2 \text{ H}_2O(g)$	2	1	third
$CHCl_3(g) + Cl_2(g) \longrightarrow CCl_4(g) + HCl(g)$	1	$\frac{1}{2}$	three-halves
$2 \text{ HI}(g) \xrightarrow{\text{Au}} H_2(g) + I_2(g)$	0	—	zero

The constant of proportionality k in Equation 14.4 is referred to as the **rate constant** for the reaction. For a particular reaction, k is a function only of temperature; k is independent of the concentrations of A and B. The rate constant can of course be calculated from the rate expression if the rate is measured at known concentrations of A and B (Example 14.1).

EXAMPLE 14.1. For the reaction A → products, the rate of decomposition of A is known to be 0.020 mole/lit sec when its concentration is 1.0×10^{-3} mole/liter. Calculate the numerical value of k if the reaction is
      a. first order   b. second order

353

SOLUTION

a. If the reaction is first order, the rate expression must be

$$\text{rate} = k(\text{conc A}); \; k = \frac{\text{rate}}{(\text{conc A})} = \frac{0.020 \text{ mole/lit sec}}{1.0 \times 10^{-3} \text{ mole/lit}} = 20 \text{ sec}^{-1}$$

b. For a second order reaction,

$$\text{rate} = k(\text{conc A})^2; \; k = \frac{\text{rate}}{(\text{conc A})^2} = \frac{0.020 \text{ mole/lit sec}}{(1.0 \times 10^{-3} \text{ mole/lit})^2}$$

$$= 2.0 \times 10^4 \text{ (mole/lit)}^{-1} \text{ sec}^{-1}$$

First Order Reactions

An important example of a first order gas phase reaction is the thermal decomposition of N_2O_5

$$2 \; N_2O_5(g) \longrightarrow 4 \; NO_2(g) + O_2(g) \tag{14.5}$$

for which the rate expression is

$$\text{rate} = k(\text{conc } N_2O_5) \tag{14.6}$$

By measuring the concentration as a function of time, we can accumulate data such as those plotted on the left in Figure 14.2. From such a plot, we can evaluate the rate constant k. We can do this by drawing a tangent to the curve at a particular concentration of N_2O_5, evaluating the rate from the slope of the tangent, substituting into Equation 14.6, and solving for k. When we do this, we arrive at a value of 0.345 min^{-1} for k at 66°C.

$$\text{rate} = (0.345 \text{ min}^{-1})(\text{conc } N_2O_5) \tag{14.6a}$$

Equation 14.6a can be used to calculate the rate of reaction at any desired concentration of N_2O_5. If, for example, we wish to know the rate when the concentration of N_2O_5 is 0.600 mole/lit, we have

$$\text{rate} = (0.345 \text{ min}^{-1})(0.600 \text{ mole/lit}) = 0.207 \text{ (mole/lit)min}^{-1}$$

In other words, the concentration of N_2O_5 is decreasing at the rate of 0.207 mole/lit per minute when the concentration is 0.600 mole/liter.

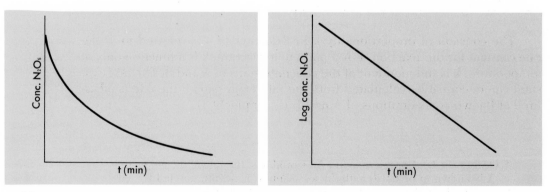

Figure 14.2 First order decomposition of N_2O_5.

Suppose, however, that we wish to know the time required for the concentration of N_2O_5 to fall to 0.600 mole/lit, starting with a concentration of 1.60 mole/lit. Equation 14.6a is of little use to us in answering this question. In principle we could obtain an answer by reading from the plot at the left of Figure 14.2 the time corresponding to a concentration of 0.600 mole/lit. In practice this is neither accurate nor particularly convenient. We would prefer to have a simple mathematical equation relating concentration and time. The form of this equation can be deduced if we plot the **logarithm** of the concentration of N_2O_5, rather than the concentration itself, versus time. This is done at the right of Figure 14.2.

It will be noted that there is a linear relationship between the logarithm of the concentration and the time. This relationship is ordinarily expressed in the form of the equation

$$\log_{10} \frac{X_0}{X} = \frac{kt}{2.30} \tag{14.7}$$

where X_0 and X represent reactant concentrations at zero time and t respectively.

Equation 14.7 is readily derived from the principles of calculus. For a first order reaction we have

$$rate = \frac{-dX}{dt} = k X$$

Rearranging,

$$\frac{dX}{X} = -kdt$$

Integrating from $t = 0$ to t,

$$\int_{X_0}^{X} \frac{dX}{X} = -k \int_0^t dt; \quad \ln \frac{X}{X_0} = -kt$$

where "$\ln X/X_0$" represents the natural logarithm (base e) of the ratio X/X_0.

Or, $\ln \frac{X_0}{X} = kt$, and finally $\log_{10} \frac{X_0}{X} = \frac{kt}{2.30}$

For a first order reaction, Equation 14.7 can be used to calculate the time required for the reactant concentration to drop to a certain level, or the concentration of that reactant remaining after a given time (Example 14.2). Alternatively, it can be employed to calculate the rate constant from concentration-time data (Example 14.3a) or the time required for a given fraction of a sample to react (Example 14.3b).

EXAMPLE 14.2. Given that the rate constant for the first order decomposition of N_2O_5 is 0.345 min^{-1} and that one starts with a concentration of 1.60 mole/lit, calculate

a. The time required for the concentration of N_2O_5 to drop to 0.600 mole/lit.

b. The concentration of N_2O_5 remaining after five minutes.

SOLUTION

a. Substituting into Equation 14.7, we have

$$\log_{10} \frac{1.60}{0.600} = \frac{0.345}{2.30 \ min} t$$

Solving,

$$t = \frac{2.30 \text{ min}}{0.345}\left(\log_{10}\frac{1.60}{0.600}\right) = \frac{2.30 \text{ min}}{0.345}\left(\log_{10}2.67\right) = \frac{(2.30)(0.426)}{0.345} \text{ min}$$

$$= 2.84 \text{ min}$$

b. $\log_{10}\dfrac{1.60}{X} = \dfrac{(0.345)(5.00 \text{ min})}{2.30 \text{ min}} = 0.750$

$$\log_{10}1.60 - \log_{10}X = 0.750;\ \log_{10}X = \log_{10}1.60 - 0.750$$

$$= 0.204 - 0.750 = -0.546$$

$$= 0.454 - 1$$

$$X = 2.84 \times 10^{-1} \text{ mole/lit}$$

EXAMPLE 14.3. For a certain first order reaction the concentration of reactant is found to drop from 2.00 mole/lit to 1.50 mole/lit in 64 minutes. Calculate

    a. The rate constant for this reaction.

    b. The time required for one-half of a sample to decompose.

SOLUTION

    a. $\log_{10}\dfrac{2.00}{1.50} = \dfrac{k(64.0 \text{ min})}{2.30}$

    Solving, $k = 4.49 \times 10^{-3} \text{ min}^{-1}$

    b. In general, when half the sample has decomposed, $X = \frac{1}{2}X_0$;

$$\log_{10}\frac{X_0}{X} = \log_{10}\frac{1}{\frac{1}{2}} = \log_{10}2 = 0.301$$

    From Equation 14.7 we have $0.301 = kt/2.30$

$$t = \frac{0.693}{k} = \frac{0.693}{4.49 \times 10^{-3}} \text{ min} = 154 \text{ minutes}$$

The analysis of Example 14.3b shows that **the time required for a given fraction (e.g., one-half) of a reactant to decompose via a first order reaction is independent of concentration**. Specifically, the time required for one-half the sample to decompose, often referred to as the **half-life** of the reaction, is

$$t_{\frac{1}{2}} = \frac{0.693}{k} \tag{14.8}$$

Note that $t_{\frac{1}{2}}$ is inversely proportional to the rate constant, k. A "fast" reaction, for which k is large, will have a short half-life; a "slow" reaction (small value of k) will be characterized by a comparatively long half-life.

The general validity of Equation 14.8 for any first order reaction suggests a simple experimental technique for determining whether or not a reaction is first order. Suppose we start with a given concentration of reactant, let us say 1.00 mole/lit, and measure the times required for the concentration to decrease by successive factors of two. If the reaction is first order, the time required for the concentration to drop from 1.00 mole/lit to 0.50 mole/lit should be equal to that required for it to decrease from 0.50 mole/lit to 0.25 mole/lit, or from 0.25 mole/lit to 0.125 mole/lit. If this behavior is not followed within experimental error the reaction must have an order other than one.

If the "half-life" is 2 hours, what fraction of reactant is left after 4 hours? one day?

Reactions of Higher Order

Among the gas phase reactions which have been shown to follow an integral order (i.e., $m + n = 0$, 1, 2, or 3), second order reactions are perhaps the most common. An example is the thermal decomposition of acetaldehyde

$$CH_3CHO(g) \longrightarrow CH_4(g) + CO(g) \tag{14.9}$$

for which the rate expression is found to be

$$rate = k(conc\ CH_3CHO)^2$$

Figure 14.3 Kinetics of a second order reaction.

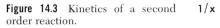

1/x

slope = k

t(min)

It can be shown, either from experiment or from the principles of calculus (cf. Problem 14.28), that for a second order reaction such as 14.9, in which there is only one reactant species, the concentration of that species is related to the elapsed time by the equation

$$\frac{1}{X} - \frac{1}{X_0} = kt \tag{14.10}$$

where X is the concentration at time t, X_0 is the concentration at $t = 0$, and k is the second order rate constant. As Equation 14.10 implies, a plot of $1/X$ vs. t will be linear for a second order reaction of this type. The slope of such a plot gives directly the numerical value of the rate constant (Figure 14.3).

EXAMPLE 14.4. For the decomposition of CH_3CHO the rate constant k is 19.8 (mole/lit)$^{-1}$ min^{-1} at 518°C. Calculate the concentration of CH_3CHO remaining after 10.0 minutes if one starts with a concentration of 0.0100 mole/lit.

SOLUTION. Substituting values of X_0, k, and t in Equation 14.10, we have

357

$$\frac{1}{X} - \frac{1}{0.0100 \text{ mole/lit}} = \frac{19.8}{(\text{mole/lit})\text{min}} \ (10.0 \text{ min})$$

$$\frac{1}{X} = \frac{100}{\text{mole/lit}} + \frac{198}{\text{mole/lit}} = \frac{298}{\text{mole/lit}}$$

$$X = \frac{1}{298} \text{ mole/lit} = 3.34 \times 10^{-3} \text{ mole/lit}$$

An equation for the half-life of a second order reaction involving a single reactant is readily derived from Equation 14.10. Remembering that, by definition, $t = t_{\frac{1}{2}}$ when $X = \frac{1}{2}X_0$, we have

$$\frac{1}{X_0/2} - \frac{1}{X_0} = kt_{\frac{1}{2}}; \quad \frac{2}{X_0} - \frac{1}{X_0} = kt_{\frac{1}{2}}; \quad \frac{1}{X_0} = kt_{\frac{1}{2}}$$

(14.11)

$$t_{\frac{1}{2}} = \frac{1}{kX_0}$$

This suggests a simple way of telling whether a reaction is first or second order.

Note that for a second order reaction, unlike the first order case, the half-life depends upon the initial concentration. The greater the initial concentration, the shorter the time required for one-half the reactant to decompose.

Many second order reactions involve two reactant species; an example is the hydrogenation of ethylene

$$C_2H_4(g) + H_2(g) \longrightarrow C_2H_6(g)$$

(14.12)

for which the rate expression has been shown experimentally to be

$$\text{rate} = k(\text{conc } C_2H_4)(\text{conc } H_2)$$

For second order reactions of this type, Equations 14.10 and 14.11 apply if the reactants have the same initial concentrations. Otherwise, the rate expressions are considerably more complicated.

Reactions with orders greater than two are comparatively rare. Many of them involve reactions of nitric oxide with the elements:

$$2 \text{ NO}(g) + O_2(g) \longrightarrow 2 \text{ NO}_2(g); \text{ rate} = k(\text{conc NO})^2(\text{conc } O_2) \quad (14.13)$$

$$2 \text{ NO}(g) + 2 \text{ H}_2(g) \longrightarrow N_2(g) + 2 \text{ H}_2O(g); \text{ rate} = k(\text{conc NO})^2(\text{conc } H_2) \ (14.14)$$

Many reactions of higher order can be treated as first order processes when one of the reactants is present in large excess. These are referred to as "pseudo–first order reactions" and are frequently encountered in aqueous solutions when water itself is involved in the reaction. An example is the hydrolysis of sucrose:

$$\underset{\text{sucrose}}{C_{12}H_{22}O_{11}} + H_2O \longrightarrow \underset{\text{glucose}}{C_6H_{12}O_6} + \underset{\text{fructose}}{C_6H_{12}O_6}$$

(14.15)

Kinetic data obtained when this reaction is carried out in nonaqueous solvents indicate that it is first order in both sucrose and water, giving an overall order of two. In water solution, the number of water molecules is so much greater than the number of sugar molecules that one cannot detect any change in water concentration as the reaction proceeds, and the kinetics are adequately described by the simple first order expression:

$$\text{rate} = k(\text{conc } C_{12}H_{22}O_{11})$$

14.3 DEPENDENCE OF REACTION RATE UPON TEMPERATURE

The rates of chemical reactions are profoundly influenced by the temperature at which the reaction is carried out. An increase in temperature ordinarily increases the rate; a reaction that requires months at room temperature may occur in hours at 100°C. As a general and very approximate rule, it is often stated that an increase in temperature of 10°C can be expected to double the reaction rate. If this rule holds, one can expect the rate to increase by a factor of 2^{10}, or approximately 1000, if the temperature is increased by 100°C.

This is one very practical reason for using refrigerators.

To derive a quantitative relationship between reaction rate and temperature, let us consider a specific reaction, that between CO and NO_2. To study the effect of temperature upon the rate of this reaction, we might measure the rate at different temperatures but at the same reactant concentrations. An equivalent but more convenient approach would be to determine the rate constant k as a function of temperature. Data obtained from such a study are presented in Table 14.3.

TABLE 14.3 TEMPERATURE DEPENDENCE OF THE RATE CONSTANT FOR THE REACTION $CO(g) + NO_2(g) \longrightarrow CO_2(g) + NO(g)$

TEMPERATURE (°K)	k (MOLE/LIT)$^{-1}$ SEC^{-1}
600	0.028
650	0.22
700	1.3
750	6.0
800	23

If one plots directly the data in Table 14.3 (k on the vertical axis, T along the horizontal axis), the curve shown at the left of Figure 14.4 is obtained. We discover a more useful way to represent this data if we choose to plot $\log_{10}k$ vs. $1/T$; in this case, the straight-line plot shown at the right of Figure 14.4 is obtained.

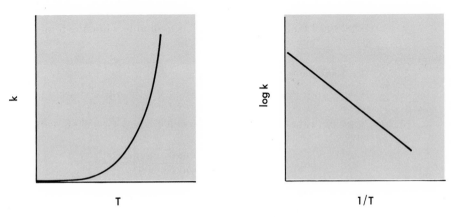

Figure 14.4 Temperature dependence of rate constant.

359

The linear plot of Figure 14.4 tells us that the relationship between $\log_{10}k$ and $1/T$ is of the form

$$\log_{10}k = A - \frac{B}{T} \tag{14.16}$$

where A and B are constants that can be determined from the intercept and slope of the plot. The validity of Equation 14.16 for representing the variation of the rate constant with temperature was first demonstrated by the Swedish chemist Svandte Arrhenius in 1887.

The constant B in Equation 14.16 is directly related to an important parameter known as the **activation energy** E_a for the reaction. We will discuss the physical significance of E_a shortly. It turns out that $B = E_a/2.30\,R$; making this substitution in Equation 14.16, we obtain

$$\log_{10}k = A - \frac{E_a}{2.30\,RT} \tag{14.17}$$

Equation 14.17 can be manipulated (cf. Equations 10.3 and 10.4, Chapter 10) to obtain a rather simple relation which facilitates the calculation of the rate constant at one temperature (T_2) if its value at another temperature (T_1) is known. This equation is

Compare with equation 13.23, Chapter 13.

$$\log_{10} \frac{k_2}{k_1} = \frac{E_a}{2.30\,R} \left(\frac{T_2 - T_1}{T_2 T_1} \right) \tag{14.18}$$

where k_2 and k_1 are the rate constants at temperatures T_2 and T_1 (°K) respectively, E_a is the activation energy in calories, and R is the gas law constant, 1.99 cal/°K. Some of the applications of Equation 14.18 are illustrated in Example 14.5.

EXAMPLE 14.5

a. If the slope of a plot of $\log_{10}k$ vs. $1/T$ is $-5000°K$, calculate E_a.

b. If the rate constant for a certain reaction is 2.0×10^{-3} sec^{-1} at 27°C and E_a is known to be 10,000 calories, calculate k at 127°C.

c. What must be the value of E_a if k is to double when the temperature increases from 27 to 37°C?

SOLUTION

a. $B = 5000°K = E_a/2.30\,R$

$E_a = (2.30)(R)(5000°K) = 2.30 \times 1.99 \text{ cal/°K} \times 5000°K = 22,900 \text{ cal}$

b. Applying Equation 14.18 with $T_2 = 400°K$ and $T_1 = 300°K$, we have

$$\log_{10} \frac{k_2}{k_1} = \frac{10,000}{(2.30)(1.99)} \left(\frac{400 - 300}{400 \times 300} \right) = 1.82$$

Taking antilogs, $k_2/k_1 = 66$; but since $k_1 = 2.0 \times 10^{-3}$ sec^{-1}, we have

$k_2 = 66 \times 2.0 \times 10^{-3}$ sec^{-1} = 0.13 sec^{-1}

c. If $k_2/k_1 = 2.00$, then $\log_{10}(k_2/k_1) = \log_{10}2 = 0.301$

Substituting in Equation 14.18, $0.301 = \dfrac{E_a (310 - 300)}{2\ 30(1.99)(310)(300)}$

$E_a = \dfrac{0.301 \times 2.30 \times 1.99 \times 310 \times 300}{10}$ cal $= 12{,}800$ cal

Note that if E_a were appreciably greater than 12.8 kcal, k would more than double for a 10° rise in temperature; if E_a were considerably smaller than 12.8 kcal, k would increase by less than a factor of two. Clearly the empirical rule that a temperature increase of 10°C doubles the reaction rate is a rather rough approximation.

The energy of activation associated with a reaction can be given a simple physical interpretation. In order for a reaction to occur between stable molecules, a certain amount of energy must be absorbed to break (or at least to weaken) the bonds holding the reactant molecules together. The activation energy E_a represents the energy required to bring the reactants to the point where they can rearrange to form products. From this point of view, one might expect a large activation energy to lead to a small rate constant, other factors being equal. Equation 14.16 tells us that this is the case; remembering that the constant B in this equation is directly proportional to the activation energy $(B = E_a/2.3\ R)$, it is clear that a large value of E_a will tend to make $\log_{10}k$ more negative and hence make k smaller.

A large E_a corresponds to a small k, but one which increases rapidly with T.

Expanding upon this physical picture, we might say that reactant molecules, before being converted to products, must pass through an unstable, high-energy intermediate. This transient, highly reactive species is referred to as an **activated complex**. Its exact nature is difficult to determine; in the reaction between CO and NO_2 it might be a "pseudomolecule" made up of one CO molecule and one NO_2 molecule in close contact. The path of reaction might be somewhat as follows:

$$O{\equiv}C + O{-}N{\diagdown}_O \quad \longrightarrow \quad O{\equiv}C{-}O{-}N{\diagdown}_O \quad \longrightarrow \quad O{=}C{=}O + N{=}O$$

reactants   activated complex   products

In general, one can interpret the activation energy for a reaction as representing the difference in energy between the reactant molecules and the less stable activated complex. This concept is shown schematically in Figure 14.5.

Just as one cannot estimate the height of a mountain by measuring the difference in elevation of the valleys that surround it, so one cannot calculate the activation energy E_a from the energy difference between reactants and products ΔE. However, we can see from Figure 14.5 that ΔE represents the difference between the activation energies of the forward and reverse reactions:

$$\Delta E = E_a - E_a' \tag{14.19}$$

If there is a net evolution of energy ($\Delta E < 0$), the activation energy for the

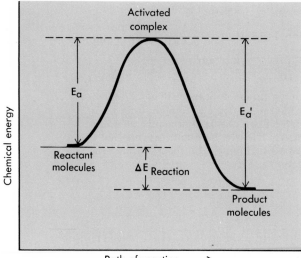

Figure 14.5 Energy changes during a reaction.

reverse reaction E_a' must be greater than that for the forward reaction, E_a. If, on the other hand, the products are in a higher energy state than the reactants ($\Delta E > 0$), then $E_a > E_a'$.

14.4 CATALYSIS

Relatively early in the development of chemistry it was discovered that small amounts of certain substances can drastically alter reaction rates. Such substances, called **catalysts**, are not consumed in the reaction but participate in it in such a way as to increase the rate. Since the catalyst is not a reactant or product in the overall reaction, it cannot change the equilibrium constant or the final equilibrium state of the system; what it does is to decrease the time required to reach equilibrium.

An example of a gas phase reaction whose rate is extremely sensitive to the presence of certain catalysts is that between sulfur dioxide and oxygen:

$$SO_2(g) + \tfrac{1}{2} O_2(g) \longrightarrow SO_3(g) \tag{14.20}$$

Could a catalyst be found to bring about the reverse reaction at room T and 1 atm?

Even though this reaction is thermodynamically spontaneous at room temperature and atmospheric pressure ($\Delta G = -16.7$ kcal), it occurs at an infinitesimal rate under these conditions. Raising the temperature increases the rate of reaction but decreases the yield of sulfur trioxide, since the reaction is exothermic ($\Delta H = -23.5$ kcal).

An effective catalyst for this reaction is a gaseous mixture of nitric oxide and nitrogen dioxide. The exact mechanism of the catalyzed reaction is difficult to establish, but a plausible reaction path would involve the two-step sequence

$$SO_2(g) + NO_2(g) \longrightarrow SO_3(g) + NO(g) \tag{14.20a}$$

$$NO(g) + \tfrac{1}{2} O_2(g) \longrightarrow NO_2(g) \tag{14.20b}$$

362 Addition of Equations 14.20a and 14.20b yields the equation for the overall reaction, 14.20. Neither NO nor NO_2 is consumed in the reaction, yet they participate in it intimately.

The use of oxides of nitrogen to speed up the reaction between sulfur dioxide and oxygen is an example of **homogeneous catalysis**, in which the catalyzed reaction occurs within a single phase. Among the most important reactions of this type are those which take place in the body under the influence of certain catalysts known as enzymes. These substances, which are themselves rather complicated protein molecules, are responsible for the digestion of carbohydrates in the mouth and of proteins in the stomach and intestines. Enzymes are highly specific in their action; the enzyme maltase, for example, is an effective catalyst for the hydrolysis of maltose to glucose:

$$\underset{\text{maltose}}{C_{12}H_{22}O_{11}} + H_2O \xrightarrow{\text{maltase}} 2 \underset{\text{glucose}}{C_6H_{12}O_6} \qquad (14.21)$$

but lactose, a sugar with the same molecular formula as maltose, requires a different enzyme, lactase, for its hydrolysis.

The reaction between sulfur dioxide and oxygen can also be speeded up by certain finely divided solid catalysts, notably platinum and vanadium pentoxide. This is an example of **heterogeneous catalysis**, where the catalyzed reaction takes place at a phase boundary, usually at a solid surface. A classical application of heterogeneous catalysis occurs in the Haber process for the synthesis of ammonia from the elements, in which a mixture of iron and metal oxides serves as the catalyst. Elementary nickel is used as a catalyst in the commercial hydrogenation of liquid fats to yield margarine or shortening. Solid catalysts are used extensively in the petroleum industry to rearrange hydrocarbon molecules, thereby improving the quality of motor fuels and increasing the amount of fuel that can be obtained from a given amount of petroleum (Chapter 24).

Heterogeneous catalysis comes about as the result of *chemical adsorption* of reactant molecules at the solid surface. These molecules are held to the surface by forces comparable in strength to chemical bonds. The bonds within the molecule are thereby weakened, and the molecule becomes more susceptible to decomposition or reaction with other molecules.

A reaction of this type that has been studied extensively is the decomposition of nitrous oxide, N_2O, on metal surfaces:

$$N_2O(g) \xrightarrow{M} N_2(g) + \tfrac{1}{2} O_2(g)$$

This reaction is believed to proceed by the following mechanism:

$$N\equiv N-O + M \longrightarrow N\equiv N \cdots O \cdots M \longrightarrow N\equiv N + O-M$$

Adsorption of the N_2O molecule at a metal surface results in the formation of a bond between oxygen and a metal atom, M. This weakens the bond holding nitrogen to oxygen; the molecule breaks apart to form molecular nitrogen. The last step in the process involves the desorption of adsorbed oxygen atoms to form molecular oxygen.

The effect of certain "poisons" in inhibiting heterogeneous catalysis is readily explained in terms of an adsorption mechanism. A foreign substance which is more strongly adsorbed on the surface of a catalyst greatly reduces the fraction of surface available for adsorption of reactant molecules. In extreme cases, traces of inhibitors can completely cover the surface, reducing

363

the rate of the catalyzed reaction to zero. Frequently, a product of a reaction is strongly enough adsorbed to poison the catalyst. This is the case in the thermal decomposition of ammonia on a platinum surface,

$$2 \text{ NH}_3(g) \xrightarrow{\text{Pt}} \text{N}_2(g) + 3 \text{ H}_2(g)$$

What would you expect to be the reaction order with respect to H_2?

where the rate of reaction slows down as the hydrogen formed covers the platinum catalyst and prevents further adsorption of ammonia.

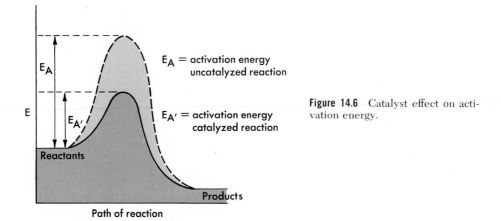

E_A = activation energy
uncatalyzed reaction

$E_{A'}$ = activation energy
catalyzed reaction

Reactants

Products

Path of reaction

Figure 14.6 Catalyst effect on activation energy.

The effectiveness of a catalyst in increasing the rate of a reaction may be attributed to its ability to lower the activation energy. For example, the homogeneous gas-phase decomposition of hydrogen iodide to the elements has an activation energy of 44.0 kcal/mole; in the presence of a platinum catalyst, the activation energy for the same overall reaction is only 33.8 kcal. The use of a gold catalyst further reduces the activation energy for this reaction, to 25.0 kcal. The general case is shown schematically in Figure 14.6. Note that the addition of a catalyst reduces the activation energies of both forward and reverse reactions. This means that the rate constants of the two opposing reactions are increased in the same proportion. The equilibrium constant remains unchanged.

14.5 THE COLLISION THEORY OF REACTION RATES

A successful theory of reaction rates should explain the experimental observations discussed in the previous section. In particular, it should account quantitatively for the dependence of reaction rate upon concentration and temperature. Hopefully, it should go further and predict the numerical value of the rate constant. We shall confine our discussion to one of the simpler theories of reaction rates which is based on the kinetic theory of gases. The **collision theory** of gas reactions, although unsatisfactory in certain respects, is physically reasonable and contributes to our understanding of chemical kinetics.

Molecules are like people; it is the small amount of high energy ones that get the job done.

According to the collision theory, reactions occur as a result of collision between molecules. However, not all collisions result in reaction; only those collisions involving molecules with high enough kinetic energies to supply the activation energy are fruitful. The rate of reaction, expressed in terms of the number of molecules reacting per second in a given volume, is equal to the number of collisions per second (Z) times the fraction (f) of those collisions that occur with sufficient energy to result in reaction.

$$\text{rate} = Z \times f \tag{14.22}$$

In principle, Z and f can be calculated exactly from the kinetic theory of gases, given the concentrations and diameters of the reactant molecules, the temperature, and the activation energy for the reaction. Since these calculations are rather involved, we shall consider only the predictions of collision theory concerning the dependence of reaction rate upon concentration and temperature.

For a given reaction, the collision number Z depends primarily upon the concentration of reactant molecules. The relationship between Z and concentration is readily deduced if we consider a specific reaction:

$$A + B \longrightarrow \text{products} \tag{14.23}$$

Clearly, the number of collisions between A and B molecules per unit volume in a given time will increase with increasing concentration of both A and B. If the concentration of A were doubled, there would be twice as many possibilities for collision and we would expect the collision number Z to be doubled. By the same token, doubling the concentration of B should double Z. We conclude that the collision number Z will be directly proportional to the concentrations of both A and B,

$$Z = C \text{ (conc A)(conc B)} \tag{14.24}$$

where C is a proportionality constant having only a small temperature dependence.

The main effect of temperature upon reaction rate arises in connection with f, the fraction of collisions which occur with sufficient energy to result in reaction. It seems reasonable to assume that f will be approximately equal to the fraction of molecules which have kinetic energies of motion equal to or greater than the activation energy, E_a. You may recall from our discussion of the kinetic theory of gases in Chapter 5 that many years ago James Clerk Maxwell derived an expression for the relative number of molecules having a given speed at a given temperature. It is possible to cast this equation in a form which states the relative number of molecules having any given energy. In Figure 14.7, we have shown how the relative number of molecules having a given energy depends upon the energy selected. Two different curves are drawn, one at a relatively low temperature, T_1, and the other at a substantially higher temperature, T_2.

From the graphs, it is clear that at a given temperature most of the molecules have energies near the maximum of the curve, with only a small number having energies equal to or greater than E_a. It is also clear that the likelihood of a molecule having an energy of at least E_a is enormously greater at high temperatures than at low temperatures. This means that at high temperatures the fraction of effective collisions and hence the reaction rate will be much greater.

Using the equation for the curves shown in Figure 14.7, it is possible to

365

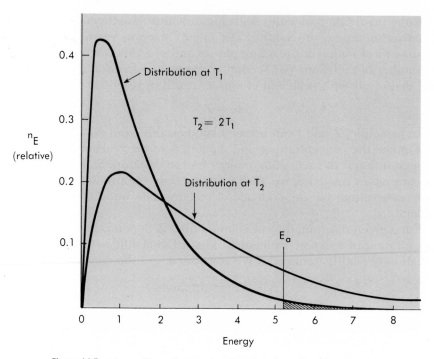

Figure 14.7 Maxwellian distribution function for molecular energies.

solve mathematically for the fraction of molecules having energies equal to or greater than E_a. It is found that, to a satisfactory degree of approximation,

$$f = e^{-E_a/RT} \tag{14.25}$$

where e is the base of natural logarithms, E_a is the activation energy, R is the gas constant, and T is the absolute temperature.

Having found values for Z and f, we can substitute them in Equation 14.22 to obtain the collision theory expression for the rate of reaction:

$$\text{rate} = Z \times f = C \text{ (conc A)(conc B) } e^{-E_a/RT} \tag{14.26}$$

Equation 14.26 gives us immediately the concentration dependence of the reaction rate. It predicts that the reaction between A and B will be first order in both reactants, second order overall. As we have seen, many gas phase reactions are second order, so in this respect the collision theory gives a plausible explanation for the dependence of reaction rates on concentration.

Equation 14.26 can be used to estimate how the rate constant of a reaction would be expected to vary with temperature according to the collision theory. Since the rate constant k is equal to the rate at unit concentration of reactants, it follows that

$$k = Ce^{-E_a/RT} \tag{14.27}$$

To compare Equation 14.27, obtained from collision theory, with experiment, we take natural logarithms of both sides, obtaining

$$\ln k = \ln C - \frac{E_a}{RT}; \quad \log_{10} k = \log_{10} C - \frac{E_a}{2.30\ RT} \qquad (14.28)$$

This equation becomes identical with 14.16 if we take $A = \log_{10}C$ and $B = E_a/2.30\ R$. Since Equation 14.16 was obtained from experiment, we conclude that the collision theory explains satisfactorily the dependence of reaction rate upon temperature as well as upon concentration.

In order to go one step further and see whether collision theory can predict the actual magnitude of the rate constants for bimolecular reactions, it is necessary to calculate the quantity C in Equation 14.27. This can be accomplished with the aid of the kinetic theory of gases; the results of such calculations are compared with experiment in Table 14.4.

TABLE 14.4

| | | k IN $(MOLE/LIT)^{-1}$ SEC^{-1} | |
REACTION	T(°K)	OBSERVED	CALCULATED
$2\ HI(g) \longrightarrow H_2(g) + I_2(g)$	556	3.5×10^{-7}	5.2×10^{-7}
$H_2(g) + I_2(g) \longrightarrow 2\ HI(g)$	700	6.4×10^{-2}	14×10^{-2}
$NO_2(g) + CO(g) \longrightarrow NO(g) + CO_2(g)$	500	0.55	1.7
$2\ NOCl(g) \longrightarrow 2\ NO(g) + Cl_2(g)$	300	1.5×10^{-5}	9×10^{-5}

Considering the various approximations that are necessary to calculate rate constants from collision theory, the deviations between observed and calculated values of k are not too surprising. It is interesting, however, that the rate constants predicted by collision theory are invariably too large (Table 14.4). Qualitatively, this can be explained by refining the theory to take account of the fact that two molecules must be favorably oriented with respect to each other if they are to react upon collision. If, for example, the collision of a CO molecule with an NO_2 molecule brings the carbon and nitrogen atoms into contact with each other, it is unlikely that the transfer of an oxygen atom, which is necessary for reaction, will take place.
Taking this factor into account, we could write Equation 14.22 as

$$\text{rate} = P \times Z \times f \qquad (14.29)$$

where P, a fraction less than one, is the probability that the colliding molecules will be favorably oriented for reaction. Unfortunately, this fraction is difficult to estimate accurately; in the few cases where P has been calculated, the agreement between theory and experiment has not improved markedly.

We have seen that the collision theory gives a reasonably satisfactory explanation of the kinetics of second order reactions. However, the great majority of gaseous reactions are not second order; reactions of other integral orders (1, 3) or fractional orders ($\frac{1}{2}$, $\frac{3}{2}$, and so forth) are common. In order to

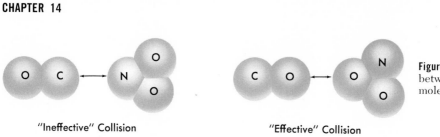

"Ineffective" Collision   "Effective" Collision

Figure 14.8 Collision between CO and NO_2 molecules.

explain data of this type, it is necessary to postulate a series of steps rather than a single, bimolecular collision. Consider, for example, a first order reaction:

$$A \longrightarrow B; \quad \text{rate} = \text{constant (conc A)}$$

The observed kinetics can be explained in terms of the collision theory if we assume that the reaction occurs in three steps which comprise what is called the mechanism of the reaction:

1. As the result of a bimolecular collision, a high-energy, unstable species A^* is formed.

$$A + A \longrightarrow A^* + A$$

2. The activated molecule A^* may lose its energy by colliding with an ordinary molecule of A.

$$A^* + A \longrightarrow A + A$$

(Note that this is the reverse of the reaction by which A was formed).

3. Alternatively, A^* may decompose to products.

$$A^* \longrightarrow B$$

If reactions 1 and 2 are very rapid compared to 3, A^* will be essentially in a state of equilibrium with A:

$$A + A \rightleftharpoons A^* + A; \quad K = \frac{(\text{conc } A^*)(\text{conc } A)}{(\text{conc } A)^2} = \frac{(\text{conc } A^*)}{(\text{conc } A)}$$

The overall rate of reaction will be governed by that of the slow **(rate-determining)** step

$$\text{rate} = \frac{\Delta \text{conc B}}{\Delta t} = k \, (\text{conc } A^*)$$

But, from the expression for the equilibrium constant, we see that

$$(\text{conc } A^*) = K \, (\text{conc } A)$$

so that rate = k K (conc A) = constant (conc A). This is, of course, the rate equation for a first order reaction.

A similar type of stepwise mechanism can be invoked to explain third order kinetics (cf. Problem 14.13). As we shall see in the next section, stepwise mechanisms can also account for reactions that have a fractional order with respect to one or more of the reactants.

14.6 CHAIN REACTIONS

Certain gaseous reactions are initiated by the formation of an extremely reactive species at very low concentrations which sets off a series of reactions resulting in the formation of products. Such reactions are referred to as **chain reactions**. An example is the formation of hydrogen chloride from the elements. A mixture of hydrogen and chlorine stored at room temperature in the dark shows no evidence of reaction over long periods of time. However, if the mixture is exposed to ultraviolet light or heated to 200°C, a vigorous reaction occurs. The first step in this reaction, referred to as **chain initiation**, involves the dissociation of a chlorine molecule into atoms:

$$Cl_2 \longrightarrow 2\ Cl \qquad\qquad (14.30a)$$

The chlorine atoms formed are extremely reactive towards hydrogen molecules:

$$Cl + H_2 \longrightarrow HCl + H \qquad\qquad (14.30b)$$

This reaction forms another highly reactive species, a hydrogen atom, which attacks a chlorine molecule:

$$H + Cl_2 \longrightarrow HCl + Cl \qquad\qquad (14.30c)$$

At any time, the concentrations of Cl and H are extremely small.

In this way, the chlorine atoms are regenerated; the **chain propagation**, represented by Equations 14.30b and 14.30c, occurs over and over again until the H_2 and Cl_2 are almost completely converted to HCl.

The hydrogen and chlorine atoms, which act as **chain carriers**, can be consumed by reaction with each other:

$$H + Cl \longrightarrow HCl; \quad Cl + Cl \longrightarrow Cl_2; \quad H + H \longrightarrow H_2 \quad (14.30d)$$

These processes represent **chain termination**, since they break the chain mechanism.

Chain termination seldom occurs by a direct, two-body collision as Equation 14.30d might imply. Species such as H and Cl ordinarily have energies that are too high to permit the formation of a stable molecule by direct collision. Instead, chain termination may occur via a simultaneous, three-body collision involving a third species such as H_2 or Cl_2 to which some of the extra energy can be transferred. Alternatively, reactions such as 14.30d may occur on the surface of the vessel in which the reaction is taking place. In the reaction of hydrogen with chlorine, it is found that the reaction rate is decreased by the addition of powdered glass or silica, which acts as a surface catalyst for the recombination of hydrogen and chlorine atoms.

The chain carriers in a gaseous reaction are ordinarily present in very small quantities. In the reaction of hydrogen with chlorine, the concentrations of hydrogen or chlorine atoms may be less than 10^{-10} moles/lit and yet serve to continue the chain. Throughout most of the reaction the concentrations of these species remain nearly constant, since they are regenerated at approximately the same rate at which they are consumed.

The highly reactive species that act as chain carriers commonly contain an unpaired electron and are referred to as radicals. They may be individual

369

atoms such as

$$H\cdot, \quad :\overset{..}{\underset{..}{Cl}}:, \quad or \quad :\overset{..}{\underset{..}{O}}:$$

or polyatomic species such as the methyl radical

Why do traces of
NO often have a
drastic effect on
the rate of chain
reactions?

which is a chain carrier in many reactions involving organic molecules, as in
the decomposition of acetaldehyde:

$$CH_3CHO \longrightarrow CH_3 + H + CO \quad \text{(chain initiation)} \quad \text{(14.31a)}$$

$$CH_3 + CH_3CHO \longrightarrow CH_4 + CH_3 + CO \quad \text{(chain propagation)} \quad \text{(14.31b)}$$

$$\left.\begin{array}{l} CH_3 + CH_3 \longrightarrow C_2H_6 \\ CH_3 + H \longrightarrow CH_4 \\ H + H \longrightarrow H_2 \end{array}\right\} \quad \text{(chain termination)} \quad \text{(14.31c)}$$

Chain reactions are often inhibited by traces of substances which stop
the chain propagation by reacting with the carriers. The addition of very
small amounts of alcohol to chloroform stabilizes it against oxidation, presumably by interfering with the chain reaction involved in the reaction of
$CHCl_3$ with oxygen. The effectiveness of tetraethyl lead as an anti-knock addi-
tive is apparently related to its ability to slow down certain of the chain reactions involved in the oxidation of hydrocarbons. The ethyl radicals produced
when $Pb(C_2H_5)_4$ decomposes are believed to inhibit the formation of the chain
reaction carriers which are responsible for detonation.

The rate expressions for chain reactions are frequently quite complicated,
primarily because of the many steps involved in the reaction. An example of
a chain reaction with a fractional kinetic order is the chlorination of chloroform,

$$CHCl_3(g) + Cl_2(g) \longrightarrow CCl_4(g) + HCl(g) \quad \text{(14.32)}$$

for which the observed rate expression is

$$\text{rate} = k \, (\text{conc } Cl_2)^{\frac{1}{2}}(\text{conc } CHCl_3)$$

A chain mechanism which leads to the rate expression is

$$Cl_2 \rightleftharpoons 2\,Cl \qquad \text{(fast)} \qquad \text{(14.32a)}$$

$$Cl + CHCl_3 \overset{k}{\longrightarrow} CCl_4 + H \quad \text{(slow, rate determining)} \quad \text{(14.32b)}$$

$$H + Cl_2 \longrightarrow HCl + Cl \qquad \text{(14.32c)}$$

From 14.32b,

$$\text{rate} = \frac{\Delta\text{conc } CCl_4}{\Delta t} = k \, (\text{conc } Cl)(\text{conc } CHCl_3)$$

But the concentration of atomic chlorine is related to the concentration of Cl_2

by the equation

$$\frac{(\text{conc Cl})^2}{(\text{conc Cl}_2)} = K,$$

where K is the equilibrium constant for 14.32a. Therefore

$$(\text{conc Cl}) = K^{\frac{1}{2}} (\text{conc Cl}_2)^{\frac{1}{2}}$$

Substituting in the rate expression, we obtain

$$\text{rate} = k\, K^{\frac{1}{2}} (\text{conc Cl}_2)^{\frac{1}{2}} (\text{conc CHCl}_3)$$

which is identical in form to the experimentally observed rate equation.

PROBLEMS

14.1 For the reaction $2\ N_2O_5(g) \rightarrow 4\ NO_2(g) + O_2(g)$ it is found that the N_2O_5 is decomposing at the rate of 0.0100 mole/lit sec. Calculate the "rate of the reaction," defined as

a. $-\Delta\text{conc } N_2O_5/\Delta t$, with Δt in seconds
b. $-\Delta\text{conc } N_2O_5/\Delta t$, with Δt in minutes
c. $\Delta\text{conc } NO_2/\Delta t$, with Δt in seconds
d. $\Delta\text{conc } O_2/\Delta t$, with Δt in seconds

14.2 In the decomposition of N_2O_5 at 45°C the following concentration data were obtained.

t	conc N_2O_5	t	conc N_2O_5
0 sec	0.250 mole/lit	800 sec	0.153 mole/lit
200 sec	0.222 mole/lit	1000 sec	0.136 mole/lit
400 sec	0.196 mole/lit	1200 sec	0.120 mole/lit
600 sec	0.173 mole/lit		

a. Plot the data (concentration on the vertical axis).
b. From the plot determine the rate of reaction $(-\Delta\text{conc } N_2O_5/\Delta t)$ at $t = 500$ seconds.
c. Calculate from the data the average rate between $t = 400$ and $t = 600$ seconds, and compare this to your answer in b.
d. Estimate from your plot the concentration of N_2O_5 at 1500 seconds; at 2000 seconds.

14.3 a. Show that the data in Problem 14.2 correspond to a first order reaction.
b. What is the first order rate constant for the decomposition of N_2O_5?
c. Using the first order rate law, find the concentration of N_2O_5 at $t = 500$ seconds, $t = 1500$ seconds, and $t = 2000$ seconds.
d. Calculate the rate of reaction at $t = 500$ seconds, using the first order rate law.

14.4 Consider the reaction $A + B \rightarrow$ products. The following table gives the initial rates of reaction at various concentrations of A and B.

conc A	conc B	rate
0.10 mole/lit	0.10 mole/lit	0.0090 mole/lit sec
0.20 mole/lit	0.10 mole/lit	0.036 mole/lit sec
0.10 mole/lit	0.20 mole/lit	0.018 mole/lit sec
0.10 mole/lit	0.30 mole/lit	0.027 mole/lit sec

What is the order of the reaction with respect to A? with respect to B? the overall order?

14.5   For the reaction A + B → products, the rate of reaction is 2.0×10^{-4} mole/lit sec when the concentrations of A and B are each 0.10 mole/lit. Determine the rate constant and the rate to be expected when the concentrations are doubled if the reaction is

a. First order in both A and B
b. First order in A, second order in B

14.6   For the decomposition of NOCl, the following data are obtained at 100°C

t	conc NOCl
0 sec	0.500 mole/lit
10 sec	0.357 mole/lit
20 sec	0.278 mole/lit
30 sec	0.227 mole/lit

From these data, decide whether the reaction is first or second order, and find the rate constant.

14.7   For the first order decomposition of CH_3NO_2 at 200°C, the rate constant is 0.0170 min^{-1}. Starting with a concentration of 0.200 mole/lit, calculate

a. The concentration of CH_3NO_2 after ten minutes; twenty minutes; two days.
b. The time required for the concentration of CH_3NO_2 to drop to 0.150 mole/liter.
c. The half-life.

14.8   The reaction HI(g) + CH_3I(g) → CH_4(g) + I_2(g) is known to be first order in both reactants. The rate constant at 50°C is 2.30×10^{-3} (mole/lit)$^{-1}$ min^{-1}. Starting with equal concentrations of HI and CH_3I, 0.200 mole/lit, calculate

a. The concentration of HI remaining after 100 minutes.
b. The concentration of I_2 after 200 minutes.
c. The half-life.

14.9   The half-life of the reaction A → products is observed to be 160 seconds, when the original concentration of A is 1.0×10^{-3} mole/lit. How long will it take for conc A to drop to 1.0×10^{-4} mole/lit if the reaction is

a. first order   b. second order

14.10   For the thermal decomposition of $C_2H_5NO_2$, the following data are obtained for the first order rate constant k as a function of temperature.

k (sec^{-1})	0.0047	0.029	0.15	0.66	2.6
T (°K)	500	525	550	575	600

What is the activation energy for this reaction?

14.11   The rate constant for a certain first order reaction is 1.0×10^{-3} sec^{-1} at 25°C. Calculate the rate constant at 50°C if the activation energy is

a. 0   b. 10.0 kcal   c. 20.0 kcal

14.12   For a certain reaction, the rate constant at 100°C is 0.025 min^{-1}. The activation energy is 24.0 kcal. When the temperature is raised to 110°C, calculate

a. The per cent increase in the collision number Z. (Z = constant × $T^{1/2}$.)
b. The per cent increase in the fraction, f, of molecules having an energy greater than E_a.
c. The per cent increase in the rate constant k.

14.13   Consider the reaction 2 NO(g) + O_2(g) → 2 NO_2(g). Assume that this reaction occurs via the following mechanism:

$$2 \text{ NO} \rightleftharpoons N_2O_2, \text{ followed by: } N_2O_2 + O_2 \xrightarrow{\text{k}} 2 \text{ NO}_2$$

Show that if N_2O_2 is in equilibrium with NO, the reaction will be second order in NO and first order in O_2 (consider the rate of formation of NO_2).

14.14 Consider the following mechanism for a chain reaction:

$$X_2 \rightleftharpoons 2\,X$$

$$X + H_2 \xrightarrow{k_1} HX + H$$

$$H + X_2 \xrightarrow{k_2} HX + X$$

Obtain the rate expression for the formation of HX in terms of the concentrations of X_2 and H_2. Assume that X is in equilibrium with X_2 and that the concentration of H does not change with time (i.e., Δconc $H/\Delta t = 0$).

14.15 For the reaction $2\,NH_3(g) \rightarrow N_2(g) + 3\,H_2(g)$, it is found that NH_3 is decomposing at the rate of 0.160 moles/lit sec. Determine the "rate of reaction," defined as

a. $-\Delta$conc $NH_3/\Delta t$, with t in seconds.
b. Δconc $H_2/\Delta t$, with t in seconds.
c. Δconc $N_2/\Delta t$, with t in seconds.
d. Δconc $N_2/\Delta t$, with t in minutes.

14.16 For the decomposition of diazomethane CH_2N_2 at 600°C, the following data were obtained.

t	conc CH_2N_2	t	conc CH_2N_2
0 min	0.100 mole/lit	15 min	0.044 mole/lit
5 min	0.076 mole/lit	20 min	0.033 mole/lit
10 min	0.058 mole/lit	25 min	0.025 mole/lit

a. Plot these data (concentration on the vertical axis).
b. From the plot, determine the rate of decomposition at t = 10 minutes.
c. Show graphically that the reaction is first order, and determine the rate constant.

14.17 For the reaction $A + B \rightarrow$ products, the initial rates of reaction at various concentrations of A and B are

conc A	conc B	rate
0.20 mole/lit	0.020 mole/lit	1.6×10^{-3} mole/lit sec
0.30 mole/lit	0.020 mole/lit	2.4×10^{-3} mole/lit sec
0.30 mole/lit	0.030 mole/lit	3.6×10^{-3} mole/lit sec

Calculate the order with respect to A and B, the overall order, and the rate constant.

14.18 For the reaction $A \rightarrow$ products, the rate of reaction is 1.5×10^{-3} mole/lit min when the concentration of A is 0.16 mole/lit. Calculate the rate when the concentration of A is 0.25 mole/lit if the reaction is

a. Zero order.   b. One-half order.   c. First order.   d. Second order.

14.19 For the reaction $C_2H_4O(g) \rightarrow CH_4(g) + CO(g)$, the following data are obtained at 415°C.

Time (min)	0	10	20	30	40
Conc (mole/lit)	0.200	0.177	0.156	0.138	0.123

From these data, decide whether the reaction is first or second order, and determine the rate constant.

14.20 For the first order decomposition of CH_3NO_2 at 300°C, the rate constant is 0.23 sec^{-1}. Starting with a concentration of 0.60 mole/lit, calculate

a. The concentration of CH_3NO_2 after 10 seconds; after one minute.
b. The time required for the concentration of CH_3NO_2 to drop to 0.10 mole/lit.
c. The half-life.

14.21 The dimerization of ethylene, $C_2H_4(g) \rightarrow \frac{1}{2} C_4H_8(g)$, is known to be second order with a rate constant of 1.6 $(mole/lit)^{-1} sec^{-1}$ at 500°C. If one starts with a concentration of 0.020 mole/lit, calculate

a. The concentration of C_2H_4 remaining after 20 seconds.
b. The concentration of C_4H_8 after 20 seconds.
c. The half-life.

14.22 The half-life of the reaction A → products is found to be 12 minutes when the initial concentration of A is 0.100 mole/lit. How long will it take for the concentration of A to drop to 0.020 mole/lit if the reaction is

a. First order?   b. Second order?

14.23 Using appropriate data from Problems 14.7 and 14.20, calculate the activation energy for the decomposition of CH_3NO_2 and the rate constant at 250°C.

14.24 According to kinetic theory, the fraction of molecules having kinetic energy greater than E is given by the expression

$$f = e^{-E_a/RT}$$

Find f at 100°C and 120°C for

a. $E_a = 10$ kcal   b. $E_a = 30$ kcal

14.25 Assuming that the decomposition of ozone occurs by the following mechanism,

$$O_3 \rightleftharpoons O_2 + O$$
$$O + O_3 \xrightarrow{k} 2 O_2$$

and that the first reaction is at equilibrium, derive an expression for the formation of O_2 by the second reaction.

14.26 It is found that, under certain conditions, the rate of formation of CH_4 from CH_3CHO is 3/2 order in CH_3CHO. Show that this is consistent with the mechanism

$$CH_3CHO \xrightarrow{k_1} CH_3 + CHO$$
$$CH_3 + CH_3CHO \xrightarrow{k_2} CH_4 + CH_3 + CO$$
$$2 CH_3 \xrightarrow{k_3} C_2H_6$$

where the concentration of CH_3 radicals does not change with time (i.e., Δconc $CH_3/\Delta t = 0$.

°14.27 The rate of a certain heterogeneous reaction is given by the expression

$$\text{rate} = 2.0 \times 10^{-3}\ \theta$$

where θ is the fraction of surface covered by reactant molecules A. The quantity θ is given by

$$\theta = b\ (\text{conc A})/[1 + b\ (\text{conc A})] \text{ with } b = 0.10\ (mole/lit)^{-1}$$

Calculate the rate of reaction at (conc A) = 0.10, 0.20, 0.50, 1.00, and 2.00 mole/lit respectively. Plot a graph of the data (rate vs. concentration). What is the limiting value of the reaction order as (conc A) → 0?

°14.28 Using the principles of calculus,

a. Derive Equation 14.10.
b. Derive the rate expression for a third order reaction, 3A → products.

°14.29 For a certain reaction, $2A(g) \rightarrow B(g) + 3C(g)$, the total pressure of the gaseous mixture is measured as a function of time at 200°C, starting with pure A.

p_{tot} (mm Hg)	200	241	274	300	320
t(min)	0	10	20	30	40

Show that the reaction is first order and determine the rate constant.

°14.30 A substance A undergoes two different reactions when heated:

$$A \xrightarrow{k_1} B; \quad \text{1st order,} \quad k_1 = 2.0 \times 10^{-2} \text{ sec}^{-1}$$

$$A \xrightarrow{k_2} C; \quad \text{2nd order,} \quad k_2 = 2.0 \times 10^{-2} \text{ (mole/lit)}^{-1} \text{ sec}^{-1}$$

a. Write an expression for the rate of reaction, $-\Delta \text{conc A}/\Delta t$.
b. Calculate the rate of reaction at concentrations of 0.2, 0.4, 0.6, 0.8, and 1.0 mole/lit.
c. Estimate the apparent order of the reaction between 0.2 and 0.4 mole/lit; between 0.8 and 1.0 mole/lit.

°14.31 The decomposition of N_2O_5 is believed to occur by the following mechanism:

$$2 N_2O_5 \rightleftharpoons N_2O_5^* + N_2O_5 \quad \text{equilibrium constant} = K_1$$

$$N_2O_5^* \rightleftharpoons NO_2 + NO_3 \quad \text{equilibrium constant} = K_2$$

$$NO_2 + NO_3 \xrightarrow{k_1} NO_2 + O_2 + NO$$

$$NO + NO_3 \xrightarrow{k_2} 2 NO_2$$

Show that the rate of formation of O_2 is directly proportional to the concentration of N_2O_5.

15

Reactions of Elements
with Each Other

Throughout the remainder of this text, we shall be concerned with various kinds of chemical reactions. In this chapter we shall survey one particular type of reaction, that between pairs of elementary substances. Specifically, we shall discuss the reactions of hydrogen and oxygen with a variety of other elements, both metals and nonmetals.

Our approach to the chemistry of reactions between elements will be primarily descriptive. We shall, however, draw heavily on the principles of chemical kinetics (Chapter 14), equilibrium (Chapter 12), and covalent and ionic bonding (Chapter 8). In addition we shall find it convenient to introduce certain new concepts, among them a generalized concept of oxidation and reduction.

15.1   OXIDATION AND REDUCTION

Loss and Gain of Electrons

The formation of the ionic compound sodium fluoride from sodium and fluorine atoms can be represented in electron dot notation as

$$\text{Na} \cdot + \cdot \ddot{\text{F}} : \longrightarrow \text{Na}^+ + (: \ddot{\text{F}} :)^- \tag{15.1}$$

It is clear that in this process a sodium atom has transferred an electron to a fluorine atom.

$$\text{Na} \cdot \longrightarrow \text{Na}^+ + e^- \tag{15.1a}$$

$$\cdot \ddot{\text{F}} : + e^- \longrightarrow (: \ddot{\text{F}} :)^- \tag{15.1b}$$

Any reaction between atoms which leads to the formation of ions may be analyzed similarly. For example, the reaction between magnesium and oxygen atoms

$$\cdot \text{Mg} \cdot + : \dot{\text{O}} : \longrightarrow \text{Mg}^{2+} + (: \ddot{\text{O}} :)^{2-} \tag{15.2}$$

may be broken down into two half-reactions:

$$\cdot \text{Mg} \cdot \longrightarrow \text{Mg}^{2+} + 2 \ e^- \qquad (15.2a)$$

and
$$: \overset{\cdot}{\underset{\cdot}{O}} : + 2 \ e^- \longrightarrow (: \overset{..}{\underset{..}{O}} :)^{2-} \qquad (15.2b)$$

Processes such as 15.1a and 15.2a, which involve the **loss of electrons**, are referred to as **oxidation** half-reactions; atoms of sodium and magnesium lose electrons to become positively charged ions. The processes represented by 15.1b and 15.2b, which involve the **gain of electrons**, are referred to as **reduction** half-reactions; in gaining electrons atoms of fluorine and oxygen are said to be reduced. The overall Reactions 15.1 and 15.2, in which oxidation and reduction occur simultaneously, are called **oxidation-reduction** reactions or simply "redox" reactions. Since no electrons are created or destroyed in the process, in any oxidation-reduction reaction there can be no net gain or loss of electrons. For example, when lithium reacts with oxygen, two lithium atoms are oxidized to Li^+ ions for every oxygen atom reduced to an O^{2-} ion.

oxidation:
$$2 \ \text{Li} \cdot \longrightarrow 2 \ \text{Li}^+ + 2 \ e^- \qquad (15.3a)$$

reduction:
$$: \overset{\cdot}{\underset{\cdot}{O}} : + 2 \ e^- \longrightarrow (: \overset{..}{\underset{..}{O}} :)^{2-} \qquad (15.3b)$$

overall reaction:
$$2 \ \text{Li} \cdot + : \overset{\cdot}{\underset{\cdot}{O}} : \longrightarrow 2 \ \text{Li}^+ + (: \overset{..}{\underset{..}{O}} :)^{2-} \qquad (15.3)$$

We shall shortly find that it is possible to broaden the meaning of the terms oxidation and reduction so that they can be applied to a wide variety of reactions, including many in which no ionic species are involved. To accomplish this, it is necessary to introduce a new concept — oxidation number.

Oxidation Number

The chemical equation written for the reaction between hydrogen and fluorine

$$\tfrac{1}{2} \ H_2(g) + \tfrac{1}{2} \ F_2(g) \longrightarrow HF(g) \qquad (15.4)$$

bears at least a superficial resemblance to that for the reaction of sodium with fluorine

$$Na(s) + \tfrac{1}{2} \ F_2(g) \longrightarrow NaF(s)$$

Indeed, the two reactions themselves have much in common. In both there is an exchange of electrons between atoms. The major difference between the two reactions lies in the extent to which electron transfer takes place. In the case of sodium and fluorine, an electron is donated by a sodium atom to a fluorine atom, forming the pair of ions Na^+ and F^-. When hydrogen reacts with fluorine, the valence electron of hydrogen is shared with fluorine to form the covalently bonded HF molecule. This distinction is one of degree rather than kind. The electrons in the covalent bond are displaced strongly toward the fluorine atom. The electronic environment of the fluorine atom in the HF molecule is not markedly different from that in the F^- ion:

$$: \overset{..}{\underset{..}{F}} :^- \quad \text{vs.} \quad H \ : \overset{..}{\underset{..}{F}} :$$

So far as "electron bookkeeping" is concerned, it is reasonable to assign both bonding electrons in the HF molecule to the fluorine atom. This is equivalent to assigning a -1 charge to the fluorine, which now has one more valence electron (eight) than the neutral fluorine atom (seven). The hydrogen atom, deprived of its valence electron by this assignment, could then be said to have a charge of $+1$.

Many years ago, chemists adopted the accounting system that we have just outlined to assign electrons in covalently bonded substances. The concept of **oxidation number** was introduced to refer to the charge an atom would have if the bonding electrons were assigned arbitrarily to the more electronegative element. In the HF molecule, hydrogen is said to have an oxidation number of $+1$, fluorine an oxidation number of -1. In water, the bonding electrons are assigned to the more electronegative oxygen atom:

There are no ions in HF; H does not have a $+1$ charge.

$$\text{H} \bigg| \ :\ddot{\underset{..}{\text{O}}}:\ \bigg| \text{H}$$

This gives oxygen an oxidation number of -2 (8 valence e^- vs. 6 e^- in the neutral atom) and hydrogen an oxidation number of $+1$ (0 e^- vs. 1 e^- in the neutral atom). In a nonpolar covalent bond, the bonding electrons are split evenly between the two atoms:

$$:\ddot{\underset{..}{\text{F}}} \cdot \bigg| \cdot \ddot{\underset{..}{\text{F}}}: \qquad \text{oxidation no. F} = 0$$

It should be emphasized that the oxidation number of an atom in a covalently bonded substance is an artificial concept. Unlike the charge of an ion, the oxidation number of an element cannot be determined experimentally. The hydrogen atom in the HF or H_2O molecule does not carry a full positive charge; its oxidation number of $+1$ in these molecules may be regarded as a "pseudocharge."

In principle, one could obtain oxidation numbers for the various atoms in any covalently bonded compound whose electronic structure is known by simply assigning valence electrons to the more electronegative atom. In practice, such a method of assignment is neither convenient nor necessary. Instead, oxidation numbers are assigned according to certain arbitrary rules which, while consistent with the scheme we have described, are much simpler to apply. These rules make it possible for us to assign oxidation numbers to atoms in any substance, regardless of the structure or type of bonding involved.

1. **The oxidation number* of an element in an elementary substance is 0.** For example, the oxidation number of chlorine in Cl_2 or of phosphorus in P_4 is zero.

2. **The oxidation number of an element in a monatomic ion is equal to the charge of that ion.** In the ionic compound NaCl, sodium has an oxidation number of $+1$, chlorine an oxidation number of -1. The oxidation numbers of aluminum and oxygen in Al_2O_3 (Al^{3+}, O^{2-} ions) are $+3$ and -2 respectively.

3. **Certain elements have the same oxidation number in all or almost all their compounds.** The 1A metals always exist as $+1$ ions in their compounds and hence are assigned an oxidation number of $+1$. By the same token, the 2A elements always have oxidation numbers of $+2$ in their compounds. Fluorine,

---

* Many authors prefer to speak of the **oxidation state** of an element in contrast to the **oxidation number** of an atom. We shall follow the common practice of using these terms interchangeably.

the most electronegative of all elements, has an oxidation number of -1 in almost all its compounds. Oxygen, second only to fluorine in electronegativity, is ordinarily assigned an oxidation number of -2 (certain exceptions will be pointed out later).

Hydrogen, in its compounds with metals (as in NaH and CaH$_2$), exists as a -1 ion and therefore has an oxidation number of -1. In its compounds with the nonmetals, hydrogen is assigned an oxidation number of $+1$.

4. **The sum of the oxidation numbers of all the atoms in the formula of a substance is 0.** To illustrate the application of this rule, consider the ionic compound sodium selenide, Na$_2$Se. Knowing the oxidation number of sodium in this compound to be $+1$, we have

$$2(+1) + \text{oxidation no. Se} = 0, \quad \text{oxidation no. Se} = -2$$

Again, for the compound V$_2$O$_5$, in which the oxidation number of oxygen is -2,

$$2(\text{oxidation no. V}) + 5(-2) = 0, \quad \text{oxidation no. V} = +5$$

As an example of a somewhat more complex situation, consider the ternary compound potassium dichromate, K$_2$Cr$_2$O$_7$:

$$2(\text{oxidation no. K}) + 2(\text{oxidation no. Cr}) + 7(\text{oxidation no. O}) = 0$$

$$2(+1) + 2(\text{oxidation no. Cr}) + 7(-2) = 0$$

Solving, we have $\qquad$ oxidation no. Cr $= +6$

A corollary to this rule states that the sum of the oxidation numbers of the atoms in a polyatomic ion is equal to the charge of that ion. Consider, for example, the permanganate ion, MnO$_4^-$:

What is the oxidation number of N in NH$_3$OH$^+$?

$$\text{oxidation no. Mn} + 4(-2) = -1, \quad \text{oxidation no. Mn} = +7$$

Oxidation States of the Elements

The common oxidation states of the elements are tabulated in Figure 15.1. The oxidation states shown by different elements can be correlated with their position in the Periodic Table as follows:

1. The metals in groups 1A and 2A show only one oxidation state (1A $= +1$; 2A $= +2$).

2. The transition metals commonly show two or more oxidation states. For example, iron forms one series of compounds (FeCl$_2$, FeSO$_4$, . . .) in which it has an oxidation state of $+2$, and another series (FeCl$_3$, Fe$_2$(SO$_4$)$_3$, . . .) in which it has an oxidation state of $+3$.

The lowest common oxidation state of the transition metals is $+2$, corresponding to the loss of two outer "s" electrons. The major exception to this rule occurs with the 1B elements, where the $+1$ state is common (CuCl, AgNO$_3$, and so forth). A possible explanation for this anomaly lies in the electronic structures of the 1B atoms, which are believed to have only one outer "s" electron. Another apparent exception is the $+1$ state of mercury, which corresponds to the unique Hg$_2^{2+}$ ion in which there is a metal-metal bond

$$(\text{Hg} : \text{Hg})^{2+}$$

Among the transition metals, the highest stable oxidation state is ordinarily given by the group number. Thus, titanium shows a maximum oxidation state of +4 (TiO_2, $TiCl_4$), and chromium a maximum oxidation state of +6 (CrO_3, $K_2Cr_2O_7$). Again the 1B elements fail to follow this rule; they show oxidation numbers of +2 ($CuCl_2$, $CuSO_4$) and +3 ($AuCl_3$, $HAuCl_4$) as well as +1.

As we move down a given group of transition metals, the higher oxidation states become more stable. For example, among the 6B elements, the +6 state becomes more stable as we move from chromium to molybdenum to tungsten, while the +3 state becomes less stable in the same order. Among the elements in the iron subgroup, only the two higher members, ruthenium and osmium, show an oxidation state of +8; iron forms few compounds in which its oxidation state is greater than +3.

Can you find other examples of this trend in Table 15.1?

3. The nonmetals can exist in a variety of oxidation states. Here again the maximum oxidation state is given by the group number (+4 for 4A, +5 for 5A, +6 for 6A, +7 for 7A). The minimum oxidation number is determined by the number of electrons that must be acquired to form an octet. The 7A elements, which need to gain one electron to attain a noble-gas structure, show a minimum oxidation number of −1; the 6A elements, −2.

The positive oxidation states of nonmetals do not, of course, correspond to positively charged ions. They occur when the nonmetal atom is covalently bonded to a different, more electronegative atom, frequently oxygen. For example, chlorine has a +1 oxidation state in the ClO^- ion,

$$(: \ddot{\underset{..}{C}}l \,\Big|\, : \ddot{\underset{..}{O}} :)^-$$

where the bonding electrons are arbitrarily assigned to oxygen, leaving chlorine with only six valence electrons, one less than the seven associated with a neutral chlorine atom. The addition of another oxygen to give the ClO_2^- ion

$$(: \ddot{\underset{..}{O}} : \,\Big|\, \ddot{\underset{..}{C}}l \,\Big|\, : \ddot{\underset{..}{O}} :)^-$$

takes two more electrons "away" from chlorine, increasing its oxidation number to +3. This process can be continued to form the ClO_3^- ion

$$(: \ddot{\underset{..}{O}} : |\ddot{\underset{..}{C}}l|: \ddot{\underset{..}{O}} :)^-$$
$$: \ddot{\underset{..}{O}} :$$

in which the oxidation number of chlorine is +5, and finally the ClO_4^- ion

$$\left[\begin{array}{c} : \ddot{\underset{..}{O}} : \\ : \ddot{\underset{..}{O}} : |\ddot{\underset{..}{C}}l| : \ddot{\underset{..}{O}} : \\ : \ddot{\underset{..}{O}} : \end{array} \right]^-$$

in which the chlorine atom has "lost" all its valence electrons, giving it an oxidation number of +7. In terms of this sequence, it is clear why the maximum oxidation number of chlorine must be +7; it has no more electrons to "lose" by sharing them with oxygen. Moreover, one can readily see why the successive positive oxidation states of chlorine (and those of many other nonmetals) differ from each other by two units; every time the chlorine forms another electron-pair bond, it "loses" two electrons according to the arbitrary conventions that are used to assign oxidation numbers.

1A	2A	3B	4B	5B	6B	7B	8B	8B	8B	1B	2B	3A	4A	5A	6A	7A	8A
H −1 +1																H +1 −1	He
Li +1	Be +2											B +3	C +4 +2 −4	N +5 +4 +3 +2 +1 −3	O −1 −2	F −1	Ne
Na +1	Mg +2											Al +3	Si +4 +2 −4	P +5 +3 −3	S +6 +4 +2 −2	Cl +7 +5 +3 +1 −1	Ar
K +1	Ca +2	Sc +3	Ti +4 +3 +2	V +5 +4 +3 +2	Cr +6 +3 +2	Mn +7 +4 +3 +2	Fe +3 +2	Co +3 +2	Ni +3 +2	Cu +2 +1	Zn +2	Ga +3	Ge +4 −4	As +5 +3 −3	Se +6 +4 −2	Br +5 +1 −1	Kr +4 +2
Rb +1	Sr +2	Y +3	Zr +4 +3	Cb +5 +3	Mo +6 +4 +3 +2	Tc +7	Ru +8 +6 +4 +3 +2	Rh +4 +3 +2	Pd +4 +2	Ag +1	Cd +2	In +3	Sn +4 +2	Sb +5 +3 −3	Te +6 +4 −2	I +7 +5 +1 −1	Xe +6 +4 +2
Cs +1	Ba +2		Hf +4	Ta +5	W +6 +4 +3 +2	Re +7 +6 +4	Os +8 +4	Ir +4 +3	Pt +4 +2	Au +3 +1	Hg +2 +1	Tl +3 +1	Pb +4 +2	Bi +3	Po +2	At −1	Rn

Figure 15.1 Oxidation states of the elements.

Oxidation and Reduction: General Definition

The concept of oxidation number leads directly to a working definition of the terms oxidation and reduction. **Oxidation** is defined as an **increase in oxidation number, reduction** as a **decrease in oxidation number**. Reactions in which one element increases in oxidation number at the expense of another are referred to as oxidation-reduction reactions. Two simple examples follow.

$$2 \ Al(s) + 3 \ Cl_2(g) \longrightarrow 2 \ AlCl_3(s) \quad \text{Al oxidized (oxidation no. } 0 \longrightarrow +3)$$
$$\text{Cl reduced (oxidation no. } 0 \longrightarrow -1) \tag{15.5}$$

$$4 \ As(s) + 5 \ O_2(g) \longrightarrow 2 \ As_2O_5(s) \quad \text{As oxidized (oxidation no. } 0 \longrightarrow +5)$$
$$\text{O reduced (oxidation no. } 0 \longrightarrow -2) \tag{15.6}$$

These definitions are compatible, of course, with the earlier interpretation of oxidation and reduction in terms of the loss and gain of electrons. An element which loses electrons inevitably increases in oxidation number; the gain of electrons always results in a decrease in oxidation number. By defining oxidation and reduction in terms of changes in oxidation number, the scope of oxidation-reduction reactions is greatly increased. In particular, any

381

reaction in which an elementary substance participates, either as a reactant or product, falls into this category. Furthermore, the use of oxidation number simplifies the electron bookkeeping in oxidation-reduction reactions. For example, analysis of the reaction

The balancing of
redox equations is
discussed in Chap-
ter 20.

$$HCl(g) + HNO_3(l) \longrightarrow NO_2(g) + \tfrac{1}{2} Cl_2(g) + H_2O(l) \qquad (15.7)$$

in terms of oxidation numbers reveals immediately that chlorine is oxidized (oxidation no. $= -1$ in HCl, 0 in Cl_2) while nitrogen is reduced (oxidation no. $= +5$ in HNO_3, $+4$ in NO_2). It is much more difficult to decide precisely which atoms are "losing" or "gaining" electrons.

In discussing oxidation-reduction reactions, the phrases **oxidizing agent** and **reducing agent** are frequently used to designate the species responsible for oxidation and reduction. We speak of chlorine in Reaction 15.5 and oxygen in 15.6 as being oxidizing agents, since they bring about the oxidation of aluminum and arsenic respectively. In these reactions, aluminum and arsenic act as reducing agents, being responsible for the reduction of chlorine and oxygen. In the more complex reaction represented by 15.7, the species which is oxidized, HCl, acts as a reducing agent; HNO_3, which undergoes reduction, is the oxidizing agent.

15.2 REACTIONS OF HYDROGEN

Hydrogen reacts with a wide variety of elements to form three different types of compounds. With the highly electropositive elements (those in Group 1A and 2A of the Periodic Table), hydrogen forms ionic compounds, called **saline** or **saltlike hydrides**, containing H^- ions. With the nonmetals, including oxygen, nitrogen, and the halogens, hydrogen forms volatile **molecular hydrides**. The reaction of hydrogen with the transition metals leads to the formation of so-called **metallic hydrides**, whose electronic structures are not well established. We shall now consider the preparation and properties of these three types of hydrogen compounds.

Saline Hydrides

Hydrogen reacts with all of the 1A metals and with the heavier members of the 2A group (Ca, Sr, Ba) to form compounds referred to as "saline" hydrides because of their resemblance to the halides of these metals. Lithium hydride can be formed by passing hydrogen over molten lithium at 600°C:

$$2 Li(l) + H_2(g) \longrightarrow 2 LiH(s) \qquad (15.8)$$

Calcium reacts readily with hydrogen at considerably lower temperatures (150 to 300°C):

$$Ca(s) + H_2(g) \longrightarrow CaH_2(s) \qquad (15.9)$$

The crystal structures of these compounds resemble those of the corresponding halides. Their ionic character is demonstrated by their low volatility, their insolubility in common organic solvents, and their ability to conduct an electric current when melted. Upon electrolysis, hydrogen gas is formed at the positive electrode, indicating the presence of H^- ions.

TABLE 15.1   PROPERTIES OF THE SALINE HYDRIDES

PROPERTY	LiH	NaH	KH	RbH	CsH	CaH$_2$	SrH$_2$	BaH$_2$
ΔH_f (kcal/mole)	−21.6	−13.8	−14.5	−12	−19.9	−46.6	−42.4	−41.0
Melting Point (°C)	680	d.500	d.400	d-	d-	d.1000	–	1200
Density (g/ml)	0.8	1.40	1.42	2.59	3.42	1.90	3.27	4.15

Of the saline hydrides, those of lithium and the 2A metals are most stable with respect to decomposition into the elements, reflecting their higher heats of formation (Table 15.1). All these compounds evolve hydrogen if exposed to oxygen or water; the reactions with lithium hydride are typical.

$$2 \text{ LiH(s)} + \tfrac{1}{2} \text{ O}_2(g) \longrightarrow \text{Li}_2\text{O(s)} + \text{H}_2(g) \tag{15.10}$$

$$\text{LiH(s)} + \text{H}_2\text{O(l)} \longrightarrow \text{LiOH(s)} + \text{H}_2(g) \tag{15.11}$$

It will be noted that in both of these reactions lithium hydride acts as a reducing agent; the hydride ion (oxidation no. $= -1$), in being oxidized to molecular hydrogen, brings about the reduction of the other reactant ($O_2 \rightarrow O^{2-}$; $H_2O \rightarrow H_2$). The saline hydrides in general and lithium hydride in particular are sometimes used as reducing agents in organic chemistry. A related compound, lithium aluminum hydride, $LiAlH_4$, is widely used for this purpose.

If CaH$_2$ is exposed to moist air, what is the final product?

Metallic (Interstitial) Hydrides

The behavior of hydrogen toward certain of the transition metals is quite different from that described with the 1A and 2A metals. Many of the heavy metals, notably platinum and palladium, take up large quantities of hydrogen when exposed to the gas at room temperature and atmospheric pressure. For many years, it was supposed that only a physical interaction was involved; hydrogen was believed to form a solid solution in the metal. It is indeed difficult to explain the wide deviations from stoichiometry observed in these hydrides in terms of a chemical reaction between metal and hydrogen. Formulas such as $PdH_{0.7}$ and $ZrH_{1.9}$ are difficult to rationalize in terms of definite compounds.

Nevertheless, a great deal of evidence accumulated over the past two decades suggests that hydrogen does interact chemically with the transition metals. For one thing, the heat effects observed are much greater than one would expect if only a solid solution were involved. When 1 mole of hydrogen is taken up by palladium, 17.1 kcal of heat are evolved; the heat of solution of hydrogen in various liquid solvents is of the order of 0.2 kcal. Moreover, x-ray studies reveal that the interaction of hydrogen with a transition metal almost always leads to a distortion or rearrangement of the crystal structure of the metal. This is hard to reconcile with the simple picture of hydrogen molecules entering randomly into holes in the metal lattice to form a solid solution. Even more significant is the fact that the introduction of hydrogen frequently alters the magnetic properties of the transition metal. Certain metals with unpaired electrons lose their paramagnetism when they take up hydrogen. This requires an electronic interaction and hence, by definition, a reaction between hydrogen and the transition metal.

383

The particle structure of the transition metal hydrides is not definitely established. It has been suggested that hydrogen enters the metal lattice as individual atoms, each with an unpaired electron. If this is the case, one might visualize a transition metal hydride as a network of close-packed metal atoms with a variable number of hydrogen atoms located at interstitial positions within the metal lattice.

Molecular Hydrides

Very few of the hydrogen compounds of the nonmetals listed in Table 15.2 can be prepared by the direct reaction of the elements with each other. Indeed, only nitrogen, oxygen, and the halogens react with hydrogen at a reasonable rate to give products in good yield.

$$N_2(g) + 3\ H_2(g) \longrightarrow 2\ NH_3(g) \tag{15.12}$$

$$O_2(g) + 2\ H_2(g) \longrightarrow 2\ H_2O(g) \tag{15.13}$$

$$X_2(g) + H_2(g) \longrightarrow 2\ HX(g) \tag{15.14}$$

$$(X = F,\ Cl,\ Br,\ I)$$

The failure of many nonmetals to react with hydrogen can be attributed to unfavorable thermodynamic factors. This is particularly true of those elements which are close to hydrogen in electronegativity; in these cases the bonds within the molecular hydride are relatively weak. An example is arsenic (electronegativity of H = 2.1, As = 2.0); the compound arsine, AsH_3, has a positive heat of formation ($\Delta H_f = +41.0$ kcal) and free energy of formation ($\Delta G_f^\circ = +38$ kcal). It cannot be formed directly from the elements; if prepared by the reduction of a compound such as As_2O_3, it readily decomposes to the elements (Figure 15.2).

TABLE 15.2 HYDROGEN COMPOUNDS OF THE NONMETALS

3B	4B	5B	6B	7B
B_2H_6	CH_4	NH_3	H_2O	HF
		N_2H_4	H_2O_2	
	SiH_4	PH_3	H_2S	HCl
	GeH_4	AsH_3	H_2Se	HBr
	SnH_4	SbH_3	H_2Te	HI
	PbH_4	BiH_3		

Even if the compound formed by the reaction of hydrogen with a nonmetal is thermodynamically stable, its rate of formation may be so slow as to make the reaction useless from a practical standpoint. The activation energies for reactions involving hydrogen are frequently very large, primarily because of the great strength of the H—H bond (bond energy = 104 kcal). If the nonmetal with which hydrogen is reacting is also held together by strong bonds, as is the case with carbon (C—C bond energy = 83 kcal), one can expect to find very high activation energies and correspondingly small rate constants.

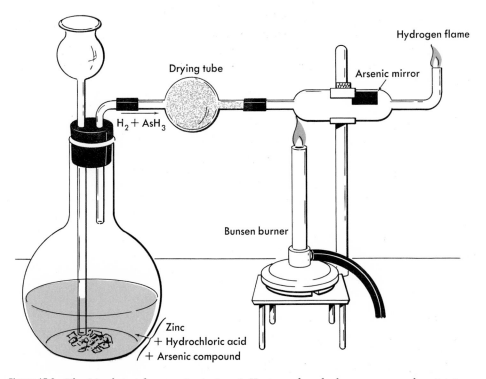

Figure 15.2 The Marsh test for arsenic. Arsine, AsH_3, is produced when a compound containing arsenic is added to hydrochloric acid in the presence of zinc. The arsine decomposes to the metal when heated, forming a mirror which serves as a test for the presence of arsenic.

Nitrogen $\qquad$ $N_2(g) + 3\ H_2(g) \rightleftharpoons 2\ NH_3(g)$ $\qquad\qquad$ (15.15)

Over six million tons of ammonia are produced annually in this country by the reaction represented by Equation 15.15. The industrial process used today to convert atmospheric nitrogen to ammonia is essentially similar to that worked out by the German chemist Fritz Haber in 1913. The success of Haber and his co-workers in making this reaction feasible on a large scale made Germany independent of outside sources of nitrogen compounds, particularly nitrates, which are used to make explosives, and undoubtedly had a great deal to do with Germany's ability to wage World War I.

The problem faced by Haber was that of finding experimental conditions under which nitrogen would react rapidly with hydrogen to give a reasonably high yield of ammonia. Calculations show that at room temperature and atmospheric pressure, the position of the equilibrium strongly favors the formation of ammonia ($K_c = 4 \times 10^9$). Unfortunately, under these conditions the rate of reaction is virtually zero. The rate at which equilibrium is reached can, of course, be increased by raising the temperature. However, since Reaction 15.15 is exothermic ($\Delta H = -22.0$ kcal), high temperatures reduce the yield of ammonia (Table 15.3). High pressures, on the other hand, have a favorable influence on both the rate of the reaction and the position of the equilibrium.

385

An increase in pressure makes it possible for the system to attain equilibrium more rapidly because it increases the concentrations of the gases involved; it also increases the relative amount of ammonia present at equilibrium, since the forward reaction represented by 15.15 results in a decrease in the number of moles of gas.

TABLE 15.3 EFFECT OF TEMPERATURE AND PRESSURE ON THE YIELD OF AMMONIA IN THE HABER PROCESS ($[H_2] = 3\ [N_2]$)

		MOLE % NH_3 IN EQUILIBRIUM MIXTURE				
°C	K_c	10 atm	50 atm	100 atm	300 atm	1000 atm
200	650	51	74	82	90	98
300	9.5	15	39	52	71	93
400	0.5	4	15	25	47	80
500	0.08	1	6	11	26	57
600	0.014	0.5	2	5	14	13

On the basis of data such as that in Table 15.3, Haber concluded that direct synthesis of ammonia from the elements was industrially feasible. The reaction was to be carried out at a moderately high temperature with a catalyst, to hasten the attainment of equilibrium, and at high pressures to increase the yield of ammonia. Much of Haber's research was devoted to finding a suitable catalyst; he also had to deal with the metallurgical problem of developing alloy steels capable of withstanding pressures up to 300 atm at temperatures in the vicinity of 400°C. His success in overcoming these difficulties won him the Nobel Prize in 1918.

In a modern ammonia synthesis plant, nitrogen and hydrogen are reacted at a temperature of 400 to 450°C and pressures of 200 to 600 atm. Various catalysts have been used; perhaps the most common is a specially prepared mixture of iron, potassium oxide, and aluminum oxide. Ammonia is separated from the gaseous reaction mixture by being condensed out as a liquid (bp = −33°C); the unreacted hydrogen and nitrogen are recycled to raise the yield of ammonia.

Ammonia is a major component of commercial fertilizers.

A portion of the ammonia produced in this country is used to prepare nitric acid, a fundamental starting material for the manufacture of almost all high explosives. The so-called Ostwald process for the preparation of nitric acid makes use of the following series of reactions:

$$4\ NH_3(g) + 5\ O_2(g) \longrightarrow 4\ NO(g) + 6\ H_2O(g) \quad \text{(1000°C, Pt catalyst)}$$
$$2\ NO(g) + O_2(g) \longrightarrow 2\ NO_2(g) \quad \text{(NO allowed to cool in air)}$$
$$3\ NO_2(g) + H_2O(l) \longrightarrow 2\ HNO_3(l) + NO(g) \quad \text{(NO}_2 \text{ passed into water)}$$

Halogens $\qquad\qquad H_2(g) + X_2(g) \longrightarrow 2\ HX(g) \qquad\qquad$ (15.16)

The reactions of hydrogen with the halogens lend themselves particularly well to an analysis of the factors that govern the ease with which a reaction occurs — the rate of reaction and the stability of the products. Experimentally, it is observed that both these factors become less favorable as one moves from fluorine to iodine.

The reaction of hydrogen with chlorine was discussed in Chapter 14 where it was pointed out that it occurs via a chain mechanism, the first step of which is the dissociation of a chlorine molecule into atoms

1. $Cl_2(g) \longrightarrow 2\ Cl(g)$, $\qquad\qquad$ $\Delta H = +58$ kcal

followed by the two-step chain

2. $Cl(g) + H_2(g) \longrightarrow HCl(g) + H(g)$, $\qquad$ $\Delta H = +1$ kcal
3. $H(g) + Cl_2(g) \longrightarrow HCl(g) + Cl(g)$, $\qquad$ $\Delta H = -45$ kcal

Steps 2 and 3, occurring over and over again in the gaseous mixture, account for the formation of hydrogen chloride and the net evolution of energy:

$$H_2(g) + Cl_2(g) \longrightarrow 2\ HCl(g), \qquad \Delta H = -44\ \text{kcal}$$

The slow step of the chain mechanism is the dissociation of chlorine molecules into atoms; once a few chlorine atoms (about one in 10^{12}) are formed, Steps 2 and 3 occur rapidly. The energy required to break the Cl—Cl bond (58 kcal/mole = 4.0×10^{-12} ergs/molecule) can be supplied **photochemically** by the absorption of a photon of light. Using Planck's equation:

$$E = h\nu = \frac{hc}{\lambda}; \quad \begin{array}{l} h = 6.6 \times 10^{-27}\ \text{erg sec} \\ c = 3.0 \times 10^{10}\ \text{cm/sec} \end{array} \qquad (15.17)$$

one can calculate that light of wavelength 4900 Å or less should be able to initiate this reaction:

$$\lambda = \frac{hc}{E} = \frac{(6.6 \times 10^{-27}\ \text{erg sec})(3.0 \times 10^{10}\ \text{cm/sec})}{4.0 \times 10^{-12}\ \text{erg}} = 4.9 \times 10^{-5}\ \text{cm} = 4900\ \text{Å}$$

This is the maximum wavelength capable of breaking the Cl—Cl bond.

This wavelength falls in the blue region of the spectrum. Sunlight, which is rich in high energy radiation in the blue, violet, and ultraviolet ($\lambda < 4000$ Å) range, is capable of supplying this energy.

Fluorine and bromine react with hydrogen in much the same way as does chlorine. The chain reaction is more readily initiated with fluorine, since the F—F bond energy is only 37 kcal/mole as compared to 58 kcal/mole for Cl—Cl. Consequently, fluorine and hydrogen react explosively at temperatures as low as 20°K. Bromine reacts more slowly with hydrogen than does chlorine, despite the lower dissociation energy (46 kcal/mole) of the Br_2 molecule. The inhibiting factor here is the increased amount of energy required in the second step of the reaction:

$$Br(g) + H_2(g) \longrightarrow HBr(g) + H(g), \quad \Delta H = +16\ \text{kcal}$$

The amount of energy absorbed in this step is 15 kcal greater than for the corresponding step in the reaction of chlorine with hydrogen, reflecting the weakness of the H—Br bond (bond energy = 88 kcal/mole) as compared to the H—Cl bond (103 kcal/mole).

In the case of iodine, the step

$$I(g) + H_2(g) \longrightarrow HI(g) + H(g), \quad \Delta H = +33\ \text{kcal}$$

is so difficult to accomplish that a chain reaction takes place only at extremely

387

high temperatures. Under ordinary conditions, reaction results from collisions between H_2 and I_2 molecules and the rate of reaction is quite slow.

Turning from rate of reaction to stability of products, we find that the hydrogen halides become less stable as we pass from hydrogen fluoride to hydrogen iodide. As Table 15.4 indicates, hydrogen iodide is the only member of the series which is appreciably dissociated below 1000°C. This means that the reactions of hydrogen with fluorine, chlorine, and bromine are essentially irreversible at ordinary temperatures. Of the four hydrogen halides, HF, HCl, and HBr can be prepared in good yield by the direct reaction of the elements with each other; HI prepared in this way contains appreciable quantities of unreacted hydrogen and iodine.

HI is prepared by a quite different reaction (see Chapter 22).

TABLE 15.4 PERCENTAGE OF HX DISSOCIATED INTO H_2 AND X_2

TEMPERATURE (°C)	HF	HCl	HBr	HI
200	—	—	—	13
600	—	—	0.04	22
1000	—	0.01	0.4	25
2000	0.00006	0.4	4	30

The thermal stabilities of the hydrogen halides can be correlated with the free energy and enthalpy changes for their formation from the gaseous elements.

	$\Delta G°_{25°C}$	ΔH
$\frac{1}{2} H_2(g) + \frac{1}{2} F_2(g) \longrightarrow HF(g)$	−64,700 cal	−64,200 cal
$\frac{1}{2} H_2(g) + \frac{1}{2} Cl_2(g) \longrightarrow HCl(g)$	−22,800 cal	−22,100 cal
$\frac{1}{2} H_2(g) + \frac{1}{2} Br_2(g) \longrightarrow HBr(g)$	−13,100 cal	−12,300 cal
$\frac{1}{2} H_2(g) + \frac{1}{2} I_2(g) \longrightarrow HI(g)$	− 2,000 cal	− 1,200 cal

For all these reactions, except that for the formation of HI, the equilibrium constant at 25°C is a very large number. In the case of HCl, for example, we have

Why is $K_c = K_p$? (Recall Chap. 13.)

$$\log_{10} K_c = \log_{10} K_p = -\frac{\Delta G°}{2.30\ RT} = \frac{22,800}{(2.30)(1.99)(298)} = 16.7$$

$$K_c \text{ at } 25°C = 5 \times 10^{16}$$

Since this reaction is exothermic ($\Delta H = -22,100$ cal), K_c decreases with temperature. However, the calculated value of K_c for the formation of HCl from the elements at 1000°C is still quite a large number, about 2×10^4, indicating that even at this elevated temperature, the per cent dissociation of HCl will be quite small.

The situation is quite different with HI, where $\Delta G°$ at 25°C is a much smaller negative number. At 25°C we calculate a value of about 30 for the equilibrium constant,

$$\log K_c = \frac{2000}{(2.30)(1.99)(298)} = 1.48; \quad K_c = 30$$

which indicates significant dissociation at this temperature. Since ΔH for the formation of HI from gaseous hydrogen and iodine is very nearly zero, the equilibrium constant and therefore the per cent dissociation will be relatively independent of temperature (cf. Table 15.4).

15.3 REACTIONS OF OXYGEN

Metals

1A and 2A Metals. Peroxides and Superoxides. The alkali metals react readily with oxygen, even at room temperature and atmospheric pressure. In the case of lithium, the product is, as one might expect, Li_2O. However, on exposure to dry oxygen, sodium gives a product which can be shown by analysis to have the empirical formula NaO. The oxygen in this compound is present as a diatomic ion with a -2 charge, i.e., $O_2{}^{2-}$. This anion, called the **peroxide ion**, has the electronic structure

$$(: \ddot{O}\!-\!\ddot{O} :)^{2-} \qquad \text{oxidation no. of oxygen} = -1$$

The formula of sodium peroxide is written Na_2O_2 to indicate that there are two Na^+ ions present for every $O_2{}^{2-}$ ion. The reaction between sodium and oxygen can be represented by the equation

$$2\ Na\cdot + : \ddot{O}\!-\!\ddot{O} : \longrightarrow 2\ Na^+ + (: \ddot{O}\!-\!\ddot{O} :)^{2-}$$

or $\qquad\qquad 2\ Na(s) + O_2(g) \longrightarrow Na_2O_2(s)$ (15.18)

Peroxides are known for several of the 1A and 2A metals (Table 15.5). Of these, only the peroxides of sodium and barium are readily prepared by direct reaction of the metal with oxygen.

When potassium burns in oxygen, the principal product formed is a compound of empirical formula KO_2. The potassium in this compound is present as K^+ ions; the oxygen is in the form of a diatomic ion with a unit negative charge. This ion, called the **superoxide ion**, has the electronic structure

$$(: \ddot{O}\!-\!\ddot{O} :)^- \qquad \text{oxidation no. of oxygen} = -\tfrac{1}{2}$$

It is formed by the transfer of an electron from a potassium atom to an oxygen molecule:

$$K\cdot + : \ddot{O}\!-\!\ddot{O} : \longrightarrow K^+ + (: \ddot{O}\!-\!\ddot{O} :)^-$$

or $\qquad\qquad K(s) + O_2(g) \longrightarrow KO_2(s)$ (15.19)

From a kinetic standpoint, it should be easier to form $O_2{}^{2-}$ than O^{2-}, since no bond need be broken.

TABLE 15.5 OXYGEN COMPOUNDS OF THE 1A AND 2A METALS*

	Li	Na	K	Rb	Cs	Be	Mg	Ca	Sr	Ba
Oxide	Li₂O	Na₂O	K₂O	Rb₂O	Cs₂O	BeO	MgO	CaO	SrO	BaO
Peroxide	Li₂O₂	Na₂O₂	K₂O₂	Rb₂O₂	Cs₂O₂			CaO₂	SrO₂	BaO₂
Superoxide		NaO₂	KO₂	RbO₂	CsO₂					

* The compounds underlined represent the principal product of the direct reaction of the metal with oxygen.

The higher members of the 1A group, rubidium and cesium, also form super-oxides as principal products when they react with oxygen.

Metal peroxides decompose on heating to give off oxygen and form the normal oxide. Barium peroxide, for example, begins to give off oxygen at 700°C.

$$BaO_2(s) \longrightarrow BaO(s) + \tfrac{1}{2} O_2(g) \tag{15.20}$$

Superoxides decompose to give first the peroxide and finally, at higher temperatures the normal oxide.

$$2 \ KO_2(s) \longrightarrow K_2O_2(s) + O_2(g) \tag{15.21}$$

$$K_2O_2(s) \longrightarrow K_2O(s) + \tfrac{1}{2} O_2(g) \tag{15.22}$$

All the binary oxygen compounds of the 1A and 2A metals except BeO react exothermically with water to give hydroxides. With the peroxides, a second product, hydrogen peroxide, H_2O_2, is formed. The superoxides yield elementary oxygen as well.

oxides:
$$Li_2O(s) + H_2O(l) \longrightarrow 2 \ LiOH(s) \tag{15.23}$$

$$CaO(s) + H_2O(l) \longrightarrow Ca(OH)_2(s) \tag{15.24}$$

peroxides:
$$Na_2O_2(s) + 2 \ H_2O(l) \longrightarrow 2 \ NaOH(s) + H_2O_2(l) \tag{15.25}$$

$$BaO_2(s) + 2 \ H_2O(l) \longrightarrow Ba(OH)_2(s) + H_2O_2(l) \tag{15.26}$$

superoxide:
$$2 \ KO_2(s) + 2 \ H_2O(l) \longrightarrow 2 \ KOH(s) + H_2O_2(l) + O_2(g) \tag{15.27}$$

Can you show, in electron dot notation, what happens to the ions and molecules in 15.27?

Transition Metals. Most of the transition and post-transition metals form more than one stable oxide (Table 15.6). Among the oxides of a given metal, the metal-to-oxygen bond becomes more covalent as the oxidation state of the metal increases. Consider, for example, the oxides of manganese. The +2 oxide, MnO, appears to be a typical, high-melting (mp = 1650°C), ionic solid. Manganese(IV) oxide, MnO_2, is a high-melting, macromolecular solid similar in structure to silicon dioxide. The highest oxide of manganese, Mn_2O_7, is a reddish liquid that freezes below −20°C, indicating a molecular structure.

TABLE 15.6 OXIDES OF THE METALS OF THE FIRST TRANSITION SERIES

OXIDATION NO. METAL	3B Sc	4B Ti	5B V	6B Cr	7B Mn	8B Fe	8B Co	8B Ni	1B Cu	2B Zn
7					Mn_2O_7					
6				CrO_3						
5			V_2O_5							
4		TiO_2	VO_2		MnO_2					
3	Sc_2O_3	Ti_2O_3	V_2O_3	Cr_2O_3	Mn_2O_3	Fe_2O_3	Co_2O_3			
$\!*\tfrac{8}{3}$					Mn_3O_4	Fe_3O_4	Co_3O_4			
2		TiO	VO	CrO	MnO	FeO	CoO	NiO	CuO	ZnO
1									Cu_2O	

* Two +3 ions, one +2.

When a metal forms more than one oxide, it is found that low temperatures and excess oxygen favor the formation of the higher oxide. Powdered copper, exposed to air at room temperature, is slowly converted to black copper(II) oxide:

$$Cu(s) + \tfrac{1}{2} O_2(g) \longrightarrow CuO(s) \quad \text{at} \quad 25°C \qquad (15.28)$$

Lower oxides such as red copper(I) oxide can be formed by heating the higher oxide to a high temperature in air,

$$2\, CuO(s) \longrightarrow Cu_2O(s) + \tfrac{1}{2} O_2(g) \quad \text{at} \quad 900°C \qquad (15.29)$$

or, more readily, by heating the higher oxide with an equivalent amount of metal in the absence of oxygen.

$$CuO(s) + Cu(s) \longrightarrow Cu_2O(s) \qquad \ll 900°C \qquad (15.30)$$

Stability of Metal Oxides. The reactions of oxygen with metals are invariably exothermic ($\Delta H < 0$) and thermodynamically spontaneous at room temperature and atmospheric pressure ($\Delta G°$ at $25°C < 0$). The amount of heat evolved per gram atomic weight of oxygen is a reliable measure of the thermal stability of the metal oxide formed. Referring to Table 15.7, we find that the oxides that decompose most readily to the elements are those whose formation evolves the least amount of heat. Silver oxide, Ag_2O, breaks down to the elements at 190°C; mercury(II) oxide, HgO, gives off oxygen readily at 450°C; indeed, this is the way Joseph Priestley discovered oxygen in 1774. Oxides of the elements above mercury in Table 15.7 are progressively more resistant to thermal decomposition. For example, copper(I) oxide, Cu_2O, decomposes to the elements at 2000°C; Fe_2O_3 decomposes at 2800°C. The oxides located near the top of the table (CaO, MgO, and so forth) cannot be broken down by heat alone.

TABLE 15.7 THERMODYNAMIC QUANTITIES FOR THE FORMATION OF METAL OXIDES FROM ONE GAW OF OXYGEN AT 25°C

	ΔH (KCAL)	$\Delta G°$ (KCAL)	$\Delta S°$ (CAL/°K)			
CaO	−152	−144	−25			
MgO	−144	−136	−26			
Al₂O₃	−133	−126	−25			
TiO₂	−109	−102	−23			
Cr₂O₃	−90	−83	−23			
ZnO	−83	−76	−25			
Fe₂O₃	−66	−59	−22			
CdO	−61	−54	−24			
NiO	−58	−52	−22			
CoO	−57	−51	−21	Reduced	Reduced	
PbO	−52	−45	−23	by	by	
Cu₂O	−40	−35	−16	Al	H₂	Reduced
HgO	−22	−14	−26			by
Ag₂O	−7	−3	−16			heat

The correlation between ΔH and the decomposition temperature has a simple thermodynamic explanation. The decomposition temperature, T_d, can be taken to be that temperature at which $\Delta G°$ for the reaction becomes zero (i.e., the reaction is at equilibrium at an oxygen pressure of one atmosphere).

Applying the Gibbs-Helmholtz equation (Chapter 13), we have

$$\Delta G^\circ = 0 = \Delta H - T_d \Delta S^\circ; \quad T_d = \frac{\Delta H}{\Delta S^\circ}$$

Can you suggest why ΔS° should be nearly constant?

But from Table 15.7 we note that ΔS° is very nearly the same for all of the oxides. This means that the value of the decomposition temperature is governed almost entirely by the magnitude of ΔH.

Many metal oxides that are not readily decomposed by heat can be reduced to the corresponding metals by reaction with an element that has a strong affinity for oxygen, such as hydrogen, carbon, or aluminum. As one might expect, the extent to which such reactions occur is related to the position of the metal oxide in Table 15.7. The oxides at the bottom of the table (Ag_2O, HgO) react so readily that heating these compounds with hydrogen frequently results in explosions. Oxides such as CoO and NiO react with hydrogen at relatively low temperatures, and so are easily reduced in the laboratory. Iron(III) oxide must be heated to quite a high temperature, approximately 500°C, before its reaction with hydrogen becomes spontaneous.

$$Fe_2O_3(s) + 3\ H_2(g) \xrightarrow{500°C} 2\ Fe(s) + 3\ H_2O(g) \qquad (15.31)$$

Even at temperatures as high as 1600°C, the reduction of zinc oxide, ZnO, by hydrogen is incomplete.

Commercially, the reduction of metal oxides is most often accomplished by heating with carbon, usually in the form of coke. Iron and tin are prepared from their ores Fe_2O_3 (hematite) and SnO_2 (cassiterite) by this process; many other metals can be reduced in the same way, but some (like titanium, zirconium, and tungsten) form stable carbides under these conditions. The actual reducing agent appears to be carbon monoxide, produced by the incomplete combustion of carbon. The overall reaction, however, may be written

What are the advantages and disadvantages of using C rather than H_2?

$$3\ C(s) + 2\ Fe_2O_3(s) \longrightarrow 3\ CO_2(g) + 4\ Fe(s) \qquad (15.32)$$

This reaction is spontaneous only above 700°C, indicating that carbon is a somewhat weaker reducing agent than hydrogen. Blast furnace operating temperatures are ordinarily much higher (1300 to 1500°C); at these temperatures iron is obtained in molten form and can readily be separated from the glassy silicate slag with which it is contaminated.

From the position of aluminum oxide in Table 15.7, one can deduce that aluminum metal should be an effective agent for the reduction of metal oxides. Powdered aluminum is capable of reducing all of the oxides listed up to and including TiO_2. The reaction of aluminum with oxides of elements such as chromium and iron is highly exothermic. A mixture of aluminum powder and finely divided iron(III) oxide, ignited by means of a fuse, produces a temperature above 3000°C. The iron formed is readily separated from the aluminum oxide slag that adheres loosely to its surface. At one time, this reaction, known as the Thermit process, was used to weld steel rails and repair broken machinery.

$$2\ Al(s) + Fe_2O_3(s) \longrightarrow Al_2O_3(s) + 2\ Fe(l)$$

Nonmetals

Table 15.8 lists the oxides of the nonmetals that are known to be thermo-dynamically stable with respect to decomposition to the elements at room temperature and atmospheric pressure. Curiously enough, several well-known nonmetal oxides do not meet this criterion. In particular, all of the oxides of nitrogen (N_2O_5, N_2O_4, NO_2, N_2O_3, NO, and N_2O) have positive free energies of formation (i.e., ΔG_f° at $25°C = +30$, $+23.5$, $+12.4$, $+33.5$, $+20.7$, and $+24.8$ kcal respectively), and are potentially unstable with respect to decomposition to nitrogen and oxygen. The fact that compounds such as N_2O, NO, and NO_2 can exist for long periods of time must be attributed to kinetic factors. The activation energy for the reaction

$$N_2O(g) \longrightarrow N_2(g) + \tfrac{1}{2} O_2(g) \qquad (15.33)$$

is relatively high (60 kcal for the homogeneous gas phase reaction) and the rate constant is correspondingly small at ordinary temperatures. In the presence of a metal catalyst the activation energy is lowered, and nitrous oxide decomposes rapidly at relatively low temperatures.

TABLE 15.8 STABLE OXIDES OF THE NONMETALS (ΔG_f° NEGATIVE)

3A	4A	5A	6A	7A
B_2O_3	CO_2			
	CO			
	SiO_2	P_4O_{10}	SO_3	
		P_4O_6	SO_2	
	GeO_2	As_2O_5	SeO_3	
		As_4O_6	SeO_2	
		Sb_2O_5	TeO_3	I_2O_5
		Sb_4O_6	TeO_2	

The kinetic stability of the oxides of nitrogen has long intrigued chemists interested in the fixation of atmospheric nitrogen. Even though K_c for the reaction

$$\tfrac{1}{2} N_2(g) + \tfrac{1}{2} O_2(g) \longrightarrow NO(g) \qquad (15.34)$$

is very small at room temperature (about 1×10^{-15}), it increases with temperature, since the reaction is endothermic ($\Delta H = +21.6$ kcal). At 2000°C, for example, K_c has increased to about 10^{-1}. It should then be possible, by heating nitrogen and oxygen to very high temperatures, to produce an equilibrium mixture containing significant amounts of nitric oxide. If this mixture could be cooled quickly to room temperature, the nitric oxide should be kinetically stable even though thermodynamically unstable. In practice, the problem of rapidly quenching the high temperature gas mixture has never been solved satisfactorily.

Perhaps the most important of the reactions of oxygen with the nonmetals are those with carbon and sulfur. We shall now examine these reactions in some detail.

Carbon. The combustion of elementary carbon or carbon compounds in excess oxygen gives the higher oxide CO_2.

$$C(s) + O_2(g) \longrightarrow CO_2(g) \qquad (15.35)$$

Solid carbon dioxide (dry ice) is widely used as a refrigerant. Carbon dioxide has an unusually high triple-point pressure, 5.2 atm at −58°C. This is, of course, the lowest pressure at which the liquid state of carbon dioxide is stable. The sublimation pressure of carbon dioxide reaches one atmosphere at a considerably lower temperature (−78°C); consequently, in an open container the solid passes directly to the gaseous state without melting.

A common type of fire extinguisher contains liquid carbon dioxide. The pressure within such a cylinder is the equilibrium vapor pressure of carbon dioxide, which increases rapidly with temperature from about 59 atmospheres at 20°C to 73 atmospheres at 31°C, the critical temperature of CO_2. When the cylinder is opened, the liquid boils; the amount of heat absorbed in this process is frequently sufficient to lower the temperature of the escaping gas to the point where it forms a finely divided "snow" of solid carbon dioxide. Since CO_2 is a nonconductor of electricity, extinguishers of this type are particularly useful in dealing with electrical fires. They cannot be used to put out fires of burning magnesium or aluminum, since these elements take oxygen away from carbon.

The lower oxide of carbon, carbon monoxide, can be formed by the combustion of hydrocarbons in a limited amount of oxygen or by the reduction of carbon dioxide by carbon at elevated temperatures in the absence of oxygen.

Note the similarity to Equation 15.30.

$$C(s) + CO_2(g) \longrightarrow 2\ CO(g) \tag{15.36}$$

Since the free energy change for this reaction at 25°C and 1 atm is a large positive number ($\Delta G°$ at 25°C = +28.7 kcal), the equilibrium constant at ordinary temperatures is small and the concentration of CO is minute. However, since Reaction 15.36 is endothermic (ΔH = +41.3 kcal), K_c increases with temperature (Table 15.9). At temperatures above 400°C, significant amounts of CO are present in the equilibrium mixture.

TABLE 15.9 EFFECT OF TEMPERATURE ON THE EQUILIBRIUM
$C(s) + CO_2(g) \rightleftharpoons 2\ CO(g)$

T(°C)	25	200	400	600	800	1000
K_c	1×10^{-22}	2×10^{-12}	1×10^{-6}	1×10^{-3}	0.07	1
Mole % CO at 1 atm	3×10^{-9}	1×10^{-3}	0.7	23	87	99

The conditions favorable to the formation of CO by means of Reaction 15.36 (high temperature, insufficient oxygen to burn the CO to CO_2) prevail at the surface of the fuel bed in a coal furnace. In principle, the carbon monoxide should be converted to carbon dioxide as it comes in contact with air at higher levels in the chimney.

$$CO(g) + \tfrac{1}{2}\ O_2(g) \longrightarrow CO_2(g) \tag{15.37}$$

However, Reaction 15.37 occurs very slowly under ordinary conditions; the gases coming out of industrial smokestacks often contain appreciable concentrations of carbon monoxide.

Carbon monoxide is highly toxic; a concentration of 0.2 per cent brings about unconsciousness, while concentrations of 1 per cent or greater, if inhaled for only a few minutes, are fatal. Gas masks that are designed to protect against carbon monoxide poisoning contain a mixture of metal oxides which will catalyze Reaction 15.37.

Carbon monoxide is an important constituent of several industrial fuel gases. One of these is water gas, made by passing steam over hot coal:

$$C(s) + H_2O(g) \longrightarrow CO(g) + H_2(g) \qquad (15.38)$$

Before natural gas was available, gas stoves and furnaces were fired with water gas.

The resulting gas mixture contains the two highly combustible gases, carbon monoxide and hydrogen, in about equal amounts, contaminated by small amounts of carbon dioxide and nitrogen.

Sulfur. The reaction of sulfur with oxygen is important from an industrial standpoint, since it forms the basis for the preparation of sulfuric acid. When sulfur is heated in an open container above its melting point, it catches fire, forming a choking gas made up principally of sulfur dioxide.

$$S(l) + O_2(g) \longrightarrow SO_2(g) \qquad (15.39)$$

Up to 2 per cent of sulfur trioxide is formed simultaneously. To prepare this compound in good yield, sulfur dioxide is reacted further with oxygen.

$$SO_2(g) + \tfrac{1}{2} O_2(g) \rightleftharpoons SO_3(g); \quad \Delta H = -23 \text{ kcal} \qquad (15.40)$$

Recall the discussion of this reaction in Chapter 14.

The rate of this reaction and the equilibrium yield of sulfur trioxide depend upon temperature and pressure. The principles involved here are the same as those discussed previously in connection with the Haber process. At low temperatures, the equilibrium constant for the formation of sulfur trioxide is large, but equilibrium is reached very slowly. As the temperature is raised, the rate increases, but since the reaction is exothermic, the yield of sulfur trioxide drops off. High pressures would tend to increase both the yield of sulfur trioxide and the rate of reaction.

TABLE 15.10 EFFECT OF TEMPERATURE ON THE EQUILIBRIUM
$SO_2(g) + \tfrac{1}{2} O_2(g) \rightleftharpoons SO_3(g)$

T(°C)	400	500	600	700	800
K_c	2300	400	70	20	7
Mole % SO_3°	96	88	66	40	20

° Under the condition that $[O_2] = \tfrac{1}{2} [SO_2]$, and $p_{tot} = 1$ atm

In the so-called contact process for the manufacture of sulfuric acid, the equilibrium represented by Equation 15.40 is reached rapidly by passing sulfur dioxide and oxygen, at atmospheric pressure, over a solid catalyst at a temperature of about 650°C. The equilibrium mixture is then recycled at a lower temperature, 400 to 500°C, to increase the yield of sulfur trioxide. The two catalysts which have proved most effective are vanadium pentoxide, V_2O_5, and finely divided platinum. The sulfur trioxide produced can be converted to sulfuric acid by reaction with water.

$$SO_3(g) + H_2O(l) \longrightarrow H_2SO_4(l) \qquad (15.41)$$

This reaction is ordinarily carried out in sulfuric acid solution; if sulfur trioxide is added directly to water, sulfuric acid is formed as a fog of tiny particles which are difficult to recover.

395

PROBLEMS

15.1   Give the oxidation numbers of each atom in

    a. MoO_2 b. $MgCO_3$ c. C_2H_6O d. $Ca(ClO_3)_2$ e. MnO_4^-

15.2   For each of the following reactions

$$2\ Cr(s) + 3\ Cl_2(g) \longrightarrow 2\ CrCl_3(s)$$
$$ZnO(s) + H_2(g) \longrightarrow Zn(s) + H_2O(g)$$
$$H_2O_2(l) + NO_2^- \longrightarrow H_2O + NO_3^-$$

indicate

    a. The element that is oxidized.
    b. The element that is reduced.
    c. The substance that acts as an oxidizing agent.
    d. The substance that acts as a reducing agent.

15.3   Using Figure 15.1, predict the formulas of

    a. Yttrium sulfide.
    b. Four different compounds containing the elements sodium, chlorine, and oxygen.
    c. Two different chlorides of phosphorus.

15.4   Suggest an explanation for the fact that the maximum oxidation state of a transition metal is ordinarily given by its group number.

15.5   Give the formulas of six different compounds that contain no elements other than potassium, hydrogen, and oxygen.

15.6   Write balanced equations for reactions in which

    a. A salt is formed by a reaction involving elementary hydrogen.
    b. Hydrogen is formed by the reaction of a salt with water.
    c. Hydrogen acts as an oxidizing agent.
    d. Hydrogen acts as a reducing agent.

15.7   For the reaction $PH_3(g) \rightarrow P(s) + \frac{3}{2}\ H_2(g)$, ΔH is -2.2 kcal and $\Delta G°$ at 25°C is -4.4 kcal.

    a. Explain why PH_3 cannot be formed by the direct reaction of the elements with each other.
    b. Explain how it is possible for PH_3 to exist in the laboratory when prepared indirectly (for example, by the reaction of elementary phosphorus with sodium hydroxide).

15.8   For the reaction $2\ NO(g) + O_2(g) \rightarrow 2\ NO_2(g)$, $\Delta H = -27.0$ kcal. Discuss the effect of temperature and pressure upon the rate of this reaction and the position of the equilibrium.

15.9   Consider the data in Table 15.3.

    a. From the table, determine the number of moles of NH_3, H_2, and N_2 in an equilibrium mixture containing a total of one mole of gas at 300°C and 100 atm. Note that $[H_2] = 3[N_2]$.
    b. Using the Ideal Gas Law, calculate the volume occupied by one mole of gas under these conditions.
    c. Combining your answers to a and b, calculate the equilibrium concentrations, in moles/lit, of NH_3, H_2, and N_2.
    d. From your answer to c, calculate K_c at 300°C and 100 atm and compare to the value given in the table.

15.10   Consider the following mechanism for a chain reaction between hydrogen and iodine:

$$I_2 \longrightarrow 2\ I \qquad\qquad \text{(Step 1)}$$

$$I + H_2 \longrightarrow HI + H \qquad\qquad \text{(Step 2)}$$

$$H + I_2 \longrightarrow HI + I \qquad\qquad \text{(Step 3)}$$

a. From the data in Table 8.8 (page 185) estimate the energy change for each step.

b. Would this reaction be more or less difficult to *initiate* than that between hydrogen and chlorine?

c. We can find the overall energy change for the reaction by summing the energy changes for Steps 2 and 3. Why is Step 1 not included in taking this sum?

d. Suggest several different processes that could break the chain of reaction.

15.11 Discuss briefly how the following compounds can be prepared from the elements.

a. NH_3 b. HBr c. CaH_2 d. AsH_3

15.12 Calculate the wavelength of light required to initiate the reaction of hydrogen with bromine (H—H bond energy = 104 kcal, Br—Br bond energy = 46 kcal, 1 kcal = 7.0×10^{-14} ergs).

15.13 Write balanced equations to represent reactions in which

a. Lithium metal is converted to lithium hydroxide (two steps).

b. Oxygen is converted to a solid containing the peroxide ion.

c. Potassium combines with oxygen.

d. Oxygen is formed by the reaction of an ionic compound with water.

15.14 Describe briefly how the following conversions might be accomplished.

a. $CuO \longrightarrow Cu_2O$ e. $S \longrightarrow SO_3$

b. $O_2 \longrightarrow H_2O_2$ f. $Fe_2O_3 \longrightarrow Fe$

c. $CO_2 \longrightarrow CO$ g. $Ag_2O \longrightarrow Ag$

d. $CO \longrightarrow CO_2$

15.15 Give the formulas of

a. Three metal oxides which can be reduced by magnesium but not by hydrogen.

b. A metal oxide that could be formed by reacting tin with tin(IV) oxide.

c. The normal oxide, peroxide, and superoxide of francium (element 87).

15.16 Using the K_c value given in Table 15.9, calculate the concentration of CO in equilibrium with a CO_2 concentration of 0.10 mole/lit at 800°C.

15.17 Give the oxidation numbers of each atom in

a. P_4O_{10} b. KNO_3 c. $C_2H_4O_2$ d. $Cr_2O_7{}^{2-}$

15.18 For the reaction

$$4\ NH_3(g) + 5\ O_2(g) \longrightarrow 4\ NO(g) + 6\ H_2O(g)$$

indicate

a. The element that is oxidized.

b. The element that is reduced.

c. The reducing agent.

d. The oxidizing agent.

15.19 Using Figure 15.1, predict the formulas of

a. Three possible oxides of osmium.

b. The fluoride of germanium.

c. Several possible oxyanions of nitrogen.

15.20 Give the formulas of five different compounds containing no elements other than barium, hydrogen, and oxygen.

397

15.21 Write balanced equations for reactions in which

    a. Hydrogen is formed by heating an ionic compound.
    b. Hydrogen is formed by heating a molecular compound.
    c. Oxygen is formed by heating an ionic compound.
    d. Oxygen is formed by heating a molecular compound.

15.22 For the reaction $\frac{1}{2} H_2(g) + \frac{1}{2} Br_2(g) \rightarrow HBr(g)$, $\Delta H = 12.3$ kcal and $\Delta G°$ at 25°C = -13.1 kcal. Find K_c at 25°C.

15.23 For the reaction $H_2(g) + I_2(g) \rightarrow 2 HI(g)$, $\Delta H = -2400$ cal.
Discuss the effect of temperature and pressure on the rate of this reaction and the position of the equilibrium.

15.24 Consider the data in Table 15.10.

    a. Given that the mole per cent of SO_3 at 700°C is 40 and that $[O_2] = \frac{1}{2} [SO_2]$, calculate the mole per cents of all three gases in the equilibrium mixture at this temperature.
    b. Using the Ideal Gas Law, find the number of moles of gas in a one-liter container at 1 atm and 700°C.
    c. Combining your results in a and b, find the equilibrium concentrations of all gases in a one-liter container at 1 atm and 700°C.
    d. From your answers to c, find K_c at 700°C and compare to the value given in the table.

15.25 Determine the wavelength of light required to break

    a. The F—F bond (bond energy = 37 kcal).
    b. The H—Br bond (bond energy = 88 kcal).

15.26 Discuss briefly how each of the following compounds can be prepared from the elements (more than one step may be required).

    a. LiH   b. NaOH   c. H_2O_2 d. Al_2O_3 e. CO

15.27 Write balanced equations for each of the following conversions.

    a. $HgO \longrightarrow Hg$ e. $NH_3 \longrightarrow NO$
    b. $TiO_2 \longrightarrow Ti$ f. $SO_3 \longrightarrow H_2SO_4$
    c. $Na_2O_2 \longrightarrow NaOH$ g. $CaH_2 \longrightarrow CaO$
    d. $C \longrightarrow CO$ h. $SrO_2 \longrightarrow SrO$

15.28 Give the electronic structures (electron-dot notation) for

    a. The peroxide ion.   d. The hydroxide ion.
    b. The hydride ion.   e. The PH_3 molecule.
    c. The sulfur dioxide molecule.   f. The oxide ion.

15.29 Using the data in Table 15.4, calculate K_c for the reaction $\frac{1}{2} H_2(g) + \frac{1}{2} Br_2(g) \rightarrow HBr(g)$ at 600°C, 1000°C, and 2000°C.

°15.30 Using the data in Table 15.7, calculate ΔH and $\Delta G°$ at 25°C for the reaction

$$Fe_2O_3(s) + 3 Zn(s) \longrightarrow 3 ZnO(s) + 2 Fe(s)$$

Could this reaction be reversed by increasing the temperature?

°15.31 When 0.100 g of a certain metal is heated with an excess of chlorine, it reacts to form 0.290 g of a volatile metal chloride. When the same amount of this metal is added to an excess of dilute hydrochloric acid, it goes into solution with the evolution of hydrogen; evaporation of this solution at 100°C gives 0.227 g of a deliquescent metal chloride. Identify the metal.

°15.32 Using the data in Table 15.3, calculate the mole percentage of NH_3 in an equilibrium mixture at 600°C and 500 atm. Take $[H_2] = 3[N_2]$.

°15.33 A chemist heats cesium superoxide in air to produce a material which analysis shows to contain 10.74% O and 89.26% Cs. On this basis, he reports the preparation of cesium peroxide, Cs_2O_2. What other possible reaction might explain his results? How would you distinguish chemically between the two possibilities?

°15.34 Using an appropriate reference source, look up the formulas of the known oxides of the metals in the second and third transition series. Use this information to test the principle that as one moves down in a given group, higher oxidation states tend to become more stable. Note particularly any exceptions to this rule.

16

Precipitation Reactions

Almost all the reactions considered up to this point have involved pure substances as reactants and products. While such reactions are of great importance to the layman and chemist alike, reactions taking place in solution are of perhaps even greater significance and are certainly more widespread. Life itself depends upon the intricate processes occurring in solution within our bodies, in the blood, in the digestive fluids, and in glandular secretions.

Having considered some of the properties of solutions in Chapter 11, we are now in a position to study reactions taking place between particles in solution. Such a study will occupy our attention for the next several chapters. We shall be concerned primarily with reactions occurring in water solution, since water is by far the most common solvent. The majority of the reactions considered will be those taking place between dissolved ions.

In this chapter, we shall consider what is perhaps the simplest type of reaction that can occur between ions in water solution, that of precipitation. When certain pairs of solutions such as nickel chloride and sodium hydroxide or barium bromide and aluminum sulfate are mixed, ions combine to form an insoluble solid. We shall develop principles to predict when such reactions will occur, what their products will be, and how they can be represented by chemical equations.

16.1 NET IONIC EQUATIONS

If 0.1 M solutions of nickel chloride, $NiCl_2$, and sodium hydroxide, $NaOH$, are mixed, a green, gelatinous precipitate forms. In principle, the green precipitate might be either sodium chloride, $NaCl$, resulting from the interaction of Na^+ ions of the NaOH solution with Cl^- ions of the $NiCl_2$ solution, or nickel hydroxide, $Ni(OH)_2$, formed when Ni^{2+} ions of the $NiCl_2$ solution come in contact with OH^- ions of the NaOH solution. Experience enables us to make a choice between these two possibilities: we know that sodium chloride, ordinary table salt, is neither green nor insoluble. By elimination, we deduce that nickel hydroxide, $Ni(OH)_2$, must be the product of the reaction. To confirm this reasoning, we might filter off the precipitate and conduct various tests upon it, all of which would give the results to be expected for this particular compound. The green color of the precipitate can be ascribed to the presence of the Ni^{2+} ion, which is also responsible for the green color of the original $NiCl_2$ solution.

Having deduced the nature of the reaction, it is now possible to write an equation for it. The product is solid nickel hydroxide, $Ni(OH)_2$; the reactants are Ni^{2+} and OH^-. Consequently, the balanced equation must be

$$Ni^{2+} + 2 OH^- \longrightarrow Ni(OH)_2(s) \qquad (16.1)$$

Such an equation, representing a reaction between ions in water solution, is often referred to as a **net ionic equation**. Many such equations will be encountered throughout the remainder of this text. We shall use them to describe many types of reactions other than precipitation reactions. It is important to keep in mind that whenever a formula such as Ni^{2+} or OH^- appears alone in a chemical equation, **it is understood that the corresponding species is in aqueous solution**. This is sometimes emphasized by writing (aq) after the formula. Thus, we might write

$$Ni^{2+}(aq) + 2 OH^-(aq) \longrightarrow Ni(OH)_2(s) \qquad (16.1')$$

In writing net ionic equations, the same conventions are followed as in writing any equation. Only those species which actually take part in the reaction are included in the equation. In Equation 16.1, neither the Na^+ nor the Cl^- ions are included, since they do not take part in the reaction. By the same token, an equation for the combustion of aluminum in air does not include the N_2 molecules present in the reaction mixture, since they do not enter into the reaction.

Net ionic equations will be used extensively from now on.

To illustrate further how net ionic equations can be deduced from experimental observations, let us consider what happens when dilute solutions of aluminum sulfate, $Al_2(SO_4)_3$, and barium bromide, $BaBr_2$, are mixed. In this case, a white, granular precipitate forms. To decide upon the nature of this precipitate, we note that the ions originally present are

$$Al_2(SO_4)_3 \text{ solution:} \quad Al^{3+}, SO_4^{2-}$$

$$BaBr_2 \text{ solution:} \quad Ba^{2+}, Br^-$$

Logically, one could obtain either barium sulfate, $BaSO_4$, or aluminum bromide, $AlBr_3$, as a precipitate. In order to choose between these possibilities, one could obtain bottles of the two solids from the storeroom and test their solubilities. Alternatively, the identity of the precipitate could be decided upon by testing other pairs of solutions. Suppose, for example, solutions of $Al_2(SO_4)_3$ and NaBr are mixed. If this is done, it is found that no precipitate forms, indicating that aluminum bromide is soluble in water. On the other hand, mixing solutions of $Ba(NO_3)_2$ and $Al_2(SO_4)_3$ gives a precipitate. Indeed, whenever two solutions, one containing Ba^{2+} ions and the other containing SO_4^{2-} ions, are mixed, a white precipitate is formed, identical in its properties to that resulting from the reaction under study. We deduce, therefore, that the process occurring when solutions of $Al_2(SO_4)_3$ and $BaBr_2$ are mixed must be the reaction of Ba^{2+} ions with SO_4^{2-} ions to form solid barium sulfate, $BaSO_4$. Hence, the net ionic equation must be

$$Ba^{2+} + SO_4^{2-} \longrightarrow BaSO_4(s) \qquad (16.2)$$

In certain cases, two precipitation reactions occur when a pair of solutions is mixed. Suppose, for example, a solution of barium hydroxide, $Ba(OH)_2$, is added to a solution of nickel sulfate, $NiSO_4$. It has been pointed out that both of the possible products, $Ni(OH)_2$ and $BaSO_4$, are insoluble in water. Hence, we would predict that both of these compounds should precipitate. Experiment confirms this deduction; if one examines the precipitate under a microscope, it is possible to distinguish white crystals of $BaSO_4$ from green particles of $Ni(OH)_2$. In representing this double precipitation reaction, we write two net ionic equations, since two entirely different reactions are taking place:

$$Ba^{2+} + SO_4^{2-} \longrightarrow BaSO_4(s)$$

and

$$Ni^{2+} + 2\ OH^- \longrightarrow Ni(OH)_2(s)$$

Net ionic equations such as these can be given a quantitative meaning in precisely the manner outlined in Chapter 3 for reactions between pure substances. Example 16.1 illustrates how this is done.

EXAMPLE 16.1. 400 ml of 0.30 M $Al_2(SO_4)_3$ is added to 300 ml of 0.50 M $BaBr_2$ solution. Calculate

    a. The number of moles of $BaSO_4$ precipitated.

    b. The number of moles and concentration of each ion remaining in solution after precipitation.

SOLUTION. Before making any calculations of this type, one must determine the number of moles of each ion originally present:

$$\text{no. moles } Al_2(SO_4)_3 = 0.400 \text{ lit} \times 0.30 \frac{\text{mole}}{\text{lit}} Al_2(SO_4)_3 = 0.12 \text{ mole } Al_2(SO_4)_3$$

$$\text{no. moles } BaBr_2 = 0.300 \text{ lit} \times 0.50 \frac{\text{mole}}{\text{lit}} BaBr_2 = 0.15 \text{ mole } BaBr_2$$

Noting that *one* mole of $Al_2(SO_4)_3$ gives *two* moles of Al^{3+} and *three* moles of SO_4^{2-}, while *one* mole of $BaBr_2$ yields *one* mole of Ba^{2+} and *two* moles of Br^-, we have

    no. moles $Al^{3+} = 2 \times 0.12 = 0.24$ no. moles $Ba^{2+} = 1 \times 0.15 = 0.15$

    no. moles $SO_4^{2-} = 3 \times 0.12 = 0.36$ no. moles $Br^- = 2 \times 0.15 = 0.30$

    a. It may be noted from Equation 16.2 that *one* mole of Ba^{2+} is required to react with *one* mole of SO_4^{2-}, producing *one* mole of $BaSO_4$. The limiting factor here is clearly the number of moles of Ba^{2+} available:

$$0.15 \text{ mole } Ba^{2+} + 0.15 \text{ mole } SO_4^{2-} \longrightarrow 0.15 \text{ mole } BaSO_4$$

    b. Subtracting the 0.15 mole of Ba^{2+} and 0.15 mole of SO_4^{2-} reacting from the number of moles originally present gives

        no. moles Ba^{2+} = 0.15 mole − 0.15 mole = 0.00 mole
        no. moles SO_4^{2-} = 0.36 mole − 0.15 mole = 0.21 mole
        no. moles Al^{3+} = 0.24 mole − 0   = 0.24 mole
        no. moles Br^- = 0.30 mole − 0   = 0.30 mole

    To obtain the concentrations of these ions in moles per liter, we need only divide by the total volume of the final solution. Assuming no volume change on mixing, the total volume must be 0.400 lit + 0.300 lit = 0.700 lit.

$$\text{conc. } Ba^{2+} = \frac{0.00 \text{ mole}}{0.70 \text{ lit}} = 0.00 \text{ M}$$

$$\text{conc. } SO_4{}^{2-} = \frac{0.21 \text{ mole}}{0.70 \text{ lit}} = 0.30 \text{ M}$$

$$\text{conc. } Al^{3+} = \frac{0.24 \text{ mole}}{0.70 \text{ lit}} = 0.34 \text{ M}$$

$$\text{conc. } Br^{-} = \frac{0.30 \text{ mole}}{0.70 \text{ lit}} = 0.43 \text{ M}$$

Summarizing, we have

| | Number of Moles | | | | | Concentration (M) |
	original	−	reacted	=	final	final
Ba^{2+}	0.15	−	0.15	=	0.00	0.00 M
$SO_4{}^{2-}$	0.36	−	0.15	=	0.21	0.30 M
Al^{3+}	0.24	−	0.00	=	0.24	0.34 M
Br^{-}	0.30	−	0.00	=	0.30	0.43 M

16.2 SOLUBILITIES OF IONIC COMPOUNDS

It is always possible to decide by direct experimental observation whether or not a precipitate forms when two solutions are mixed. More conveniently, solubility data obtained from a limited number of precipitation reactions can be used to predict the results of a great many other reactions. For example, having established the fact that nickel hydroxide is insoluble in water, one can predict that mixing solutions of $Ni(NO_3)_2$ and NaOH, $NiSO_4$ and KOH, or $NiBr_2$ and $Ca(OH)_2$ will result in the formation of a precipitate of $Ni(OH)_2$, according to Equation 16.1. Again, if it is established through experiment that sodium chloride and potassium nitrate are both soluble in water, it follows that no precipitation reaction will occur when solutions of $NaNO_3$ and KCl are mixed. It might then appear that all one needs to predict the results of possible precipitation reactions is a table of solubilities in which every ionic solid is neatly classified as soluble or insoluble. Unfortunately, there are difficulties associated with any attempt to set up such a simple classification scheme.

Ionic solids do not fall clearly into the categories soluble and insoluble with a sharp dividing line between them. Instead, they cover an enormous range of solubility. One of the most soluble salts known is lithium chlorate, $LiClO_3$, which dissolves to the extent of 35 moles/lit at room temperature. One can safely predict that this compound will not be the product of a precipitation reaction. At the other extreme is mercury(II) sulfide, HgS; one can calculate that a saturated solution should contain only about 10^{-26} mole/lit of Hg^{2+} and S^{2-} ions. Quite clearly, we can expect to get a precipitate of mercuric sulfide whenever water solutions containing Hg^{2+} and S^{2-} ions are mixed, even if the solutions are extremely dilute. Lead chloride, $PbCl_2$, is an example of a compound of intermediate solubility. One can calculate that if equal volumes of 0.1 M $Pb(NO_3)_2$ and 0.1 M NaCl are mixed, the solubility of $PbCl_2$ will be exceeded and it will precipitate. If, on the other hand, the solutions

403

mixed are somewhat more dilute, say 0.04 M, no precipitate is formed. Compounds such as lead chloride are difficult to classify as soluble or insoluble; it is perhaps begging the question to classify them as slightly soluble.

In attempting to decide whether or not a precipitate will form when two solutions are mixed, one must always consider the effect of temperature on solubility. This factor is particularly important when the compound in question is in the slightly soluble class. Consider, for example, lead chloride: its solubility in boiling water is about four times as great as at room temperature. As a result, although lead chloride precipitates when equal volumes of solutions 0.1 M in Pb^{2+} and Cl^- are mixed at room temperature, no precipitate forms when the same reaction is attempted at 100°C. Because of this behavior, lead chloride is sometimes described as being soluble in hot water and insoluble in cold water.

TABLE 16.1 SOLUBILITY RULES

NO_3^-, ClO_3^-	All nitrates and chlorates are soluble.
Cl^-, Br^-, I^-	All chlorides are soluble except AgCl, Hg_2Cl_2 and $PbCl_2$.* The same rule applies to bromides and iodides except that $HgBr_2$* and HgI_2 are also insoluble.
F^-	All fluorides are soluble except those of the 2A elements, FeF_3, PbF_2, and AlF_3.
SO_4^{2-}	All sulfates are soluble except $CaSO_4$,* $SrSO_4$, $BaSO_4$, Hg_2SO_4, $HgSO_4$, $PbSO_4$, and Ag_2SO_4.
CO_3^{2-}, PO_4^{3-}	All carbonates are insoluble except those of the 1A elements and NH_4^+. The same rule applies to phosphates except that Li_3PO_4 is also insoluble.
OH^-	All hydroxides are insoluble except those of the 1A elements, $Sr(OH)_2$ and $Ba(OH)_2$. ($Ca(OH)_2$ is slightly soluble.)
S^{2-}	All sulfides except those of the 1A and 2A elements and NH_4^+ are insoluble.

* Insoluble compounds are those which precipitate upon mixing equal volumes of solutions 0.1 M in the corresponding ions. Compounds which fail to precipitate at concentrations slightly below 0.1 M are starred.

Bearing these qualifications in mind, we can classify the more common ionic solids on the basis of their solubility behavior according to the rules outlined in Table 16.1. The use of solubility rules to predict the results of precipitation reactions is illustrated by Example 16.2.

EXAMPLE 16.2. Write balanced equations for the reactions, if any, that occur when equal volumes of 0.1 M solutions of the following compounds are mixed:

a. $Hg(NO_3)_2$ and $AlBr_3$ b. Na_2SO_4 and KOH

c. $CuSO_4$ and $Ba(OH)_2$

SOLUTION. In each case, we first deduce the formulas of the two possible precipitates and then, on the basis of the solubility rules, decide whether one or both of these compounds will precipitate. Having made this decision, the net ionic equation for the precipitation reaction is readily derived.

  a. Possible precipitates: $HgBr_2$, $Al(NO_3)_3$. Table 16.1 indicates that $HgBr_2$ is insoluble, while $Al(NO_3)_3$ is soluble (all nitrates are soluble). Consequently, we have

$$Hg^{2+} + 2\ Br^- \longrightarrow HgBr_2(s)$$

  b. Possible precipitates: $NaOH$, K_2SO_4. From the solubility rules, it is clear that both of these compounds are soluble. No precipitation reaction occurs.

  c. Possible precipitates: $Cu(OH)_2$, $BaSO_4$. Of these compounds, $BaSO_4$ is specifically listed as insoluble; it can be deduced that $Cu(OH)_2$ is also insoluble. Two precipitation reactions occur simultaneously:

$$Ba^{2+} + SO_4^{2-} \longrightarrow BaSO_4(s) \text{ and } Cu^{2+} + 2\ OH^- \longrightarrow Cu(OH)_2(s)$$

Unfortunately, there is no simple theory which enables us to explain the solubility rules listed in Table 16.1 or even to predict the relative solubilities of different ionic compounds. The difficulties involved in predicting relative solubilities of ionic compounds are inherent in the nature of the solution process. When an ionic solute is added to water, a balance is achieved between the attractive forces of the ions for each other, which tend to hold them in the solid state, and the attractive forces of the ions for water, which tend to bring the compound into solution. These two attractive forces are essentially similar in origin. The ions are held together in the crystal lattice by electrostatic attraction; the attraction of a polar water molecule for ions is also largely electrostatic. As a result, the factors which tend to increase the attractive forces between ions simultaneously increase the attraction of these ions for water molecules.

Many of the general trends in solubility of ionic compounds can be explained in terms of the strength of the forces holding the ions together in the solid. If these forces are particularly strong, one expects, and ordinarily finds, that the compound has low solubility in water. If the forces are particularly weak, the compound is likely to be readily soluble.

From this point of view, one would expect to find a general relation between the solubility of an ionic compound and the **charge density** (charge to size ratio) of the ions of which it is composed. Compounds containing ions of low charge density (low charge, large size) should be more soluble than those of high charge density, since the electrostatic forces holding them in the crystal lattice are weaker. Experimentally, we find this to be generally true. For example, the salts of the 1A metals, where we are dealing with a relatively large cation with a +1 charge, are soluble in water, almost without exception. In contrast, many of the salts of the 2A and transition metals are insoluble. Again, salts containing large anions of low charge, such as NO_3^- and ClO_4^-, tend to be more soluble than those containing anions of higher charge (CO_3^{2-}, PO_4^{3-}) or smaller size (F^-, OH^-).

Certain compounds which, on the basis of charge density considerations alone, might be expected to be quite insoluble in water, show abnormally

405

high solubilities. An example is magnesium sulfate, $MgSO_4$, which dissolves in water to the extent of about two moles per liter at 20°C, despite the high charges $(+2, -2)$ of its ions. The high solubility of this compound can be explained on the basis of the relative sizes of its ions. The Mg^{2+} ion (radius = 0.65 Å) is so much smaller than the SO_4^{2-} ion (S—O distance = 1.44 Å) that there is actually anion-anion contact in a crystal of anhydrous magnesium sulfate. The repulsion between adjacent ions of like charge weakens the crystal lattice of magnesium sulfate and makes it much more soluble than one might otherwise expect. In barium sulfate, $BaSO_4$, where the ions are of comparable size (radius Ba^{2+} = 1.35 Å), this effect disappears and the solubility drops to a very low value, about 3×10^{-5} moles/liter.

We can explain relative solubilities of salts but we can not predict them with confidence.

16.3 SOLUBILITY EQUILIBRIA

Qualitative Aspects: Common Ion Effect

When silver acetate is dissolved in pure water to form a saturated solution, an equilibrium is established between the solid and its ions in solution.

$$AgC_2H_3O_2(s) \rightleftharpoons Ag^+ + C_2H_3O_2^- \qquad (16.3)$$

The concentrations of Ag^+ and $C_2H_3O_2^-$ ions under these conditions will, of course, be equal to each other and will have a fixed value, independent of the amount of silver acetate or water used in preparing the saturated solution. At 20°C the equilibrium concentrations of Ag^+ and $C_2H_3O_2^-$ in a solution prepared by dissolving silver acetate in pure water are found experimentally to be 0.045 mole/lit.

It is possible to change the relative concentrations of Ag^+ and $C_2H_3O_2^-$ ions in equilibrium with solid silver acetate in any of various ways. Specifically, one might accomplish this by adding to the solution:

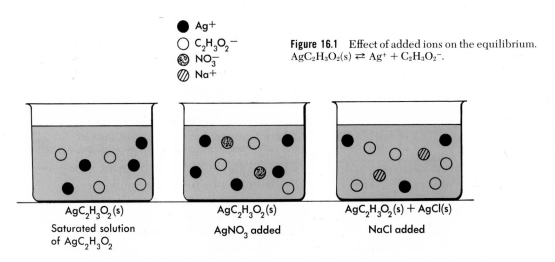

● Ag^+
○ $C_2H_3O_2^-$
◉ NO_3^-
◍ Na^+

Figure 16.1 Effect of added ions on the equilibrium. $AgC_2H_3O_2(s) \rightleftharpoons Ag^+ + C_2H_3O_2^-$.

$AgC_2H_3O_2$(s)
Saturated solution of $AgC_2H_3O_2$

$AgC_2H_3O_2$(s)
$AgNO_3$ added

$AgC_2H_3O_2$(s) + AgCl(s)
NaCl added

1. *An electrolyte containing a common ion, Ag^+ or $C_2H_3O_2^-$.* Suppose, for example, that we add a concentrated solution of a more soluble silver salt, such as silver nitrate. Some of the added Ag^+ ions will combine with the

$C_2H_3O_2^-$ ions in solution to precipitate silver acetate. This behavior is often referred to as the **common ion** effect; addition of another silver salt (which has the Ag^+ ion in common with $AgC_2H_3O_2$) to a saturated solution of silver acetate $AgC_2H_3O_2$, disturbs the solubility equilibrium so as to bring about precipitation. When equilibrium is reestablished, it is found that the concentration of $C_2H_3O_2^-$ ion is reduced from its original value while that of Ag^+ ion is increased (Figure 16.1).

2. *An electrolyte which forms a precipitate with one of the ions.* If one adds sodium chloride to a saturated solution of silver acetate, it is observed that the very insoluble salt, silver chloride, precipitates. As Ag^+ ions are removed from solution as AgCl, more silver acetate dissolves to restore equilibrium between Ag^+ and $C_2H_3O_2^-$ ions. The net effect is to increase the concentration of $C_2H_3O_2^-$ ions in solution and simultaneously decrease the concentration of Ag^+ ions (Figure 16.1).

Another way to decrease the conc. of $C_2H_3O_2^-$ is to add strong acid (Chap. 17).

Quantitative Treatment: Solubility Product

As the foregoing discussion implies, there is an inverse relationship between the equilibrium concentrations of ions in contact with the electrolyte from which they are derived. Any process which shifts the equilibrium so as to increase the concentration of one of these ions simultaneously decreases the concentration of the other. The quantitative relationship between these concentrations can be derived in a manner analogous to that used in Chapter 12 in dealing with gaseous equilibria. This derivation, outlined as follows, is strictly applicable only in very dilute solution in which ionic interactions of the type described in Chapter 11, Section 11.7, may be neglected.

Let us consider a slightly soluble 1:1 electrolyte of general formula MX in equilibrium with its ions M^+ and X^- in aqueous solution.

$$MX(s) \rightleftharpoons M^+ + X^- \qquad (16.4)$$

For this equilibrium, we can write

$$K_{sp} = [M^+] \times [X^-] \qquad (16.5)$$

where the square brackets are used to denote equilibrium concentrations, in moles/liter, of the ions M^+ and X^- in the solution. The concentration of solid MX does not appear in the equilibrium expression. The situation here is completely analogous to that discussed in Chapter 12, Section 12.6, in connection with heterogeneous equilibria involving gases, where it was pointed out that for the equilibrium

$$NH_4Cl(s) \rightleftharpoons NH_3(g) + HCl(g) \qquad (16.6)$$

the equilibrium constant expression has the form

$$K = [NH_3] \times [HCl]$$

In both systems (16.4 and 16.6), equilibrium is established as long as some solid is present; adding or removing solid does not affect the position of the equilibrium.

The quantity K_{sp} appearing in Equation 16.5 is a particular type of equilibrium constant, referred to as the solubility product. For any given elec-

407

trolyte at a particular temperature, K_{sp} should have a fixed value, independent of the concentrations of the individual ions in equilibrium with the solid.

The concept of the solubility product is readily extended to electrolytes of any valence type. For an electrolyte MX_2, for which the solubility equilibrium may be expressed by the equation

$$MX_2(s) \rightleftharpoons M^{2+} + 2\,X^- \tag{16.7}$$

we can write

$$K_{sp} = [M^{2+}] \times [X^-]^2$$

Thus, we have, for example,

$$K_{sp} \text{ of } PbCl_2 = [Pb^{2+}] \times [Cl^-]^2 = 1.7 \times 10^{-5}$$

Again, for a compound such as arsenic(III) sulfide, As_2S_3,

$$As_2S_3(s) \rightleftharpoons 2\,As^{3+} + 3\,S^{2-} \tag{16.8}$$

$$K_{sp} = [As^{3+}]^2 \times [S^{2-}]^3 = 5 \times 10^{-27}$$

In general, the solubility product principle may be stated as follows: **In any water solution in equilibrium with a slightly soluble ionic compound, the product of the ion concentrations, each raised to a power equal to the coefficient of the ion in the solubility equation, has a constant value.**

In principle, at least, the numerical value of the solubility product can be calculated from the measured solubility, or vice-versa (Example 16.3). In practice, values of K_{sp}, particularly for electrolytes of very low solubility, are ordinarily calculated by less direct methods, one of which is mentioned in Chapter 21 in connection with voltage measurements on electrical cells.

EXAMPLE 16.3

a. The measured solubility of $AgC_2H_3O_2$ at 20°C is 0.045 mole/lit. Calculate K_{sp} for this salt.

b. Use the known solubility product of BaF_2, 2×10^{-6}, to estimate the solubility, in mole/lit of this compound.

SOLUTION

a. From the equation,

$$AgC_2H_3O_2(s) \rightleftharpoons Ag^+ + C_2H_3O_2^-$$

it is evident that for every mole of silver acetate that dissolves, a mole of Ag^+ and a mole of $C_2H_3O_2^-$ enter the solution. It follows that if 0.045 mole of solid dissolves per liter of solution, the equilibrium concentrations of Ag^+ and $C_2H_3O_2^-$ must be 0.045 M.

$$K_{sp} = [Ag^+] \times [C_2H_3O_2^-] = (0.045)(0.045) = 2.0 \times 10^{-3}$$

b. In this case, we know that $[Ba^{2+}] \times [F^-]^2 = 2 \times 10^{-6}$

From the equation $BaF_2(s) \longrightarrow Ba^{2+} + 2\,F^-$,

we note that for every mole of BaF_2 that dissolves, *one* mole of Ba^{2+} and *two* moles of F^- are formed. Consequently, if we let   s = solubility BaF$_2$ (mole/lit),

then: $[Ba^{2+}] = s$, and $[F^-] = 2s$

Substituting for these concentrations in the expression for K_{sp}

$$s\,(2s)^2 = 2 \times 10^{-6}; \quad 4\,s^3 = 2 \times 10^{-6}$$

Solving, $s = 8 \times 10^{-3}$ mole/lit

The solubility product principle, in the form stated, is strictly applicable only in very dilute solution. We find that if we study the solubility of silver chloride in potassium nitrate solutions of different concentrations (Table 16.2) that the product,

$$[Ag^+] \times [Cl^-]$$

which has the limiting value of 1.6×10^{-10} in pure water, has increased by about 50 per cent by the time we get to 0.04 M KNO$_3$.

The failure of the ion concentration product to behave as a true constant can be attributed to the interionic attractions discussed in Chapter 11 in connection with the colligative properties of electrolytes. The effect of these forces can be taken into account by writing the expression for K_{sp} of silver chloride in the form

$$\gamma_{Ag^+} \times \gamma_{Cl^-} \times [Ag^+] \times [Cl^-] = 1.6 \times 10^{-10} \tag{16.9}$$

The quantities γ_{Ag^+} and γ_{Cl^-} are known as **activity coefficients** for the Ag$^+$ and Cl$^-$ ions. In very dilute solution, where the forces between ions play a minor role, the activity coefficients approach unity, and we obtain the simple expression

$$[Ag^+] \times [Cl^-] = 1.6 \times 10^{-10}$$

As the total concentration of ions increases, the effect of the interionic forces becomes more significant, and the activity coefficients, whose magnitudes reflect the strength of these forces, deviate from unity. In moderately dilute solution, the activity coefficients of the ions are ordinarily less than one. In 0.04 M KNO$_3$, for example, knowing that $[Ag^+] \times [Cl^-] = 2.4 \times 10^{-10}$, it follows that $\gamma_{Ag^+} \times \gamma_{Cl^-} = 0.67$. That is,

$$\gamma_{Ag^+} \times \gamma_{Cl^-} \times 2.4 \times 10^{-10} = 1.6 \times 10^{-10}$$

$$\gamma_{Ag^+} \times \gamma_{Cl^-} = 1.6/2.4 = 0.67$$

Unfortunately, we do not have a theory which tells us exactly how the product $\gamma_{Ag^+} \times \gamma_{Cl^-}$ changes with concentration. If such a quantitative theory were available, the accuracy of predictions based upon K_{sp} or other equilibrium constants dealing with ions in solution would be greatly improved.

TABLE 16.2 SOLUBILITY OF AgCl IN KNO$_3$ SOLUTIONS

CONCENTRATION KNO$_3$	SOLUBILITY AgCl	$[Ag^+] \times [Cl^-]$
0	1.28×10^{-5}	1.64×10^{-10}
0.0010	1.33×10^{-5}	1.77×10^{-10}
0.0050	1.38×10^{-5}	1.90×10^{-10}
0.010	1.43×10^{-5}	2.04×10^{-10}
0.020	1.49×10^{-5}	2.22×10^{-10}
0.040	1.55×10^{-5}	2.40×10^{-10}

We shall be interested in using solubility products primarily to determine (1) whether or not a precipitate will form under a given set of conditions (Example 16.4) and (2) the extent to which a precipitation reaction occurs (Example 16.5).

TABLE 16.3 SOLUBILITY PRODUCTS

Acetate	$AgC_2H_3O_2$	2×10^{-3}	Iodides	AgI	1×10^{-16}
				PbI_2	1×10^{-8}
Bromides	AgBr	1×10^{-13}			
	$PbBr_2$	5×10^{-6}	Sulfates	$BaSO_4$	1.5×10^{-9}
				$CaSO_4$	3×10^{-5}
Carbonates	$BaCO_3$	1×10^{-9}		$PbSO_4$	1×10^{-8}
	$CaCO_3$	5×10^{-9}			
	$MgCO_3$	2×10^{-8}	Sulfides	Ag_2S	1×10^{-49}
	$PbCO_3$	1×10^{-13}		CdS	1×10^{-26}
				CoS	1×10^{-21}
Chlorides	AgCl	1.6×10^{-10}		CuS	1×10^{-25}
	Hg_2Cl_2	1×10^{-18}		FeS	2×10^{-17}
	$PbCl_2$	1.7×10^{-5}		HgS	1×10^{-52}
				MnS	1×10^{-13}
Chromates	Ag_2CrO_4	1×10^{-12}		NiS	1×10^{-22}
	$BaCrO_4$	2×10^{-10}		PbS	1×10^{-27}
	$PbCrO_4$	2×10^{-14}		ZnS	1×10^{-23}
Fluorides	BaF_2	2×10^{-6}			
	CaF_2	2×10^{-10}			
	PbF_2	4×10^{-8}			
Hydroxides	$Al(OH)_3$	5×10^{-33}			
	$Cr(OH)_3$	1×10^{-30}			
	$Fe(OH)_2$	1×10^{-15}			
	$Fe(OH)_3$	5×10^{-38}			
	$Mg(OH)_2$	1×10^{-11}			
	$Zn(OH)_2$	5×10^{-17}			

EXAMPLE 16.4

a. If enough $CaCl_2$ is added to a solution of Na_2SO_4 to make the concentrations of Ca^{2+} and $SO_4{}^{2-}$ 0.01 M and 0.001 M respectively, will a precipitate of $CaSO_4$ form ($K_{sp} = 3 \times 10^{-5}$)?

b. If 400 ml of 0.09 M $CaCl_2$ is added to 800 ml of 0.009 M Na_2SO_4, will a precipitate of $CaSO_4$ form?

SOLUTION

a. Let us calculate the product of the concentrations of Ca^{2+} and $SO_4{}^{2-}$ and compare to K_{sp} for $CaSO_4$, 3×10^{-5}.

conc. $Ca^{2+} \times$ conc. $SO_4{}^{2-} = (1 \times 10^{-2}) \times (1 \times 10^{-3}) = 1 \times 10^{-5}$

If the conc. product is less than K_{sp} we do not get a precipitate.

Since the concentration product is less than K_{sp} ($1 \times 10^{-5} < 3 \times 10^{-5}$), equilibrium is not established and no calcium sulfate forms.

b. This problem is completely analogous to a, except for the addition of one step. We must first calculate the concentrations of Ca^{2+} and $SO_4{}^{2-}$ in the solution formed by adding 400 ml of a solution 0.09 M in Ca^{2+} to 800 ml of a solution 0.009 M in $SO_4{}^{2-}$.

To obtain the concentration of Ca^{2+}, we note that the 400 ml of $CaCl_2$ solution has, in effect, been diluted to 1200 ml by the addition of the Na_2SO_4 solution. Consequently, the concentration of Ca^{2+} has decreased by a factor of 400/1200:

$$\text{conc. } Ca^{2+} = 0.09 \text{ M} \times \frac{400}{1200} = 0.03 \text{ M}$$

Similarly, conc. $SO_4^{2-} = 0.009 \text{ M} \times \frac{800}{1200} = 0.006 \text{ M}$

Consequently,

$(\text{conc. } Ca^{2+})(\text{conc. } SO_4^{2-}) = (3 \times 10^{-2})(6 \times 10^{-3}) = 18 \times 10^{-5}$

Since the concentration product 18×10^{-5} is greater than K_{sp} (3×10^{-5}), a precipitate will form. Solid $CaSO_4$ will precipitate until the concentrations of Ca^{2+} and SO_4^{2-} have dropped to the point at which their product has become equal to 3×10^{-5}.

If the conc. product exceeds K_{sp} a precipitate forms.

EXAMPLE 16.5. Sufficient CrO_4^{2-} ion is added to a solution originally 0.010 M in Ag^+ to make the equilibrium concentration of CrO_4^{2-} 2×10^{-3} M.
  a. What is the equilibrium concentration of Ag^+ at this point?

  b. What percentage of the Ag^+ originally present remains in solution?

SOLUTION
  a. From the equation

$$Ag_2CrO_4(s) \rightleftharpoons 2 Ag^+ + CrO_4^{2-}$$
$$K_{sp} = [Ag^+]^2 \times [CrO_4^{2-}] = 1 \times 10^{-12}$$

But $[CrO_4^{2-}]$ is given as 2×10^{-3} M.

Therefore $\quad [Ag^+]^2 (2 \times 10^{-3}) = 1 \times 10^{-12}$

$$[Ag^+]^2 = \frac{1 \times 10^{-12}}{2 \times 10^{-3}} = 5 \times 10^{-10}$$

$[Ag^+] = (5 \times 10^{-10})^{\frac{1}{2}} = 2 \times 10^{-5}$ M (1 significant figure)

  b. Originally, the concentration of Ag^+ was 1×10^{-2} M; it is now 2×10^{-5} M.

$$\% \ Ag^+ \text{ remaining} = \frac{2 \times 10^{-5}}{1 \times 10^{-2}} \times 100 = 0.2\%$$

In other words, 99.8 per cent of the Ag^+ has been precipitated as Ag_2CrO_4.

16.4 PRECIPITATION REACTIONS IN ANALYTICAL CHEMISTRY

Quantitative Analysis

In Chapter 3, precipitation reactions were cited as a method of elementary analysis. For example, one can find the percentage of chloride ion in a solid mixture by determining the weight of silver chloride produced when excess silver nitrate is added to a weighed sample of the mixture. Similarly, it is possible to analyze for barium by adding sulfate ions to precipitate barium sulfate. Analyses of this type, based on weights of products and reactants, constitute a branch of quantitative analysis known as gravimetric analysis. The calculations involved in analyses of this type are illustrated in Example 16.6.

EXAMPLE 16.6. A sample consisting of a mixture of Na_2SO_4 and $NaCl$ is analyzed by adding an excess of Ba^{2+} ions to precipitate $BaSO_4$. A 1.000 g sample of the mixture yields 0.846 g of $BaSO_4$. Calculate the percentage of Na_2SO_4 in the mixture.

SOLUTION. From the net ionic equation for the precipitation reaction

$$Ba^{2+} + SO_4^{2-} \longrightarrow BaSO_4(s)$$

it should be obvious that 1 mole of Na_2SO_4, containing 1 mole of SO_4^{2-} ions, yields 1 mole of $BaSO_4$.

$$1 \text{ mole } Na_2SO_4 \longrightarrow 1 \text{ mole } BaSO_4$$

But, since the gram formula weights of Na_2SO_4 and $BaSO_4$ are, respectively, 142.1 g and 233.4 g, it follows that

$$142.1 \text{ g } Na_2SO_4 \longrightarrow 233.4 \text{ g } BaSO_4$$

Consequently, the weight of Na_2SO_4 required to produce 0.846 g of $BaSO_4$ must be

$$0.846 \text{ g } BaSO_4 \times \frac{142.1 \text{ g } Na_2SO_4}{233.4 \text{ g } BaSO_4} = 0.515 \text{ g } Na_2SO_4$$

The percentage of Na_2SO_4 in the one-gram sample must then be

$$\frac{0.515 \text{ g}}{1.000 \text{ g}} \times 100 = 51.5\%$$

We can also determine the composition of a mixture by measuring the volume of a reagent of known concentration required to react exactly with one of the components of the mixture. Analyses based on volume measurements of this type fall in the general area of volumetric analysis. The principles involved are illustrated in Example 16.7.

EXAMPLE 16.7. A chloride sample weighing 0.208 g is found to require 15.0 ml of 0.184 M $AgNO_3$ for complete reaction. Calculate the percentage of Cl^- in the mixture.

SOLUTION. To obtain the percentage of Cl^-, we must know the weight of Cl^- in the sample. This is readily determined from the number of moles of Cl^- present, which can be calculated directly from the data:

no. moles Cl^- = no. moles Ag^+ (1 mole Cl^- reacts with 1 mole Ag^+)

$$= 0.184 \frac{\text{mole}}{\text{lit}} \times 0.0150 \text{ lit} = 0.00276 \text{ mole } Cl^-$$

$$\text{no. g Cl}^- = 0.00276 \text{ mole Cl}^- \times \frac{35.45 \text{ g Cl}^-}{1 \text{ mole Cl}^-} = 0.0978 \text{ g Cl}^-$$

$$\% \text{ Cl}^- = \frac{\text{g Cl}^-}{\text{g sample}} \times 100 = \frac{0.0978}{0.208} \times 100 = 47.0\% \text{ Cl}^-$$

In this method for the determination of chloride, as in all volumetric analyses, it is essential to know the exact point at which reaction is complete. In principle, we could do this by noting the point at which the precipitate of silver chloride stops forming. In practice, it is more convenient to add a substance known as an *indicator*, which changes color when the equivalence point is reached, that is, when chemically equivalent quantities of precipitating agent and sample are present. A few drops of potassium chromate solution added to a chloride sample serves as a suitable indicator for titration with silver ions. At the equivalence point, when essentially all of the chloride ion has been removed as silver chloride, a dark red precipitate of silver chromate, Ag_2CrO_4, forms. Since this compound requires a somewhat higher concentration of Ag^+ to precipitate than does silver chloride, it does not form, as long as an appreciable amount of chloride ion remains in solution. See Problem 16.24.

The experimental setup for the quantitative determination of Cl^- ion by this method, referred to as a Mohr titration, is shown in Figure 16.2. Silver nitrate solution of accurately known molarity is added from a buret until a permanent red color develops. Many other ions can be determined in a simi-

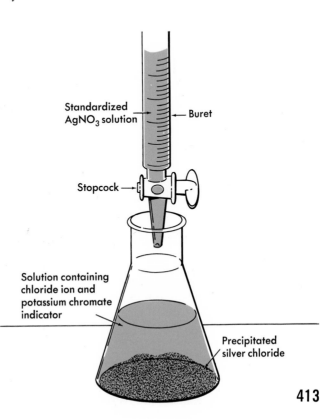

Standardized AgNO$_3$ solution — Buret

Stopcock →

Figure 16.2 Chloride titration (Mohr).

Solution containing chloride ion and potassium chromate indicator

Precipitated silver chloride

413

lar manner. Bromides and iodides as well as chlorides can be determined by titrating with Ag^+.

Qualitative Analysis

The qualitative detection of ions in a mixture is commonly accomplished by a scheme of analysis in which precipitation reactions play a major role. The general objectives of such a scheme are illustrated in Example 16.8.

EXAMPLE 16.8. Develop a scheme, based on the information given in Table 16.1, to separate and identify the ions Ag^+, Ca^{2+} and Ni^{2+} in a water solution. You may assume that no other cations are present in the solution.

SOLUTION. To carry out this analysis, we look first for a reagent that will precipitate one ion, leaving the other two in solution. Clearly the Cl^- ion is such a species, since it precipitates Ag^+ but not Ca^{2+} or Ni^{2-}. The first step in the analysis might then be the addition of a solution of sodium chloride or hydrochloric acid. Formation of a precipitate would indicate the presence of Ag^+; if no precipitate forms, Ag^+ must be absent.

Having removed any Ag^+ ions present, we next choose a reagent that will distinguish between Ca^{2+} and Ni^{2+}. Table 16.1 reveals that this can be done by adding sulfide ions (for example, a solution of sodium sulfide). This precipitates any Ni^{2+} present in the solution but has no effect on the Ca^{2+} ions.

Can you suggest an entirely different scheme based on Table 16.1?

At this point, the solution remaining need only be tested for Ca^{2+}. A convenient method of accomplishing this is to add a solution of sodium carbonate, which will precipitate Ca^{2+} as $CaCO_3$. Failure to obtain a precipitate at this point would show Ca^{2+} to be absent. In summary,

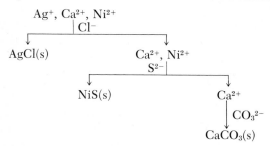

Obviously, this is not the only scheme that could be used to test for these ions. It does, however, illustrate two important points concerning precipitation analyses:

1. The order in which precipitating agents are added is crucial. It would have been useless to have added CO_3^{2-} to the original solution, since it gives a precipitate with all three cations.

2. All precipitations must be quantitative. Any Ag^+ left after the first step would interfere with the tests for Ni^{2+} and Ca^{2+}, since both Ag_2S and Ag_2CO_3 are insoluble.

Several different analytical schemes have been devised to separate and identify the 25 or more ions commonly included in a laboratory course in qualitative analysis. Needless to say, these schemes are a great deal more complex than that outlined in Example 16.8. Many other types of reactions in addition to precipitation are used. We shall consider later how acid-base reactions (Chapter 18), complex-ion formation reactions (Chapter 19), and oxidation-reduction reactions (Chapter 22) are utilized in qualitative analysis. At this stage, it may be useful to indicate the scope of qualitative analysis by giving a brief outline of one of the simpler schemes of cation analysis. The experimental procedures pertinent to this and other such schemes are described in various laboratory manuals dealing with qualitative analysis.

The cations are first separated into five major groups on the basis of the solubilities of their compounds.

Group 1. (Ag^+, Pb^{2+}, Hg_2^{2+}) Separated as the insoluble chlorides, usually by the addition of dilute hydrochloric acid. All other cations remain in solution, since their chlorides are soluble.

Group 2. (Cu^{2+}, Bi^{3+}, Cd^{2+}, Hg^{2+}, As^{3+}, Sb^{3+}, Sn^{4+}) These ions, which form extremely insoluble sulfides, are precipitated by adding S^{2-} in low concentration. This can be done by saturating with hydrogen sulfide the acidic solution remaining after precipitation of Group 1.

Group 3. (Al^{3+}, Cr^{3+}, Fe^{3+}, Zn^{2+}, Ni^{2+}, Co^{2+}, Mn^{2+}) The sulfides of these metals are somewhat more soluble than those in Group 2; the ions may be precipitated by adding H_2S in basic solution. Al^{3+} and Cr^{3+} come down as the hydroxides, the other ions as the sulfides.

Group 4. (Mg^{2+}, Ca^{2+}, Sr^{2+}, Ba^{2+}) May be precipitated as the carbonates.

Group 5. (Na^+, K^+, NH_4^+) Special reagents are ordinarily used to identify these ions. In the case of K^+, the violet color imparted to a flame by potassium salts is a useful test.

From the equilibrium $H_2S \rightleftharpoons 2H^+ + S^{2-}$, we deduce that the conc. of H^+ and S^{2-} must be inversely related.

Once the cations have been separated into groups, the problem becomes one of further subdividing the groups. We shall not attempt at this point to describe the chemistry involved in these separations. A single example may, however, be instructive. Lead can be separated from the other group 1 ions by boiling the chloride precipitate with water, filtering the hot mixture, and testing the hot filtrate with CrO_4^{2-}. The formation of a yellow precipitate of lead chromate, $PbCrO_4$, which is much less soluble than $PbCl_2$, indicates the presence of lead.

$$PbCl_2(s) \longrightarrow Pb^{2+} + 2\ Cl^-$$
$$\underline{Pb^{2+} + CrO_4^{2-} \longrightarrow PbCrO_4(s)}$$
$$PbCl_2(s) + CrO_4^{2-} \longrightarrow 2\ Cl^- + PbCrO_4(s) \qquad (16.10)$$

16.5 PRECIPITATION REACTIONS IN INORGANIC PREPARATIONS

Precipitation reactions serve as a convenient source of insoluble ionic compounds. Barium sulfate, for example, is readily prepared by the addition of a slight excess of a solution containing sulfate ions, SO_4^{2-}, to a solution of a barium salt. The precipitate is filtered, washed to remove foreign ions, and dried at 100°C to give a product of high purity. Similar techniques can be used to' convert silver nitrate to any one of the insoluble halides of silver, AgCl, AgBr, or AgI.

In addition to their use in preparing insoluble inorganic compounds, precipitation reactions form the basis of a general method of converting one type of soluble ionic compound to another. Let us suppose, for example, that we desire to convert a soluble metal chloride to the corresponding nitrate. If we add silver nitrate to a solution of the chloride, a precipitate of AgCl will form; Cl⁻ ions in solution will be replaced by NO₃⁻ ions. If just enough Ag⁺ ions are added to react with all the Cl⁻ ions originally present, the solution remaining after the AgCl is filtered off will contain only two kinds of ions; the metal ions originally present and the NO₃⁻ ions introduced with the silver nitrate.

$$M^+ + Cl^- + Ag^+ + NO_3^- \longrightarrow AgCl(s) + M^+ + NO_3^- \qquad \text{(16.11)}$$

Evaporation of this solution gives the pure metal nitrate:

$$M^+ + NO_3^- \longrightarrow MNO_3(s) \qquad \text{(16.12)}$$

Quite clearly, the quantities of reagents used in preparations of this type are critical. If too little silver nitrate is added as a precipitant, some of the chloride ions will remain in solution; evaporation will give a mixture of the metal nitrate and chloride. If too much precipitant is added, the product will be contaminated with the excess silver nitrate. Only if exactly equivalent quantities of the two solutions are mixed will a pure product be obtained on evaporation.

The method of preparation just described is applicable to a wide variety of inorganic syntheses. Metal bromides and iodides as well as chlorides are readily converted to the corresponding nitrates, since AgBr and AgI, like AgCl, are insoluble in water. We can convert soluble sulfates to other salts by taking advantage of the insolubility of BaSO₄. For example, we can prepare cesium chloride, CsCl, from cesium sulfate by adding an equivalent amount of barium chloride, filtering off the precipitated BaSO₄, and evaporating:

This equation, like 16.11, is not a net ionic equation.

precipitation:

$$2 \; Cs^+ + SO_4^{2-} + Ba^{2+} + 2 \; Cl^- \longrightarrow BaSO_4(s) + 2 \; Cs^+ + 2 \; Cl^- \qquad \text{(16.13)}$$

evaporation: $\qquad\qquad 2 \; Cs^+ + 2 \; Cl^- \longrightarrow 2 \; CsCl(s) \qquad\qquad\qquad \text{(16.14)}$

16.6 WATER SOFTENING

A great many important industrial processes involve a precipitation reaction as an essential step. One of these is the "softening" of water. "Hardness" in water is brought about by certain cations, notably Ca²⁺ and Mg²⁺. Hard water has many undesirable characteristics. Among these is its tendency to form a precipitate with soap, which we recognize as a slimy bathtub deposit. The most common type of soap has as its major ingredient sodium stearate, NaC₁₈H₃₅O₂, the sodium salt of an organic acid. The calcium and magnesium salts of this acid are insoluble in water; the reaction taking place when soap is used in hard water may be represented as

$$M^{2+} + 2 \; C_{18}H_{35}O_2^- \longrightarrow M(C_{18}H_{35}O_2)_2(s) \qquad M = Mg \quad \text{or} \quad Ca \quad \text{(16.15)}$$

This reaction continues until nearly all the Ca^{2+} or Mg^{2+} ions are removed from the water; only then does soap become effective as a cleansing agent.

If the only disadvantage of hard water were its tendency to form a precipitate with soap, hardness would be a relatively minor problem. Synthetic detergents do not form a precipitate with Ca^{2+} and Mg^{2+} ions and hence are more effective sudsing agents than soap in hard water. From an industrial standpoint, a more serious disadvantage of hard water is its tendency to form precipitates when heated or partially evaporated, as in a boiler. The nature of the precipitate depends upon the anions present in the hard water. One of the most common anions in surface water is the HCO_3^- ion, formed by the reaction of atmospheric carbon dioxide with water. When a solution containing Ca^{2+} and HCO_3^- ions is heated, a precipitate of calcium carbonate forms as a result of the reaction sequence

$$2\ HCO_3^- + \text{heat} \longrightarrow CO_3^{2-} + CO_2(g) + H_2O$$
$$\underline{Ca^{2+} + CO_3^{2-} \longrightarrow CaCO_3(s)}$$
$$2\ HCO_3^- + Ca^{2+} \longrightarrow CaCO_3(s) + CO_2(g) + H_2O \qquad (16.16)$$

The scale formed on tea kettles in the home is frequently of this type; it can be removed without too much trouble if treated with a weakly acidic solution such as vinegar.

More serious problems arise if significant quantities of sulfate ion are present in hard water. In this case, heating may produce a precipitate of calcium sulfate, which, curiously enough, is less soluble at high than at low temperatures:

$$Ca^{2+} + SO_4^{2-} \longrightarrow CaSO_4(s) \qquad (16.17)$$

Crystals of calcium sulfate are particularly likely to form in a steam boiler in which water is constantly being vaporized, thereby increasing the concentrations of Ca^{2+} and SO_4^{2-} ions. The calcium sulfate deposits as a tightly adherent scale which lowers the heat conductivity of the boiler. Eventually, the deposition of calcium sulfate in boiler tubes may block the flow of water and lead to rupture of the pipes.

The various methods used to soften water have as their common objective the removal of Ca^{2+} and, to a lesser extent, Mg^{2+} ions. We shall consider two methods of water softening.

Lime-Soda Method

One of the oldest methods of softening water, still used in a great many localities, involves the addition of two chemicals, slaked lime, $Ca(OH)_2$, and soda ash, Na_2CO_3. The purpose of adding the sodium carbonate is rather obvious; it acts as a source of carbonate ions for the removal of Ca^{2+} ions through the reaction

$$Ca^{2+} + CO_3^{2-} \longrightarrow CaCO_3(s) \qquad (16.18)$$

The function served by the lime is less apparent. Indeed, it might seem that the addition of $Ca(OH)_2$ to water in an attempt to remove Ca^{2+} ions would be a step in the wrong direction. As a matter of fact, lime is an effective water softener only when there are HCO_3^- ions in the hard water. To understand

417

why this is the case, consider what happens when one mole of $Ca(OH)_2$ is added to water containing two moles of HCO_3^- ions:

dissolving of $Ca(OH)_2$:
$$Ca(OH)_2(s) \longrightarrow Ca^{2+} + 2\,OH^-$$

conversion of
$HCO_3^- \longrightarrow CO_3^{2-}$:
$$2\,HCO_3^- + 2\,OH^- \longrightarrow 2\,CO_3^{2-} + 2\,H_2O$$

precipitation of
added Ca^{2+}:
$$Ca^{2+} + CO_3^{2-} \longrightarrow CaCO_3(s)$$

overall reaction:
$$Ca(OH)_2(s) + 2\,HCO_3^- \longrightarrow CaCO_3(s) + CO_3^{2-} + 2\,H_2O$$
$$(16.19)$$

We note from this reaction sequence that the lime furnishes the OH^- ions required to convert HCO_3^- ions in the water to CO_3^{2-} ions. The net effect of the addition of the lime, as represented by Equation 16.19, is to form an extra mole of free CO_3^{2-} ions. These ions are then capable of removing a mole of Ca^{2+} ions originally present in the hard water. Thus, for every mole of $Ca(OH)_2$ added, an extra mole of Ca^{2+} ions is removed from the hard water. It is, of course, extremely important not to add too much $Ca(OH)_2$ to the water; any excess over that required for Reaction 16.19 will tend to increase the Ca^{2+} ion concentration in the water.

The procedure used in softening hard water may be summarized as follows:

Explain, in your own words, why $Ca(OH)_2$ is added in this ratio.

1. The water is first analyzed for Ca^{2+} and HCO_3^- ions.
2. Sufficient lime is added to give one mole of $Ca(OH)_2$ for every two moles of HCO_3^-. Every mole of $Ca(OH)_2$ added removes one mole of Ca^{2+} from the water.
3. Any Ca^{2+} ions remaining in the hard water are removed by adding soda ash, Na_2CO_3, in a 1:1 mole ratio.

EXAMPLE 16.9. A sample of water is found on analysis to contain 0.0030 mole/lit of Ca^{2+}, 0.0040 mole/lit of HCO_3^- and 0.0010 mole/lit of SO_4^{2-}. Calculate the number of moles of $Ca(OH)_2$ and Na_2CO_3 that should be added to one liter of this water to soften it.

SOLUTION. The first step is to add 1 mole of $Ca(OH)_2$ for every 2 moles of HCO_3^- present:

no. moles $Ca(OH)_2 = 0.0040/2 = 0.0020$ mole $Ca(OH)_2$

This removes an equal amount, 0.0020 mole, of Ca^{2+} from the water, leaving in one liter of the water

$(0.0030 - 0.0020)$ mole $= 0.0010$ mole of Ca^{2+}

Consequently, it is necessary to add 0.0010 mole of Na_2CO_3 to the water.

Ion Exchange

The lime-soda method of water softening has gradually lost favor because of its many economic disadvantages. Before treatment the water must be analyzed for Ca^{2+} and HCO_3^- ions; the amounts of lime and soda added must be carefully controlled, and the calcium carbonate formed must be removed from the water by filtration or settling. All of these processes are time-consuming and expensive.

The so-called *ion exchange* method of water softening suffers from none of these drawbacks. The water need not be analyzed, reagents need not be added in specified amounts, and there is no precipitate to be removed. Moreover, the method is applicable to large or small-scale operations; ion exchangers can be used to treat the water supply of a home, a factory, or an entire city.

The ion-exchange process makes use of a class of substances known collectively as zeolites. A typical zeolite particle, which looks much like a grain of sand, contains the four elements sodium, silicon, aluminum, and oxygen. The atoms of aluminum, silicon, and oxygen are bonded into a vast anionic network similar in structure to such macromolecular solids as quartz. The negative charge of the network is compensated for by a large number of Na^+ ions, located in "corridors" or "cages" within the anionic lattice. Many zeolites of different and rather complicated empirical formulas ($NaAlSi_3O_8$, $NaAlSi_2O_6$, and so on) occur in nature; others have been synthesized in the laboratory.

If pure water is passed through a zeolite column, nothing spectacular happens. The crystals themselves contain far too many atoms to dissolve in water; the small Na^+ ions cannot dissolve since this would leave behind an

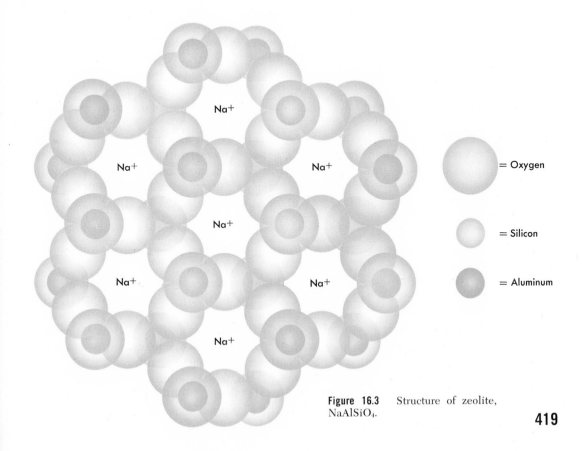

Figure 16.3 Structure of zeolite, $NaAlSiO_4$.

419

unstable anionic network carrying a large negative charge. However, if the water coming in contact with the zeolite contains Ca^{2+} ions, the situation is quite different. One Ca^{2+} ion replaces two Na^+ ions in the lattice. The reaction that takes place may be represented by the equation

$$Ca^{2+} + 2\ NaZ(s) \rightleftharpoons CaZ_2(s) + 2\ Na^+ \tag{16.20}$$

in which Z represents a small portion of the anionic lattice of the zeolite. The net effect of this process is the replacement of Ca^{2+} ions responsible for hardness with unobjectionable Na^+ ions. This, of course, is precisely what happens in the lime-soda process when Na_2CO_3 is added to hard water; the difference is that in the ion-exchange process there is no precipitate to filter off since the Ca^{2+} ions are incorporated into the zeolite lattice.

Since the equilibrium constant for Reaction 16.20 is comparatively small (usually less than ten), it is ordinarily necessary to pass the water through a column several feet long to achieve a quantitative replacement of Ca^{2+} ions by Na^+ ions. Moreover, the column slowly loses its effectiveness as the Na^+ ions in the network are replaced by Ca^{2+} ions. To rejuvenate the column, Reaction 16.20 is reversed by passing a concentrated solution of sodium chloride through the column. With a properly designed column, this is the only type of maintenance required—an occasional flushing with saturated brine.

The type of ion-exchange column we have described, which replaces positive ions in water with Na^+ ions, is referred to as a cation exchanger. Zeolites can be developed to replace positive ions with any desired cation. For example, if the zeolite framework contains K^+ ions, these ions will be substituted for the cations in the water passed through the column. By the proper use of cation-exchangers, it is possible to convert salts of one metal to the corresponding salts of a different metal. If one runs a solution of NaCl through a zeolite containing K^+ ions, a solution of KCl results. By reversing the procedure, i.e., running a KCl solution through a sodium ion exchanger, potassium chloride can be converted to sodium chloride.

Synthetic zeolites have been developed for anion exchange. A column packed with particles in which Cl^- ions are trapped in a cationic network can be used to substitute Cl^- ions for the anions originally present in a water solution. Suppose, for example, that a solution containing nitrate ions is passed through a chloride exchanger. As a result of the reaction

How might one convert $AgNO_3$ to AgF? $CaCl_2$ to $NiCl_2$?

$$NO_3^- + RCl(s) \longrightarrow RNO_3(s) + Cl^- \quad (R = \text{portion of cationic network.}) \tag{16.21}$$

Cl^- ions are substituted for NO_3^- ions and a metal nitrate is transformed into the corresponding chloride.

By combining cationic and anionic exchangers containing special types of synthetic ion-exchange resins, it is possible to develop an apparatus known as a **deionizer**, which completely removes the ions from a sample of water. The water is first passed through a column containing a cation exchanger, HZ, which replaces cations with H^+ ions. The water leaving this column is then pumped through a second column filled with an anion exchange resin whose formula may be written as ROH. In this column, the negative ions present in the water are replaced by OH^- ions. The hydroxide ions produced by the anion exchanger react with the hydrogen ions formed in the first column to give water. If a solution of sodium chloride is passed through a two-stage exchanger of this type, the reactions taking place may be represented as

cation exchange: $$Na^+ + HZ(s) \longrightarrow NaZ(s) + H^+$$

anion exchange: $$Cl^- + ROH(s) \longrightarrow RCl(s) + OH^-$$

$$H^+ + OH^- \longrightarrow H_2O$$

overall reaction: $$Na^+ + Cl^- + HZ(s) + ROH(s) \longrightarrow NaZ(s) + RCl(s) + H_2O$$

(16.22)

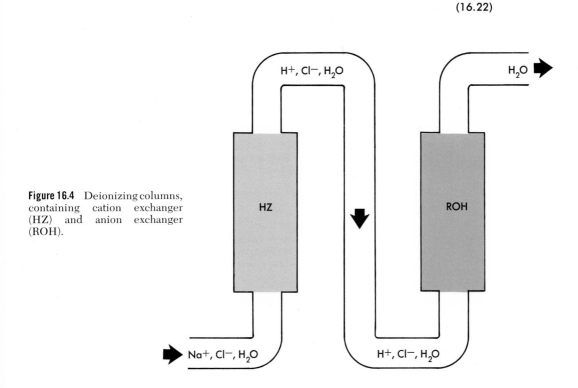

Figure 16.4 Deionizing columns, containing cation exchanger (HZ) and anion exchanger (ROH).

Such a combination of ion-exchange columns produces water comparable in purity to distilled water. Indeed, such deionizers have replaced stills in a great many teaching and industrial laboratories. It is necessary to distill water for certain biological or medicinal applications to remove nonionic impurities which sometimes cause fever ("pyrogenic substances").

PROBLEMS

16.1 Write net ionic equations for the reactions, if any, that occur when 0.1 M solutions of the following ionic compounds are mixed.

a. $Pb(NO_3)_2$ and $NaCl$
b. $Al(NO_3)_3$ and $NaOH$
c. $NiSO_4$ and KCl

d. $NiSO_4$ and BaS
e. $Ca(NO_3)_2$ and NaF
f. KNO_3 and $SrCl_2$

16.2 200 ml of 0.20 M $AgNO_3$ solution is added to 300 ml of 0.10 M Na_2CrO_4.

a. Calculate the number of moles of each ion present before the solutions are mixed.

421

b. Calculate the number of moles of Ag_2CrO_4 formed.

c. Calculate the number of moles and concentration of each ion remaining in solution.

16.3 Given the following solubilities, find the corresponding values of K_{sp}.

a. $MgCO_3$: 1.4×10^{-4} mole/lit   b. Ag_2CrO_4: 6×10^{-5} mole/lit

16.4 From the K_{sp} values given in Table 16.3, determine the solubility, in moles/liter of

a. $CaCO_3$ b. PbF_2

16.5 State whether or not precipitates will form under the following conditions:

a. Enough Ca^{2+} is added to a solution 1×10^{-4} M in F^- to make the concentration of Ca^{2+} 0.1 M.

b. 0.10 M $Pb(NO_3)_2$ is added to tap water in which the concentration of Cl^- is 0.001 M.

c. Equal volumes of 0.008 M solutions of $Ca(NO_2)_2$ and Na_2SO_4 are mixed.

16.6 If Ba^{2+} is added to a solution of 0.1 M in F^-,

a. At what concentration of Ba^{2+} does a precipitate start to form?

b. When the concentration of Ba^{2+} is 0.001 M, what per cent of the F^- remains in solution?

16.7 A solution is 0.10 M in Br^- and 0.01 M in CrO_4^{2-}. If Ag^+ ions are slowly added,

a. What is the precipitate that forms first?

b. What is the concentration of the first ion when the second ion starts to precipitate?

c. On the basis of your answers to a and b, suggest why CrO_4^{2-} is a suitable indicator for the titration of Br^- with Ag^+.

16.8 How many moles of AgCl will dissolve in two liters of 0.10 M NaCl solution?

16.9 Neither of the following statements is entirely valid in the form given. Explain why this is the case and rephrase the statements to make them correct.

a. Of two slightly soluble electrolytes, the one with the larger K_{sp} value is the more soluble in water.

b. In dilute solution, the product (conc. Ag^+) × (conc. Cl^-) is approximately equal to 1.6×10^{-10}.

16.10 A student analyzes a solid mixture for Ba^{2+} ions by adding excess Na_2SO_4 solution and weighing the precipitated $BaSO_4$. He finds that a 1.200 g sample yields 0.900 g of $BaSO_4$. What is the per cent of Ba^{2+} in the mixture?

16.11 A student determines the per cent of Cl^- in a mixture by weighing out 0.500 g, dissolving in 20 cc of water, and titrating with 24.0 ml of 0.150 M $AgNO_3$ solution. What is the per cent of Cl^- in the mixture?

16.12 Develop a scheme, based on Table 16.1, for analyzing qualitatively for the following ions.

a. Ca^{2+}, Cu^{2+}, Pb^{2+} b. I^-, SO_4^{2-}, F^-

16.13 Before analyzing a Group 1 unknown, a student runs through the separation scheme on a solution prepared by mixing equal volumes of 0.10 M $AgNO_3$, $Pb(NO_3)_2$, and $Hg_2(NO_3)_2$. To precipitate the Group 1 cations, he adds enough HCl to give a Cl^- ion concentration of 0.02 M. Later on in the analysis, he obtains positive tests for Ag^+ and Hg_2^{2+} but not for Pb^{2+}. Explain what happened, using the K_{sp} values for AgCl, $PbCl_2$, and Hg_2Cl_2.

16.14 Describe, in some detail, how the following conversions could be accomplished by means of precipitation reactions.

a. $SrCl_2 \longrightarrow Sr(NO_3)_2$ b. $Na_2SO_4 \longrightarrow NaCl$

c. $KF \longrightarrow KCl$ d. $KCl \longrightarrow KF$

16.15 Water-softening consists essentially of the removal of Ca^{2+} and Mg^{2+}. Yet, in the lime-soda process, more Ca^{2+} ions are added to the water in the form of $Ca(OH)_2$. Explain why this is done.

16.16 A water supply contains the following ions in the indicated concentrations. Determine the number of moles of Na_2CO_3 and $Ca(OH)_2$ that must be added to one liter of the water to soften it most economically.

    a. 2.0×10^{-4} M Ca^{2+}, 4.0×10^{-4} M HCO_3^-
    b. 2.0×10^{-4} M Ca^{2+}, 4.0×10^{-4} M Cl^-
    c. 1.5×10^{-4} M Ca^{2+}, 1.0×10^{-4} M HCO_3^-, 2.0×10^{-4} M Cl^-

16.17 Write net ionic equations for the reactions, if any, that occur when 0.1 M solutions of the following ionic compounds are mixed.

    a. $AgNO_3$ and KCl f. $SbCl_3$ and Na_2S
    b. $Pb(NO_3)_2$ and Na_2SO_4 g. $CuSO_4$ and KOH
    c. KNO_3 and $NaClO_3$ h. Na_2CO_3 and $Ba(NO_3)_2$
    d. NH_4Cl and $Hg_2(NO_3)_2$ i. $ZnSO_4$ and BaS
    e. $FeCl_3$ and NaF j. $Pb(NO_3)_2$ and KI

16.18 Calculate the number of moles of $Fe(OH)_2$ formed and the number of moles of each ion remaining in solution when 1.20 lit of 0.380 M $FeCl_2$ is added to 1.50 lit of 0.460 M $NaOH$.

16.19 Calculate the number of moles and concentration of each ion remaining in solution when 212 cc of 0.352 M KOH is mixed with 616 cc of 0.219 M $NaCl$ solution.

16.20 Given the following solubilities, calculate the corresponding values of K_{sp}.

    a. $BaCO_3$: 3×10^{-5} mole/lit   b. SrF_2: 1.0×10^{-3} mole/lit

16.21 From the K_{sp} values given in Table 16.3, calculate

    a. The solubility of $Mg(OH)_2$, in moles per liter.
    b. The solubility of $PbCO_3$, in grams per liter.

16.22 State whether or not precipitates will form under the following conditions:

    a. 0.10 M $AgNO_3$ is added to tap water in which the concentration of Cl^- is 1.0×10^{-5} M.
    b. Enough $AgNO_3$ is added to a 0.10 M solution of Na_2CrO_4 to make $[Ag^+] = 1 \times 10^{-3}$ M.
    c. 100 ml of 0.10 M $Pb(NO_3)_2$ is added to 900 ml of 0.010 M $NaCl$.
    d. 10 ml of 0.010 M $Ca(NO_3)_2$ is added to 40 ml of 0.010 M Na_2SO_4.

16.23 If Ag^+ is slowly added to a solution 0.10 M in Na_2CrO_4,

    a. At what concentration of Ag^+ does a precipitate start to form?
    b. When $[Ag^+] = 1 \times 10^{-4}$ M, what percentage of the CrO_4^{2-} is left in solution?
    c. What concentration of Ag^+ will be required to precipitate 99% of the CrO_4^{2-}?

16.24 A solution is 0.10 M in Cl^- and 0.010 M in CrO_4^{2-}. If Ag^+ ions are slowly added,

    a. What precipitate forms first?
    b. What is the concentration of the first ion when the second ion starts to precipitate?
On the basis of your answers to a and b, suggest why CrO_4^{2-} ion might serve as a suitable indicator for the titration of Cl^- with Ag^+.

16.25 How many moles of $AgBr$ will dissolve in 5 lit of 0.02 M KBr solution?

16.26 A student analyzes a solid mixture for S^{2-} ions by adding excess $Cu(NO_3)_2$ solution and weighing the precipitated CuS. He finds that a 0.500 g sample yields 0.618 g of CuS. Find the percentage of S^{2-} in the mixture.

423

16.27 A student determines the percentage of Br^- in a mixture by weighing out 0.582 g, dissolving in 15 cc of water, and titrating with 0.154 M $AgNO_3$ solution. 32.0 ml of this solution are required. What is the percentage of Br^- in the mixture?

16.28 Develop a scheme, based on Table 16.1, for analyzing qualitatively for the following ions (cf. Example 16.8).

a. Ag^+, Ca^{2+}, Na^+ b. Pb^{2+}, Ni^{2+}, Mg^{2+} c. Cl^-, SO_4^{2-}

16.29 Explain why, in the scheme of cation analysis described in Section 16.4, Pb^{2+} often precipitates in Group 2, even though an excess of HCl has been added to precipitate the Group 1 ions.

16.30 Describe in some detail how the following conversions could be accomplished by means of precipitation reactions.

a. $RbCl \rightarrow RbNO_3$ b. $CuSO_4 \rightarrow CuCl_2$ c. $LiCl \rightarrow LiClO_3$ d. $RaI_2 \rightarrow RaCl_2$

16.31 Write balanced equations for

a. The removal of Mg^{2+} salts from water, using the zeolite NaZ.
b. The removal of dissolved $CaCl_2$ by passing water through successive columns containing an anion exchanger (ROH) and a cation exchanger (HZ).
c. The regeneration of a cation exchange column packed with a resin of formula HZ which has been used to remove K^+ salts from water.

16.32 A municipal water supply contains the following ions in the indicated concentrations. Calculate the number of grams of Na_2CO_3 and $Ca(OH)_2$ that must be added to ten liters of the water to soften it most economically.

a. 1.0×10^{-3} M Ca^{2+} and 2.0×10^{-3} M HCO_3^-
b. 1.0×10^{-3} M Ca^{2+}, 1.0×10^{-3} M HCO_3^-, 1.0×10^{-3} M Cl^-
c. 2.0×10^{-4} M Ca^{2+}, 2.0×10^{-4} M SO_4^{2-}

°16.33 A certain zeolite column is designed to remove 10 per cent of the Ca^{2+} ions in hard water per foot length of column. If the water originally contains a Ca^{2+} concentration of 1.0×10^{-3} M, through how long a column must it pass to lower the $[Ca^{2+}]$ to 5.0×10^{-4} M?

°16.34 In testing for Group 1 cations, a student works with a precipitate containing approximately one-tenth of a gram of $PbCl_2$.

a. How much water must be added to this precipitate if it is to be completely dissolved by heating to 100°C (solubility $PbCl_2$ at 100°C = 33 g/lit)?
b. What volume of 0.10 M Na_2CrO_4 solution must be added to the solution formed in a to precipitate 99% of the Pb^{2+}?

°16.35 A student finds that 20.0 ml of 0.200 M $AgNO_3$ is required to react with a solid sample weighing 0.300 g. The sample consists of a mixture of NaCl and NaBr. Calculate the per cent by weight of NaCl.

°16.36 According to the Debye-Hückel theory, the product of the activity coefficients of Ag^+ and Cl^-, $\gamma_+ \gamma_-$, in a solution of a 1:1 electrolyte of molarity M should be given by the expression

$$\log_{10}(\gamma_+ \gamma_-) = -1.0 \ M^{\frac{1}{2}}$$

Using this equation, and taking K_{sp} for AgCl to be 1.64×10^{-10}, determine the solubility of AgCl in KNO_3 solutions of the concentrations listed in Table 16.2 and compare to the observed values.

Acids and Bases

In earlier chapters, we have referred to the properties and structures of acids and bases. In Chapter 11, for example, the electrical conductivities of water solutions of acids and bases were touched on briefly.

We shall now expand the discussion of acids and bases, paying particular attention to the properties of acidic and basic water solutions, the nature of the ions responsible for these properties, and the methods by which such solutions are formed. We shall find that quantitative study of the equilibria involved in acidic and basic water solutions enables us to distinguish between strong and weak acids or bases. Furthermore, from structural considerations we can predict the relative acidic or basic strengths of different compounds.

17.1 PROPERTIES OF ACIDIC AND BASIC WATER SOLUTIONS

Acidic solutions may be identified in various ways. Their characteristic sour taste reveals itself in vinegar, lemon juice, and rhubarb pie. It has been reliably reported that solutions of nitric, sulfuric, and hydrochloric acids also have a sour taste. A safer way to determine whether or not a solution is acidic is to test its behavior with zinc or magnesium; when added to acidic water solutions, these metals evolve hydrogen. A more sensitive test involves the addition of carbonate ions to the solution; sodium carbonate or calcium carbonate, in the presence of an acid, react to give off bubbles of carbon dioxide.

Perhaps the simplest experimental test for acidity utilizes the fact that certain organic dyes change color when placed in acidic solution. One of the most familiar of these acid-base indicators is litmus, which turns red in acidic solution. Interestingly enough, acidic solutions bring about the same color change in grape juice. For various reasons, litmus is more widely used than grape juice to test for acidity in the general chemistry laboratory.

Basic water solutions, like acidic solutions, possess certain characteristic properties. Anyone who uses liquid detergents or household ammonia in cleaning is aware of the soapy feel of these solutions. This property of basic solutions, which results from dissolving a surface layer of skin, is possessed to an even greater extent by a solution of lye, sodium hydroxide. A more satisfactory way of finding out whether or not a water solution is basic is to add to it a small amount of a solution of a magnesium salt such as $MgCl_2$. The formation of a white precipitate of magnesium hydroxide indicates the presence of a base.

The color of indicators is affected by basic as well as acidic solutions; litmus turns blue in a basic solution. The organic dye phenolphthalein, which is colorless in acid or neutral solution, takes on a bright red color in the presence of a base.

17.2 IONS PRESENT IN ACIDIC AND BASIC WATER SOLUTIONS

Basic Solution: OH^-

The characteristic properties of basic water solutions are those of a specific ion, common to all such solutions, the hydroxide ion, OH^-. It is this species which is responsible for the soapy feel, the effect on indicators, and the formation of a precipitate with magnesium salts.

$$Mg^{2+} + 2\ OH^- \longrightarrow Mg(OH)_2(s) \tag{17.1}$$

Acidic Solution: the Hydrated Proton or Hydronium Ion

The ion responsible for the properties common to acidic solutions may be described most simply as a proton or H^+ ion. It is this ion which accounts for the reaction of acidic water solutions with zinc or with carbonate ions.

$$Zn(s) + 2\ H^+ \longrightarrow Zn^{2+} + H_2(g) \tag{17.2}$$

$$CO_3{}^{2-} + 2\ H^+ \longrightarrow CO_2(g) + H_2O \tag{17.3}$$

The sour taste of acidic solutions and their effect on the color of indicators can also be attributed to the presence of protons.

The proton, like all other positive ions of high charge density, is hydrated in water solution. Various experiments designed to determine the number of water molecules associated with a proton give hydrations numbers ranging from 1 to 25. The simplest hydrated species would be the **hydronium ion**, H_3O^+, in which the proton is joined by a coordinate covalent bond to the oxygen atom of a water molecule:

The hydronium ion bears a close resemblance to the previously discussed ammonium ion, NH_4^+. Both are derived from a neutral molecule (H_2O, NH_3) by the addition of a proton. The existence of the hydronium ion has been demonstrated in the solid $HClO_4 \cdot H_2O$, which is made up of H_3O^+ and ClO_4^- ions.

In discussing the properties of acidic water solutions, we shall often find it convenient to ignore the hydration of the proton and write it as the simple, unhydrated ion, H^+. On the other hand, in writing equations to describe the way in which acidic solutions are formed, we shall often find it helpful to write the hydrated proton as H_3O^+.

Equilibrium Between H+ and OH−: Concept of K_w

In any water solution, regardless of what other species may be present, there will always be an equilibrium between H_2O molecules, H^+ ions and OH^- ions expressed most simply as

$$H_2O \rightleftharpoons H^+ + OH^- \tag{17.4}$$

The corresponding equilibrium constant takes the form

$$K = \frac{[H^+] \times [OH^-]}{[H_2O]} \quad \text{or} \quad K \times [H_2O] = [H^+] \times [OH^-] \tag{17.5}$$

Equation 17.5 may be simplified by taking account of the fact that in any dilute water solution, the concentration of water molecules greatly exceeds that of any other species and hence remains nearly constant. In other words, the entire left-hand side of Equation 17.5 is a constant, independent of the concentration of H^+ or OH^- ions, and we may write

$[H_2O] \approx 56$ moles/lit

$$K_w = [H^+] \times [OH^-]$$

in which K_w, referred to as the ionization constant of water, is written in place of the expression $K \times [H_2O]$ in Equation 17.5.

The numerical value of the ionization constant of water may be determined experimentally in any of several ways. At 25°C, it has the value 1.0×10^{-14}; that is,

$$K_w = [H^+] \times [OH^-] = 1.0 \times 10^{-14} \tag{17.6}$$

Equation 17.6 indicates that in any water solution the concentrations of H^+ and OH^- are inversely proportional to each other. Accordingly, a 100-fold increase in the concentration of one of these ions produces a 100-fold decrease in the concentration of the other (Table 17.1).

It is possible to distinguish between neutral, basic, and acidic solutions on the basis of the relative concentrations of H^+ and OH^-. A **neutral solution** may be defined as one in which the concentrations of these two ions are equal:

$$[H^+] = [OH^-] = (10^{-14})^{\frac{1}{2}} = 10^{-7} \text{ M}$$

Chemically pure water, of course, is exactly neutral.

An **acidic solution** may be defined as one in which the concentration of H^+ ions exceeds that of OH^- ions:

$$[H^+] > [OH^-], \qquad [H^+] > 10^{-7} \text{ M}, \qquad [OH^-] < 10^{-7} \text{ M}$$

The three solutions listed at the left of Table 17.1 are acidic; in each of these solutions, litmus takes on a red color.

In **basic solution** there are more OH^- ions than H^+ ions:

$$[OH^-] > [H^+], \qquad [OH^-] > 10^{-7} \text{ M}, \qquad [H^+] < 10^{-7} \text{ M}$$

427

TABLE 17.1

SOLUTION	No. 1	No. 2	No. 3	No. 4	No. 5	No. 6	No. 7
H^+	10^{-1}	10^{-3}	10^{-5}	10^{-7}	10^{-9}	10^{-11}	10^{-13}
OH^-	10^{-13}	10^{-11}	10^{-9}	10^{-7}	10^{-5}	10^{-3}	10^{-1}
pH	1	3	5	7	9	11	13

Solutions 5 to 7 in Table 17.1 are basic; they show to a steadily increasing extent the properties characteristic of basic water solutions.

17.3 pH

We have seen that it is possible to describe quantitatively the acidity or basicity of a solution by specifying the concentration of H^+ ion. In 1909 Sörenson proposed an alternative method of accomplishing this purpose, making use of a term known as pH, defined as follows:

$$pH \equiv -\log_{10} [H^+] = \log_{10} 1/[H^+] \tag{17.7}$$

A great many important solutions have a pH between 1 and 14, corresponding to a variation in hydrogen ion concentration of 10^{-1} to 10^{-14} M. For such solutions, it is perhaps more convenient to express acidity in terms of pH rather than H^+ ion concentration, thereby avoiding the use of either small fractions or negative exponents.

The term $pOH = -\log_{10}[OH^-]$ is also used.

It follows from the definition that the lower the pH, the more acidic a solution is. A solution of pH 1 has an H^+ ion concentration 100 times greater than one of pH 3. Conversely, a solution of pH 13 has an OH^- ion concentration 100 times greater than one of pH 11 (Table 17.1). Just as it is possible to differentiate between neutral, basic, and acidic solutions on the basis of the concentrations of H^+ or OH^- ions, so one can make this distinction in terms of pH:

neutral solution: $[H^+] = 10^{-7}$ M, pH = 7.0

acidic solution: $[H^+] > 10^{-7}$ M, pH < 7.0

basic solution: $[H^+] < 10^{-7}$ M, pH > 7.0

Example 17.1 illustrates the calculation of the pH of water solutions and the interrelationships between pH, $[H^+]$, and $[OH^-]$.

TABLE 17.2 pH OF COMMON MATERIALS

Vinegar	2.2	Human urine	7.4
Apples	3	Sea water	8.3
Sauerkraut	3.5	0.1 M $NaHCO_3$	8.4
Tomatoes	4.2	Satd $Mg(OH)_2$	10.5
Carrots	5	0.1 M NH_3	11.2
Saliva	5.7–7.1	0.1 M Na_2CO_3	11.6
Milk	6.6	Satd $Ca(OH)_2$	12.1

EXAMPLE 17.1. Calculate

   a. The pH of a solution 0.020 M in H^+.

   b. The pH of a solution 2.5×10^{-3} M in OH^-.

   c. The $[H^+]$ in sea water.

SOLUTION. Before working this problem, you may wish to refer to the material on the use of logarithms in Appendix 1, p. 651.

   a. $[H^+] = 0.020$ M $= 2.0 \times 10^{-2}$ M

$$pH = -\log_{10} [H^+] = -\log_{10} (2.0 \times 10^{-2}) =$$
$$-(0.3 - 2.0) = -(-1.7) = 1.7$$

   b. Before calculating the pH, we need to know $[H^+]$. We can obtain this by making use of the relation $[H^+] \times [OH^-] = 1.0 \times 10^{-14}$

$$[H^+] = \frac{1.0 \times 10^{-14}}{[OH^-]} = \frac{1.0 \times 10^{-14}}{2.5 \times 10^{-3}} = 0.40 \times 10^{-11} = 4.0 \times 10^{-12}$$

$$pH = -\log_{10} (4.0 \times 10^{-12}) = -(0.6 - 12.0) = 11.4$$

   c. From Table 17.2 we note that the pH of sea water is 8.3.

$$-\log_{10} [H^+] = 8.3$$
$$\log_{10} [H^+] = -8.3 = 0.7 - 9.0$$

Taking antilogs, $[H^+] = 5 \times 10^{-9}$ M

The pH, $[H^+]$, or $[OH^-]$ of a water solution may be determined experimentally in a number of different ways, one of which involves the use of acid-base indicators, which undergo a color change over a rather narrow pH range. By the judicious use of one or more of the indicators listed in Table 17.3, it is possible to bracket quite accurately the pH or $[H^+]$ of a solution. For example, if one finds that a solution gives a red (basic) color with phenolphthalein but a yellow (acidic) color with alizarine yellow, its pH must be approximately 10.

A universal indicator made by combining several acid-base indicators may be used to determine, within about one unit, the pH of any water solution. This mixture of indicators shows an entire spectrum of colors ranging from deep red in strongly acidic solution to deep blue in strongly basic solution. A similar principle is used to prepare pH paper, widely used to test the acidity of biological fluids. Strips of paper impregnated with a mixture of in-

If an indicator contained phenolphthalein and methyl red, what would be its color at pH 10? 7? 3?

TABLE 17.3 COLOR CHANGE INTERVALS OF INDICATORS

NAME	pH INTERVAL	ACID COLOR	BASE COLOR
Methyl violet	0–2	yellow	violet
Methyl yellow	2.9–4.0	red	yellow
Methyl orange	3.1–4.4	red	yellow
Methyl red	4.4–6.2	red	yellow
Bromthymol blue	6.0–8.0	yellow	blue
Thymol blue	8.0–9.6	yellow	blue
Phenolphthalein	8.5–10.0	colorless	red
Alizarine yellow	10.1–12.0	yellow	red

429

dicators can be designed to give gradations of color over a wide or narrow pH range.

17.4 FORMATION OF ACIDIC WATER SOLUTIONS

The **transfer of a proton from a solute particle to a water molecule** leads to the formation of an acidic water solution. The hydronium ion, H_3O^+, formed by such a transfer, gives the solution its characteristic acidic properties. The proton donor may be a neutral molecule, a negative ion, or a positive ion. We shall now consider examples of each of these three types of proton transfer.

Transfer of a Proton from a Neutral Solute Molecule

The molecular compounds which react with water to produce acidic solutions can be subdivided into three major categories:

1. The binary hydrogen compounds of the 6A and 7A elements. The reaction of gaseous hydrogen chloride with water is representative:

$$HCl(g) + H_2O \longrightarrow Cl^- + H_3O^+ \tag{17.8}$$

In electron dot notation, this equation takes the form

$$H : \ddot{\underset{\cdot\cdot}{C}}l : + \; H : \ddot{\underset{\cdot\cdot}{O}} : H \longrightarrow (: \ddot{\underset{\cdot\cdot}{C}}l :)^- + \left[H : \underset{\underset{H}{\cdot\cdot}}{\overset{\cdot\cdot}{O}} : H \right]^+$$

from which it can be seen that a proton has been transferred from the chlorine atom of an HCl molecule to the oxygen atom of a water molecule.

2. Certain molecular compounds in which a hydrogen atom is joined by a covalent bond to an oxygen atom that is in turn bonded to a nonmetal atom. The electronic structures of several such compounds, known collectively as **oxyacids**, are shown in Table 17.4. Three of the most important oxyacids are perchloric acid, $HClO_4$, nitric acid, HNO_3, and sulfuric acid, H_2SO_4. The reactions of these compounds with water are analogous to that of HCl (compare Equations 17.9 to 17.11 with Equation 17.8).

The formula of H_2SO_4 could be written $SO_2(OH)_2$.

$$HClO_4(l) + H_2O \longrightarrow ClO_4^- + H_3O^+ \tag{17.9}$$

$$HNO_3(l) + H_2O \longrightarrow NO_3^- + H_3O^+ \tag{17.10}$$

$$H_2SO_4(l) + H_2O \longrightarrow HSO_4^- + H_3O^+ \tag{17.11}$$

In each case, a proton is transferred from an oxygen atom of the oxyacid molecule to the oxygen atom of a water molecule. The products are an oxyanion (ClO_4^-, NO_3^-, HSO_4^-) and a hydronium ion.

3. Certain nonmetal oxides, which react with water to form oxyacids. Typical of these compounds is sulfur trioxide, which reacts with water to form sulfuric acid:

$$SO_3(g) + H_2O(l) \longrightarrow H_2SO_4(l) \tag{17.12}$$

The sulfuric acid produced by this process reacts further with water to form an acidic solution according to Equation 17.11.

Nonmetal oxides that react with water to form oxyacids are often referred to as **acid anhydrides**. Several such compounds are listed in Table 17.5, along with the corresponding oxyacids and oxyanions which they form in water

TABLE 17.4 Electronic Structures of Some Oxyacids

Acid	Formula	Structure	Acid	Formula	Structure
Perchloric acid	$HClO_4$		Nitric acid	HNO_3	
Hypochlorous acid	$HClO$		Phosphoric acid	H_3PO_4	
Sulfuric acid	H_2SO_4		Acetic acid	$HC_2H_3O_2$	
Sulfurous acid	H_2SO_3				

solution. It may be noted that in each set of related species (acid anhydride, oxyacid, and oxyanion), the central, nonmetal atom is in the same oxidation state. For example, sulfur has an oxidation number of +6 in SO_3, H_2SO_4, and HSO_4^-; chlorine has an oxidation number of +7 in Cl_2O_7, $HClO_4$, and ClO_4^-.

Transfer of a Proton from an Anion (HSO_4^-, HSO_3^-, $H_2PO_4^-$)

The hydrogen sulfate ion, HSO_4^-, produced by the reaction of sulfuric acid with water, retains a hydrogen atom attached to oxygen which can be transferred to a water molecule.

$$HSO_4^- + H_2O \longrightarrow SO_4^{2-} + H_3O^+ \tag{17.13}$$

TABLE 17.5 Acid Anhydrides, Oxyacids, and Oxyanions

Acid Anhydride	Oxyacid	Oxyanion
Cl_2O_7	$HClO_4$	ClO_4^-
Cl_2O	$(HClO)^*$	ClO^-
SO_3	H_2SO_4	HSO_4^-
SO_2	(H_2SO_3)	HSO_3^-
N_2O_5	HNO_3	NO_3^-
P_2O_5	H_3PO_4	$H_2PO_4^-$
CO_2	(H_2CO_3)	HCO_3^-

* The oxyacids enclosed in parentheses cannot be isolated from water solution as pure compounds. Indeed, in the case of carbonic acid, H_2CO_3, the evidence for its existence is fragmentary at best.

As a result of this reaction, solutions containing the HSO_4^- ion are acidic. A solution of $NaHSO_4$, for example, gives an acidic test with litmus. Solutions of sodium hydrogen sulfite, $NaHSO_3$, and sodium dihydrogen phosphate, NaH_2PO_4, are also acidic.

$$HSO_3^- + H_2O \longrightarrow SO_3^{2-} + H_3O^+ \qquad (17.14)$$

$$H_2PO_4^- + H_2O \longrightarrow HPO_4^{2-} + H_3O^+ \qquad (17.15)$$

It should be pointed out that many negative ions of this type, in which a hydrogen atom is joined to oxygen, do *not* react with water to give an acidic solution. A solution of sodium hydrogen carbonate, $NaHCO_3$, is actually basic to litmus, for reasons which we shall discuss later in this chapter.

Transfer of a Proton from a Cation (NH_4^+, Al^{3+}, Transition Metal Ions)

It is observed experimentally that concentrated solutions of NH_4Cl, $Al(NO_3)_3$ and $ZnSO_4$ are acidic. Solutions of these salts give a red color with methyl red, for example, indicating a pH of .5 or lower. Indeed, we find that solutions of a great many salts containing NH_4^+ ions, Al^{3+} ions, or transition metal ions (Zn^{2+}, Cu^{2+}, and so forth) are at least weakly acidic. The question arises as to how these ions can react with water to produce an excess of H^+ ions.

In the special case of the NH_4^+ ion, one can readily visualize a proton transfer to a water molecule:

$$NH_4^+ + H_2O \longrightarrow NH_3 + H_3O^+ \qquad (17.16)$$

We see that a proton originally bonded to nitrogen has shifted to the more strongly electronegative oxygen atom.

It is somewhat more difficult to explain the acidity of solutions containing such ions as Al^{3+} or Zn^{2+}. Perhaps the simplest way to understand their acidity is to consider the effect that such ions might be expected to have on the equilibrium:

$$H_2O \rightleftharpoons H^+ + OH^- \qquad (17.17)$$

A small cation of high charge, such as Al^{3+}, has a strong enough attraction for an OH^- ion to form with it a "complex ion," most simply represented as $Al(OH)^{2+}$:

$$Al^{3+} + OH^- \longrightarrow Al(OH)^{2+} \qquad (17.18)$$

By consuming OH^- ions, this process shifts the equilibrium in Reaction 17.17 to produce an excess of H^+ ions, thereby making the solution acidic. We may summarize the overall process by adding Equations 17.17 and 17.18 to obtain

Subsequent reactions could produce $Al(OH)_2^+$ and $Al(OH)_3$.

$$Al^{3+} + H_2O \longrightarrow Al(OH)^{2+} + H^+ \qquad (17.19)$$

The acidity of zinc salts may be explained in a similar manner:

$$Zn^{2+} + H_2O \longrightarrow Zn(OH)^+ + H^+ \qquad (17.20)$$

Equations 17.19 and 17.20 are oversimplified in that they fail to take into account the fact that bare ions such as Al^{3+} and Zn^{2+} are unlikely to exist as such in water solution. Instead, the cations are hydrated to give complex ions such as $Al(H_2O)_6^{3+}$ and $Zn(H_2O)_4^{2+}$. Proton transfer can occur from a water molecule in the complex ion to a water molecule in solution:

$$Al(H_2O)_6^{3+} + H_2O \longrightarrow Al(H_2O)_5(OH)^{2+} + H_3O^+ \qquad (17.19')$$

$$Zn(H_2O)_4^{2+} + H_2O \longrightarrow Zn(H_2O)_3(OH)^+ + H_3O^+ \qquad (17.20')$$

Comparison of Equations 17.19 and 17.19' (or 17.20 and 17.20') reveals their essential similarity. In both cases the ion responsible for acidic properties, H^+ or H_3O^+, is a product of the reaction. The only difference between the equations lies in the formulas written for the metal cations (Al^{3+} vs. $Al(H_2O)_6^{3+}$; $Al(OH)^{2+}$ vs. $Al(H_2O)_5(OH)^{2+}$).

17.5 STRONG AND WEAK ACIDS: K_a

In many of the reactions that we have been considering, the proton transfer to a water molecule is incomplete. Consider, for example, the reaction of hydrogen fluoride with water:

$$HF + H_2O \rightleftharpoons F^- + H_3O^+ \qquad (17.21)$$

It is found experimentally that only a comparatively small fraction of the HF molecules react; the concentrations of H_3O^+ and F^- ions are low compared to that of undissociated HF molecules. One can calculate (Example 17.5) that in a 0.10 M solution of hydrogen fluoride, the concentration of F^- or H_3O^+ ions is only about 0.008 mole/lit, corresponding to 8 per cent ionization.

Species such as HF, which react incompletely with water to form H_3O^+ ions, are referred to as **weak acids**. Substances which undergo complete proton transfer upon addition to water are known as **strong acids**. Relatively few strong acids are known; the common acids in this category include three hydrogen halides (HCl, HBr, HI) and three oxyacids ($HClO_4$, HNO_3, H_2SO_4). In contrast, there a large number of weak acids, including many molecular species (HF, HClO, H_3PO_4, $HC_2H_3O_2$), all the acidic anions (HSO_4^-, HSO_3^-, $H_2PO_4^-$), and all the acidic cations (NH_4^+, Al^{3+}, Zn^{2+}).

Experimentally, one can distinguish between strong and weak acids by measuring:

1. The conductivities of their water solutions. Strong acids behave as strong electrolytes; solutions of weak molecular acids such as hydrofluoric acid or acetic acid are relatively poor electrical conductors.

2. The colligative properties of their water solutions. The freezing point of a 0.10 M solution of HCl is almost exactly the same as that of 0.10 M NaCl. In contrast, the freezing point of a 0.10 M solution of acetic acid is comparable to that of 0.10 M solutions of nonelectrolytes such as sugar or urea.

3. The pH of their water solutions. A 0.10 M solution of perchloric acid has a pH of 1, indicating complete dissociation. The pH of a 0.10 M solution of HF is about 2.1, indicating that relatively few of the HF molecules are ionized.

4. The rate of reaction with metals such as zinc or magnesium (Figure 17.1).

The equilibria between species in solutions of weak acids can be handled

433

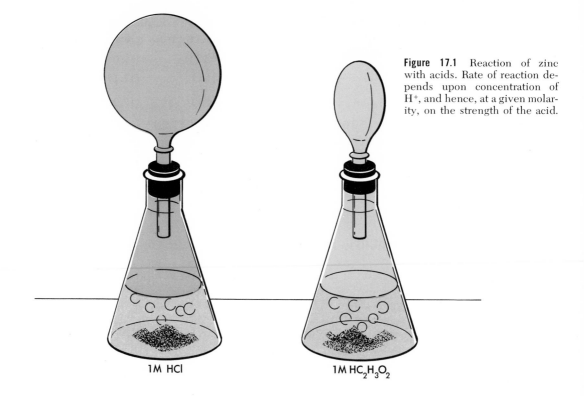

Figure 17.1 Reaction of zinc with acids. Rate of reaction depends upon concentration of H^+, and hence, at a given molarity, on the strength of the acid.

1M HCl

1M HC$_2$H$_3$O$_2$

quantitatively in a manner analogous to that developed previously in connection with gases (Chapter 13) and slightly soluble salts (Chapter 16). In particular, it is possible to derive an expression for the equilibrium constant, or the ionization constant, of a weak acid in water. We shall now consider how this equilibrium constant, symbol K_a, is expressed; how it is determined experimentally; and how it is used in practical calculations.

Expression for K_a

Consider a weak acid that reacts with water as follows:

$$HX + H_2O \rightleftharpoons H_3O^+ + X^-$$

For this equilibrium, one can write

$$K = \frac{[H_3O^+] \times [X^-]}{[HX] \times [H_2O]} \quad \text{or} \quad K \times [H_2O] = \frac{[H_3O^+] \times [X^-]}{[HX]}$$

It has been pointed out previously that the term $[H_2O]$ is itself a constant. For simplicity, $[H_3O^+]$ can be replaced by $[H^+]$. Consequently, the foregoing equation can be written

$$K_a = \frac{[H^+] \times [X^-]}{[HX]} \tag{17.22}$$

in which K_a, known as the ionization constant of the weak acid, is independent of the concentrations of H^+, X^-, or HX. Equation 17.22 is, of course, the expression one would arrive at directly by starting with the simplified equation for the dissociation of HX:

$$HX \rightleftharpoons H^+ + X^-$$

To illustrate the form taken by K_a for various weak acids, consider the three species HF, HSO_4^-, and NH_4^+:

$$HF \rightleftharpoons H^+ + F^- \qquad K_a = \frac{[H^+] \times [F^-]}{[HF]}$$

$$HSO_4^- \rightleftharpoons H^+ + SO_4^{2-} \qquad K_a = \frac{[H^+] \times [SO_4^{2-}]}{[HSO_4^-]}$$

$$NH_4^+ \rightleftharpoons H^+ + NH_3 \qquad K_a = \frac{[H^+] \times [NH_3]}{[NH_4^+]}$$

Experimental Determination of K_a

One approach to the experimental determination of the ionization constant of a weak acid is illustrated in Example 17.2.

EXAMPLE 17.2. When 1.00 mole of HF is dissolved to give one liter of solution, the equilibrium concentration of H^+ is found, through the use of indicators, to be 2.6×10^{-2} M. Calculate the ionization constant of HF.

SOLUTION. The expression for K_a is

$$HF \rightleftharpoons H^+ + F^- \qquad K_a = \frac{[H^+] \times [F^-]}{[HF]}$$

To obtain the numerical value of K_a, the equilibrium concentrations of H^+, F^-, and HF must be known. That of H^+ has been determined to be 2.6×10^{-2} M. From the expression for the dissociation of HF, it is evident that 1 mole of F^- is formed for every mole of H^+. In a solution of HF, in which the only source of H^+ and F^- ions is the HF molecule (the $[H^+]$ produced by the dissociation of water is so small in comparison that it may be neglected), the concentrations of F^- and H^+ must be equal:

$$[H^+] = [F^-] = 2.6 \times 10^{-2} \text{ M}$$

To obtain the equilibrium concentration of HF, we note that for every mole of H^+ formed, a mole of HF is consumed. The concentration of HF remaining at equilibrium can be obtained by subtracting the concentration of H^+ formed, 2.6×10^{-2} M, from the original concentration of HF, 1.00 M:

$$[HF] = 1.00 \text{ M} - 2.6 \times 10^{-2} \text{ M} = 1.00 \text{ M} - 0.026 \text{ M} = 0.97 \text{ M}$$

Substituting,

$$K_a = \frac{[H^+] \times [F^-]}{[HF]} = \frac{(2.6 \times 10^{-2}) \times (2.6 \times 10^{-2})}{(0.97)} = 7.0 \times 10^{-4}$$

The procedure illustrated by Example 17.2 can be applied to any weak acid. Two quantities must be measured—the original concentration of the weak acid and the equilibrium concentration of H^+.

A more general method for determining the ionization constant of a weak

435

acid, in which one need not know its concentration or even its chemical identity, is outlined in Example 17.3.

EXAMPLE 17.3. A student is instructed to determine the ionization constant of an acid by the following procedure: A sample of the weak acid, HA, is first dissolved in water to give 50 ml of solution. This solution is then split into two equal portions of 25 ml each. One of these portions is neutralized with sodium hydroxide and then added to the unneutralized portion. Show how the ionization constant of the acid can be calculated from the measured [H$^+$] of the final solution.

SOLUTION. By definition, $K_a = \dfrac{[H^+] \times [A^-]}{[HA]}$

By the design of the experiment, $[A^-] = [HA]$, since exactly half the HA molecules originally present were converted to A$^-$ ions by neutralization. Substituting in the expression for K_a,

$$\frac{[A^-]}{[HA]} = 1; \qquad K_a = [H^+]$$

Consequently, the ionization constant of the acid is exactly equal to the [H$^+$] in the final solution.

[H$^+$] and hence K_a could be determined by using an indicator.

Interpretation of K_a

The numerical value of the ionization constant of an acid gives a quantitative measure of its strength. The stronger the acid, the larger will be the value of K_a. From Table 17.6 (p. 443) we find that acetic acid, $HC_2H_3O_2$ ($K_a = 1.8 \times 10^{-5}$) is weaker than hydrofluoric acid ($K_a = 7.0 \times 10^{-4}$) but stronger than boric acid ($K_a = 5.8 \times 10^{-10}$). In equimolar solutions of two different acids, the concentration of H$^+$ or the fraction of the weak acid dissociated will be greater for the acid of larger ionization constant.

Comparison of the relative strengths of different acids is facilitated by the use of a term known as pK_a, defined by the equation

$$pK_a \equiv -\log_{10} K_a = \log_{10} 1/K_a \qquad (17.23)$$

The larger the value of pK_a, the weaker is the acid. Thus, for acetic acid, hydrofluoric acid, and boric acid, we have

acetic acid: $pK_a = -\log_{10} (1.8 \times 10^{-5}) = -(0.3 - 5.0) = 4.7$

hydrofluoric acid: $pK_a = -\log_{10} (7.0 \times 10^{-4}) = -(0.8 - 4.0) = 3.2$

boric acid: $pK_a = -\log_{10} (5.8 \times 10^{-10}) = -(0.8 - 10.0) = 9.2$

acid strength: $HF > HC_2H_3O_2 > H_3BO_3$

It may be noted that pK_a bears the same relation to K_a that pH bears to H$^+$ (cf. Equations 17.23 and 17.7). Just as a solution of low H$^+$ ion concentration has a high pH, so a weak acid with a small ionization constant has a high pK_a value.

Use of K_a in Calculations

Examples 17.4 and 17.5 illustrate how ionization constants can be used in calculations involving solutions of weak acids.

EXAMPLE 17.4. Calculate the $[H^+]$
    a. a 1.0 M solution of $HC_2H_3O_2$.

    b. a 0.10 M solution of NH_4Cl.

SOLUTION

    a. From Table 17.6 we have

$$HC_2H_3O_2 \rightleftharpoons H^+ + C_2H_3O_2^-$$
$$K_a = \frac{[H^+] \times [C_2H_3O_2^-]}{[HC_2H_3O_2]} = 1.8 \times 10^{-5}$$

Let x represent the unknown equilibrium concentration of H^+. It is evident from the equation for the dissociation of acetic acid that a mole of acetate ions is produced for every mole of H^+ formed. Neglecting the small amount of H^+ produced from the water, the concentration of $C_2H_3O_2^-$ must also be x. The formation of x moles of H^+ consumes an equal number of moles of acetic acid; the equilibrium concentration of $HC_2H_3O_2$ must then be $1.0 - x$. Substituting,

$$\frac{(x)(x)}{1.0 - x} = 1.8 \times 10^{-5}$$

This equation could be rearranged to the form $x^2 + bx + c = 0$ and solved for x, using the quadratic formula. Such a procedure is tedious and, in this case, unnecessary. Since $HC_2H_3O_2$ is a weak acid, only slightly dissociated in water, the equilibrium concentration of $HC_2H_3O_2$, $1.0 - x$, must be very nearly equal to its original concentration before dissociation, 1.0 M. Making this approximation, we obtain

$$\frac{x^2}{1.0} = 1.8 \times 10^{-5}; \qquad x^2 = 1.8 \times 10^{-5} = 18 \times 10^{-6}$$

Solving for x by extracting the square root of both sides, we have

$$x = [H^+] = 4.2 \times 10^{-3}$$

The fact that the concentration of H^+, 0.0042, is so much less than the original concentration of $HC_2H_3O_2$, justifies the approximation made earlier, i.e., $1.0 - x = 1.0$. In general, the expression for K_a is rarely valid to better than ±5 per cent. Consequently, in the expression

$$K_a = \frac{x^2}{a - x}$$

in which $x = [H^+]$, a = concentration of weak acid prior to dissociation, one is justified in setting $(a - x)$ equal to a, provided this approximation does not introduce an error of more than about 5 per cent. In practice, it is ordinarily simplest to make the approximation, calculate x, and compare to a. If the value of x thus obtained is less than 5 per cent of a (in this problem, x is 0.42 per cent of a), the approximation is valid. If x is greater than about 5 per cent of a, one can go back to the original equation and solve by means of the quadratic

We avoid using the quadratic formula whenever possible.

formula or, alternatively, by the method of successive approximations (Example 17.5).

b. Proceeding exactly as in a,

$$NH_4^+ \rightleftharpoons NH_3 + H^+; \qquad K_a = \frac{[NH_3] \times [H^+]}{[NH_4^+]} = 5.6 \times 10^{-10}$$

Taking x to be the equilibrium concentration of H^+,

$$\frac{(x)(x)}{(0.1 - x)} = 5.6 \times 10^{-10}$$

Since K_a is so small, it seems reasonable to assume that x will be much smaller than 0.1.

$$x^2/0.1 = 5.6 \times 10^{-10}; \quad x^2 = 5.6 \times 10^{-11} = 56 \times 10^{-12}$$

$$x = [H^+] = 7.5 \times 10^{-6}$$

Clearly, x is much less than 5% of 0.1; thus the approximation was justified.

EXAMPLE 17.5. Calculate the $[H^+]$ of a 0.100 M HF solution.

SOLUTION. Proceeding as in Example 17.4, we have

$$HF \rightleftharpoons H^+ + F^-$$

$$K_a = \frac{[H^+] \times [F^-]}{[HF]} = 7.0 \times 10^{-4}$$

Letting $[H^+] = x$,

$$\frac{(x)(x)}{0.100 - x} = 7.0 \times 10^{-4}$$

If we make the same approximation as before, i.e., $0.10 - x \approx 0.10$,

$$x^2 = 7.0 \times 10^{-5} = 70 \times 10^{-6}$$

$$x = [H^+] = 8.4 \times 10^{-3}$$

We note that in this case the calculated concentration of H^+ is greater than 5 per cent of the original concentration of undissociated acid:

$$\frac{8.4 \times 10^{-3}}{1.0 \times 10^{-1}} = 0.084 = 8.4\%$$

To refine our calculation, we can make a second approximation, more nearly valid than the first. Let us use the value of $[H^+]$ just calculated to obtain a better approximation to the true value of $[HF]$.

If $\quad [H^+] = 8.4 \times 10^{-3}, \quad$ then $\quad [HF] = 0.100 - 8.4 \times 10^{-3}$
$$= 0.100 - 0.0084 = 0.092 \text{ M}$$

Substituting this value for $[HF]$ into the expression for K_a, we have

$$\frac{x^2}{0.092} = 7.0 \times 10^{-4}$$

$$x^2 = 6.4 \times 10^{-5} = 64 \times 10^{-6}$$

$$x = [H^+] = 8.0 \times 10^{-3}$$

This value is more nearly equal to the true $[H^+]$ in the solution, since 0.092 M is a better approximation to the equilibrium concentration of HF than was 0.100 M. If we are still not satisfied, we can attempt a further refinement, using the value of $[H^+]$ just calculated to obtain a still more accurate value for $[HF]$. If we do, we find that $[HF]$ remains unchanged at 0.092 M; i.e.,

if $\quad [H^+] = 8.0 \times 10^{-3}$, then $\quad [HF] = 0.100 - 0.0080 = 0.092$ M

This means that if we were to solve again for $[H^+]$ we would get the same answer. In other words, we "have gone about as far as we can go."

The method of successive approximations just described is a very useful one for working problems of this type involving the dissociation of weak acids. In the vast majority of cases, it will be found that a first approximation [i.e., $(a - x) \approx a$] will be sufficient. Occasionally, it will be necessary, as it appeared to be in this example, to make one further refinement. Very seldom does one need to go beyond this.

This is a powerful method for solving any mathematical expression.

17.6 FORMATION OF BASIC WATER SOLUTIONS

We have seen that it is the hydroxide ion which is responsible for the basic properties of water solutions. When the concentration of OH^- ions exceeds 10^{-7} M, the solution is basic. Such a situation can arise as the result of:

1. The addition of a metal hydroxide or metal oxide to water. The hydroxides of the 1A and 2A metals, except $Be(OH)_2$, produce hydroxide ions directly when added to water. These compounds contain hydroxide ions in the solid state; when they dissolve in water, the OH^- ions are liberated to form a strongly basic solution:

$Be(OH)_2$ shows acidic properties (cf $SO_2(OH)_2$).

$$NaOH(s) \longrightarrow Na^+ + OH^- \qquad \text{(17.24)}$$

$$Ca(OH)_2(s) \longrightarrow Ca^{2+} + 2\ OH^- \qquad \text{(17.25)}$$

Magnesium hydroxide, unlike calcium hydroxide, is quite insoluble in water. A water solution or suspension of $Mg(OH)_2$, known as milk of magnesia, is only slightly basic because of the low concentration of OH^- ions. Consequently, milk of magnesia is effective in "neutralizing excess stomach acidity." Sodium hydroxide, lye, is even more effective but has the undesirable side effect of dissolving the stomach lining and other parts of the digestive tract. The hydroxides of aluminum and the transition metals, like magnesium hydroxide, are too insoluble to give high concentrations of OH^- ions in solution.

The oxides of the 1A and 2A metals, except BeO, react with water to form the corresponding hydroxides. Consequently, the addition of such compounds as CaO or Li_2O to excess water yields a basic solution. The reaction of calcium oxide may be represented as follows:

$$CaO(s) + H_2O \longrightarrow Ca(OH)_2(s)$$

$$\frac{Ca(OH)_2(s) \longrightarrow Ca^{2+} + 2\ OH^-}{CaO(s) + H_2O \longrightarrow Ca^{2+} + 2\ OH^-} \qquad \text{(17.26)}$$

With lithium oxide, the net reaction is

$$\tfrac{1}{2}\ Li_2O(s) + \tfrac{1}{2}\ H_2O \longrightarrow Li^+ + OH^- \qquad \text{(17.27)}$$

439

In electron dot notation, these reactions may be represented in terms of a proton transfer from a water molecule to an oxide ion.

$$(:\overset{..}{\underset{..}{O}}:)^{2-} + H:\overset{..}{\underset{..}{O}}:H \longrightarrow 2\,(:\overset{..}{\underset{..}{O}}:H)^-$$

Metal oxides such as CaO and Li$_2$O which react with water to give metal hydroxides are often referred to as **basic anhydrides**.

2. The addition of certain molecular compounds to water. Perhaps the most important compound in this class is ammonia, which undergoes the following reversible reaction upon addition to water:

The reagent labelled NH$_4$OH is a water solution of NH$_3$.

$$NH_3(g) + H_2O \rightleftharpoons NH_4^+ + OH^- \tag{17.28}$$

The weak base ammonia abstracts a proton from a water molecule, forming the weak acid NH$_4^+$ and the OH$^-$ ion which makes the solution basic.

Certain organic compounds, including methyl amine, CH$_3$NH$_2$, and pyridine, C$_5$H$_5$N, behave like ammonia but are even weaker bases.

$$CH_3NH_2(l) + H_2O \rightleftharpoons CH_3NH_3^+ + OH^- \tag{17.29}$$

$$C_5H_5N(l) + H_2O \rightleftharpoons C_5H_5NH^+ + OH^- \tag{17.30}$$

3. The addition to water of salts containing the anion of a weak acid.

Certain negative ions are also capable of abstracting a proton from a water molecule. Water solutions of NaF or NaC$_2$H$_3$O$_2$ are weakly basic because of the reversible reaction of the anion with water:

C$_2$H$_3$O$_2^-$ behaves like NH$_3$; HC$_2$H$_3$O$_2$ like NH$_4^+$.

$$(:\overset{..}{\underset{..}{F}}:)^- + H-\overset{..}{\underset{..}{O}}-H \rightleftharpoons H-\overset{..}{\underset{..}{F}}: + (:\overset{..}{\underset{..}{O}}-H)^-$$

$$C_2H_3O_2^- + H_2O \rightleftharpoons HC_2H_3O_2 + OH^- \tag{17.31}$$

As Equation 17.31 indicates, a molecule of the weak acid HC$_2$H$_3$O$_2$ is formed as a byproduct of the reaction. The reaction of the carbonate ion with water follows a similar path,

$$CO_3^{2-} + H_2O \rightleftharpoons HCO_3^- + OH^- \tag{17.32}$$

accounting for the fact that a water solution of sodium carbonate is basic.

The reaction of the HCO$_3^-$ ion with water is particularly interesting. Here we have two competing processes, one the transfer of a proton from a water molecule to the HCO$_3^-$ ion, which tends to make the solution basic

$$HCO_3^- + H_2O \longrightarrow H_2CO_3 + OH^- \tag{17.33}$$

and the other the transfer of a proton from the HCO$_3^-$ ion to the water molecule, which tends to make the solution acidic

$$HCO_3^- + H_2O \longrightarrow CO_3^{2-} + H_3O^+ \tag{17.34}$$

It appears that Reaction 17.33 occurs to a somewhat greater extent than Reaction 17.34, since a water solution of NaHCO$_3$ is slightly basic.

Competing reactions analogous to 17.33 and 17.34 can, of course, occur with any negative ion that has an ionizable hydrogen. In some cases (HCO$_3^-$, HS$^-$, H$_2$BO$_3^-$, HPO$_4^{2-}$, and so on), the net effect is to produce a basic solution; in others (HSO$_4^-$, HSO$_3^-$, H$_2$PO$_4^-$) proton transfer occurs predominantly in the opposite direction, from anion to water molecule, and the solution is acidic.

17.7 WEAK BASE EQUILIBRIA: K_b

Equilibria in water solutions of weak bases such as NH_3 or $C_2H_3O_2^-$ may be treated in a manner completely analogous to that used with weak acids such as NH_4^+ or HF. Consider, for example, the ionization of ammonia,

$$NH_3 + H_2O \rightleftharpoons NH_4^+ + OH^-$$

for which we can write
$$K_b = \frac{[NH_4^+] \times [OH^-]}{[NH_3]}$$

in which K_b represents the ionization constant of the weak base ammonia. In a similar manner, for the equilibrium

$$C_2H_3O_2^- + H_2O \rightleftharpoons HC_2H_3O_2 + OH^-$$

$$K_b = \frac{[HC_2H_3O_2] \times [OH^-]}{[C_2H_3O_2^-]}$$

The numerical value of K_b for the weak base $C_2H_3O_2^-$ can be determined experimentally in a manner similar to that described previously for determining K_a. Alternatively, if the value of K_a is known for a weak acid, the value of K_b for the corresponding weak base can be calculated by means of the equation

$$K_a \times K_b = K_w = 1.0 \times 10^{-14} \qquad (17.35)$$

The validity and usefulness of Equation 17.35 are demonstrated by Example 17.6.

EXAMPLE 17.6
  a. Show that the product of K_a for $HC_2H_3O_2$ and K_b for $C_2H_3O_2^-$ is equal to K_w.
  b. Determine K_b for $C_2H_3O_2^-$, given that K_a for $HC_2H_3O_2$ is 1.8×10^{-5}.

If HX is a weak acid, X^- is a weak base.

SOLUTION

a.
$$K_a \text{ of } HC_2H_3O_2 = \frac{[H^+] \times [C_2H_3O_2^-]}{[HC_2H_3O_2]} ;$$

$$K_b \text{ of } C_2H_3O_2^- = \frac{[HC_2H_3O_2] \times [OH^-]}{[C_2H_3O_2^-]}$$

$$K_a \times K_b = \frac{[H^+] \times [\cancel{C_2H_3O_2^-}]}{[\cancel{HC_2H_3O_2}]} \times \frac{[\cancel{HC_2H_3O_2}] \times [OH^-]}{[\cancel{C_2H_3O_2^-}]}$$

$$= [H^+] \times [OH^-] = K_w$$

b. K_b of $C_2H_3O_2^- = \dfrac{K_w}{K_a \text{ of } HC_2H_3O_2} = \dfrac{1.0 \times 10^{-14}}{1.8 \times 10^{-5}} = 5.6 \times 10^{-10}$

Equation 17.35 can, of course, be used to calculate K_a for a weak acid such as NH_4^+ from a known K_b value for the corresponding base, NH_3. Species such as $C_2H_3O_2^-$ and NH_3 are sometimes referred to as the **conjugate bases** of the weak acids $HC_2H_3O_2$ and NH_4^+; alternatively, $HC_2H_3O_2$ and NH_4^+ may be regarded as the **conjugate acids** corresponding to the bases $C_2H_3O_2^-$ and NH_3.

Calculations involving weak base equilibria can be handled in a manner analogous to that used with weak acids (compare Example 17.7 with Examples 17.4 and 17.5).

EXAMPLE 17.7. Calculate the $[OH^-]$ of 0.10 M solutions of

      a. NH_3 b. $NaC_2H_3O_2$

SOLUTION

a. $K_b = \dfrac{[NH_4^+] \times [OH^-]}{[NH_3]} = 1.8 \times 10^{-5}$

Taking x to be the equilibrium concentration of OH^-, it follows that $[NH_4^+] = x$, $[NH_3] = 0.10 - x$. Substituting in the expression for K_b,

$$\frac{x^2}{0.10 - x} = 1.8 \times 10^{-5}$$

Making the approximation that $0.10 - x \approx 0.10$, we obtain

$$x^2 = (1.8 \times 10^{-5})(0.10) = 1.8 \times 10^{-6}$$
$$x = [OH^-] = 1.3 \times 10^{-3}$$

The approximation is clearly valid, since $x < 5\%$ of 0.10

b. Proceeding exactly as in a,

$$K_b = \frac{[HC_2H_3O_2] \times [OH^-]}{[C_2H_3O_2^-]} = 5.6 \times 10^{-10}$$

Letting $x = [OH^-]$, we have $\dfrac{x^2}{0.10 - x} = 5.6 \times 10^{-10}$

$$x^2 \approx 5.6 \times 10^{-11} = 56 \times 10^{-12}$$
$$x = 7.5 \times 10^{-6} = [OH^-]$$

17.8 RELATIVE STRENGTHS OF ACIDS AND BASES

Up to this point in our study of acids and bases, we have been concerned with the questions "how?" and "how many?" We have considered how acidic and basic water solutions are formed and how many H^+ or OH^- ions are produced from a given amount of reactant. We shall now turn our attention to the question "why?", i.e., why do certain substances act as acids in water solution while other substances behave as bases? We shall attempt to develop principles that will make it possible to predict the relative acid or base strengths of compounds of similar structure. Three different classes of substances will be considered: nonmetal hydrides, oxyacids, and inorganic salts. Within each category, we shall be interested in relating basic or acidic properties to structure and bond type.

TABLE 17.6 IONIZATION CONSTANTS OF WEAK ACIDS AND BASES

	ACID	K_a	BASE	K_b*
Acetic acid	$HC_2H_3O_2$	1.8×10^{-5}	$C_2H_3O_2^-$	5.6×10^{-10}
Ammonium ion	NH_4^+	5.6×10^{-10}	NH_3	1.8×10^{-5}
Benzoic acid	$HC_7H_5O_2$	6.6×10^{-5}	$C_7H_5O_2^-$	1.5×10^{-10}
Boric acid	H_3BO_3	5.8×10^{-10}	$H_2BO_3^-$	1.7×10^{-5}
Carbonic acid	H_2CO_3	4.2×10^{-7}	HCO_3^-	2.4×10^{-8}
	HCO_3^-	4.8×10^{-11}	CO_3^{2-}	2.1×10^{-4}
Formic acid	$HCHO_2$	2.1×10^{-4}	CHO_2^-	4.8×10^{-11}
Hydrocyanic acid	HCN	4.0×10^{-10}	CN^-	2.5×10^{-5}
Hydrofluoric acid	HF	7.0×10^{-4}	F^-	1.4×10^{-11}
Hydrogen sulfide	H_2S	1×10^{-7}	HS^-	1×10^{-7}
	HS^-	1×10^{-15}	S^{2-}	1×10^{1}
Hypochlorous acid	$HClO$	3.2×10^{-8}	ClO^-	3.1×10^{-7}
Nitrous acid	HNO_2	4.5×10^{-4}	NO_2^-	2.2×10^{-11}
Phosphoric acid	H_3PO_4	7.5×10^{-3}	$H_2PO_4^-$	1.3×10^{-12}
	$H_2PO_4^-$	6.2×10^{-8}	HPO_4^{2-}	1.6×10^{-7}
	HPO_4^{2-}	1.7×10^{-12}	PO_4^{3-}	5.9×10^{-3}
Propionic acid	$HC_3H_5O_2$	1.4×10^{-5}	$C_3H_5O_2^-$	7.1×10^{-10}
Sulfurous acid	H_2SO_3	1.7×10^{-2}	HSO_3^-	5.9×10^{-13}
	HSO_3^-	5.6×10^{-8}	SO_3^{2-}	1.8×10^{-7}

* The base dissociation constants of the negative ions listed in this column and the acid dissociation constant of the NH_4^+ ion listed under K_a are often referred to as hydrolysis constants and given the symbol K_h.

Hydrides of the 7A, 6A, and 5A Elements

The hydrides of the 7A elements (HF, HCl, HBr, HI) form acidic water solutions. All these compounds, with the exception of hydrogen fluoride, are completely ionized in water solution. To understand why hydrogen fluoride is a much weaker acid than the other hydrogen halides, it is convenient to imagine the reaction with water as being broken up into three steps:

1. Rupture of the covalent bond in the HX molecule to form hydrogen and halogen atoms

$$H : \ddot{\underset{..}{X}} : \longrightarrow H \cdot + : \dot{\underset{..}{X}} :$$

2. Electron transfer from a hydrogen to a halogen atom

$$H \cdot + : \dot{\underset{..}{X}} : \longrightarrow H^+ + (: \ddot{\underset{..}{X}} :)^-$$

3. Proton transfer to a water molecule

$$H^+ + H : \ddot{\underset{..}{O}} : H \longrightarrow \left[H : \ddot{\underset{\overset{|}{H}}{O}} : H \right]^+$$

The third step is, of course, the same for all of the hydrogen halides; Steps 1 and 2 differ depending upon the nature of X. We see then that the relative acid strengths of the hydrogen halides should depend upon two factors—the ease with which the covalent bond in the HX molecule is broken, and the attraction of the halogen atom for electrons.

It happens that in this series of compounds, it is the first factor which predominates. Hydrogen fluoride is a weak acid because of the large amount

443

TABLE 17.7   ACID STRENGTHS OF 6A HYDRIDES

	K_1	BOND ENERGY (KCAL/MOLE)
H_2S	1×10^{-7}	H—S = 81
H_2Se	1.7×10^{-4}	H—Se = 66
H_2Te	2.3×10^{-3}	H—Te = 59

of energy (135 kcal/mole) required to break the HF bond. The bonds in HCl (103 kcal/mole), HBr (88 kcal/mole), and HI (71 kcal/mole) are considerably weaker.

The hydrogen compounds of the 6A elements (H_2S, H_2Se, H_2Te) are all weak acids in water solution. Acid strength in this series, as measured by the first ionization constant, increases from H_2S to H_2Te. This trend may be attributed to the decrease in the strength of the bond formed with hydrogen as one passes from sulfur to tellurium.

None of the binary hydrogen compounds of the elements to the left of the 6A elements in the periodic table acts as an acid in water solution. The 5A hydrides tend to accept protons from water molecules rather than to donate them. The tendency to act as a base is greatest for the first member of the series, ammonia:

$$NH_3 + H_2O \rightleftharpoons NH_4^+ + OH^-$$

The hydrides of the other 5A elements are much less basic; phosphine (PH_3), arsine (AsH_3), and stibine (SbH_3) show little or no tendency to accept a proton from a water molecule. The phosphonium ion, PH_4^+, can be formed by reacting phosphine with a strong proton donor such as hydrogen iodide:

$$PH_3(g) + HI(g) \longrightarrow PH_4I(s) \tag{17.36}$$

In water solution the phosphonium ion acts as a strong acid, decomposing quantitatively to phosphine:

$$PH_4I(s) + H_2O \longrightarrow PH_3(g) + H_3O^+ + I^- \tag{17.37}$$

Oxyacids

The ionization in water solution of a molecular oxyacid involves the separation of a proton from a molecule containing a covalent O—H bond. For example, in the case of hypochlorous acid the process is:

$$Cl—O—H \longrightarrow Cl—O^- + H^+ \tag{17.38}$$

Looking at Equation 17.38 from a structural point of view, it seems reasonable to suppose that the extent to which it occurs should depend on the magnitude of the electron density in the vicinity of the hydrogen atom in the molecule. Any factor that tends to decrease this electron density will facilitate the separation of the positively charged proton and thereby increase the acid strength.

In an oxyacid molecule the electron density in the vicinity of the ionizable hydrogen atom will depend to a considerable extent upon the electronegativity and the oxidation number of the nonmetal atom bonded to oxygen. The greater its electronegativity, the more it will tend to pull electrons away from hydrogen, thereby promoting ionization. On this basis, we would predict that hypochlorous acid, HOCl, would be a stronger acid than hypobromous acid, HOBr, since chlorine is more electronegative than bromine. Table 17.8 confirms this prediction.

H_2SO_3 is stronger than H_2SeO_3, yet H_2S is weaker than H_2Se. Why?

An increase in the oxidation number of the nonmetal atom in an oxyacid also tends to increase its acid strength. Thus we find that among the oxyacids of chlorine, acid strength increases in the order

$$\begin{array}{cccc} +1 & +3 & +5 & +7 \\ \text{HClO} < & \text{HClO}_2 < & \text{HClO}_3 < & \text{HClO}_4 \end{array}$$

This trend is readily explained when one considers that the increase in oxidation number comes about as the result of the addition of oxygen atoms to the molecule. The addition of atoms of this highly electronegative element tends to pull electrons away from the hydrogen atom, facilitating its ionization.

In summary we find that:

1. Acid strength increases with increasing electronegativity of the nonmetal atom to which the $-OH$ group is attached.

2. Acid strength increases with increasing oxidation number of the nonmetal atom. From a slightly different point of view, we might say that acid strength increases with the number of "extra" or "unprotonated" oxygen atoms in the molecule.

The strong oxyacids are all ones in which there are three ($HClO_4$) or two (HNO_3, H_2SO_4) oxygens bonded to the central nonmetal atom but not to hydrogen. Where this number is 1, as in H_3PO_4, HNO_2, or H_2SO_3, the acid is moderately weak, with an ionization constant in the vicinity of 10^{-2} to 10^{-6}. The weakest oxyacids are those such as HClO, $H_2N_2O_2$, or H_3BO_3 in which all the oxygen atoms are bonded to hydrogen.

It should be emphasized that the rule just stated applies only to *molecular* oxyacids. Oxyanions such as HSO_4^- and $H_2PO_4^-$ are much weaker acids than the molecules (H_2SO_4, H_3PO_4) from which they are derived. As one

TABLE 17.8 STRENGTH OF OXYACIDS

Variation with Oxidation Number

ACID	K_a	ACID	K_a	ACID	K_a
$HClO_4$	10^7	H_2SO_4	very large	HNO_3	very large
$HClO_3$	large	H_2SO_3	1.7×10^{-2}	HNO_2	4.5×10^{-4}
$HClO_2$	1×10^{-2}			$H_2N_2O_2$	9×10^{-8}
HClO	3.2×10^{-8}				

Variation with Electronegativity

ACID	K_a	ACID	K_a	ACID	K_a
HClO	3.2×10^{-8}	H_2SO_3	1.7×10^{-2}	HNO_3	very large
HBrO	2×10^{-9}	H_2SeO_3	3×10^{-3}	H_3PO_4	7.5×10^{-3}
HIO	5×10^{-13}	H_2TeO_3	6×10^{-6}	H_3AsO_4	5×10^{-3}

445

would expect, it is more difficult to remove a proton from a negative ion than from a neutral molecule. The successive ionization constants of such acids as H_2SO_4 and H_3PO_4 which contain more than one ionizable hydrogen atom decrease in the approximate ratio $1:10^{-5}:10^{-10}$.

Hydrolysis of Salts

We have seen that the anions of weak acids (F^-, CO_3^{2-}, and so forth) react with water to form OH^- ions; certain cations (such as NH_4^+, Al^{3+}, Zn^{2+}) react with water to form H^+ ions. These reactions are examples of a general process called hydrolysis, in which a water molecule is split by the solute. Depending upon the relative extents of hydrolysis of anion and cation, a given salt solution may be acidic, basic, or neutral.

Anions. The extent to which a negative ion hydrolyzes is inversely related to the strength of its conjugate acid. The chloride ion, derived from the strong acid HCl, does not undergo hydrolysis and hence does not affect the pH of a water solution. On the other hand, the fluoride ion reacts with water to form the weak acid HF and liberate OH^- ions, thereby producing a basic solution. One of the most strongly hydrolyzed ions is the sulfide ion, derived from the extremely weak acid HS^-. A water solution of sodium sulfide has about the same pH as a solution of sodium hydroxide of the same concentration.

$$Cl^- + H_2O \longleftarrow HCl + OH^- \qquad K_b = 0$$

$$F^- + H_2O \rightleftharpoons HF + OH^- \qquad K_b = 1.4 \times 10^{-11}$$

$$S^{2-} + H_2O \longrightarrow HS^- + OH^- \qquad K_b = 1 \times 10^1$$

Cations. The extent of hydrolysis of a metal ion is directly related to its charge density. The Al^{3+} ion (charge $= +3$, $r = 0.50$ Å) is strongly hydrolyzed to give an acidic solution. The Na^+ ion (charge $= +1$, $r = 0.95$ Å) has no detectable effect on the pH of water. The influence of charge density can be explained in terms of the attraction of a positive ion for OH^- ions. The higher the charge density of the cation, the greater will be this attraction and hence the greater the extent of hydrolysis.

Salts. The effect of a given salt upon the pH of water can be deduced if the relative extents of hydrolysis of the two ions involved are known. Four categories may be distinguished:

1. *Neither ion hydrolyzes.* Solution is *neutral.* NaCl, K_2SO_4, $Ba(NO_3)_2$.
2. *Only the cation hydrolyzes.* Solution is *acidic.* NH_4Cl, $ZnSO_4$, $Al(NO_3)_3$.
3. *Only the anion hydrolyzes.* Solution is *basic.* NaF, K_2CO_3, BaS.
4. *Both ions hydrolyze.* Solution may be acidic, neutral, or basic, depending upon the relative extents of hydrolysis of cation and anion. In the case of NH_4F, the NH_4^+ ion ($K_a = 5.6 \times 10^{-10}$) is more strongly hydrolyzed than the F^- ion ($K_b = 1.4 \times 10^{-11}$), and the solution is acidic. A solution of $NH_4C_2H_3O_2$ is neutral, since the NH_4^+ ion ($K_a = 5.6 \times 10^{-10}$) and the $C_2H_3O_2^-$ ion ($K_b = 5.6 \times 10^{-10}$) are hydrolyzed to the same extent. A solution of $(NH_4)_2S$ is distinctly basic; the S^{2-} ion ($K_b = 1 \times 10^1$) is much more strongly hydrolyzed than the NH_4^+ ion ($K_a = 5.6 \times 10^{-10}$).

Would a solution of KBr be acidic, basic, or neutral? $KC_2H_3O_2$? $AlCl_3$?

TABLE 17.9 HYDROLYSIS OF IONS

ANIONS		CATIONS	
Negligible	*Appreciable*	*Negligible*	*Appreciable*
ClO_4^-	F^-	K^+	NH_4^+
I^-	$C_2H_3O_2^-$	Na^+	Al^{3+}
Br^-	CN^-	Li^+	
Cl^-	PO_4^{3-}	Ba^{2+}	
NO_3^-	CO_3^{2-}	Ca^{2+}	transition metal ions
SO_4^{2-}	S^{2-}	Sr^{2+}	
	etc.		

17.9 GENERAL CONCEPTS OF ACIDS AND BASES

Throughout this chapter our discussion of acids and bases has centered upon water solutions. We have considered an acid to be a substance which, upon addition to water, forms hydrated protons, H_3O^+; a substance which produces hydroxide ions, OH^-, in water solution has been referred to as a base. These definitions of acids and bases represent a modernized version of the **Arrhenius acid-base concept**, first proposed in 1884.

Although the Arrhenius concept has proved to be reasonably satisfactory for a discussion of the properties of acidic and basic water solutions, we shall find it advantageous to consider other concepts of acids and bases that can be extended to reactions occurring in nonaqueous systems. Those we shall discuss are the **Brönsted-Lowry concept** and the **Lewis concept**. Curiously enough, both of these were proposed almost simultaneously in 1923. We shall place particular emphasis on the Brönsted-Lowry picture of acids and bases, in part because it serves to review much that has been said in this chapter, and because it offers a logical introduction to the topic of acid-base reactions, to be discussed in Chapter 18.

Brönsted-Lowry Concept

The Arrhenius concept defines acids and bases in terms of the species they produce upon addition to water. A more general picture, suggested independently by Brönsted in Denmark and Lowry in England in the year 1923, defines acids and bases in terms of the way in which they react *with each other*. According to the Brönsted-Lowry picture, **an acid-base reaction is one in which there is a proton transfer from one species to another**. The species which gives up **(donates)** a **proton** is referred to as an **acid**; the molecule or ion which **accepts** the **proton** is called a **base**.

To illustrate the application of the Brönsted-Lowry concept, let us examine a reaction considered previously, that of hydrogen chloride with water:

$$HCl(g) + H_2O \longrightarrow H_3O^+ + Cl^- \qquad (17.39)$$
$$\text{acid} \qquad \text{base}$$

Clearly, the HCl molecule in this reaction is acting as an acid, since it is donating a proton to a water molecule. (It is also, of course, an acid in the Arrhenius

447

sense, since it is forming H_3O^+ ions). Consider, now, the role of the water molecule in this reaction. In the Brönsted-Lowry sense, water is acting as a base, since it is accepting a proton from an HCl molecule. On the basis of the Arrhenius concept, water would not be considered to be a base, since the reaction does not result in the formation of OH^- ions. Here we see our first example of the broadening of the acid-base concept that is implied in the Brönsted-Lowry picture.

Another example of a Brönsted-Lowry acid-base reaction is the reversible ionization of hydrogen fluoride in aqueous solution:

$$HF(g) + H_2O \rightleftharpoons H_3O^+ + F^- \tag{17.40}$$
$$\text{acid} \quad\quad \text{base} \quad\quad\quad \text{acid} \quad\quad \text{base}$$

Here the two species HF and H_3O^+ are Brönsted-Lowry acids; they donate protons to the H_2O molecule and the F^- ion respectively in the forward and reverse reactions. The species H_2O and F^- are acting as bases; they accept protons in the forward and reverse reactions respectively.

At this point, it is perhaps obvious that in all the reactions considered in Section 17.4, which dealt with the formation of acidic water solutions, the water molecule is acting as a base, accepting protons from such acids as HNO_3 (Equation 17.10), HSO_4^- (Equation 17.13), NH_4^+ (Equation 17.16) and $Zn(H_2O)_4^{2+}$ (Equation 17.20'). It is also possible for water to act as an acid in the Brönsted-Lowry sense. Consider, for example its reaction with ammonia:

$$H_2O + NH_3(g) \rightleftharpoons NH_4^+ + OH^- \tag{17.41}$$
$$\text{acid} \quad\quad \text{base} \quad\quad\quad \text{acid} \quad\quad \text{base}$$

In the forward reaction the water molecule is acting as a Brönsted-Lowry acid, because it is donating a proton to the NH_3 molecule to form the NH_4^+ ion. This illustrates once again the greater generality of the Brönsted-Lowry concept as opposed to that of Arrhenius.

Reviewing the reactions discussed in Section 17.6, which had to do with the formation of basic water solutions, we recall that in Equations 17.28 to 17.34, water acts as a Brönsted-Lowry acid, since it donates a proton to a molecule such as NH_3 or CH_3NH_2 or an anion such as F^-, CO_3^{2-}, or HCO_3^-.

$$H_2O + CH_3NH_2(l) \rightleftharpoons CH_3NH_3^+ + OH^- \tag{17.42}$$
$$\text{acid} \quad\quad\quad \text{base} \quad\quad\quad\quad \text{acid} \quad\quad\quad \text{base}$$

$$H_2O + F^- \rightleftharpoons HF + OH^- \tag{17.43}$$
$$\text{acid} \quad\quad \text{base} \quad\quad \text{acid} \quad\quad \text{base}$$

We see then that water can act either as a base or an acid, depending on the species with which it reacts. When it acts as a base, it accepts a proton to form a H_3O^+ ion; when it acts as an acid, it donates a proton, leaving an OH^- ion as a residue. The water molecule is perhaps the most important example of an **amphiprotic** species, which can act as a Brönsted-Lowry acid or base, either donating or accepting a proton. Another example is the HCO_3^- ion, which acts as an acid when it is added to water

$$HCO_3^- + H_2O \rightleftharpoons H_3O^+ + CO_3^{2-} \tag{17.44}$$
$$\text{acid} \quad\quad\quad \text{base} \quad\quad\quad \text{acid} \quad\quad\quad \text{base}$$

or to a solution of sodium hydroxide

$$HCO_3^- + OH^- \rightleftharpoons H_2O + CO_3^{2-} \tag{17.45}$$
$$\text{acid} \quad\quad\quad \text{base} \quad\quad\quad \text{acid} \quad\quad\quad \text{base}$$

but acts as a base in the presence of hydrofluoric acid:

$$HF + HCO_3^- \rightleftharpoons H_2CO_3 + F^- \qquad \text{(17.46)}$$
$$\text{acid} \quad \text{base} \qquad\qquad \text{acid} \quad \text{base}$$

There is, of course, no requirement that a water molecule must be one of the participants in a Brönsted-Lowry acid base reaction (cf. Reaction 17.46). More important, the Brönsted-Lowry concept can be applied to many reactions occurring in nonaqueous solvents. Consider, for example, the reaction of sodium hydride with liquid ammonia:

$$NH_3 + H^- \longrightarrow H_2(g) + NH_2^- \qquad \text{(17.47)}$$
$$\text{acid} \quad \text{base}$$

NH$_3$, like H$_2$O or HCO$_3^-$, is amphiprotic.

Here the ammonia molecule acts as an acid in giving up a proton to form the H_2 molecule. An example of a Brönsted-Lowry acid-base reaction occurring in the gas phase is that between hydrogen chloride and ammonia:

$$HCl(g) + NH_3(g) \longrightarrow NH_4^+ + Cl^- \qquad \text{(17.48)}$$
$$\text{acid} \qquad \text{base}$$

It is possible to compare the relative strengths of different Brönsted-Lowry acids by comparing their tendency to donate protons to a given base. The most obvious choice of a base for comparison purposes is the water molecule. When we find, for example, that the reaction of HCl with H_2O proceeds almost to completion,

$$HCl(g) + H_2O \longrightarrow H_3O^+ + Cl^-$$

while that of HF with water is incomplete,

$$HF(g) + H_2O \rightleftharpoons H_3O^+ + F^-$$

we deduce that HCl is a stronger acid than HF. In another experiment we find that HCN has even less tendency than HF to donate protons to water,

$$HCN(g) + H_2O \rightleftharpoons H_3O^+ + CN^-$$

from which we conclude that HF must be a stronger acid than HCN.

By carrying out experiments such as those described in the previous paragraph, we can establish a list of common acids in order of decreasing strength, i.e., decreasing tendency to donate protons. The order of the acids in such a series will, of course, be that of decreasing K_a value in water solution. That is, acids high on the list have relatively large ionization constants, while those near the bottom of the list have very small ionization constants, of the order of 10^{-10} to 10^{-12}.

There is, however, one difficulty with such a scheme; if we restrict ourselves to water as a base, we are unable to distinguish among extremely strong acids such as HCl and HBr, which appear to be completely dissociated in water solution. The problem is that water is too strong a base to allow us to test the relative strengths of very strong acids. In this sense, water has a **leveling effect**. It is incapable of distinguishing between acids whose strengths are much greater than that of the acidic species formed by water itself, the H_3O^+ ion.

NH$_3$ would be an even poorer choice than H$_2$O. Why?

A better choice of solvent for comparing the acid strengths of such species as HCl and HBr would be one that is a weaker base than water. Several such solvents are available, one of which is hydrogen peroxide, H_2O_2, which is considerably more reluctant than water to accept a proton.

449

$$HA + H—\overset{..}{\underset{..}{O}}—\overset{..}{\underset{..}{O}}—H \rightleftharpoons \left[H—\overset{\overset{\displaystyle H}{|}}{\underset{..}{O}}—\overset{..}{\underset{..}{O}}—H \right]^+ + A^- \qquad (17.49)$$

By measuring the relative concentrations of ions formed in such solvents, it is possible to show experimentally that HBr is a stronger Brönsted-Lowry acid than HCl.

We can measure the relative strengths of different bases by observing their tendencies to accept protons from a given acid. If water is used as a solvent, the base strengths of different species are measured by their K_b values. Thus the CO_3^{2-} ion ($K_b = 2.1 \times 10^{-4}$) is a stronger base than the NH_3 molecule ($K_b = 1.8 \times 10^{-5}$), which, in turn, is stronger than the F^- ion ($K_b = 1.4 \times 10^{-11}$). Since, as pointed out previously, the K_b value for a given base is inversely related to the K_a value for the corresponding acid ($K_b = K_w/K_a$), it is obvious that the strength of a base is inversely related to that of its conjugate acid. Very strong acids, such as HBr and HCl, have very weak conjugate bases (Br^-, Cl^-); very strong bases such as H^-, OH^-, and S^{2-} have weak conjugate acids (H_2, H_2O, HS^-).

TABLE 17.10 BRÖNSTED-LOWRY ACIDS AND BASES

K_a	ACID	BASE	K_b
very large	$HClO_4$	ClO_4^-	very small
very large	HI	I^-	very small
very large	HBr	Br^-	very small
very large	HCl	Cl^-	very small
very large	HNO_3	NO_3^-	very small
very large	$H_3O_2^+$	H_2O_2	very small
very large	H_3O^+	H_2O	very small
1.7×10^{-2}	H_2SO_3	HSO_3^-	5.9×10^{-13}
7.5×10^{-3}	H_3PO_4	$H_2PO_4^-$	1.3×10^{-12}
7.0×10^{-4}	HF	F^-	1.4×10^{-11}
4.5×10^{-4}	HNO_2	NO_2^-	2.2×10^{-11}
1.8×10^{-5}	$HC_2H_3O_2$	$C_2H_3O_2^-$	5.6×10^{-10}
4.2×10^{-7}	H_2CO_3	HCO_3^-	2.4×10^{-8}
1×10^{-7}	H_2S	HS^-	1×10^{-7}
5.6×10^{-8}	HSO_3^-	SO_3^{2-}	1.8×10^{-7}
3.2×10^{-8}	HOCl	OCl^-	3.1×10^{-7}
5.6×10^{-10}	NH_4^+	NH_3	1.8×10^{-5}
4.0×10^{-10}	HCN	CN^-	2.5×10^{-5}
4.8×10^{-11}	HCO_3^-	CO_3^{2-}	2.1×10^{-4}
2.6×10^{-12}	H_2O_2	HO_2^-	3.8×10^{-3}
4.4×10^{-13}	HPO_4^{2-}	PO_4^{3-}	2.3×10^{-2}
1×10^{-15}	HS^-	S^{2-}	1×10^{1}
1.8×10^{-16}	H_2O	OH^-	5.6×10^{1}
very small	CH_3OH	CH_3O^-	very large
very small	NH_3	NH_2^-	very large
very small	OH^-	O^{2-}	very large
very small	H_2	H^-	very large

Table 17.10 lists a series of Brönsted-Lowry acids with their conjugate bases. The strength of the acids decreases as one moves down the column at the left; the strength of the bases listed in the right-hand column increases as one moves down the column. The strongest acid is the $HClO_4$ molecule, located at the top of the left-hand column; the ClO_4^- ion, located directly across from it, is the weakest base listed. At the bottom of the table we find the weakest acid, H_2, and the strongest base, H^-.

Table 17.10 can be used to predict the direction in which a Brönsted-Lowry acid-base reaction will tend to occur. Consider, for example, the reaction

$$\underset{\text{acid}}{HC_2H_3O_2} + \underset{\text{base}}{CN^-} \rightleftharpoons \underset{\text{acid}}{HCN} + \underset{\text{base}}{C_2H_3O_2^-} \tag{17.50}$$

For Reaction 17.50, $K = K_a HAc/K_a HCN = 4.5 \times 10^4$.

We note from the table that acetic acid ($K_a = 1.8 \times 10^{-5}$) is a stronger acid than HCN ($K_a = 4.0 \times 10^{-10}$). Furthermore, CN^- ($K_b = 2.5 \times 10^{-5}$) is a stronger base than $C_2H_3O_2^-$ ($K_b = 5.6 \times 10^{-10}$). It follows that the equilibrium in Reaction 17.50 will be displaced far to the right. In other words, the forward reaction will be nearly quantitative at ordinary concentrations, while the reverse reaction will occur to only a small extent. In the general case, we can say that the reaction of an acid with any base below it in the table will tend to be spontaneous in the sense that equilibrium considerations will favor that reaction.

The Lewis Concept

We have seen that the Brönsted-Lowry picture represents a considerable extension of the Arrhenius concept of acids and bases. However, the Brönsted-Lowry picture is restricted in one important respect: it can be applied only to reactions involving a proton transfer. In particular, in order to act as a Brönsted-Lowry acid, a species must contain an ionizable hydrogen atom.

The Lewis acid-base concept, first proposed by the American physical chemist G. N. Lewis in 1923, removes this restriction. **The Lewis concept considers an acid to be a species that can accept an electron pair; a base is a substance that can donate an electron pair.** According to the Lewis concept, any reaction which leads to the formation of a coordinate covalent bond is an acid-base reaction.

From a structural point of view, the Lewis concept of a base does not differ in any essential way from the Brönsted concept. In order for a species to accept a proton and thereby act as a Brönsted base it must possess an unshared pair of electrons. Consider, for example, the NH_3 molecule, the H_2O molecule, and the F^- ion, all of which can act as Brönsted bases:

$$H\!-\!\ddot{N}\!-\!H, \qquad H\!-\!\ddot{O}\!-\!H, \qquad (:\ddot{F}:)^-$$
$$\overset{|}{H}$$

Each of these species contains an unshared pair of electrons that is utilized in accepting a proton to form the NH_4^+ ion, the H_3O^+ ion, or the HF molecule.

$$\left[H:\overset{\text{\tiny ..}}{\underset{\overset{\text{\tiny ..}}{H}}{N}}:H\right]^+, \qquad \left[H:\overset{}{\underset{\overset{\text{\tiny ..}}{H}}{O}}:H\right]^+, \qquad H:\ddot{\underset{\text{\tiny ..}}{F}}:$$

451

Clearly, NH_3, H_2O, and F^- can also act as Lewis bases since they possess an unshared electron pair which can be donated to an acid. We see then that the Lewis concept does not significantly increase the number of species which can act as bases.

On the other hand, the Lewis concept greatly increases the number of species which can be considered to be acids. The substance which accepts an electron pair and therefore acts as a Lewis acid can be a proton:

$$\underset{\text{acid}}{H^+} + \underset{\text{base}}{H_2O} \longrightarrow H_3O^+ \tag{17.51}$$

$$\underset{\text{acid}}{H^+} + \underset{\text{base}}{NH_3} \longrightarrow NH_4^+ \tag{17.52}$$

It can equally well be a cation, such as Zn^{2+}, which is capable of forming a coordinate covalent bond with a Lewis base:

$$\underset{\text{acid}}{Zn^{2+}} + \underset{\text{base}}{4\ H_2O} \longrightarrow Zn(H_2O)_4^{2+} \tag{17.53}$$

$$\underset{\text{acid}}{Zn^{2+}} + \underset{\text{base}}{4\ NH_3} \longrightarrow Zn(NH_3)_4^{2+} \tag{17.54}$$

Another important class of Lewis acids comprises molecules containing an incomplete octet. An example is boron trifluoride, BF_3, which reacts readily with ammonia:

$$\tag{17.55}$$

Molecules whose octet can be expanded also can act as Lewis acids:

$$\tag{17.56}$$

From a slightly different point of view, one might say that any molecule containing a vacant orbital that can accommodate an electron pair is capable of acting as a Lewis acid. This may be a "p" orbital, as in BF_3, or a "d" orbital, as in SiF_4.

Certain molecules containing multiple bonds also are capable of acting as Lewis acids. Consider, for example, the carbon dioxide molecule, which accepts a pair of electrons from an O^{2-} ion in forming a CO_3^{2-} ion:

$$\tag{17.57}$$

The SO_3 molecule, in its reaction with an O^{2-} ion to form the SO_4^{2-} ion, is also acting as a Lewis acid by accepting a pair of electrons:

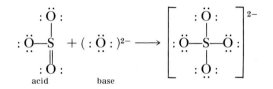

$$\text{(17.58)}$$

acid   base

Reviewing the various equations that we have written for Lewis acid-base reactions, we conclude that **any species containing an atom which can increase its coordination number through coordinate covalent bonding can act as a Lewis acid**. In Reactions 17.51 and 17.52, the proton is increasing its coordination number from 0 to 1; in 17.53 and 17.54, Zn^{2+} increases its coordination number from 0 to 4. In 17.55, the coordination number of boron increases from 3 to 4; in 17.56, the coordination number of silicon goes from 4 to 6. Finally, in 17.57 and 17.58, the elements carbon and sulfur go from coordination numbers of 2 and 3, respectively, to 3 and 4.

Other potential Lewis acids include PCl_3 and C_2H_2. Explain.

PROBLEMS

17.1   Explain exactly what is meant by each of the following terms.

a. pH   e. Conjugate acid.   i. Lewis acid.
b. Strong acid.   f. Conjugate base.   j. Lewis base.
c. Weak base.   g. Brönsted acid
d. Acid anhydride.   h. Brönsted base.

17.2   Determine the pH of solutions with the following [H^+].

a. 10^{-8} b. 10^1 c. 2.0×10^{-5} d. 6.0×10^{-9}

17.3   Calculate [H^+] in solutions with the following pH.

a. 7.0   b. −1.0   c. 12.4   d. 10.7

17.4   Calculate [H^+], [OH^-], and the pH of the following (assuming complete ionization).

a. 0.020 M HCl
b. 0.10 M CsOH
c. A solution made by diluting 10 ml of 0.1 M HNO_3 to 500 ml.
d. A solution made by adding 200 cc of 0.10 M HCl to 100 cc of 0.20 M $Ca(OH)_2$.

17.5   Write formulas for

a. An oxyacid of sulfur in the +6 oxidation state.
b. The anhydride of H_3PO_4.
c. A hydrogen halide that is a weak acid.
d. An anion that reacts with water to form an acidic solution.
e. An anion that reacts with water to form a basic solution.

17.6   Write balanced net ionic equations to explain why water solutions of the following species are acidic.

a. $HClO_3$ b. I_2O_5 c. NH_4^+ d. $H_2PO_4^-$ e. Cu^{2+}

17.7   Write balanced net ionic equations to explain why water solutions of the following species are basic.

a. LiOH   b. BaO   c. $C_2H_5NH_2$ d. CN^-

453

17.8 Write balanced net ionic equations for the reactions, if any, of the following ions with water.

a. $C_2H_3O_2^-$ c. HSO_4^- e. Zn^{2+}
b. NH_4^+ d. Cl^- f. CO_3^{2-}

17.9 A solution containing 4.7 g of HNO_2 in one liter of solution has a pH of 2.2. What is the ionization constant of HNO_2?

17.10 A student measures the ionization constant of an acid of unknown molecular weight by the following procedure: He dissolves 1.0 g of the acid in 50 ml of water and splits the resulting solution into two exactly equal parts. One of these portions is neutralized with NaOH and added to the other portion. The pH of the resulting solution is found to be 5.7. Calculate K_a.

17.11 Calculate the $[H^+]$, pH, per cent dissociation, and $[OH^-]$ in

a. 0.20 M HCl   c. 0.20 M HF
b. 0.20 M $HC_2H_3O_2$ d. 0.20 M NH_4Cl

17.12 Calculate the $[OH^-]$, per cent dissociation, $[H^+]$, and pH in

a. 0.10 M NaOH   b. 0.10 M NH_3 c. 0.10 M NaF

17.13 Given that K_a for a certain weak acid HA is 1.0×10^{-6}, find K_b for the hydrolysis of NaA.

17.14 Which would you expect to be the stronger acid in water solution?

a. H_2Se or H_2Te c. HNO_2 or HNO_3 e. Al^{3+} or Mg^{2+}
b. H_2SeO_3 or H_2TeO_3 d. H_2SO_4 or HSO_4^- f. Rb^+ or Ag^+

17.15 Predict whether water solutions of the following compounds will be acidic, basic, or neutral. Where appropriate, write net ionic equations to explain your answers.

a. HClO   d. NH_4Cl g. BaO   j. KF
b. N_2O_5 e. NH_3 h. $BaCl_2$ k. $NaNO_3$
c. KOH   f. NaCl   i. $Zn(NO_3)_2$ l. NaCN

17.16 Write equations for reactions in which

a. $H_2PO_4^-$ acts as a Brönsted acid.   d. HF acts as a Brönsted acid.
b. $H_2PO_4^-$ acts as a Brönsted base.   e. HSO_4^- acts as a Brönsted acid.
c. O^{2-} acts as a Brönsted base.   f. HSO_4^- acts as a Brönsted base.

17.17 For each of the following reactions, underline the Brönsted acids and bases.

a. $HCO_3^- + H_2O \rightleftharpoons H_2CO_3 + OH^-$
b. $C_2H_3O_2^- + HCl \rightleftharpoons HC_2H_3O_2 + Cl^-$
c. $H_2O + H^- \rightleftharpoons H_2(g) + OH^-$

17.18 Using Table 17.10, predict the direction in which each of the reactions in Problem 17.17 will be displaced at ordinary concentrations.

17.19 Write balanced equations for reactions in which

a. BF_3 acts as a Lewis acid.
b. NH_3 acts as a Lewis base.
c. SO_2 acts as a Lewis acid.

17.20 Which of the following species can act as Brönsted acids? Brönsted bases? Lewis acids? Lewis bases?

a. H_2O b. NH_3 c. BeF_2 d. F^-

17.21 Calculate the pH of solutions with the following $[H^+]$.

a. 10^{-7} b. 10^{-2} c. 10^1 d. 2.5×10^{-6} e. 1.6×10^{-9} f. 2.0×10^{-15}

17.22 Calculate the [H^+] in solutions of the following pH.

a. 6   b. 12   c. 1.5   d. 9.6   e. -1.5 f. 13.9

17.23 Calculate the [H^+], [OH^-], and pH of each of the following (assume complete ionization).

a. 0.010 M HNO_3
b. 0.090 M KOH
c. A solution made by diluting 20 ml of 0.1 M HCl to one liter.
d. A solution made by dissolving 32.0 g of KOH in enough water to make 200 ml of solution.
e. A solution made by mixing 20 cc of 0.2 M HCl with 60 cc of 0.1 M $Ca(OH)_2$.

17.24 If you were given 10 cc of 1.0 M HCl, how would you prepare from it solutions of the following pH?

a. 1.0   b. 2.0   c. 3.0   d. 7.0

17.25 Complete the following table.

Acid Anhydride	Oxyacid	Oxyanion
Cl_2O_7	—	—
SeO_3	—	—
—	H_3PO_4	—
—	H_2TeO_3	—
—	—	NO_2^-

17.26 Write balanced net ionic equations to explain why solutions prepared by adding the following species to water are acidic.

a. HBr   b. H_2S c. $HClO_4$ d. HSO_4^- e. NH_4^+

17.27 Write balanced net ionic equations to explain why solutions prepared by adding the following species to water are basic.

a. $Ca(OH)_2$ b. BaO   c. NH_3 d. CH_3NH_2

17.28 Write balanced net ionic equations for the reactions, if any, of the following ions with water.

a. Na^+ c. CO_3^{2-} e. $Cu(H_2O)_4^{2+}$ g. HCO_3^-
b. $Cr(H_2O)_6^{3+}$ d. F^- f. Cl^-

17.29 A student dissolves 105 g of an acid of formula HN_3 to form one liter of solution. The pH of this solution is 2.2. Calculate the ionization constant of HN_3.

17.30 A student measures the ionization constant of an unknown acid by dissolving a sample in 20 ml of water. He draws off 5 ml of the solution, neutralizes it with base, and adds it to the other 15 ml. The pH of the resulting solution is found to be 4.5. What is the ionization constant of the acid?

17.31 Determine the pH, percentage ionization, [H^+], and [OH^-] of the following solutions.

a. 0.12 M $HC_2H_3O_2$ c. 0.20 M HF
b. 1.0 M HCN   d. Pure water

17.32 Calculate the [OH^-] in a 0.10 M solution of a salt NaY, given that K_a for HY is 1.0×10^{-7}.

17.33 Write net ionic equations for reactions in which

a. HF acts as a Brönsted acid.   d. H_2O acts as a Brönsted acid.
b. HCO_3^- acts as a Brönsted acid.   e. H_2O acts as a Brönsted base.
c. HCO_3^- acts as a Brönsted base.   f. $C_2H_3O_2^-$ acts as a Brönsted base.

455

17.34 In each of the following reversible reactions, underline the Brönsted acids and bases.

$$NH_4^+ + H_2O \rightleftharpoons NH_3 + H_3O^+$$

$$HCO_3^- + HF \rightleftharpoons H_2CO_3 + F^-$$

$$CH_3OH + OH^- \rightleftharpoons CH_3O^- + H_2O$$

17.35 Write an equation for a reaction, other than those listed in the text, which would be considered an acid-base reaction by the Lewis but not by the Brönsted-Lowry definition.

17.36 Show that for an acid H_2X, which ionizes in two steps,

$$\frac{[H^+]^2 \times [X^{2-}]}{[H_2X]} = K_1K_2$$

in which K_1 and K_2 are the first and second ionization constants respectively.

*17.37 List, in order of decreasing concentration, all the species (molecules and ions) present in a 0.1 M solution of H_2S. Calculate the approximate concentration of each species.

*17.38 It is found that 0.1 M solutions of three sodium salts NaX, NaY, and NaZ have pH's of 7.0, 9.0, and 11.0 respectively. Arrange the acids HX, HY, and HZ in order of increasing strength. Where possible, calculate the ionization constants of the acids.

*17.39 When a certain solution of acetic acid is made 1.0 M in Ag^+, a precipitate of $AgC_2H_3O_2$ just starts to form. What is the concentration of $HC_2H_3O_2$ in this solution?

*17.40 a. Show that the equilibrium constant for the reaction

$$HA + B^- \rightleftharpoons HB + A^-$$

is given by the expression $K = \dfrac{K_a \text{ of HA}}{K_a \text{ of HB}} = \dfrac{K_b \text{ of B}^-}{K_b \text{ of A}^-}$

b. Show that for

$$HA + OH^- \rightleftharpoons H_2O + A^-, \quad K = (1.0 \times 10^{14})(K_a \text{ of HA})$$

c. Show that for

$$B^- + H^+ \rightleftharpoons HB, \quad K = (1.0 \times 10^{14})(K_b \text{ of B}^-)$$

Acid-Base Reactions

We pointed out in Chapter 17 that in terms of the Brönsted-Lowry or the Lewis concepts of acids and bases, a wide variety of reactions can be classified as acid-base reactions. In this chapter we shall be concerned with the applications of certain acid-base reactions to analytical chemistry, synthetic inorganic chemistry, and industrial processes.

18.1 TYPES OF ACID-BASE REACTIONS

The acid-base reactions we shall consider are ones which take place in water solution and go virtually to completion. They can be divided into three general categories.

1. The reaction of a solution of a strong acid ($HClO_4$, HNO_3, H_2SO_4, HCl, HBr, HI) **with a solution of a strong base** (1A or 2A hydroxide).

$$H^+ + OH^- \longrightarrow H_2O, \quad \Delta H = -13.4 \text{ kcal}, \tag{18.1}$$

This reaction, commonly referred to as **neutralization**, occurs when solutions of NaOH and HCl, $Ca(OH)_2$ and $HClO_4$, or any other strong acid–strong base combination are mixed. The other ions present in these solutions (Na^+, Cl^-, Ca^{2+}, ClO_4^-) take no part in the reaction. Perhaps the best proof of this statement lies in the experimental observation that the amount of heat evolved per mole of H^+ or OH^- reacting is the same, 13.4 kcal, regardless of the identity of the strong acid or strong base.

Since the equilibrium constant for Reaction 18.1 is a very large number ($K = 1/K_w = 10^{14}$), the neutralization process goes essentially to completion. If, for example, one neutralizes a solution of hydrochloric acid with sodium hydroxide, a solution of pure sodium chloride is produced. This solution is, of course, neutral; i.e., $[H^+] = [OH^-] = 1 \times 10^{-7}$ M.

Explain why $K = 1/K_w$.

2. The reaction of a solution of a weak acid with a solution of a strong base.
Typical reactions of this type include the following:

$$\Delta H$$

$$HCN + OH^- \longrightarrow H_2O + CN^-, \quad -2.5 \text{ kcal} \tag{18.2}$$

$$HF + OH^- \longrightarrow H_2O + F^-, \quad -16.5 \tag{18.3}$$

$$NH_4^+ + OH^- \longrightarrow H_2O + NH_3, \quad -1.0 \tag{18.4}$$

It will be noted that ΔH for these reactions differs significantly from that for the simple neutralization reaction. A reaction such as 18.2 can be broken up into two steps:

$$HCN \longrightarrow H^+ + CN^-; \qquad \Delta H = +10.9 \text{ kcal} \qquad (18.2a)$$

$$H^+ + OH^- \longrightarrow H_2O; \qquad \Delta H = -13.4 \text{ kcal} \qquad (18.2b)$$

$$HCN + OH^- \longrightarrow H_2O + CN^-; \qquad \Delta H = -2.5 \text{ kcal} \qquad (18.2)$$

The first step in the reaction is endothermic, indicating that the amount of energy absorbed in breaking the bond holding hydrogen to carbon in HCN is greater than that evolved when the H^+ and CN^- ions are formed in aqueous solution. Consequently, the evolution of energy in the overall reaction is considerably smaller than the 13.4 kcal given off upon neutralization.

The equilibrium constant for Reaction 18.2 can be obtained by multiplying together the equilibrium constants for Reactions 18.2b (1×10^{14}) and 18.2a (K_a of HCN = 4×10^{-10}).

If two equations, 1 and 2, add to a third equation, 3, then: $K_3 = K_1 \times K_2$.

$$K = 1 \times 10^{14} \times 4 \times 10^{-10} = 4 \times 10^4$$

Since K is a very large number (40,000) it follows that Reaction 18.2 goes virtually to completion under ordinary conditions. Indeed, for any reaction of a weak acid, HA, with OH^- ions, we can write:

$$K = (1 \times 10^{14})(K_a \text{ of HA}) \qquad (18.5)$$

Since K_a of a weak acid is ordinarily greater than 10^{-14}, it follows that the equilibrium constant for a reaction such as 18.2 to 18.4 is greater than one. Unless the acid is extremely weak, the reaction will go virtually to completion at ordinary concentrations (Example 18.1).

With weak acids containing more than one acidic hydrogen, reaction with OH^- ions can occur stepwise. An important example is the reaction of car-

---

EXAMPLE 18.1

a. Using Equation 18.5, calculate K for the reaction of HF with OH^-:

$$HF + OH^- \rightleftharpoons H_2O + F^-$$

b. Calculate the concentration of HF remaining at equilibrium when one liter of a 0.10 M solution of this acid is reacted with 0.10 of a mole of OH^-.

SOLUTION

a. $K = (1.0 \times 10^{14}) K_a$; for HF, $K = (1.0 \times 10^{14})(7.0 \times 10^{-4})$
   $= 7.0 \times 10^{10}$.

b. The equilibrium constant for the reaction of HF with OH^- may be written:

$$K = 7.0 \times 10^{10} = \frac{[F^-]}{[HF] \times [OH^-]}$$

Since K is so large, the reaction will go virtually to completion and it is reasonable to suppose that $[F^-]$ will be almost 0.10 M. If we choose our unknown x to be the equilibrium concentration of HF, it follows that $[OH^-]$ must also be x, since the two reactants started out at equal concentrations and combined in a 1:1 ratio. Therefore,

$$7.0 \times 10^{10} = \frac{0.10}{(x)(x)}; \quad x^2 = \frac{1.0 \times 10^{-1}}{7.0 \times 10^{10}} = 1.4 \times 10^{-12}$$

$$x = 1.2 \times 10^{-6}$$

We see that, as predicted, very little HF is left at equilibrium.

bonic acid (a water solution of carbon dioxide) with a solution of a strong base such as sodium hydroxide:

$$H_2CO_3 + OH^- \longrightarrow H_2O + HCO_3^- \tag{18.6}$$

$$HCO_3^- + OH^- \longrightarrow H_2O + CO_3^{2-} \tag{18.7}$$

The ultimate product depends upon the relative quantities of acid and base used. If one mole of OH^- ions is added per mole of H_2CO_3, the product is the HCO_3^- ion. Addition of another mole of OH^- removes the second proton to give the carbonate ion, CO_3^{2-}. Reaction 18.7 can be brought about directly by the addition of a strong base to a solution prepared by dissolving sodium hydrogen carbonate, $NaHCO_3$, in water.

3. The reaction of a solution of a weak base with a solution of a strong acid. Examples of reactions of this type include the reaction of a strong acid (such as $HClO_4$, HCl) with ammonia,

$$H^+ + NH_3 \longrightarrow NH_4^+ \tag{18.8}$$

or with an anion that is a weak base:

$$H^+ + C_2H_3O_2^- \longrightarrow HC_2H_3O_2$$

A reaction of this type, like that between a weak acid and a strong base, can be visualized as occurring in two steps:

This reaction occurs when hydrochloric acid is added to sodium acetate.

$$NH_3 + H_2O \longrightarrow NH_4^+ + OH^-; \qquad \Delta H = +1.0 \text{ kcal} \tag{18.8a}$$

$$\underline{H^+ + OH^- \longrightarrow H_2O; \qquad\qquad \Delta H = -13.4 \text{ kcal}} \tag{18.8b}$$

$$NH_3 + H^+ \longrightarrow NH_4^+; \qquad\qquad \Delta H = -12.4 \text{ kcal}$$

The amount of heat evolved in the reaction ordinarily differs from the heat of neutralization, 13.4 kcal; it may be smaller or larger than this amount. The equilibrium constant for the reaction of a weak base with a strong acid may be expressed as

$$K = (1.0 \times 10^{14})(K_b \text{ for weak base}) \tag{18.9}$$

Again, since K_b is ordinarily much greater than 10^{-14}, the equilibrium constant will be greater than unity and one can expect the reaction to go nearly to completion. In the case of the reaction of ammonia with a strong acid, we have

$$NH_3 + H^+ \longrightarrow NH_4^+$$

$$K = \frac{[NH_4^+]}{[NH_3] \times [H^+]} = (1.0 \times 10^{14})(1.8 \times 10^{-5}) = 1.8 \times 10^9$$

If the anion is capable of acquiring two protons, a two-step reaction takes place.

$$H^+ + CO_3^{2-} \longrightarrow HCO_3^- \tag{18.10}$$

Calculate K for this reaction (see Table 17.6).

TABLE 18.1   CHARACTERISTICS OF ACID-BASE REACTIONS

TYPE	EXAMPLE	EQUILIBRIUM CONSTANT
Strong acid + strong base	$H^+ + OH^- \longrightarrow H_2O$	1×10^{14}
Weak acid + strong base	$HF + OH^- \longrightarrow H_2O + F^-$	7×10^{10}
Strong acid + weak base	$H^+ + NH_3 \longrightarrow NH_4^+$	1.8×10^9

$$H^+ + HCO_3^- \longrightarrow H_2CO_3 \longrightarrow CO_2(g) + H_2O \qquad (18.11)$$

Addition of excess acid to a solution containing the CO_3^{2-} ion leads to the formation of CO_2 via Reactions 18.10 and 18.11. Carbon dioxide may also be formed by adding sodium hydrogen carbonate to a strongly acidic solution. This is, of course, what happens in the stomach when bicarbonate of soda, $NaHCO_3$, is taken to relieve acid indigestion.

18.2   ACID-BASE TITRATIONS

Acid-base reactions are commonly used in quantitative analysis to determine the concentration or amount of an acidic or basic constituent of a mixture. The experimental setup resembles that shown in Figure 16.2. The calculations involved are illustrated by Examples 18.2 and 18.3.

EXAMPLE 18.2.   It is found that 22.3 ml of 0.240 M NaOH is required to react with a 50.0 ml sample of vinegar, a solution of acetic acid in water. Calculate the concentration of acetic acid in the vinegar.

SOLUTION.   From the equation for the reaction of OH^- ions with acetic acid

$$HC_2H_3O_2 + OH^- \longrightarrow C_2H_3O_2^- + H_2O$$

it is evident that the reactants are consumed in a 1:1 mole ratio. That is,

no. moles OH^- = no. moles $HC_2H_3O_2$

But   no. moles of OH^- = (M NaOH)(volume NaOH)

no. moles $HC_2H_3O_2$ = (M $HC_2H_3O_2$)(volume $HC_2H_3O_2$ solution)

Hence

(M NaOH)(volume NaOH) = (M $HC_2H_3O_2$)(volume $HC_2H_3O_2$ solution)

$$\left(0.240 \frac{\text{mole}}{\text{lit}}\right)(0.0223 \text{ lit}) = (\text{M } HC_2H_3O_2)(0.0500 \text{ lit})$$

Solving,   M $HC_2H_3O_2 = 0.240 \frac{\text{mole}}{\text{lit}} \times \frac{0.0223}{0.0500} = 0.107$ M

EXAMPLE 18.3.   A solid mixture of $Ca(OH)_2$ and $CaCl_2$ is analyzed by titration with HCl. It is found that a sample weighing 0.500 g requires 24.0 ml of 0.200 M HCl. Determine the percentage of $Ca(OH)_2$ in the mixture.

SOLUTION.   We shall first calculate the number of moles of HCl used in the titration and equate this to the number of moles of OH^- ion in the sample. From the latter quantity we can determine the number of moles of $Ca(OH)_2$, convert this to grams and finally to percentage by weight.

$$\text{no. moles HCl} = 0.200 \, \frac{\text{mole}}{\text{lit}} \times 0.0240 \text{ lit} = 0.00480$$

Since 1 mole of HCl reacts with 1 mole of OH^-,

$$\text{no. moles OH}^- = 0.00480$$

Since 1 mole $Ca(OH)_2$ contains 2 moles of OH^-,

$$\text{no. moles Ca(OH)}_2 = \tfrac{1}{2} \times 0.00480 = 0.00240$$

Since 1 mole of $Ca(OH)_2$ weighs 74.0 g,

$$\text{no. g Ca(OH)}_2 = 0.00240 \text{ mole} \times \frac{74.0 \text{ g}}{1 \text{ mole}} = 0.178 \text{ g}$$

$$\% \text{ Ca(OH)}_2 = \frac{\text{wt. Ca(OH)}_2}{\text{wt. sample}} \times 100 = \frac{0.178 \text{ g}}{0.500 \text{ g}} \times 100 = 35.6\%$$

Normality: Gram Equivalent Weights of Acids and Bases

The concentrations of solutions used in acid-base titrations are frequently expressed in terms of normality (cf. Chapter 11, Section 11.3).

$$\text{Normality} = \frac{\text{no. of GEW solute}}{\text{no. lit solution}} \qquad (18.12)$$

In acid-base reactions, the gram equivalent weight of an acid is defined as the weight that reacts with one mole of OH^- ions; the gram equivalent weight of a base is defined as the weight that reacts with one mole of H^+ ions. Since OH^- and H^+ ions react in a 1:1 mole ratio, it follows that one gram equivalent weight of any acid must react exactly with one gram equivalent weight of any base. Stated another way, in any acid-base reaction,

GEW is always defined so that 1 GEW of one species reacts exactly with 1 GEW of the other.

$$\text{no. of GEW acid} = \text{no. of GEW base} \qquad (18.13)$$

Combining Equations 18.12 and 18.13, we conclude that in any acid-base titration,

$$(\text{N acid}) \, (\text{V acid}) = (\text{N base}) \, (\text{V base}) \qquad (18.14)$$

in which the symbols N and V stand for normality and volume respectively.

The simplicity and general validity of Equation 18.14 explains the usefulness of normality as a concentration unit in analytical chemistry. It tells us, for example, that acidic and basic solutions of equal normality will react in a 1:1 volume ratio, regardless of the particular acid or base involved. Furthermore, it makes it possible to determine the normality of an acidic or basic solution directly from an acid-base titration. If it is found, for example, that 20 cc of 0.40 N NaOH are required to titrate 30 cc of a solution of a certain acid, it can be established immediately from Equation 18.14 that

$$\text{N acid} = \text{N base} \times \frac{\text{V base}}{\text{V acid}} = 0.40 \text{ N} \times \frac{20}{30} = 0.27 \text{ N}$$

The normality of an acidic or basic solution is readily calculated from the molarity by the equation

$$N = x \, M \qquad (18.15)$$

in which x is the number of gram equivalent weights furnished by one mole. For an acid such as HCl, one mole of which always neutralizes one mole of OH^- ions, the gram equivalent weight is identical with the mole, x in Equation 18.15 is one, and the normality is equal to the molarity.

For acids such as H_2SO_4 or H_3PO_4, which are capable of reacting with more than one mole of OH^- ions, the value of x in Equation 18.15 will depend upon the number of protons that react. In the case of phosphoric acid, x may be one, two, or three:

$$H_3PO_4 + OH^- \longrightarrow H_2PO_4^- + H_2O; \qquad x = 1, N = M \qquad \text{(18.16)}$$

$$H_3PO_4 + 2\ OH^- \longrightarrow HPO_4^{2-} + 2\ H_2O; \quad x = 2, N = 2\ M \qquad \text{(18.17)}$$

$$H_3PO_4 + 3\ OH^- \longrightarrow PO_4^{3-} + 3\ H_2O; \qquad x = 3, N = 3\ M \qquad \text{(18.18)}$$

Equations 18.16 to 18.18 illustrate the ambiguity involved in the use of normality as a concentration unit. A solution containing one mole (98 g) of H_3PO_4 per liter is always 1 M; its normality may be 1, 2, or 3, depending upon which of these three reactions it undergoes. In general, the normality of a given laboratory reagent can be stated only when one specifies the reaction in which it is to be used.

If we titrate H_3PO_4 with NaOH to an end point with methyl red, only one proton reacts.

Acid-Base Indicators

Acid-base titrations are ordinarily carried out in the presence of a small amount of indicator. To understand how an indicator works, consider a specific example, bromthymol blue. This organic dye is a weak acid in which the undissociated molecule, which we shall represent simply as HIn, has a color (yellow) different from that of its conjugate base, In^- (blue). For the equilibrium involved, we can write

$$\underset{\text{yellow}}{HIn} \rightleftharpoons H^+ + \underset{\text{blue}}{In^-} \qquad K_a = \frac{[H^+] \times [In^-]}{[HIn]} = 1 \times 10^{-7}$$

The color of a solution of bromthymol blue depends upon the relative concentrations of the two colored species, HIn and In^-. When these concentrations are equal, the indicator has a green color intermediate between the yellow of the HIn molecule and the blue of the In^- ion. It is evident from the foregoing equation that this condition is fulfilled, that is, $[HIn] = [In^-]$, when

$$[H^+] = K_a = 1 \times 10^{-7}\ M$$

In other words, if bromthymol blue is used in an acid-base titration, the **end point** (the point at which a color change occurs) is reached at a H^+ ion concentration of about 10^{-7} M or a pH of 7. In more strongly acidic solution (pH < 7), bromthymol blue will exist primarily in the form of the yellow HIn molecule; above a pH of 7, the blue color of the In^- ion predominates.

In practice, an acid-base indicator such as bromthymol blue does not undergo an abrupt color change at a particular pH. Instead, it shows a gradual color change over a range of about 2 pH units. If a strongly acidic solution containing a drop of bromthymol blue is slowly neutralized, one can begin to detect a color change when the concentration of In^- is about one tenth that of HIn, that is, at a pH of 6. Not until the concentration of In^- is 10 times that

Show that when $[In^-] = 0.1[HIn]$, pH = 6.

of HIn (pH = 8) is the yellow color of the latter species completely obscured by the blue of the In⁻ ion.

It should be evident from this discussion that the end point observed with a particular indicator depends upon the magnitude of its ionization constant, K_a. Methyl red ($K_a = 1 \times 10^{-5}$) shows a color change from red to yellow at pH 5; phenolphthalein ($K_a = 1 \times 10^{-9}$) turns red at pH 9. In general, the stronger the acid used as an indicator, the lower the pH at which a color change occurs.

In carrying out an acid-base titration, we try to use an indicator which changes color at or near the **equivalence point**, the point at which equivalent quantities of acid and base have been added. The pH at the equivalence point (and hence the most suitable indicator for an acid-base titration) depends primarily upon the relative strengths of the acid and base involved.

1. Titration of a strong acid with a strong base (pH at equivalence point = 7). A typical example of such a titration is that involved in the reaction of HCl with NaOH. At the equivalence point, there is present a solution of NaCl, a neutral salt. It would seem that the most suitable indicator would be one, such as bromthymol blue or litmus, which changes color at pH 7. In practice, almost any indicator is suitable for such a titration, because the pH changes very rapidly near the equivalence point. The addition of less than a drop of reagent at that point can change the pH by as much as 6 units (e.g., from 4 to 10, Example 18.4). Methyl red (end point at pH 5) or phenolphthalein (end point at pH 9) are quite satisfactory indicators for the titration of a strong acid with a strong base.

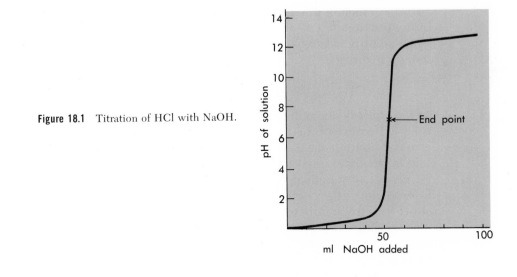

Figure 18.1   Titration of HCl with NaOH.

EXAMPLE 18.4.   50.00 ml of 1.00 M HCl is titrated with 1.00 M NaOH. Calculate the pH

  a. After 49.99 ml of 1.00 M NaOH has been added.

  b. After 50.00 ml of 1.00 M NaOH has been added.

  c. After 50.01 ml of 1.00 M NaOH has been added.

463

SOLUTION

a. At this point, 0.01 ml of 1.00 M HCl remains unreacted in a total volume of almost exactly 100 ml. Consequently,

$$\text{conc. HCl} = \text{conc. H}^+ = 1.00 \text{ M} \times \frac{0.01}{100} = 1 \times 10^{-4} \text{ M}$$

$$pH = -\log_{10} [\text{H}^+] = 4$$

b. At the equivalence point, we have a neutral solution of sodium chloride

$$pH = 7$$

c. We now have an excess of 0.01 ml of 1.00 M NaOH in a total of 100 ml.

$$\text{conc. NaOH} = \text{conc. OH}^- = 1.00 \text{ M} \times \frac{0.01}{100} = 1 \times 10^{-4} \text{ M}$$

$$\text{conc. H}^+ = \frac{1 \times 10^{-14}}{1 \times 10^{-4}} = 1 \times 10^{-10} \text{ M}$$

$$pH = 10$$

We see that near the equivalence point the addition of only 0.02 ml of base ($\approx \frac{1}{2}$ drop) changes the pH by six units. If the acid and base are more dilute, the change is slightly less abrupt, but the general conclusions remain valid.

2. Titration of a weak acid with a strong base (pH at equivalence point > 7). In the titration of a weak acid with a strong base, the solution at the equivalence point is basic. Consider, for example, the titration of acetic acid with sodium hydroxide:

$$\text{HC}_2\text{H}_3\text{O}_2 + \text{OH}^- \longrightarrow \text{C}_2\text{H}_3\text{O}_2{}^- + \text{H}_2\text{O}$$

At the equivalence point, one is left with a solution of sodium acetate which is basic because of the hydrolysis of the acetate ion.

The titration of a weak acid with a strong base differs from that of a strong acid in one other important respect; the pH changes much more slowly in the vicinity of the equivalence point (compare Figure 18.2 with Figure 18.1). Consequently, the choice of indicator is critical; phenolphthalein (end point at pH 9) is frequently used in titrations of this type.

The slope of the titration curve for a weak acid near the equivalence point depends upon the value of its ionization constant. If the acid is extremely weak ($K_a = 10^{-9}$ or less) the vertical portion of the curve virtually disappears, the change in pH near the equivalence point is very gradual, and titration becomes extremely difficult or even impossible. An example of an acid that is very difficult to titrate by the use of an indicator is HCN ($K_a = 4 \times 10^{-10}$; cf. Figure 18.2).

3. Titration of a weak base with a strong acid (pH at equivalence point < 7). In the titration of ammonia with hydrochloric acid, a solution of ammonium chloride is formed.

See also Problem 18.33.

$$\text{NH}_3 + \text{H}^+ \longrightarrow \text{NH}_4{}^+$$

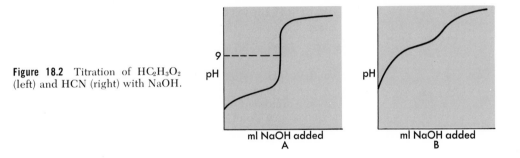

Figure 18.2 Titration of $HC_2H_3O_2$ (left) and HCN (right) with NaOH.

The solution at the equivalence point is acidic, with a pH of approximately 5, because of the hydrolysis of the NH_4^+ ion (Example 18.5a). Moreover, the pH changes rather slowly near the equivalence point. Consequently, one should use an indicator such as methyl red (end point at pH 5) which undergoes a color change in weakly acidic solution.

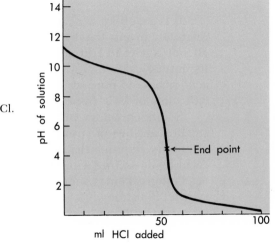

Figure 18.3 Titration of NH_3 with HCl.

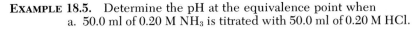

EXAMPLE 18.5. Determine the pH at the equivalence point when
   a. 50.0 ml of 0.20 M NH_3 is titrated with 50.0 ml of 0.20 M HCl.

   b. 50.0 ml of 0.20 M $HC_2H_3O_2$ is titrated with 50.0 ml of 0.20 M NaOH.

SOLUTION
   a. Referring to the equation for the titration, we see that a solution of NH_4^+ ions is formed. To determine the concentration of NH_4^+, we note that we started with

   $$(0.0500 \text{ lit})(0.20 \text{ mole/lit}) = 0.0100 \text{ mole of } NH_3$$

   The NH_3 was converted to NH_4^+ in the titration; at the equivalence point, the volume of solution is 100 ml. Consequently, the concentration of NH_4^+ must be

   $$(0.0100 \text{ mole})/(0.100 \text{ lit}) = 0.100 \text{ M}$$

   Now you may recall that in Example 17.4b, we calculated

465

the H^+ in a 0.10 M solution of NH_4^+. Taking the answer from page 438,

$$[H^+] = 7.5 \times 10^{-6}$$

so that $\quad pH = -\log_{10}(7.5 \times 10^{-6}) = -(0.9 - 6.0) = 5.1$

b. By the argument outlined in part a, we conclude that, at the equivalence point, we have a solution 0.10 M in $C_2H_3O_2^-$. In Example 17.7b, page 442, we found that in such a solution

$$[OH^-] = 7.5 \times 10^{-6}$$

but $\quad [H^+] = \dfrac{1.0 \times 10^{-14}}{[OH^-]} = \dfrac{1.0 \times 10^{-14}}{7.5 \times 10^{-6}} = 1.3 \times 10^{-9}$

$$pH = 8.9$$

18.3 BUFFERS

It is a well-known fact that the pH of blood and many other body fluids is relatively insensitive to the addition of acid or base. If one adds 0.01 mole of HCl or NaOH to 1 lit of blood, its pH changes by less than 0.1 of a unit from its normal value of 7.4. In contrast, the addition of these same quantities of acid or base to pure water changes its pH by about 5 units (from 7 to 2 with 0.01 mole of HCl, from 7 to 12 with 0.01 mole of NaOH).

Solutions whose pH remains virtually unchanged upon the addition of small quantities of strong acid or base are said to be **buffered**. Buffer action ordinarily requires that the solution contain two different components, one capable of reacting with H^+ ions, the other with OH^- ions. A simple example of a buffer system is one prepared by adding acetic acid to a solution of sodium acetate. If a strong base such as sodium hydroxide is added to the system, the OH^- ions react with acetic acid molecules.

$$OH^- + HC_2H_3O_2 \longrightarrow H_2O + C_2H_3O_2^- \qquad (18.19)$$

Since the equilibrium constant for this reaction is large, the OH^- ions are almost completely removed.

Addition of a strong acid to the buffer results in a reaction of the H^+ ions of the acid with the acetate ions of the sodium acetate:

$$H^+ + C_2H_3O_2^- \longrightarrow HC_2H_3O_2 \qquad (18.20)$$

In either case, the OH^- ions or H^+ ions added are consumed and hence fail to produce the drastic pH change observed when a strong acid or base is added to pure water or to an unbuffered solution.

The addition of acid or base to a buffered solution such as the $HC_2H_3O_2$-$C_2H_3O_2^-$ system does result in a small pH change, the magnitude of which can be calculated as illustrated in Example 18.6.

EXAMPLE 18.6. A buffered system contains 0.10 mole of $HC_2H_3O_2$ and 0.10 mole of $C_2H_3O_2^-$ per liter.
a. Calculate the pH of the buffer.

b. Calculate the pH after addition of 0.01 mole/lit of HCl.

c. Calculate the pH after addition of 0.01 mole/lit of NaOH.

SOLUTION

a. To obtain the pH for the buffered system, we set up the expression for the ionization constant of acetic acid:

$$K_a = 1.8 \times 10^{-5} = \frac{[H^+] \times [C_2H_3O_2^-]}{[HC_2H_3O_2]}$$

Substituting $[HC_2H_3O_2] = [C_2H_3O_2^-] = 0.10$ M and solving for the hydrogen ion concentration gives $[H^+] = 1.8 \times 10^{-5}$

$$pH = -\log_{10}[H^+] = -\log(1.8 \times 10^{-5}) = -(0.26 - 5.00) = 4.74$$

b. When 0.01 mole of H^+ is added, Reaction 18.20 occurs, thereby consuming 0.01 mole of $C_2H_3O_2^-$ and forming 0.01 mole of $HC_2H_3O_2$. This means that $(0.10 - 0.01) = 0.09$ mole of $C_2H_3O_2^-$ and $(0.10 + 0.01) = 0.11$ mole of $HC_2H_3O_2$ remain. Therefore,

$$[H^+] = 1.8 \times 10^{-5} \times \frac{[HC_2H_3O_2]}{[C_2H_3O_2^-]} =$$
$$1.8 \times 10^{-5} \frac{(0.11)}{(0.09)} = 2.2 \times 10^{-5}, \quad pH = 4.66$$

In other words, the addition of 0.01 mole of HCl to this buffered system reduces the pH by only 0.08 unit.

c. Here, Reaction 18.19 takes place, consuming 0.01 mole of $HC_2H_3O_2$ and producing 0.01 mole of $C_2H_3O_2^-$.

$$\text{no. moles } HC_2H_3O_2 = 0.10 - 0.01 = 0.09$$
$$\text{no. moles } C_2H_3O_2^- = 0.10 + 0.01 = 0.11$$
$$[H^+] = 1.8 \times 10^{-5} \times \frac{[HC_2H_3O_2]}{[C_2H_3O_2^-]} = 1.8 \times 10^{-5} \frac{(0.09)}{(0.11)} = 1.5 \times 10^{-5}$$
$$pH = 4.82$$

Again, we find that the pH changes by less than 0.1 unit.

In preparing buffer solutions, it is important to take into account two quantitative principles of buffer action:

1. The capacity of a buffer to absorb H^+ and OH^- ions without an appreciable change in pH is limited by the concentrations of the weak acid and weak base making up the buffer. To illustrate this principle, let us consider the buffer system discussed in Example 18.5. Suppose that instead of adding 0.01 mole/lit of HCl, we were to add five times as much, or 0.05 mole/lit. We now calculate the concentrations of $C_2H_3O_2^-$ and $HC_2H_3O_2$ to be 0.05 M and 0.15 M respectively. Consequently,

$$[H^+] = K_a \times \frac{[HC_2H_3O_2]}{[C_2H_3O_2^-]} = 1.8 \times 10^{-5} \times \frac{0.15}{0.05} = 5.4 \times 10^{-5}$$

$$pH = 4.26$$

We see that the pH has changed by 0.48 units (from 4.74 to 4.26). This contrasts unfavorably with the pH change of only 0.08 units (from 4.74 to 4.66) which was calculated for the addition of 0.01 mole of HCl.

If the capacity of the buffer is exceeded by the addition of an amount of a strong acid (or base) greater than the amount of weak base (or acid) making up the buffered system, a drastic change in pH occurs, comparable to that taking place in an unbuffered system. Suppose, for example, that we were to add 0.15 mole of HCl to one liter of a buffered solution containing 0.10 mole of $C_2H_3O_2^-$. In this case, 0.05 mole of the strong acid would remain unreacted, and we would have

$$[H^+] = 0.050 \text{ M}; \quad pH = 1.70$$

At what pH would an $NH_4^+ - NH_3$ buffer be effective?

2. A buffer system is capable of maintaining a relatively constant pH in a rather narrow range, about one unit greater or less than the pK_a of the weak acid ($pK_a = -\log_{10} K_a$). Consider, once again, the $HC_2H_3O_2$-$C_2H_3O_2^-$ ststem, where $pK_a = -\log_{10} (1.8 \times 10^{-5}) = 4.74$. If these two species are present in equal concentrations, let us say 1.0 M, the pH will be about 4.7:

$$[H^+] = K_a \times \frac{[HC_2H_3O_2]}{[C_2H_3O_2^-]} = 1.8 \times 10^{-5} \times \frac{1.0}{1.0} = 1.8 \times 10^{-5}$$

$$pH = 4.74$$

In order to obtain a pH of 3.7, it would be necessary to use 10 times as much $HC_2H_3O_2$ as $NaC_2H_3O_2$. We might, for example, make the concentrations of $HC_2H_3O_2$ and $C_2H_3O_2^-$ 1.0 M and 0.10 M respectively.

$$[H^+] = 1.8 \times 10^{-5} \times \frac{1.0}{0.1} = 1.8 \times 10^{-4}; \quad pH = 3.74$$

Consider, now, what would have to be done to reduce the pH by another unit, to 2.7. The concentration of $HC_2H_3O_2$ would have to be about 100 times that of $C_2H_3O_2^-$. If we hold $[HC_2H_3O_2]$ at 1.0 M (in general we cannot make this concentration much higher without exceeding the solubility of a weak acid), this would require that $[C_2H_3O_2^-]$ be reduced to 0.01 M.

$$[H^+] = 1.8 \times 10^{-5} \times \frac{1.0}{0.01} = 1.8 \times 10^{-3}; \quad pH = 2.74$$

But the capacity of a buffer system containing only 0.01 mole/lit of $C_2H_3O_2^-$ would be very limited. The addition of more than 0.01 mole/lit of a strong acid such as HCl would consume all the $C_2H_3O_2^-$ ions and thereby produce a sharp decrease in pH.

Quite clearly, the $HC_2H_3O_2$-$C_2H_3O_2^-$ buffer system will be of little practical use if we wish to work at a pH lower than about 3.7 (or higher than about 5.7). Many other buffer systems are, of course, available. Phosphate buffers, widely used in biological work, are made by neutralizing phosphoric acid, H_3PO_4, with a strong base such as sodium hydroxide. Three different buffer combinations are possible: H_3PO_4-$H_2PO_4^-$, $H_2PO_4^-$-HPO_4^{2-}, and HPO_4^{2-}-PO_4^{3-}, effective at pH 2, 7, and 12 respectively.

Buffered solutions are used whenever it is important to maintain constant pH throughout a physical or chemical change. They are particularly useful in studying rates of chemical reactions in which H^+ ions are produced or consumed. Suppose, for example, one wishes to study the rate of oxidation of iodide ions with hydrogen peroxide:

$$2 I^- + H_2O_2 + 2 H^+ \longrightarrow I_2 + 2 H_2O \tag{18.21}$$

Since this reaction involves H^+ ions, its rate may be expected to be strongly influenced by pH. If the pH is maintained at a constant value, it becomes possible to eliminate this variable and determine more accurately the effect of such factors as the concentration of I^-, the temperature, or the presence of a catalyst upon the reaction rate. Almost all biochemical reactions, particularly those involving enzymes, are pH dependent. Buffers are an invaluable tool in the study of such reactions.

Buffered solutions have many applications in analytical and physical chemistry. In particular, it may be recalled that one of the methods described in Chapter 17 for the determination of the ionization constant of a weak acid involves its partial neutralization with base. Since the resulting solution is buffered, its pH can be determined quite accurately even in the presence of basic or acidic impurities. This feature is perhaps the outstanding advantage of this particular method of determining ionization constants.

18.4 APPLICATION OF ACID-BASE REACTIONS IN INORGANIC SYNTHESIS

Acid-base reactions are frequently used to prepare inorganic compounds. They may be adapted to the preparation of almost any salt; in addition, acid-base reactions may be used to prepare certain volatile acids and bases.

Preparation of Salts

Let us suppose that we wish to prepare a sample of pure cesium nitrate, $CsNO_3$, from cesium hydroxide, $CsOH$. This may be done by a two-step process, in which the hydroxide is first neutralized by nitric acid:

$$Cs^+ + OH^- + H^+ + NO_3^- \longrightarrow Cs^+ + NO_3^- + H_2O \qquad \text{(18.22a)}$$

This is not a net ionic equation.

and the resulting solution, which should contain only Cs^+ and NO_3^- ions, is then evaporated:

$$Cs^+ + NO_3^- \longrightarrow CsNO_3(s) \qquad \text{(18.22b)}$$

In a completely analogous manner, it is possible to convert ammonia to ammonium salts by reaction with an equivalent amount of acid and evaporation of the solution:

$$NH_3 + H^+ + Cl^- \longrightarrow NH_4^+ + Cl^- \qquad \text{(18.23a)}$$

$$NH_4^+ + Cl^- \longrightarrow NH_4Cl(s) \qquad \text{(18.23b)}$$

Acids such as HF are readily converted to their salts in the same manner:

$$Rb^+ + OH^- + HF \longrightarrow Rb^+ + F^- + H_2O \qquad \text{(18.24a)}$$

$$Rb^+ + F^- \longrightarrow RbF(s) \qquad \text{(18.24b)}$$

When acid-base reactions such as 18.22a, 18.23a, and 18.24a are to be used to prepare pure inorganic salts, it is highly desirable that one use exactly equivalent quantities of acid and base. Suppose, for example, that in Reaction 18.22a, one were to use an excess of cesium hydroxide; upon evaporation the cesium nitrate would be contaminated with unreacted cesium hydroxide.

When an acid contains more than one ionizable hydrogen, the product obtained depends upon the relative quantities of acid and base used. An example is carbonic acid, H_2CO_3. Either $NaHCO_3$ or Na_2CO_3 can be prepared by bubbling carbon dioxide through a solution of sodium hydroxide and evaporating. To make $NaHCO_3$, 1 mole of H_2CO_3 is needed for every mole of NaOH:

$$H_2CO_3 + Na^+ + OH^- \longrightarrow Na^+ + HCO_3^- + H_2O \qquad (18.25a)$$

$$Na^+ + HCO_3^- \longrightarrow NaHCO_3(s) \qquad (18.25b)$$

If the desired product is Na_2CO_3, 2 moles of NaOH should be used:

$$H_2CO_3 + 2\ Na^+ + 2\ OH^- \longrightarrow 2\ Na^+ + CO_3^{2-} + 2\ H_2O \qquad (18.26a)$$

$$2\ Na^+ + CO_3^{2-} \longrightarrow Na_2CO_3(s) \qquad (18.26b)$$

In practice, $NaHCO_3$ and Na_2CO_3 are not ordinarily prepared in this way, but rather are made by the Solvay process (Section 18.6).

Preparation of Volatile Acids or Bases

A convenient way to prepare small quantities of anhydrous ammonia is to heat a solution containing NH_4^+ ions with a strong base such as sodium hydroxide:

$$NH_4^+ + OH^- \longrightarrow NH_3(g) + H_2O \qquad (18.27)$$

By passing the gas evolved over a drying agent, it is possible to remove any water present and thereby obtain relatively pure ammonia.

Several volatile weak acids are readily prepared by treating a solution of one of their salts with a strong, nonvolatile acid such as sulfuric acid. Thus, carbon dioxide is produced by adding sulfuric acid to sodium carbonate or sodium hydrogen carbonate:

$$CO_3^{2-} + 2\ H^+ \longrightarrow (H_2CO_3) \longrightarrow CO_2(g) + H_2O \qquad (18.28)$$

$$HCO_3^- + H^+ \longrightarrow (H_2CO_3) \longrightarrow CO_2(g) + H_2O \qquad (18.29)$$

In a similar manner, hydrogen sulfide can be prepared from a soluble sulfide:

$$S^{2-} + 2\ H^+ \longrightarrow H_2S(g) \qquad (18.30)$$

TABLE 18.2 VOLATILITY OF ACIDS FROM WATER SOLUTION

ACID	BOILING POINT (°C)	
Hydrogen sulfide	100	H_2S escapes on warming
Carbonic acid	100	CO_2 escapes on warming
Sulfurous acid	100	SO_2 escapes on warming
Hydrochloric acid	110	constant boiling mixture (20% HCl)
Hydrofluoric acid	120	constant boiling mixture (35% HF)
Nitric acid	121	constant boiling mixture (68% HNO_3)
Hydrobromic acid	126	constant boiling mixture (47% HBr)
Hydriodic acid	127	constant boiling mixture (57% HI)
Phosphoric acid	213	$2\ H_3PO_4 \longrightarrow H_4P_2O_7 + H_2O$
Sulfuric acid	330	$H_2SO_4 \longrightarrow SO_3 + H_2O$

The usefulness of the type of reaction illustrated by Equations 18.28 to 18.30 depends upon the weak acid being volatile, not on the relative strengths of the acids. Indeed, one can prepare volatile strong acids by reactions entirely analogous to these. For example, hydrogen chloride can be formed by allowing sulfuric acid to drop on sodium chloride (Figure 18.4):

$$NaCl(s) + H_2SO_4(l) \longrightarrow NaHSO_4(s) + HCl(g) \qquad (18.31)$$

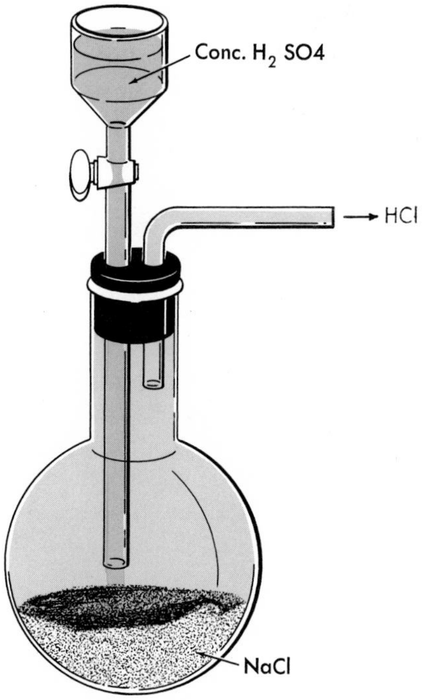

Figure 18.4 Preparation of HCl(g) from NaCl(s).

To prepare hydrogen bromide or hydrogen iodide, phosphoric acid is ordinarily used rather than sulfuric acid to avoid oxidation of the hydrogen halide to the free halogen (cf. Chapter 22, Section 22.5).

What is the equation for the reaction of sodium bromide with phosphoric acid?

18.5 APPLICATION OF ACID-BASE REACTIONS IN QUALITATIVE ANALYSIS

Acid-base reactions are widely used in qualitative analysis for either of two different purposes. An acid-base reaction may be used to test for a specific ion by converting that ion into a volatile, easily detectable product. Alternatively, such a reaction may be used to separate one ion from another. In this case, advantage is taken of the ability of acidic solutions to dissolve certain water-insoluble salts while leaving others unaffected.

Tests for Specific Ions

The reaction that occurs when an ammonium salt is heated with a strong base (Equation 18.27) is frequently used as a qualitative test for the NH_4^+ ion.

The ammonia formed may be detected either by its odor or its effect on litmus; of all the gases commonly encountered in the analytical laboratory, ammonia is the only one that is basic to litmus.

The reactions represented by Equations 18.28 to 18.30 are useful for the qualitative detection of CO_3^{2-}, HCO_3^-, or S^{2-} ions. Carbon dioxide, formed by Reaction 18.28 or 18.29, is detected by passing it through a solution of calcium hydroxide, which gives a white precipitate of calcium carbonate:

$$Ca^{2+} + 2\ OH^- + CO_2(g) + H_2O \longrightarrow CaCO_3(s) + 2\ H_2O \qquad (18.32)$$

Sketch an apparatus that could be used to detect NH_4^+, CO_3^{2-}, and S^{2-}.

Hydrogen sulfide, produced by Reaction 18.31, may be recognized either by its foul odor or its ability to form a black precipitate with Pb^{2+} ions. A piece of filter paper moistened with lead nitrate solution turns black if exposed to a gas mixture containing hydrogen sulfide:

$$Pb^{2+} + H_2S(g) \longrightarrow PbS(s) + 2\ H^+ \qquad (18.33)$$

Separation of Ions

The use of an acid-base reaction to separate one ion from another is illustrated by a procedure commonly used in anion analysis to distinguish between CO_3^{2-} and SO_4^{2-}. These ions are first precipitated as the barium salts. The mixed precipitate of $BaCO_3$ and $BaSO_4$ is then treated with hydrochloric acid. The carbonate is brought into solution by Reaction 18.28, while the sulfate is unaffected.

$$CO_3^{2-},\ SO_4^{2-}$$
$$\downarrow Ba^{2+}$$
$$BaCO_3,\ BaSO_4$$
$$\downarrow H^+$$

$$Ba^{2+},\ CO_2(g) \qquad\qquad BaSO_4(s)$$

A more subtle application of this principle is involved in the reaction of hydrochloric acid with a mixed precipitate of CuS and CoS. The cobalt sulfide dissolves in acid while copper sulfide does not. This separation is more commonly carried out in reverse. That is, by adding hydrogen sulfide in acidic solution to a mixture of Co^{2+} and Cu^{2+} ions, CuS is precipitated while Co^{2+} remains in solution:

$$Co^{2+},\ Cu^{2+}$$
$$\downarrow H_2S,\ H^+(.3\ M)$$

$$Co^{2+} \qquad\qquad CuS(s)$$

As noted in Chapter 16, ions that form water-insoluble sulfides can be separated into two groups in this manner. The ions which form extremely insoluble sulfides (group 2 = Cu^{2+}, Bi^{3+}, As^{3+}, Sb^{3+}, Sn^{2+}, Sn^{4+}, Cd^{2+}) are precipitated by hydrogen sulfide in acidic solution, while those sulfides which are somewhat more soluble (Co^{2+}, Fe^{2+}, Fe^{3+}, Ni^{2+}, Zn^{2+}, Mn^{2+} in group 3) are precipitated only after the solution is made basic.

To understand the principles behind these separations, one must consider the factors that influence the solubility of a solid in a dilute, nonoxidizing acid. These factors may be deduced by studying the equilibria involved. Consider, for example, the process by which a metal sulfide, general formula MS, dissolves in acid:

$$MS(s) \rightleftharpoons M^{2+} + S^{2-} \qquad (18.34a)$$

$$S^{2-} + 2\ H^+ \rightleftharpoons H_2S(g) \qquad (18.34b)$$

$$\overline{MS(s) + 2\ H^+ \rightleftharpoons M^{2+} + H_2S(g)} \qquad (18.34)$$

The solubility of a metal sulfide in hydrochloric acid depends upon the position of the equilibrium in 18.34, which is determined by:

1. *The solubility of the solid in water.* The greater the solubility in water, the farther to the right equilibrium (18.34a) will be displaced, and consequently the greater will be the solubility of the solid in acid. Cobalt sulfide ($K_{sp} = 1 \times 10^{-21}$) is more soluble in water than is copper sulfide ($K_{sp} = 1 \times 10^{-25}$); the same order of solubility is maintained in acid. It so happens that in a solution 0.3 M in H^+, the solubility range is such that CuS precipitates while CoS remains in solution (Example 18.7).

2. *The concentration of the acid.* Since H^+ ions appear on the left side of Equation 18.34b, it is clear that an increase in their concentration shifts the solubility equilibrium to the right, thereby bringing more of the solid into solution. Conversely, if the concentration of acid is lowered, the solid shows less tendency to dissolve. When an acidic solution containing Co^{2+} and S^{2-} ions is partially neutralized with base, cobalt sulfide precipitates.

3. *The strength of the weak acid formed.* The weaker the acid (H_2S, H_2CO_3, and so on) formed, the more soluble the salt is in strong acid. All water-insoluble carbonates are readily dissolved by dilute acid to form the very weak acid H_2CO_3 ($K_1 = 4 \times 10^{-7}$, $K_2 = 5 \times 10^{-11}$).

$$MCO_3(s) + 2\ H^+ \longrightarrow M^{2+} + CO_2(g) + H_2O \qquad (18.35)$$

In contrast, water-insoluble fluorides (K_a HF $= 7 \times 10^{-4}$) can be brought into solution only by the use of concentrated acid. The comparison is particularly striking for the two compounds $BaCO_3$ and BaF_2:

$$BaCO_3(s) + 2\ H^+ \longrightarrow Ba^{2+} + CO_2(g) + H_2O \qquad (18.36)$$

$$BaF_2(s) + 2\ H^+ \rightleftharpoons Ba^{2+} + 2\ HF \qquad (18.37)$$

Even though $BaCO_3$ is considerably less soluble in water than BaF_2 (K_{sp} $BaCO_3 = 1 \times 10^{-9}$, K_{sp} $BaF_2 = 2 \times 10^{-6}$), it dissolves readily in very dilute acid, while BaF_2 can be brought into solution only at high acid concentrations. Water-insoluble salts of strong acids, such as $BaSO_4$, PbI_2, and $AgCl$, show no tendency to dissolve in dilute acid.

The quantitative application of these principles is illustrated in Example 18.7.

Only those sulfides with $K_{sp} < 10^{-23}$ will precipitate in 0.3 M acid.

EXAMPLE 18.7. A solution 0.1 M in Cu^{2+} and Co^{2+} and 0.3 M in H^+ is saturated with H_2S so as to make $[H_2S] = 0.1$ M.

   a. Show that under these conditions CuS but not CoS precipitates.

b. To what value must [H⁺] be reduced to start to precipitate CoS?

SOLUTION

a. In order to determine whether or not a precipitate is formed under these conditions, we must first find out how the concentration products, conc. Cu^{2+} × conc. S^{2-}, and conc. Co^{2+} × conc. S^{2-} compare to the solubility products of these salts (K_{sp} CuS $= 1 \times 10^{-25}$, K_{sp} CoS $= 1 \times 10^{-21}$). The concentrations of Cu^{2+} and Co^{2+} are given as 0.1 M; the concentration of S^{2-} must be calculated from that of H^+ and H_2S. To do this, we multiply the first and second ionization constants of H_2S together to give

$$\frac{[H^+] \times [HS^-]}{[H_2S]} \times \frac{[H^+] \times [S^{2-}]}{[HS^-]} = \frac{[H^+]^2 \times [S^{2-}]}{[H_2S]}$$
$$= (1 \times 10^{-7})(1 \times 10^{-15})$$

or $\qquad \dfrac{[H^+]^2 \times [S^{2-}]}{[H_2S]} = 1 \times 10^{-22}$

Substituting, $[H^+] = 0.3$ M, $[H_2S] = 0.1$ M, and solving, gives

$$[S^{2-}] = 1 \times 10^{-22} \times \frac{0.1}{0.09} = 1 \times 10^{-22} \qquad \text{(1 significant figure)}$$

Therefore,

conc. Cu^{2+} × conc. $S^{2-} = (10^{-1})(10^{-22})$
$$= 10^{-23} > K_{sp} \text{ CuS } (1 \times 10^{-25})$$

CuS precipitates.

conc. Co^{2+} × conc. $S^{2-} = (10^{-1})(10^{-22})$
$$= 10^{-23} < K_{sp} \text{ CoS } (1 \times 10^{-21})$$

CoS does not precipitate.

b. A precipitate of CoS will form when $[Co^{2+}] \times [S^{2-}] = K_{sp}$ CoS $= 10^{-21}$. Since $[Co^{2+}] = 10^{-1}$ M, this requires that $[S^{2-}] = 10^{-20}$ M.

We must now calculate what concentration of H^+ will give this value for $[S^{2-}]$ in a solution 0.1 M in H_2S:

$$\frac{[H^+]^2 \times [S^{2-}]}{[H_2S]} = 1 \times 10^{-22}$$

$$[H^+]^2 = \frac{1 \times 10^{-22} \times 0.1}{10^{-20}} = 1 \times 10^{-3} = 10 \times 10^{-4}; \; [H^+] = 3 \times 10^{-2} \text{ M}$$

In other words, if the concentration of H^+ is reduced from 0.3 M to 0.03 M by the addition of a base, a precipitate of CoS will start to form.

18.6 AN INDUSTRIAL APPLICATION OF ACID-BASE REACTIONS: THE SOLVAY PROCESS

A commercially important process in which acid-base reactions play an important part is the so-called Solvay Process for the manufacture of sodium hydrogen carbonate, $NaHCO_3$, and sodium carbonate, Na_2CO_3. This process is economically superior to the method described earlier for the preparation of these compounds, i.e., the reaction of carbon dioxide with a solution of sodium hydroxide. The principal economic advantage of the Solvay Process is that it uses as a raw material sodium chloride, which is considerably less expensive than sodium hydroxide (which has to be made from the chloride by electrolysis).

Preparation of NaHCO₃

The first product formed in the Solvay Process is sodium hydrogen carbonate, $NaHCO_3$. To prepare this compound, advantage is taken of its comparatively low solubility at temperatures near the freezing point of water. Carbon dioxide is bubbled through a concentrated brine solution saturated with ammonia and maintained at a temperature of approximately 0°C. Under these conditions, finely divided crystals of sodium hydrogen carbonate precipitate. The equation for the overall reaction may be written:

$$CO_2(g) + H_2O + NH_3(g) + Na^+ + Cl^- \longrightarrow NaHCO_3(s) + NH_4^+ + Cl^-$$

$$(18.38)$$

The sodium hydrogen carbonate is filtered from the solution of ammonium chloride. It is possible to prepare $NaHCO_3$ in this manner because its solubility at 0°C (0.82 mole/lit) is considerably less than that of any of the other possible products, NH_4Cl (5.5 moles/lit), $NaCl$ (6.1 moles/lit) or NH_4HCO_3 (1.5 moles/lit).

Note that NaHCO₃ is by no means "insoluble" at 0°C.

A reaction such as 18.38, involving as it does a large number of species, can best be understood if we break it down into a series of relatively simple steps. In this case, three such steps may be considered:

1. Bubbling carbon dioxide through water establishes the equilibrium

$$CO_2(g) + H_2O \rightleftharpoons H_2CO_3 \rightleftharpoons H^+ + HCO_3^- \qquad (18.38a)$$

The concentration of HCO_3^- ion produced by this reaction (0.00018 M) is far below that required to precipitate $NaHCO_3$ (about 0.10 M).

2. To increase the concentration of HCO_3^- ions in solution, it is necessary to add a reagent that will shift the equilibrium in 18.38a to the right. This may be accomplished by adding ammonia, which removes H^+ ions by converting them to NH_4^+ ions:

$$NH_3(g) + H^+ \longrightarrow NH_4^+ \qquad (18.38b)$$

Adding, $$CO_2(g) + NH_3(g) + H_2O \longrightarrow NH_4^+ + HCO_3^- \qquad (18.38a + b)$$

3. Under these conditions, the HCO_3^- ions are present at a sufficiently high concentration to be precipitated by the sodium ions of the sodium chloride solution:

$$Na^+ + Cl^- + HCO_3^- \longrightarrow NaHCO_3(s) + Cl^- \qquad (18.38c)$$

Adding all three equations,

$$CO_2(g) + H_2O + NH_3(g) + Na^+ + Cl^- \longrightarrow NaHCO_3(s) + NH_4^+ + Cl^-$$

$$(18.38a + b + c)$$

475

Preparation of Na_2CO_3 from $NaHCO_3$

The greater part of the sodium hydrogen carbonate prepared by the Solvay Process is converted to the more widely used salt, sodium carbonate. This conversion is accomplished by heating $NaHCO_3$ to about 300°C; at this temperature carbon dioxide and water vapor are given off and a white residue of sodium carbonate remains:

$$2\ NaHCO_3(s) \longrightarrow Na_2CO_3(s) + CO_2(g) + H_2O(g) \qquad (18.39)$$

The carbon dioxide formed is recycled to prepare more $NaHCO_3$ by Reaction 18.38.

Preparation of CO_2 and Recovery of NH_3

For every mole of $NaHCO_3$ or Na_2CO_3 produced by the Solvay Process, one mole of CO_2 is consumed. The carbon dioxide is produced by heating limestone:

$$CaCO_3(s) \longrightarrow CaO(s) + CO_2(g) \qquad (18.40)$$

The economics of the Solvay Process, like those of all industrial operations, depend upon the effective use of all the products. In particular, it is important that the calcium oxide produced by Reaction 18.40 be utilized. Furthermore, the solution of ammonium chloride remaining after the precipitation of sodium hydrogen carbonate is far too valuable to be discarded. Indeed, if the ammonia consumed in Reaction 18.38 were not recovered, the Solvay Process would be economically impossible, since the ammonia costs more than the sodium hydrogen carbonate is worth. Fortunately, these two problems—the utilization of the CaO produced in Reaction 18.40 and the recovery of the NH_3 used in 18.38—can be solved simultaneously. It was pointed out earlier that ammonia can be formed from ammonium salts by heating with a base:

$$2\ NH_4^+ + 2\ OH^- \longrightarrow 2\ NH_3(g) + 2\ H_2O$$

Why not use CaO instead of NH_3 to bring about Reaction 18.38?

Calcium oxide, upon addition to water, gives a strongly basic solution:

$$CaO(s) + H_2O \longrightarrow Ca^{2+} + 2\ OH^-$$

Consequently, if the solution of NH_4Cl remaining after the precipitation of $NaHCO_3$ is heated with CaO, the reaction

$$CaO(s) + 2\ NH_4^+ + 2\ Cl^- \longrightarrow Ca^{2+} + 2\ Cl^- + 2\ NH_3(g) + H_2O \quad (18.41)$$

occurs, liberating ammonia and leaving, as a final by-product, a solution of calcium chloride.

Summary of the Solvay Process

The raw materials used in the Solvay Process are sodium chloride, water, and limestone. The first product formed is sodium hydrogen carbonate, $NaHCO_3$, which precipitates at 0°C from a sodium chloride solution saturated with CO_2 (produced by the thermal decomposition of $CaCO_3$) and NH_3 (recovered by heating the solution remaining with CaO). The $NaHCO_3$ may be

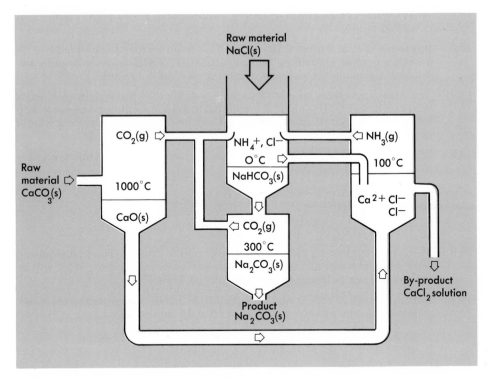

Figure 18.5 Flow diagram of the Solvay process.

sold for use in medicine, as a baking powder, and so on, or it may be converted to Na_2CO_3 by heating. More than six million tons of sodium carbonate are used annually in this country in the manufacture of glass, paper, soap, and other chemicals. Calcium chloride, formed as a by-product, is used as a drying agent and in ice removal from highways.

PROBLEMS

18.1 Write balanced net ionic equations for the reactions that occur when the following solutions are mixed (do not include "spectator" ions).

  a. NaOH and HNO_3 d. NH_3 + $HClO_4$
  b. NaOH and HF   e. $NaC_2H_3O_2$ and HI
  c. NaOH and NH_4Cl f. $Ca(OH)_2$ and HCl

18.2 Consider the reaction $HC_2H_3O_2 + OH^- \rightarrow C_2H_3O_2^- + H_2O$.

  a. Calculate the equilibrium constant for this reaction.
  b. Calculate the concentration of $HC_2H_3O_2$ at equilibrium when 0.20 of a mole of NaOH is added to a liter of 0.20 M $HC_2H_3O_2$.

18.3 Calculate equilibrium constants for the following reactions.

  a. $H^+ + OH^- \longrightarrow H_2O$
  b. $H^+ + C_2H_3O_2^- \longrightarrow HC_2H_3O_2$

18.4 Two hundred ml of 0.100 M HNO_3 is added to 300 ml of 0.100 M $Ba(OH)_2$. Calculate the number of moles of each ion and its concentration remaining in solution after neutralization.

18.5 What volume of 0.20 M NaOH will be required to react completely with 32 ml of 0.15 M H_2SO_4?

18.6 It is found that 20.0 ml of 0.100 M NaOH is required to react completely with a 1.000 g sample containing benzoic acid C_6H_5COOH (one replaceable hydrogen). Calculate the per cent of benzoic acid in the sample.

18.7 The gram equivalent weight of an acid is defined as the weight that reacts with one mole of OH^- ion. It is found that 0.466 g of a solid acid requires 22.6 ml of 0.254 M NaOH to neutralize it. Calculate the gram equivalent weight of the acid.

18.8 What is the ratio of normality to molarity for each of the underlined reagents in the following acid-base reactions?

a. $\underline{HC_2H_3O_2} + OH^- \longrightarrow C_2H_3O_2^- + H_2O$
b. $\underline{H_3PO_4} + 3\ OH^- \longrightarrow PO_4^{3-} + 3\ H_2O$
c. $\underline{H_3PO_4} + OH^- \longrightarrow H_2PO_4^- + H_2O$
d. $2\ H^+ + \underline{Zn(OH)_2(s)} \longrightarrow Zn^{2+} + 2\ H_2O$

18.9 The ionization constant of a certain weak acid is 1×10^{-6}. The undissociated acid has a red color; the anion derived from it has a yellow color. What will be the color of the indicator in a solution of pH 5? 6? 7? 8?

18.10 Fifty ml of 0.10 M HCl is titrated with 0.10 M NaOH. Calculate the pH of the solution after the following volumes of NaOH are added.

a. 49.00 ml   c. 49.99 ml   e. 50.01 ml   g. 51.00 ml
b. 49.90 ml   d. 50.00 ml   f. 50.10 ml

From your data, construct a titration curve similar to that shown in Figure 18.1.

18.11 Of the three indicators, methyl red (end point at pH 5), bromthymol blue (end point at pH 7), and phenolphthalein (end point at pH 9), which would you suggest for the titration of

a. HCl with KOH?   b. $HC_2H_3O_2$ with KOH?   c. $NaC_2H_3O_2$ with HCl?
Explain your choices.

18.12 A buffer solution is prepared by adding 0.10 mole of formic acid ($K_a = 2 \times 10^{-4}$) and 0.10 mole of sodium formate to one liter of water.

a. What is the pH of this solution?
b. What is the pH after 0.02 mole of HCl is added?
c. What is the pH after 0.02 mole of NaOH is added?
d. What is the pH after 0.20 mole of HCl is added?

18.13 Describe, in some detail, how one could prepare

a. NaBr from NaOH   d. NH_3 from NH_4Cl
b. Na_2CO_3 from NaOH   e. H_2S from Na_2S
c. NH_4Cl from NH_3 f. $CaBr_2$ from HBr

18.14 Suggest ways of separating the following ions, making use of acid-base reactions.

a. CO_3^{2-}, SO_4^{2-} c. NH_4^+, Na^+
b. Cu^{2+}, Co^{2+} d. F^-, CO_3^{2-}

18.15 Indicate briefly how you would distinguish between the following, all of which are white solids.

a. NaCl and NaOH   d. $BaCl_2$ and $CaCl_2$
b. Na_2SO_4 and Na_2CO_3 e. $Ca(OH)_2$ and CaO
c. Na_2S and $NaNO_3$ f. $NaHCO_3$ and Na_2CO_3

18.16 A solution is 0.10 M in H_2S, Ni^{2+}, and Pb^{2+}.

    a. At what $[H^+]$ will NiS start to precipitate?
    b. At what $[H^+]$ will PbS start to precipitate?
    c. In view of your answers to a and b, suggest a method of separating Ni^{2+} and Pb^{2+}.

18.17 State whether or not a precipitate will form when enough OH^- is added to a solution 0.10 M in Zn^{2+} and H_2S to make the pH = 7. Consider both ZnS and $Zn(OH)_2$ as possible precipitates.

18.18 In the Solvay process,

    a. What is the purpose of adding ammonia?
    b. Why not use $Ca(OH)_2$ instead of ammonia?
    c. Why not use NaOH instead of ammonia?

18.19 Write balanced net ionic equations for the reactions that occur when the following solutions are mixed.

    a. $HClO_4$ and KOH   c. NH_3 and HCl
    b. $HC_2H_3O_2$ and NaOH   d. HNO_3 and $Ca(OH)_2$

18.20 Consider the reaction $NO_2^- + H^+ \rightarrow HNO_2$.

    a. Calculate the equilibrium constant for this reaction.
    b. Calculate the pH of a solution formed by adding 0.10 mole of HCl to one liter of a 0.10 M solution of sodium nitrite.

18.21 22.0 ml of 0.150 M HCl reacts exactly with 18.0 ml of a certain NaOH solution. What is the concentration of NaOH in this solution?

18.22 A solid mixture contains only Na_2CO_3 and NaCl. It is found that 20.0 ml of 0.180 M HCl is required to react completely with a 0.324 g sample of the mixture. Calculate the percentage of Na_2CO_3.

18.23 Explain why Equation 18.14 is valid for all acid-base reactions. If, in this equation, normality were replaced by molarity, would the equation remain generally valid? Explain, citing examples.

18.24 The ionization constant of a certain weak acid is 1×10^{-9}. The undissociated acid is yellow; the anion derived from it has a blue color. What will be the color of this compound in a solution of pH 7? pH 8? pH 9? pH 10?

18.25 Explain why the choice of indicator in the titration of a weak acid with a strong base is much more critical than is the case when both acid and base are strong.

18.26 A buffer solution is made up by adding one mole of NH_4Cl to one lit of a 1.00 M solution of ammonia.

    a. What is the pH of this solution?
    b. What is the pH when 0.20 mole of HCl is added?
    c. What is the pH when 0.20 mole of NaOH is added?

18.27 Describe how the following conversions might be accomplished.

    a. $NaCl \longrightarrow NaHCO_3$ e. $HF \longrightarrow KF$
    b. $NaHCO_3 \longrightarrow Na_2CO_3$ f. $KF \longrightarrow HF$
    c. $CoCl_2 \longrightarrow CoBr_2$ g. $Na_2CO_3 \longrightarrow NaF$
    d. $NaCl \longrightarrow NaHSO_4$ h. $Na_2S \longrightarrow NaCl$

18.28 Suggest ways of separating the following series of ions, using acid-base and/or precipitation reactions.

    a. CO_3^{2-}, SO_4^{2-}, Cl^- c. F^-, SO_4^{2-} e. Ag^+, Cu^{2+}
    b. HCO_3^-, NO_3^- d. Cu^{2+}, Zn^{2+}, Ca^{2+}

18.29 A solution is 0.10 M in H_2S and Co^{2+} and 0.01 M in Pb^{2+}.

    a. At what $[H^+]$ will CoS start to precipitate?
    b. At what $[H^+]$ will PbS start to precipitate?
    c. In view of your answers to a and b, suggest a procedure for separating Co^{2+} and Pb^{2+}.

18.30 A solution 0.01 M in Ni^{2+} and Zn^{2+} and 1.0 M in H^+ is saturated with H_2S so as to make $[H_2S] = 0.10$ M.

    a. Will a precipitate form under these conditions?
    b. How much base must be added to a liter of such solution to start to precipitate one of these ions?

18.31 The solubility product of $Fe(OH)_2$ is 1×10^{-15}. What must be the maximum pH of a solution in which $[Fe^{2+}] = 0.1$ M? Would you expect $Fe(OH)_2$ to be soluble in strongly acidic solution?

°18.32 In general, one can say that a compound is soluble in acid if 0.1 mole dissolves in 1 lit of 1 M acid. What is the smallest value that K_{sp} for a +2 hydroxide, $M(OH)_2$, can have if it is to be soluble in acid? A +3 hydroxide, $M(OH)_3$? Referring to the K_{sp} values of the various hydroxides given in Table 16.3, explain why one can make the general statement that "All hydroxides are soluble in acid."

°18.33 Consider the titration of the weak acid $HC_2H_3O_2$ with NaOH. Suppose one starts with 50.0 ml of 1.00 M $HC_2H_3O_2$ and adds the following volumes of 1.00 M NaOH. Calculate the pH in each case.

    a. 40.0 ml   b. 49.0 ml   c. 50.0 ml   d. 51.0 ml   e. 60.0 ml

    In part c note that you are calculating the pH of a 0.50 M solution of $NaC_2H_3O_2$. In parts a and b calculate the ratio of $C_2H_3O_2^-$ to $HC_2H_3O_2$ and use K_a for acetic acid to obtain the pH.
    From your results, construct a titration curve for the reaction. Note particularly how such a curve differs from that for the titration of a strong acid with NaOH.

°18.34 A buffer solution is made up by adding 0.10 mole of $NaC_2H_3O_2$ to one lit of 0.10 M $HC_2H_3O_2$. What is the maximum amount of HCl that can be added to the solution without changing the pH by more than 0.5 unit?

°18.35 The overall reaction in the Solvay Process for the preparation of Na_2CO_3 is

$$2\ Na^+ + CaCO_3(s) \longrightarrow Na_2CO_3(s) + Ca^{2+}$$

Write balanced equations for the various steps of the process and add them to give this equation as the net result.

Complex Ions

If a powdered sample of white, anhydrous copper(II) sulfate is exposed to ammonia gas, a deep blue crystalline product is formed. Analysis reveals that this product contains four moles of ammonia for every mole of copper sulfate. X-ray studies indicate that the positive ion in this compound consists of a Cu^{2+} ion bonded to four ammonia molecules, i.e., $Cu(NH_3)_4^{2+}$. The reaction between the Cu^{2+} ion and the NH_3 molecules is represented in electron dot notation as

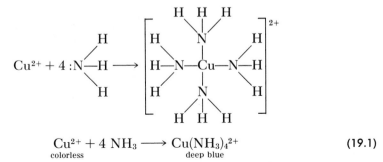

$$Cu^{2+} + 4 NH_3 \longrightarrow Cu(NH_3)_4^{2+} \qquad (19.1)$$
colorless   deep blue

The nitrogen atom of each ammonia molecule contributes a pair of unshared electrons to form a coordinate covalent bond between the Cu^{2+} ion and an NH_3 molecule. In this reaction the ammonia is acting as a Lewis base, the Cu^{2+} ion as a Lewis acid. In this sense, Reaction 19.1 resembles that between an ammonia molecule and a proton, which combine with each other to form an ammonium ion.

Addition of aqueous ammonia to a water solution of copper sulfate gives the same deep blue color. If the solution is carefully evaporated, blue crystals of hydrated $Cu(NH_3)_4SO_4$ can be obtained. From these observations it would appear that the $Cu(NH_3)_4^{2+}$ ion may exist in aqueous solution as well as in the solid state. Indeed, whenever Cu^{2+} ions and NH_3 molecules are brought into contact with each other, they show a strong tendency to combine in this manner.

The $Cu(NH_3)_4^{2+}$ ion is an example of a type of particle commonly referred to as a **complex ion**. In the broadest sense, a complex ion may be considered to be a charged aggregate consisting of more than one atom. Strictly speaking, such oxyanions as nitrate (NO_3^-) and sulfate (SO_4^{2-}) can be classified as complex ions. However, we shall use the term in a more restricted sense to refer to a **charged aggregate in which a metal atom is joined by coordinate covalent bonds to neutral molecule(s) and/or negative ions.** Species such as $Cu(H_2O)_4^{2+}$ and $Zn(H_2O)_3(OH)^+$, encountered in previous chapters, are properly considered to be complex ions.

The metal atom in a complex ion (Cu, Zn, Al, and so on) is often referred to as the **central atom**. The molecules (such as NH_3, H_2O) or anions (such as OH^-, Cl^-) attached to the central atom by coordinate covalent bonds are known as **coordinating groups** or **ligands**. In order to act as a coordinating group, a molecule or ion must ordinarily have a pair of unshared electrons that can be shared with the central atom to form a covalent bond. The number of bonds (2, 4, 6) formed by the central atom is called its **coordination number**. The nomenclature of complex ions and their salts is discussed in Appendix 2.

19.1 CHARGES OF COMPLEX IONS: NEUTRAL COMPLEXES

We can obtain the charge of a complex ion in a formal way by taking the algebraic sum of the oxidation number of the central atom and the charges of the ligands. The application of this simple rule is shown in Table 19.1 for certain complex ions of Pt(II).

As Table 19.1 indicates, a complex ion may carry either a positive or a negative charge. When a water solution of $[Pt(NH_3)_4]Cl_2$ is electrolyzed, the platinum, in the form of the $Pt(NH_3)_4^{2+}$ ion, migrates to the negative electrode. If the compound used is $K_2[PtCl_4]$, the platinum, present as the $PtCl_4^{2-}$ anion, moves to the positive electrode. The chlorine atoms in these two compounds also behave quite differently. In $[Pt(NH_3)_4]Cl_2$, chlorine is present as monatomic Cl^- ions which are immediately precipitated as AgCl by addition of silver nitrate. Addition of Ag^+ ions to a solution of $K_2[PtCl_4]$, in which the chlorine is covalently bonded to platinum, fails to give a precipitate.

TABLE 19.1 CHARGES OF COMPLEX IONS

COMPLEX ION	OXIDATION NO. PT	COORDINATING GROUPS	CHARGE OF ION
$Pt(NH_3)_4^{2+}$	+2	4 NH_3	$+2 + 4(0) = +2$
$Pt(NH_3)_3Cl^+$	+2	3 NH_3, 1 Cl^-	$+2 + 3(0) - 1 = +1$
$Pt(NH_3)Cl_3^-$	+2	1 NH_3, 3 Cl^-	$+2 + 0 - 3 = -1$
$PtCl_4^{2-}$	+2	4 Cl^-	$+2 + 4(-1) = -2$

Would [Pt(NH₃)₂Cl₂] give a precipitate with Ag⁺?

Species such as $[Pt(NH_3)_2Cl_2]$ or $[Zn(H_2O)_2(OH)_2]$, in which the oxidation number of the central atom (+2) is exactly balanced by the total charge of the ligands (−2), are examples of neutral complexes. Compounds of this type are usually quite insoluble in polar solvents such as water. If sufficient base is added to a solution of a zinc salt to give an appreciable concentration of the neutral species $[Zn(H_2O)_2(OH)_2]$, a precipitate of zinc hydroxide is obtained.

Among the most interesting of the neutral complexes are the *metal carbonyls*, in which a transition metal atom in the zero oxidation state is bonded to molecules of carbon monoxide. One of the most stable of these compounds is nickel carbonyl, $[Ni(CO)_4]$, formed by passing carbon monoxide over elementary nickel at 60°C:

$$Ni(s) + 4\ CO(g) \rightleftharpoons Ni(CO)_4(l) + heat \qquad (19.2)$$

Nickel carbonyl, a colorless liquid that boils at 43°C, has the characteristic properties of molecular compounds. It is insoluble in water but soluble in organic solvents such as ether. Reaction 19.2 is reversed by heating to 200°C. The Mond process for the refining of nickel takes advantage of this reversible complex formation to remove impurities such as cobalt and iron.

19.2 COMPOSITION OF COMPLEX IONS

The determination of the exact composition of complex ions is by no means a simple task. If a complex ion can be isolated in the solid state in the form of a salt, chemical analysis can often be used to suggest its composition. For example, when we find that the compound formed by the reaction of ammonia with copper(II) sulfate contains the elements copper, nitrogen, hydrogen, sulfur, and oxygen in the atom ratio

$$1Cu : 4N : 12H : 1S : 4O$$

we have reason to believe that the solid consists of equal numbers of $SO_4{}^{2-}$ and $Cu(NH_3)_4{}^{2+}$ ions. The existence in this compound of the $Cu(NH_3)_4{}^{2+}$ ion is confirmed by a great deal of other evidence, in particular by x-ray diffraction data which shows that each copper atom is surrounded by four nitrogen atoms.

The formulas of complex ions in water solution are much more difficult to establish experimentally. What we frequently do is to assume the existence of species in solution which are known to exist in the solid state. Knowing, for example, that evaporation of an ammoniacal solution of copper(II) sulfate gives the compound $[Cu(NH_3)_4]SO_4$, we may assume the presence in the solution of the $Cu(NH_3)_4{}^{2+}$ ion. Going one step further, one might postulate that in a solution prepared by dissolving a copper(II) salt in pure water, the species $Cu(H_2O)_4{}^{2+}$, analogous to $Cu(NH_3)_4{}^{2+}$, should be present. Such a process of extrapolation can and often does lead to erroneous conclusions. In particular, one can seldom be certain of the exact number of water molecules present as ligands in an aquo-complex ion in solution.

Central Atom

The metals which show the greatest tendency to form stable complex ions are the transition metals falling in groups 6B through 2B of the periodic table (e.g., in the first transition series, the elements from chromium through zinc). The cations of these metals are ordinarily complexed both in water solution and in the solid state. Elements in other regions of the periodic

483

table, notably aluminum in group 3A and tin and lead in 4A, form a more limited number of stable complex ions.

The metals which are most active in complex-ion formation are those which have small atomic and ionic radii. In the first transition series, for example, the atomic radii of the elements from chromium through zinc fall in the range 1.25 to 1.33 Å, being significantly smaller than the radii of the first three elements of this period, potassium (2.31 Å), calcium (1.97 Å), and scandium (1.60 Å), all of which are poor complex formers. The correlation of atomic size with complexing tendency is understandable if one considers the driving force behind complex-ion formation to be the electrostatic attraction of a metal ion for the unshared electron pair of a ligand. The smaller the cation and the larger its positive charge, the greater should be its attraction for electrons.

Many metals exhibit a variety of oxidation states in the complex ions they form. Iron forms the neutral complex $[Fe(CO)_5]$, in which its oxidation number is zero, the $Fe(CN)_6^{4-}$ ion (oxid. no. $Fe = +2$) and the $Fe(CN)_6^{3-}$ ion (oxid. no. $Fe = +3$). Metals which form more than one cation frequently show a greater tendency toward complex-ion formation in the higher oxidation state. The complex ions of Cr^{3+}, Fe^{3+}, and Co^{3+} are both more stable and more numerous than those of Cr^{2+}, Fe^{2+}, and Co^{2+}. This effect can be explained in terms of the effect of oxidation number or charge upon the electrostatic attraction between cation and ligand. The higher the oxidation number or charge of the central atom, the greater should be its attraction for a negative ion or a polar molecule.

Coordinating Group: Chelating Agents

In principle, any molecule or anion possessing an unshared pair of electrons can donate them to a metal to form a complex ion. In practice, the atom within the ligand which furnishes these electrons is ordinarily derived from one of the more electronegative elements (C, N, O, S, F, Cl, Br, I). Hundreds of different ligands containing one or more of these atoms are known. Among those most frequently encountered in general chemistry are the NH_3 and H_2O molecules and the OH^-, Cl^-, and CN^- ions (Table 19.2).

TABLE 19.2 A Few Complex Ions of the Transition Metals

H_2O	NH_3	OH^-	Cl^-	CN^-
$Cu(H_2O)_4^{2+}$	$Ag(NH_3)_2^+$	$Zn(OH)_4^{2-}$	$AgCl_2^-$	$Ag(CN)_2^-$
$Zn(H_2O)_4^{2+}$	$Cu(NH_3)_4^{2+}$	$Cr(OH)_6^{3-}$	$CuCl_4^{2-}$	$Zn(CN)_4^{2-}$
$Ni(H_2O)_6^{2+}$	$Zn(NH_3)_4^{2+}$		$HgCl_4^{2-}$	$Ni(CN)_4^{2-}$
$Co(H_2O)_6^{3+}$	$Ni(NH_3)_6^{2+}$		$PtCl_4^{2-}$	$Au(CN)_4^-$
$Cr(H_2O)_6^{3+}$	$Co(NH_3)_6^{3+}$		$AuCl_4^-$	$Fe(CN)_6^{3-}$
			$PtCl_6^{2-}$	$Fe(CN)_6^{4-}$

The relative abilities of different ligands to coordinate with metal ions depend upon a great many factors. One of the most important of these is the basicity of the ligand. It is perhaps not too surprising to find that molecules or ions which have a strong attraction for a proton are the best coordinating

How would Al^{3+} compare to Mg^{2+} as a complex-former?

agents. The species NH_3, OH^-, and CN^-, all of which are strong bases in the Brönsted-Lowry or Lewis sense, form stable complexes with a wide variety of transition metal ions. The ClO_4^- ion, which shows no tendency to acquire a proton in water solution, is a notoriously poor coordinating agent; the NO_3^- and HSO_4^- ions, both derived from strong acids, form relatively few stable complexes. It should be pointed out, however, that the Cl^- ion, which does not act as a base in water, forms stable complexes with many transition metal ions (cf. Table 19.2).

One of the first coordinating groups to be studied extensively was the ethylenediamine molecule:

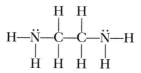

This molecule, containing two nitrogen atoms, each with an unshared pair of electrons, forms two coordinate covalent bonds with metal atoms. These bonds are extremely stable, as shown by the fact that the addition of ethylenediamine to a solution containing the $Cu(NH_3)_4^{2+}$ ion results in the displacement of the four NH_3 molecules by two $H_2N\text{---}CH_2\text{---}CH_2\text{---}NH_2$ molecules.

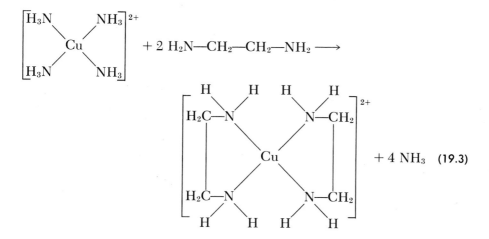

The fact that ethylenediamine is a powerful coordinating agent appears to be due at least in part to the high stability of the five-membered rings which it forms with metal ions.

Another factor which tends to make reactions such as 19.3 spontaneous is the increase in entropy that almost invariably accompanies such processes. The positive value of ΔS can be explained most simply by noting that the substitution of two ethylenediamine molecules for four ammonia molecules results in an increase in the number of particles in solution and, consequently, an increase in randomness.

A great many molecules and anions in addition to ethylenediamine form more than one coordinate covalent bond with a metal atom. Coordinating groups which behave in this manner are known as chelating agents; the complexes which they form are called **chelates**. Two important chelating agents are the oxalate ion, $C_2O_4^{2-}$, and the carbonate ion, CO_3^{2-}, both of which are

Ligands are called monodentate (one bond to central atom), bidentate (two bonds), etc.

capable of forming two oxygen-to-metal bonds. (The numbers (1) and (2) in the diagram indicate the atoms involved in chelate formation.)

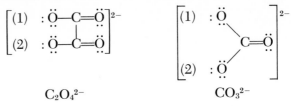

$$C_2O_4^{2-} \qquad\qquad CO_3^{2-}$$

Chlorophyll, the green coloring matter of plants, and hemoglobin, the pigment responsible for the red color of blood, are both chelates. In each of these substances, metal atoms are coordinated to four nitrogen atoms, which in turn form part of an intricate organic ring system. In chlorophyll, the central atom is magnesium; in hemoglobin, it is iron. The structure of the chlorophyll molecule is shown schematically in Figure 19.1.

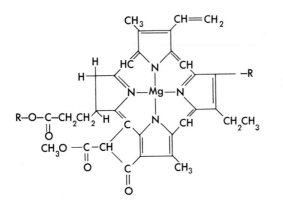

Figure 19.1   Structure of chlorophyll.

In hemoglobin, it appears that the central iron atom is coordinated to six ligands rather than four. One of these "extra" ligands is a water molecule which can be replaced reversibly by a molecule of oxygen to give a derivative known as oxyhemoglobin.

$$\text{Hemoglobin} + O_2 \rightleftharpoons \text{Oxyhemoglobin} + H_2O$$

The position of this equilibrium is sensitive to the pressure of oxygen. In the lungs, where the blood is saturated with air (partial pressure $O_2 = 0.2$ atm), the hemoglobin is almost completely converted to the oxidized form. In the tissues serviced by arterial blood, the partial pressure of oxygen drops and the oxyhemoglobin breaks down to release the elementary oxygen essential for the combustion of food. By this reversible process, hemoglobin acts as an oxygen carrier, absorbing oxygen in the lungs and liberating it to the tissues.

Unfortunately, hemoglobin forms a complex with carbon monoxide that is considerably more stable than oxyhemoglobin. The carbon monoxide complex is formed preferentially in the lungs even at CO concentrations as low as one part per thousand. When this happens, the flow of oxygen to the tissues is cut off, resulting eventually in muscular paralysis and death.

Coordination Number

As may be seen from Table 19.3, the most common coordination numbers shown by metal atoms in complex ions are 6 and 4. A coordination number of 2 occurs less frequently, being restricted for the most part to the complex ions formed by the 1B metals in the +1 oxidation state. Odd coordination numbers (1, 3, 5, 7) occur very rarely.

A metal in a given oxidation state usually shows only one coordination number, regardless of the particular complex ion in which it is present. For example, platinum in the +2 oxidation state invariably forms four bonds with

TABLE 19.3 Coordination Number and Geometry of Complex Ions

Coordination No.	Geometry	Examples
2	linear	Cu^+, Ag^+, Au^+
4	square planar	Cu^{2+}, Ni^{2+}, Pd^{2+}, Pt^{2+}, Au^{3+}
4	tetrahedral	Cu^+, Zn^{2+}, Cd^{2+}, Hg^{2+}, Al^{3+}
6	octahedral	Co^{3+}, Cr^{3+}, Fe^{2+}, Fe^{3+}, Pt^{4+}, Cd^{2+}, Al^{3+}, Ni^{2+}

coordinating groups, giving complex ions such as $Pt(H_2O)_4{}^{2+}$, $Pt(NH_3)_4{}^{2+}$, and $PtCl_4{}^{2-}$. Similarly, chromium(III) and cobalt(III) in their complex ions always show a coordination number of 6. Certain metals can show more than one coordination number in a given oxidation state. The Ni^{2+} ion, for example, is known to have two different coordination numbers, 4 and 6, shown respectively in the complex ions $Ni(CN)_4{}^{2-}$ and $Ni(H_2O)_6{}^{2+}$. Aluminum also shows coordination numbers of 4 or 6 depending upon the nature of the attached ligands.

19.3 GEOMETRY OF COMPLEX IONS

Coordination Number = 2

Complex ions in which two ligands are coordinated to a central metal atom are invariably linear. The structures of the $Ag(NH_3)_2{}^+$, $Ag(CN)_2{}^-$ and $Au(CN)_2{}^-$ ions may be represented as follows:

$$(H_3N—Ag—NH_3)^+, \quad (N\equiv C—Ag—C\equiv N)^-, \quad (N\equiv C—Au—C\equiv N)^-$$

Coordination Number = 4

For a complex ion in which four ligands are arranged about the central atom, two symmetrical geometric arrangements are possible. The four coordinating groups may be located at the corners of a square, giving what is

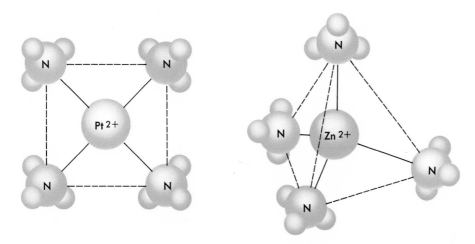

Figure 19.2 Geometry of ammonia complex ions formed by platinum(II) and zinc(II).

487

known as a **square planar complex**, or at the corners of a regular tetrahedron (**tetrahedral complex**). Both arrangements are known; x-ray studies show that the complexes of platinum(II) are of the square planar type, while the four-coordinated complexes of zinc(II) are tetrahedral. As we shall see later, the electronic structure of the central metal atom plays a major role in determining which of these two structures a particular metal forms.

It has been found experimentally that certain square complexes can exist in two different forms with quite different properties. Consider, for example, the neutral complex $[Pt(NH_3)_2Cl_2]$. Two forms of this compound, differing in color, water solubility, and chemical reactivity have been prepared. One of these, made by reacting ammonia with the $PtCl_4^{2-}$ ion, has a structure in which the two ammonia molecules are located at adjacent corners of a square. In the other form, prepared by reacting the $Pt(NH_3)_4^{2+}$ ion with hydrochloric acid, the two ammonia molecules are located at opposite corners of the square:

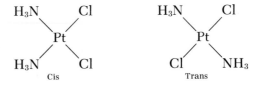

The two forms of $Pt(NH_3)_2Cl_2$ are called **geometrical isomers**. From a structural standpoint, they differ only in the spatial arrangement of the groups coordinated about the central atom. The form in which like groups are as close together as possible is called the **cis** isomer; the form in which like groups are far apart is referred to as the **trans** isomer. Geometrical isomerism can occur with any square planar complex of general formula Ma_2b_2 or Ma_2bc, in which M refers to the central atom and a, b, c represent ligands.

The assignment of a **cis** or **trans** configuration to a particular isomer of a coordination compound is by no means a simple experimental problem. X-ray diffraction studies have been used successfully in a few cases, but are generally difficult to apply. Another approach involves studying the reactivity of the coordination compound toward certain chelating agents. When the **cis** isomer of $[Pt(NH_3)_2Cl_2]$ is reacted with oxalate ions in solution, two Cl^- ions are displaced by an $C_2O_4^{2-}$ ion:

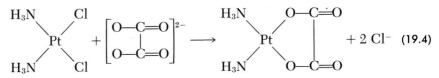

The **trans** isomer does not react readily since the oxalate ion cannot be sufficiently distorted to become attached to platinum at two points trans to each other.

Geometric isomers cannot exist for tetrahedral complexes, since each ligand is equidistant from the other three. However, if four *different* ligands are present, another type of isomerism is possible. A tetrahedral complex of general formula Mabcd, where a, b, c, and d represent four different ligands, can exist in two different forms, known as **optical isomers**, which have the ability to rotate the plane of polarized light in different directions. The "d-form" (dextrorotatory) rotates the plane to the right; the 1-form (levorota-

tory) to the left. In all other respects, these two isomers show identical properties.

Optical isomerism was discovered first and has been most extensively
studied among organic compounds (cf. Chapter 24). It arises when a molecule
or ion has no element of symmetry, as is the case with a tetrahedral species
Mabcd. The two isomers shown in Figure 19.3 cannot be superimposed upon
each other in such a way that all like atoms coincide. In this respect, they
resemble the right and left hands (or gloves), which are not superimposable
(try fitting a left-handed rubber glove on your right hand!).

Would a square
planar complex
Mabcd have op-
tical isomers?
geometrical iso-
mers?

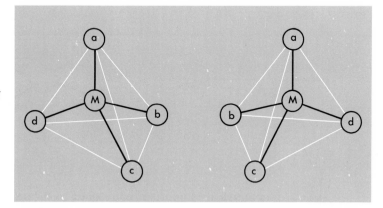

Figure 19.3 Optical isomers of Mabcd.

Coordination Number = 6

The six groups surrounding the central atom in such complexes as
$Fe(CN)_6{}^{3-}$ and $Co(NH_3)_6{}^{3+}$ are located at the corners of a regular octahedron,
a geometric figure with eight sides, all of which are equilateral triangles, and
six apices (Figure 19.4). The metal atom is located at the center of the octahedron.

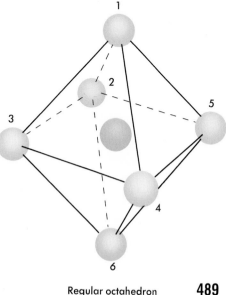

Figure 19.4 Regular octahedron.

Regular octahedron

The spatial distribution of ligands in octahedral complexes is often shown by skeleton structures such as those in Figure 19.5. The drawing at the left emphasizes that the six ligands are located along three axes at right angles to each other (x, y, and z axes), equidistant from the metal atom at the center. The skeleton structure at the right of Figure 19.5 is easier to draw and serves to emphasize another characteristic of an octahedral complex; it can be visualized as a derivative of a square planar complex in which the two additional ligands are located along a line perpendicular to the square at its center.

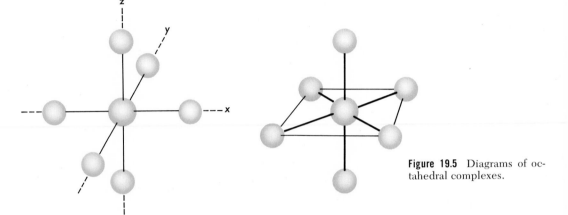

Figure 19.5 Diagrams of octahedral complexes.

Geometrical isomerism can occur in octahedral as well as square complexes. It may be noted from Figure 19.4 that for any given position of a ligand in an octahedral complex, there are four equivalent positions equidistant from the first, and one at a greater distance. If, for example, we choose position 1 as a point of reference, groups located at 2, 3, 4, and 5 will be equidistant from it, while a group at position 6 will be farther away. We may refer to positions 1 and 2, 1 and 3, 1 and 4, or 1 and 5 as being **cis** to each other while positions 1 and 6 are **trans**. Consequently, an ion such as $Co(NH_3)_4Cl_2^+$ can exist in two isomeric forms, one in which the two chloride ions are in a **cis** relationship to each other and another in which they are in a **trans** configuration (Figure 19.6).

How many isomers for $[Co(NH_3)_3Cl_3]$?

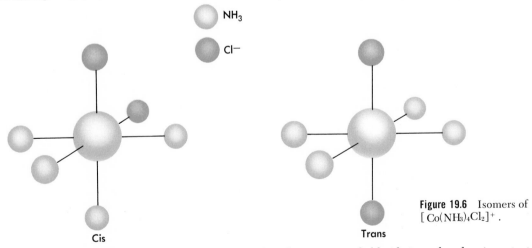

Figure 19.6 Isomers of $[Co(NH_3)_4Cl_2]^+$.

Cis

Trans

The complexes formed by cobalt(III) with ammonia and chloride ions played an important part in the development of the structural theory of coordination compounds. By the end of the nineteenth century, five different compounds containing trivalent cobalt, chloride ions and ammonia were known:

1. $CoCl_3 \cdot 6\ NH_3$ orange-yellow
2. $CoCl_3 \cdot 5\ NH_3$ violet
3. $CoCl_3 \cdot 4\ NH_3$ green

4. $CoCl_3 \cdot 4\ NH_3$ violet
5. $CoCl_3 \cdot 3\ NH_3$ green

Treatment of any of these salts with hydrochloric acid fails to remove the ammonia, indicating that the NH_3 molecules must be strongly bonded to cobalt. When silver nitrate is added to a water solution of 1, all the chlorine is precipitated immediately as AgCl. With compound 2, on the other hand, only two thirds of the chlorine is precipitated by Ag^+ at room temperature, while compounds 3 and 4 give up only one third of their chlorine under these conditions. Compounds 3 and 4 differ from each other in chemical reactivity as well as color. Compound 5 fails to react with Ag^+.

Alfred Werner, in 1893, reflecting on the properties of these and analogous compounds of chromium, platinum, and palladium, proposed the basic structural theory of coordination complexes which is accepted today and has been presented in the foregoing discussion. His ideas, reputed to have come to him in a dream, were revolutionary at the time, opening up a whole new area of inorganic chemistry. Following Werner, the compounds just listed may be assigned the structures:

1. $[Co(NH_3)_6]\ Cl_3$
2. $[Co(NH_3)_5Cl]\ Cl_2$
3. $[Co(NH_3)_4Cl_2]\ Cl$ **(trans)**

4. $[Co(NH_3)_4Cl_2]\ Cl$ **(cis)**
5. $[Co(NH_3)_3Cl_3]$

Only those chlorides which are outside the coordination sphere are precipitated by silver nitrate.

On the basis of his theory, Werner predicted the existence of geometrical isomers for certain square planar complexes. Many years later, he was able to isolate cis and trans isomers of $[Pt(NH_3)_2Cl_2]$ and analogous platinum complexes. In 1913, in recognition of his outstanding contributions to inorganic chemistry, Werner received the Nobel Prize in chemistry.

An interesting example of a type of optical isomerism that can arise in octahedral complexes is afforded by the compound $[Co(en)_2Cl_2]Cl$, where cobalt(III) is coordinated to two ethylenediamine molecules (abbreviated "en") and two chloride ions. The **cis** form of this compound, shown at the left of Figure 19.7, can be resolved into two optical isomers, since it possesses no element of symmetry. On the other hand, the **trans** form cannot show optical isomerism, since it has a plane of symmetry; the two chloride ions are equidistant from the plane passing through the two ethylenediamine molecules and the cobalt(III) ion. Werner used this property to distinguish between the **cis** and **trans** forms of this compound.

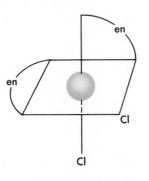

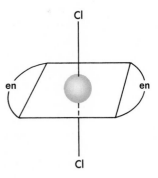

Figure 19.7 Isomerism in $[Co(en)_2Cl_2]^+$.

Cis (2 optical isomers)   Trans (no optical isomers)

19.4   ELECTRONIC STRUCTURE OF COMPLEX IONS

One of the first attempts to explain the stability of complex ions in terms of the electronic structure of the central metal atom was that of Sidgwick, who

491

suggested in 1927 that metal atoms tend to achieve a noble-gas structure by accepting electron pairs from ligands. This approach, which proved so successful in explaining the formulas and properties of the molecular compounds formed by the nonmetals of the first and second periods of the periodic table, can be applied to many complex ions.

Among the complex ions in which the central metal atom has a noble gas structure are those in which four ligands are attached to a Zn^{2+} ion, e.g., $Zn(H_2O)_4^{2+}$, $Zn(NH_3)_4^{2+}$, and so on. To show that this is the case, let us count the electrons surrounding the zinc nucleus. A Zn^{2+} ion has 28 electrons, two less than a zinc atom (AN = 30). Each ligand donates two electrons; four ligands must then contribute a total of eight electrons. Adding these eight electrons to the 28 associated with the bare Zn^{2+} ion gives a total of 36, which is the number of electrons in an atom of the noble gas krypton.

TABLE 19.4   NUMBER OF ELECTRONS SURROUNDING CENTRAL ATOM IN COMPLEXES OF FIRST SERIES OF TRANSITION METAL

ATOM	No. e⁻	MONATOMIC ION	No. e⁻	COMPLEX ION	No. e⁻
Cr	24	Cr^{3+}	21	$Cr(H_2O)_6^{3+}$	$21 + 12 = 33$
Mn	25	Mn^{2+}	23	$Mn(H_2O)_6^{2+}$	$23 + 12 = 35$
Fe	26	Fe^{3+}	23	$Fe(CN)_6^{3-}$	$23 + 12 = 35$
		Fe^{2+}	24	$Fe(CN)_6^{4-}$	$24 + 12 = 36^*$
Co	27	Co^{3+}	24	$Co(NH_3)_6^{3+}$	$24 + 12 = 36^*$
		Co^{2+}	25	$Co(NH_3)_6^{2+}$	$25 + 12 = 37$
Ni	28	Ni^{2+}	26	$Ni(CN)_4^{2-}$	$26 + 8 = 34$
				$Ni(NH_3)_6^{2+}$	$26 + 12 = 38$
Cu	29	Cu^{2+}	27	$Cu(NH_3)_4^{2+}$	$27 + 8 = 35$
		Cu^+	28	$Cu(CN)_4^{3-}$	$28 + 8 = 36^*$
Zn	30	Zn^{2+}	28	$Zn(NH_3)_4^{2+}$	$28 + 8 = 36^*$

* Complexes in which the central atom has a noble gas structure.

Several other complex ions are known in which the central metal has a noble gas structure (Table 19.4). However, exceptions to this rule are far more common than examples. Among these exceptions are all of the complex ions formed by chromium(III), manganese(II), iron(III), cobalt(II), nickel(II), and copper(II). Quite clearly, any attempt to predict the stoichiometry of complex ions based on the tendency of the central atom to acquire a noble gas structure is doomed to failure. In the case of nickel(II) complexes, for example, one would predict a coordination number of 5, since this would put $26 + 10 = 36$ electrons around the nickel. In practice nickel(II) shows a coordination number of 4 or 6 rather than 5.

For one particular series of complex compounds, the metal carbonyls, predictions based on the tendency to achieve a noble gas electronic configuration agree remarkably well with the observed stoichiometry. Considering the first transition series, we find that the metals of even atomic number, $_{24}Cr$, $_{26}Fe$, and $_{28}Ni$, form stable carbonyl complexes $Cr(CO)_6$, $Fe(CO)_5$, and $Ni(CO)_4$, in each of which the central atom has a noble-gas electronic configuration. Single atoms of the elements of odd atomic number, $_{25}Mn$, $_{27}Co$, and $_{29}Cu$ cannot reach a noble gas structure by accepting pairs of electrons from CO molecules. Experimentally, no carbonyl complex of copper(0) has ever been isolated; the carbonyls of cobalt and manganese are dimeric species in which two metal atoms are linked together, pairing their odd electrons.

In the past three decades, several more sophisticated models of bonding in complex ions have been developed. One of these, the **atomic orbital** or **valence bond** approach, was introduced in Chapter 8 to explain the bonding in relatively simple molecules of the nonmetals. A quite different approach, known as the **crystal field theory**, has been developed to explain certain of the chemical and physical properties of coordination compounds which were difficult to understand in terms of the valence bond theory.

The molecular orbital approach (Chap. 8) can also be used.

Valence Bond (Atomic Orbital) Approach

The valence bond treatment presented in Chapter 8 is readily extended to explain the electronic structures of complex ions in which various ligands are bonded to a central metal atom. It is assumed that the electron pairs donated by the ligands enter hybrid orbitals associated with the metal atom to form coordinate covalent bonds. In discussing the hybrid orbitals occupied by bonding electrons in complex ions, it will be convenient to organize our discussion according to the coordination number shown by the central metal atom (2, 4, or 6).

Coordination Number = 2: sp hybridization. Complex ions in which the central metal atom has a coordination number of 2 can be assigned electronic structures analogous to that postulated for the BeF_2 molecule (p. 198). The two pairs of bonding electrons are assumed to occupy two hybrid orbitals formed by combining an s and a p orbital. These hybrid orbitals should be oriented at angles of 180° to each other, in agreement with the experimental observation that complex ions of this type have a linear structure.

As an illustration of the formation of sp hybrid bonds, consider the electronic structure of the complex ions of copper(I) in which the coordination number is 2.

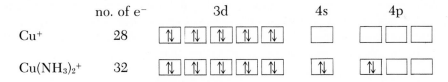

(In this structure and those that follow, the 18 electrons of the argon shell are not shown; heavy horizontal lines are drawn above and below the bonding orbitals.)

Coordination Number = 4: sp³ hybridization. Complexes in which zinc or aluminum have a coordination number of 4 show the tetrahedral geometry characteristic of such compounds as CH_4 and CCl_4. In the $Zn(NH_3)_4^{2+}$ ion, as in the CH_4 molecule, the four pairs of bonding electrons can be accommodated in four equivalent sp³ hybrid orbitals.

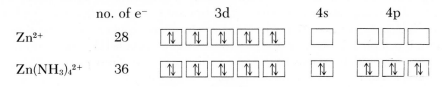

493

Coordination Number = 4: dsp² hybridization. The square planar complexes formed, for example, by nickel(II) differ in electronic structure as well as in geometry from the tetrahedral complexes of zinc. In the valence bond approach, the four bonding orbitals occupied in square planar complexes are described as **dsp²** hybrids, formed by combining a d, an s, and two p orbitals.

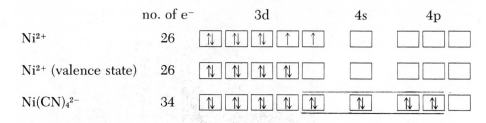

Derive the electronic structure of $Cu(NH_3)_4{}^{2+}$.

Coordination Number = 6: d²sp³ hybridization and Inner and Outer Complexes. In the octahedral complex ion $Fe(CN)_6{}^{4-}$, the six pairs of electrons contributed by the ligands can be located in six hybrid **d²sp³** bonding orbitals, formed by combining two 3d, one 4s and three 4p orbitals.

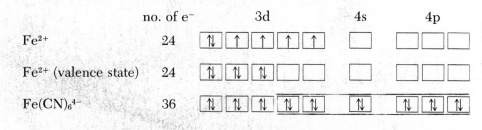

What complex ions are isoelectronic with $Fe(CN)_6{}^{4-}$?

The fact that compounds such as $K_4Fe(CN)_6$ are diamagnetic tends to confirm a structure such as this in which there are no unpaired electrons.

Certain octahedral complexes of iron(II) are known to be paramagnetic; the $Fe(H_2O)_6{}^{2+}$ ion, for example, has been shown from magnetic studies to have four unpaired electrons. One way to explain this in terms of the valence bond picture is to postulate that the d orbitals involved in bonding are those in the fourth principal energy level rather than the third.

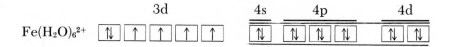

Bonds of this type are sometimes referred to as **sp³d²** hybrids (recall the discussion of the bonding in SF_6, p. 199), to distinguish them from the **d²sp³** hybrid bonds found in the $Fe(CN)_6{}^{4-}$ ion. Perhaps more frequently, the terms **outer complex** and **inner complex** are used to distinguish between these two types of octahedral complexes. In general, it is found that ligands such as the CN^- ion, which are strong bases, tend to form inner complexes. Ligands such as H_2O and, in particular, the F^- ion, which are weaker bases, tend to form outer complexes. Among the complex ions of cobalt(III), for example, the only ones which appear to be outer complexes are those such as $CoF_6{}^{3-}$, in which several fluoride ions are bonded to cobalt.

Crystal Field Theory

Although the valence bond approach has proved extremely useful in explaining and correlating the geometries, electronic structures, and many of the properties of complex ions, it is deficient in certain important respects. For example, it cannot explain the wide variety of brilliant colors characteristic of so many coordination compounds. Again, although the valence bond model can rationalize the existence of two different kinds of octahedral complexes of iron(II), it cannot explain why the CN^- ion forms one type of complex and the H_2O molecule another, nor can it predict in advance which type of complex a particular ligand will form.

It has long been recognized that many of the properties of complex ions can best be explained in terms of an electrostatic rather than a covalent model of bonding between metal and ligand. The fact that the ability of a metal ion to form complexes seems to be directly related to its charge density implies that it forms ionic or ion-dipole bonds with anions or molecules acting as ligands. In many cases, the strength of the metal-ligand bonds follows the same trend and lends itself to the same explanation.

The crystal field model starts with the assumption that the attractive forces holding a complex ion together are primarily electrostatic rather than covalent. The major part of the stabilization energy holding metal to ligand is attributed to coulombic interactions between the positively charged metal ion and the electrons associated with the ligand. While the valence bond model assumes that electrons donated by the ligand enter the orbitals of the central metal ion to form covalent bonds, crystal field theory considers that the modification of the electronic structure of the metal ion is due only to coulombic interactions between the ligands and the d electrons of the metal. The physical and chemical properties of the complex are explained largely in terms of these modifications.

To illustrate the changes that ligands can produce in the electronic structure of a metal ion, let us consider a specific example, the formation of the $Fe(CN)_6^{4-}$ complex ion.

$$Fe^{2+} + 6\ CN^- \longrightarrow Fe(CN)_6^{4-} \qquad (19.5)$$

In the uncomplexed Fe^{2+} ion, there are six electrons distributed among five 3d orbitals, all of which have the same energy. However, when CN^- ions approach the Fe^{2+} ion to form an octahedral complex, geometric considerations suggest that these orbitals should no longer have the same energy. Two of the five 3d orbitals (the $d_{x^2-y^2}$ and d_{z^2} orbitals) are oriented along the axes in the direction in which the CN^- ions are approaching (Figure 19.8). The other three orbitals (d_{xy}, d_{yz} and d_{xz}) are concentrated in the areas between the axes.

Electrons in the d_{xy}, d_{yz} and d_{xz} orbitals are repelled less strongly by the approaching negative ions than are those in the other two orbitals. In other words, in the "octahedral field" created by the approach of CN^- ions, the five 3d orbitals of Fe^{2+} are split into two groups. One group, comprising the d_{xy}, d_{yz}, and d_{xz} orbitals, is lower in energy than the other group of two orbitals, d_{z^2} and $d_{x^2-y^2}$. Schematically, we have the situation shown in Figure 19.9. Since the total energy of the d orbitals is unchanged in the process, the two higher orbitals are displaced to a greater extent than the lower three. As a

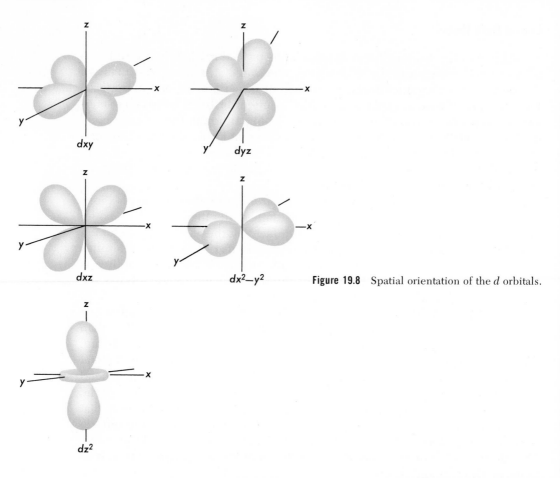

Figure 19.8 Spatial orientation of the *d* orbitals.

result, one might expect to find a rearrangement of the electronic structure of the Fe^{2+} ion; the six 3d electrons will tend to pair in the three lower energy levels. In other words, as a result of the formation of the $Fe(CN)_6^{4-}$ ion, the electronic structure of the Fe^{2+} ion changes from

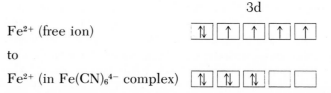

Fe^{2+} (free ion)

to

Fe^{2+} (in $Fe(CN)_6^{4-}$ complex)

The extent to which the energies of the d orbitals are modified by complex ion formation will depend upon how strongly the ligands interact with the electrons within these orbitals. The H_2O molecule, which is a weaker Lewis base than the CN^- ion, will repel electrons in the d orbitals to a lesser extent, resulting in a smaller energy separation. It is not unreasonable to suppose that in the weaker crystal field exerted by the water molecules, the electron distribution of the uncomplexed Fe^{2+} ion, in which there are a maximum number of unpaired electrons (Hund's rule), will be retained. In other words water molecules do not interact strongly enough with the d electrons of the Fe^{2+} ion to overcome their tendency to remain unpaired insofar as possible.

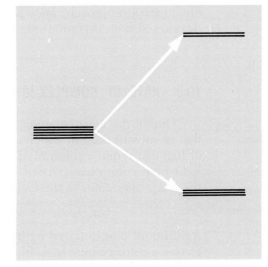

Figure 19.9 Splitting of *d* orbitals in an octahedral field.

Fe^{2+} (in $Fe(H_2O)_6^{2+}$) ⇅ ↑ ↑ ↑ ↑

Looking at the structures just written for the Fe^{2+} ion in the $Fe(CN)_6^{4-}$ and $Fe(H_2O)_6^{2+}$ complex ions, we see that the crystal field theory explains why the first complex is diamagnetic (no unpaired electrons) while the second is paramagnetic (with four unpaired electrons). Indeed, in many cases, it is possible to calculate in advance which of the two possible complexes, "high-spin" or "low-spin," will be more favored energetically. To be sure, valence bond theory also explains these observations, in terms of inner and outer complexes. The advantage of the crystal field theory is that it suggests *why* the CN^- ion should form one type of complex and the H_2O molecule another. In general, the crystal field theory predicts that ligands that are weak bases (H_2O, F^-) tend to form complexes in which there is a maximum number of unpaired electrons, while strongly basic ligands (CN^-) tend to produce complexes in which the d electrons are paired.

The crystal field theory also offers a rational explanation for the fact that most complexes in which the metal ion has unfilled d orbitals available are colored. The splitting of the d orbitals by ligand interactions creates a situation in which electrons can move into higher energy levels by absorbing light. The extent to which the d orbitals are split will determine the amount of energy required to promote an electron from a lower to a higher orbital and hence the wavelength of the light which is absorbed. In some cases it is possible to predict quite accurately from the crystal field theory the wavelengths at which absorption will occur.

Lest it be supposed that the crystal field theory can explain all the properties of complex ions, it may be advisable to point out at least one of its deficiencies. If one thinks of the bonding in complex ions as being primarily electrostatic, it is hard to explain why certain molecules that have very small dipole moments, such as CO, can be effective coordinating agents. In order to explain this, it is necessary to modify the crystal field theory to take into account covalent as well as ionic bonding. A more sophisticated version of the

Would there be "low spin" and "high spin" structures for Cr^{3+}? Cr^{2+}?

497

electrostatic approach, known as the **ligand field theory**, has recently been developed with this in mind.

19.5 RATE OF COMPLEX-ION FORMATION

One of the most common reactions shown by complex ions in solution is that of substitution, in which a new complex is formed through an exchange of ligands. When a solution of a copper(II) salt is treated with ammonia, the light blue color characteristic of the $Cu(H_2O)_4^{2+}$ complex ion is replaced by the deep blue color of the $Cu(NH_3)_4^{2+}$ ion.

$$Cu(H_2O)_4^{2+} + 4\ NH_3 \longrightarrow Cu(NH_3)_4^{2-} + 4\ H_2O \qquad (19.6)$$

Addition of concentrated hydrochloric acid to a solution of a copper(II) salt causes a color change from blue to green, reflecting the formation of species such as $CuCl_4^{2-}$, in which Cl^- ions rather than H_2O molecules are bonded to copper:

$$Cu(H_2O)_4^{2+} + 4\ Cl^- \longrightarrow CuCl_4^{2-} + 4\ H_2O \qquad (19.7)$$

Reactions such as 19.6 and 19.7 are, of course, reversible. One obtains an equilibrium mixture of products whose relative concentrations depend upon the relative (thermodynamic) stabilities of the various complex ions. We shall consider some of the equilibrium relations involved in such reactions later in this chapter. At the moment, we are concerned with the **rate** at which equilibrium is established in substitution reactions involving complex ions.

In the case of copper(II) complexes, the establishment of equilibria involving species such as $Cu(H_2O)_4^{2+}$ and $Cu(NH_3)_4^{2+}$ is virtually instantaneous. If one adds ammonia to a water solution of a copper(II) salt, the deep blue color of the $Cu(NH_3)_4^{2+}$ ion appears immediately. The color quickly changes to green if hydrochloric acid is added to replace ammonia molecules in the complex by chloride ions. Complexes such as these, in which the rate of ligand exchange is too rapid to be measured by ordinary techniques, are said to be **labile**. Species in this category include, in addition to copper(II) complexes, all the known complexes of silver(I) and zinc(II), most of those of nickel(II), and all the outer **(sp^3d^2)** octahedral complexes of the transition metals.

Complexes which undergo substitution reactions in solution at a measurable rate are said to be **nonlabile** or inert. If the purple compound $[Co(NH_3)_5Cl]Cl_2$ is added to water, a slow reaction occurs in which the chloride ion inside the complex is replaced by a water molecule:

$$Co(NH_3)_5Cl^{2+} + H_2O \longrightarrow Co(NH_3)_5H_2O^{3+} + Cl^- \qquad (19.8)$$

The rate of this reaction can be determined by following the color change or by measuring the increase in conductivity brought about by the formation of free chloride ions. It is found that at room temperature only about six out of every 1000 $Co(NH_3)_5Cl^{2+}$ ions have reacted with water in an hour's time. Even after a day has passed, better than 85 per cent of the starting material remains unreacted. This is true despite the fact that the aquo-complex $Co(NH_3)_5H_2O^{3+}$ is inherently the more stable species under these conditions. That is, the equilibrium constant for Reaction 19.8 is so large that, given sufficient time, virtually all of the cobalt(III) is converted to the $Co(NH_3)_5H_2O^{3+}$ complex. Here we see again the importance of distinguishing between kinetic considerations, as described in the terms *lability* or inertness, and thermodynamic factors, which can predict only the ultimate state of a system in terms of the *stabilities* of the species involved.

Almost all the complex ions formed by cobalt(III), chromium(III), and platinum(II and IV) can be classified as inert. These ions vary greatly, however, in the rate at which they participate in substitution reactions. The rate of substitution depends not only upon the nature of the central atom but also upon the ligands bonded to it. For example, if one adds the complex compound $[Co(NH_3)_5I]Cl_2$ to water, a reaction analogous to 19.8

$$Co(NH_3)_5I^{2+} + H_2O \longrightarrow Co(NH_3)_5H_2O^{3+} + I^-$$

occurs at a rate such that about 3 per cent of the $Co(NH_3)_5I^{2+}$ ions are consumed in one hour. On the other hand, the $[Co(NH_3)_5NO_2]^{2+}$ complex ion is so inert that no detectable reaction with water takes place within a week.

By studying the kinetics of reactions involving nonlabile complex ions, it is possible to learn a great deal about their mechanisms. Consider, for example, the reaction

$$[Co(NH_3)_5Cl]^{2+} + NO_2^- \longrightarrow [Co(NH_3)_5NO_2]^{2+} + Cl^-$$

This is an example of a **nucleophilic substitution** reaction, in which an electron donating group, the NO_2^- ion, substitutes for another such group in the coordination sphere of cobalt(III). One might visualize two quite different mechanisms for this reaction: *This type of reaction is discussed further in Chapter 24.*

1. A mechanism which involves as a first step the slow, rate-determining dissociation of a $[Co(NH_3)_5Cl]^{2+}$, to give an intermediate in which the central ion has a coordination number of 5:

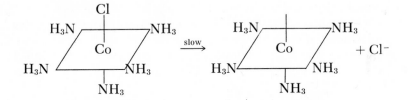

followed by rapid addition of a NO_2^- ion to this intermediate:

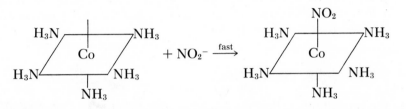

2. The direct displacement of a Cl^- ion in the coordination sphere by a NO_2^- ion. Such a reaction would presumably pass through a 7-coordinated species:

499

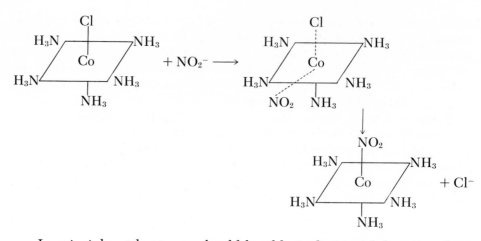

In principle, at least, one should be able to distinguish between these two possible mechanisms. The first mechanism would be expected to show first order kinetics

$$\text{rate} = k \, (\text{conc. } Co(NH_3)_5Cl^{2+})$$

since the NO_2^- ion does not participate in the rate determining step. Reactions of this type are often referred to as S_N1 reactions (substitution, nucleophilic, first order). The second mechanism would result in second order kinetics

$$\text{rate} = k \, (\text{conc. } Co(NH_3)_5Cl^{2+}) \, (\text{conc. } NO_2^-)$$

and would be described as an S_N2 reaction (substitution, nucleophilic, second order).

Unfortunately, if this reaction is carried out in water solution a complication occurs: a water molecule rather than NO_2^- ion attacks the $[Co(NH_3)_5Cl]^{2+}$ ion to give as a first product the aquo-complex ion $[Co(NH_3)_5H_2O]^{3+}$, which then undergoes further reaction with NO_2^- ions. This complication can be avoided by carrying out the reaction in methyl alcohol, which shows little tendency to undergo substitution reactions. Under these conditions, it is found that the reaction follows second order kinetics, which indicates that it occurs via an S_N2 mechanism. On the other hand, the reaction

$$Co(NH_3)_5Cl^{2+} + Br^- \longrightarrow Co(NH_3)_5Br^{2+} + Cl^-$$

in methanol solution follows first order kinetics (rate independent of conc. of Br^-), implying an S_N1 mechanism. Presumably, the difference in mechanism in these two reactions reflects the fact that the NO_2^- ion is a much stronger Lewis base than the Br^- ion.

19.6 COMPLEX-ION EQUILIBRIA

In the preceding section we compared the (kinetic) labilities of different complex ions. It is, of course, equally important to be able to compare their (thermodynamic) stabilities. One way to do this is to formulate and measure the equilibrium constant for the reaction that occurs when a compound con-

500

taining a particular complex ion is added to water. In doing this, it is customary to treat the reaction as if it were a simple dissociation, ignoring the water molecules involved. Thus for the equilibrium which is set up when a compound containing $Ag(NH_3)_2^+$ ions is added to water, we write

$$Ag(NH_3)_2^+ \rightleftharpoons Ag^+ + 2\ NH_3 \qquad K_c = \frac{[Ag^+] \times [NH_3]^2}{[Ag(NH_3)_2^+]}$$

The fact that the equilibrium constant for this reaction is a very small number, 4×10^{-8}, means that the $Ag(NH_3)_2^+$ ion dissociates to only a very slight extent when added to water. Looking at it another way, the addition of ammonia in low concentrations to a solution containing Ag^+ ions should convert most of them to the $Ag(NH_3)_2^+$ ion (Example 19.1).

EXAMPLE 19.1. Using the value of 4×10^{-8} for the dissociation constant of the $Ag(NH_3)_2^+$ ion, calculate

a. The ratio of the concentrations of Ag^+ and $Ag(NH_3)_2^+$ in a solution 1 M in NH_3.

b. The equilibrium concentration of NH_3 necessary to produce a 50 per cent conversion of Ag^+ to $Ag(NH_3)_2^+$.

SOLUTION

a. All that is required is to substitute for the concentration of NH_3 in the equilibrium expression:

$$K_c = \frac{[Ag^+] \times [NH_3]^2}{[Ag(NH_3)_2^+]} = 4 \times 10^{-8}; \qquad \frac{[Ag^+]\ (1)^2}{[Ag(NH_3)_2^+]} = 4 \times 10^{-8}$$

$$\frac{[Ag^+]}{[Ag(NH_3)_2^+]} = 4 \times 10^{-8}$$

The fact that this ratio is extremely small means that in a solution 1 M in ammonia, nearly all the silver is in the form of the $Ag(NH_3)_2^+$ complex. If, for example, the concentration of $Ag(NH_3)_2^+$ is 1 M, that of uncomplexed Ag^+ is only 4×10^{-8} M.

b. The conditions of part b, 50 per cent conversion of Ag^+ to $Ag(NH_3)_2^+$, require that the equilibrium concentrations of these two ions be equal to each other:

$$\frac{[Ag^+] \times [NH_3]^2}{[Ag(NH_3)_2^+]} = 4 \times 10^{-8}; [NH_3]^2 = 4 \times 10^{-8}; [NH_3] = 2 \times 10^{-4}$$

It is evident that a very low concentration of NH_3 is sufficient to convert half the Ag^+ ions to the ammine complex. At concentrations of NH_3 above 2×10^{-4}, better than half the silver ions will be complexed.

The relative stabilities of different complex ions of silver can be estimated from a knowledge of their dissociation constants. For example, when we find that the equilibrium constant for the reaction

$$Ag(CN)_2^- \rightleftharpoons Ag^+ + 2\ CN^- \qquad K_c = \frac{[Ag^+] \times [CN^-]^2}{[Ag(CN)_2^-]}$$

is only about 1×10^{-21}, we deduce that the $Ag(CN)_2^-$ complex is even more stable than $Ag(NH_3)_2^+$. We would expect that the addition of CN^- ions to a solution containing $Ag(NH_3)_2^+$ would convert most of these ions to the more stable $Ag(CN)_2^-$ complex. This is shown experimentally by the fact that AgI dissolves more readily in a solution of NaCN than in aqueous ammonia.

The experimental determination of the dissociation constants of complex ions is a very tricky problem. Complex ions which are relatively non-labile may take a long time to reach equilibrium. In such cases, one can either find a catalyst for the reaction, raise the temperature, or be very, very patient. A more serious obstacle to the direct determination of dissociation constants of complex ions is the fact that these dissociations seldom occur in a single step. In studying the equilibrium

$$Cu(NH_3)_4^{2+} \rightleftharpoons Cu^{2+} + 4\ NH_3$$

Calculations involving K_c for $Cu(NH_3)_4^{2+}$ are inaccurate for this same reason.

we can expect to find, in addition to $Cu(NH_3)_4^{2+}$ and (aquated) Cu^{2+} ions, several intermediate species containing three, two, and one NH_3 molecules as ligands. The presence of such species makes it difficult to determine the concentrations of $Cu(NH_3)_4^{2+}$ and Cu^{2+}, which we need to know to calculate the equilibrium constant. For this reason, among others, dissociation constants of complex ions are usually obtained indirectly by observing the effect of a complexing agent on the solubility of a metal salt (Section 19.7) or on the voltage of an electrical cell (Chapter 21).

TABLE 19.5 Dissociation Constants of Complex Ions

MA₂		MA₄		MA₆	
$AgCl_2^-$	1×10^{-6}	$CdCl_4^{2-}$	4×10^{-3}	$Cr(OH)_6^{3-}$	1×10^{-38}
$Ag(NH_3)_2^+$	4×10^{-8}	$Cd(NH_3)_4^{2+}$	1×10^{-7}	$Co(NH_3)_6^{3+}$	1×10^{-35}
$Ag(SCN)_2^-$	1×10^{-10}	$Cd(CN)_4^{2-}$	1×10^{-19}	$Co(CN)_6^{3-}$	1×10^{-64}
$Ag(S_2O_3)_2^{3-}$	1×10^{-13}	$Cu(NH_3)_4^{2+}$	2×10^{-13}	$Fe(CN)_6^{3-}$	1×10^{-31}
$Ag(CN)_2^-$	1×10^{-21}	$Cu(CN)_4^{2-}$	1×10^{-25}	$Ni(NH_3)_6^{2+}$	2×10^{-9}
$CuCl_2^-$	3×10^{-6}	$Ni(CN)_4^{2-}$	1×10^{-14}		
$Cu(NH_3)_2^+$	1×10^{-7}	$PdCl_4^{2-}$	1×10^{-13}		
		$Zn(NH_3)_4^{2+}$	3×10^{-10}		
		$Zn(OH)_4^{2-}$	3×10^{-16}		
		$Zn(CN)_4^{2-}$	1×10^{-17}		

19.7 COMPLEX IONS IN ANALYTICAL CHEMISTRY

Qualitative Analysis

Reactions involving the formation of complex ions are widely used in qualitative analysis for either of two purposes. The ability of a metal ion to form a colored complex or a precipitate with a particular complexing agent may be used as a specific test for that ion. Alternatively, two ions may be separated from one another by adding a complexing agent that forms a complex with only one of them. Not infrequently, these two purposes are achieved simultaneously by adding the proper complexing agent at a particular stage in an analysis.

Tests for Specific Ions. An extremely sensitive test for the Cu^{2+} ion in water solution involves its ability to form a deep blue complex with ammonia. The color of the $Cu(NH_3)_4^{2+}$ ion is much more intense than that of the light blue $Cu(H_2O)_4^{2+}$ ion; it can be detected at concentrations of Cu^{2+} as low as 10^{-4} M. Certain other ions interfere with this test; nickel, for example, also forms a deep blue complex ion with ammonia.

Iron(III) ions are readily detected by adding a solution of potassium thiocyanate, KSCN. A blood-red color develops as a result of the formation of a complex ion whose exact composition is difficult to determine; the formula of the complex is often written simply as $Fe(SCN)^{2+}$. Iron(III) ions also give a precipitate known as Prussian blue upon addition of a solution of potassium hexacyanoferrate.

$$Fe^{3+} + K^+ + Fe(CN)_6^{4-} \longrightarrow KFe[Fe(CN)_6](s) \qquad (19.9)$$

The same precipitate is obtained when solutions of $FeCl_2$ and $K_3[Fe(CN)_6]$ are mixed.

Chelating agents, because of their strong complexing ability, are widely used in qualitative analysis. Dimethyl glyoxime

$$\begin{array}{c} H_3C-C-C-CH_3 \\ \parallel \quad \parallel \\ HO-\underset{\cdot\cdot}{N} \quad \underset{\cdot\cdot}{N}-OH \end{array}$$

is one such chelating agent. It uses the unshared pairs of electrons on the two nitrogen atoms to form chelates with many metal ions. The complexes formed with nickel(II) (red) and palladium(II) (yellow) are both insoluble in water.

Separation of Ions. Metal ions are often separated from each other by taking advantage of differences in their tendencies to form complex ions with a particular coordinating agent. To illustrate the method, consider the separation of the two ions Fe^{3+} and Al^{3+}, which are ordinarily precipitated in the same group in cation analysis. If one adds sodium hydroxide to a solution containing these ions, they both precipitate as the hydroxides:

$$Al^{3+} + 3\ OH^- \longrightarrow Al(OH)_3(s) \qquad (19.10)$$

$$Fe^{3+} + 3\ OH^- \longrightarrow Fe(OH)_3(s) \qquad (19.11)$$

However, as more sodium hydroxide is added to increase the OH^- concentration, it is found that the aluminum hydroxide dissolves to form a complex ion which may be represented most simply as $Al(OH)_4^-$

$$Al(OH)_3(s) + OH^- \longrightarrow Al(OH)_4^- \qquad (19.12)$$

Iron(III) hydroxide fails to dissolve and is thus separated from the aluminum.

$$Al^{3+},\ Fe^{3+}$$
$$\downarrow\ OH^-\ \text{(dilute)}$$
$$Al(OH)_3(s),\ Fe(OH)_3(s)$$
$$\downarrow\ OH^-\ \text{(conc.)}$$
$$Al(OH)_4^- \qquad\qquad Fe(OH)_3(s)$$

Many of the transition metals act like aluminum and form sufficiently

503

stable complex ions with OH^- to bring their hydroxides into solution in concentrated sodium hydroxide. Among the hydroxides which dissolve in a strong base are $Zn(OH)_2$, $Cr(OH)_3$ and, to a lesser extent, $Cu(OH)_2$. Compounds such as these, which are capable of reacting with OH^- ions as well as H^+ ions are said to be **amphoteric**.

Another complexing agent which is frequently used to separate metal ions is the ammonia molecule, NH_3. In the analysis of the group 1 cations, advantage is taken of the stability of the $Ag(NH_3)_2{}^+$ complex to separate silver from mercury. Treatment of a precipitate containing AgCl with dilute ammonia leads to the reaction

$$AgCl(s) + 2\ NH_3 \longrightarrow Ag(NH_3)_2{}^+ + Cl^- \qquad (19.13)$$

bringing the silver into solution in the form of the complex ion. To confirm the presence of Ag^+, one can add nitric acid to the solution. The hydrogen ions from the acid destroy the complex by converting NH_3 molecules to $NH_4{}^+$ ions.

$$Ag(NH_3)_2{}^+ + Cl^- + 2\ H^+ \longrightarrow AgCl(s) + 2\ NH_4{}^+ \qquad (19.14)$$

The ability of silver(I) salts to form a complex ion with ammonia may be used in a more subtle way to separate the three anions Cl^-, Br^-, and I^- from each other. Addition of silver nitrate to a solution containing these three ions gives a mixed precipitate of silver chloride, silver bromide, and silver iodide. Silver chloride dissolves in dilute ammonia to give the $Ag(NH_3)_2{}^+$ complex. Silver bromide goes into solution only in concentrated ammonia, while silver iodide remains insoluble even at very high ammonia concentrations. Consequently, the addition of silver nitrate, followed first by dilute and then concentrated ammonia, serves to separate Cl^-, Br^-, and I^- ions from each other.

Cl^-, Br^-, I^-

| Ag^+

$AgCl(s)$, $AgBr(s)$, $AgI(s)$

| dilute NH_3 (6 M)

$Ag(NH_3)_2{}^+$, Cl^- $AgBr(s)$, $AgI(s)$

| conc. NH_3 (15 M)

$Ag(NH_3)_2{}^+$, Br^- $AgI(s)$

To understand the principles underlying the separation of metal ions by complex formation, one must consider the factors which determine the solubility of a solid in a complexing agent. These factors are very similar to those enumerated in Chapter 18 in discussing the solubility of solids in strong acids. They may be deduced from a consideration of the equilibria involved. Take, for example, the process by which a silver halide, AgX, dissolves in ammonia:

$$AgX(s) \rightleftharpoons Ag^+ + X^- \qquad (19.15a)$$

$$Ag^+ + 2\ NH_3 \rightleftharpoons Ag(NH_3)_2{}^+ \qquad (19.15b)$$

$$\overline{AgX(s) + 2\ NH_3 \rightleftharpoons Ag(NH_3)_2{}^+ + X^-} \qquad (19.15)$$

The solubility of a silver halide in ammonia or, indeed, the solubility of any solid in a complexing agent will depend upon:

1. *The solubility of the solid in water* (Equilibrium 19.15a). The greater the solubility in water, the greater will be the solubility in a complexing agent. The three silver halides, AgCl, AgBr, and AgI, are successively less soluble in water as shown by their solubility products (K_{sp} AgCl $= 1.6 \times 10^{-10}$, AgBr $= 1 \times 10^{-13}$, AgI $= 1 \times 10^{-16}$). Consequently, one can predict that the solubility of these three compounds in ammonia or in any complexing agent will decrease in the same order: AgCl > AgBr > AgI. It so happens that the solubilities of the three silver halides in ammonia cover a convenient range insofar as their separation is concerned.

2. *The concentration of complexing agent* (Equilibrium 19.15b). The greater the concentration of complexing agent, the greater will be the tendency of the solid to dissolve to form a complex ion. As pointed out earlier, silver bromide is soluble in concentrated ammonia but insoluble in dilute ammonia. An increase in the concentration of ammonia from 6 M to 15 M exerts a sufficient influence on the equilibrium in Reaction 19.15b to bring a significant amount of silver bromide into solution.

Another example of the effect of concentration of complexing agent on solubility is furnished by the reaction with ammonia of solutions of copper(II) salts. Addition of a small amount of ammonia gives a precipitate of $Cu(OH)_2$. As more ammonia is added to increase the concentration of NH_3 molecules relative to OH^- ions, the hydroxide dissolves to form the $Cu(NH_3)_4^{2+}$ complex ion.

The per cent dissociation of NH_3 decreases with its concentration.

3. *The stability of the complex ion formed.* In many cases, a solid which does not dissolve in one complexing agent can be brought into solution by using a reagent which forms a more stable complex. Silver iodide, which is insoluble in ammonia (K_c Ag(NH$_3$)$_2{}^+ = 4 \times 10^{-8}$), dissolves readily in potassium cyanide solution to form the $Ag(CN)_2{}^-$ complex (K_c Ag(CN)$_2{}^- = 1 \times 10^{-21}$) or in sodium thiosulfate to form the $Ag(S_2O_3)_2{}^{3-}$ complex (K_c Ag(S$_2$O$_3$)$_2{}^{3-} = 1 \times 10^{-13}$).

Solubility in a complexing agent may be treated quantitatively by using the solubility product principle in combination with the dissociation constant of the complex ion involved. Example 19.2 illustrates how this is done.

EXAMPLE 19.2. Calculate the minimum concentration of NH_3 necessary to dissolve:

    a. 0.01 mole/lit of AgCl.   b. 0.01 mole/lit of AgBr.

    c. 0.01 mole/lit of AgI.

SOLUTION

a. We note from the expression

$$K_c = \frac{[Ag^+] \times [NH_3]^2}{[Ag(NH_3)_2{}^+]} = 4.0 \times 10^{-8}$$

that in order to calculate the concentration of NH_3, the concentrations of Ag^+ and $Ag(NH_3)_2{}^+$ must first be determined. From the equation

$$AgCl(s) + 2\ NH_3 \longrightarrow Ag(NH_3)_2{}^+ + Cl^-$$

it is evident that in a solution formed by dissolving 0.01

505

mole/lit of silver chloride, the concentrations of both $Ag(NH_3)_2^+$ and Cl^- must be 0.01 M. But the concentrations of Cl^- and Ag^+ are related by the expression

$$K_{sp}\ AgCl = [Ag^+] \times [Cl^-] = 1.6 \times 10^{-10}$$

Consequently, $[Ag^+] \times 0.01 = 1.6 \times 10^{-10}$ $[Ag^+] = 1.6 \times 10^{-8}$

Substituting these values for the concentrations of $Ag(NH_3)_2^+$ and Ag^+ in the expression for the dissociation constant of the $Ag(NH_3)_2^+$ complex, we have

$$\frac{(1.6 \times 10^{-8}) \times [NH_3]^2}{10^{-2}} = 4.0 \times 10^{-8}$$

Solving, $[NH_3]^2 = \dfrac{4.0 \times 10^{-10}}{1.6 \times 10^{-8}} = 0.025;$ $[NH_3] = 0.16\ M$

b. Following the same reasoning process, we arrive at a value of 0.01 M for the concentration of $Ag(NH_3)_2^+$ and, for the concentration of Ag^+,

$$[Ag^+] = \frac{K_{sp}\ AgBr}{[Br^-]} = \frac{1 \times 10^{-13}}{10^{-2}} = 1 \times 10^{-11}$$

Substituting and solving as before,

$$\frac{(1 \times 10^{-11}) \times [NH_3]^2}{10^{-2}} = 4 \times 10^{-8}; \qquad [NH_3]^2 = \frac{4 \times 10^{-10}}{1 \times 10^{-11}} = 40$$

$$[NH_3] = 6\ M$$

c. Following the procedure outlined in b, using 1×10^{-16} for K_{sp} of AgI, one can calculate that

$$[NH_3] = 200\ M$$

In summary, to dissolve 0.01 mole/lit of AgCl, the concentration of ammonia need be only 0.16 M, while to accomplish the same result with AgBr, an NH_3 concentration of about 6 M is required. The answer obtained with AgI, an impossibly high value of 200 M, means that it is not possible to dissolve significant quantities of AgI even at high ammonia concentrations.

Quantitative Analysis

The reaction of a metal ion with a complexing agent bears at least a superficial resemblance to the reaction of an H^+ ion with an OH^- ion. Comparing the two equations

$$Cu^{2+} + 4\ NH_3 \longrightarrow Cu(NH_3)_4^{2+}$$

$$H^+ + OH^- \longrightarrow H_2O$$

we note that in both cases a positive ion is converted to an extremely stable, covalently bonded species ($K_w = 1 \times 10^{-14}$, $K_c Cu(NH_3)_4^{2+} = 2 \times 10^{-13}$). One might suppose, then, that one could determine the concentration of a metal ion by titrating with a complexing agent in much the same way that an acid is titrated with a base.

In practice, it is seldom possible to use ordinary complexing agents to analyze quantitatively for metal ions in solution. The difficulty is that, as previously noted, the formation of a metal complex is a stepwise process. If one adds ammonia to a solution of a copper(II) salt, the hydrated Cu^{2+} ion is not converted directly to the $Cu(NH_3)_4^{2+}$ ion. Instead, intermediate species containing one, two, or three ammonia molecules are formed. As the concentration of ammonia increases, the various equilibria gradually shift to lower the Cu^{2+} ion concentration. There is no sharp change in "free" Cu^{2+} ion concentration analogous to the abrupt change in H^+ ion concentration that one observes at the equivalence point of an acid-base titration. The end point, instead of being sharp and precise, is drawn out and diffuse.

Within the past few years, analytical chemists have developed a series of reagents that react with metal ions to give extremely stable 1:1 complexes and hence are suitable for metal-ion titrations. These substances are chelating agents; the best known is the sodium salt of ethylenediaminetetracetic acid, commonly called **EDTA**. The anion of this salt has the following structure:

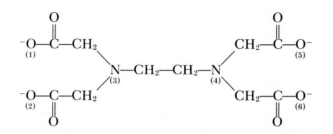

A single EDTA anion can attach itself to a metal ion through as many as six different atoms (numbered 1 to 6 in the foregoing structural formula), filling all of its coordination requirements. Difficulties inherent in stepwise complex formation are thereby avoided; EDTA titrations yield a sharp, easily observed end point.

EDTA is a hexadentate ligand.

EDTA is among the most effective complexing agents known; it forms stable 1:1 chelates with a wide variety of metals. One of its earliest applications was in the determination of the alkaline-earth metals calcium and magnesium found in hard water. Several hundred papers, appearing over the past 20 years, have described the use of EDTA titrations in the determination of over 60 different elements.

To illustrate how an EDTA titration may be carried out, let us consider the use of this chelating agent in the determination of iron(III) salts. The reaction that occurs may be represented as follows:

$$Fe^{3+} + EDTA^{4-} \longrightarrow Fe(EDTA)^- \qquad (19.16)$$

The SCN^- ion, which forms a blood-red complex with Fe^{3+}, can be used as an indicator. As EDTA is added, the thiocyanate complex of iron(III) is converted to the more stable EDTA complex. At the equivalence point, the SCN^- ions attached to iron are quantitatively displaced. The color change from deep red to yellow yields a sharp end point. Knowing the concentration of the EDTA solution and the volume which must be added to reach the end point, one can readily calculate the amount of Fe^{3+} present.

507

PROBLEMS

19.1 Determine the charges of platinum(IV) complexes in which the ligands are

a. 6 NH_3 molecules.
b. 4 NH_3 molecules, 2 Cl^- ions.
c. 2 ethylenediamine molecules, 2 SCN^- ions.
d. 2 $C_2O_4^{2-}$ ions, 2 Cl^- ions.

19.2 Draw sketches to show the geometry of

a. $Cu(NH_3)_2^+$ b. $Fe(CN)_6^{4-}$ c. $Zn(H_2O)_4^{2+}$ d. $Cu(NH_3)_2Cl_2$

19.3 Draw structures for each of the following octahedral complexes.

a. $Co(NH_3)_4Cl_2^+$ b. Co en NH_3Cl_3 c. $Co(en)_2ClBr^+$ d. $Co(H_2O)_4NH_3Cl^{2+}$

19.4 Which of the compounds listed in Problems 19.2 and 19.3 would be expected to show geometrical isomerism? optical isomerism?

19.5 Calculate the total number of electrons associated with the central metal atom in

a. $Ni(CN)_4^{2-}$ b. $Cr(H_2O)_6^{3+}$ c. $PtCl_4^{2-}$ d. $Cu(en)_2^{2+}$

19.6 Using the valence bond model, give the electronic structures of

a. $Ag(NH_3)_2^+$ b. $Zn(OH)_4^{2-}$ c. $Cr(NH_3)_6^{3+}$ d. $PtCl_4^{2-}$

19.7 Indicate how one might experimentally

a. Distinguish between the cis and trans isomers of $[Cr(NH_3)_4Cl_2]Cl$.
b. Determine whether an ion $[MCl_2Br_2]^{2-}$ is square planar or tetrahedral.
c. Determine which of the two ions $Cr(NH_3)_5Cl^{2+}$ or $Cr(NH_3)_5NO_2^{2+}$ is the more labile.
d. Determine which of the two ions $Zn(OH)_4^{2-}$ or $Zn(NH_3)_4^{2+}$ is the more stable.

19.8 Using the crystal field model, write two plausible structures for an octahedral complex of Fe(III). How would you determine experimentally which of these structures is correct for a particular complex of Fe(III)?

19.9 Using Table 19.5, find

a. The ratio of the concentrations of Ag^+ and $Ag(CN)_2^-$ in a solution 10^{-5} M in CN^-.
b. The concentration of CN^- required to bring about a 20 per cent conversion of Zn^{2+} to $Zn(CN)_4^{2-}$.
c. The ratio of the concentration of Cu^{2+} to that of Cd^{2+} in a solution in which there are equal amounts of $Cu(CN)_4^{2-}$ and $Cd(CN)_4^{2-}$.

19.10 Outline separation schemes for

a. Ag^+, Pb^{2+}, Ca^{2+} b. Cl^-, I^-, NO_3^- c. Ni^{2+}, Al^{3+}

19.11 Suggest at least two reagents capable of bringing the following compounds into solution.

a. AgBr   b. Al_2O_3 c. $NiCO_3$

19.12 Write balanced net ionic equations for the reactions that occur when

a. Silver chloride is treated with sodium thiosulfate.
b. Excess ammonia is added to zinc hydroxide.
c. Excess sodium hydroxide is added to zinc hydroxide.
d. Excess sodium cyanide is added to an ammoniacal solution of copper sulfate.

19.13 Consider the reaction

$$[Co(NH_3)_5A]^{2+} + H_2O \longrightarrow [Co(NH_3)_5H_2O]^{3+} + A^-$$

    a. How could you follow experimentally the rate of this reaction?

    b. It is found experimentally that the concentration of $[Co(NH_3)_5A]^{2+}$ decreases at the rate of 10 per cent per hour over a one day period. Show that the reaction must be first order, and calculate the rate constant.

    c. Does the fact that this reaction is first order prove that an S_N1 mechanism is involved? Explain your answer.

19.14 Calculate the minimum concentration of SCN^- required to dissolve 0.1 mole/lit of AgCl.

19.15 A sample weighing 0.200 g and containing Fe^{3+} is titrated with 20.0 ml of 0.100 M EDTA solution. Calculate the per cent of iron in the sample.

19.16 Determine the charges of chromium(III) complexes in which the ligands are

    a. $6 NH_3$ molecules.
    b. $3 NH_3$ molecules, $3 Cl^-$ ions.
    c. $2 NH_3$ molecules, $4 Cl^-$ ions.
    d. 3 ethylenediamine molecules.
    e. $5 H_2O$ molecules, $1 OH^-$ ion.

19.17 Write structural formulas to indicate the geometry of

    a. $AgCl_2^-$
    b. $Cu(NH_3)_4^{2+}$
    c. $Zn(NH_3)_4^{2+}$
    d. $Zn(H_2O)_3(OH)^+$
    e. $Pt(NH_3)_2Br_2$ (two forms)

19.18 Write structural formulas for the following ions (en = ethylenediamine).

    a. $Co(en)_2NH_3Cl^{2+}$
    b. $Co(NH_3)_4CO_3^+$
    c. $Cr(C_2O_4)_3^{3-}$
    d. $Cu(en)_2^{2+}$

19.19 Write structural formulas for all of the compounds having the empirical formula $CrN_4H_{12}Cl_2Br$. Suggest how one might distinguish experimentally between these various compounds.

19.20 Calculate the total number of electrons associated with the central metal atom in

    a. $Zn(OH)_4^{2-}$
    b. $Au(CN)_2^-$
    c. $Fe(CN)_6^{4-}$
    d. $Fe(CN)_6^{3-}$
    e. $Ni(H_2O)_6^{2+}$
    f. $PdCl_4^{2-}$
    g. $Cd(en)_2^{2+}$
    h. $Co(en)_3^{3+}$

19.21 Consider the $AuCl_4^-$ complex ion.

    a. Give the electronic structure of the gold atom in this complex, assuming it to be square planar.

    b. Repeat a, assuming the ion to be tetrahedral.

    c. How could you determine experimentally whether this ion is square planar or tetrahedral?

19.22 Give the electronic structures of the following complex ions, using the valence bond model.

    a. $Ag(CN)_2^-$
    b. $Pt(NH_3)_4^{2+}$
    c. $Al(H_2O)_6^{3+}$
    d. $Cd(NH_3)_4^{2+}$
    e. $Ni(CN)_4^{2-}$
    f. $PtCl_6^{2-}$

19.23 In terms of the crystal field model, explain why a nickel(II) octahedral complex would be expected to have two unpaired electrons.

19.24 Using the value of 1×10^{-17} for the dissociation constant of the $Zn(CN)_4^{2-}$ ion, calculate

    a. The ratio of the concentrations of Zn^{2+} and $Zn(CN)_4^{2-}$ in a solution 0.01 M in CN^-.

    b. The concentration of CN^- necessary to produce a 10 per cent conversion of Zn^{2+} to $Zn(CN)_4^{2-}$.

509

19.25 Outline a separation scheme for

   a. Ag^+, Ni^{2+}, Al^{3+} b. Cl^-, Br^-, SO_4^{2-} c. Br^-, I^-, CO_3^{2-}

19.26 Suggest at least two reagents, including perhaps a complexing agent, which could be used to bring each of the following compounds into solution.

   a. $CaCO_3$ b. $AgCl$ c. AgI d. $Fe(OH)_3$ e. $Al(OH)_3$ f. $Zn(OH)_2$

19.27 What is the minimum concentration of $S_2O_3^{2-}$ necessary to dissolve 0.10 mole/lit of AgBr; 0.10 mole/lit of AgI?

19.28 Write balanced net ionic equations for the reactions that occur when

   a. Excess sodium hydroxide is added to a solution of zinc chloride.
   b. Aluminum hydroxide dissolves in concentrated potassium hydroxide.
   c. Excess ammonia is added to a solution of copper(II) nitrate.
   d. Nitric acid is added to a solution prepared by dissolving silver chloride in ammonia.
   e. Cis $Pd(NH_3)_2Cl_2$ is treated with a solution of sodium oxalate.

19.29 Suggest how one might determine experimentally whether the reaction

$$Co(NH_3)_5Cl^{2+} + H_2O \longrightarrow Co(NH_3)_5H_2O^{3+} + Cl^-$$

proceeds via an S_N1 or an S_N2 mechanism.

°19.30 A solution containing Ni^{2+} and Al^{3+} ions is treated with aqueous ammonia. A bluish precipitate forms at first; as more ammonia is added, part of the precipitate dissolves to form a deep blue solution. The precipitate that remains is white. The solution is separated and treated with dimethylglyoxime to form a red precipitate. The white precipitate previously referred to, upon treatment with excess OH^-, forms a clear solution. If acid is slowly added to this solution, a white precipitate forms, which dissolves as more acid is added. Write balanced net ionic equations for each reaction that took place.

°19.31 A sample of 1 g of silver chloride is shaken with 1 M NH_3. Assuming equilibrium is reached when half of the precipitate has dissolved, calculate the concentrations of Ag^+, $Ag(NH_3)_2^+$, Cl^-, H^+, and OH^- in this solution. (Assume that the concentration of NH_3 does not change.)

°19.32 Solid $Cu(OH)_2$ ($K_{sp} = 1 \times 10^{-19}$) is in equilibrium with a solution which is 1 M in NH_3 and 0.1 M in NH_4^+. Calculate the concentrations of Cu^{2+}, OH^- and $Cu(NH_3)_4^{2+}$ in this solution.

*19.33 Suggest how the d orbitals would be split in a tetrahedral crystal field. You may wish to consult a reference text in coordination chemistry.

°19.34 A certain coordination compound analyzes as follows:

   23.4% Co,   22.3% N,   5.6% H,   6.4% O,   42.3% Cl

It is found that the conductivity of a water solution of this compound corresponds to that of a 1:1 electrolyte such as NaCl. Write a structural formula for the compound consistent with this information. Suggest at least two further experiments which one might perform to check this structural formula.

Electrolytic Cells: Balancing Oxidation-Reduction Equations

An electrolytic cell is a device for converting electrical energy into chemical energy. The process going on within such a cell, known as electrolysis, employs a direct electric current to bring about an oxidation-reduction reaction. To understand how an electrolytic cell operates, consider the generalized cell diagram shown in Figure 20.1.

The portion of the circuit labeled B at the top of the diagram represents a battery whose positive and negative terminals are indicated by + and − signs. The battery is connected by means of two wires to the electrolytic cell, which consists of two electrodes, A and C, dipping into a liquid in which ions M^+ and X^- are free to move.

By a mechanism which will be considered in Chapter 21, the battery acts as an electron pump, pushing electrons into the electrode shown at the left

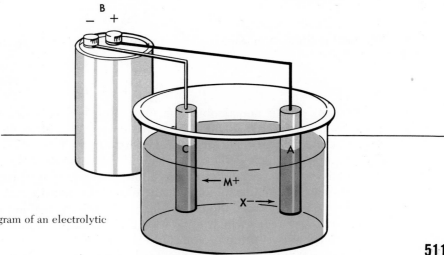

Figure 20.1 Diagram of an electrolytic cell.

in Figure 20.1 and withdrawing them from the electrode at the right. In order to maintain electrical neutrality, some process must take place within the cell so as to consume electrons at C and liberate them at A. This process is an oxidation-reduction reaction. At electrode C, known as the **cathode**, an ion or molecule undergoes **reduction** by accepting electrons. At the **anode**, A, electrons are produced by the **oxidation** of an ion or molecule. The overall cell reaction is the sum of the two half-reactions occurring at the electrodes. While electrolysis is proceeding, there is a steady flow of ions to the two electrodes. Positive ions (*cations*) move towards the *cathode*; negative ions (*anions*) move toward the *anode*.

In *any* cell, reduction occurs at the cathode, oxidation at the anode.

Electrolyses may be carried out by passing a direct current through a water solution of an electrolyte or through a molten salt or oxide. The cell reactions are somewhat easier to visualize in the latter case.

20.1 ELECTROLYSIS OF MOLTEN IONIC COMPOUNDS

In principle, any ionic compound can be decomposed to the elements by melting and electrolyzing it. To illustrate the principles involved, we shall consider three electrolyses of this type. These involve the production of elementary sodium from sodium chloride, aluminum from aluminum oxide, and fluorine from a potassium fluoride–hydrogen fluoride mixture.

Na from NaCl

The so-called Downs cell, used commercially to electrolyze molten sodium chloride, is shown in Figure 20.2. The half-reactions occurring in this cell are particularly simple. At the circular iron *cathode*, sodium ions are reduced to metallic sodium:

$$Na^+ + e^- \longrightarrow Na(l) \qquad (20.1a)$$

For every sodium ion reduced at the cathode, a chloride ion is oxidized to chlorine gas at the graphite *anode*:

$$Cl^- \longrightarrow \tfrac{1}{2} Cl_2(g) + e^- \qquad (20.1b)$$

The total cell reaction, obtained by summing 20.1a and 20.1b, is

$$Na^+ + Cl^- \longrightarrow Na(l) + \tfrac{1}{2} Cl_2(g)$$

or $\qquad NaCl(l) \longrightarrow Na(l) + \tfrac{1}{2} Cl_2(g) \qquad (20.1)$

Reaction 20.1 is nonspontaneous. One can calculate, for example, that at the operating temperature, 600°C, the free energy change for the reaction as written is +77,200 cal. This quantity of energy, in the form of electrical work, must be supplied to make the reaction go. If the products of the cell reaction, elementary sodium and chlorine, are allowed to come in contact with each other, they will combine spontaneously to give the starting material, sodium chloride. To prevent this, the electrodes in the Downs cell are separated by a circular iron screen, which allows for the migration of ions but prevents direct contact between the products of electrolysis.

Some 10,000 tons of sodium are made annually in the United States by the electrolysis of molten sodium chloride. The chlorine formed simultaneously is a valuable by-product. There is, however, a cheaper way of producing chlorine electrolytically (Section 20.2). The relatively high cost of the sodium produced by the Downs process (20 cents a pound) reflects the large amount of energy that must be expended to carry out the electrolysis and maintain the sodium chloride in the liquid state. In practice, the cell is operated at a temperature of about 600°C, some 200°C below the melting point of pure sodium chloride. The lower temperature is made possible by adding a small amount of calcium chloride, thereby forming an ionic solution with a melting point lower than that of the pure "solvent," sodium chloride.

Al from Al$_2$O$_3$

Aluminum is the third most abundant element in the earth's crust. Its importance as a structural material is indicated by the fact that about 2,000,000 tons of aluminum are produced annually in the United States, an amount greater than that of any other metal except copper and iron. Yet, from 1828, when aluminum was first isolated by Wöhler, until 1886, when the Hall electrolytic process for its manufacture was developed, the metal remained little more than a scientific curiosity. In this 58-year period, the price of aluminum never fell below $8 a pound; today it sells for about 30 cents a pound.

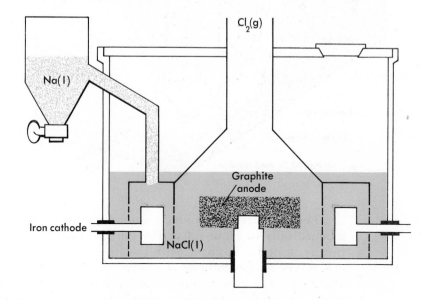

Figure 20.2 Electrolysis of molten sodium chloride.

The long time lag between the isolation of aluminum and its commercial utilization reflected the difficulties of extracting the metal from its ores. Aluminum occurs in such common minerals as feldspar, granite, and clay, but unfortunately the aluminum in these materials is tightly bound in a network of silicon and oxygen atoms from which its extraction is extremely difficult. The principal source of aluminum has always been bauxite ore, in which the element occurs as the hydrated oxide. Prior to 1886, it was necessary to first convert the oxide to the chloride and then reduce the latter with sodium. The

513

high cost of the sodium in this two-step process made the aluminum prohibitively expensive.

The electrolytic process by which aluminum is produced today from
aluminum oxide was worked out by Charles Hall, a graduate student at
Oberlin College. After experimenting with a great many materials, he found
that the mineral cryolite, Na_3AlF_6, could be used in the molten state as a
solvent for Al_2O_3. The use of cryolite makes it possible to reduce the temperature of electrolysis from 2000°C, the melting point of pure Al_2O_3, to about
1000°C. Curiously enough, within a few weeks of the time that Hall produced
his first aluminum, a young Frenchman, Heroult, independently worked out
an almost identical process for its manufacture.

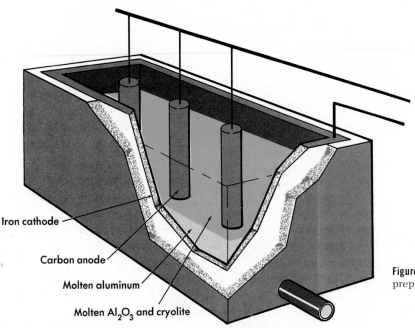

Iron cathode

Carbon anode

Molten aluminum

Molten Al$_2$O$_3$ and cryolite

Figure 20.3 Electrolytic preparation of aluminum.

How might Al_2O_3
be purified? (See
Section 19.7,
Chap. 19).

The cell used to produce aluminum from aluminum oxide is shown
schematically in Figure 20.3. The purified bauxite ore is placed in the electrically heated cell, mixed with cryolite,* and melted. The iron wall of the
cell serves as the cathode at which Al^{3+} ions are reduced to form molten
aluminum. The anodes, retractable carbon rods, are attacked by the oxygen
produced in the cell to form a mixture of carbon dioxide and carbon monoxide.
The two half-reactions occurring at the electrodes may be represented most
simply as

cathode: $\qquad\qquad 2\ Al^{3+} + 6\ e^- \longrightarrow 2\ Al(l)$ \hfill **(20.2a)**

---

* A mixture of the fluorides of aluminum, sodium, and calcium is now used in place of
cryolite. This mixture gives a solution with aluminum oxide which has a lower melting point
and density than that obtained with cryolite. The lower density facilitates the separation of the
molten aluminum, which sinks to the bottom of the cell.

anode:

$$3\ O^{2-} \longrightarrow \tfrac{3}{2}\ O_2(g) + 6\ e^- \qquad \text{(20.2b)}$$

$$2\ Al^{3+} + 3\ O^{2-} \longrightarrow 2\ Al(s) + \tfrac{3}{2}\ O_2(g) \qquad \text{(20.2)}$$

The production of 1 lb of aluminum consumes about 2 lb of aluminum oxide, 0.6 lb of anodic carbon, 0.1 lb of cryolite and 10 kilowatt-hours (kwhr) of electrical energy.

F$_2$ from KF-HF Mixture

The fluoride ion has such a strong attraction for electrons that it cannot be oxidized to fluorine chemically. However, it is possible to oxidize F^- ions in an electrolytic cell at a sufficiently high voltage. Commercially, fluorine is prepared by electrolysis of a solution of potassium fluoride in anhydrous hydrogen fluoride. Nickel electrodes are used. The cell itself is made of nickel, one of the few metals that withstands attack by fluorine. The products are two gases, hydrogen and fluorine.

cathode:
$$2\ H^+ + 2\ e^- \longrightarrow H_2(g) \qquad \text{(20.3a)}$$

anode:
$$2\ F^- \longrightarrow F_2(g) + 2\ e^- \qquad \text{(20.3b)}$$

$$2\ H^+ + 2\ F^- \longrightarrow H_2(g) + F_2(g) \qquad \text{(20.3)}$$

The electrode reactions are actually somewhat more complex than those indicated by Equations 20.3a and b. In a solution of hydrogen fluoride in potassium fluoride, the principal anionic species is the HF_2^- ion, formed by the interaction of an HF molecule with a fluoride ion:

$$\left(:\ddot{\underset{..}{F}}: \right)^- + H : \ddot{\underset{..}{F}}: \longrightarrow \left(:\ddot{\underset{..}{F}} : H : \ddot{\underset{..}{F}}: \right)^-$$

This hydrogen-bonded species is reduced to elementary hydrogen at the cathode and is oxidized to elementary fluorine at the anode.

20.2 ELECTROLYSIS OF WATER SOLUTIONS

It is ordinarily less expensive and more convenient to carry out electrolytic reactions in water solution rather than in the molten salt. However, the presence of water multiplies the number of possible reactions that can take place in an electrolytic cell. We shall first consider what the possible electrode reactions are and then illustrate the principles involved by describing what happens when water solutions of a few typical ionic compounds are electrolyzed.

Cathode Reactions (Reduction)

When a direct electric current is passed through a water solution of an electrolyte, either of two reduction processes may occur at the cathode.

1. The cation may be reduced to the corresponding metal.

$$M^{n+} + ne^- \longrightarrow M(s) \qquad \text{(20.4)}$$

(n = charge of cation)

This is, of course, the reaction which occurs at the cathode when a molten ionic compound is electrolyzed (cf. $Na^+ \rightarrow Na$; $Al^{3+} \rightarrow Al$).

2. Water molecules may be reduced to elementary hydrogen.

515

In acidic solution
we would write
$2H^+ + 2e^- \rightarrow H_2(g)$.

$$2 H_2O + 2 e^- \longrightarrow H_2(g) + 2 OH^- \qquad (20.5)$$

Which of these two reactions will occur with a particular salt depends upon several different factors. One of the most important of these is the relative ease of reduction of the cation as opposed to a water molecule. With salts containing transition metal cations (e.g., Cu^{2+}, Ag^+, Ni^{2+}), which are relatively easy to reduce, Reaction 20.4 ordinarily occurs at the cathode. On the other hand, if the cation present is derived from an A group metal (e.g., Na^+, Mg^{2+}, Al^{3+}), the water molecules are reduced.

Anode Reactions (Oxidation)

The oxidation process that occurs at the anode of an electrolytic cell operating in aqueous solution may be:

1. The oxidation of the anion to the corresponding nonmetal. An example is the reaction that occurs when a water solution containing iodide ions is electrolyzed:

$$2 I^- \longrightarrow I_2(s) + 2 e^- \qquad (20.6)$$

2. The oxidation of water molecules to elementary oxygen.

$$H_2O \longrightarrow \tfrac{1}{2} O_2(g) + 2 H^+ + 2 e^- \qquad (20.7)*$$

Relative ease of
oxidation or reduc-
tion is discussed
in Chapter 21.

Again, the relative ease of oxidation of the anion compared to that of a water molecule is the major factor in determining which of these two reactions will take place. With salts containing iodide, bromide, or chloride ions, it is usually found that the nonmetal (I_2, Br_2, Cl_2) is formed at the anode. When the anion present is F^- or SO_4^{2-}, both of which are very difficult to oxidize, elementary oxygen is ordinarily the product at the anode.

Having listed some of the possible electrode reactions which can occur in the electrolysis of a water solution, we shall now consider what happens when solutions of particular salts are electrolyzed.

Electrolysis of a Solution of CuCl₂

Here the electrode reactions are the same as those to be expected in the absence of water:

cathode: $\qquad\qquad Cu^{2+} + 2 e^- \longrightarrow Cu(s)$

anode: $\qquad\qquad\dfrac{2 Cl^- \longrightarrow Cl_2(g) + 2 e^-}{Cu^{2+} + 2 Cl^- \longrightarrow Cu(s) + Cl_2(g)} \qquad (20.8)$

Electrolysis of a Solution of NaCl

As one would expect, the anode reaction is the oxidation of Cl^- ions:

$$2 Cl^- \longrightarrow Cl_2(g) + 2 e^- \qquad (20.9a)$$

---

* In basic solution, we would write

$$2 OH^- \longrightarrow \tfrac{1}{2} O_2(g) + H_2O + 2 e^-$$

At the cathode, bubbles of hydrogen gas form, and the solution surrounding the electrode becomes basic. This evidence indicates that it is a water molecule rather than a sodium ion which is reduced.

$$2 H_2O + 2 e^- \longrightarrow H_2(g) + 2 OH^- \qquad (20.9b)$$

The overall cell reaction for the electrolysis of an aqueous solution of sodium chloride is obtained by summing the two half-reactions.

$$2 H_2O + 2 Cl^- \longrightarrow Cl_2(g) + H_2(g) + 2 OH^- \qquad (20.9)$$

It will be noted that one effect of this cell reaction is the replacement of the chloride ions originally present by an equal number of hydroxide ions. Consequently, evaporation of the solution remaining after electrolysis yields a residue of sodium hydroxide. The greater part of the sodium hydroxide and almost all the chlorine made in the United States is prepared by the electrolysis of aqueous sodium chloride; hydrogen is an important by-product.

Electrolysis of a Solution of $CuSO_4$

cathode: $Cu^{2+} + 2 e^- \longrightarrow Cu(s)$

anode: $\underline{H_2O \longrightarrow \frac{1}{2} O_2(g) + 2 H^+ + 2 e^-}$

$$Cu^{2+} + H_2O \longrightarrow Cu(s) + \frac{1}{2} O_2(g) + 2 H^+ \qquad (20.10)$$

These reactions reflect the fact that the Cu^{2+} ion is more readily reduced than the H_2O molecule, which, in turn, is easier to oxidize than the sulfate ion. The products of the reaction include copper metal at the cathode and oxygen gas at the anode. As the electrolysis proceeds, the solution becomes strongly acidic; eventually the electrolyte is converted to a solution of sulfuric acid when all the Cu^{2+} is gone.

EXAMPLE 20.1. Predict what reactions will occur at the anode and cathode and write an equation for the overall reaction in the electrolysis of aqueous solutions of

      a. KI   b. Na_2SO_4

SOLUTION

    a. The electrolysis of a KI solution should follow a path analogous to that described for NaCl:

        cathode: $2 H_2O + 2 e^- \longrightarrow H_2(g) + 2 OH^-$

        anode: $\underline{2 I^- \longrightarrow I_2(s) + 2 e^-}$

            $2 H_2O + 2 I^- \longrightarrow H_2(g) + 2 OH^- + I_2(s)$

            (compare Equation 20.9)

    b. In this case, neither ion participates in an electrode reaction. The Na^+ ion is too difficult to reduce, the SO_4^{2-} too difficult to oxidize.

        cathode: $2 H_2O + 2 e^- \longrightarrow H_2(g) + 2 OH^-$

        anode: $H_2O \longrightarrow \frac{1}{2} O_2(g) + 2 H^+ + 2 e^-$

What would be the product if the salt were $Ni(NO_3)_2$?

517

Taking account of the fact that the OH⁻ and H⁺ ions produced at the electrodes will react with each other, i.e.,

$$2 \ H^+ + 2 \ OH^- \longrightarrow 2 \ H_2O$$

we obtain for the overall reaction,

$$H_2O \longrightarrow H_2(g) + \tfrac{1}{2} \ O_2(g)$$

which, of course, represents the electrolysis of water. The sodium sulfate takes no part in the electrolysis reaction, but does increase the conductivity and so allows us to pass more current through the cell.

Electroplating

In discussing the electrolysis of aqueous solutions, we have tacitly assumed that the electrodes themselves do not participate in the reaction. In practice, it is often found that the anode reaction involves the oxidation of the metal used to form that electrode. Indeed, the electrolytic cells used in the commercial process of electroplating are designed with precisely that purpose in mind. Consider, for example, the cell used in copper plating, shown in Figure 20.4.

The anode of this cell consists of a bar or strip of copper metal. When a direct current is passed through the cell, copper atoms at this electrode are oxidized to Cu^{2+} ions. The electrolyte used in this cell is a water solution of a copper(II) salt, usually $CuSO_4$. At the cathode, Cu^{2+} ions are reduced to copper metal, which forms a uniform coating on the object to be plated. The reactions at the two electrodes are precisely the reverse of one another:

cathode: $$Cu^{2+} + 2 \ e^- \longrightarrow Cu(s)$$

anode: $$Cu(s) \longrightarrow Cu^{2+} + 2 \ e^-$$

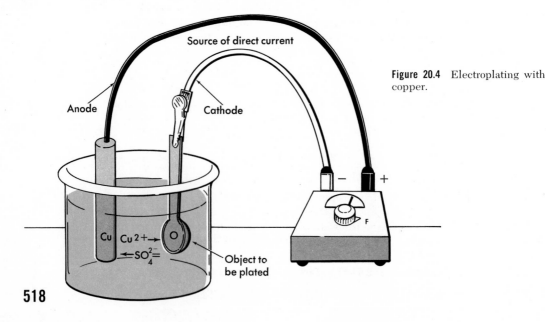

Figure 20.4 Electroplating with copper.

The net result of the cell reaction is simply the transfer of copper atoms from anode to cathode. The concentration of Cu^{2+} ions in solution remains constant, ensuring the formation of a smooth, adherent plate.

In many electroplating processes, the electrolyte contains a complexing agent that keeps the concentration of metal ions at a low, constant value. It is particularly important to do this when the cation is easily reduced; otherwise it may plate too rapidly, giving a rough or pitted surface or a nonadherent mass. In silver-plating, the bath used contains CN^- ions which react with Ag^+ ions to form the very stable complex $Ag(CN)_2^-$. Here, as in copper plating, the anode is made from the metal to be plated, in this case silver; the object to be plated is made the cathode; the electrode reactions are

Nickel plating is often carried out in NH_3 solution. Explain.

anode: $$Ag(s) + 2\ CN^- \longrightarrow Ag(CN)_2^- + e^-$$

cathode: $$Ag(CN)_2^- + e^- \longrightarrow Ag(s) + 2\ CN^-$$

20.3 FARADAY'S LAW OF ELECTROLYSIS

From an economic standpoint, one of the most important aspects of an electrochemical process is the relationship between the quantity of electricity passed through the cell and the amounts of substances produced by oxidation and reduction at the electrodes. Before discussing this relationship, it may be helpful to review the units used to express it.

In electrochemistry, one of the most convenient units for expressing amount of electrical charge is the mole of electrons (6.02×10^{23} electrons). This quantity is so important that it is given a special name—the **faraday**. Another widely used unit is the **coulomb**. It can be shown experimentally that one faraday, to three significant figures, is equal to 96,500 coulombs:

$$1 \text{ mole of electrons} = 1 \text{ faraday} = 96,500 \text{ coulombs}$$

We shall also have occasion to refer to a unit of current flow, the **ampere**; an ampere (amp) is a rate of flow of electricity of such magnitude that one coulomb passes a given point in the circuit in one second. The number of coulombs flowing through a cell can be calculated from the amperage and time using the relation

$$\text{no. of coulombs} = \text{no. of amperes} \times \text{no. of seconds}$$

Of the three units of mass that we shall use in electrochemistry, the gram, the mole, and the gram equivalent weight, only the third requires comment. It will be recalled that in Chapter 2 the gram equivalent weight of an element was defined as the weight that reacts with or is otherwise equivalent to eight grams of oxygen. This definition is useful for a particular type of oxidation-reduction reaction, that between elements. We are now in a position to frame a more general definition of gram equivalent weight valid for any oxidation-reduction reaction. The **gram equivalent weight** of a substance is the weight in grams which reacts with, is produced by, or is otherwise equivalent to **one mole of electrons (one faraday)**. The use of this definition to calculate the gram equivalent weights of species taking part in oxidation-reduction reactions is illustrated by Example 20.2.

519

EXAMPLE 20.2.   Calculate the gram equivalent weights of

    a. Aluminum and oxygen in the electrolysis of Al_2O_3 (Reaction 20.2).

    b. Zinc and Ag^+ in the reaction $Zn(s) + 2 Ag^+ \rightarrow Zn^{2+} + 2 Ag(s)$.

SOLUTION

    a. Breaking the overall reactions into two half-reactions, we have

cathode: $\qquad 2 Al^{3+} + 6 e^- \longrightarrow 2 Al(l)$

anode $\qquad\qquad 3 O^{2-} \longrightarrow 3 O + 6 e^-$

Clearly, 6 moles of electrons (6 faradays) produces 2 moles of aluminum and 3 moles of (monatomic) oxygen. It follows that 1 mole of electrons (1 faraday) will yield $\frac{2}{6}$ or $\frac{1}{3}$ of a mole of Al and $\frac{3}{6}$ or $\frac{1}{2}$ of a mole of O:

$$6 \text{ moles } e^- \simeq 2 \text{ moles Al} \simeq 3 \text{ moles O}$$

$$1 \text{ mole } e^- \simeq \tfrac{1}{3} \text{ mole Al} \simeq \tfrac{1}{2} \text{ mole O}$$

Hence, from the definition,

$$\text{GEW O} = \tfrac{1}{2} \text{ mole O} = \tfrac{1}{2} \times 16.00 \text{ g O} = 8.00 \text{ g O}$$

$$\text{GEW Al} = \tfrac{1}{3} \text{ mole Al} = \tfrac{1}{3} \times 26.97 \text{ g Al} = 8.99 \text{ g Al}$$

8.99 g of Al reacts with 8.00 g of O.

    b. Proceeding exactly as in a:

$$Zn(s) \longrightarrow Zn^{2+} + 2 e^- \qquad 2 Ag^+ + 2 e^- \longrightarrow 2 Ag(s)$$

$$2 \text{ moles } e^- \simeq 1 \text{ mole Zn} \simeq 2 \text{ moles } Ag^+$$

$$1 \text{ mole } e^- \simeq \tfrac{1}{2} \text{ mole Zn} \simeq 1 \text{ mole } Ag^+$$

$$\text{GEW Zn} = \tfrac{1}{2} \text{ mole Zn} = \tfrac{1}{2} \times 65.4 \text{ g Zn} = 32.7 \text{ g Zn}$$

$$\text{GEW Ag} = 1 \text{ mole } Ag^+ = 1 \times 107.9 \text{ g } Ag^+ = 107.9 \text{ g } Ag^+$$

Going back to the overall equation, it may be noted that 2 GEW of Ag^+ (2 moles Ag^+) react exactly with 2 GEW of Zn (1 mole Zn). In any oxidation-reduction reaction, the number of gram equivalent weights of oxidizing agent (Ag^+, and so on) will always be exactly equal to the number of gram equivalent weights of reducing agent (Zn, and so on). This is, of course, a necessary consequence of the way in which gram equivalent weight is defined (cf. the definitions of GEW for acid-base reactions, Chapter 18).

    The foregoing definition of gram equivalent weight leads immediately to the fundamental law of electrolysis, which was first discovered by Michael Faraday more than a century ago. This law can be stated as follows: **The number of faradays of electricity (moles of electrons) passed through an electrical cell is exactly equal to the number of gram equivalent weights of the substances produced at each of the two electrodes.**

$$\text{no. of faradays} = \text{no. of GEW} \qquad (20.11)$$

    The basic Equation 20.11, expressing the relationship between quantity of electricity and quantity of matter produced by electrolysis, can be applied to a great many practical problems that arise. Example 20.3 illustrates the general approach.

EXAMPLE 20.3. Calculate

a. The number of grams of aluminum produced when 7200 coulombs of electricity pass through aluminum oxide dissolved in molten cryolite.

b. The time required to plate a spoon 14.0 cm² in area to a depth of 0.010 cm with silver, using an $Ag(CN)_2^-$ plating bath with a current of 0.0120 amp. (d. Ag = 10.5 g/cm³).

SOLUTION

a. We start with the fundamental relationship

$$\text{no. of faradays} = \text{no. of GEW}$$

The number of faradays in 7200 coulombs is readily calculated, knowing that 1 faraday = 96,500 coulombs:

$$\text{no. of faradays} = 7200 \text{ coulombs} \times \frac{1 \text{ faraday}}{96,500 \text{ coulombs}} = 0.0746$$

The number of gram equivalent weights of aluminum must then be 0.0746. But since one gram equivalent weight of aluminum is 8.99 g (Example 20.2),

$$\text{no. of g Al} = 0.0746 \text{ GEW Al} \times \frac{8.99 \text{ g Al}}{1 \text{ GEW Al}} = 0.671 \text{ g Al}$$

b. Here, we shall proceed along a well-defined path. We shall first calculate the volume of silver to be plated; then, using the density of silver, we can find the weight required. We shall then find the number of gram equivalent weights and hence the number of faradays required, from which the number of coulombs and finally the time can be calculated.

$$\text{Vol. Ag} \rightarrow \text{wt. Ag} \rightarrow \text{no. GEW Ag} \rightarrow$$
$$\quad (1) \qquad\quad (2) \qquad\qquad (3)$$
$$\text{no. faradays} \rightarrow \text{no. coulombs} \rightarrow \text{no. secs.}$$
$$\quad (4) \qquad\qquad (5) \qquad\qquad (6)$$

(1) Volume Ag = 14.0 cm² × 0.0100 cm = 0.140 cm³

(2) no. g Ag = 0.140 cm³ × 10.5 g/cm³ = 1.47 g Ag

(3) To calculate the number of GEW of Ag, we note from the equation

$$Ag(CN)_2^- + e^- \longrightarrow Ag(s) + 2 \text{ CN}^-$$

that one mole of electrons gives one mole of Ag and hence

$$\text{GEW Ag} = \text{GAW Ag} = 108 \text{ g}$$

$$\text{no. of GEW Ag} = 1.47 \text{ g Ag} \times \frac{1 \text{ GEW Ag}}{108 \text{ g}} = 0.0136 \text{ GEW Ag}$$

(4) no. faradays = no. GEW = 0.0136

(5) no. coulombs = 0.0136 faradays × 96,500 $\frac{\text{coulombs}}{\text{faraday}}$

$$= 1310 \text{ coulombs}$$

521

$$(6) \ \text{no. sec} = \frac{\text{no. coulombs}}{\text{no. amp}} = \frac{1310 \text{ coulombs}}{0.0120 \text{ amp}}$$
$$= 109{,}000 \text{ sec } (30.3 \text{ hrs})$$

In working these problems, it has been tacitly assumed that all the electrons passing into the cell are used in forming the products, aluminum and silver. In practice, this is never exactly true of any electrolytic process. There are always side reactions which consume at least a small fraction of the current. In the Hall process for the preparation of aluminum, the **current efficiency**, that is, the fraction of the current used to produce aluminum, is only about 80 per cent. We calculated that 7200 coulombs should yield 0.671 g of aluminum, assuming 100 per cent current efficiency. The amount of aluminum actually formed is closer to 80 per cent of this amount, or about 0.54 g. In silver plating, with a properly designed cell and a plating bath free of impurities, it is possible to approach very closely a current efficiency of 100 per cent.

Faraday's Law offers a convenient experimental means of determining the gram equivalent weight of a substance involved in an oxidation-reduction reaction. All that one has to do is to determine the weight of the substance produced or consumed when 1 faraday of electricity passes through a cell. Going one step further, it is possible to combine the gram equivalent weight, as determined by electrolysis, with the gram formula weight to calculate the number of moles of electrons involved in the oxidation-reduction reaction. In the simplest case, in which the product is an elementary substance formed by oxidation or reduction of a monatomic ion, this information serves to establish the charge of that ion. Example 20.4 illustrates the calculations involved.

EXAMPLE 20.4. A sample of gadolinium metal (AW = 157) is dissolved in hydrochloric acid and the resulting solution is electrolyzed. It is found that when 3216 coulombs pass through the cell, 1.74 g of gadolinium is formed at the cathode. Calculate the charge on the gadolinium ion.

SOLUTION. Let us first calculate the gram equivalent weight of gadolinium. Noting that 3216 coulombs pass through the cell, the number of faradays must be

$$\text{no. faradays} = 3216 \text{ coulombs} \times \frac{1 \text{ faraday}}{96{,}500 \text{ coulombs}} = 0.0333 \text{ faraday}$$
$$= 0.0333 \text{ GEW Gd}$$

It follows that 1.74 g of gadolinium must represent 0.0333 GEW

$$1.74 \text{ g Gd} \simeq 0.0333 \text{ GEW Gd}$$

$$\text{no. of g/GEW} = 1 \text{ GEW Gd} \times \frac{1.74 \text{ g Gd}}{0.0333 \text{ GEW}} = 52.3 \text{ g}$$

Since the gram equivalent weight of gadolinium (52.3 g) is one third of its gram atomic weight (157 g), it is clear that three moles of electrons are required to form one mole of the element. This means that the gadolinium cation must carry a charge of $+3$; the equation for the reduction half-reaction is

$$Gd^{3+} + 3\ e^- \longrightarrow Gd(s)$$

The electrolytic method illustrated by Example 20.4 was one of the first employed to determine the charges of ions in solution. It has the advantage that the compound used for electrolysis need not be highly pure, provided the impurities present do not take part in the electrode reaction.

How could you show that the chromate ion has a charge of -2 $(CrO_4{}^{2-})$?

20.4 BALANCING OXIDATION-REDUCTION EQUATIONS

Throughout this chapter we have had frequent occasion to represent the overall reaction going on in an electrolytic cell by means of a balanced equation. We have arrived at such an equation by first writing balanced half-equations for the half-reactions of oxidation and reduction occurring at the electrodes. These two half-equations are then combined into one equation representing the entire electrolysis reaction. To recall one example, in arriving at the balanced equation for the electrolysis of a water solution of sodium chloride, we first wrote an equation for the oxidation reaction at the anode (20.9a) and the reduction reaction at the cathode (20.9b). Combining these in such a way that there was no net loss or gain of electrons, we arrived at the final equation, 20.9.

This method of analyzing an oxidation-reduction reaction and the corresponding equation helps one to understand what is taking place within an electrolytic cell. We shall find it equally applicable when we consider voltaic cells in Chapter 21. Indeed, whenever one is faced with the problem of arriving at an equation for an oxidation-reduction reaction, whether it is occurring in an electrical cell, a test tube, or elsewhere, it is convenient to first break the equation down into two half-equations, balance these separately, and then combine them so as to arrive at an overall equation involving no net loss or gain of electrons.

To illustrate the application of this method to the balancing of a simple oxidation-reduction equation, consider the reaction that occurs when a water solution of iron(III) chloride is electrolyzed. The products are observed to be metallic iron and chlorine gas. The unbalanced equation may be written

$$Fe^{3+} + Cl^- \longrightarrow Fe(s) + Cl_2(g)$$

To balance this equation, we proceed as follows:

1. Split the equation into two half-equations, one oxidation and one reduction:

reduction: $\qquad\qquad\qquad Fe^{3+} \longrightarrow Fe(s) \qquad\qquad\qquad$ (1a)

oxidation: $\qquad\qquad\qquad Cl^- \longrightarrow Cl_2(g) \qquad\qquad\qquad$ (1b)

2. Balance these half-equations, first with respect to mass and then with respect to charge:

Equation 1a is balanced insofar as mass is concerned, since there is one

523

atom of Fe on both sides. The charges, however, are unbalanced: the Fe atom on the right has 0 charge while the Fe^{3+} ion on the left has a charge of $+3$. To correct this, we add three electrons to the left of 1a, arriving at

$$Fe^{3+} + 3\ e^- \longrightarrow Fe(s) \tag{2a}$$

Equation 1b must first be balanced with respect to mass by providing two Cl^- ions to give one molecule of Cl_2:

$$2\ Cl^- \longrightarrow Cl_2(g)$$

To balance charges, two electrons must be added to the right, giving a charge of -2 on both sides:

$$2\ Cl^- \longrightarrow Cl_2(g) + 2\ e^- \tag{2b}$$

3. Having arrived at two balanced half-equations, combine them so as to make the number of electrons gained in reduction equal to the number lost in oxidation.

In Equation 2a, three electrons are gained; in 2b, two electrons are given off. To arrive at a final equation in which no electrons appear, multiply 2a by 2, 2b by 3, and add:

$2 \times 2a$:
$$2\ Fe^{3+} + 6\ e^- \longrightarrow 2\ Fe(s) \tag{3a}$$

$3 \times 2b$:
$$6\ Cl^- \longrightarrow 3\ Cl_2(g) + 6\ e^- \tag{3b}$$

$$\overline{2\ Fe^{3+} + 6\ Cl^- \longrightarrow 2\ Fe(s) + 3\ Cl_2(g)} \tag{3}$$

The equation just balanced corresponds to the simplest type of oxidation-reduction reaction, in which only two elements are involved. One of these elements (chlorine, in the form of Cl^- ions) underwent oxidation, the other (iron, in the form of Fe^{3+} ions) was reduced. Equations for such reactions can ordinarily be balanced by inspection. The reaction itself becomes more complicated, the equation more difficult to balance, and the method just described more pertinent when atoms of elements other than those being oxidized or reduced are involved in the overall reaction. The method we have outlined, with minor modifications, can be applied regardless of the number of elements participating in the reaction or the complexity of the resulting equation.

For oxidation-reduction reactions in water solution, the two most common "extra" elements (elements whose atoms undergo no change in oxidation number) are hydrogen and oxygen. Compounds containing these elements take part in a great many oxidation-reduction reactions, often without any change in the oxidation number of H $(+1)$ or oxygen (-2).

To illustrate the balancing of an oxidation-reduction equation in which "extra" elements appear along with the elements being oxidized or reduced, consider the reaction that occurs between chloride and permanganate ions in acidic solution. Experimental evidence indicates that this reaction can best be represented by the equation

$$MnO_4^- + H^+ + Cl^- \longrightarrow Mn^{2+} + Cl_2(g) + H_2O$$

Note that the two elements that undergo a change in oxidation number are manganese $(+7 \rightarrow +2)$ and chlorine $(-1 \rightarrow 0)$. Neither hydrogen nor oxygen change oxidation number, yet atoms of these elements participate in the reaction. The oxygen atoms tied up originally in the MnO_4^- ion end up as H_2O molecules; the H^+ ions meet the same fate.

To balance this equation, we proceed as follows.

1. Separate into two half-equations:

Oxidation: $$Cl^- \longrightarrow Cl_2(g) \tag{1a}$$

Reduction: $$MnO_4^- + H^+ \longrightarrow Mn^{2+} + H_2O \tag{1b}$$

2. Half-equation 1a is readily balanced as before, giving us

$$2\ Cl^- \longrightarrow Cl_2(g) + 2\ e^- \tag{2a}$$

To balance 1b, we first make sure that there are the same number of Mn atoms on both sides, one. Next, the oxygen is balanced by writing a coefficient of 4 in front of the H_2O on the right to account for the four oxygens in the MnO_4^- ion:

$$MnO_4^- + H^+ \longrightarrow Mn^{2+} + 4\ H_2O$$

To complete the mass balance, the number of hydrogen atoms on the two sides must be equalized. The four H_2O molecules on the right contain eight hydrogen atoms; there must then be eight H^+ ions on the left:

$$MnO_4^- + 8\ H^+ \longrightarrow Mn^{2+} + 4\ H_2O$$

Finally, the charges must be balanced; at the moment there is a charge of $+2$ on the right and $+7$ on the left $(-1+8)$. To balance, five electrons are added to the left:

$$MnO_4^- + 8\ H^+ + 5\ e^- \longrightarrow Mn^{2+} + 4\ H_2O \tag{2b}$$

It may be seen from this discussion that balancing a half-equation containing hydrogen and oxygen in addition to the elements undergoing oxidation or reduction involves, as one would expect, two extra steps. In obtaining the mass balance for the half-equation, it is ordinarily simplest to deal first with the element being oxidized or reduced, then with the oxygen, and finally the hydrogen.

3. Half-equations 2a and 2b are now combined as usual so as to eliminate electrons from the final equation. To do this, multiply 2a by 5 and 2b by 2, producing 10 e^- on both sides:

$5 \times 2a$: $$10\ Cl^- \longrightarrow 5\ Cl_2(g) + 10\ e^- \tag{3a}$$

$2 \times 2b$: $$2\ MnO_4^- + 16\ H^+ + 10\ e^- \longrightarrow 2\ Mn^{2+} + 8\ H_2O \tag{3b}$$

$$2\ MnO_4^- + 16\ H^+ + 10\ Cl^- \longrightarrow 2\ Mn^{2+} + 8\ H_2O + 5\ Cl_2(g) \tag{3}$$

We frequently have occasion to write balanced equations for oxidation-reduction reactions taking place in basic solution. For such reactions, it would be inappropriate to write equations in which H^+ ions appear, since this ion is present in only very small concentrations in basic solution. Instead, the equations should contain hydrogen in the form of OH^- ions or H_2O molecules. A simple way to accomplish this is to eliminate any H^+ ions appearing in the half-equations, "neutralizing" them by adding an equal number of OH^- ions to both sides. To illustrate, consider the oxidation, in basic solution, of iodide

It is also possible to balance the equation directly, using OH^- and H_2O.

525

by permanganate ions:

$$I^- + MnO_4^- \longrightarrow I_2 + MnO_2(s) \text{ (basic solution)}$$

One can proceed exactly as in the foregoing example, to obtain the half-equations

Oxidation: $\qquad\qquad\qquad 2\ I^- \longrightarrow I_2 + 2\ e^- \qquad\qquad\qquad$ **(2a)**

Reduction: $\qquad MnO_4^- + 4\ H^+ + 3\ e^- \longrightarrow MnO_2(s) + 2\ H_2O \qquad$ **(2b)**

The H^+ ions appearing in the reduction half-equation must now be removed to obtain an equation valid in basic solution. To do this, four OH^- ions are added to both sides:

$$MnO_4^- + 4\ H^+ + 3\ e^- \longrightarrow MnO_2(s) + 2\ H_2O$$
$$\underline{+\ 4\ OH^- \qquad\qquad \longrightarrow \qquad\qquad\qquad\qquad +\ 4\ OH^-}$$
$$MnO_4^- + 4\ H_2O + 3\ e^- \longrightarrow MnO_2(s) + 2\ H_2O + 4\ OH^-$$

Eliminating two water molecules from each side, we arrive at

$$MnO_4^- + 2\ H_2O + 3\ e^- \longrightarrow MnO_2(s) + 4\ OH^- \qquad\qquad \textbf{(2b′)}$$

for the reduction half-reaction in basic solution. To obtain the overall equation, we proceed as before, combining 2a and 2b′ in such a way as to make the electron gain equal the electron loss:

$3 \times 2a$: $\qquad\qquad\qquad 6\ I^- \longrightarrow 3\ I_2 + 6\ e^-$

$2 \times 2b′$: $\quad \underline{2\ MnO_4^- + 4\ H_2O + 6\ e^- \longrightarrow 2\ MnO_2(s) + 8\ OH^-}$

$\qquad\qquad 6\ I^- + 2\ MnO_4^- + 4\ H_2O \longrightarrow 3\ I_2 + 2\ MnO_2(s) + 8\ OH^- \qquad$ **(3)**

Oxidation Number Method of Balancing Equations

The **half-equation** method just described is by no means the only or even the simplest method of balancing oxidation-reduction equations. We have stressed this particular method because it involves techniques that are valuable in studying other aspects of oxidation-reduction reactions. Of the various other methods which can be used to balance redox equations, we shall discuss only one—the **oxidation number** method.

To illustrate the application of this method, consider the equation referred to earlier:

$$MnO_4^- + Cl^- + H^+ \longrightarrow Mn^{2+} + Cl_2(g) + H_2O$$

To balance this equation by the oxidation number method, we proceed as follows:

1. Determine the oxidation number of each element on both sides of the equation, thereby determining which elements have undergone oxidation and reduction:

	Oxid. No. Reactants	Oxid. No. Products	
Mn	+7	+2	reduced
O	−2	−2	
Cl	−1	0	oxidized
H	+1	+1	

2. By adjusting the coefficients of the species being oxidized and reduced, make the total increase in oxidation number equal to the total decrease.

In this case, each Mn atom undergoes a decrease in oxidation number of five units; each Cl atom increases in oxidation number by one unit. To make the increase in oxidation number equal to the decrease, there must be five Cl atoms oxidized for every Mn reduced. Thus

$$MnO_4^- + 5\ Cl^- \longrightarrow Mn^{2+} + \tfrac{5}{2}\ Cl_2(g)$$

or, multiplying through by two to eliminate fractional coefficients,

$$2\ MnO_4^- + 10\ Cl^- \longrightarrow 2\ Mn^{2+} + 5\ Cl_2(g) \qquad\qquad \text{(a)}$$

Note that for Equation a, the total increase in oxidation number of Cl is $10 \times 1 = 10$, the total decrease in oxidation number of Mn $= 2 \times 5 = 10$.

3. Having determined the coefficients of the species being oxidized and reduced, balance the number of atoms of the remaining elements in the usual manner.

Here, starting with Equation a, the oxygen is balanced first. The presence of 2 MnO_4^- ions on the left, containing a total of eight oxygen atoms, requires that there be eight H_2O molecules, each with one oxygen atom, on the right:

$$2\ MnO_4^- + 10\ Cl^- \longrightarrow 2\ Mn^{2+} + 5\ Cl_2(g) + 8\ H_2O \qquad\qquad \text{(b)}$$

Finally, to balance the hydrogen, 16 H^+ ions must be added to the left:

$$2\ MnO_4^- + 10\ Cl^- + 16\ H^+ \longrightarrow 2\ Mn^{2+} + 5\ Cl_2(g) + 8\ H_2O \qquad\qquad \text{(c)}$$

The final balanced equation is, of course, identical to that previously derived by the half-equation method.

Calculations Involving Balanced Equations

A balanced oxidation-reduction equation, like any other balanced equation, can be used to calculate the relative amounts of reactants and products involved in a reaction. The way in which this is done was discussed in Chapter 3. Just in case you have forgotten the principles involved, it may be helpful to review them in Example 20.5.

EXAMPLE 20.5. Consider the reaction between potassium permanganate and hydrochloric acid, for which we may write the balanced equation

$$2\ MnO_4^- + 16\ H^+ + 10\ Cl^- \longrightarrow 2\ Mn^{2+} + 8\ H_2O + 5\ Cl_2(g)$$

Calculate

a. The number of moles of MnO_4^- required to form 16.0 g of Cl_2.

b. The number of grams of $KMnO_4$ that one must start with to obtain one liter of Cl_2, measured at 25°C and one atmosphere pressure.

c. The theoretical yield, in moles, of Cl_2 if 12.64 g of $KMnO_4$ reacts with 40.0 ml of 12.0 M HCl.

SOLUTION

a. From the balanced equation, 2 moles $MnO_4^- \approx 5$ moles Cl_2

But $\qquad\qquad\qquad\qquad$ 1 mole $Cl_2 = 70.9$ g Cl_2.

Hence, the no. moles MnO_4^-

$$= 16.0\ g\ Cl_2 \times \frac{1\ mole\ Cl_2}{70.9\ g\ Cl_2} \times \frac{2\ moles\ MnO_4^-}{5\ moles\ Cl_2}$$

$$= 0.0903\ moles\ MnO_4^-$$

b. Here, we shall first calculate the number of moles of Cl_2, using the Ideal Gas Law. From the coefficients of the balanced equation, we can then calculate the number of moles of MnO_4^- required. This must be equal to the number of moles of $KMnO_4$. Knowing the gram formula weight of $KMnO_4$, we can finally calculate the number of grams of $KMnO_4$.

Remember the Ideal Gas Law? (Chap. 5).

$$(1)\ n\ Cl_2 = \frac{PV}{RT} = \frac{(1\ atm)(1\ lit)}{\left(0.0821\ \dfrac{lit\ atm}{mole\ °K}\right)(298°\ K)} = 0.0409\ moles\ Cl_2$$

527

(2) no. moles MnO_4^-

$$= 0.0409 \text{ moles } Cl_2 \times \frac{2 \text{ moles } MnO_4^-}{5 \text{ moles } Cl_2} = 0.0164$$

(3) no. moles $KMnO_4$ = no. moles MnO_4^- = 0.0164

(4) no. g $KMnO_4$

$$= 0.0164 \text{ mole } KMnO_4 \times \frac{158 \text{ g } KMnO_4}{1 \text{ mole } KMnO_4}$$
$$= 2.59 \text{ g } KMnO_4$$

c. We must first decide which reagent to base our calculations upon. To do this, let us calculate the number of moles of MnO_4^-, H^+, and Cl^-.

Remember theoretical yields? (Section 3.6, Chap. 3).

$$\text{no. moles } MnO_4^- = \text{no. moles } KMnO_4 = \frac{12.64 \text{ g}}{158 \text{ g/mole}} = 0.0800$$

$$\text{no. moles } H^+ = \text{no. moles } HCl = (0.0400 \text{ lit})(12.0 \text{ mole/lit})$$
$$= 0.480$$

no. moles Cl^- = no. moles HCl = 0.480

Now, the equation tells us that

$$1 \text{ mole } MnO_4^- \doteq 8 \text{ moles } H^+ \doteq 5 \text{ moles } Cl^-$$

A moment's reflection should convince you that the critical reagent is the H^+ ion. We do not have enough H^+ to react with all the MnO_4^- ($8 \times 0.0800 = 0.640$ mole of H^+ would be required to do this). Hence, the yield of Cl_2 is limited by the amount of H^+ present. But,

$$16 \text{ moles } H^+ \doteq 5 \text{ moles } Cl_2$$

Theoretical yield:

$$\text{no. moles } Cl_2 = 0.480 \text{ moles } H^+ \times \frac{5 \text{ moles } Cl_2}{16 \text{ moles } H^+}$$

$$= 0.150 \text{ mole } Cl_2$$

PROBLEMS

20.1 Explain precisely what is meant by each of the following terms.

a. Electrolysis
b. Cathode
c. Anode
d. Gram equivalent weight (in electrolysis)
e. Faraday
f. Coulomb
g. Ampere

20.2 Describe in some detail the commercial preparations of the following elements.

a. Sodium   b. Aluminum   c. Chlorine   d. Fluorine

20.3 Explain why

a. When a water solution of sodium chloride is electrolyzed, the solution becomes basic.
b. The electrolysis of a water solution of KI gives products different from those formed when the molten salt is electrolyzed.
c. Aluminum cannot be prepared by the electrolysis of a water solution of an aluminum salt.

d. In the electrolysis of a water solution, O_2 may be formed at the anode but never at the cathode.

20.4 Write balanced net ionic equations for the reactions that occur when the following water solutions are electrolyzed.

    a. NiI_2 b. KI c. NaF d. CuF_2

20.5 Describe in words how one might prepare

    a. Br_2 from $NaBr$ c. KOH from KCl
    b. H_2SO_4 from $CuSO_4$ d. O_2 from water

20.6 Describe how the following transformations might be accomplished with the aid of an electrolytic process.

    a. $Al_2O_3 \longrightarrow Al$ b. $CaCl_2 \longrightarrow Ca(OH)_2$ c. $NaBr \longrightarrow NaCl$

20.7 When a solution of $NiSO_4$ is electrolyzed, the products are Ni and O_2. If 38,200 coulombs are passed through the solution,

    a. How many gram equivalent weights of nickel and oxygen are produced?
    b. How many grams of Ni and O_2 are produced?

20.8 In the electrolysis of a water solution of $CoCl_2$,

    a. If a current of 0.600 amp is used, how many grams of cobalt are plated in one hour?
    b. How long does it take to form 10.0 g of Cl_2 using a current of 12.0 amp?
    c. What current is required to plate cobalt at the rate of one gram per minute?

20.9 When a solution of a certain lanthanum salt is electrolyzed, it is found that a current of 1.93 amperes forms 3.33 g of La in one hour. Calculate the charge of the lanthanum cation.

20.10 Calculate the gram equivalent weight of the metal in each of the following (unbalanced) half-equations.

    a. $PtCl_6^{2-} \longrightarrow PtCl_4^{2-} + 2\ Cl^-$
    b. $MnO_4^- \longrightarrow Mn^{2+}$
    c. $Cr_2O_7^{2-} \longrightarrow 2\ Cr^{3+}$

20.11 Balance the following oxidation-reduction equations.

    a. $Zn(s) + H^+ + NO_3^- \longrightarrow Zn^{2+} + NH_4^+ + H_2O$
    b. $HSO_3^- + NO_3^- \longrightarrow HSO_4^- + NO(g)$ (acidic solution)
    c. $AuCl_4^- + H_2O_2 \longrightarrow Au(s) + O_2(g)$ (acidic solution)
    d. $MnO_4^{2-} \longrightarrow MnO_4^- + MnO_2(s)$ (basic solution)
    e. $N_2H_4 + ClO^- \longrightarrow N_2(g) + Cl^-$ (basic solution)

20.12 When air is bubbled through a water solution of hydrogen sulfide, the solution becomes cloudy because of the formation of colloidal sulfur.

    a. Write a balanced equation for this reaction.
    b. Calculate the number of liters of air (21.0% O_2) at 20°C and 1 atm required to oxidize one liter of 0.100 M H_2S.

20.13 Consider the reaction

$$MnO_2(s) + I^- \longrightarrow Mn^{2+} + I_2(s) \text{ (acidic solution)}$$

If one adds 6.00 g of MnO_2 to 250 ml of 2.00 M HI, what is the theoretical yield of I_2?

20.14 Explain why

    a. In the electrolysis of molten sodium chloride, the electrodes are separated by a screen.
    b. It is cheaper to produce chlorine by the electrolysis of a water solution of sodium chloride than by electrolyzing molten sodium chloride.

c. Cryolite is used in the electrolysis of aluminum oxide.

d. The carbon electrodes used in the Hall process have to be replaced frequently.

e. Fluorine is produced by the electrolysis of a molten KF-HF mixture rather than by electrolyzing a water solution of KF.

20.15 A paper company located in Maine is producing Cl_2 by the electrolysis of NaCl. In order to make use of the H_2 obtained as by-product, the company buys land in South Carolina and raises peanuts on it. They intend to crush the peanuts and hydrogenate the resulting peanut oil. Where would you advise them to build the hydrogenation plant?

20.16 Write balanced net ionic equations for the reactions that occur when the following water solutions are electrolyzed.

a. NaCl   b. $CuBr_2$ c. KF   d. HCl   e. $Cu(NO_3)_2$ f. K_2SO_4

20.17 Describe in some detail how one might prepare

a. I_2 from KI
b. Cu from $CuSO_4$
c. Pure Cu from impure Cu
d. H_2 from H_2O
e. KOH from KBr

20.18 Describe how the following transformations might be carried out with the aid of electrolytic reactions.

a. NaCl $\longrightarrow$ NaBr   b. NaBr $\longrightarrow$ Br_2 c. $Cu(NO_3)_2 \longrightarrow HNO_3$

20.19 When a solution of $ZnBr_2$ is electrolyzed, the products are Zn and Br_2. If 20,200 coulombs are passed through the solution,

a. How many gram equivalent weights of zinc and bromine are produced?
b. How many gram atomic weights of Zn and Br are produced?
c. How many grams of Zn and Br_2 are produced?

20.20 In the electrolysis of Al_2O_3, using a current of 25.0 amp,

a. What is the rate of production of Al in grams per hour?
b. The O_2 liberated at the positive carbon electrode reacts with it to form CO_2. How many grams of CO_2 are produced per hour?

20.21 A sample of brass weighing 12.02 g and containing 62.3% Cu is dissolved in concentrated nitric acid, converting the copper to Cu^{2+} ions. The resulting solution is electrolyzed. If it is desired to plate out all of the copper using a current of 0.542 amp, how long will it take?

20.22 A dry cell has an outer coating of zinc that weighs 55 g. If the cell fails when 20 per cent of the zinc is consumed, how long can it be used to supply a current of 2.0 amp?

20.23 A student finds that 16.8 g of a certain metal is plated in the same time as 12.9 g of copper, using the same current. What is the gram equivalent weight of the metal?

20.24 In the electrolysis of a potassium dichromate solution, it is found that 0.90×10^{-4} g of chromium are produced per coulomb.

a. How many electrons are required to produce one atom of chromium?
b. How many electrons are required to reduce one $Cr_2O_7^{2-}$ ion?
c. Write a balanced half-equation for the reduction of $Cr_2O_7^{2-}$ to Cr(s).

20.25 Balance the following oxidation-reduction equations.

a. $Cu(s) + H^+ + NO_3^- \longrightarrow Cu^{2+} + NO_2(g) + H_2O$
b. $SO_2(g) + NO_3^- + H_2O \longrightarrow SO_4^{2-} + NO(g) + H^+$
c. $MnO_2(s) + H^+ + Cl^- \longrightarrow Cl_2(g) + Mn^{2+} + H_2O$
d. $Cu(s) + H^+ + SO_4^{2-} \longrightarrow Cu^{2+} + SO_2(g) + H_2O$

20.26 Balance the following equations, all of which correspond to reactions occurring in acidic solution.

a. $Cr_2O_7^{2-} + Fe^{2+} \longrightarrow Cr^{3+} + Fe^{3+}$
b. $CuS(s) + NO_3^- \longrightarrow Cu^{2+} + SO_4^{2-} + NO(g)$
c. $PbO_2(s) + Mn^{2+} \longrightarrow MnO_4^- + Pb^{2+}$
d. $NH_4^+ + NO_3^- \longrightarrow N_2O(g)$

20.27 Balance the following equations for reactions occurring in basic solution.

a. $Cr(OH)_3(s) + ClO^- \longrightarrow CrO_4^{2-} + Cl^-$
b. $Fe(OH)_2(s) + O_2(g) \longrightarrow Fe(OH)_3(s)$
c. $Cl_2(g) \longrightarrow Cl^- + ClO_3^-$

20.28 Write balanced equations to correspond to the following reactions.

a. A solution of nitric acid and hydrochloric acid is heated to give nitric oxide and chlorine.
b. Permanganate ions (MnO_4^-) and sulfide ions in acidic solution react to give Mn^{2+} and sulfur.
c. Mn^{2+} ions are oxidized to manganate ions (MnO_4^{2-}) by nitrate ions in basic solution (NO_2^- ions are also produced).

20.29 What weight of copper can be oxidized to Cu^{2+} ions by .0500 lit of 12.0 M H_2SO_4 (cf. Problem 20.25d)?

20.30 What volume of chlorine at 25°C and 750 mm Hg can be produced by the reaction of 20.0 g of MnO_2 with excess hydrochloric acid (cf. Problem 20.25c)?

°20.31 Explain how Avogadro's number might be determined by an electrolysis experiment. What assumptions would have to be made in the calculations? How could these assumptions be checked experimentally?

°20.32 A brass cylinder, open only at the top, is to be plated with silver. The cylinder is 12.0 cm high and its circular base is 2.56 cm in diameter.

a. What weight of silver is required to plate the cylinder (inside and outside) to a thickness of 0.10 mm? (The density of silver is 10.54 g/cc.)
b. If the plating is carried out from $Ag(CN)_2^-$, using a current of 0.15 amp, how much time is required to give a coating this thick?

°20.33 Write a balanced equation for the reaction of sulfuric acid with sodium iodide. All the iodide ions are oxidized to iodine. Of the SO_4^{2-} ions, it is found that 50 per cent are reduced to SO_3^{2-} ions, 30 per cent to elementary sulfur and the remainder to H_2S.

°20.34 A mixture of $CuCl_2$, NaCl, and $NaNO_3$ is analyzed. It is found that a 1.000 g sample, on treatment with excess silver nitrate, gives 1.628 g of AgCl. An electrolysis of a 1.000 g sample gives 0.148 g of copper. Calculate the percentages of the three constituents of the mixture.

21

Voltaic Cells: Spontaneity of Oxidation-Reduction Reactions

Electrolytic cells, to which Chapter 20 was devoted, use a direct electric current to bring about a non-spontaneous oxidation-reduction reaction. Voltaic cells, which will be discussed in this chapter, perform the reverse function: a spontaneous oxidation-reduction reaction taking place within the cell generates a direct electric current. Voltaic cells have long been used for such mundane purposes as starting an automobile or operating a flashlight. More recently, they have served as a source of power for hearing aids and satellite communications systems. In both the industrial and teaching laboratory, voltaic cells are commonly used to provide electrical energy to operate electrolytic cells of the type discussed in Chapter 20.

To the chemist, the most important application of voltaic cells is their use in determining the spontaneity of oxidation-reduction reactions. By measuring cell voltages, it is possible to predict whether a given reaction will take place in the laboratory and, if so, the extent to which it will occur.

21.1 A SIMPLE VOLTAIC CELL: THE Zn–Cu CELL

If a piece of zinc is added to a solution of copper sulfate, the following, spontaneous oxidation-reduction reaction takes place:

$$Zn(s) + Cu^{2+} \longrightarrow Zn^{2+} + Cu(s) \tag{21.1}$$

Electron transfer from Zn atoms to Cu^{2+} ions takes place directly at the surface of the zinc. A spongy, reddish-brown deposit of copper forms on the surface of the zinc; the blue color of the solution fades as Cu^{2+} ions are replaced by Zn^{2+} ions. Energy is liberated as heat; the temperature of the solution rises several degrees. The spontaneity of the reaction is commonly explained by saying that zinc loses electrons more readily than copper or, alternatively, that Cu^{2+} ions gain electrons more readily than Zn^{2+} ions.

532

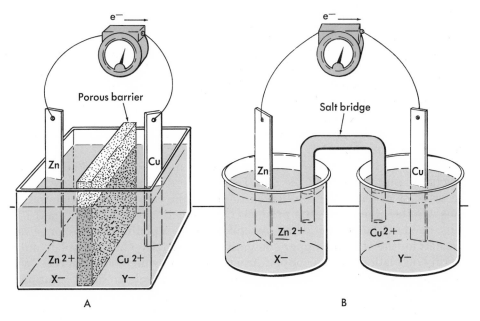

Figure 21.1 Zinc-copper voltaic cells.

To design a cell that uses Reaction 21.1 as a source of electrical energy, the electron transfer must be made to occur indirectly. That is, the electrons given up by the zinc atoms must pass through an electrical circuit and do some work before they reduce Cu^{2+} ions to copper atoms. A schematic diagram of a cell in which this is achieved is shown in Figure 21.1A.

It will be instructive to trace the path of electrical current through this cell.

1. At the zinc **anode**, electrons are produced by the **oxidation** half-reaction

anode: $$Zn(s) \longrightarrow Zn^{2+} + 2\ e^-$$ (21.1a)

This electrode, which "pushes" electrons into the external circuit, is ordinarily considered to be the negative pole of the cell.

2. The electrons generated by Reaction 21.1a move through the external circuit, which may be a simple resistance wire, a motor, or an electrolytic cell.

3. Electrons pass through the external circuit into the copper **cathode**, where they are consumed in **reducing** Cu^{2+} ions in the solution to copper atoms:

cathode: $$Cu^{2+} + 2\ e^- \longrightarrow Cu(s)$$ (21.1b)

The copper electrode, which "pulls" electrons out of the external circuit, is considered to be the positive pole of the cell.

4. The circuit is completed by the passage of ions through the cell. As Reactions 21.1a and 21.1b proceed, a surplus of positive ions (Zn^{2+}) is created around the zinc electrode. The region around the copper electrode becomes deficient in positive ions because of the discharge of Cu^{2+} ions. To maintain electrical neutrality, cations must move into the region around the copper

533

cathode or, alternatively, anions must flow toward the zinc anode. In practice, both of these migrations occur.

The net result of the half-reactions occurring within the Zn-Cu cell is, of course, Reaction 21.1. The cell is so designed that electron transfer from Zn atoms to Cu^{2+} ions occurs indirectly through an external circuit. In this way, the energy liberated by the reaction is used to do electrical work rather than being dissipated as heat.

Two essential points concerning the design of a zinc-copper cell are implied by Equation 21.1:

1. The only species which need be present in the cell initially are the reactants, zinc atoms and Cu^{2+} ions. In other words, the anode of the cell must be made of zinc and the solution surrounding the cathode must contain Cu^{2+} ions. On the other hand, the products of the reaction, copper atoms and Zn^{2+} ions, need not be present when the cell is set up. A platinum wire can be substituted for the strip of copper shown in Figure 21.1A; Cu^{2+} ions plate out as readily on platinum as on copper. Any positive ion which does not react with zinc can replace the Zn^{2+} ions surrounding the zinc electrode. A solution of Na_2SO_4, or KNO_3 works as well as a solution of $ZnSO_4$.

2. It is imperative that Cu^{2+} ions not come in contact with the zinc electrode. If this happens, Reaction 21.1 will occur directly at the surface of the zinc, with no electrons going through the external circuit. One way of preventing the Cu^{2+} ions formed at the cathode from diffusing over to the zinc anode is to interpose a porous partition between the two halves of the cell (Figure 21.1A). The partition permits the migration of ions while the cell is in use but minimizes the random diffusion of ions that otherwise takes place when the cell is not connected.* Another way of accomplishing this same purpose is to connect the two halves of the cell by means of a U-shaped tube referred to as a salt bridge (Figure 21.1B). The tube is filled with a solution of an electrolyte such as potassium nitrate to which gelatin or agar-agar has been added to give a semisolid paste. When current is drawn from the cell, ions move slowly through the salt bridge to complete the circuit. When the cell is not in use, diffusion of ions through the bridge is so slow that it may take several days for the Cu^{2+} ions to reach the zinc electrode.

Still another way of preventing Cu^{2+} ions from coming in contact with the zinc electrode is illustrated in the gravity cell shown in Figure 21.2. To form this cell, enough copper sulfate solution is added to the jar to cover the copper electrode. A more dilute, less dense solution of zinc sulfate is then carefully poured over the copper sulfate. So long as the cell is not subjected to vibrations, the boundary between the layers may be maintained over long periods of time. Cells of this design were once used extensively to operate telegraph relays, doorbells, and other stationary electrical apparatus. Since their internal resistance is much lower than that of porous-partition or salt-bridge cells, much larger currents can be drawn from them.

Could we use H^+ ions?

Why not use a gravity cell in an automobile?

---

* It will be recalled that when the zinc-copper cell is operating, Cu^{2+} ions move away from the zinc electrode. This means that so long as current is being drawn from the cell, there is no tendency for Cu^{2+} ions to come in contact with the zinc electrode. For this reason, it is common practice, with cells of this type which are used intermittently, to set up an auxiliary circuit to draw a small amount of current when the cell is not in use. This procedure is followed in the commercial model of the zinc-copper cell (Figure 21.2).

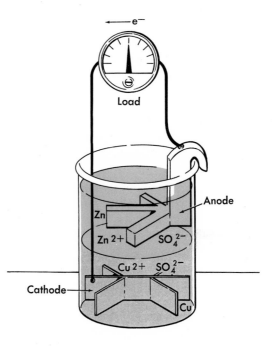

Figure 21.2 Gravity (Daniell) cell.

21.2 OTHER VOLTAIC CELLS: COMMERCIAL CELLS

A great many simple voltaic cells can be set up in a manner analogous to the zinc-copper cell shown in Figure 21.1. One can, for example, devise cells (Figure 21.3) in which the following spontaneous oxidation-reduction reactions serve as a source of electrical energy:

$$Ni(s) + Cu^{2+} \longrightarrow Ni^{2+} + Cu(s) \tag{21.2}$$

$$Zn(s) + 2\,H^+ \longrightarrow Zn^{2+} + H_2(g) \tag{21.3}$$

In each case, the apparatus consists of two half-cells, each containing an elec-

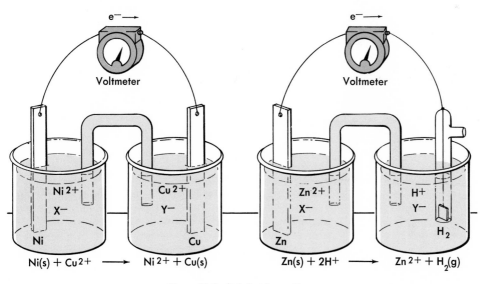

Figure 21.3 Salt-bridge cells.

535

trode dipping into a solution of an appropriate electrolyte, separated by a salt bridge or similar device. Atoms of the element having the greatest tendency to lose electrons (Ni, Zn) are oxidized at the anode, giving up electrons which travel through the external circuit to the cathode, where they reduce the cation (Cu^{2+}, H^+) in that half-cell.

All the cells discussed to this point, with the single exception of the gravity cell shown in Figure 21.2, are unsuitable for commercial use because of their high internal resistance. There are on the market today several dozen different cells, all of which are capable of supplying a comparatively large current, at least for a short time. Two of these, the dry cell and the lead storage battery, were as familiar to our grandparents as they are to us. Another type of voltaic cell, now in the development stage, gives promise of becoming a major source of electrical energy, perhaps within this decade. This is the so-called fuel cell, which has received so much attention during the past few years.

Dry Cell (Leclanché Cell)

The construction of the ordinary dry cell used in flashlights, portable radios, and similar appliances is shown in Figure 21.4. The zinc wall of the cell serves as the anode; the graphite rod passing through the center of the cell is the cathode. The space between the electrodes is filled with a moist paste containing manganese dioxide, carbon black, and ammonium chloride. When the cell is being used to generate energy, the half-reaction at the anode is

$$Zn(s) \longrightarrow Zn^{2+} + 2\ e^- \qquad\qquad (21.4a)$$

At the cathode, manganese dioxide is reduced to a variety of species in which manganese is in the $+3$ oxidation state. These include Mn_2O_3, $MnO(OH)$, and $Mn_2O_4^{2-}$. For simplicity, we shall write the half-reaction as:

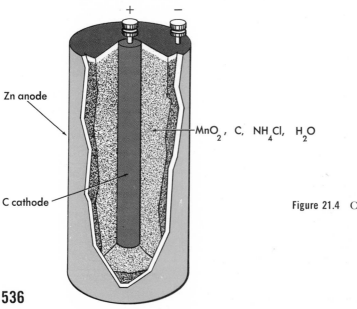

Zn anode

C cathode

MnO_2, C, NH_4Cl, H_2O

Figure 21.4 Cross section of Leclanché dry cell.

$$MnO_2(s) + 4 \ NH_4^+ + e^- \longrightarrow Mn^{3+} + 2 \ H_2O + 4 \ NH_3 \qquad \text{(21.4b)}$$

For the overall reaction,

$$Zn(s) + 2 \ MnO_2(s) + 8 \ NH_4^+ \longrightarrow Zn^{2+} + 2 \ Mn^{3+} + 8 \ NH_3 + 4 \ H_2O \quad \text{(21.4)}$$

If too large a current is drawn from the cell, the ammonia formed by Reaction 21.4b forms a gaseous, insulating layer around the carbon electrode. In normal operation, this condition is prevented by the migration of Zn^{2+} ions to the cathode where they react with ammonia molecules to form complex ions such as $Zn(NH_3)_4^{2+}$, $Zn(NH_3)_2(H_2O)_2^{2+}$, and so on.

Can you suggest why a dry cell is sometimes revived by heating?

Lead Storage Battery

The 12-volt storage battery commonly used in automobiles consists of six voltaic cells of the type shown in Figure 21.5 connected in series. A group of lead plates, the grids of which are filled with spongy, gray lead, forms the anode of the cell. The multiple cathode consists of a group of plates of similar design filled with lead dioxide. These two series of plates, alternating with each other throughout the cell, are immersed in a water solution of sulfuric acid, which acts as the electrolyte.

When a lead storage battery is supplying current, the lead in the anode grids is oxidized to Pb^{2+} ions, which immediately precipitate on the plates as lead sulfate, $PbSO_4$. At the cathode, the lead dioxide is reduced to Pb^{2+} ions, which also precipitate as $PbSO_4$.

$$Pb(s) + SO_4^{2-} \longrightarrow PbSO_4(s) + 2 \ e^- \qquad \text{(21.5a)}$$

$$PbO_2(s) + 4 \ H^+ + SO_4^{2-} + 2 \ e^- \longrightarrow PbSO_4(s) + 2 \ H_2O \qquad \text{(21.5b)}$$

$$\overline{Pb(s) + PbO_2(s) + 4 \ H^+ + 2 \ SO_4^{2-} \longrightarrow 2 \ PbSO_4(s) + 2 \ H_2O} \qquad \text{(21.5)}$$

It is important that the PbSO$_4$ adhere to the plates (Equation 21.6).

Deposits of lead sulfate formed by Reactions 21.5a and 21.5b slowly build up on the plates, partially covering and replacing the lead and lead dioxide. As the cell discharges, the concentration of sulfuric acid decreases; for every mole of lead reacting, two moles of H_2SO_4 ($4 \ H^+$, $2 \ SO_4^{2-}$) are replaced by two moles of water. The state of charge of a storage battery can be checked by measuring the density of the electrolyte. A low density indicates a low sulfuric acid concentration and hence a partially discharged cell.

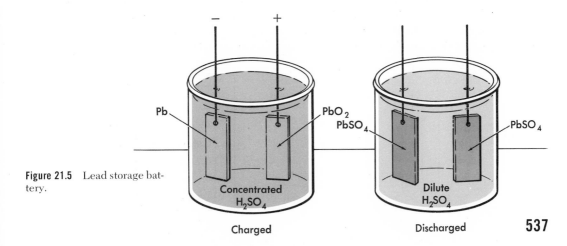

Figure 21.5 Lead storage battery.

A lead storage battery, unlike an ordinary dry cell, can be restored to its original condition by passing a direct current through it in the reverse direction. While a storage battery is being charged, it acts as an electrolytic cell; the half-reactions represented by equations 21.5a and 21.5b are reversed:

$$2 \; PbSO_4(s) + 2 \; H_2O \longrightarrow Pb(s) + PbO_2(s) + 4 \; H^+ + 2 \; SO_4^{2-} \qquad \textbf{(21.6)}$$

The electrical energy required to bring about Reaction 21.6 may be furnished by a direct-current generator, as in older automobiles, or by an alternator equipped with a rectifier (in modern cars).

Fuel Cells

By far the major portion of our electrical energy is obtained indirectly from the combustion of fuels. The thermal energy produced by burning coal, oil, or natural gas is converted to mechanical energy, which in turn is used to run electrical generators. Although this method of producing electrical energy is more economical than a conventional voltaic cell, it is relatively inefficient. A steam power plant using coal as a fuel converts only about 35 per cent of the available chemical energy of the coal into electrical energy.

Scientists for generations have speculated on the possibility of converting the chemical energy of fuels directly to electrical energy in a type of voltaic cell known as a fuel cell. In principle, there is no reason why this cannot be done. The combustion of a fuel is a spontaneous oxidation-reduction reaction and hence should serve as the basis for a voltaic cell. In practice, it turns out to be extremely difficult to design cells in which such apparently simple reactions as

$$C(s) + O_2(g) \longrightarrow CO_2(g)$$

and
$$H_2(g) + \tfrac{1}{2} \; O_2(g) \longrightarrow H_2O(l)$$

will occur. Recently, however, fuel cells have been built which convert up to 80 per cent of the energy available from combustion reactions to electrical energy.

A prototype of the modern fuel cell, built in Germany in 1937, is shown in Figure 21.6A. The anode consists of a layer of carbon granules packed around a central, inert electrode. Air or oxygen is blown through magnetite (Fe_3O_4) particles which surround the cathode. The electrolyte consists of a molten mixture of several different metal oxides; oxide ions (O^{2-}) carry the current through the cell. The reaction taking place may be represented most simply as

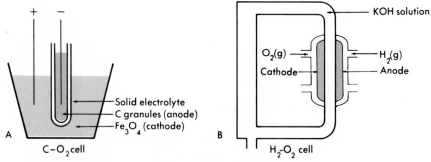

Figure 21.6 Fuel cells.

anode: $$C(s) + 2\ O^{2-} \longrightarrow CO_2(g) + 4\ e^- \qquad \text{(21.7a)}$$

cathode: $$O_2(g) + 4\ e^- \longrightarrow 2\ O^{2-} \qquad \text{(21.7b)}$$

$$\overline{\quad C(s) + O_2(g) \longrightarrow CO_2(g) \qquad \text{(21.7)}}$$

An inherent disadvantage of this cell is the high operating temperature, approximately 1000°C, required to maintain the electrolyte in the molten state.

A modern fuel cell with attractive commercial possibilities is sketched in Figure 21.6B. This cell uses the spontaneous reaction between hydrogen and oxygen to produce electrical energy. The operating temperature of the cell is about 250°C under a pressure of 50 atm. A water solution of sodium or potassium hydroxide serves as the electrolyte. In the presence of certain catalysts, the reactions

anode: $$H_2(g) + 2\ OH^- \longrightarrow 2\ H_2O + 2\ e^- \qquad \text{(21.8a)}$$

cathode: $$\tfrac{1}{2}\ O_2(g) + H_2O + 2\ e^- \longrightarrow 2\ OH^- \qquad \text{(21.8b)}$$

$$\overline{\quad H_2(g) + \tfrac{1}{2}\ O_2(g) \longrightarrow H_2O \qquad \text{(21.8)}}$$

occur at graphite electrodes to give a maximum voltage of about 1 V. Either air or pure oxygen can be used at the cathode. At present, pure hydrogen appears to be required for the anode reaction.

21.3   STANDARD REDUCTION AND OXIDATION POTENTIALS

It was stated earlier that the principal interest of the chemist in voltaic cells lies in the information these cells provide concerning the spontaneity of oxidation-reduction reactions. This information can be deduced from a measurement of cell voltages. With a properly designed cell, it is found that the voltage at a given temperature, let us say 25°C, depends upon two factors: the nature of the cell reaction and the concentrations of the various species (atoms, ions, or molecules) participating in the cell reaction. The **standard voltage** corresponding to a given cell reaction is that obtained when all such species have an activity of one. This condition is approximately fulfilled when all ions or molecules in solution are at a concentration of 1 M and all gases at a partial pressure of 1 atm. To illustrate, consider the Zn-H_2 cell shown in Figure 21.3; it is found experimentally that when the pressure of hydrogen is 1 atm and the concentrations of both Zn^{2+} and H^+ are 1 M, the cell voltage is +0.76 V. This quantity, +0.76 V, is referred to as the standard cell voltage, corresponding to the cell reaction represented by Equation 21.3.

Just as one can split a cell reaction such as 21.3 into two half-reactions of reduction and oxidation, so it is possible to divide a standard cell voltage into two parts, one corresponding to the reduction half-reaction, the other to the oxidation half-reaction. The potential corresponding to the reduction half-reaction is referred to as a **standard reduction potential**; associated with the oxidation half-reaction is the quantity which we shall call a **standard oxidation potential**.*

Activities of species in solution were discussed in Chap. 16.

---

* The International Union of Pure and Applied Chemistry, at a meeting in Stockholm in 1953, suggested that the term "standard electrode potential" be used to refer to what we have called the "standard reduction potential" and that the use of the term "standard oxidation potential" be abandoned. These conventions have not received wide acceptance in this country, but will be found in some textbooks.

For Reaction 21.3, we have

Redn. $2 H^+(1 M) + 2 e^- \longrightarrow H_2(g, 1 atm);$ SRP $H^+ = x$

Oxid. $Zn(s) \longrightarrow Zn^{2+}(1 M) + 2 e^-;$ SOP $Zn = y$

$Zn(s) + 2 H^+(1 M) \longrightarrow Zn^{2+}(1 M) + H_2(g, 1 atm);$ std. voltage $= x + y = +0.76 V$

By methods which we shall now consider, it is possible to establish a table of standard electrode potentials corresponding to a wide variety of oxidation and reduction half-reactions. We shall find such a table to be of value in predicting the results of reactions occurring in voltaic or electrolytic cells and, even more important, the spontaneity and extent of oxidation-reduction reactions in general, wherever they take place.

Assignment of Potentials

To illustrate how standard potentials can be established, let us consider once again the Zn-H$_2$ cell. It has been pointed out that at standard concentrations, the voltage of this cell is $+0.76$ V and further, that this quantity is the sum of the two half cell potentials involved. That is,

$$+0.76 V = SOP\ Zn + SRP\ H^+ \qquad (21.9)$$

It is experimentally impossible to determine the absolute value of either of the two quantities on the right-hand side of Equation 21.9. To establish a value for the standard oxidation potential of zinc, one must arbitrarily assign a value to the standard reduction potential of the H$^+$ ion. It was agreed many years ago to take this value to be 0.00 V. That is,

$$2 H^+(1 M) + 2 e^- \longrightarrow H_2(g, 1 atm) \qquad SRP\ H^+ = 0.00\ V \qquad (21.10)$$

Substituting in Equation 21.9, it is clear that, on the basis of this convention, the standard oxidation potential of zinc must be $+0.76$ V.

Suppose we took SRP H$^+$ to be 1.00 V. What would be SOP Zn?

$$Zn(s) \longrightarrow Zn^{2+}(1 M) + 2 e^- \qquad SOP = +0.76\ V \qquad (21.11)$$

To obtain the standard reduction potential of the Zn^{2+} ion, we simply reverse the sign of the standard oxidation potential. That is,

$$Zn^{2+}(1 M) + 2 e^- \longrightarrow Zn(s) \qquad SRP = -0.76\ V \qquad (21.12)$$

In general, for any oxidation-reduction couple, **the standard potentials for the forward and reverse half-reactions (oxidation and reduction) are equal in magnitude but opposite in sign**.

As soon as a few potentials have been established, it becomes relatively easy to determine others. Suppose, for example, we wish to determine the standard reduction potential of the Cu^{2+} ion. One way to do this is to set up the cell shown in Figure 21.1, using 1 M solutions of Zn^{2+} and Cu^{2+}. When it is found that the voltage of this cell is 1.10 volts, it follows that the standard reduction potential of Cu^{2+} must be $+0.34$ V. That is,

$$SOP\ Zn + SRP\ Cu^{2+} = 1.10\ V$$

$$+0.76\ V + SRP\ Cu^{2+} = 1.10\ V$$

$$SRP\ Cu^{2+} = +0.34\ V$$

In Table 21.1, we have listed standard reduction potentials for a series of species (Li^+, K^+, etc.) which are capable of being reduced. The standard oxidation potentials for the corresponding oxidation half-reactions can be obtained, of course, by reversing the sign. For example,

$$Li^+ + e^- \longrightarrow Li(s) \qquad SRP = -3.05 \text{ V}$$

$$Li(s) \longrightarrow Li^+ + e^- \qquad SOP = +3.05 \text{ V}$$

Calculation of Cell Voltages from Standard Potentials

As the foregoing discussion implies, **the standard voltage of any cell (symbol $E°$) is the algebraic sum of the standard oxidation potential of the species being oxidized in the cell reaction and the standard reduction potential of the species being reduced**. This simple relationship makes it possible, using Table 21.1, to calculate the standard voltages of more than 3000 different voltaic cells. Example 21.1 illustrates how this may be done for three specific cases.

EXAMPLE 21.1. Calculate the voltages of cells in which the following reactions occur:

    a. $Cl_2(g, 1 \text{ atm}) + 2 I^-(1 \text{ M}) \longrightarrow 2 Cl^-(1 \text{ M}) + I_2(s)$

    b. $MnO_4^-(1 \text{ M}) + 8 H^+(1 \text{ M}) + 5 Cl^-(1 \text{ M}) \longrightarrow Mn^{2+}(1 \text{ M}) + 4 H_2O + \frac{5}{2} Cl_2(g, 1 \text{ atm})$

    c. $Sn^{2+}(1 \text{ M}) + 2 Ag(s) \longrightarrow Sn(s) + 2 Ag^+(1 \text{ M})$

SOLUTION. In each case we shall split the reaction into two half-reactions, tabulate the proper electrode potentials, and add to obtain the cell voltage.

    a. reduction: $Cl_2(g) + 2 e^- \longrightarrow 2 Cl^-$ $SRP = +1.36$ V
        oxidation: $2 I^- \longrightarrow I_2(s) + 2 e^-$ $SOP = -0.53$ V
                                              $E° = +0.83$ V

    b. reduction:
        $MnO_4^- + 8 H^+ + 5 e^- \longrightarrow Mn^{2+} + 4 H_2O$ $SRP = +1.52$ V
        oxidation: $2 Cl^- \longrightarrow Cl_2(g) + 2 e^-$ $SOP = -1.36$ V
                                              $E° = +0.16$ V

    c. reduction: $Sn^{2+} + 2 e^- \longrightarrow Sn(s)$ $SRP = -0.14$ V
        oxidation: $Ag(s) \longrightarrow Ag^+ + e^-$ $SOP = -0.80$ V
                                              $E° = -0.94$ V

The negative voltage means that the reaction as written is not spontaneous and cannot occur in a voltaic cell. If Ag-Ag^+ and Sn-Sn^{2+} half-cells are connected, the reverse reaction

$$2 Ag^+ + Sn(s) \longrightarrow 2 Ag(s) + Sn^{2+}$$

occurs, producing a voltage of $+0.94$ V.

Standard potentials can be used in connection with electrolytic as well as voltaic cells. By adding the proper potentials, one can calculate the minimum applied voltage necessary, at standard concentrations, to bring about a nonspontaneous oxidation-reduction reaction in an electrolytic cell. Con-

TABLE 21.1 STANDARD REDUCTION POTENTIALS

REDUCTION HALF REACTION		STANDARD REDUCTION POTENTIAL (VOLTS)
$Li^+ + e^-$	$\longrightarrow Li(s)$	-3.05
$K^+ + e^-$	$\longrightarrow K(s)$	-2.93
$Ba^{2+} + 2\ e^-$	$\longrightarrow Ba(s)$	-2.90
$Ca^{2+} + 2\ e^-$	$\longrightarrow Ca(s)$	-2.87
$Na^+ + e^-$	$\longrightarrow Na(s)$	-2.71
$Mg^{2+} + 2\ e^-$	$\longrightarrow Mg(s)$	-2.37
$Al^{3+} + 3\ e^-$	$\longrightarrow Al(s)$	-1.66
$Mn^{2+} + 2\ e^-$	$\longrightarrow Mn(s)$	-1.18
$Zn^{2+} + 2\ e^-$	$\longrightarrow Zn(s)$	-0.76
$Cr^{3+} + 3\ e^-$	$\longrightarrow Cr(s)$	-0.74
$Fe^{2+} + 2\ e^-$	$\longrightarrow Fe(s)$	-0.44
$Cr^{3+} + e^-$	$\longrightarrow Cr^{2+}$	-0.41
$Cd^{2+} + 2\ e^-$	$\longrightarrow Cd(s)$	-0.40
$PbSO_4(s) + 2\ e^-$	$\longrightarrow Pb(s) + SO_4^{2-}$	-0.36
$Tl^+ + e^-$	$\longrightarrow Tl(s)$	-0.34
$Co^{2+} + 2\ e^-$	$\longrightarrow Co(s)$	-0.28
$Ni^{2+} + 2\ e^-$	$\longrightarrow Ni(s)$	-0.25
$AgI(s) + e^-$	$\longrightarrow Ag(s) + I^-$	-0.15
$Sn^{2+} + 2\ e^-$	$\longrightarrow Sn(s)$	-0.14
$Pb^{2+} + 2\ e^-$	$\longrightarrow Pb(s)$	-0.13
$2\ H^+ + 2\ e^-$	$\longrightarrow H_2(g)$	0.00
$AgBr(s) + e^-$	$\longrightarrow Ag(s) + Br^-$	0.10
$S(s) + 2\ H^+ + 2\ e^-$	$\longrightarrow H_2S$	0.14
$Sn^{4+} + 2\ e^-$	$\longrightarrow Sn^{2+}$	0.15
$Cu^{2+} + e^-$	$\longrightarrow Cu^+$	0.15
$SO_4^{2-} + 4\ H^+ + 2\ e^-$	$\longrightarrow SO_2(g) + 2\ H_2O$	0.20
$Cu^{2+} + 2\ e^-$	$\longrightarrow Cu(s)$	0.34
$Cu^+ + e^-$	$\longrightarrow Cu(s)$	0.52
$I_2(s) + 2\ e^-$	$\longrightarrow 2\ I^-$	0.53
$Fe^{3+} + e^-$	$\longrightarrow Fe^{2+}$	0.77
$Hg_2^{2+} + 2\ e^-$	$\longrightarrow 2\ Hg(l)$	0.79
$Ag^+ + e^-$	$\longrightarrow Ag(s)$	0.80
$2\ Hg^{2+} + 2\ e^-$	$\longrightarrow Hg_2^{2+}$	0.92
$NO_3^- + 4\ H^+ + 3\ e^-$	$\longrightarrow NO(g) + 2\ H_2O$	0.96
$AuCl_4^- + 3\ e^-$	$\longrightarrow Au(s) + 4\ Cl^-$	1.00
$Br_2(l) + 2\ e^-$	$\longrightarrow 2\ Br^-$	1.07
$O_2(g) + 4\ H^+ + 4\ e^-$	$\longrightarrow 2\ H_2O$	1.23
$MnO_2(s) + 4\ H^+ + 2\ e^-$	$\longrightarrow Mn^{2+} + 2\ H_2O$	1.23
$Cr_2O_7^{2-} + 14\ H^+ + 6\ e^-$	$\longrightarrow 2\ Cr^{3+} + 7\ H_2O$	1.33
$Cl_2(g) + 2\ e^-$	$\longrightarrow 2\ Cl^-$	1.36
$ClO_3^- + 6\ H^+ + 5\ e^-$	$\longrightarrow \frac{1}{2}\ Cl_2(g) + 3\ H_2O$	1.47
$Au^{3+} + 3\ e^-$	$\longrightarrow Au(s)$	1.50
$MnO_4^- + 8\ H^+ + 5\ e^-$	$\longrightarrow Mn^{2+} + 4\ H_2O$	1.52
$H_2O_2 + 2\ H^+ + 2\ e^-$	$\longrightarrow 2\ H_2O$	1.77
$Co^{3+} + e^-$	$\longrightarrow Co^{2+}$	1.82
$F_2(g) + 2\ e^-$	$\longrightarrow 2\ F^-$	2.87

BASIC SOLUTION

$Zn(OH)_4^{2-} + 2\ e^-$	$\longrightarrow Zn(s) + 4\ OH^-$	-1.22
$Fe(OH)_2(s) + 2\ e^-$	$\longrightarrow Fe(s) + 2\ OH^-$	-0.88
$2\ H_2O + 2\ e^-$	$\longrightarrow H_2(g) + 2\ OH^-$	-0.83
$Fe(OH)_3(s) + e^-$	$\longrightarrow Fe(OH)_2(s) + OH^-$	-0.56

TABLE 21.1 STANDARD REDUCTION POTENTIALS (*continued*)

BASIC SOLUTION (*continued*)

$S(s) + 2\ e^-$	$\longrightarrow S^{2-}$	-0.48
$Cu(OH)_2(s) + 2\ e^-$	$\longrightarrow Cu(s) + 2\ OH^-$	-0.36
$CrO_4^{2-} + 4\ H_2O + 3\ e^-$	$\longrightarrow Cr(OH)_3(s) + 5\ OH^-$	-0.12
$NO_3^- + H_2O + 2\ e^-$	$\longrightarrow NO_2^- + 2\ OH^-$	0.01
$Ag_2O(s) + H_2O + 2\ e^-$	$\longrightarrow 2\ Ag(s) + 2\ OH^-$	0.34
$ClO_4^- + H_2O + 2\ e^-$	$\longrightarrow ClO_3^- + 2\ OH^-$	0.36
$O_2(g) + 2\ H_2O + 4\ e^-$	$\longrightarrow 4\ OH^-$	0.40
$ClO_3^- + 3\ H_2O + 6\ e^-$	$\longrightarrow Cl^- + 6\ OH^-$	0.62
$ClO^- + H_2O + 2\ e^-$	$\longrightarrow Cl^- + 2\ OH^-$	0.89

sider, for example, the electrolysis of a water solution of copper(II) chloride, discussed in Chapter 20. Adding the standard potentials for the two half-reactions, we have

$$Cu^{2+} + 2\ e^- \longrightarrow Cu(s) \qquad\qquad \text{SRP } Cu^{2+} = +0.34 \text{ V}$$

$$\underline{2\ Cl^- \qquad\qquad \longrightarrow Cl_2(g) + 2\ e^- \qquad\qquad \text{SOP } Cl^- = -1.36 \text{ V}}$$

$$Cu^{2+} + 2\ Cl^- \longrightarrow Cu(s) + Cl_2(g) \qquad\qquad -1.02 \text{ V}$$

We deduce that in order to operate the cell, a potential of at least 1.02 V must be applied across the electrodes. A similar calculation for the electrolysis of a sodium chloride solution shows that, at standard concentrations, a potential of at least 2.19 V must be applied to bring about the nonspontaneous reaction

$$2\ Cl^- + 2\ H_2O \longrightarrow Cl_2(g) + 2\ OH^- + H_2(g)$$

In practice, it is ordinarily found that the voltage required to operate an electrolytic cell is somewhat higher than that calculated from electrode potentials. The excess voltage, referred to as **overvoltage**, may be 1 V or more; it is particularly large when one of the products of the cell reaction is a gas such as hydrogen or oxygen. In the evolution of hydrogen, the magnitude of the overvoltage is found to vary greatly with the metal used as a cathode. Zinc, tin, or cadmium electrodes give hydrogen overvoltages in the neighborhood of 1 V. On the other hand, at platinum or palladium cathodes, hydrogen can be generated with zero overvoltage.

Although the exact mechanism of overvoltage is poorly understood, it is known to arise from kinetic effects. Electrode processes, involving the transfer of electrons between a metal electrode and ions in solutions, are inherently slow. By using a voltage higher than that theoretically required, one supplies the activation energy necessary to make the reaction proceed at a finite rate.

In the evolution of hydrogen, it appears that the most difficult step is the combination of hydrogen atoms to give H_2 molecules. This process occurs on the surface of the metal electrode. It is significant that the two metals, platinum and palladium, which show the lowest hydrogen overvoltages, are precisely the metals which adsorb hydrogen atoms most strongly. They provide an effective catalytic surface for the formation of diatomic molecules. Traces of impurities can poison the electrode surfaces; 0.001 cc of carbon monoxide spread out over an electrode 1 cm² in area can raise the hydrogen overvoltage by as much as half a volt.

Qualitative Interpretation of Standard Potentials

The standard potentials listed in Table 21.1 may be regarded as a measure of the relative tendencies of the different species to be reduced or oxidized.

The more positive the potential, the more readily the corresponding half-reaction occurs. A large negative potential signifies a half-reaction that is difficult to accomplish.

Of the species listed in the left-hand column of Table 21.1, the more difficult to reduce are those located at the top. The cations of the 1A and 2A metals (Li^+, K^+, Ba^{2+}, and so forth), whose standard reduction potentials are large negative numbers (-3.05 V, -2.93 V, -2.90 V), cannot be reduced in water solution. Moving down the column, we come to species which are successively easier to reduce. The most readily reduced species is the fluorine molecule (SRP $= +2.87$ V), located at the bottom of the column. Substances which are capable of being oxidized (Li, K, Ba, and so forth) are listed in the right-hand column in order of decreasing ease of oxidation. The most easily oxidized species is the lithium atom (SOP $= +3.05$ V). The zinc atom (SOP $= +0.76$ V) is less readily oxidized than lithium; the F^- ion, located at the bottom of the column (SOP $= -2.87$ V), is so reluctant to lose electrons that it cannot be oxidized chemically.

Which is more readily oxidized, Ni or Cd?

The potentials listed in Table 21.1 enable us to explain the electrode reactions in the electrolysis of aqueous salt solutions (Chapter 20, Section 20.2). Comparing, for example, the standard reduction potentials of the three species

$$Al^{3+} + 3\ e^- \longrightarrow Al(s) \qquad -1.66\ V$$

$$2\ H_2O + 2\ e^- \longrightarrow H_2(g) + 2\ OH^- \qquad -0.83\ V$$

$$Cu^{2+} + 2\ e^- \longrightarrow Cu(s) \qquad +0.34\ V$$

we see that Cu^{2+} is more easily reduced than H_2O, which in turn is more readily reduced than Al^{3+}. This explains why it is possible to prepare copper but not aluminum by electrolysis of a water solution of the appropriate salt. As another example, if we compare the standard oxidation potentials of the three species

$$2\ I^- \longrightarrow I_2(s) + 2\ e^- \qquad -0.53\ V$$

$$H_2O \longrightarrow \tfrac{1}{2} O_2(g) + 2\ H^+ + 2\ e^- \qquad -1.23\ V$$

$$2\ F^- \longrightarrow F_2(g) + 2\ e^- \qquad -2.87\ V$$

we see why iodine can be prepared by the electrolysis of a water solution of sodium iodide (I^- more readily oxidized than H_2O), but fluorine cannot be prepared by electrolyzing a water solution of sodium fluoride (H_2O oxidized rather than F^-).

From a slightly different point of view, Table 21.1 tells us the relative strengths of various oxidizing and reducing agents. The stronger oxidizing agents are located at the lower left ($F_2 > Co^{3+} > H_2O_2 > MnO_4^-$). The halogens listed above fluorine (chlorine, bromine and iodine) are successively less powerful oxidizing agents in that order. The species listed in the right-hand column are all capable of acting as reducing agents, at least in principle. The stronger reducing agents are those at the upper right, including the 1A and 2A metals, Al, and a few of the transition metals (Mn, Zn and Fe, among others). Species such as Ag, Au and Cl^-, which are located well down in the right-hand column hold on to their electrons too tightly to be very effective as reducing agents.

Which is the better reducing agent, Ni or Cd?

21.4 SPONTANEITY AND EXTENT OF OXIDATION-REDUCTION REACTIONS

In the first paragraph of this chapter it was pointed out that a voltaic cell is one in which a spontaneous oxidation-reduction reaction occurs. The converse of this statement is also true: any reaction that can occur in a voltaic cell to produce a positive voltage must be spontaneous. To decide whether a given reaction is capable of taking place under a particular set of conditions, all one has to do is to calculate the voltage associated with it. If the calculated voltage is positive, the reaction must be spontaneous. If the voltage is negative, the reaction cannot go by itself; the reverse reaction will be spontaneous.

These criteria can readily be applied to determine whether or not an oxidation-reduction reaction can occur at standard concentrations. All one has to do is add the appropriate standard electrode potentials to obtain the E° values corresponding to the reaction and note whether this quantity is positive or negative. Referring to Example 21.1, we recall that the E° values associated with the reactions

$$\text{Cl}_2(\text{g, 1 atm}) + 2 \text{ I}^-(1 \text{ M}) \longrightarrow 2 \text{ Cl}^-(1 \text{ M}) + \text{I}_2(\text{s}) \qquad \textbf{(21.13)}$$

$$\text{MnO}_4^-(1 \text{ M}) + 8 \text{ H}^+(1 \text{ M}) + 5 \text{ Cl}^-(1 \text{ M}) \longrightarrow$$
$$\text{Mn}^{2+}(1 \text{ M}) + 4 \text{ H}_2\text{O} + \tfrac{5}{2} \text{ Cl}_2(\text{g, 1 atm}) \qquad \textbf{(21.14)}$$

$$\text{Sn}^{2+}(1 \text{ M}) + 2 \text{ Ag}(\text{s}) \longrightarrow \text{Sn}(\text{s}) + 2 \text{ Ag}^+(1 \text{ M}) \qquad \textbf{(21.15)}$$

are +0.83 V, +0.16 V, and −0.94 V respectively. We deduce that Reactions 21.13 and 21.14 will occur spontaneously in the laboratory, while Reaction 21.15 will not. In other words, at standard concentrations, chlorine gas will oxidize iodide ions to elementary iodine, and permanganate ions will oxidize hydrochloric acid to chlorine. Under the same conditions, Sn^{2+} ions will not react with silver metal; instead, the reverse reaction will be spontaneous.

Example 21.2 illustrates the application of this simple principle to a somewhat more complex situation.

EXAMPLE 21.2. Assuming standard concentrations, will a reaction occur between nitric acid and a solution of iron(II) chloride?

SOLUTION. Before we can answer this question, we must decide what the possible reactions are; only then can we calculate whether or not a reaction will occur. To be sure that we do not neglect any possibilities, let us list separately, using Table 21.1, every possible half-reaction. Noting that the species present are H^+, NO_3^-, Fe^{2+}, and Cl^- ions, we have:

possible oxidations:

$$\text{O}_1 \quad \text{Fe}^{2+}(1 \text{ M}) \longrightarrow \text{Fe}^{3+}(1 \text{ M}) + e^- \qquad \text{SOP} = -0.77 \text{ V}$$
$$\text{O}_2 \quad 2 \text{ Cl}^-(1 \text{ M}) \longrightarrow \text{Cl}_2(\text{g, 1 atm}) + 2 \text{ } e^- \qquad \text{SOP} = -1.36 \text{ V}$$

possible reductions:

$$\text{R}_1 \quad 2 \text{ H}^+(1 \text{ M}) + 2 \text{ } e^- \longrightarrow \text{H}_2(\text{g, 1 atm}) \qquad\qquad\qquad \text{SRP} = 0.00 \text{ V}$$
$$\text{R}_2 \quad \text{NO}_3^-(1 \text{ M}) + 4 \text{ H}^+(1 \text{ M}) + 3 \text{ } e^- \longrightarrow \text{NO}(\text{g, 1 atm}) + 2 \text{ H}_2\text{O} \quad \text{SRP} = +0.96 \text{ V}$$
$$\text{R}_3 \quad \text{Fe}^{2+}(1 \text{ M}) + 2 \text{ } e^- \longrightarrow \text{Fe}(\text{s}) \qquad\qquad\qquad\qquad \text{SRP} = -0.44 \text{ V}$$

545

In principle, the two possible oxidations, O_1 and O_2, taken in combination with the three possible reductions, R_1, R_2, and R_3, could lead to six different overall reactions ($O_1 + R_1$, $O_1 + R_2$, $O_1 + R_3$; $O_2 + R_1$, $O_2 + R_2$, $O_2 + R_3$). In practice, it is readily seen that only one of these combinations, $O_1 + R_2$, will give a positive voltage. We deduce that the spontaneous reaction is

oxidation: $\qquad\qquad$ $3\ Fe^{2+}(1\ M) \longrightarrow 3\ Fe^{3+}(1\ M) + 3\ e^-$ $\qquad$ $-0.77\ V$

reduction:
$$NO_3^-(1\ M) + 4\ H^+(1\ M) + 3\ e^- \longrightarrow NO(g,\ 1\ atm) + 2\ H_2O \quad +0.96\ V$$

$3\ Fe^{2+}(1\ M) + NO_3^-(1\ M) + 4\ H^+(1\ M) \longrightarrow 3\ Fe^{3+}(1\ M)$
$\qquad\qquad\qquad\qquad\qquad\qquad\qquad\qquad + NO(g,\ 1\ atm) + 2\ H_2O \quad +0.19\ V$

In other words, nitric acid, at standard concentrations, will oxidize a solution of iron(II) chloride to a solution of iron(III) chloride, forming NO as a reduction product.

While it is, of course, important to know whether or not an oxidation-reduction reaction will occur at standard concentrations, this information alone is hardly sufficient for the chemist who is interested in carrying out the reaction in the laboratory.

In the first place, it is highly unlikely that he will wish to run the reaction under such conditions that all reactants and products are at standard concentrations. Moreover, he would like to know, not only whether a particular reaction will take place, but also the *extent* to which it will occur.

In order to decide whether a reaction will take place under a particular set of conditions and, if so, the extent to which it will occur, one must know the numerical value of the equilibrium constant for the reaction. Fortunately, it is possible to calculate the equilibrium constant for an oxidation-reduction reaction from standard electrode potentials.

Relation between E° and ΔG°

You will recall from Chapter 13 that a spontaneous reaction is characterized by a *negative* free energy change. We have seen in this chapter that a spontaneous oxidation-reduction reaction is always associated with a *positive* voltage. These two statements, taken together, imply a direct relationship between the free energy decrease ($-\Delta G$) and the voltage (E). This relationship is of the form

$$\Delta G = -nFE \qquad\qquad (21.16)$$

where ΔG is the free energy change in calories, n is the number of moles of electrons transferred in the reaction, F is the value of the faraday in calories per volt (23,060 cal/volt), and E is the voltage.*

---

* The quantity nFE represents the electrical work which is obtained from a voltaic cell. It can be identified with the maximum amount of useful work (W') that can be extracted from the spontaneous redox reaction taking place within the cell. It will be recalled from Chapter 13 that $\Delta G = -W'$.

Equation 21.16 can be interpreted qualitatively in a very simple manner. We may distinguish three cases:

$$E > 0, \quad \Delta G < 0; \quad \text{reaction is spontaneous}$$
$$E = 0, \quad \Delta G = 0; \quad \text{reaction is at equilibrium}$$
$$E < 0, \quad \Delta G > 0; \quad \text{reverse reaction is spontaneous}$$

We are particularly interested in applying this relationship when reactants and products are at their standard concentrations (conc. of 1 M for species in solution, partial pressure of gases 1 atmosphere). Under these conditions we can write

$$\Delta G^\circ = -nFE^\circ = -23{,}060\ nE^\circ \qquad (21.17)$$

where ΔG° is the standard free energy change in calories and E° is the standard voltage.

EXAMPLE 21.3. Using Equation 21.17, calculate
a. ΔG° for the reaction $Zn(s) + 2\ H^+ \longrightarrow Zn^{2+} + H_2(g)$

b. E° for the reaction $Ag(s) + \frac{1}{2}\ Cl_2(g) \longrightarrow AgCl(s)$, using Table 13.2.

SOLUTION
a. We have previously shown that E° for this reaction is $+0.76$ V. The number of moles of electrons transferred is clearly two (2 moles of e^- required to reduce 2 moles of H^+; 2 moles of electrons released by the oxidation of one mole of Zn). Consequently,

$$\Delta G^\circ = -23{,}060(2)\ (+0.76)\ cal = -35{,}000\ cal$$

b. We first note that ΔG° for this reaction is simply the standard free energy of formation of AgCl. This quantity is found from Table 13.2 to be $-26{,}200$ cal. In this reaction, n is 1: one mole of electrons is given off when one mole of Ag is oxidized to Ag^+.

$$-26{,}200 = -23{,}060(1)E^\circ$$
$$E^\circ = +1.13\ V$$

We see from Example 21.3 that Equation 21.17 is convenient for calculating standard free energy changes from measured cell voltages. Alternatively, if ΔG° is known, E° can be predicted in advance.

Relation Between E° and K

As we pointed out in Chapter 13, the standard free energy change, ΔG°, is related to the equilibrium constant by the relation

$$\Delta G^\circ = -2.30\ RT \log_{10} K \qquad (21.18)$$

where R is the gas constant and T the absolute temperature. Combining Equations 21.17 and 21.18, we obtain

$$nFE^\circ = 2.30\ RT \log_{10} K$$

or
$$E° = 2.30 \frac{RT}{nF} \log_{10} K$$

Substituting the values for the constants R and F in the proper units (1.99 cal/°K, 23,060 cal/volt) and taking T to be 298°K (25°C), we obtain

$$E° = \frac{2.30(1.99)(298)}{n(23,060)} \log_{10} K = \frac{0.059}{n} \log_{10} K$$

If E° is +, K > 1;   or
if E° is −, K < 1.

$$\log_{10} K = \frac{nE°}{0.059} \qquad (21.19)$$

Equation 21.19 enables us to determine* the equilibrium constant K for an oxidation-reduction reaction from the value of E° as calculated from standard electrode potentials. To illustrate the procedure involved, let us consider three reactions having quite different E° values.

(1) $$Zn(s) + Cu^{2+} \longrightarrow Zn^{2+} + Cu(s)$$

$$E° = SOP\ Zn + SRP\ Cu^{2+} = 0.76\ V + 0.34\ V = 1.10\ V$$

n = 2 (2 moles of e⁻ required to reduce 1 mole of Cu²⁺ ions)

$$\log_{10} K = \frac{2(1.10)}{0.059} = 37 \qquad K = \frac{[Zn^{2+}]}{[Cu^{2+}]} = 10^{37}$$

The fact that the equilibrium constant for this reaction is such a large number, 10^{37}, means that at ordinary concentrations, the reaction between zinc and Cu²⁺ ions goes to completion. If excess zinc is added to a 1 M solution of CuSO₄, nearly all the Cu²⁺ ions will be replaced by Zn²⁺ ions; the equilibrium concentration of Zn²⁺ will be 1 M, while that of Cu²⁺ will be 10^{-37} M.

(2) $$Sn(s) + Pb^{2+} \longrightarrow Sn^{2+} + Pb(s)$$

$$E° = +0.14\ V - 0.13\ V = 0.01\ V, \qquad n = 2$$

$$\log_{10} K = \frac{(2)(0.01)}{0.059} = 0.3 \qquad K = \frac{[Sn^{2+}]}{[Pb^{2+}]} = 2$$

Here, the situation is quite different from that in (1). Since the equilibrium constant is relatively small (K = 2), we can expect to find both products and reactants present at equilibrium in significant amounts (see Example 21.4). This situation will arise whenever the E° value for a reaction is small. The direction in which such reactions will proceed is particularly sensitive to changes in concentrations of reactants or products (Example 21.5).

(3) $$Ag(s) + Tl^+ \longrightarrow Ag^+ + Tl(s)$$

$$E° = -0.80\ V - 0.34\ V = -1.14\ V, \qquad n = 1$$

$$\log_{10} K = \frac{(1)(-1.14)}{0.059} = -19 \qquad K = \frac{[Ag^+]}{[Tl^+]} = 10^{-19}$$

The equilibrium constant for this reaction is so small (K = 10^{-19}) that it is impossible to form significant amounts of Ag⁺ by reacting elementary silver with

---

* In the expression for K, concentrations of species in water solution are expressed in moles per liter. For gases participating in the reaction, partial pressures in atmospheres are used. Concentrations of solids or solvent water molecules do not appear in the equilibrium expression.

Tl^+ ions. Conversely, one can predict that the reverse reaction, that of thallium metal with Ag^+ ions, will proceed almost to completion.

Calculations Involving K

The equilibrium constant for a redox reaction can be used to calculate the extent to which that reaction will occur if one starts with pure reactants at known concentrations (Example 21.4).

EXAMPLE 21.4. Calculate the equilibrium concentrations of Sn^{2+} and Pb^{2+} if excess tin is added to a solution originally 4.5 M in Pb^{2+}.

SOLUTION. The reaction involved is $Sn(s) + Pb^{2+} \rightarrow Sn^{2+} + Pb(s)$, for which we have shown that

$$K = \frac{[Sn^{2+}]}{[Pb^{2+}]} = 2$$

If we let $[Sn^{2+}] = x$, then $[Pb^{2+}] = 4.5 - x$, since 1 mole of Pb^{2+} reacts for every mole of Sn^{2+} formed. Substituting in the expression for K,

$$\frac{x}{4.5 - x} = 2$$

Solving, $x = [Sn^{2+}] = 3$; $4.5 - x = [Pb^{2+}] = 1.5$

We see that the reaction gives an equilibrium mixture in which there are two Sn^{2+} ions for every Pb^{2+} ion.

The equilibrium constant K can also be used to predict the direction in which a redox reaction will occur spontaneously under specified conditions (Example 21.5).

EXAMPLE 21.5. Will the following reaction occur in the direction indicated?

$$Ni(s) + Co^{2+}(1\ M) \longrightarrow Ni^{2+}(0.01\ M) + Co(s)$$

SOLUTION. To answer this question, let us first calculate K and then determine which "side" of the equilibrium we are on.

$$E° = 0.25\ V - 0.28\ V = -0.03\ V, \qquad n = 2$$

$$\log_{10} K = \frac{2(-0.03)}{0.059} = -1 \qquad K = \frac{[Ni^{2+}]}{[Co^{2+}]} = 0.1$$

From the statement of the problem, the concentration of Ni^{2+} starts out at 0.01 M, only 1/100 that of Co^{2+}. In order to reach equilibrium, the concentration of Ni^{2+} must increase until it becomes 1/10 that of Co^{2+} (K = 0.1). To accomplish this, the reaction must proceed from left to right. In so doing, the concentration of Ni^{2+} increases and that of Co^{2+} decreases until their ratio becomes 0.1.

It is worth pointing out that this reaction will not take place at standard concentrations; i.e., conc. Ni^{2+} = conc. Co^{2+} = 1. This is immediately obvious from the fact that E° is negative. From a different point of view, one

549

can say that the ratio, conc. Ni²⁺/conc. Co²⁺ is too high (1 > 0.1); the reaction must proceed in the reverse direction to reach equilibrium. Finally, if we were to add nickel metal to a solution 1 M in Co²⁺ and 0.1 M in Ni²⁺, no net change would occur; we would already be at equilibrium and there would be no tendency for the reaction to proceed in either direction. In summary, if

Recall the discussion in Chapter 12, Section 12.3.

$$\frac{\text{conc. Ni}^{2+}}{\text{conc. Co}^{2+}} < K, \text{ reaction } \longrightarrow$$

$$\frac{\text{conc. Ni}^{2+}}{\text{conc. Co}^{2+}} > K, \text{ reaction } \longleftarrow$$

$$\frac{\text{conc. Ni}^{2+}}{\text{conc. Co}^{2+}} = K, \text{ reaction is at equilibrium}$$

21.5 EFFECT OF CONCENTRATION ON VOLTAGE: THE NERNST EQUATION

Our discussion of cell voltages up to this point has been limited to cases in which all species participating in the cell reaction are at standard concentrations (more precisely, unit activities). We shall now consider how, for a given reaction, the voltage is affected by changes in the concentrations of reactants and products. We shall proceed from a qualitative discussion of the direction of this effect to a quantitative treatment of its magnitude.

Direction of Effect

If the voltage of a cell is taken to be a measure of the spontaneity of the reaction occurring, it follows that any change in conditions which makes that reaction take place more readily should increase the voltage. We know from equilibrium considerations that a reaction becomes more spontaneous when the concentrations of *reactants* are *increased* or those of the *products* are *decreased*. Concentration changes in this direction should then *increase* the cell voltage. Conversely, the reaction should become less spontaneous and the voltage less positive if the concentrations of reactants are decreased or those of the products increased. To illustrate, consider the familiar reaction

$$\text{Zn(s)} + \text{Cu}^{2+} \longrightarrow \text{Zn}^{2+} + \text{Cu(s)}$$

At standard concentrations, this reaction, going on in a cell such as that shown in Figure 21.1, produces a voltage of 1.10 V. From the argument just outlined, one would expect the cell voltage to rise above 1.10 V if the concentration of Cu²⁺ ions were increased above 1 M, or if that of the Zn²⁺ ions produced by the reaction were decreased below 1 M. A decrease in the concentration of Cu²⁺ ions or an increase in the concentration of Zn²⁺ ions would be expected to have the opposite effect, decreasing the cell voltage below 1.10 V. The data in Table 21.2 indicate that this reasoning is in accord with experiment.

From Table 21.2 we can see that a change of as much as a power of 10 in the concentration of a reactant or product produces only a small change in

TABLE 21.2 EFFECT OF CHANGE IN CONCENTRATION ON VOLTAGE OF Zn-Cu CELL

CONC. Cu^{2+}	CONC. Zn^{2+}	CONC. Cu^{2+}/CONC. Zn^{2+}	VOLTAGE
10	0.10	100	1.16
10	1.0	10	1.13
1.0	0.10	10	1.13
1.0	1.0	1	1.10
0.10	0.10	1	1.10
1.0	10	0.1	1.07
0.10	1.0	0.1	1.07
0.10	10	0.01	1.04

the cell voltage — about 0.03 V with this particular cell. It follows that as long as the concentrations of the ions used in setting up voltaic cells do not differ too greatly from 1 M, the cell voltage will be very nearly that calculated from standard electrode potentials. In practice, with simple metal-metal ion cells, large voltage changes are observed only when the concentration of one of the species is drastically lowered by the addition of a complexing or precipitating agent. Suppose, for example, one were to add a large excess of ammonia or sulfide ions to the Cu-Cu^{2+} half-cell of Figure 21.1. This would remove all but a tiny fraction of the free Cu^{2+} ions from solution and hence would materially reduce the cell voltage. If sufficient sulfide ions were added to make $[S^{2-}] = 1$ M, then since K_{sp} of CuS $= 1 \times 10^{-25}$, the concentration of Cu^{2+} would be reduced to 1×10^{-25}. One can calculate that under these conditions, the cell voltage would be only 0.01 V.

It is possible to design voltaic cells whose driving force depends solely on concentration differences of the type just discussed. An example of such a *concentration cell* is shown in Figure 21.7. Each of the half-cells consists of a copper electrode dipping into a solution of copper sulfate. They differ only in the concentration of $CuSO_4$: one solution is more dilute than the other. The cell produces a small but measurable voltage whose magnitude depends on the ratio of the concentrations of Cu^{2+} ions in the two solutions.

When current is drawn from the cell, it is found that the copper electrode dipping into the more dilute solution decreases in weight. At the same time, the blue color of this solution intensifies, indicating an increase in Cu^{2+} ion concentration. Exactly the reverse process takes place in the half-cell containing the more concentrated $CuSO_4$ solution; copper plates out on the electrode, while the concentration of Cu^{2+} ions decreases. Clearly, the half-reactions must be

anode: $$Cu(s) \longrightarrow Cu^{2+} \text{ (dilute)} + 2 \text{ e}^-$$

cathode: $$Cu^{2+} \text{ (conc.)} + 2 \text{ e}^- \longrightarrow Cu(s)$$

The net effect of this process is to cause the two solutions to approach each other in concentration. The cell produces a voltage because this process is spontaneous; matter always tends to go from a region of high concentration to one of low concentration. If the cell is operated long enough, the concentrations of copper sulfate in the two half-cells become equal, equilibrium is reached, and the voltage drops to zero.

What would be the sign of ΔS for this process? ΔH? ΔG?

551

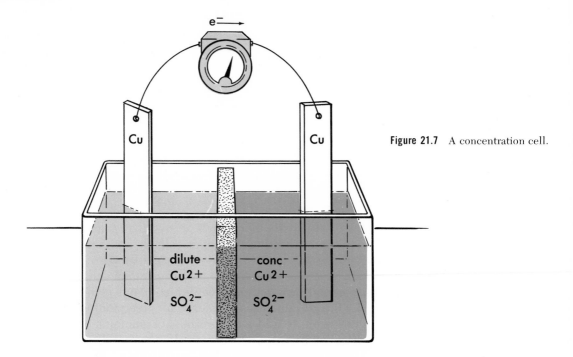

Figure 21.7 A concentration cell.

Magnitude of Effect

A quantitative study of the effect of concentration upon the voltage of the Zn-Cu cell leads to the following equation:

$$E = 1.10 \text{ V} - 0.030 \log_{10} \frac{\text{conc. Zn}^{2+}}{\text{conc. Cu}^{2+}} \qquad (21.20)$$

It is easily shown that this equation is in agreement with the observations made earlier. For example, if the concentration of Zn^{2+} ions is increased above that of Cu^{2+}, the second term on the right of Equation 21.20 is negative and the voltage drops below 1.10 V. The same effect can be achieved by reducing the concentration of Cu^{2+} ions. Conversely, an increase in the ratio of Cu^{2+} ions to Zn^{2+} ions makes the concentration term in the equation positive and therefore increases the voltage. Finally, it may be seen that a change of a power of 10 in the concentration ratio changes the voltage by 0.03 V, as pointed out earlier.

Equation 21.20 is a special form of a more general relation known as the **Nernst equation**. For the general oxidation-reduction reaction at 25°C

$$aA + bB \longrightarrow cC + dD$$

in which A, B, C, and D are species whose concentrations can be varied and a, b, c, and d are the corresponding coefficients of the balanced equation, the Nernst equation has the form

$$E = E° - \frac{0.059}{n} \log_{10} \frac{(\text{conc. C})^c (\text{conc. D})^d}{(\text{conc. A})^a (\text{conc. B})^b} \qquad (21.21)$$

(E = cell voltage, E° = standard voltage, n = no. moles e⁻ transferred in reaction.)

Example 21.6 illustrates the calculations involved in using the Nernst equation.

EXAMPLE 21.6. Calculate the voltage of a cell in which the following reaction occurs:

$$Zn(s) + 2 H^+(0.001 \text{ M}) \longrightarrow Zn^{2+}(1 \text{ M}) + H_2(g, 1 \text{ atm})$$

SOLUTION. For this reaction, n is 2 (2 moles of electrons are required to reduce 2 moles of H^+ ions). Hence, the Nernst equation becomes

$$E = E° - \frac{0.059}{2} \log_{10} \frac{(\text{conc. } Zn^{2+})(P_{H_2})}{(\text{conc. } H^+)^2}$$

Substituting $E° = +0.76$ V, conc. $Zn^{2+} = 1$ M, $P_{H_2} = 1$ atm, conc. $H^+ = 10^{-3}$ M, we have

$$E = 0.76 \text{ V} - \frac{0.059}{2} \log_{10} \frac{1}{(\text{conc. } H^+)^2} = +0.76 \text{ V} - \frac{0.059}{2} \log_{10} \frac{1}{(10^{-3})^2}$$

$$E = 0.76 \text{ V} - \frac{0.059}{2} (6) = +0.58 \text{ V}$$

The Nernst equation can also be used to determine the effect of changes in concentration on the potential of an individual half-cell (see Problem 21.35 and Example 22.1).

Use of the Nernst Equation to Determine Concentrations of Ions in Solution

One of the most important applications of the Nernst equation is in the experimental determination of concentrations of solutions from measured cell voltages. For example, the relation derived in Example 21.6

$$E = +0.76 \text{ V} - \frac{0.059}{2} \log_{10} \frac{1}{(\text{conc. } H^+)^2}$$

enables one to determine the concentration of H^+ ions in a solution by measuring the voltage of a properly constructed Zn-H_2 cell. From cell voltages it is possible to determine concentrations of ions present in amounts so small that more conventional analytical techniques are inapplicable. One could, for example, use a Zn-H_2 cell to determine the concentration of H^+ in a 0.01 M NaOH solution. Under these conditions, the measured cell voltage is 0.05 V, corresponding to an H^+ ion concentration of 10^{-12} M. That is,

$$+0.05 = +0.76 - \frac{0.059}{2} \log_{10} \frac{1}{(\text{conc. } H^+)^2}$$

$$\log_{10} \frac{1}{(\text{conc. } H^+)^2} = \frac{2(0.71)}{0.059} = 24$$

$$(\text{conc. } H^+)^2 = 10^{-24}, \quad \text{conc. } H^+ = 10^{-12} \text{ M}$$

How could you design a cell to measure the conc. of Cl⁻?

553

The sensitivity of this method at very low ion concentrations makes it ideal for the measurement of ionic equilibrium constants. Electrochemical techniques are readily applicable to the measurement of ionization constants of weak acids, stability constants of complex ions, or solubility products of slightly soluble salts. Example 21.7 illustrates the procedure and calculations involved.

EXAMPLE 21.7. In order to determine the solubility product of zinc sulfide, a student sets up a voltaic cell consisting of a standard Cu-Cu^{2+} half-cell (conc. Cu^{2+} = 1 M) joined to a Zn-Zn^{2+} half-cell. The concentration of Zn^{2+} is adjusted by adding enough sodium sulfide to precipitate almost all of the Zn^{2+} ions; the equilibrium concentration of S^{2-} is 1 M. Under these conditions, the measured cell voltage is +1.78 V. Calculate K_{sp} of ZnS.

SOLUTION. The cell reaction is

$$Zn(s) + Cu^{2+} \longrightarrow Zn^{2+} + Cu(s)$$

For this reaction, the Nernst equation takes the form

$$E = E^\circ - \frac{0.059}{n} \log_{10} \frac{(\text{conc. } Zn^{2+})}{(\text{conc. } Cu^{2+})}$$

Or, since E° = 1.10 V, n = 2 and conc. Cu^{2+} = 1

$$E = 1.10 - \frac{0.059}{2} \log_{10} (\text{conc. } Zn^{2+})$$

Substituting the measured voltage, 1.78 V, and solving for the concentration of Zn^{2+},

$$1.78 = 1.10 - \frac{0.059}{2} \log_{10} (\text{conc. } Zn^{2+})$$

$$\log (\text{conc. } Zn^{2+}) = \frac{0.68 \times 2}{-0.059} = -23 \qquad \text{conc. } Zn^{2+} = 10^{-23} \text{ M}$$

Knowing that the concentration of S^{2-} is 1 M, we have

$$K_{sp} \text{ ZnS} = \text{conc. } Zn^{2+} \times \text{conc. } S^{2-} = 10^{-23} \times 1 = 1 \times 10^{-23}$$

One of the most useful of analytical instruments, the pH meter, shown in Figure 21.8, utilizes the dependence of cell voltage upon concentration to measure pH. The instrument incorporates three elements: a vacuum tube voltmeter for accurate potential measurements, a reference half-cell of known potential, and another half-cell whose potential depends upon the concentration of H^+ ions. This half-cell consists of a metal electrode dipping into a solution of known pH separated by a thin glass membrane from the solution whose pH is to be determined. The potential across this *glass electrode* and consequently the cell voltage itself is a linear function of the pH of the solution outside the membrane. The mechanism by which the glass electrode operates is believed to involve conduction by transport of unipositive ions (Na^+ and H^+) through the ionic network of the glass, plus some exchange of H^+ for Na^+ ions at the surface.

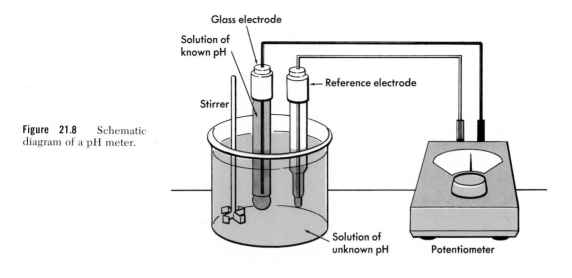

Glass electrode

Solution of
known pH

Stirrer

← Reference electrode

Figure 21.8 Schematic
diagram of a pH meter.

Solution of
unknown pH

Potentiometer

Use of the Nernst Equation to Determine Reaction Spontaneity

We pointed out in Section 21.4 that one can use the equilibrium constant for a redox reaction to determine whether or not the reaction will be spontaneous at specified concentrations. Alternatively, one can use the Nernst equation to calculate the voltage under these conditions. If the calculated voltage is positive, the reaction will be spontaneous; if it is negative, the reverse reaction will occur. If, perchance, E turns out to be zero, the reaction must be at equilibrium at the specified concentrations.

To illustrate this application of the Nernst equation, let us refer to the reaction considered in Example 21.5,

$$Ni(s) + Co^{2+}(1\ M) \longrightarrow Ni^{2+}(0.01\ M) + Co(s)$$

for which $E° = -0.03$ V, $n = 2$, and the Nernst equation takes the form

$$E = -0.03 - \frac{0.059}{2} \log_{10} \frac{(conc.\ Ni^{2+})}{(conc.\ Co^{2+})}$$

Substituting for the concentrations of Ni^{2+} and Co^{2+}, we obtain

$$E = -0.03 - \frac{0.059}{2} \log_{10} 0.01$$

$$= -0.03 + 0.059 = +0.03\ V$$

and we come to the conclusion that at these concentrations, the reaction will be spontaneous. It will be recalled that we arrived at the same conclusion in Example 21.5 by comparing the ratio (conc. Ni^{2+})/(conc. Co^{2+}) to the equilibrium constant, K.

555

PROBLEMS

21.1 A voltaic cell is made up of two half-cells, one consisting of a Cd electrode dipping into a solution 1 M in Cd^{2+}, the other a Ni electrode in a solution 1 M in Ni^{2+}.

    a. Write a balanced equation for the cell reaction.
    b. Draw a diagram of the cell. Label anode and cathode. Indicate the direction in which ions move through the cell and the direction of electron flow through the external circuit.
    c. Calculate the voltage of the cell.

21.2 Consider a cell in which the following reaction occurs:

$$Ni(s) + 2\ H^+ \longrightarrow Ni^{2+} + H_2(g)$$

    a. What would be a suitable cathode for this cell?
    b. List three positive ions which could be present in the anode compartment.
    c. List two positive ions which could not be present around the anode without short-circuiting the cell.

21.3 Explain why

    a. The fuel cell shown in Figure 21.6A is operated at a high temperature.
    b. The voltage of a gravity cell (Figure 21.2) drops to zero if the electrolyte solutions are thoroughly mixed.
    c. A "dry cell" actually contains a considerable amount of water.
    d. NH_4Cl rather than HCl is used as an electrolyte in the dry cell.

21.4 Write equations for

    a. The reaction that occurs at the anode of a lead storage battery when it is discharging.
    b. The reaction that occurs at the cathode of a lead storage battery when it is charged.
    c. The overall reaction that occurs in a fuel cell in which carbon monoxide is used as the fuel.

21.5 Assuming standard concentrations in each case, calculate

    a. The voltage developed in a cell in which a Pb electrode dipping into a $Pb(NO_3)_2$ solution is connected to an Fe electrode in a $FeSO_4$ solution.
    b. The voltage of a cell in which the following reaction occurs:

$$NO_3^- + S^{2-} \longrightarrow NO_2^- + S(s) \quad \text{(basic solution)}$$

    c. The voltage required to electrolyze molten KI.
    d. The voltage required to electrolyze a water solution of KI.

21.6 Which of the following reactions could serve as a source of electrical energy in a voltaic cell? Which would have to be carried out in an electrolytic cell?

    a. $Cu(s) + 2\ Ag^+(1\ M) \longrightarrow Cu^{2+}(1\ M) + 2\ Ag(s)$
    b. $2\ Br^-(1\ M) + 2\ H_2O \longrightarrow Br_2(l) + H_2(g) + 2\ OH^-(1\ M)$
    c. $2\ AuCl_4^-(1\ M) + 3\ Sn^{2+}(1\ M) \longrightarrow 2\ Au(s) + 8\ Cl^-(1\ M) + 3\ Sn^{4+}(1\ M)$
    d. $ClO_3^-(1\ M) + 5\ Cl^-(1\ M) + 6\ H^+(1\ M) \longrightarrow 3\ Cl_2(g) + 3\ H_2O$

21.7 a. Arrange the following reducing agents in order of increasing strength (acidic solution): Ni(s), Mg(s), Fe^{2+}, F^-, Co(s).
    b. Arrange the following oxidizing agents in order of increasing strength (acidic solution): $AuCl_4^-$, ClO_3^-, SO_4^{2-}, Sn^{4+}, H_2O_2.
    c. Using the species in a and b, write balanced equations for three different spontaneous oxidation-reduction reactions.

21.8 Decide what reaction, if any, will occur when the following are mixed (standard concentrations).

a. SO_4^{2-}, H^+, Fe^{2+} b. $O_2(g)$, OH^-, S^{2-}
c. $Cr_2O_7^{2-}$, H^+, Sn^{2+} d. I^-, MnO_4^-, H^+, Sn^{2+}
e. Cu^{2+}, Ag^+

21.9 Calculate $\Delta G°$ for the following reactions.

a. $2\ H_2S(g) + O_2(g) \longrightarrow 2\ S(s) + 2\ H_2O$
b. $Cl_2(g) + 2\ Br^- \longrightarrow 2\ Cl^- + Br_2(l)$
c. $Cu(s) + SO_4^{2-} + 4\ H^+ \longrightarrow Cu^{2+} + SO_2(g) + 2\ H_2O$

21.10 Calculate the equilibrium constants for the reactions listed in 21.9.

21.11 Consider the reaction

$$Sn(s) + Pb^{2+} \longrightarrow Sn^{2+} + Pb(s)$$

a. Calculate the equilibrium ratio of the concentrations of Sn^{2+} to Pb^{2+}.
b. If 0.10 M Pb^{2+} is added to excess Sn, what will be the equilibrium concentration of Sn^{2+}?
c. Will this reaction occur spontaneously if the concentration of Pb^{2+} is 0.10 M and that of Sn^{2+} is 1.0 M?

21.12 Consider a cell in which the following reaction occurs:

$$Cu^{2+} + Ni(s) \longrightarrow Cu(s) + Ni^{2+}$$

State qualitatively what will happen to the voltage of this cell if

a. The anode compartment is saturated with H_2S.
b. Ammonia is added to the cathode compartment.
c. The half-cell containing Ni^{2+} is diluted with water.
d. The Ni^{2+} ions are replaced by Na^+ ions.

21.13 Using the Nernst equation, calculate the voltages of the following cells.

a. $Cd(s) + Co^{2+}(0.01\ M) \longrightarrow Cd^{2+}(1\ M) + Co(s)$
b. $Ni(s) + Sn^{4+}(2 \times 10^{-3}\ M) \longrightarrow Ni^{2+}(0.10\ M) + Sn^{2+}(0.01\ M)$
c. $2\ MnO_4^-(2\ M) + 16\ H^+(0.10\ M) + 10\ Cl^-(0.10\ M) \longrightarrow 2\ Mn^{2+}(0.001\ M) + 5\ Cl_2(g, 1\ atm) + 8\ H_2O$

21.14 A cell consisting of a Zn electrode dipping into a 1 M solution of $ZnSO_4$ and a hydrogen electrode (1 atm) has a voltage of 0.64 V. Calculate the pH of the solution surrounding the hydrogen electrode.

21.15 The voltage of a cell consisting of a Mn electrode in a solution 1 M in Mn^{2+} and a hydrogen electrode in 1 M NaOH is 0.34 V. Use these data to calculate the ion product for water, i.e., $[H^+] \times [OH^-]$.

21.16 A voltaic cell is made up of two half-cells, one consisting of a silver electrode dipping into a solution of 1 M $AgNO_3$, the other a copper electrode in a solution 1 M in $Cu(NO_3)_2$.

a. Write a balanced equation for the cell reaction.
b. Draw a diagram of the cell. Label anode and cathode. Indicate the direction of electron flow through the external circuit and the direction in which ions move through the cell.
c. Calculate the voltage of the cell.

21.17 Consider a cell in which the following reaction occurs:

$$Zn(s) + 2\ Ag^+ \longrightarrow Zn^{2+} + 2\ Ag(s)$$

a. Could a platinum electrode be used as the anode in this cell? The cathode?
b. Could a copper electrode be used as the anode in this cell? The cathode?

557

   c. Could Cl⁻ ions be present in the solution surrounding the anode? The cathode?

   d. Could one use a solution of pure KNO_3 to surround the zinc electrode? The silver electrode?

21.18 Consider cells in which the following reactions occur.

   a. $Ni(s) + 2 H^+(1 M) \longrightarrow Ni^{2+}(1 M) + H_2 (g, 1 atm)$
   b. $3 Sn(s) + 2 Au^{3+}(1 M) \longrightarrow 3 Sn^{2+}(1 M) + 2 Au(s)$
   c. $Cl_2 (g, 1 atm) + 2 I^-(1 M) \longrightarrow 2 Cl^-(1 M) + I_2(s)$
   In each case, draw a diagram of the cell, label the + and − electrodes, and calculate the voltage.

21.19 Calculate the voltages of cells made up as follows:

   a. Zn electrode in 1 M $ZnSO_4$; C electrode in solution 1 M in Cl⁻, surrounded by Cl_2 at 1 atm.
   b. Pt wire dipping into pool of Hg, surrounded by 1 M $Hg_2(NO_3)_2$ solution; Ag electrode in 1 M $AgNO_3$ solution.
   c. Sn electrode in 1 M $SnCl_2$ solution; Au electrode in solution 1 M in $AuCl_4^-$ and Cl⁻.

21.20 Write balanced equations for the reactions that occur when

   a. A dry cell is used to supply current.
   b. A lead storage battery discharges.
   c. A lead storage battery is charged.

21.21 A lead storage battery is to be charged by connecting it to another battery. To which terminal of the discharged battery should the negative terminal of the fresh battery be connected? Explain.

21.22 Explain why

   a. The ammonia liberated by Reaction 21.4 does not ordinarily build up a gas pressure within a dry cell.
   b. The density of the electrolyte in a lead storage battery decreases as it is discharged.
   c. The fuel cell shown in Figure 21.6B is operated at a high pressure.
   d. Fuel cells, at least in principle, are a cheaper source of energy than ordinary voltaic cells.

21.23 Calculate the minimum voltage necessary to electrolyze a solution of $NiCl_2$ so as to produce nickel at one electrode and chlorine gas at the other.

21.24 Referring to Table 21.1,

   a. Which is the stronger oxidizing agent, Ni^{2+} or I_2?
   b. Which is the stronger reducing agent, Ni or H_2S?
   c. Will oxygen react spontaneously, at standard concentrations, with a water solution of hydrogen iodide? Hydrogen chloride?

21.25 From the table of standard potentials, decide whether or not each of the following reactions will go.

   a. $2 Cr(s) + 3 Ni^{2+}(1 M) \longrightarrow 2 Cr^{3+}(1 M) + 3 Ni(s)$
   b. $Sn(s) + Cl_2(g, 1 atm) \longrightarrow Sn^{2+}(1 M) + 2 Cl^-(1 M)$
   c. $I_2(s) + 2 Co^{2+}(1 M) \longrightarrow 2 Co^{3+}(1 M) + 2 I^-(1 M)$
   d. $MnO_4^-(1 M) + 8 H^+(1 M) + 5 F^-(1 M) \longrightarrow \frac{5}{2} F_2(g, 1 atm) + Mn^{2+}(1 M) + 4 H_2O$
   e. $Cl_2 + Cr_2O_7^{2-} \longrightarrow ClO_3^- + Cr^{3+}$ (acidic soln., std. conc.)

21.26 Decide what reaction, if any, will occur when the following are mixed (standard concentrations).

   a. $H_2(g)$ and Ag^+ c. I^-, H^+, Fe^{3+} e. Cl_2, Br^-, and F^-
   b. $Cr_2O_7^{2-}$, H^+ and Fe^{2+} d. H_2S, Br^-, Mg^{2+} f. NO_3^-, H^+, Fe^{2+}

21.27   The standard free energy of formation of $PbBr_2(s)$ is -62.1 kcal/mole. Calculate the standard voltage of a cell in which the following reaction occurs:

$$Pb(s) + Br_2(l) \longrightarrow PbBr_2(s)$$

21.28   Consider the reaction

$$Cd(s) + Fe^{2+} \longrightarrow Cd^{2+} + Fe(s)$$

If excess cadmium is added to a solution 1 M in Fe^{2+}, what will be the equilibrium concentration of Cd^{2+}?

21.29   Will the reaction cited in Problem 21.28 occur spontaneously if the ratio (conc. Cd^{2+})/(conc. Fe^{2+}) is 1×10^{-4}?

21.30   Using the Nernst equation, calculate the voltages of the following cells.

a. $Zn(s) + Cu^{2+}(1\ M) \longrightarrow Zn^{2+}(10^{-5}\ M) + Cu(s)$
b. $Cl_2(g, 1\ atm) + 2\ I^-(0.01\ M) \longrightarrow 2\ Cl^-(10^{-3}\ M) + I_2(s)$
c. $MnO_2(s) + 4\ H^+(10\ M) + 2\ Cl^-(10\ M) \longrightarrow Mn^{2+}(1\ M) + Cl_2(g, 1\ atm) + 2\ H_2O$

21.31   Explain how the zinc-hydrogen cell shown in Figure 21.3 could be used to determine

a. The pH of a certain water solution.
b. The ionization constant of water.
c. The solubility product of $Zn(OH)_2$.

21.32   A voltaic cell consists of two half-cells. One contains a zinc electrode dipping into a solution 1 M in $Zn(NO_3)_2$. The other is made up of a lead electrode dipping into a solution 1 M in Cl^- and saturated with lead chloride. The observed voltage of the cell under these conditions is 0.49 V. Estimate the solubility product of $PbCl_2$.

°21.33   Starting with the Nernst equation (21.21) and noting that the voltage of a cell is zero when all species are at their equilibrium concentrations, derive Equation 21.19, relating $\log_{10} K$ to $E°$.

°21.34   Excess MnO_2 is added to one liter of a solution originally 2 M in HCl. Assuming chlorine gas is produced at 1 atm pressure, calculate

a. The equilibrium concentrations of H^+, Cl^-, and Mn^{2+}.
b. The volume of Cl_2 produced at 1 atm and 25°C.

°21.35   By applying the Nernst equation to the half-reaction

$$H_2(g) \longrightarrow 2\ H^+ + 2\ e^-$$

show that the standard oxidation potential for hydrogen in basic solution, i.e., the voltage corresponding to the half-reaction

$$H_2(g) + 2\ OH^-(1\ M) \longrightarrow 2\ H_2O + 2\ e^-$$

is $+0.83$ V.

*21.36   From the values given in Table 21.1 for the standard reduction potentials of $PbSO_4$ and Pb^{2+}, calculate the solubility product for this salt and compare to the value given in Table 16.3.

22

Oxidizing Agents in Water Solution

In Chapter 21, we considered how standard electrode potentials can be used to determine whether a given redox reaction will occur and, if so, the extent to which it will take place. In this chapter, we shall apply the principles that we have developed to a descriptive study of some of the more common oxidation-reduction reactions occurring in water solution. It is convenient to organize this material according to the oxidizing agents that participate in the reaction.

An oxidizing agent is, by definition, an atom, ion, or molecule that is capable of accepting electrons in an oxidation-reduction reaction. In principle, any species in which an element is in an oxidation state above its minimum value can serve as an oxidizing agent. For example, the ClO_3^- ion, the ClO^- ion, and the Cl_2 molecule are all potential oxidizing agents, since, in each of these species, chlorine is in an oxidation state $(+5, +1, 0)$ higher than the minimum shown by the element, -1. We shall consider only a limited number of oxidizing agents which commonly participate in aqueous reactions. These species can be classified for convenience into four major categories:

1. The **H$^+$ ion** which, in acting as an oxidizing agent, is reduced to hydrogen gas.

2. **Metal cations**, such as Ag^+, Cu^{2+}, and Fe^{3+}. These ions may be reduced to the free metal or, in certain cases, to another cation in a lower oxidation state.

3. **Molecules of nonmetals.** Two of the more powerful oxidizing agents in this category are elementary chlorine, Cl_2, and oxygen, O_2.

4. **Oxyanions** such as NO_3^-, SO_4^{2-}, $Cr_2O_7^{2-}$, and MnO_4^- in which the central atom is in its highest oxidation state $(+5$ for N, $+6$ for S, $+6$ for Cr, $+7$ for Mn$)$.

The species which is oxidized (i.e., the reducing agent) in an oxidation-reduction reaction must contain an element in a state of oxidation below its maximum. The NH_4^+ ion, the N_2 molecule, and the NO_2^- ion can all be oxidized; the nitrogen atom in these species is in an oxidation state $(-3, 0, +3)$ lower than its maximum value of $+5$. In our study of oxidizing agents, we shall deal primarily with their action on three different types of reducing agents:

1. *Metals*, which can be oxidized to metal cations.
2. *Metal cations* such as Fe^{2+} and Cu^+ which can be oxidized to a higher state (Fe^{3+}, Cu^{2+}).
3. *Anions*, such as I^- and S^{2-}, which are oxidized to the corresponding nonmetals.

22.1 H⁺ ION

$$2 H^+ + 2 e^- \longrightarrow H_2(g), \qquad SRP = 0.00 \text{ V}$$

Reaction of Metals with Acids

A very important type of reaction in which the H^+ ion participates as an oxidizing agent is that which occurs when a metal above hydrogen in the activity series is added to an acidic solution. The metal atom is oxidized to the corresponding cation; the H^+ ion is reduced to elementary hydrogen.

$$Mg(s) + 2 H^+ \longrightarrow Mg^{2+} + H_2(g) \tag{22.1}$$

$$Zn(s) + 2 H^+ \longrightarrow Zn^{2+} + H_2(g) \tag{22.2}$$

$$Al(s) + 3 H^+ \longrightarrow Al^{3+} + \tfrac{3}{2} H_2(g) \tag{22.3}$$

With many active metals such as sodium (SOP = +2.71 V) and calcium (SOP = +2.87 V), this reaction is violently exothermic; the hydrogen formed often ignites or explodes. Reaction proceeds more smoothly with less active metals such as magnesium, zinc, and aluminum, which are often used with dilute hydrochloric or sulfuric acid to produce small quantities of hydrogen in the general chemistry laboratory. Metals such as copper and silver, which have negative standard oxidation potentials (Cu = −0.34 V, Ag = −0.80 V), do not react with dilute acids to produce hydrogen.

In principle, any strong acid can serve as a source of H^+ ions for the oxidation of a metal. Hydrochloric acid is frequently used; evaporation of the solution remaining after reaction with excess metal gives the corresponding metal chloride. It is possible, for example, to convert nickel to nickel chloride by the following two-step process:

reaction with HCl: $\quad Ni(s) + 2 H^+ + 2 Cl^- \longrightarrow Ni^{2+} + H_2(g) + 2 Cl^-$

evaporation: $\qquad\qquad\quad Ni^{2+} + 2 Cl^- \longrightarrow NiCl_2(s)$

$$\overline{Ni(s) + 2 H^+ + 2 Cl^- \longrightarrow NiCl_2(s) + H_2(g)} \tag{22.4}$$

Cu and Ag do react with conc. HNO_3 (Section 22.5).

At 25°C, $NiCl_2 \cdot 6H_2O$ is produced on evaporation.

Bromides and iodides may also be prepared by this method. To obtain cobalt(II) iodide, CoI_2, from cobalt, one can react the metal with hydriodic acid, HI, and evaporate the resulting solution. Dilute sulfuric acid reacts smoothly with metals above hydrogen in the activity series to give solutions of metal sulfates. With concentrated sulfuric acid, the SO_4^{2-} ion rather than the H^+ ion may act as the oxidizing agent, forming sulfur dioxide as a reduction product rather than hydrogen. The reaction of metals with nitric acid, unless carried out in very dilute solution, does not evolve hydrogen; the NO_3^- ion rather than the H^+ ion is reduced (Section 22.5).

The reaction of metals with weak acids is both less spontaneous and

561

slower than with strong acids. A 1 N solution of acetic acid ($[H^+] = 0.004$ M) reacts much more slowly with zinc or magnesium than does a 1 N solution of HCl or H_2SO_4. Solutions of extremely weak acids such as H_3BO_3 (Ka = 6 × 10^{-10}) or HCN (Ka = 4 × 10^{-10}) fail to react with such metals as zinc or magnesium.

Reaction of Metals with Water

The metals above magnesium in the activity series (Na, Ca, Ba, K, Li) are so readily oxidized by H^+ ions that they react even at extremely low H^+ ion concentrations. In particular, these metals react with pure water ($[H^+] = 10^{-7}$ M) to evolve hydrogen. One may consider the reaction of water with a metal such as sodium to occur in two steps:

$$H_2O \longrightarrow H^+ + OH^-$$

The heat produced here is often sufficient to ignite the H_2.

$$\underline{Na(s) + H^+ \longrightarrow Na^+ + \tfrac{1}{2} H_2(g)}$$

$$Na(s) + H_2O \longrightarrow Na^+ + OH^- + \tfrac{1}{2} H_2(g) \tag{22.5}$$

Calcium reacts similarly, forming a solution of calcium hydroxide:

$$Ca(s) + 2 H_2O \longrightarrow Ca^{2+} + 2 OH^- + H_2(g) \tag{22.6}$$

With sodium, the exothermic reaction generates heat so rapidly that the hydrogen ignites; with calcium, the reaction is less vigorous.

It might be supposed that aluminum, which has a standard oxidation potential (+1.66 V) nearly as great as those of the 1A and 2A metals, would react with water in a similar manner:

$$Al(s) + 3 H_2O \longrightarrow Al^{3+} + 3 OH^- + \tfrac{3}{2} H_2(g) \tag{22.7}$$

In practice, no reaction can be detected when aluminum is added to pure water, presumably because the Al^{3+} and OH^- ions formed combine to give an insoluble product, $Al(OH)_3$ (more exactly, hydrated aluminum oxide, $Al_2O_3 \cdot x\ H_2O$), which adheres tightly to the metal surface. If the aluminum is coated with mercury (amalgamated) so that no oxide can adhere, the metal reacts with water very rapidly, in keeping with its high potential. Aluminum can also be oxidized in strongly basic solution; $Al(OH)_3$, being amphoteric, does not precipitate in the presence of excess OH^- ions:

$$Al(s) + 3 H_2O \longrightarrow Al^{3+} + 3 OH^- + \tfrac{3}{2} H_2(g)$$

$$\underline{Al^{3+} + 4 OH^- \longrightarrow Al(OH)_4^-}$$

Al metal "dissolves" in NaOH as well as HCl.

$$Al(s) + 3 H_2O + OH^- \longrightarrow Al(OH)_4^- + \tfrac{3}{2} H_2(g) \tag{22.8}$$

22.2 METAL CATIONS

Any metal cation, in undergoing reduction to the corresponding metal, can serve as an oxidizing agent. The oxidizing power of the cation depends upon:

1. **The magnitude of its standard reduction potential.** Ions such as Al^{3+} (SRP = −1.66 V) or Zn^{2+} (SRP = −0.76 V) are extremely weak oxidizing agents. The Ag^+ ion (SRP = +0.80 V), on the other hand, is quite a powerful oxidizing

agent, considerably more effective than the H^+ ion. A piece of copper added to a solution of silver nitrate is oxidized to Cu^{2+} ions as a result of the spontaneous reaction

$$2\ Ag^+ + Cu(s) \longrightarrow 2\ Ag(s) + Cu^{2+} \tag{22.9}$$

$$E° = SRP\ Ag^+ + SOP\ Cu = +0.80\ V - 0.34\ V = +0.46\ V$$

This reaction, carried out with a copper wire bent into an appropriate shape, leads to the formation of a "chemical Christmas tree," in which shiny needles of metallic silver appear to grow out of the surface of the wire.

2. **The concentration of "free" (hydrated) cation.** If the concentration of a metal cation is reduced below the standard value of 1 M, its reduction potential drops and it becomes a less effective oxidizing agent. This effect is ordinarily small unless one adds a precipitant or complexing agent so as to drastically lower the concentration of free metal ions. Addition of excess CN^- ions to a solution of a silver salt, resulting in the formation of the very stable $Ag(CN)_2^-$ complex, reduces the concentration of Ag^+ to the point where it becomes a very weak oxidizing agent (Example 22.1).

EXAMPLE 22.1. Enough sodium cyanide is added to a solution of silver nitrate to make $[CN^-] = [Ag(CN)_2^-] = 1$ M. Calculate the reduction potential of Ag^+ under these conditions.

SOLUTION. Let us first calculate the concentration of Ag^+ in this solution, using the fact that the dissociation constant for the $Ag(CN)_2^-$ ion is 1×10^{-21} (Table 19.5). We can then use the Nernst equation to determine the reduction potential of Ag^+ at this concentration.

$$Ag(CN)_2^- \rightleftharpoons Ag^+ + 2\ CN^- \qquad K_c = \frac{[Ag^+] \times [CN^-]^2}{[Ag(CN)_2^-]} = 1 \times 10^{-21}$$

Substituting $[CN^-] = [Ag(CN)_2^-] = 1$, we obtain

$$[Ag]^+ = 1 \times 10^{-21}$$

For the half-reaction $Ag^+ + e^- \rightarrow Ag(s)$, the Nernst equation has the form

$$RP = SRP - \frac{0.059}{n} \log_{10} \frac{1}{[Ag^+]}$$

But,

$$SRP\ Ag^+ = +0.80\ V,\ n = 1,\ [Ag]^+ = 1 \times 10^{-21}$$

$$RP = +0.80\ V - 0.059 \log_{10} \frac{1}{10^{-21}}$$

$$= +0.80\ V - 0.059(21)\ V = -0.44\ V$$

The Nernst equation can be applied to half reactions.

The information obtained in Example 22.1 is sometimes expressed by saying that the standard reduction potential of the $Ag(CN)_2^-$ ion, i.e., the reduction potential corresponding to the half-reaction

$$Ag(CN)_2^-(1\ M) + e^- \longrightarrow Ag(s) + 2\ CN^-(1\ M)$$

is -0.44 V. The fact that this potential is so much less positive than the standard reduction potential of Ag^+ explains, at least in part, why silver plating is ordinarily conducted from a cyanide bath. The presence of excess CN^- ions prevents the spontaneous formation of a rough, porous coating of silver which is obtained at high Ag^+ ion concentrations.

When the Ag^+ ion acts as an oxidizing agent, it must, of course, be reduced to the metal. However, with many transition and post-transition metal ions of higher charge such as Fe^{3+} or Cu^{2+}, another possibility arises. Instead of being reduced to the metal (Fe, Cu), they may be reduced to an ion in a lower oxidation state (Fe^{2+}, Cu^+). The path which the reduction takes will depend upon several factors. If we limit ourselves to situations in which there is an excess of oxidizing agent (Fe^{3+}, Cu^{2+}) available, it is a relatively simple task to decide between the two possibilities. To illustrate the principle involved, let us consider a redox reaction involving excess Fe^{3+} ions. Let us suppose, for the moment, that this reaction were to produce elementary iron. If this happened, one can calculate that the iron would be re-oxidized to the Fe^{2+} state by the excess Fe^{3+} ions, since the reaction

$$2\ Fe^{3+} + Fe(s) \longrightarrow 3\ Fe^{2+} \tag{22.10}$$

$$E° = SRP\ (Fe^{3+} \longrightarrow Fe^{2+}) + SOP\ (Fe \longrightarrow Fe^{2+})$$

$$= +0.77\ V + 0.44\ V = 1.21\ V$$

Would Sn^{4+} be reduced to Sn^{2+} or Sn? (cf. Table 21.1).

is spontaneous. Regardless of what the reducing agent may be, so long as excess Fe^{3+} ions are available, the product will be Fe^{2+} ions rather than Fe atoms.

When the Cu^{2+} ion acts as an oxidizing agent, the situation is quite different. The reaction, analogous to 22.10,

$$Cu^{2+} + Cu(s) \longrightarrow 2\ Cu^+$$

$$E° = SRP\ (Cu^{2+} \longrightarrow Cu^+) + SOP\ (Cu \longrightarrow Cu^+)$$

$$= +0.15\ V - 0.52\ V = -0.37\ V$$

is nonspontaneous. Any Cu^+ ions formed by the reduction of Cu^{2+} undergo a reaction known as **disproportionation** (simultaneous oxidation and reduction), precisely the reverse of that just stated.

$$2\ Cu^+ \longrightarrow Cu^{2+} + Cu(s) \qquad E° = +0.37\ V \tag{22.11}$$

We deduce that when the Cu^{2+} ion is used as an oxidizing agent, the product is elementary copper rather than Cu^+. The spontaneity of Reaction 22.11 explains why such salts as copper(I) sulfate, Cu_2SO_4, are unstable in water solution; they decompose to copper metal and the corresponding copper(II) salt. The only copper(I) compounds which are stable in contact with water are those which are only very slightly soluble such as CuI ($K_{sp} = 5 \times 10^{-12}$).

22.3 CHLORINE

$$Cl_2(g) + 2\ e^- \longrightarrow 2\ Cl^- \qquad SRP = +1.36\ V$$

Elementary chlorine is a powerful oxidizing agent, as shown by the magnitude of its standard reduction potential. Perhaps the most familiar redox reactions in which chlorine acts as an oxidizing agent are those involving bromide and iodide ions.

$$Cl_2(g) + 2 Br^- \longrightarrow 2 Cl^- + Br_2 \qquad E° = +1.36 V - 1.07 V = +0.29 V \quad (22.12)$$

$$Cl_2(g) + 2 I^- \longrightarrow 2 Cl^- + I_2 \qquad E° = +1.36 V - 0.53 V = +0.83 V \quad (22.13)$$

These reactions are frequently used to test for the presence of Br^- or I^- ions. Addition of chlorine to a solution containing either of these ions gives the free halogens Br_2 or I_2. If the water solution is then shaken with a small amount of a nonpolar organic solvent such as carbon disulfide or carbon tetrachloride, the free halogens enter the organic layer, to which they impart their characteristic colors, reddish-brown (bromine) or violet (iodine).

Bromine is prepared commercially from sea water, in which it occurs as Br^- ions, by oxidation with chlorine (Reaction 22.12). About 100,000 tons of bromine are produced annually in the United States by this method. The concentration of iodide ions in sea water is so low (conc. $I^- = 4 \times 10^{-7}$ M vs. conc. $Br^- = 8 \times 10^{-4}$ M) that it is not feasible to produce iodine in this way.

Chlorine, unlike oxygen, forms many compounds in which it has a positive oxidation number. These compounds are ordinarily formed by a disproportionation reaction in which elementary chlorine is simultaneously oxidized and reduced. An example of one such reaction is that which occurs when chlorine is added to water.

$$Cl_2(g) + H_2O \rightleftharpoons HOCl + H^+ + Cl^- \qquad (22.14)$$

The resulting solution, called *chlorine water*, contains equimolar amounts of the weak acid HOCl (hypochlorous acid) and the strong acid HCl. Half the chlorine (oxid. state = 0) is reduced to Cl^- ions (oxid. state = -1) while the remainder is oxidized to HOCl (oxid. state Cl = $+1$).

The hypochlorous acid formed by Reaction 22.14 is stable in the dark but slowly decomposes on exposure to sunlight, to give oxygen and a solution of hydrochloric acid:

$$HOCl \longrightarrow \tfrac{1}{2} O_2(g) + H^+ + Cl^- \qquad (22.15)$$

Why is Cl_2 water a useful disinfectant?

To prevent this reaction, chlorine water is stored in bottles made of brown or amber-colored glass.

The position of the equilibrium in Reaction 22.14 is strongly affected by the concentration of H^+ ions. In basic solution, in which the concentration of H^+ ions is low, chlorine is much more soluble than in pure water. The overall reaction that takes place when chlorine is bubbled through a solution of sodium hydroxide maintained at room temperature is

$$Cl_2(g) + 2 OH^- \longrightarrow ClO^- + Cl^- + H_2O \qquad (22.16)$$

One may regard this as a two-step process:

oxidation-reduction: $\qquad Cl_2(g) + H_2O \rightleftharpoons HOCl + H^+ + Cl^-$

neutralization: $\qquad \underline{HOCl + H^+ + 2 OH^- \longrightarrow ClO^- + 2 H_2O}$

$$Cl_2(g) + 2 OH^- \longrightarrow ClO^- + Cl^- + H_2O$$

The solution formed by the reaction of chlorine with sodium hydroxide via 22.16 is sold under various trade names as a household bleach and disinfectant. It is prepared commercially by electrolyzing a stirred water solution of sodium chloride. Recall that the electrolysis of an NaCl solution gives Cl_2 molecules and OH^- ions (Chapter 20); stirring ensures that these species react with each other. The active ingredient of the resulting solution is the hypochlorite ion, a relatively potent oxidizing agent:

$$ClO^- + H_2O + 2\ e^- \longrightarrow Cl^- + 2\ OH^- \qquad SRP = +0.89\ V$$

The reaction of chlorine with a hot, concentrated solution of sodium or potassium hydroxide is quite different from that observed at room temperature. Any ClO^- ions formed decompose on heating to ClO_3^- and Cl^- ions; the net reaction is

$$3\ Cl_2(g) + 6\ OH^- \longrightarrow ClO_3^- + 5\ Cl^- + 3\ H_2O \qquad (22.17)$$

Here, as in 22.16, chlorine acts as both an oxidizing and a reducing agent: $\frac{5}{6}$ of the chlorine atoms (oxid. state $= 0$) are reduced to Cl^- (oxid. state $= -1$) while $\frac{1}{6}$ of them are oxidized to ClO_3^- (oxid. state $Cl = +5$). Potassium chlorate, a powerful oxidizing agent is made commercially by Reaction 22.17. The chlorine and potassium hydroxide required are made *in situ* by electrolyzing a hot, stirred solution of potassium chloride. On cooling, $KClO_3$ crystallizes out first, since it is less soluble than KCl (0.85 M vs. 5.0 M at 30°C). Sodium chlorate, $NaClO_3$, is more difficult to prepare in this manner, since its solubility exceeds that of sodium chloride.

22.4 OXYGEN

Of all oxidizing agents, elementary oxygen is the most abundant and, in many ways, the most important. Its presence in air insures that all water supplies will contain dissolved oxygen. Water solutions used in the laboratory are ordinarily saturated with atmospheric oxygen. Finally, and most important, whenever we carry out a reaction in an open container, we must consider the possibility of elementary oxygen entering into the reaction.

As implied by its standard reduction potential,

$$\tfrac{1}{2}\ O_2(g) + 2\ H^+ + 2\ e^- \longrightarrow H_2O \qquad SRP = +1.23\ V$$

oxygen is a comparatively powerful oxidizing agent, at least in acidic solution. Since H^+ ions are involved as a reactant in this half-reaction, the reduction potential and consequently the oxidizing strength of elementary oxygen decreases as one moves from acidic to neutral to basic solution (Example 22.2). In basic solution, the half-equation for the reduction of oxygen is more properly written as

$$\tfrac{1}{2}\ O_2(g) + H_2O + 2\ e^- \longrightarrow 2\ OH^- \qquad SRP = +0.40\ V$$

EXAMPLE 22.2. Calculate the reduction potential of oxygen, at 1 atm pressure, in

    a. Neutral solution, i.e., $[H^+] = 10^{-7}$ M

    b. A solution 1 M in OH^-, $[H^+] = 10^{-14}$ M

SOLUTION. Here, as in Example 22.1, we can apply the Nernst equation to determine the effect of the concentration of H^+ on the potential for the half-reaction

$$\tfrac{1}{2} O_2(g) + 2\ H^+ + 2\ e^- \longrightarrow H_2O \qquad SRP = +1.23\ V$$

$$RP = +1.23\ V - \frac{0.059}{n} \log_{10} \frac{1}{[H^+]^2(P_{O_2})^{\frac{1}{2}}}$$

Noting that $n = 2$, $P_{O_2} = 1$ atm we have

$$RP = +1.23\ V - \frac{0.059}{2} \log_{10} \frac{1}{[H^+]^2}$$

Simplifying, $\qquad RP = +1.23\ V + 0.059 \log_{10} [H^+]$

a. $[H^+] = 10^{-7}\ M \quad RP = +1.23\ V + 0.059(-7)\ V = +0.82\ V$

b. $[H^+] = 10^{-14}\ M \quad RP = +1.23\ V + 0.059(-14)\ V = +0.40\ V$

Note that the potential just calculated is the standard reduction potential for oxygen in basic solution, i.e., for the half-reaction

$$\tfrac{1}{2} O_2\ (g,\ 1\ atm) + H_2O + 2\ e^- \longrightarrow 2\ OH^-(1\ M) \quad SRP = +0.40\ V$$

Can you explain why O_2 does not oxidize Br^- or Cl^- in neutral solution?

Reaction with Ions in Solution

Laboratory reagents are frequently contaminated by products formed from the oxidation of an anion or cation by elementary oxygen. Of the many positive and negative ions which are susceptible to oxidation by dissolved air, we shall consider only four: I^-, S^{2-}, Sn^{2+} and Fe^{2+}.

Iodide Ions. When a solution of hydriodic acid is exposed to air, it slowly takes on a yellow and finally a brown color as a result of the redox reaction,

$$\tfrac{1}{2} O_2(g) + 2\ H^+ + 2\ I^- \longrightarrow H_2O + I_2 \qquad (22.18)$$

$$E° = SRP\ O_2 + SOP\ I^- = +0.70\ V$$

The same reaction occurs, somewhat less readily, in a neutral solution of an alkali metal iodide, such as NaI or KI.

Sulfide Ions. A solution of sodium sulfide in contact with air turns cloudy because of the oxidation of sulfide ions to colloidal sulfur. Since the sodium sulfide solution is strongly basic as a result of hydrolysis, the equations for its reaction with oxygen are best written

oxidation: $\qquad\qquad\qquad S^{2-} \longrightarrow S(s) + 2\ e^- \quad SOP = +0.48\ V$

reduction: $\quad \tfrac{1}{2} O_2(g) + H_2O + 2\ e^- \longrightarrow 2\ OH^- \qquad SRP = +0.40\ V$

$$\tfrac{1}{2} O_2(g) + H_2O + S^{2-} \longrightarrow 2\ OH^- + S(s) \qquad +0.88\ V \quad (22.19)$$

Solutions of hydrogen sulfide undergo a similar reaction, taking on a milky appearance.

Cations. Many transition metal cations in intermediate oxidation states are oxidized by the oxygen of the air. Among these is the Sn^{2+} ion; a freshly prepared solution of tin(II) chloride gradually turns cloudy, due to the formation of a finely divided precipitate of tin(IV) oxide, SnO_2.

$$\tfrac{1}{2} O_2(g) + H_2O + Sn^{2+} \longrightarrow SnO_2(s) + 2 H^+ \qquad (22.20)$$

The Fe^{2+} ion, like the Sn^{2+} ion, is readily oxidized. If one adds hydroxide ions to a solution of an iron(II) salt, the $Fe(OH)_2$ first formed (green) rapidly turns brown as a result of its oxidation to $Fe(OH)_3$.

$$2 Fe(OH)_2(s) + \tfrac{1}{2} O_2(g) + H_2O \longrightarrow 2 Fe(OH)_3(s) \qquad (22.21)$$

Reaction with Metals. Corrosion of Iron

Reaction with oxygen can take place at room temperature if a metal is exposed to a water solution containing dissolved air. The "water solution" may be a body of fresh or salt water or a thin film of water on the surface of a metal exposed to the atmosphere. Attack by oxygen under these conditions is referred to as **corrosion**. The corrosion of iron is a particularly serious problem. It has been estimated that the annual cost to this country of the corrosion of iron and steel exceeds five billion dollars.

In order to understand the mechanism by which iron corrodes, let us consider what happens when a sheet of iron is exposed to a neutral water solution containing an electrolyte such as sodium chloride. The iron tends to oxidize according to the half-reaction

$$Fe(s) \longrightarrow Fe^{2+} + 2 e^- \qquad SOP = +0.44 \text{ V} \qquad (22.22a)$$

For this reaction to take place, some other species must be reduced simultaneously. Quite clearly, the sodium ion cannot pick up electrons (SRP $Na^+ = -2.71$ V). A more reasonable possibility would be the reduction of H^+ ions to elementary hydrogen (SRP $H^+ = 0.00$ V). This does occur in strongly acidic solution, but cannot take place in neutral solution, in which the concentration of H^+ ions is only 10^{-7} M. Instead, oxygen molecules dissolved in the solution are reduced:

$$\tfrac{1}{2} O_2(g) + H_2O + 2 e^- \longrightarrow 2 OH^- \qquad SRP = +0.40 \text{ V} \qquad (22.22b)$$

Adding these two half-equations and noting that iron(II) hydroxide is insoluble in water, we obtain for the primary corrosion reaction,

$$Fe(s) + \tfrac{1}{2} O_2(g) + H_2O \longrightarrow Fe(OH)_2(s) \qquad (22.22)$$

There is a great deal of evidence to indicate that Reaction 22.22 represents the first and most important step in the corrosion of iron or steel. However, as pointed out earlier, iron(II) hydroxide, exposed to air or dissolved oxygen, is further oxidized to iron(III) hydroxide, $Fe(OH)_3$.

The overall reaction for the corrosion process is obtained by multiplying Equation 22.22 (oxidation of iron to $Fe(OH)_2$) by two and adding Equation 22.21 (oxidation of $Fe(OH)_2$ to $Fe(OH)_3$).

$$2 Fe(s) + O_2(g) + 2 H_2O \longrightarrow 2 Fe(OH)_2(s)$$

$$\underline{2 Fe(OH)_2(s) + \tfrac{1}{2} O_2(g) + H_2O \longrightarrow 2 Fe(OH)_3(s)}$$

$$2 Fe(s) + \tfrac{3}{2} O_2(g) + 3 H_2O \longrightarrow 2 Fe(OH)_3(s) \qquad (22.23)$$

The final product, the loose, flaky deposit that we call rust, has the reddish-brown color of iron(III) hydroxide.

One of the most significant clues to the mechanism of the corrosion of iron emerges from the experimental observation that the oxidation half-reaction (22.22a) and the reduction half-reaction (22.22b) do not occur at the same location. If one examines a nail extracted from an old building, it is commonly found that the rust is concentrated near the head of the nail, which has been in contact with moist air. The most serious pitting, often amounting to disintegration, is found along the shank of the nail, which is embedded in the wood. These observations lead us to believe that oxidation (22.22a) is occurring along a surface that may be an inch or more away from the point at which oxygen is being reduced (22.22b).

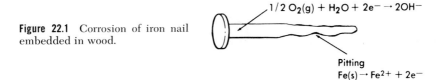

Figure 22.1 Corrosion of iron nail embedded in wood.

$$1/2 \ O_2(g) + H_2O + 2e^- \rightarrow 2OH^-$$

Pitting
$$Fe(s) \rightarrow Fe^{2+} + 2e^-$$

The fact that oxidation and reduction half-reactions take place at different locations suggests that corrosion occurs by an electrochemical mechanism. The surface of a piece of corroding iron may be visualized as consisting of a series of localized voltaic cells. At the anode **(anodic area)**, iron is oxidized to Fe^{2+} ions; at the cathode **(cathodic area)**, elementary oxygen is reduced to OH^- ions. Electrons are transferred through the iron, which acts like the external conductor of an ordinary voltaic cell. The electrical circuit is completed by the flow of ions through the water solution or film covering the iron. The fact that rust ordinarily accumulates at cathodic areas suggests that it is primarily Fe^{2+} ions which move through the solution, from anode to cathode. When these ions arrive at a cathodic area, they are precipitated, first as $Fe(OH)_2$ and ultimately, through further reaction with oxygen, as $Fe(OH)_3$.

Many of the characteristics of corrosion are most readily explained in terms of an electrochemical mechanism. A perfectly dry metal surface is not attacked by oxygen; iron exposed to dry air does not corrode. This seems plausible if corrosion occurs through a voltaic cell, which requires a water solution through which ions can move to complete the circuit. The fact that corrosion occurs more readily in sea water than in fresh water has a similar explanation.* The dissolved salts in sea water supply the ions necessary for the conduction of current.

The existence of discrete cathodic and anodic areas on a piece of corroding iron implies that adjacent surface areas differ chemically from each other. There are several ways in which one small area on a piece of iron or steel can

---

* Certain salts, unlike sodium chloride, inhibit corrosion rather than enhance it. An example is zinc sulfate, $ZnSO_4$. The Zn^{2+} ions react with the OH^- ions produced by Reaction 22.22b to form an adherent, protective film of $Zn(OH)_2$. Salts containing the chromate ion, CrO_4^{2-}, are also effective in inhibiting corrosion. Indeed, the high resistance to corrosion of stainless steel (13 to 27 per cent chromium) is believed to be due in large measure to the formation of a surface layer of metal chromate.

569

become anodic or cathodic with respect to an adjacent area. Two of the most important are:

1. *The presence of impurities at scattered locations along the metal surface.* A tiny crystal of a less active metal such as copper or tin embedded in the surface of the iron acts as a cathode at which oxygen molecules are reduced. The iron atoms in the vicinity of these impurities are anodic and undergo oxidation to Fe^{2+} ions. This effect can be demonstrated on a large scale by immersing in water an iron plate which has been partially copper plated (Figure 22.2). At the interface between the two metals, a voltaic cell is set up in which the iron is anodic and the copper cathodic. A thick deposit of rust forms at the interface. The formation of rust inside an automobile bumper where the chromium plate stops is another example of this phenomenon.

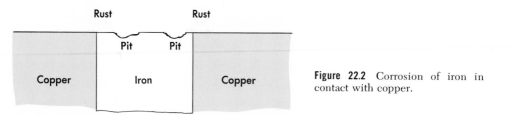

Figure 22.2 Corrosion of iron in contact with copper.

The corrosion resistance of alloy steels depends to a large extent upon their degree of homogeneity. A steel in which the carbon is present in the form of a solid solution has a nearly homogeneous structure and is extremely resistant to corrosion. On the other hand, steels in which the carbon is present in a separate phase as iron carbide, Fe_3C, are sufficiently heterogeneous to corrode readily. The tiny flakes of iron carbide act as cathodic areas; the surrounding phase of pure iron is anodic.

2. *Differences in oxygen concentration along the metal surface.* To illustrate this effect, consider what happens when a drop of water adheres to the surface of a piece of iron exposed to the air (Figure 22.3). The metal around the edges of the drop is in contact with water containing a high concentration of dissolved oxygen. The water touching the metal beneath the center of the drop is depleted in oxygen, since it is cut off from contact with air. As a result, a small oxygen-concentration cell is set up. The area around the edge of the drop, where the oxygen concentration is high, becomes cathodic; oxygen molecules are reduced there. Directly beneath the drop is an anodic area where the iron is oxidized. A particle of dirt on the surface of an iron object can act in much the same way as a drop of water to cut off the supply of oxygen to the area beneath it and thereby establish anodic and cathodic areas. This explains why garden tools left covered with soil are particularly susceptible to corrosion.

Iron or steel objects may be protected from corrosion in either of two ways:

1. The surface may be covered with a protective coating. This may be a layer of paint which cuts off access to moisture and oxygen. Under more severe conditions, it may be desirable to cover the surface of the iron or steel object with a layer of another metal. Metallic plates, applied electrically (Cr, Ni, Cu, Ag, Zn, Sn) or by immersion at high temperatures (Zn, Sn), are ordi-

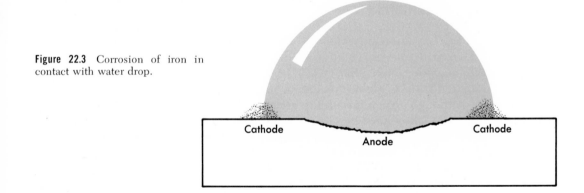

Figure 22.3 Corrosion of iron in contact with water drop.

narily more resistant to heat and chemical attack than the organic coating left when paint dries. However, if the plating metal is less active than iron, there is a danger that cracks in its surface may enhance the corrosion of the iron or steel. This problem can arise with "tin cans," which are made by applying a layer of tin over a steel base. If the food in the can contains citric acid, some of the tin plate may dissolve, exposing the steel beneath.[*] When the can is opened, exposing the interior to the air, rust forms spontaneously on the iron surrounding the breaks in the tin surface. A thin coating of lacquer is ordinarily applied over the tin to prevent corrosive effects of this type.

2. The object may be brought into electrical contact with a bar of a more active metal such as magnesium or zinc. The iron becomes cathodic and hence is protected against rusting; the more active metal serves as a sacrificial anode in a large-scale corrosion cell. This method of combating corrosion, known as *cathodic protection* is particularly useful for steel objects such as cables or pipelines that are buried under soil or water. (Figure 22.4.)

It would hardly be practical to protect a hammer from corrosion this way.

---

[*] Tin forms an extremely stable complex with citrate ion and hence is attacked more readily by citric acid than by many stronger inorganic acids.

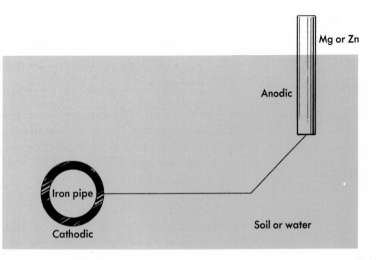

Figure 22.4 Cathodic protection.

22.5   OXYANIONS

Any oxyanion in which the central atom is in a positive oxidation state can, in principle, serve as an oxidizing agent. The four oxyanions which we shall consider (NO_3^-, SO_4^{2-}, $Cr_2O_7^{2-}$, and MnO_4^-) are ones in which the central atom is in its highest oxidation state ($N = +5$, $S = +6$, $Cr = +6$, $Mn = +7$).

Although oxyanions differ greatly from one another in their oxidizing properties, they have certain characteristics in common. In particular:

1. **Oxyanions are stronger oxidizing agents in acidic than in neutral or basic solution.** Thus, we find that whereas concentrated nitric and sulfuric acids are relatively powerful oxidizing agents, neutral salts containing NO_3^- or SO_4^{2-} ions such as KNO_3 or K_2SO_4 are ineffective oxidizing agents in water solution. A solution prepared by adding sulfuric acid to potassium dichromate is frequently used to clean laboratory glassware; it is particularly effective in oxidizing greases and oils which are impervious to a solution containing only $K_2Cr_2O_7$. Finally, it may be mentioned that a strongly acidic solution containing the MnO_4^- ion is a dangerously powerful oxidizing agent; neutral or weakly acidic solutions of potassium permanganate are much safer to work with.

One can readily explain the direct relationship between oxidizing strength and acidity by examining the half-equation for the reduction of an oxyanion. Consider, for example, the NO_3^- ion. The half-reaction for the reduction of NO_3^- to NO_2 involves H^+ as a reactant.

$$NO_3^- + 2\ H^+ + e^- \longrightarrow NO_2(g) + H_2O$$

An increase in the concentration of H^+ ions makes this reaction more spontaneous. In terms of potentials, one can calculate that at a H^+ ion concentration of 1 M, the reduction potential for the foregoing reaction is $+0.78$ V; in neutral solution ($[H^+] = 10^{-7}$ M), it drops to -0.05 V, while in a solution 1 M in OH^- ($[H^+] = 10^{-14}$ M), it is a large negative number, -0.87 V.

These potentials can be calculated from the Nernst equation.

The principle we have used to explain the effect of H^+ ion concentration on the oxidizing strength of the nitrate ion can be applied to other oxyanions as well. Thus, we find that H^+ ions are involved as a reactant in the half-reactions

$$SO_4^{2-} + 4\ H^+ + 2\ e^- \longrightarrow SO_2(g) + 2\ H_2O$$

$$Cr_2O_7^{2-} + 14\ H^+ + 6\ e^- \longrightarrow 2\ Cr^{3+} + 7\ H_2O$$

$$MnO_4^- + 8\ H^+ + 5\ e^- \longrightarrow Mn^{2+} + 4\ H_2O$$

In general, whenever an oxyanion acts as an oxidizing agent, one or more of the oxygen atoms bonded to the central atom is converted to a water molecule. This means that at least two H^+ ions will be required for the reduction of an oxyanion, causing the potential to increase with increasing H^+ ion concentration.

A corollary to this rule is that in preparing oxyanions such as NO_3^-, SO_4^{2-}, $Cr_2O_7^{2-}$, or MnO_4^- from species in which the central atom is in a lower oxidation state, it is ordinarily best to work in a strongly basic medium, in which half-reactions such as those just written are most easily reversed. This principle is applied in the commercial preparation of potassium permanganate from pyrolusite (MnO_2). The first step in the preparation involves fusing

MnO_2 with potassium hydroxide in the presence of air:

$$2\ MnO_2(s) + 4\ OH^- + O_2(g) \longrightarrow 2\ MnO_4^{2-} + 2\ H_2O \qquad (22.24)$$

The product of this reaction is a green salt, potassium manganate, K_2MnO_4, which disproportionates in water to give a solution of potassium permanganate and a precipitate of MnO_2.

$$3\ MnO_4^{2-} + 2\ H_2O \longrightarrow 2\ MnO_4^- + MnO_2(s) + 4\ OH^- \qquad (22.25)$$

2. **Oxyanions can be reduced to a variety of species, depending upon the experimental conditions.** Table 22.1 indicates some of the species to which such ions as NO_3^-, SO_4^{2-}, $Cr_2O_7^{2-}$, and MnO_4^- can be reduced when they act as oxidizing agents.

TABLE 22.1 OXIDATION STATES OF N, S, Cr, Mn

| | NITROGEN | | | SULFUR | |
	Acidic Solution	Basic Solution		Acidic Solution	Basic Solution
+5	NO_3^-	NO_3^-	+6	SO_4^{2-}, HSO_4^-	SO_4^{2-}
+4	$NO_2(g)$	$NO_2(g)$	+4	$SO_2(g)$, H_2SO_3	SO_3^{2-}
+3	HNO_2	NO_2^-	+2	–	$S_2O_3^{2-}$
+2	$NO(g)$	$NO(g)$	0	$S(s)$	$S(s)$
+1	$N_2O(g)$	$N_2O(g)$	−2	$H_2S(g)$	S^{2-}
0	$N_2(g)$	$N_2(g)$			
−1	NH_3OH^+	NH_2OH			
−2	$N_2H_5^+$	N_2H_4			
−3	NH_4^+	NH_3			

| | CHROMIUM | | | MANGANESE | |
	Acidic Solution	Basic Solution		Acidic Solution	Basic Solution
+6	$Cr_2O_7^{2-}$	CrO_4^{2-}	+7	MnO_4^-	MnO_4^-
+3	$Cr(H_2O)_6^{3+}$	$Cr(H_2O)_2(OH)_4^-$	+6	–	MnO_4^{2-}
+2	$Cr(H_2O)_6^{2+}$	$Cr(OH)_2(s)$	+4	$MnO_2(s)$	$MnO_2(s)$
			+2	$Mn(H_2O)_6^{2+}$	$Mn(OH)_2(s)$

It may be noted from Table 22.1 that the stable species in a given oxidation state often depends upon the acidity or basicity of the solution. In the majority of cases, what is involved is an equilibrium between a weak base and its conjugate acid. Consider, for example, the +3 state of nitrogen. The nitrite ion, NO_2^-, which is stable in basic solution, acquires a proton in strongly acidic solution to form a molecule of the weak acid HNO_2. An analogous situation applies to the −3, −2, and −1 states of nitrogen. Molecules of the weak bases ammonia (NH_3), hydrazine (N_2H_4), and hydroxylamine (NH_2OH) are capable of adding a proton in acidic solution to form the ions NH_4^+, $N_2H_5^+$, and NH_3OH^+ respectively.

The existence of different species in acidic and basic solutions of chromium(III) may be explained by the tendency for OH^- ions to replace H_2O molecules in the coordination sphere as the solution becomes more basic. What is actually involved here is a complex equilibrium with a series of species such as $Cr(H_2O)_6^{3+}$ (strongly acidic solution), $Cr(H_2O)_5(OH)^{2+}$ (weakly

573

acidic solution), $Cr(H_2O)_3(OH)_3$ (weakly basic solution), and $Cr(H_2O)_2(OH)_4^-$ (strongly basic solution).

Chromium in the $+6$ oxidation state forms two different oxyanions, the chromate ion, CrO_4^{2-}, stable in basic solution, and the dichromate ion, $Cr_2O_7^{2-}$, stable in acidic solution. If a water solution of a metal chromate is acidified, the yellow color of the CrO_4^{2-} ion changes to the red color of the dichromate ion:

This is not a redox reaction.

$$2\ CrO_4^{2-} + 2\ H^+ \rightleftharpoons Cr_2O_7^{2-} + H_2O \qquad \text{(22.26)}$$
yellow   red

This reaction is readily reversed by adding base. At a pH lower than about 7, the $Cr_2O_7^{2-}$ ion is the principal species present; at a higher pH the CrO_4^{2-} ion predominates.

Of the oxidation states listed in Table 22.1, it will be noted that in two cases ($+6$ manganese, $+2$ sulfur) there is no stable species in acidic solution. In both instances, the ion which is stable in basic solution disproportionates in acid. The reaction of the manganate ion (MnO_4^{2-}) with water was mentioned earlier (Equation 22.25). The thiosulfate ion ($S_2O_3^{2-}$) is considerably more stable; solutions prepared by dissolving in water such salts as sodium thiosulfate, $Na_2S_2O_3$, show little if any tendency to decompose. However, the addition of a strong acid to such solutions brings about the spontaneous redox reaction

$$S_2O_3^{2-} + 2\ H^+ \longrightarrow S(s) + SO_2(g) + H_2O \qquad E° = +0.10\ V \qquad \text{(22.27)}$$

in which half the sulfur atoms are reduced to elementary sulfur ($+2 \rightarrow 0$) and half are oxidized to sulfur dioxide ($+2 \rightarrow +4$). The small positive voltage associated with this reaction suggests that it should readily be reversed in basic solution. Indeed, sodium thiosulfate is prepared commercially by bubbling sulfur dioxide through a suspension of sulfur in concentrated sodium hydroxide.

In practice, the problem of deciding what species an oxyanion such as NO_3^- will be reduced to is not as difficult as the large number of possible reduction products might imply. The mechanisms by which redox reactions occur are poorly understood, but it is generally agreed that reduction and oxidation ordinarily take place in a series of one-electron steps. Consequently, it would be expected that as one moves down the oxidation scale, the first stable species formed would be the principal product. In the case of the NO_3^- ion (oxid. state $N = +5$), this would lead us to predict that the NO_2 molecule (oxid. state $N = +4$) would be a likely reduction product. This is indeed the case; brown fumes of nitrogen dioxide are almost invariably observed in reactions in which concentrated nitric acid is used as an oxidizing agent. In more dilute solutions (low H^+ ion concentrations), some of the NO_2 disproportionates to NO and HNO_3,

$$3\ NO_2(g) + H_2O \longrightarrow NO(g) + 2\ H^+ + 2\ NO_3^- \qquad \text{(22.28)}$$

and significant quantities of nitric oxide can be detected among the reduction products.

It is found experimentally that the species to which an oxyanion is reduced in a redox reaction depends to some extent upon the strength of the reducing agent used. Consider, for example, the reaction of concentrated sulfuric acid with bromide ions. If one allows sulfuric acid to drop on solid sodium bromide, the reaction

Why use H_3PO_4 rather than H_2SO_4 to make HBr from NaBr?

$$SO_4^{2-} + 4\ H^+ + 2\ Br^- \longrightarrow SO_2(g) + Br_2 + 2\ H_2O \qquad \text{(22.29)}$$

occurs. Here, the only sulfur-containing species produced in significant amounts is sulfur dioxide (oxid. state S $= +4$). On the other hand, the addition of sulfuric acid to sodium iodide gives detectable quantities of sulfur (oxid. state S $= 0$) and hydrogen sulfide (oxid. state S $= -2$). One might explain this difference in behavior of Br^- and I^- ions in terms of the greater reducing strength of the iodide ion (SOP $= -0.53$ V) as compared to the bromide ion (SOP $= -1.07$ V); the I^- ion should be capable of reducing any SO_2 formed to lower oxidation states such as S and H_2S. It seems quite likely, however, that the difference between the two reactions should be ascribed to a difference in mechanism and that any argument involving electrode potentials may be of dubious validity.

Having considered some of the general principles governing the use of oxyanions as oxidizing agents, we shall now describe some of the more important redox reactions involving these ions.

Reactions of Metals with Nitric Acid

When a metal reacts with nitric acid, the products differ markedly from those obtained with dilute HCl or H_2SO_4. Unless the acid is very dilute, it is the NO_3^- ion rather than the H^+ ion which is reduced; no hydrogen is evolved. Since the nitrate ion in acidic solution is a much stronger oxidizing agent than the H^+ ion, many metals, including copper and silver, which do not react with dilute hydrochloric or sulfuric acids are brought into solution by nitric acid.

The species to which nitric acid is reduced by a metal depends on the concentration of the acid and the activity of the metal. With the very unreactive metal silver, nitrogen dioxide is the principal product:

$$Ag(s) + NO_3^- + 2\ H^+ \longrightarrow Ag^+ + NO_2(g) + H_2O \qquad \text{(22.30)}$$

Copper reacts with concentrated nitric acid (16 M) to give nitrogen dioxide:

$$Cu(s) + 2\ NO_3^- + 4\ H^+ \longrightarrow Cu^{2+} + 2\ NO_2(g) + 2\ H_2O \qquad \text{(22.31)}$$

Dilute acid (6 M) gives nitric oxide as a principal product:

$$3\ Cu(s) + 2\ NO_3^- + 8\ H^+ \longrightarrow 3\ Cu^{2+} + 2\ NO(g) + 4\ H_2O \qquad \text{(22.32)}$$

Zinc, a strong reducing agent, may give any of a series of reduction products depending upon the concentration of the acid used:

$$Zn(s) + 2\ NO_3^- +\ \ 4\ H^+ \longrightarrow\ \ Zn^{2+} + 2\ NO_2(g) + 2\ H_2O$$

$$2\ Zn(s) + 2\ NO_3^- +\ \ 6\ H^+ \longrightarrow 2\ Zn^{2+} + 2\ HNO_2\ \ + 2\ H_2O$$

$$3\ Zn(s) + 2\ NO_3^- +\ \ 8\ H^+ \longrightarrow 3\ Zn^{2+} + 2\ NO(g)\ \ + 4\ H_2O$$

$$4\ Zn(s) + 2\ NO_3^- + 10\ H^+ \longrightarrow 4\ Zn^{2+} + N_2O(g)\ \ + 5\ H_2O$$

$$5\ Zn(s) + 2\ NO_3^- + 12\ H^+ \longrightarrow 5\ Zn^{2+} + N_2(g)\ \ \ \ + 6\ H_2O$$

$$8\ Zn(s) + 2\ NO_3^- + 20\ H^+ \longrightarrow 8\ Zn^{2+} + 2\ NH_4^+\ \ + 6\ H_2O$$

incr. conc. HNO_3

Curiously enough, concentrated nitric acid fails to react with certain active metals including aluminum and chromium, which are readily attacked by dilute hydrochloric or sulfuric acid. The inertness of these metals toward nitric acid is poorly understood; it has been suggested that an insoluble oxide film forms on the metal surface, thereby preventing further reaction. In support of this idea, it is found that the addition of a few drops of hydrofluoric acid, in which Al_2O_3 and Cr_2O_3 are soluble, enables the oxidation of aluminum or chromium by nitric acid to proceed smoothly.

A few very inactive metals including platinum and gold are not attacked by nitric acid. Both of these metals are dissolved by **aqua regia**, a mixture of concentrated nitric and hydrochloric acids. The oxidizing agent here is the nitrate ion; the hydrochloric acid serves as a source of chloride ions to form the stable complexes $AuCl_4^-$ and $PtCl_6^{2-}$:

$$Au(s) + 3\ NO_3^- + 6\ H^+ \longrightarrow Au^{3+} + 3\ NO_2(g) + 3\ H_2O$$

$$Au^{3+} + 4\ Cl^- \longrightarrow AuCl_4^-$$

$$\overline{Au(s) + 3\ NO_3^- + 6\ H^+ + 4\ Cl^- \longrightarrow AuCl_4^- + 3\ NO_2(g) + 3\ H_2O} \quad \textbf{(22.33)}$$

Use of MnO$_4^-$ in Volumetric Analysis

A species that is readily oxidized can be determined quantitatively by titration with an oxidizing agent in much the same way that a base is titrated with an acid. An oxidizing agent which is frequently used in redox titrations is potassium permanganate, $KMnO_4$.

To illustrate the use of MnO_4^- ion as an oxidizing agent in volumetric analysis, let us consider a specific redox titration, the determination of Fe^{2+} ions with MnO_4^-.

$$MnO_4^- + 8\ H^+ + 5\ Fe^{2+} \longrightarrow Mn^{2+} + 4\ H_2O + 5\ Fe^{3+} \quad \textbf{(22.34)}$$

$$E° = SRP\ MnO_4^- + SOP\ Fe^{2+} = (+1.52 - 0.77)\ V = 0.75\ V$$

The large positive E° value for this reaction means that the equilibrium constant is large enough ($K = 10^{64}$) to make the reaction go essentially to completion. What one does in the titration is to start with a known volume of an acidified solution containing Fe^{2+} ions and add from a buret a solution of potassium permanganate of known concentration. At the instant the MnO_4^- ions are added, the solution takes on the pink or purple color characteristic of that ion. As the MnO_4^- ions are used up by Reaction 22.34, the color fades almost immediately. However, when all the Fe^{2+} ions have been titrated, i.e., at the equivalence point, the addition of one or two drops of excess MnO_4^- produces a permanent pink color. The volume of titrant necessary to reach this end point is recorded and the concentration of Fe^{2+} ions calculated as indicated in Example 22.3.

EXAMPLE 22.3. A 20.0 ml sample containing Fe^{2+} ions requires 18.0 ml of 0.100 M $KMnO_4$ solution for complete reaction. Calculate the concentration of Fe^{2+} ions in the solution.

SOLUTION. Let us first calculate the number of moles of MnO_4^- added. Then, using Equation 22.34, we can calculate the number of moles of Fe^{2+} ion in the sample. Finally, knowing the volume of the sample, we can calculate the concentration of Fe^{2+}.

$$\text{no. moles MnO}_4^- = 0.100 \, \frac{\text{mole}}{\text{lit}} \times 0.0180 \, \text{lit} = 0.00180 \, \text{mole}$$

According to Equation 22.34,

$$1 \text{ mole MnO}_4^- \simeq 5 \text{ moles Fe}^{2+}$$

Hence,

$$\text{no. moles Fe}^{2+} = 0.00180 \text{ mole MnO}_4^- \times \frac{5 \text{ moles Fe}^{2+}}{1 \text{ mole MnO}_4^-} = 0.00900 \text{ mole Fe}^{2+}$$

$$\text{conc. Fe}^{2+} = \frac{0.00900 \text{ mole}}{0.0200 \text{ lit}} = 0.450 \text{ M}$$

The MnO_4^- ion is a powerful enough oxidizing agent to react with a wide variety of oxidizable substances. Thus, potassium permanganate can be used to titrate such species as I^-, NO_2^-, and $C_2O_4^{2-}$ ions.

$$MnO_4^- + 8 \text{ H}^+ + 5 \text{ I}^- \longrightarrow Mn^{2+} + 4 \text{ H}_2O + \tfrac{5}{2} \text{ I}_2 \qquad \textbf{(22.35)}$$

$$2 \text{ MnO}_4^- + 6 \text{ H}^+ + 5 \text{ NO}_2^- \longrightarrow 2 \text{ Mn}^{2+} + 3 \text{ H}_2O + 5 \text{ NO}_3^- \qquad \textbf{(22.36)}$$

$$2 \text{ MnO}_4^- + 16 \text{ H}^+ + 5 \text{ C}_2O_4^{2-} \longrightarrow 2 \text{ Mn}^{2+} + 8 \text{ H}_2O + 10 \text{ CO}_2(g) \qquad \textbf{(22.37)}$$

Reaction 22.37 is particularly useful, since it can be adapted to the determination of metals that form insoluble oxalates. The concentration of Ca^{2+} ions in a solution can be determined by adding excess $C_2O_4^{2-}$ to precipitate calcium oxalate, CaC_2O_4, and then titrating this precipitate with a standard solution of $KMnO_4$.

See Problem 22.14.

Use of Oxyanions in Qualitative Analysis

Many of the reactions involved in the standard schemes of cation and anion analysis are of the oxidation-reduction type. Oxidizing agents are commonly used in qualitative analysis for three different purposes:

1. **To bring a sample into solution so it can be analyzed.** Solid samples which are insoluble in both water and dilute nonoxidizing acids can frequently be brought into solution by an oxidizing agent. Concentrated nitric acid is most often used for this purpose; it reacts with inactive metals such as silver or copper (Equations 22.30, 22.31) or insoluble metal sulfides such as CuS.

$$CuS(s) + 2 \text{ NO}_3^- + 4 \text{ H}^+ \longrightarrow Cu^{2+} + S(s) + 2 \text{ NO}_2(g) + 2 \text{ H}_2O \qquad \textbf{(22.38)}$$

to form a solution containing the corresponding metal ion.

It should be emphasized that a solution prepared by treating a sample with concentrated nitric acid or other powerful oxidizing agent can hardly be expected to contain readily oxidizable ions such as Fe^{2+} or Sn^{2+}. In the presence of nitric acid, for example, Fe^{2+} ions are spontaneously oxidized to Fe^{3+}:

$$Fe^{2+} + NO_3^- + 2 \text{ H}^+ \longrightarrow Fe^{3+} + NO_2(g) + H_2O \qquad E° = +0.04 \text{ V} \qquad \textbf{(22.39)}$$

2. **To test for the presence of a particular ion.** Chromium(III) salts are ordi-

narily detected by oxidizing them to chromium(VI) compounds. One way to do this is to use sodium peroxide, Na_2O_2, as an oxidizing agent in basic solution.

$$Cr(OH)_3(s) + \tfrac{3}{2}\,O_2{}^{2-} \longrightarrow CrO_4{}^{2-} + OH^- + H_2O \qquad (22.40)$$

This reaction is often used in qualitative analysis to test for the presence of Cr^{3+}; the existence of $CrO_4{}^{2-}$ ions in the final solution may be demonstrated either by acidifying to obtain the orange color of the $Cr_2O_7{}^{2-}$ ion* or by adding Ba^{2+} to give yellow, insoluble barium chromate, $BaCrO_4$.

3. **To separate ions from one another.** Mercuric sulfide, because of its very low solubility in water, is not as readily oxidized by nitric acid as are the sulfides of the other ions in group 2 (Cu^{2+}, Bi^{3+}, Cd^{2+}, and so on). Advantage is taken of this difference in separating the Hg^{2+} ion from the other ions of this group. The sulfide precipitate obtained by saturating the solution of the unknown with H_2S is heated with moderately dilute nitric acid. All of the sulfides except HgS react under these conditions (cf. Equation 22.38); the residue consists of a mixture of mercuric sulfide and sulfur.

Would NiS dissolve in HNO_3? (K_{sp} NiS > K_{sp} CuS).

22.6 THE PHOTOGRAPHIC PROCESS

A commercial process which involves, at a critical stage, an oxidation-reduction reaction in water solution, is photography. We shall consider the chemistry of the three steps by which a negative is produced on a photographic film: exposure, development, and fixing.

Exposure

The thin, light-sensitive layer which coats a photographic film or plate consists of an emulsion of a silver halide, usually AgBr, in gelatin. The silver bromide is dispersed in the gelatin as tiny crystals or **grains**, which are clearly visible under the microscope. The grains range in diameter from 10^{-5} to 10^{-4} cm (1000 to 10,000 Å); on the average, a silver bromide grain is made up of about 10^9 Ag^+ ions and an equal number of Br^- ions.

When a film is exposed momentarily to light, a few of the Ag^+ ions in each grain of silver bromide are reduced to Ag atoms. Aggregates of from 10 to 500 silver atoms form at various points within each grain. The number of aggregates produced depends upon the amount of light striking the film. Under ordinary conditions, the amount of silver constituting the so-called **latent image** is so small as to be invisible under a microscope. Long exposure to bright light reduces all the silver bromide to silver, darkening the entire film.

---

* When an alkaline solution containing $CrO_4{}^{2-}$ and $O_2{}^{2-}$ ions is acidified, there is obtained a fleeting blue color which quickly changes to red. The blue color has been attributed to the triperoxy chromate ion, $CrO_7{}^{2-}$:

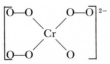

in which three of the four oxide ions of the $CrO_4{}^{2-}$ anion have been replaced by peroxide ions.

Increasing the particle size of the grains facilitates the reaction; high-speed and x-ray film contain comparatively large crystals of silver bromide, approximately 10^{-4} cm in diameter.

Despite a great deal of research devoted to the subject, the precise mechanism of latent image formation is not well-established. Several steps appear to be involved:

1. A photon of light ejects an electron from a bromide ion:

$$Br^- + h\nu \longrightarrow Br + e^-$$

2. Electrons produced in the first step migrate through the crystal until they encounter an **electron-trap**. The simplest type of trap is an anion vacancy, that is, a defect lattice position from which a Br^- ion is missing. Defects of this type are particularly abundant at the surface of the crystal or along the edges of dislocations within the crystal where tiny microcrystals intersect each other. High concentrations of vacant anion sites may also be found in the vicinity of impurities in the AgBr crystal. The substitution of a S^{2-} ion for a Br^- ion requires the simultaneous formation of a Br^- ion vacancy in order to maintain electroneutrality. The catalytic effect on latent image formation of small quantities of silver sulfide, Ag_2S, has been attributed to this phenomenon.

3. Trapped electrons are not free to migrate, so they tend to participate in local chemical reactions. For example, they can reduce local Ag^+ ions to silver. Since Ag^+ ions are much smaller than Br^- ions, they can move around in the lattice with more freedom. Any that reach an electron trap are reduced and the speck of silver grows:

$$Ag^+ + e^- \longrightarrow Ag$$

Clusters of silver atoms formed in this way constitute the latent image.

Development

After a film has been exposed briefly to light, it is treated, in the dark, with a solution of a weak reducing agent, referred to as a developer. The silver bromide grains which have been sensitized by exposure are reduced to metallic silver. The development reaction may be represented as

$$AgBr + e^- \longrightarrow Ag + Br^- \text{ (e}^- \text{ furnished by reducing agent)}$$

The relatively small number of silver atoms constituting the latent image act as a catalyst for this reaction. They facilitate the electron transfer from reducing agent to Ag^+ ion, perhaps by adsorbing from solution the active ingredient of the developing solution.

In regions of the film that were not struck by light during exposure, no silver atoms are available and reaction occurs more slowly. If the film is left in the developing solution too long, the unexposed as well as the sensitized silver bromide is reduced to silver. The same undesirable effect can result from the use of a developer whose reducing strength is too great.

A wide variety of reducing agents may be used as developers. Iron(II) salts, often in the form of complexes, have been used successfully:

579

$$Fe^{2+} \longrightarrow Fe^{3+} + e^-$$

Hydroxylamine, in basic solution, is effective:

$$H_2NOH + OH^- \longrightarrow 2\ H_2O + \tfrac{1}{2}\ N_2(g) + e^-$$

Most commercial developers employ organic reducing agents such as hydroquinone:

$$\underset{\text{hydroquinone}}{C_6H_6O_2} \longrightarrow \underset{\text{quinone}}{C_6H_4O_2} + 2\ H^+ + 2\ e^-$$

Can you suggest another possible reducing agent?

Fixing

After development, a film shows dark areas of metallic silver where it was exposed to light and light areas of unchanged silver bromide in the regions where light did not reach the film. The remaining silver bromide must be removed so that the finished negative will not be light-sensitive. This is accomplished by dipping the negative into a **fixing bath**, a water solution of sodium thiosulfate, $Na_2S_2O_3$. The silver bromide is dissolved by complex formation:

$$AgBr(s) + 2\ S_2O_3{}^{2-} \longrightarrow Ag(S_2O_3)_2{}^{3-} + Br^- \qquad \text{(22.41)}$$

This reaction is, of course, reversible. Care must be taken not to allow the concentrations of products to rise to the point where they reprecipitate silver bromide on the film.

Preparation of the Positive

To prepare a positive print, the negative is placed over a piece of printing paper coated with a photographic emulsion. Light is then passed through the negative to the printing paper. The amount of light reaching the print is inversely related to the thickness of the silver deposit on the negative. Subsequent development and fixing of the print gives a picture in which the light and dark areas of the negative are reversed, giving an accurate reproduction of the article being photographed.

One of the most interesting developments in photography in recent years has been an increased emphasis on processes for preparing a positive print directly without going through a negative. One method of doing this involves as a first step the overexposure of the film. Curiously enough, if photographic film is exposed for longer than usual, a **reverse image** appears on development. The grains of silver bromide which have been overexposed are less readily reduced than those in the underexposed portions of the film. In development, the underexposed silver bromide is reduced to black metallic silver, while the overexposed portions remain white. This effect is enhanced by the use of special developers. In this way, a phenomenon once regarded as a nuisance has been converted into a commercially valuable technique of amateur photography. The mechanism of the reverse image effect is not well understood. It is believed, however, that overexposure liberates bromine, which acts as an oxidizing agent to inhibit the reduction of silver bromide to silver.

PROBLEMS

In answering the following problems, use Table 21.1 to obtain the necessary potentials.

22.1   Give an example of

a. A metal that is a more powerful reducing agent than magnesium.
b. A nonmetal whose strength as an oxidizing agent depends upon the concentration of H^+.
c. An oxidizing agent that can convert Fe^{2+} to Fe^{3+}.
d. A metal that reacts with 1 M HCl but not with water.
e. An ion derived from the element chlorine that cannot act as an oxidizing agent.
f. A reducing agent capable of converting Zn^{2+} to Zn.

22.2   Write balanced net ionic equations for the reaction, if any, of

a. Aluminum with hydrochloric acid.
b. Calcium with water.
c. Copper with dilute sulfuric acid.
d. Nickel with dilute sulfuric acid.

22.3   Describe, in words, how one can prepare

a. Aluminum sulfate from aluminum.
b. Cobalt(II) bromide from cobalt.
c. Silver from silver nitrate.
d. Tin(IV) oxide from tin(II) chloride.
e. Sodium nitrate from sodium nitrite.
f. Calcium hydroxide from calcium.

22.4   Give the formulas of two ions which

a. Are reduced by elementary hydrogen.
b. Are formed by the reaction of metals with water.
c. Are oxidized by chlorine.
d. Can be formed by the electrolysis of a sodium chloride solution.
e. Are oxidized to cations by oxygen.
f. Are oxidized to nonmetallic elements by oxygen.
g. Are oxidized to anions by oxygen.

22.5   Explain why

a. Aluminum is oxidized more readily in basic than in neutral solution.
b. Ni^{2+} is more difficult to reduce in the presence of ammonia.
c. Solutions of $SnCl_2$ and H_2S become cloudy on exposure to air.
d. A solution of $FeCl_2$ can be protected from oxidation by adding iron filings.
e. Solutions of $NaNO_2$ invariably give a test for NO_3^-.
f. The compound $Sn(NO_3)_2$ is unstable.

22.6   Using the Nernst equation, calculate

a. The reduction potential of O_2 in neutral solution.
b. The reduction potential of O_2 in a solution 1 M in OH^-.
c. The reduction potential of $NO_3^- \rightarrow NO(g)$ in neutral solution.
d. The reduction potential of Ag^+ in a solution 1 M in CN^- and $Ag(CN)_2^-$ (cf. Table 19.5).

22.7   Write balanced equations to describe the observations of Problem 22.5.

581

22.8 Describe, in words and by equations, how the following conversions can be accomplished.

a. $Fe^{2+} \longrightarrow Fe^{3+}$ d. $Fe(OH)_2 \longrightarrow Fe(OH)_3$
b. $MnO_4^- \longrightarrow Mn^{2+}$ e. $Cr(OH)_3 \longrightarrow CrO_4^{2-}$
c. $MnO_2 \longrightarrow MnO_4^-$ f. $I^- \longrightarrow I_2$

22.9 Write balanced redox equations for

a. The disproportionation of Cu_2SO_4.
b. The disproportionation of NO_2^- in acidic solution.
c. The disproportionation of Cl_2 in basic solution.

22.10 Explain why

a. It is believed that corrosion proceeds by an electrochemical mechanism.
b. Rust often forms at an area removed from that at which iron is being oxidized.
c. Corrosion can occur in acidic solution in the absence of oxygen.
d. Steel objects corrode very slowly in the desert.
e. Magnesium bars are often driven into the soil in the vicinity of pipelines.
f. Differences in oxygen concentration on an iron surface promote corrosion.

22.11 Explain the following color changes in terms of balanced equations.

a. A solution of NaI turns yellow on standing.
b. A solution of K_2CrO_4 turns red on addition of acid.
c. A solution of $KMnO_4$ fades in color when added to hydrochloric acid.

22.12 Suppose you wished to prepare $K_2Cr_2O_7$ from $CrCl_3$.

a. Suggest a reason for carrying out this oxidation in basic rather than in acidic solution.
b. Referring to the table of oxidation potentials, what would you select as a suitable oxidizing agent?
c. What would be the product of the oxidation in basic solution? How would you convert this to $K_2Cr_2O_7$?
d. On the basis of your answers to a, b, and c, write balanced net ionic equations for each step involved in the preparation of $K_2Cr_2O_7$ from $CrCl_3$.

22.13 A sample containing I^- ions is titrated with $KMnO_4$. It is found that 20.0 ml of 0.100 M $KMnO_4$ is required to react with 2.000 g of the sample. Calculate the per cent of I^- in the sample.

22.14 A sample containing Ca^{2+} is precipitated as CaC_2O_4. It is found that 18.0 ml of 0.110 M $KMnO_4$ is required to react with the calcium oxalate formed from a sample weighing 1.650 g. Calculate the per cent of Ca^{2+} in the sample.

22.15 Write balanced net ionic equations for the reactions, if any, that occur when the following metals are treated with concentrated nitric acid.

a. Ag   b. Cu   c. Zn   d. Al   e. Au

22.16 Using balanced equations, explain how one can use oxidizing agents to separate

a. Bi^{3+} from Hg^{2+} b. Ag from Zn   c. Cr^{3+} from Al^{3+}

22.17 Explain why

a. Exposed AgBr is reduced preferentially by a developer.
b. The developer used in photography must not be too strong a reducing agent.
c. A fixer is used in photography.
d. A fixer solution becomes ineffective on repeated use.
e. A photographic developer should be stored in a closed bottle.

22.18 State whether the following species, when taking part in a redox reaction, are capable of acting *only* as oxidizing agents, *only* as reducing agents, or can act as either oxidizing or reducing agents depending upon the circumstances.

a. Cl_2 b. Cl^- c. ClO_4^- d. N_2 e. NO_3^- f. NH_3

22.19 Give an example of

a. A metal cation which can act only as an oxidizing agent in a redox reaction.
b. A metal cation which can act as either an oxidizing or reducing agent.
c. An oxyanion which can act only as an oxidizing agent in a redox reaction.
d. A nonmetal which can act only as an oxidizing agent in a redox reaction.

22.20 Indicate which of the following metals will react with dilute HCl and which will react directly with water.

a. Mg   b. Au   c. Zn   d. Ba   e. Cu

22.21 Describe, in some detail, a suitable method for preparing the following compounds, starting with the corresponding metals.

a. $ZnCl_2$ b. $Co(NO_3)_2$ c. $MgBr_2$ d. $Ca(OH)_2$ e. $Zn(OH)_2$

22.22 List three metals with which Cu^{2+} ions will react spontaneously.

22.23 In the reaction of excess Sn^{4+} with Cr^{2+},

a. Would you expect to find $Cr_2O_7^{2-}$ among the reaction products?
b. Would you expect to find $Sn(s)$ among the products?
c. Would your answer to a or b differ if Cr^{2+} were in excess?

22.24 Solutions containing Sn^{2+}, on exposure to air, slowly form a precipitate of SnO_2.

a. Write a balanced equation for the redox reaction involved.
b. Explain why this reaction is inhibited by the addition of tin to the solution.

22.25 Which of the following species would you expect to have a pronounced effect on the oxidizing power of the Ag^+ ion? Explain your answers.

a. CN^- b. NH_3 c. NO_3^- d. $S_2O_3^{2-}$ e. H_2O

22.26 Calculate the reduction potential of the Ag^+ ion in a solution 1 M in NH_3 and $Ag(NH_3)_2^+$.

22.27 Explain why

a. Oxygen is a more powerful oxidizing agent in acidic than in neutral solution.
b. Solutions of hydrogen sulfide become cloudy on standing.
c. Solutions of sodium iodide exposed to air slowly develop a yellow color.
d. Solutions of sodium sulfite frequently give a positive test for sulfate ion.

22.28 Explain the following observations regarding corrosion.

a. Corrosion occurs more readily in acidic than in neutral solution.
b. The presence of dissolved salts increases the rate of corrosion.
c. The presence of dissolved air makes corrosion occur more readily.
d. A tin can rusts rapidly when punctured while a piece of galvanized iron does not.
e. A piece of steel in contact with a piece of nickel plated steel rusts rapidly at the point of contact.
f. Aluminum, although more active chemically than iron, does not corrode as rapidly as iron.
g. Corrosion of a cast iron pipe passing through acid soil can be prevented by giving it a slight negative charge.
h. A steel bridge support rusts more rapidly at the water line than at any other point.

583

22.29 Describe, with the aid of balanced equations, how the following compounds can be made from potassium chloride.

a. Cl_2 b. KOH   c. HOCl   d. $KClO_3$ e. $KClO_4$

22.30 Calculate the change in reduction potential caused by increasing the pH by one unit in the following half-reactions.

a. $\frac{1}{2} O_2(g) + 2 H^+ + 2 e^- \longrightarrow H_2O$
b. $NO_3^- + 6 H^+ + 5 e^- \longrightarrow \frac{1}{2} N_2(g) + 3 H_2O$
c. $SO_4^{2-} + 8 H^+ + 6 e^- \longrightarrow S(s) + 4 H_2O$
d. $MnO_4^- + 8 H^+ + 5 e^- \longrightarrow Mn^{2+} + 4 H_2O$

22.31 Explain why

a. A color change occurs when an acidic solution of MnO_4^- is treated with Fe^{2+}.
b. A color change occurs when a solution of K_2CrO_4 is acidified.
c. Silver reacts with nitric but not with sulfuric acid.
d. Sodium thiosulfate decomposes in acidic solution.

Write balanced equations to represent the reactions involved in a through d.

22.32 A sample containing Fe^{2+} ions is titrated with $KMnO_4$. It is found that a 2.000 g sample requires 16.4 ml of 0.100 M $KMnO_4$ for complete reaction. Calculate the percentage of Fe^{2+} in the sample.

22.33 A solution containing $C_2O_4^{2-}$ ions is titrated with $KMnO_4$. It is found that 12.0 ml of 0.124 M $KMnO_4$ is required to react with 26.2 ml of this solution. Calculate the concentration of $C_2O_4^{2-}$ in the solution.

22.34 Suggest how the following pairs of ions might be separated with the aid of an oxidizing agent.

a. I^-, Cl^- b. Hg^{2+}, Cu^{2+} c. Cr^{3+}, Fe^{3+}

22.35 Write balanced equations to represent

a. The reaction which occurs when silver bromide which has been exposed to light is treated with a solution of hydroquinone.
b. The reaction that occurs when silver bromide is treated with a solution of sodium thiosulfate.

°22.36 The gram equivalent weight of an oxidizing agent is defined as the weight that reacts with 1 mole of electrons. Calculate the normality of

a. 0.100 M $K_2Cr_2O_7$ when it is reduced to Cr(III).
b. 0.100 M $KMnO_4$ when it is reduced to Mn^{2+}.
c. 0.100 M $KMnO_4$ when it is reduced to MnO_2.

°22.37 Calculate E° for the reaction represented by Equation 22.22 and compare it to the E° value for the same reaction in acidic solution.

°22.38 Consider the following ions: Ag^+, Fe^{2+}, Sn^{4+}, I^-, MnO_4^-. Which of these ions could not be present simultaneously, at ordinary concentrations, in acidic solution?

°22.39 Calculate the minimum concentration of HNO_3 required to dissolve HgS, which is in equilibrium with 0.1 M Hg^{2+} (K_{sp} HgS $= 1 \times 10^{-52}$). Assume that the products include NO(g) and S(s).

Nuclear Reactions

In Chapter 7 the composition of atomic nuclei was discussed briefly in terms of two fundamental particles, the proton and the neutron. The proton carries a unit positive charge; the neutron has zero charge. Both particles have a mass of approximately one on the atomic weight scale. The charge of a nucleus is determined by the number of protons (the **atomic number**).

The nucleus of a fluorine atom is made up of 9 protons and 10 neutrons, giving it an atomic number of 9 and a mass number of 19. In nuclear symbolism, the atomic number is indicated as a subscript at the lower left of the symbol of the element; the mass number is given as a superscript at the upper left. Thus, the fluorine nucleus is designated as $^{19}_{9}F$. The symbol $^{23}_{11}Na$ is written to represent the nucleus of a sodium atom in which there are 11 protons (atomic number $= 11$) and 12 neutrons (mass number $= 11 + 12 = 23$).

Atoms of a given element can differ from one another in mass number. For example, the nucleus of a "light" hydrogen atom consists of a single proton ($^{1}_{1}H$); a "heavy" hydrogen (deuterium) nucleus contains one proton and one neutron ($^{2}_{1}H$). Species having the same atomic number but different mass number are referred to as **isotopes**. Examples, in addition to $^{1}_{1}H$ and $^{2}_{1}H$, include $^{235}_{92}U$ and $^{238}_{92}U$, isotopes of uranium. Both of these nuclei contain the same number of protons (92) but differ in the number of neutrons (143, 146) and consequently in mass number (235, 238).

Throughout this chapter, we shall discuss several different types of nuclear reactions. It is important to point out that these reactions differ from ordinary chemical reactions in many important respects. In particular:

1. In ordinary reactions, the different isotopes of an element show virtually identical chemical properties; in nuclear reactions they behave quite differently. Consider, for example, the two isotopes of carbon, $^{12}_{6}C$ and $^{14}_{6}C$. The chemical properties of these isotopes are very similar. Their nuclear properties differ considerably; the $^{12}_{6}C$ nucleus is extremely stable while the $^{14}_{6}C$ nucleus decomposes spontaneously.

2. The nuclear reactivity of an element is essentially independent of its state of chemical combination. In the nuclear chemistry of radium, it makes little difference whether we deal with the element itself or one of its compounds. The radium atom in elementary radium and the Ra^{2+} ion in $RaCl_2$ behave similarly from a nuclear standpoint.

Ra salts are used as a radioactive source.

585

In discussing nuclear reactions or writing equations to represent them, we shall not ordinarily be concerned with what happens to the electrons outside the nucleus. Even though the species taking part in these reactions are atoms, molecules, or ions, the reactions themselves occur within the nucleus.

3. Nuclear reactions frequently involve the conversion of one element to another. Whenever a nuclear process results in a change in the number of protons in the nucleus, a new element of different atomic number is formed. In contrast, elements taking part in ordinary chemical reactions retain their identity.

4. Nuclear reactions are accompanied by energy changes which exceed, by several orders of magnitude, those associated with ordinary chemical reactions. The energy evolved when one gram of radium undergoes radioactive decay (Section 23.1) is about 500,000 times as great as that given off when an equal amount of radium reacts with chlorine to form radium chloride. Still larger amounts of energy are given off in nuclear fission (Section 23.4) and nuclear fusion (Section 23.6).

23.1 NATURAL RADIOACTIVITY

The first type of nuclear reaction to be studied was that of natural radioactivity, in which the nucleus of an unstable, naturally occurring isotope spontaneously decomposes. This phenomenon was discovered by a French scientist, Henri Becquerel, as a result of some curious observations encountered in a study of the fluorescence of uranium salts. Becquerel thought that by exposing these salts to sunlight, he might be able to produce high-energy radiation similar to x-rays. In February of 1896, he wrapped several photographic plates with black paper, covered the paper with a thin layer of potassium uranyl sulfate, $K_2UO_2(SO_4)_2$, and exposed the setups to sunlight for a few hours. Upon developing the plates, he found them to be blackened, exactly as he had expected. However, Becquerel soon discovered to his surprise that plates exposed on cloudy days when there was very little sunlight gave images just as intense as those exposed for an equal amount of time to bright sunlight. Further experiments showed that plates covered with uranium salts were darkened even when they were sealed within an opaque cardboard box put away in a closed locker. From these observations, Becquerel concluded that uranium must spontaneously emit a powerful type of radiation whose existence had not previously been observed or even suspected.

Becquerel, in further studies on the radioactivity of uranium salts, was able to show that the rate at which radiation was emitted from a sample was directly proportional to the amount of uranium present. There was one apparent exception to this rule: a certain uranium ore known as pitchblende gave off radiation at a rate nearly four times as great as one would calculate on the basis of its uranium content. In July of 1898, Marie and Pierre Curie, colleagues of Becquerel at the Sorbonne, were able to isolate from a ton of pitchblende ore a fraction of a gram of a new element, which was much more intensely radioactive than uranium. They named this element polonium, after Marie Curie's native country. Six months later, the Curies isolated still another, intensely radioactive, previously unknown element, radium. The Nobel Prize for physics in 1903 was awarded jointly to Henri Becquerel and

Marie and Pierre Curie; eight years later, Madame Curie received an un-precedented second Nobel Prize, this time in chemistry.

Nature of Radiation

The radiation emitted by naturally radioactive elements can be split by an electrical or magnetic field into three distinct parts (Figure 23.1):

1. **Alpha rays**, which consist of a stream of positively charged particles (alpha particles) that carry a charge of +2 and have a mass of 4 on the atomic weight scale. These particles are identical with the nuclei of ordinary helium atoms (at. no. = 2, mass no. = 4).

When an alpha particle is ejected from the nucleus, there is a decrease of two units in atomic number and a decrease of four in mass number. For example, the loss of an alpha particle by the nucleus of an ordinary uranium atom (at. no. 92, mass no. 238) gives an isotope of thorium with an atomic number of 90 and a mass number of 234. This nuclear reaction may be represented by the equation

$$^{238}_{92}\text{U} \longrightarrow {}^{4}_{2}\text{He} + {}^{234}_{90}\text{Th} \tag{23.1}$$

Note that here, as in all nuclear equations, there is a balance of both atomic number ($90 + 2 = 92$) and mass number ($4 + 234 = 238$) on the two sides.

2. **Beta rays**, which are made up of a stream of negatively charged particles (beta particles) that have all the properties of electrons. The ejection of a beta particle (mass $\cong 0$, charge $= -1$) results from the transformation of a neutron (mass = 1, charge = 0) at the surface of the nucleus into a proton

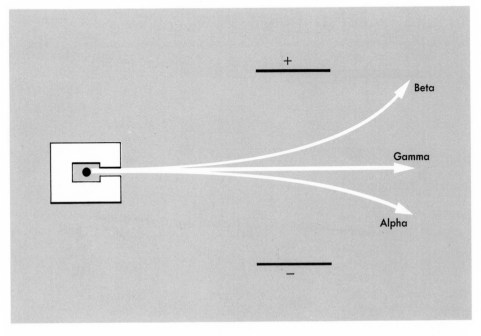

Why are β rays bent more than α rays?

Figure 23.1 Deflection in an electric field of rays emitted by naturally radioactive materials.

(mass $= 1$, charge $= +1$). Consequently, the emission of a beta particle leaves the mass number unchanged but increases the atomic number by one unit. An example of beta-emission is the spontaneous radioactive decay of thorium-234 (90 protons, 144 neutrons) to protactinium-234 (91 protons, 143 neutrons):

$$^{234}_{90}\text{Th} \longrightarrow {}^{\ 0}_{-1}\text{e} + {}^{234}_{91}\text{Pa} \tag{23.2}$$

The symbol $_{-1}^{\ 0}\text{e}$ is written to represent a beta particle (electron).

3. **Gamma rays**, which consist of electromagnetic radiation of very short wavelength ($\lambda = 0.005$ to 1 Å), i.e., high-energy photons. The emission of gamma rays accompanies virtually all nuclear reactions, as the result of an energy change within the nucleus, whereby an unstable, excited nucleus resulting from alpha- or beta-emission gives off a photon and drops to a lower, more stable energy state. Since gamma-emission changes neither the atomic number nor the mass number, we shall frequently neglect it in writing nuclear equations.

Radioactive Series

Natural radioactivity produces many isotopes which are themselves unstable and undergo further decay. To illustrate, consider what happens when uranium-238 decays (Equation 23.1): the thorium-234 isotope produced by this reaction is itself radioactive, decaying via Equation 23.2 to protactinium-234. This isotope, like those which preceded it, is unstable, spontaneously decaying by beta-emission:

$$^{234}_{91}\text{Pa} \longrightarrow {}^{\ 0}_{-1}\text{e} + {}^{234}_{92}\text{U} \tag{23.3}$$

TABLE 23.1   URANIUM-238 RADIOACTIVE SERIES°

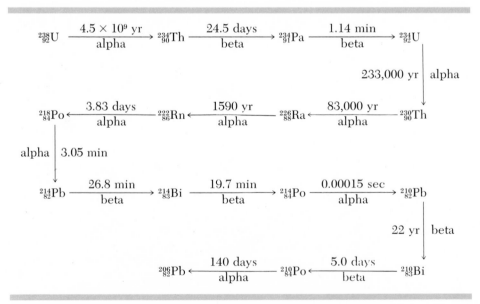

° The times listed in this table are half-lives, discussed in Section 23.3.

The product of this reaction, $^{234}_{92}U$, again decomposes, this time by alpha-emission:

$$^{234}_{92}U \longrightarrow ^4_2He + ^{230}_{90}Th \qquad (23.4)$$

The successive decompositions represented by Equations 23.1 to 23.4 continue until one finally arrives at a stable isotope, $^{206}_{82}Pb$. A total of 14 separate steps is involved in this series of reactions; eight of these occur by alpha-emission, six by beta-emission. The isotopes in this chain (Table 23.1) compose what is known as a **radioactive series**. For the overall process, one can write the net nuclear equation:

$$^{238}_{92}U \longrightarrow ^{206}_{82}Pb + 8\ ^4_2He + 6\ _{-1}^0e \qquad (23.5)$$

The natural radioactive series just described was the first to be discovered; two other series were later found. One of these starts with the less abundant isotope of uranium, $^{235}_{92}U$, and ends with another stable lead isotope, $^{207}_{82}Pb$. The other starts with $^{232}_{90}Th$ and ends with $^{208}_{82}Pb$.

When $^{238}_{92}U$ goes to $^{206}_{82}Pb$, how many α particles are lost? β particles?

Interaction of Radiation with Matter

The alpha, beta, and gamma rays given off during radioactive decay lose their energy in passing through matter by transferring it to the atoms, molecules, or ions with which they collide. These collisions may be **elastic**, in the sense that only kinetic energy is transferred, thereby increasing the temperature of the exposed material. The increased kinetic energy of the target particles is eventually translated into heat, which appears in the surroundings.

More frequently, the interaction of high energy radiation with matter leads to **inelastic** collisions, in which the potential energy of the target particle is raised. A common type of inelastic collision is one in which an electron is excited to a higher energy level. When the electron drops back to a lower level, energy is evolved in the form of electromagnetic radiation which may be visible light (λ = 3000 to 8000 Å), ultraviolet light (λ = 100 to 3000 Å), or x-rays (λ = 0.05 to 100 Å). Radium salts give off an intense luminescence, which is easily visible in the dark and can even be detected in broad daylight with a sample containing more than 0.1 g of radium. The dials of luminous watches and many other instruments are painted with a mixture containing a tiny amount of a radium salt and a substance such as zinc sulfide or anthracene, which fluoresces on exposure to radiation.

Perhaps the most important type of inelastic collision is one in which an electron is completely removed from an atom or molecule to form a positively charged ion. Alpha, beta, and gamma rays are all capable of producing ionization. This characteristic was discovered by Becquerel, who found that an electroscope could be discharged by bringing it near a radioactive sample. Many of the instruments used to detect and study radiation utilize its ability to produce ionization in gaseous atoms or molecules. In the Wilson cloud chamber (Figure 23.2), the ions produced serve as nuclei upon which tiny water droplets condense. The path of an alpha or beta particle through the chamber can be observed by following the condensation trail produced in the vapor. The Geiger-Müller counter (Figure 23.3) amplifies the electric current produced by a cascading flow of electrons and positive ions to oppositely charged electrodes. The current pulse resulting from each ionization

589

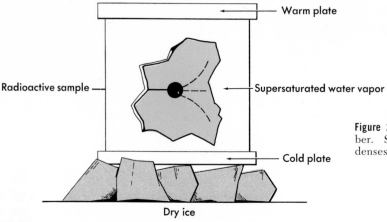

Figure 23.2 Diffusion cloud chamber. Supersaturated vapor condenses on ions formed by radiation.

Figure 23.3 Geiger-Müller counter. Ions produced by radiation cause electrical discharge.

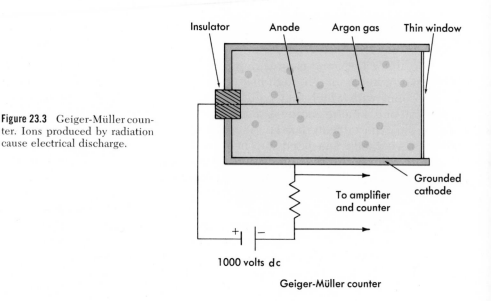

is amplified and detected by means of an electrically activated counting device.

The harmful effect of high-energy radiation on human tissue is caused by its ability to ionize the organic molecules of which body cells are composed. Table 23.2 lists some of the effects to be expected when a human being is exposed to a single dose of radiation of successively higher energies. Exposure to about 100 rads (a rad corresponds to the absorption of 100 ergs of energy per gram of tissue) brings about the typical symptoms of radiation sickness. Doses of 500 rads or more are almost certain to result in death; the destruction of white blood cells is virtually complete and the body becomes highly susceptible to bacterial infection. At this level, death usually occurs within 10 to 15 days after exposure. At 500 to 1500 rads survival time drops to 3 to 5 days; death results from severe damage to the intestinal tract.

Small doses of radiation repeated over long periods of time can also produce extremely serious consequences. Many of the early workers in the field

TABLE 23.2 EFFECT OF EXPOSURE TO A SINGLE DOSE OF RADIATION

DOSE (RADS)	PROBABLE EFFECT
0 to 25	?
25 to 50	Small decrease in white blood cell count
50 to 100	Lesions, marked decrease in white blood cells
100 to 200	Nausea, vomiting, loss of hair
200 to 500	Hemorrhaging, ulcers, possible death
500 +	Fatal

of radioactivity developed cancer as a result of chronic overexposure to radiation. Cases are known in which cancers developed as long as 40 years after initial exposure. Recent studies have shown an abnormally large number of cases of leukemia among the survivors of the nuclear bombs detonated at Hiroshima and Nagasaki.

It has long been known that radiation can produce mutations in plants and animals by bringing about changes in chromosomes. There is every reason to suppose that similar genetic effects can arise in human beings as well. Statistical surveys of the children of American radiologists indicate an increased frequency of congenital defects. This is confirmed by studies which have been made of children born to the survivors of Nagasaki and Hiroshima. Perhaps the most disturbing aspect of this problem is that there appears to be no lower limit or "tolerance level" below which the genetic effects of radiation become negligible. Even a small increase in the background level of radiation can be expected to produce a proportional increase in undesirable mutations. Considerations of this type played a major role in influencing world leaders to ratify the Nuclear Test Ban Treaty of 1963.

23.2 RATE OF RADIOACTIVE DECAY

The rate at which a radioactive sample decays can be measured by counting the number of particles emitted in a given time. Modern measuring instruments perform this counting automatically. In determining the rate at which a particular isotope decays, one must, of course, be careful not to include counts arising from radioactive "daughter" isotopes produced from the parent. Fortunately, it is usually possible to surmount this problem, either by cleverly designing the experiment so that the daughter isotope is removed as it is formed or by modifying the counting device so that it becomes sensitive to only one of the emitted particles.

One of the most striking generalizations that emerges from rate studies with radioactive isotopes is that the rate of decay is completely independent of temperature. From an experimental standpoint, this eliminates any need for temperature control in rate studies. From a theoretical point of view, it implies that the activation energy for radioactive decay is zero.

Logarithmic Rate Law

Radioactive decay is one of the most important examples of the first order rate process discussed in Chapter 14. It will be recalled that for first order reactions, the rate equation is

$$\log_{10} \frac{X_0}{X} = \frac{k\,t}{2.30} \qquad (23.6)$$

where X_0 is the amount of radioactive material at zero time (i.e., when the counting process starts), and X is the amount remaining after time t. The first order rate constant, k, is characteristic of the isotope undergoing radioactive decay.

The application of Equation 23.6 to calculations involving the rate of radioactive decay is illustrated in Example 23.1.

EXAMPLE 23.1. It is found that if one starts with a 1.00 g sample of $^{210}_{83}$Bi, 0.250 g is left after 10.0 days. Calculate

a. The rate constant k for the decay of this isotope.

b. The amount left after one day.

c. The time required for one-half the sample to decay.

SOLUTION

a. From the data given, it is possible to calculate k by direct substitution into Equation 23.6. We see that $X_0 = 1.00$ g, $X = 0.250$ g, $t = 10.0$ days

$$\log_{10} \frac{1.00}{0.250} = \frac{k}{2.30} \,(10.0 \text{ days})$$

$$\log_{10} 4.00 = 0.602 = \frac{k}{2.30} \,(10.0 \text{ days})$$

$$k = \frac{2.30}{10.0 \text{ days}} (0.602) = 0.138 \text{ (day)}^{-1}$$

b. Here $X_0 = 1.00$ g, $t = 1.00$ day, $k = 0.138$ day^{-1}

$$\log_{10} \frac{1.00 \text{ g}}{X} = \frac{0.138}{2.30} \text{ (day)}^{-1} \times 1.00 \text{ day} = 0.0600$$

Taking antilogs, $\frac{1.00 \text{ g}}{X} =$ antilog of $0.0600 = 1.15$

$$X = 1.00 \text{ g}/1.15 = 0.87 \text{ g}$$

c. $X_0 = 1.00$ g, $X = 0.500$ g, $k = 0.138$ day^{-1}, $t = ?$

$$\log_{10} \frac{1.00}{0.500} = \frac{0.138}{2.30} \text{ (day)}^{-1} \times t$$

$$\log_{10} 2.00 = 0.302 = 0.0600 \text{ (day)}^{-1} \times t$$

$$t = \frac{0.302}{0.0600} \text{ day} = 5.00 \text{ days}$$

Half-Life

From Chapter 14 you will recall that the time required for one-half of a sample to decompose via a first order reaction is independent of the amount of sample. In other words, the "half-life" of a first order reaction is character-istic of that particular reaction. Decay rates of radioactive isotope are often expressed in terms of their half-lives (cf. Table 23.1). The half-life of an iso-tope may be extremely long, as is the case with $^{238}_{92}U$ ($t_{\frac{1}{2}} = 4.5 \times 10^9$ years); alternatively, it may be a small fraction of a second, as with $^{214}_{84}Po$ ($t_{\frac{1}{2}} = 0.00015$ sec).

Qualitatively, half-lives can be interpreted in terms of the relative stabili-ties of isotopes. The fact that uranium-238 has a very long half-life means that it is comparatively stable as far as radioactive decay is concerned. It gives off alpha particles very slowly, corresponding to a very low level of radiation whose intensity remains nearly constant with time. Polonium-214, on the other hand, is extremely unstable; half of a sample of this isotope has decom-posed after 0.00015 sec. Within a few seconds, virtually all of the radiation associated with polonium-214 has been dissipated. Isotopes such as polo-nium-214, which have very short half-lives, produce a tremendously high level of radiation during their brief existence.

The most danger-ous isotopes are those with inter-mediate half-lives.

Table 23.3 illustrates how the concept of half-life can be used to estimate the fraction of a radioactive sample remaining after a given number of half-life periods have elapsed.

TABLE 23.3   RATE OF DECAY OF $^{210}_{83}Bi$ ($t_{\frac{1}{2}} = 5$ DAYS)

TIME (DAYS)	NO. OF HALF-LIVES	FRACTION LEFT	FRACTION DECAYED
0	0	1	0
5	1	$\frac{1}{2}$	$\frac{1}{2}$
10	2	$\frac{1}{4}$	$\frac{3}{4}$
15	3	$\frac{1}{8}$	$\frac{7}{8}$
20	4	$\frac{1}{16}$	$\frac{15}{16}$
$\vdots$	$\vdots$	$\vdots$	$\vdots$
5n	n	$(\frac{1}{2})^n$	$1 - (\frac{1}{2})^n$

It is possible to calculate the rate constant for radioactive decay from the half-life, using the equation derived in Chapter 14:

$$k = \frac{0.693}{t_{\frac{1}{2}}} \qquad\qquad (23.7)$$

EXAMPLE 23.2.   Calculate

a. The rate constant for the decay of $^{226}_{88}Ra$, using the data in Table 23.1.

b. The fraction of a sample of radium-226 which has decayed in 100 years.

SOLUTION

a. From Table 23.1 we note that the half-life of this isotope is 1590 years.

$$k = \frac{0.693}{1690 \text{ yrs}} = 4.36 \times 10^{-4} \text{ (yrs)}^{-1}$$

b. Knowing k, we can calculate the fraction of the sample that is left after 100 years. Then, by subtraction, we can obtain the fraction that has decayed.

$$\log_{10} \frac{X_0}{X} = \frac{4.36 \times 10^{-4} \ (yr)^{-1}}{2.30} \times 100 \ yr = 0.0190$$

$$\frac{X_0}{X} = 1.04$$

$$\text{Fraction left} = \frac{X}{X_0} = \frac{1}{1.04} = 0.96; \qquad \text{fraction decayed} = 0.04$$

Age of Rocks

A knowledge of the rate of decay of certain radioactive isotopes makes it possible to estimate the time at which various rock deposits solidified, or, in other words, to estimate their age. To understand how this can be done, consider a uranium-bearing rock, formed billions of years ago at time "zero." The uranium present immediately started to decay, establishing the uranium-238, lead-206 radioactive series. Since the half-lives of all of the intermediate members of that series are very short compared to that of uranium-238 (4.5×10^9 years), virtually all the uranium decaying over a long period of time was converted to the stable, end product of the series, lead-206. It should then be possible, by comparing the quantity of lead-206 produced to the quantity of uranium-238 remaining, to estimate the age of the rock. If, for example, analysis shows that equal numbers of atoms of these two isotopes are present, it follows that one half-life must have passed and that the rock must be 4.5×10^9 (4.5 billion) years old.

See Problems 23.9 and 23.28.

This method of estimating the age of mineral deposits assumes, among other things, that none of the $^{206}_{82}Pb$ has become separated from its parent, $^{238}_{92}U$. If atmospheric weathering or some natural process were to remove one or the other of these elements preferentially, the $^{238}_{92}U/^{206}_{82}Pb$ ratio would, of course, be misleading. One way to correct for such effects is to measure simultaneously the $^{235}_{92}U/^{207}_{82}Pb$ ratio; you may recall that these two isotopes form the beginning and end products of a second radioactive series, with a half-life of 7.1×10^8 years. Although the results of these two methods are not in complete agreement, both indicate the age of the oldest rocks to be in the vicinity of 3.5 billion years. This period of time is often taken as an approximate lower limit for the age of the earth.

Age of Organic Material

During the 1950's, Professor W. F. Libby and others worked out a method based upon the decay rate of a naturally occurring isotope, carbon-14, for determining the age of organic matter. This method can be applied to objects from a few hundred up to 50,000 years old. It has been used, for example, to check the authenticity of canvases of Renaissance painters and to determine the age of relics left by prehistoric cavemen.

Carbon-14 is produced in the atmosphere by the interaction of neutrons from cosmic radiation with ordinary nitrogen atoms:

$$^{14}_{7}N + ^{1}_{0}n \longrightarrow ^{14}_{6}C + ^{1}_{1}H \qquad \text{(23.8)}$$

The carbon-14 produced by this nuclear reaction is eventually incorporated into the carbon dioxide of the air. A steady-state concentration, amounting to about one atom of carbon-14 for every 10^{12} atoms of carbon-12, is established in atmospheric CO_2. A living plant, taking in carbon dioxide, has this same $^{14}C/^{12}C$ ratio in the organic compounds that make up its tissues. By the same token, the $^{14}C/^{12}C$ ratio in plant-eating animals or human beings has this same equilibrium value of about $1/10^{12}$.

This assumes a constant neutron density over a long period.

When a plant or animal dies, the intake of radioactive carbon stops. Consequently, the radioactive decay of carbon-14

$$^{14}_{6}C \longrightarrow ^{14}_{7}N + ^{0}_{-1}e \text{ (half-life = 5760 years)} \qquad \text{(23.9)}$$

takes over and the ratio of $^{14}C/^{12}C$ drops. By measuring this ratio and comparing it to that in living plants, one can estimate the time at which the plant or animal died (Example 23.3).

EXAMPLE 23.3. A piece of wood claimed to have been taken from the cross of Christ, is found to have a $^{14}C/^{12}C$ ratio 0.785 times that in a living plant. Estimate the age of the wood.

SOLUTION. Using the half-life of ^{14}C ($t_{\frac{1}{2}} = 5760$ yrs) we can calculate the first order rate constant from Equation 23.7. Then, using Equation 23.6, we can calculate the elapsed time.

$$k = \frac{0.693}{5760 \text{ yrs}} = 1.20 \times 10^{-4} \text{ (yr)}^{-1}$$

$$\log_{10} \frac{X_0}{X} = \frac{1.20 \times 10^{-4} \text{ (yr)}^{-1} \times t}{2.30}$$

$$\log_{10} \frac{1.000}{0.785} = \log 1.274 = 0.105 = \frac{1.20 \times 10^{-4} \text{ (yr)}^{-1} \times t}{2.30}$$

$$t = \frac{(2.30)(0.105) \text{ yr}}{1.20 \times 10^{-4}} = 2010 \text{ yr}$$

23.3 BOMBARDMENT REACTIONS. ARTIFICIAL RADIOACTIVITY

Prior to 1933, the study of radioactivity was limited to the relatively few radioisotopes which occur in nature. In that year Irène (daughter of Marie and Pierre) Curie and her husband, Frédéric Joliot, discovered the process of inducing artificial radioactivity. By bombarding certain light isotopes with alpha particles, they produced several isotopes of low mass number which were unstable toward radioactive decay. One of the reactions they studied was

This was one of the first neutron sources.

$$^{27}_{13}Al + ^{4}_{2}He \longrightarrow ^{30}_{15}P + ^{1}_{0}n \qquad \text{(23.10)}$$

The product of this nuclear reaction, phosphorus-30, is radioactive. It decays by emitting a particle called a *positron*, which has the same mass as the electron but the opposite charge:

$$\ce{^{30}_{15}P} \longrightarrow \ce{^{30}_{14}Si} + \ce{^{0}_{1}e} \tag{23.11}$$

Hundreds of different radioactive isotopes have now been produced in the laboratory by bombardment reactions. At least one such isotope has been prepared for every element that occurs in nature; in addition, small quantities of isotopes of at least 15 previously unknown elements have been prepared.

Bombarding Particles

A large number of different types of bombardment reactions have been carried out in the laboratory; examples of some of the more important ones are listed in Table 23.4. The particles which are most frequently used include:

Positively charged particles: protons ($\ce{^{1}_{1}H}$), deuterons ($\ce{^{2}_{1}H}$), alpha particles ($\ce{^{4}_{2}He}$), and sometimes $\ce{^{12}_{6}C}$ ions. For a positively charged particle to penetrate an atomic nucleus, which itself carries a positive charge, the particle must be moving at a very high velocity. One can calculate that if protons, deuterons, or alpha particles approaching a nucleus are to overcome the normal coulombic repulsion, they must be accelerated to energies of several million electron-volts. (An electron-volt is the amount of energy acquired by a particle

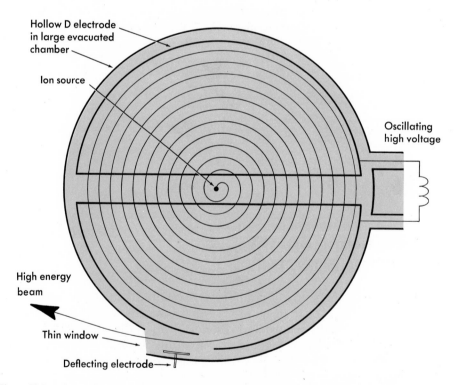

Figure 23.4 The cyclotron consists of two oppositely charged, evacuated "dees" placed between the poles of a powerful electromagnet. Positive ions, originating at the center, enter the upper dee, which is originally at a negative potential. They pass through this dee in a curved path. At the instant they reenter the central corridor, the polarity of the dees is reversed, and the particles enter the lower dee at an increased velocity. This procedure is repeated over and over; the particles move at higher and higher velocities in paths of greater and greater radius. Eventually they are deflected from the periphery of one of the dees to strike the target.

TABLE 23.4 TYPICAL BOMBARDMENT REACTIONS

BOMBARDING PARTICLE	EXPELLED PARTICLE	EXAMPLE
Proton	$_0^1n$, $_2^4He$, γ	$_{15}^{31}P + _1^1H \longrightarrow _{16}^{31}S + _0^1n$
Deuteron	$_1^1H$, $_2^4He$, $_0^1n$, γ	$_{33}^{75}As + _1^2H \longrightarrow _{33}^{76}As + _1^1H$
Alpha	$_0^1n$, $_1^1H$, $_1^2H$	$_9^{19}F + _2^4He \longrightarrow _{11}^{22}Na + _0^1n$
Photon	$_0^1n$, $_1^1H$	$_{35}^{81}Br + \gamma \longrightarrow _{35}^{80}Br + _0^1n$
Neutron	γ, $_2^4He$, $_1^1H$	$_{34}^{82}Se + _0^1n \longrightarrow _{34}^{83}Se + \gamma$

of unit charge passing through a potential gradient of one volt.) Nuclear physicists and engineers have built several different kinds of instruments for this purpose. One of these, the cyclotron, designed by E. O. Lawrence at the University of California, is shown schematically in Figure 23.4.

Photons (gamma or x-rays). The gamma radiation emitted by radioactive isotopes is too low in energy to initiate more than a handful of nuclear reactions. More effective photons can be produced by allowing high-energy electrons, accelerated in an instrument known as a betatron, to impinge on a tungsten target. In this manner, photons with energies as high as 300 million electron-volts can be produced.

Neutrons. Since a neutron experiences no coulombic repulsion when it approaches a nucleus, it need have only a very small kinetic energy to initiate a nuclear reaction. So-called "slow" or "thermal" neutrons, with energies of the order of 0.03 to 0.04 eV, are most effective. Every known nucleus save one, the proton, is unstable toward neutron bombardment.

Transuranium Elements

Bombardment reactions have proved particularly useful in synthesizing isotopes of elements which do not occur in nature. Thirty years ago, the periodic table ended with uranium (at. no. 92). Above uranium, there were four gaps in the table, corresponding to elements 43, 61, 85, and 87. To be sure, various groups of scientists had reported the discovery of these elements and assigned names to them, but none of these claims had been substantiated. Within a period of about five years, between 1937 and 1942, radioactive isotopes of these elements (technetium, at. no. 43; promethium, at. no. 61; astatine, at. no. 85; francium, at. no. 87) were synthesized in the laboratory. The past quarter-century has seen the preparation by Glenn Seaborg and his colleagues at the University of California of a series of transuranium elements (actinides) extending from neptunium (at. no. 93) to lawrencium (at. no. 103).

The first transuranium elements to be discovered, neptunium (at. no. 93) and plutonium (at. no. 94) were prepared by McMillan and Seaborg in 1940 by bombarding uranium-238 with low-energy neutrons:

$$_{92}^{238}U + _0^1n \longrightarrow _{92}^{239}U + \gamma \tag{23.12}$$

The product of this reaction, uranium-239, decays by beta-emission to give an isotope of neptunium,

$$_{92}^{239}U \longrightarrow _{93}^{239}Np + _{-1}^0e \tag{23.13}$$

597

which in turn decomposes, again by beta-emission:

$$^{239}_{93}\text{Np} \longrightarrow {}^{239}_{94}\text{Pu} + {}_{-1}^{0}e \qquad (23.14)$$

In principle, neutron bombardment can be used to prepare a wide variety of transuranium elements. The two elements einsteinium (at. no. 99) and fermium (at. no. = 100) were first discovered in uranium that had been exposed to the very high neutron density which accompanies a thermonuclear explosion (Section 23.4). It is postulated that these isotopes arise from the sudden absorption of a large number of neutrons by uranium-238, followed by successive beta-emissions. The overall processes could be represented as:

$$^{238}_{92}\text{U} + 15\ {}_{0}^{1}\text{n} \longrightarrow [{}^{253}_{92}\text{U}] \longrightarrow {}^{253}_{99}\text{Es} + 7\ {}_{-1}^{0}e \qquad (23.15)$$

and

$$^{238}_{92}\text{U} + 17\ {}_{0}^{1}\text{n} \longrightarrow [{}^{255}_{92}\text{U}] \longrightarrow {}^{255}_{100}\text{Fm} + 8\ {}_{-1}^{0}e \qquad (23.16)$$

In practice, the yield of transuranium isotopes formed by this process falls off exponentially with increasing atomic number. A more practical method of forming isotopes of elements of high atomic number involves bombarding appropriate targets with high-energy, positively charged particles accelerated in a cyclotron. In this way, the element mendelevium has been prepared by bombarding $^{253}_{99}\text{Es}$ with alpha particles:

$$^{253}_{99}\text{Es} + {}_{2}^{4}\text{He} \longrightarrow {}^{256}_{101}\text{Mv} + {}_{0}^{1}\text{n} \qquad (23.17)$$

By using bombarding particles containing a larger number of protons, it is possible to employ target isotopes of lower atomic number, which are more readily available. In this way, element 102 (nobelium) was prepared from curium (at. no. = 96), using carbon-12 nuclei as bombarding particles:

$$^{244}_{96}\text{Cm} + {}_{6}^{12}\text{C} \longrightarrow {}^{254}_{102}\text{No} + 2\ {}_{0}^{1}\text{n} \qquad (23.18)$$

Element 103 was first prepared in 1961 by a similar reaction in which californium (at. no. 98) was bombarded by $^{11}_{5}\text{B}$ nuclei.

With few exceptions, the isotopes of the transuranium elements have very short half-lives. Moreover, most of them, especially those of very high atomic number, have been formed in extremely minute quantities, amounting, in some cases, to only a few atoms. One of the greatest achievements of the scientists working in this field has been their ability to work out methods of studying the chemical and physical properties of submicrogram amounts of these elements. Both chemical and physical evidence indicate that the transuranium elements up to atomic number 103 are filling out a second rare-earth series by completing the 5f subshell.

The prospects for extending the number of transuranium elements beyond those presently known appear much brighter than they did a few years ago. Progress along this line will depend primarily upon the ability to develop more powerful accelerators so that larger and larger bombarding particles can be used. In this way, Dr. Seaborg has suggested, it should be possible to synthesize elements of considerably higher atomic number than those now known.

Decay of Artificially Produced Radioactive Isotopes

Over a thousand radioactive isotopes have been prepared by the various types of bombardment reactions just discussed. Each of these decays to one or another of the approximately 270 stable isotopes of elements ranging from hydrogen (at. no. 1) to bismuth (at. no. 83). The ratio of neutrons to protons required for stability increases with atomic number (Figure 23.5) but in any given region of the periodic table is restricted within very narrow limits. The way in which an artificially radioactive isotope decays depends upon whether its neutron-to-proton ratio is greater or less than that required for stability.

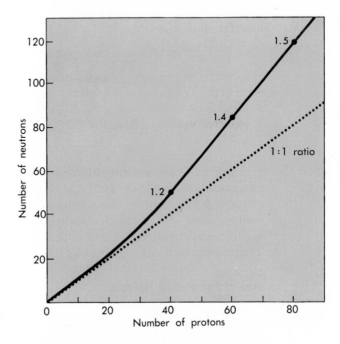

Figure 23.5 Neutron-to-proton ratio in stable isotopes.

Neutron-to-proton ratio too high. An isotope whose nucleus has too many neutrons, i.e., a neutron-to-proton ratio above that required for stability, can become more stable if one of the neutrons in its nucleus decays to a proton. Such a conversion results in the formation of an electron, which is ejected from the nucleus:

$$\text{neutron} \longrightarrow \text{proton} + \text{electron}$$

Beta-emission can be expected to occur whenever the neutron-to-proton ratio is too high; this is almost always the case when the *mass number of the radio-active isotope is greater than the average atomic weight of the element.* An example is the beta-decay of $^{14}_{6}C$ (at. wt. C = 12):

$$^{14}_{6}C \longrightarrow {}^{14}_{7}N + {}_{-1}^{0}e \qquad (23.19)$$

In this type of reaction, beta-emission from the nucleus converts a neutron into a proton, leaving the mass number unchanged but increasing the atomic number by one unit.

Neutron-to-proton ratio too low. A neutron-deficient nucleus tends to change in such a way that a proton is replaced by a neutron. This ordinarily happens with a radioactive isotope whose *mass number is less than the aver-age atomic weight of the element.* This may come about in either of two ways —by emission of a positron or by orbital electron capture.

Neutrons or pro-
tons are seldom
emitted directly.

1. emission of a positron: proton $\longrightarrow$ neutron + positron

This mode of decay is prevalent with the light isotopes of elements of low atomic number. An example is

$$^{13}_{7}N \longrightarrow {}^{13}_{6}C + {}^{0}_{1}e \qquad (23.20)$$

In this reaction, a proton is converted to a neutron, leaving the mass number unchanged but decreasing the atomic number by one unit. The particle emitted, called a **positron**, has a charge opposite to that of an electron but has the same mass.

2. orbital electron capture: proton + electron $\longrightarrow$ neutron

Nuclei of too light isotopes of elements of relatively high atomic number tend to decay by capturing orbital electrons. The electron which falls into the nucleus is ordinarily one in the K shell, closest to the nucleus, hence the name **K electron capture** to describe this type of reaction. An example is

$$^{82}_{37}Rb + {}^{0}_{-1}e \longrightarrow {}^{82}_{36}Kr \qquad (23.21)$$

The nuclear transformation resulting from K-capture is the same as that observed with positron emission; in both cases, a proton in the nucleus is converted to a neutron. In K-capture, as soon as a vacancy is created in the energy level closest to the nucleus, electrons move in from successively higher levels to fill this vacancy. The excess energy involved in this electronic transition is ordinarily given off in the form of x-rays.

Uses of Radioactive Isotopes

The large number of radioactive isotopes which are now available have been used to study a variety of problems in basic and applied research. Techniques involving radioactive isotopes have proved extremely useful in four major areas.

Medicine.   The high energy radiation given off by radium has been used for many years in the treatment of cancer to destroy or arrest the growth of malignant tissue. More recently, a radioactive isotope of cobalt, cobalt-60, which is cheaper than radium and gives off even more powerful radiation, has been used for this purpose. Certain types of cancer can be treated internally with the aid of radioactive isotopes. If a patient suffering from cancer of the thyroid (malignant goiter) drinks a solution of sodium iodide containing radioactive iodine (^{131}I or ^{128}I), the radioactive iodine moves preferentially to the thyroid gland, where the radiation destroys the malignant cells without affecting the rest of the body.

Trace amounts of radioactive samples injected into the blood stream can be used to detect circulatory disorders. For example, by injecting a sodium chloride solution containing a small amount of radioactive sodium into the leg of a patient and measuring the build-up of radiation in the foot, a physician can quickly find out whether or not the circulation in that area is abnormal. This may help him to decide whether amputation is necessary and, if so, where it should be done. It is even possible, by this method, to detect certain types of heart malfunctions.

Industry.   The frictional wear of piston rings can be monitored by making a test ring slightly radioactive by neutron bombardment, and measuring the activity of the iron dust in the lubricating oil that circulates around the piston. The rate of corrosion of steel and the locations at which it is most severe can also be measured by a similar technique.

The thickness of very thin sheets of metal, paper, or plastic can be determined by interposing them between a radioactive source and a detector such as a Geiger counter. If the fraction of the radiation which passes through the sheet is known, it is possible to estimate the thickness of the sheet quite accurately. One can also quickly detect the presence of thin spots where the sheet might break down in use.

Analytical Chemistry. Techniques involving radioactive isotopes have proved extremely useful for determining trace amounts of certain elements. For example, it is possible to determine small amounts of Ag^+ in solution by adding sodium iodide labeled with radioactive iodine. The colloidal silver iodide formed is carried down on the surface of a precipitate such as $Fe(OH)_3$. By measuring the radioactivity of the precipitate, it is possible to calculate how much iodide ion was carried down as AgI, and hence what the concentration of Ag^+ must have been in the solution.

An extremely sensitive analytical technique involves what is known as **activation analysis.** This involves bombarding a sample containing a trace of the element to be analyzed for with particles (usually neutrons) which are known to convert that element to a radioactive isotope. The resulting radioactivity is measured; knowing the efficiency of the bombardment process it is then possible to calculate how much of the element in question was present in the sample. Activation analyses have been worked out for most of the common metals. As little as 10^{-6} g of an element is readily detected by this technique; in the most favorable cases, an impurity present to the extent of 10^{-10} g shows up.

Mechanism Determinations. Organic chemists in particular have made widespread use of radioactive isotopes in unraveling reaction mechanisms. One reaction which has been studied in this manner is the extremely important process of photosynthesis, for which we can write the overall equation

$$6\ CO_2(g) + 6\ H_2O(l) \longrightarrow C_6H_{12}O_6(s) + 6\ O_2(g) \tag{23.22}$$

Quite obviously, this reaction takes place in a series of complicated steps with a great many intermediate products. One of the questions asked by early investigators involved the origin of oxygen gas: Does it come from the CO_2 or the H_2O? This problem was solved when it was discovered from tracer work that any oxygen-18 introduced into water showed up as molecular O_2; carbon dioxide enriched in oxygen-18 gave ordinary O_2. From this evidence, it is clear that the elementary oxygen given off by green plants is formed from water, rather than from carbon dioxide. During the past 30 years, M. Calvin and his co-workers at the University of California at Berkeley have developed a several-step mechanism for the photosynthesis reaction, based in large part on experiments using minute amounts of carbon-14 as a radioactive tracer. In 1961, this work won Calvin the Nobel Prize in chemistry.

23.4 NUCLEAR FISSION

Discovery

Shortly before World War II, several groups of scientists were studying the products obtained by bombarding uranium with neutrons, in hope of discovering new elements. In 1938, two German chemists, Hahn and Strass-man, isolated from the products a compound of a group 2A element, which they originally believed to be radium (at. no. 88). Subsequent work indicated that this element was really barium (at. no. 56). Hahn's first reaction to this discovery was one of disbelief; he later stated, in January of 1939, "As chemists, we should replace the symbol Ra · · · by Ba · · · [but], as nuclear chemists, closely associated with physics, we cannot decide to take this step in contradiction to all previous experience in nuclear physics."

If Hahn was reluctant to admit the possibility of an entirely new type of nuclear reaction, a former colleague of his, Lisa Meitner, was not. In a letter published with O. R. Frisch in January of 1939, she stated: "At first sight, this result seems very hard to understand. . . . On the basis, however, of present ideas about the behavior of heavy nuclei, an entirely different picture of these new disintegration processes suggests itself. . . . It seems possible that the uranium nucleus . . . may, after neutron capture, divide itself into nuclei of roughly equal size." This revolutionary suggestion was quickly substantiated by experiments carried out in laboratories all over the world. The process by which uranium or other heavy elements split under neutron impact into smaller fragments was called **fission**, following a suggestion of Lisa Meitner.

Fissionable Isotopes

Several isotopes of the heavy elements, including platinum, gold, mercury, and lead, are capable of undergoing fission if bombarded by neutrons of sufficiently high energy. In practice, attention has centered upon two particular isotopes, $^{235}_{92}U$ and $^{239}_{94}Pu$, both of which can be split into fragments by low-energy neutrons. These are the two isotopes which have been used in the manufacture of atomic bombs. During World War II, several different processes were worked out for the separation of uranium-235 from the more abundant isotope, uranium-238, which makes up 99.3 per cent of naturally occurring uranium. The most successful separation technique was that of gaseous diffusion, described in Chapter 5. The element plutonium does not occur in nature; the 239-isotope is made from uranium-238 by the sequence of reactions described by Equations 23.12 to 23.14.

Most of the available data on fission reactions has to do with uranium-235. For this reason, our discussion from this point on will concentrate upon the nuclear reactions that take place when a $^{235}_{92}U$ nucleus interacts with a neutron:

$$^{235}_{92}U + ^{1}_{0}n \longrightarrow \text{products} + \text{energy}$$

Fission Products

When a uranium-235 atom undergoes fission, it splits into two unequal fragments and a number of neutrons and beta particles. The fission process is complicated by the fact that different uranium-235 atoms split up in many different ways. For example, while one atom of $^{235}_{92}U$ is splitting to give isotopes of rubidium (at. no. 37) and cesium (at. no. 55), another may break up to give isotopes of bromine (at. no. 35) and lanthanum (at. no. 57), while still another atom yields isotopes of zinc (at. no. 30) and samarium (at. no. 62). The fission of a macroscopic sample of uranium-235, containing billions of billions

The highest yields are of isotopes of masses around 90 and 140.

of atoms, gives a large number of products; at least 200 isotopes of 25 different elements have been identified among the fission products of uranium-235.

It will be recalled from our discussion of artificial radioactivity (Section 23.3) that the stable neutron-to-proton ratio near the middle of the periodic table, where the fission products are located, is considerably smaller (~ 1.2) than that for the very heavy elements such as uranium (~ 1.5). Consequently, the immediate products of the fission process, such as $_{37}^{90}\text{Rb}$ and $_{55}^{144}\text{Cs}$, contain too many neutrons for stability. These isotopes are radioactive, decaying by electron emission. In the case of rubidium-90, three steps are required to reach a stable nucleus:

$$_{37}^{90}\text{Rb} \longrightarrow {}_{38}^{90}\text{Sr} + {}_{-1}^{0}\text{e} \text{ (very short half-life)} \tag{23.23}$$

$$_{38}^{90}\text{Sr} \longrightarrow {}_{39}^{90}\text{Y} + {}_{-1}^{0}\text{e} \text{ (half-life} = 25 \text{ yr)} \tag{23.24}$$

$$_{39}^{90}\text{Y} \longrightarrow {}_{40}^{90}\text{Zr} + {}_{-1}^{0}\text{e} \tag{23.25}$$

The radiation hazard associated with nuclear testing arises from the formation of radioactive isotopes such as these. One of the most dangerous of these isotopes is strontium-90, which, in the form of strontium carbonate, is readily incorporated into the bones of animals and human beings.

Neutron Emission: Nuclear Chain Reactions

Early investigators showed that considerable numbers of neutrons were emitted in the fission process. Indeed, it is found that, on the average, about two to three neutrons are liberated for every neutron absorbed in bringing about fission. Two typical fission reactions might be represented as:

$$_{92}^{235}\text{U} + {}_{0}^{1}\text{n} \begin{cases} \longrightarrow {}_{37}^{90}\text{Rb} + {}_{55}^{144}\text{Cs} + 2\ {}_{0}^{1}\text{n} & (23.26) \\ \longrightarrow {}_{35}^{87}\text{Br} + {}_{57}^{146}\text{La} + 3\ {}_{0}^{1}\text{n} & (23.27) \end{cases}$$

The neutrons which are liberated as the result of fission are capable of causing other uranium-235 nuclei to undergo fission. Since two to three neutrons are evolved for every one absorbed, the necessary condition for a chain reaction exists. Once a few atoms of uranium-235 split up, the neutrons produced can bring about the fission of many more uranium-235 atoms, which in turn yield more neutrons, capable of splitting more uranium atoms, and so on. This is, of course, precisely what happens in the atomic bomb; the energy evolved in successive fissions escalates to give, within a few seconds, a tremendous explosion.

In order to ensure that nuclear fission will occur by a chain reaction, it is important that the uranium-235 sample be large enough so that most of the neutrons will be captured internally. If the sample is too small, too many of the neutrons produced by individual fissions will escape from the surface, thereby breaking the chain. The **critical mass** of uranium-235 required to maintain a nuclear chain reaction has been variously estimated at from 2 to 200 lb; an educated guess would put it in the neighborhood of 20 lb. The problem in designing an atomic bomb is to bring together, at the time of explosion, two samples of subcritical size whose total mass exceeds the critical mass. One way to do this would be to use an ordinary explosive to fire one sample into the other.

Evolution of Energy: Nuclear Reactors

Some idea of the vast amount of energy available from nuclear fission may be obtained by comparing it to the energy evolved in ordinary chemical reactions. The fission of 1 g of uranium-235 evolves about 20,000,000 kcal of energy. The heat of combustion of coal is only about 8 kcal/g; the energy given off when 1 g of TNT explodes is still smaller, about 0.66 kcal. Putting it another way, the fission of 1 g of uranium produces as much energy as the combustion of 5500 lb of coal or the explosion of 33 tons of TNT. The enormous energy change accompanying nuclear fission is directly attributable to the change in mass that takes place in the fission process (Section 23.5).

Even before the first atomic bomb was exploded, scientists and politicians had begun to speculate on the use of nuclear fission as a source of energy for peaceful purposes. Many nuclear reactors, in which the fission reaction is made to occur at a controlled rate, have been designed to meet this need. The simplest type of reactor consists of lumps of uranium separated from each other by blocks of a neutron-moderating material such as graphite. To prevent the fission reaction from getting out of control, reactors are designed so that bars of boron or cadmium, both excellent neutron absorbers, can be inserted into the apparatus to control the neutron flux.

The energy produced by a nuclear reactor is given off primarily as heat. The construction of engines to convert this heat to mechanical or electrical energy poses special problems in engineering and metallurgy. The source of heat must be isolated from the turbogenerator that converts the heat to electricity, so that the machinery will not be exposed to radiation. This is accomplished by interposing heat-transfer loops between reactor and turbine. Despite these difficulties, we have seen in the past decade the development of nuclear ships and submarines and the establishment of many nuclear power plants in various parts of the world.

From a long-range standpoint, it seems unlikely that uranium-235 will, by itself, make an important contribution to our future power needs. It has been estimated that the fission of all the uranium-235 that is known to exist in nature could fulfill our energy needs for one century at the most. A much more attractive possibility is to use uranium-235 as a neutron source to convert the more abundant isotope, uranium-238, to plutonium-239, which can then undergo fission. A still more inviting possibility is the conversion of thorium-232 (which is more abundant than uranium) to uranium-233, by a similar process. So-called breeder reactors based on these principles have been in existence for many years; the problem is to increase the yield of ^{239}Pu and ^{233}U to the point at which the process becomes feasible on a large scale. Ultimately, the hope is to build breeder reactors capable of producing large quantities of fissionable isotopes from more abundant elements such as thorium, lead, and mercury.

Why are nuclear ships more practical than nuclear autos?

23.5 MASS-ENERGY RELATIONS

The source of the energy evolved in a nuclear reaction is the mass change that occurs simultaneously. These two quantities are related by Einstein's equation,

$$\Delta E = \Delta mc^2 \qquad\qquad (23.28)$$

where Δm is the change in mass,* ΔE is the change in energy, and c is the velocity of light. If one substitutes for c the value 3.00×10^{10} cm/sec, Equation 23.28 gives directly the relation between the energy change in ergs and the mass change in grams:

$$\Delta E \text{ (in ergs)} = 9.00 \times 10^{20} \times \Delta m \text{ (in grams)}$$

In dealing with nuclear reactions, we are frequently interested in obtaining ΔE in units other than ergs. In particular, we may wish to calculate ΔE in kilocalories or in *millions of electron volts* (MeV). Again, it is often convenient to express the mass change in moles or in *atomic mass units* (amu).† To facilitate calculations involving mass-energy conversions, it is useful to have available a series of conversion factors such as those given in Table 23.5.

TABLE 23.5 MASS-ENERGY CONVERSION FACTORS

TYPE OF CONVERSION	CONVERSION FACTOR
mass-mass	1 g = 6.02×10^{23} amu
energy-energy	1 erg = 2.39×10^{-11} kcal
	1 erg = 6.24×10^5 MeV
mass-energy	1 g $\simeq 9.00 \times 10^{20}$ ergs
	1 g $\simeq 2.15 \times 10^{10}$ kcal
	1 amu $\simeq$ 931 MeV

Calculations Involving Mass-Energy Conversions

Using the appropriate conversion factors given in Table 23.5 in conjunction with the appropriate isotopic masses (Table 23.6), it is possible to calculate the energy change accompanying any nuclear reaction. Example 23.4 illustrates the calculations involved.

EXAMPLE 23.4. For the nuclear reaction $^{226}_{88}\text{Ra} \rightarrow {}^{222}_{86}\text{Rn} + {}^{4}_{2}\text{He}$

    a. Calculate ΔE, in MeV, when one atom of radium decays.

    b. Calculate ΔE, in ergs, when one mole of radium decays.

    c. Calculate ΔE, in kcal when one g of radium decays.

SOLUTION

    a. We shall first calculate Δm, in amu, and then convert this to energy in MeV.

---

   * Specifically, Δm = mass of products − mass of reactants; ΔE = energy of products − energy of reactants. In most nuclear reactions, the products weigh less than the reactants (Δm negative); in this case, the energy of the products is less than that of the reactants (ΔE negative), and energy is evolved to the surroundings.

   † An atomic mass unit is defined as exactly $\frac{1}{12}$ of the mass of a carbon-12 atom. This means that the mass of a particle in atomic mass units is numerically equal to its atomic weight on the carbon-12 scale. Thus, the proton (at. wt. = 1.00728) has a mass of 1.00728 amu; the alpha particle (at. wt. = 4.00150) has a mass of 4.00150 amu, and so on.

$$\Delta m = \text{mass } {}_{2}^{4}\text{He} + \text{mass } {}_{86}^{222}\text{Rn} - \text{mass } {}_{88}^{226}\text{Ra}$$

$$= 4.0015 \text{ amu} + 221.9703 \text{ amu} - 225.9771 \text{ amu}$$

$$= -0.0053 \text{ amu}$$

(Note that Δm is extremely small; it is necessary to know the masses of the various particles very accurately to obtain an answer accurate to two significant figures.)

$$\Delta E = -0.0053 \text{ amu} \times 931 \frac{\text{MeV}}{\text{amu}} = -4.9 \text{ MeV}$$

i.e., 4.9 MeV of energy are *evolved* when an atom of radium decays.

b. Here, we calculate the mass change in grams and convert to ergs:

$$\Delta m = \text{mass 1 mole } {}_{2}^{4}\text{He}$$
$$+ \text{ mass 1 mole } {}_{86}^{222}\text{Rn} - \text{mass 1 mole } {}_{88}^{226}\text{Ra}$$

$$= 4.0015 \text{ g} + 221.9703 \text{ g} - 225.9771 \text{ g} = -0.0053 \text{ g}$$

(Note that Δm in grams for the decay of a mole of radium is numerically equal to Δm in amu for the decay of one atom of radium.)

$$\Delta E = -0.0053 \text{ g} \times 9.0 \times 10^{20} \text{ ergs/g} = -4.8 \times 10^{18} \text{ ergs}$$

i.e., 4.8×10^{18} ergs of energy are *evolved* when a mole of radium decays.

c. Perhaps the simplest way to analyze this problem is to recall from b that the decay of 1 mole (~ 226 g) of ${}_{88}^{226}\text{Ra}$ resulted in the loss of 0.0053 g of mass; i.e., $\Delta m = -0.0053$ g when 226 g of Ra decays.

Accordingly, when 1 g of Ra decays,

$$\Delta m = \frac{-0.0053 \text{ g}}{226} = -2.3 \times 10^{-5} \text{ g}$$

$$\Delta E = -2.3 \times 10^{-5} \text{ g} \times 2.15 \times 10^{10} \text{ kcal/g} = -4.9 \times 10^{5} \text{ kcal}$$

(Note that while this amount of energy is vastly greater than that evolved in an ordinary chemical reaction, it is significantly less than the energy evolved in nuclear fission (about 2×10^{7} kcal/g).)

Nuclear Stability: Binding Energy

In Chapter 4 we pointed out that the energy change in an ordinary chemical reaction reflects the difference in stability between products and reactants. If the products are more stable than the reactants, energy is evolved. If a reaction absorbs energy, we deduce that the products are less stable than the reactants. We can, of course, apply the same interpretation to nuclear reactions. Referring to Example 23.4, the evolution of energy in the nuclear reaction

$$ {}_{88}^{226}\text{Ra} \longrightarrow {}_{86}^{222}\text{Rn} + {}_{2}^{4}\text{He}$$

means that the products, an alpha particle and a radon-222 nucleus, are more stable than the reactant, a radium-226 nucleus.

TABLE 23.6 NUCLEAR MASSES IN ATOMIC MASS UNITS*

	AT. NO.	MASS NO.	MASS			AT. NO.	MASS NO.	MASS
n	0	1	1.00867		Br	35	79	78.8992
H	1	1	1.00728			35	81	80.8971
	1	2	2.01355			35	87	86.9028
	1	3	3.01550		Rb	37	89	88.8909
He	2	3	3.01493		Sr	38	90	89.8864
	2	4	4.00150		Mo	42	99	98.8849
Li	3	6	6.01348		Ru	44	106	105.8829
	3	7	7.01436		Ag	47	109	108.8789
Be	4	9	9.00999		Cd	48	109	108.8786
	4	10	10.01134			48	115	114.8793
B	5	10	10.01019		Sn	50	120	119.8747
	5	11	11.00656		Ce	58	144	143.8816
C	6	11	11.00814			58	146	145.8865
	6	12	11.99671		Pr	59	144	143.8807
	6	13	13.00006		Sm	62	152	151.8853
	6	14	13.99995		Eu	63	157	156.8914
O	8	16	15.99052		Er	68	168	167.8941
	8	17	16.99474		Hf	72	179	178.9048
	8	18	17.99477		W	74	186	185.9107
F	9	18	17.99601		Os	76	192	191.9187
	9	19	18.99346		Au	79	196	195.9231
Na	11	23	22.98373		Hg	80	196	195.9219
Mg	12	24	23.97845		Pb	82	206	205.9295
	12	25	24.97925			82	207	206.9309
	12	26	25.97600			82	208	207.9316
Al	13	26	25.97977		Po	84	210	209.9368
	13	27	26.97439			84	218	217.9628
	13	28	27.97477		Rn	86	222	221.9703
Si	14	28	27.96924		Ra	88	226	225.9771
S	16	32	31.96329		Th	90	230	229.9837
Cl	17	35	34.95952		Pa	91	234	233.9934
	17	37	36.95657		U	92	233	232.9890
Ar	18	40	39.95250			92	235	234.9934
K	19	39	38.95328			92	238	238.0003
	19	40	39.95358			92	239	239.0038
Ca	20	40	39.95162		Np	93	239	239.0019
Ti	22	48	47.93588		Pu	94	239	239.0006
Cr	24	52	51.92734			94	241	241.0051
Fe	26	56	55.92066		Am	95	241	241.0045
Co	27	59	58.91837		Cm	96	242	242.0061
Ni	28	59	58.91897		Bk	97	245	245.0129
Zn	30	64	63.91268		Cf	98	248	248.0186
	30	72	71.91128		Es	99	251	251.0255
Ge	32	76	75.90380		Fm	100	252	252.0278
As	33	79	78.90288			100	254	254.0331

* Note that these are *nuclear masses*. The masses of the corresponding atoms can be calculated by adding the mass of the extranuclear electrons (mass of electron = 0.000549 amu). For example, the mass of an *atom* of $^4_2$He is

$$4.00150 \text{ amu} + 2(0.000549) \text{ amu} = 4.00260 \text{ amu}$$

Similarly, the mass of an atom of $^{12}_6$C is

$$11.99671 \text{ amu} + 6(0.000549) \text{ amu} = 12.00000 \text{ amu}$$

It is of considerable interest to compare the relative stabilities of different nuclei. One of the most straightforward ways to do this is to compare the masses of nuclei to those of the individual protons and neutrons of which they are composed. It is found experimentally that the mass of every nucleus containing neutrons and protons is less than that of the isolated particles themselves. Consider, for example, the $^4_2$He nucleus:

$$\text{mass 2 protons } = 2(1.00728) \text{ amu} = 2.01456 \text{ amu}$$
$$\underline{\text{mass 2 neutrons} = 2(1.00867) \text{ amu} = 2.01734 \text{ amu}}$$
$$4.03190 \text{ amu}$$

$$\text{mass of } ^4_2\text{He} = 4.00150 \text{ amu}$$

In this case, there is a decrease in mass of $(4.03190 - 4.00150)$ amu $= 0.03040$ amu when a helium nucleus is formed from two protons and two neutrons. This decrease in mass, called the **mass decrement**, can be calculated for any isotope whose nuclear mass is known. A series of mass decrements, calculated for a few typical isotopes, is given in Table 23.7.

TABLE 23.7 BINDING ENERGIES OF VARIOUS NUCLEI

	MASS DECREMENT	BINDING ENERGY (MeV)	BINDING ENERGY PER NUCLEON (MeV)
$^2_1$H	0.00239	2.22	1.11
$^3_2$He	0.00829	7.72	2.57
$^4_2$He	0.0304	28.3	7.07
$^7_3$Li	0.0421	39.2	5.60
$^{10}_5$B	0.0695	64.7	6.47
$^{12}_6$C	0.0989	92.1	7.67
$^{27}_{13}$Al	0.2415	224.8	8.33
$^{59}_{27}$Co	0.5555	517.2	8.77
$^{99}_{42}$Mo	0.9146	851.6	8.60
$^{157}_{63}$Eu	1.3815	1286	8.19
$^{196}_{80}$Hg	1.6653	1550	7.91
$^{238}_{92}$U	1.9342	1801	7.57

The fact that the formation of a nucleus from protons and neutrons involves a decrease in mass means that any nucleus is stable toward decomposition into these particles. Consider, for example, the helium-4 nucleus; in order to break this up into two protons and two neutrons, one would have to add 0.03040 amu of mass. This would, of course, require the absorption of a large amount of energy. The quantity of energy which would have to be absorbed to decompose a nucleus into protons and neutrons is referred to as the **binding energy** of the nucleus. The binding energy, in MeV, is readily calculated from the mass decrement, making use of the conversion factor 931 MeV = 1 amu. For the $^4_2$He nucleus,

$$\text{binding energy (MeV)} = \text{mass decrement (amu)} \times 931 \ \frac{\text{MeV}}{\text{amu}}$$

$$= 0.0304 \text{ amu} \times 931 \ \frac{\text{MeV}}{\text{amu}} = 28.3 \text{ MeV}$$

We can interpret this binding energy to mean that the decomposition of a helium-4 nucleus (an alpha particle) into protons and neutrons would require the absorption of 28.3 MeV of energy. Conversely, 28.3 MeV would be evolved if an alpha particle were formed from two protons and two neutrons.

It will be noted from Table 23.7 that binding energy increases, with rather large fluctuations at first but then more smoothly, as the number of protons and neutrons in the nucleus increases. One might indeed expect this to be the case; the more particles there are in the nucleus, the greater should be the total amount of energy required to break the nucleus apart. One can gain a better idea of the relative stabilities of different nuclei by calculating the binding energy per nuclear particle. For the $_2^4$He nucleus, which contains a total of four nuclear particles (nucleons),

$$\text{binding energy per nucleon} = \frac{\text{binding energy}}{\text{no. nucleons}} = \frac{28.3 \text{ MeV}}{4} = 7.07 \text{ MeV}$$

Looking at the last column of Table 23.7, we note that the binding energy per nucleon is relatively small for very light isotopes such as $_1^2$H and $_2^3$He.* It rises to a maximum of about 9 MeV with isotopes of intermediate mass number such as cobalt-59 and copper-63. The binding energy per nucleon then falls off slowly to about 7.5 MeV for the very heavy elements such as uranium (Figure 23.6).

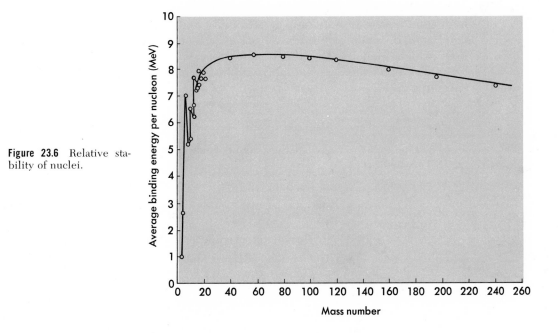

Figure 23.6 Relative stability of nuclei.

If we take the binding energy per nucleon to be a measure of the relative stability of a nucleus, it is obvious from Figure 23.6 that the most stable nuclei are those of intermediate mass such as $_{27}^{59}$Co, located near the broad

---

* It may be noted that the alpha particle $_2^4$He has an abnormally high binding energy per nucleon (7.07 MeV as compared to 5.60 MeV for $_3^7$Li). This is an indication of the unusual stability of this particular isotope.

maximum of the curve. Nuclei of very heavy elements such as uranium, which have comparatively low binding energies, should be unstable with respect to splitting into smaller nuclei. This conclusion, of course, is confirmed experimentally. The tremendous amount of energy evolved in the fission process is explained by the fact that the reactant atoms ($^{235}_{92}U$, for example) are less stable than the fragments ($^{90}_{37}Rb$, $^{144}_{55}Cs$, and so on) into which they split.

23.6   NUCLEAR FUSION

Referring again to Figure 23.6, let us concentrate our attention upon the early portion of the curve. It will be noted that the binding energy per nucleon rises very sharply from the isotopes of hydrogen to those of somewhat heavier elements such as helium, lithium, or boron. This suggests that it should be possible to obtain large amounts of energy by fusing hydrogen nuclei together to form heavier nuclei. Indeed, the energy available from nuclear fusion is considerably greater than that obtained from the fission of an equal mass of a very heavy element.

A typical fusion reaction, which has been employed in the hydrogen (thermonuclear) bomb, is

$$^{2}_{1}H + ^{3}_{1}H \longrightarrow ^{4}_{2}He + ^{1}_{0}n \qquad (23.29)$$

This reaction, unlike the fission process, has a high activation energy. It can be made to occur rapidly only at extremely high temperatures; in the hydrogen bomb, these temperatures are achieved by first carrying out a fission reaction, which then triggers the fusion process. The evolution of energy in Reaction 23.29 compares favorably with that which can be obtained by fission (Example 23.5).

EXAMPLE 23.5.   Calculate the amount of energy evolved, in kcal, per gram of reactants in

a. A fusion reaction: $^{2}_{1}H + ^{3}_{1}H \longrightarrow ^{4}_{2}He + ^{1}_{0}n$

b. A fission reaction: $^{235}_{92}U \longrightarrow ^{90}_{38}Sr + ^{144}_{58}Ce + ^{1}_{0}n + 4\,_{-1}^{0}e$

SOLUTION

a. We first calculate the mass decrement, using Table 23.6.

$$\Delta m \text{ (in amu)} = 4.00150 + 1.00867 - 2.01355 - 3.01550$$

$$= -0.01888 \text{ amu}$$

This calculation tells us that, starting with 5 amu of reactants ($^{2}H + ^{3}H$), there is a mass decrease of 0.01888 amu. It follows that if one were to start with five grams of reactants, Δm would be -0.01888 g. Stating it another way, the mass decrement per gram of reactant would be

$$\frac{-0.01888 \text{ g}}{5} = -0.00378 \text{ g}$$

Using the appropriate conversion factor from Table 23.5,

$$\Delta E = -0.00378 \text{ g} \times 2.15 \times 10^{10} \frac{\text{kcal}}{\text{g}} = -8.13 \times 10^{7} \text{ kcal}$$

b. Proceeding exactly as in a,

$$\Delta m \text{ (in amu)} = 89.8864 + 143.8816 + 1.0087$$
$$+ 4(0.00055) - 234.9934$$

$$= -0.2145 \text{ amu}$$

Mass decrement per gram of $^{235}U = \dfrac{-0.2145 \text{ g}}{235} = -0.000913 \text{ g}$

$$\Delta E = -0.000913 \text{ g} \times 2.15 \times 10^{10} \frac{\text{kcal}}{\text{g}} = -1.96 \times 10^7 \text{ kcal}$$

Comparing the answers to a and b, we conclude that nuclear fusion produces about four times as much energy per gram of starting material (8.13×10^7 kcal vs. 1.96×10^7 kcal) as the fission process.

It is believed that the energy given off by the sun and other stars results from fusion reactions in which ordinary hydrogen is eventually converted to helium. One mechanism which has been suggested for this process follows:

$$^1_1H + {}^1_1H \longrightarrow {}^2_1H + {}^0_1e; \qquad \Delta E = -0.43 \text{ MeV}$$

$$^2_1H + {}^1_1H \longrightarrow {}^3_2He; \qquad \Delta E = -4.96 \text{ MeV}$$

$$^3_2He + {}^1_1H \longrightarrow {}^4_2He + {}^0_1e; \qquad \Delta E = -19.30 \text{ MeV}$$

$$\overline{4\,{}^1_1H \qquad \longrightarrow {}^4_2He + 2\,{}^0_1e;} \qquad \Delta E = -24.69 \text{ MeV} \qquad \text{(23.30)}$$

PROBLEMS

23.1   Explain what is meant by the following terms.

a. Isotopes	f. Radioactive series	k. MeV
b. Bombardment reaction	g. Half-life	l. amu
c. Fission	h. Cyclotron	m. Mass decrement
d. Fusion	i. Cloud chamber	n. Binding energy
e. Beta ray	j. Critical mass	o. Tracer

23.2   Explain why

a. $^{14}_6C$ shows the same chemical reactivity as $^{12}_6C$.
b. So far as nuclear reactivity is concerned, radioactive Na^+ ions behave in the same way as radioactive Na atoms.
c. A nuclear reaction frequently results in the conversion of one element to another.

23.3   Write balanced nuclear equations for

a. The loss of an alpha particle by a $^{230}_{90}Th$ nucleus.
b. The loss of a beta particle by $^{214}_{82}Pb$.
c. The emission of gamma radiation from $^{203}_{82}Pb$.

23.4   Consider the radioactive series that starts with $^{235}_{92}U$ and ends with $^{207}_{82}Pb$. How many alpha particles are given off in this series? How many beta particles?

23.5   The rate constant for the decay of $^{197}_{78}Pt$ is 0.0385 hr^{-1}. Calculate

a. The number of grams of this isotope left after 12.0 hrs, starting with a sample weighing 0.106 g.
b. The fraction of a sample left after 1.00 hours.
c. The half-life of this isotope.

23.6   The half-life of $^{193}_{81}$Tl is 22.6 minutes. Calculate

a. The rate constant for the decay of this isotope.
b. The number of grams of this isotope left after one hour, starting with a one gram sample.
c. The weight of a sample which, after 12.0 min, contains 0.00160 g of this isotope.

23.7   The half-life of $^{225}_{89}$Ac is 10.0 hours. Starting with an 8.00 mg sample of this isotope,

a. How much is left after 20.0 hours?
b. How much is left after 30.0 hours?
c. How much is left after one day?

23.8   Estimate the age of an organic object in which the ratio of $^{14}_{6}$C/$^{12}_{6}$C is 0.920 that in a living plant.

23.9   The half-life of $^{238}_{92}$U is 4.5×10^9 yrs. Calculate the mole ratio of uranium to lead in a rock formed 3.0×10^9 years ago.

23.10 Explain the principle behind the radioactive method of determining the age of

a. Rocks or mineral deposits.   b. Organic material.

23.11 Write balanced nuclear equations for the following bombardment reactions.

Target Nucleus	Bombarding Particle	Expelled Particle
$^{95}_{42}$Mo	proton	neutron
$^{96}_{44}$Ru	alpha particle	neutron
$^{127}_{53}$I	proton	7 neutrons

23.12 Explain why neutrons are able to initiate nuclear reactions at much lower energies than protons or alpha particles.

23.13 Write plausible nuclear equations for the formation of an isotope of curium (at. no. = 96).

a. By beta-emission from an isotope of another element with a mass number of 242.
b. By neutron bombardment of uranium-238.
c. By a reaction analogous to Equation 23.17, starting with an isotope of another element having a mass number of 239.

23.14 Explain why the probability of decay by electron capture as compared to positron emission increases with atomic number.

23.15 Predict whether the following unstable isotopes will decay by electron or positron emission.

a. $^{24}_{11}$Na   b. $^{15}_{8}$O   c. $^{87}_{35}$Br   d. $^{26}_{14}$Si

23.16 Write a plausible nuclear equation for the neutron-induced fission of plutonium-239.

23.17 Consider the nuclear reaction

$$^{40}_{19}\text{K} + ^{1}_{0}\text{n} \longrightarrow ^{37}_{17}\text{Cl} + ^{4}_{2}\text{He}$$

Calculate

a. ΔE, in MeV, when one atom of $^{40}_{19}$K reacts.
b. ΔE, in ergs, when 1.60 moles of $^{37}_{17}$Cl is formed.
c. ΔE, in kcal, when one gram of $^{40}_{19}$K reacts.

23.18 Considering the data in Table 23.6, what can you say about the spontaneity of the following reactions?

a. $^{14}_{6}C \longrightarrow {}^{13}_{6}C + {}^{1}_{0}n$
b. $^{11}_{6}C \longrightarrow {}^{10}_{5}B + {}^{1}_{1}H$

23.19 Calculate the total binding energy and the binding energy per nucleon in $^{19}_{9}F$.

23.20 Calculate the total amount of energy evolved, in kcal, when one gram of $^{235}_{92}U$ undergoes the following fission process:

$$^{235}_{92}U + {}^{1}_{0}n \longrightarrow {}^{87}_{35}Br + {}^{146}_{58}Ce + 3\,{}^{1}_{0}n + {}_{-1}^{0}e$$

23.21 Calculate the energy evolved, in kcal/g, in the fusion reaction

$$^{3}_{1}H + {}^{3}_{1}H \longrightarrow {}^{4}_{2}He + 2\,{}^{1}_{0}n$$

23.22 Explain the principle of operation of the following instruments.

a. Wilson cloud chamber   c. Cyclotron
b. Geiger-Müller counter   d. Betatron

23.23 Write balanced nuclear equations for

a. The loss of an alpha particle by a $^{224}_{88}Ra$ nucleus.
b. The loss of a beta particle by a $^{12}_{5}B$ nucleus.
c. The emission of gamma radiation by $^{83}_{34}Se$.

23.24 A certain radioactive series starts with $^{241}_{94}Pu$ and ends with $^{209}_{83}Bi$. What is the total number of alpha and beta particles given off in this series?

23.25 The rate constant for the decay of $^{72}_{33}As$ is 0.0266 hr^{-1}. If one starts with a 0.200 g sample, calculate

a. The weight remaining after one day.
b. The time required for 70 per cent of a sample to decompose.
c. The half-life.

23.26 The half-life of $^{70}_{31}Ga$ is 20 minutes. How long will it take for $\frac{3}{4}$ of a sample of this isotope to decay? $\frac{7}{8}$?

23.27 Referring to Problem 23.26, how long will it take for 20 per cent of this sample to decay? 60 per cent? 90 per cent?

23.28 $^{238}_{92}U$ has a half-life of 4.5×10^9 years. A sample of uranium ore is found to contain 11.9 g of $^{238}_{92}U$ and 10.3 g of $^{206}_{82}Pb$. How old is the ore? What weight of helium has escaped from the ore sample? How many alpha particles does this represent?

23.29 A certain painting attributed to Raphael (1483–1520) is studied by the $^{14}_{6}C$ method. It is found that the $^{14}_{6}C$ content of a tiny piece of the canvas is 0.96 of that in living plants. On the basis of this data, can you come to any conclusion as to whether or not the painting is a forgery?

23.30 Write balanced nuclear equations for the following reactions.

Target Nucleus	Bombarding Particle	Expelled Particle
$^{63}_{29}Cu$	$^{1}_{1}H$	$^{1}_{0}n$
$^{27}_{13}Al$	$^{1}_{0}n$	$^{1}_{1}H$
$^{24}_{12}Mg$	$^{2}_{1}H$	$^{1}_{1}H$

23.31 Predict whether the following unstable isotopes will decay by electron or positron emission.

a. $^{12}_{5}B$ b. $^{11}_{6}C$ c. $^{14}_{6}C$ d. $^{14}_{8}O$

23.32 Write balanced nuclear equations for the following.

a. Loss of an alpha particle by $^{228}_{90}Th$.
b. Loss of a positron by $^{110}_{49}In$.

613

   c. K electron capture by $^{112}_{49}$In.

   d. Bombardment of $^{116}_{48}$Cd with a deuteron, resulting in the emission of two neutrons and another nucleus.

   e. Bombardment of $^{154}_{62}$Sm with a proton, resulting in the emission of a neutron and another nucleus.

   f. Fission of $^{239}_{94}$Pu to give $^{145}_{57}$La, four neutrons, and another nucleus.

23.33   Consider the reaction $^{11}_{6}$C → $^{11}_{5}$B + $^{0}_{1}$e. Using the data in Table 23.6,

   a. Calculate ΔE in MeV when one atom of $^{11}_{6}$C decays.
   b. Calculate ΔE in ergs when 1 mole of $^{11}_{6}$C decays.
   c. Calculate ΔE in kilocalories when 1 g of $^{11}_{6}$C decays.

23.34   Using the data in Table 23.6, calculate the mass decrement, the binding energy in MeV and the binding energy per nucleon for

   a. $^{3}_{2}$He   b. $^{12}_{6}$C   c. $^{59}_{27}$Co   d. $^{238}_{92}$U

   Check your values by referring to Table 23.7. How can one interpret these results in terms of the relative stabilities of the four nuclei involved?

23.35   Calculate the energy evolved per gram in the fusion of two deuterons to form an alpha particle.

°23.36   Aluminum-28 decays by β-emission to silicon-28. Using the data in Table 23.6, calculate the energy of the β-particle in MeV.

°23.37   Knowing the half-life of $^{238}_{92}$U, calculate the number of uranium atoms undergoing disintegration per second in a sample weighing 1 mg.

°23.38   One curie of a radioactive substance is the amount in which 3.7×10^{10} nuclei undergo radioactive decay during each second. Calculate the weight of 1 curie of radium (your answer should be 1 g).

°23.39   The elements francium, technetium, promethium, and astatine were discovered as products of nuclear reactions in a five-year period between 1937 and 1942. Can you suggest how the names of these elements might have been derived? Consult Samuel Glasstone, *Sourcebook on Atomic Energy*, D. Van Nostrand and Co., Princeton, N. J., 2nd edition, 1958.

Organic Chemistry

The physical and chemical properties of elemental carbon are such that there are many more organic compounds than compounds of all other elements combined. The enormous number and variety of carbon-containing compounds have made the field of organic chemistry the most extensive of all areas of chemical science.

Organic compounds are important in our daily lives in many ways. The human body, for example, consists largely of tissues composed of organic molecules of a wide variety of sizes and structures. These molecules, many of which are synthesized by our bodies, are daily utilized as ingredients in complex reactions which produce additional new molecules, release energy, and, in general, carry on the life process. Similar molecules are found in all plant and animal systems, where they selectively participate in a myriad of reactions controlling growth, maturity, and reproduction.

Among the many chemical substances man has found useful, those of organic origin are notably numerous and valuable. From petroleum and coal we obtain our major fuels and the raw materials for the production of fabrics, paints, dyes, and the great variety of plastics. From plants we obtain such important substances as sugar, cotton, alcohol, quinine, turpentine, and rubber. From animals come leather, wool, and silk. Although the many products obtained from nature have been known and used for centuries, only recently has the chemist been able to determine the actual chemical composition of such natural materials as rubber, silk, petroleum, and leather. Our understanding of the nature of these and other materials has made possible great improvement in the quality of our synthetic rubbers, fabrics, coatings, and fuels.

One particularly important area of organic chemistry is concerned with the identification and development of medicinal products. Although man has for centuries recognized certain plant and animal products as being valuable medicines, developments since 1940 have produced new chemicals that have truly revolutionized the practice of medicine. Sulfa drugs, cortisones, steroids, tranquilizers, and a variety of other medicinal agents, isolated or synthesized by researchers in the last two decades, have made it possible to deal with illness much more effectively than at any time in man's history. If one were

615

to rate the scientific areas of progress in this century which have contributed most to the well-being of mankind, it would be only fair to rank our ability to treat illness with medicines obtained through the knowledge of organic chemistry at the top of the list.

24.1 KINDS OF ORGANIC SUBSTANCES

We have pointed out that organic chemistry owes its existence to unique properties of the carbon atom. You will recall that the electron configuration of carbon, $1s^2 2s^2 2p^2$, gives the atom four valence electrons. Through hybridization, or by the principle that electron pairs in a molecule occur at maximum distances from one another, the carbon atom can form four equivalent, strong, covalent bonds, located tetrahedrally about the carbon atom. Of special significance is the fact that bonds between carbon atoms are strong, making for the possibility of long carbon chains in organic molecules. Silicon and germanium, of the same family as carbon, can also form four bonds, but the relative weakness of the Si—Si and Ge—Ge bonds restricts the chemistry of those elements severely. Because the electronegativity of carbon is near that of hydrogen and other nonmetals, such as oxygen, nitrogen, and sulfur, that commonly occur in organic compounds, the bonding in organic compounds is typically covalent in character. Organic substances are therefore primarily molecular, with physical properties determined by molecular interactions due to dispersion forces, dipole forces, or hydrogen bonding.

Organic compounds can be classified into groups and subgroups, according to the nature of the covalent bonds and the kinds of atoms present. One very large group includes those substances whose molecules contain only carbon and hydrogen atoms. These substances are called **hydrocarbons**, and, depending on the kinds of carbon bonds present, they can be further classified as paraffins, olefins, acetylenes, or aromatic substances. In this section we will first consider the hydrocarbons, then some compounds containing halogens, and finally oxygen-containing organic compounds.

Organic chemistry is the chemistry of hydrocarbons and their derivatives.

Saturated Hydrocarbons: Paraffins and Cycloparaffins

One large and structurally simple subgroup of the hydrocarbons includes those substances whose molecules contain only single carbon-carbon bonds. These substances are called **saturated** hydrocarbons or **paraffins**. In the paraffins the carbon atoms are bonded to each other in chains, which may be long or short, single or branched. In some cases the ends of a chain are bonded to each other, and **cycloparaffins** result.

The simplest paraffinic substances are methane, CH_4, and ethane, C_2H_6, whose molecules we used as models in the discussion of covalent bonding in Chapter 8.

Methane   Ethane

You will recall that in these molecules the bonding around each carbon atom is tetrahedral. Independent of the length of the carbon chain, the four bonds around each carbon atom in a saturated hydrocarbon will retain essentially tetrahedral geometry. The only exceptions occur in the small ring cycloparaffins, where the molecular structure restricts the bond angles.

Accepting the fact that carbon atoms in the paraffins occur in chains, which may or may not be branched or form rings, one can easily and correctly guess that the following higher paraffins exist:

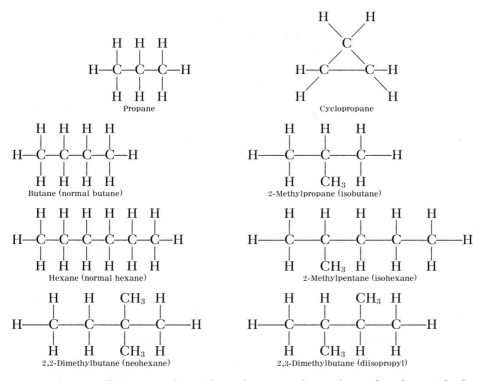

In the paraffins, since the carbon chains are bent, the molecules are fairly compact, with less atom-atom repulsion than may seem apparent in the planar diagrams. In cyclopropane, where the 60° bond angle is much smaller than the energetically favored tetrahedral angle of 109°, there is considerable ring strain in the molecule and a marked decrease in stability.

Can you explain why 5- or 6- membered rings are stable?

The molecules shown are typical of the paraffins. There are no well-defined positive and negative centers, so the molecules are relatively non-polar. They are soluble in each other and in other nonpolar solvents, but are essentially insoluble in water. The paraffins are colorless and relatively odorless. Their melting and boiling points increase with molecular weight, reflecting the fact that intermolecular bonding in the compounds is due to dispersion forces. Methane and ethane are both gases (bp − 161°C and −88°C respectively). Hexane is a liquid (bp 69°C), whereas eicosane, $C_{20}H_{42}$, a long-chain paraffin, is a solid (mp 38°C). The solids have the waxy properties of the common household paraffin used to seal jelly jars and make candles.

As a group the paraffins are relatively nonreactive chemically. The main reaction they ordinarily exhibit is burning in air. Their principal use is as fuels, both in home heating systems and internal combustion engines. Natural gas, fuel oil, and gasoline are for the most part mixtures of paraffins, selected to have the proper burning characteristics. The natural source of essentially all the paraffins is petroleum (see Section 24.3).

The preceding molecular diagrams illustrate the important general fact that in organic chemistry a given molecular formula frequently does not lead to one unique molecular structure. In the four-carbon paraffins the formula is C_4H_{10}, but the compound may be either butane or 2-methyl-propane. For the six-carbon paraffins, formula C_6H_{14}, five structures are possible, of which we show four. Compounds such as these, having the same formula but different molecular structures, are called **isomers**. Isomerism of this and other kinds is very common in organic chemistry and increases the number of possible organic compounds enormously. Although many of the 75 possible isomers having the formula $C_{10}H_{22}$ will not be known, it is important that there be available a system of nomenclature by which one can denote unambiguously any given isomer.

As you might surmise, the problem of nomenclature has plagued organic chemists and organic chemistry students for many years. In the early days of chemistry each newly discovered organic compound was given a trivial name, describing its source, use, color, or possibly the name of its discoverer. By 1900 it became apparent that some system had to be established for denoting compounds in a manner that reflected structure rather than some one of many arbitrarily chosen characteristics. By 1930 the system of nomenclature now in general use was set up by international agreement; it enables the organic chemist to name a new compound, no matter how complex it may be, in such a way as to indicate its structure to other chemists. The only real difficulty for the novice chemist is that for the most common substances, which are the ones ordinarily first encountered in the laboratory, the old trivial names persist. While the general population still speaks of muriatic acid instead of hydrochloric acid and blue vitriol instead of copper sulfate pentahydrate, the organic chemist still uses acetone instead of 2-propanone and will probably continue to do so for some time.

In this chapter we will make no attempt to discuss organic nomenclature in detail. We will name the compounds discussed in the manner of the practicing organic chemist, using the systematic notation where it is convenient and the trivial name where it is almost always employed. In some cases, as in the foregoing diagrams, both names will be given, with the common trivial name in parentheses.

Unsaturated Hydrocarbons: Olefins and Acetylenes

If one or more of the carbon-carbon bonds in a hydrocarbon is a double or triple bond, that substance is said to be **unsaturated**. If the multiple bond is double, the material is called an **olefin** (oil former). If the multiple bond is triple, the substance is an **acetylene**, after C_2H_2, the first member of that group of substances.

The olefins and acetylenes are similar to the paraffins in their number and variety. A typical paraffin, on being dehydrogenated or ruptured at a carbon-carbon bond, yields an olefin. This reaction, called *cracking*, is widely used in the petroleum industry to produce olefins from petroleum. Some cracking

C_5H_{12} would have how many isomers?

reactions that butane, C_4H_{10}, would undergo follow:

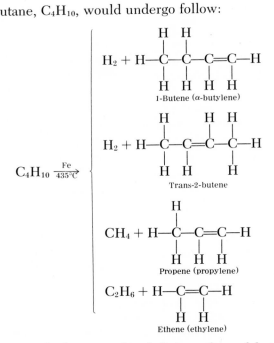

$C_4H_{10} \xrightarrow[435°C]{Fe}$

1-Butene (α-butylene)

Trans-2-butene

Propene (propylene)

Ethene (ethylene)

1-Butene, propene, and ethene are the olefinic analogs of the paraffins butane, propane, and ethane. By virtue of these names, the olefins are sometimes called **alkenes**, and the paraffins **alkanes**.

The geometry about two carbon atoms joined by a double bond is planar, with the bonded atoms at about equal angles around the carbon. There is no free rotation around a double bond, so a molecule like 2-butene (see diagram) can exist in two isomeric forms, called **geometric** isomers, in which the two hydrogen atoms attached to the double bonded carbon atoms exist either on the same side (cis form) or on opposite sides (trans form) of the double bond. The nomenclature for such isomers is analogous to that used previously in denoting the geometry in inorganic complex ions.

Olefinic hydrocarbons are more reactive chemically than are the paraffins, because of the presence of the double bond. A reaction common to these hydrocarbons is one of addition, in which a molecule such as H_2, HCl, or Br_2 adds directly to the double bond; the resulting saturated compound may contain one or more halogen atoms. Another important reaction is that of polymerization, in which an olefinic molecule adds to itself to form a long chain:

How could C_2H_5Br be made from C_2H_4?

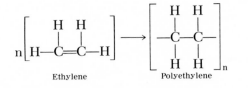

Ethylene   Polyethylene

Many modern plastics are polymers containing long-chain molecules of several thousand monomer units. Some well-known plastics produced by addition polymerization are polyvinyl chloride (CH_2=CHCl monomer), polypropylene (CH_3CH=CH_2 monomer), and polystyrene (C_6H_5CH=CH_2 monomer). These materials have the useful property of being thermoplastic,

619

softening at about 150°C, at which temperature they can be readily molded or extruded.

If more than one double bond is present in a molecule, that molecule is called a **polyene**, and the properties of the substance may reflect the relative positions of the double bonds as well as their presence. Two double bonds on the same carbon atom in a chain always produce a very unstable, highly reactive molecule. If the double bonds are relatively far apart in the carbon chain they act as single isolated double bonds. In some molecules, in which the double bonds in the chain alternate with single bonds, the carbon bonds are said to be **conjugated**. The most common example of a conjugated system is 1,3-butadiene,

$$H—C=C—C=C—H$$
$$\mid \quad \mid \quad \mid \quad \mid$$
$$H \quad H \quad H \quad H$$

1,3-Butadiene

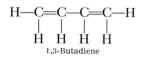

Substances like butadiene undergo rather different addition reactions than do ordinary olefins, and polymerize by addition reactions to produce polymers with one double bond per unit of monomer. This type of polymerization is typical of the process used to produce synthetic rubber (see Section 24.3).

Molecules containing a carbon-carbon triple bond are even more reactive than the olefins. Acetylene, the simplest of these substances, which are sometimes called **alkynes**, is typically unstable and chemically reactive. Its most commonly known use is in the oxyacetylene torch, where it is burned in oxygen to produce a high-temperature flame useful for welding and cutting metals. It undergoes addition reactions with hydrogen or the halogens to produce olefins or saturated substituted hydrocarbons, many of which are very important industrially. Acetylene itself is a gas. It is manufactured by reacting calcium carbide, CaC_2, with water, or by the carefully controlled oxidation of methane:

$$CaC_2(s) + 2\ H_2O(l) \longrightarrow C_2H_2(g) + Ca(OH)_2(s)$$

$$4\ CH_4(g) + 3\ O_2(g) \longrightarrow 2\ C_2H_2(g) + 6\ H_2O(g)$$

In the methane oxidation the gases are passed very quickly through an electric arc and the products are quenched in water to prevent further oxidation of the acetylene to the much more stable carbon monoxide and carbon dioxide. This reaction is thermodynamically possible, $\Delta G° < 0$, since the relatively high stability of water more than compensates for the very high instability of the acetylene produced simultaneously.

The geometric arrangement of atoms around a carbon atom on which there is a triple bond is linear. In acetylene all the atoms lie on the same straight line.

Aromatic Hydrocarbons

Relatively early in the history of organic chemistry some hydrocarbons were discovered which did not seem to behave chemically as one would have expected on the basis of their elementary composition. These substances all contained a relatively small amount of hydrogen and yet did not have the characteristic properties associated with unsaturation noted in the olefins and acetylenes. These hydrocarbons are members of a large group known as the aromatic substances.

The simplest aromatic substance, benzene, C_6H_6, was discovered by Michael Faraday in 1825. Benzene is a relatively stable liquid, which boils at about 80°C and has a characteristic odor. Though it has the same percentage by weight of hydrogen as does acetylene, it does not readily add either hydrogen or the halogens; its reaction with bromine, which must be carried out in the presence of catalysts, is one of substitution rather than the expected addition. This behavior implies that the structure of benzene must differ considerably from that of other hydrocarbons of similar composition.

Kekulé, a German chemist, was the first to suggest the structure for benzene which is used at present. Using the rather sparse data then available, he was able to conclude that, when benzene was reacted with bromine as in the following substitution reactions,

$$C_6H_6 + Br_2 \xrightarrow[50°]{Fe} C_6H_5Br + HBr$$

$$C_6H_5Br + Br_2 \xrightarrow[50°]{Fe} C_6H_4Br_2 + HBr$$

there was produced only one isomer of bromobenzene, C_6H_5Br, and three isomers of dibromobenzene, $C_6H_4Br_2$. Kekulé recognized that these facts require that benzene have a cyclic structure, with all hydrogens equivalent. In Figure 24.1 we have indicated the geometry Kekulé found necessary for the carbon skeleton and the structures of some of the substituted derivatives. On the basis of this structure, Kekulé was able to predict that there would be three isomers of the trisubstituted halogen derivative of benzene. Later ex-

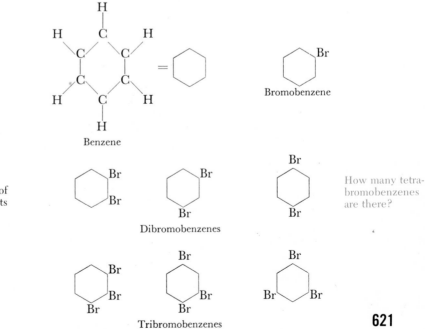

Figure 24.1 The geometry of benzene and some of its derivatives.

How many tetrabromobenzenes are there?

periments verified this and other predictions that can be made from this structure, and soon established the validity of Kekulé's reasoning. More recent x-ray diffraction and vibrational spectra results confirm that in the benzene molecule the carbon atoms are at the corners of a regular plane hexagon and are bonded to the hydrogen atoms in such a way as to maintain perfect hexagonal symmetry. The carbon-carbon bond length is 1.39 Å, intermediate between that of 1.34 Å for the double-bonded carbon atoms in ethylene and 1.55 Å, the single carbon-carbon bond length in ethane.

If one attempts to draw valence bond structures for benzene, he finds that no one structure is adequate. Kekulé suggested that, in the molecule, single and double bonds alternated in the ring:

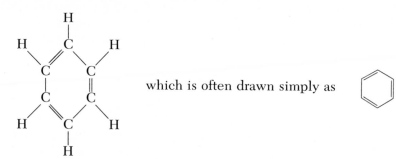

which is often drawn simply as

When it became clear that there were no disubstituted isomers of the form

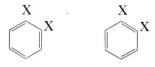

Kekulé proposed that the structure of the molecule continually oscillated between the two forms

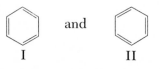

I   II

If such an oscillation were very rapid, the atomic arrangement might well appear constant at some average configuration having the observed very high symmetry and equivalence of carbon-carbon bonds.

You may recognize the rather close similarity between the two Kekulé structures for benzene and the various resonance structures which were given for some inorganic molecules in Chapter 8. A common modern description of the structure of benzene is based on the resonance concept. The benzene molecule according to this idea has only *one* structure, intermediate between structures I and II, and is called a resonance hybrid of those structures. The high relative stability of the molecule is attributed to the fact that such resonance structures can be drawn. (As a matter of fact, several other resonance structures can be drawn for benzene, but I and II are thought to be the major contributors.)

According to another form of the theory, 24 of the 30 valence electrons available for bonding in benzene are used in making up six carbon-carbon sigma (σ) bonds and six carbon-hydrogen bonds. The remaining six electrons (the π electrons) are relatively mobile and able to move around on the carbon-carbon ring, thereby furnishing the partial double bond character in equal amounts to the carbon-carbon bonds. In such a model the stability of benzene is the result of the added freedom of motion of the six electrons, a freedom which is presumably missing in ordinary molecules. As can be surmised, the story of the structure and properties of benzene is not yet complete and will probably continue to interest chemists and theorists for many years to come.

This structure is often indicated as ⬡.

Many substances derived from or related to benzene contain the carbon ring that we have been discussing. These materials, all of which have chemical properties characteristic of the benzene ring, form the group of substances we call aromatic. Many of them are found in coal tar, which is evolved when soft coal is heated to about 1200°C in the absence of air. The main part of the coal is converted to coke, a much cleaner-burning fuel than soft coal. The coal tar, a gummy black residue which is produced in about 3 per cent yield, is refined by further distillation and chemical treatment, finally yielding a fairly large group of pure aromatic substances, some of which are indicated in Figure 24.2. Many of these materials are of commercial importance, and in recent years have also been produced synthetically to meet increased demand. Benzene has some use as a solvent, but its main application is as a raw material for the manufacture of styrene, $C_6H_5CH{=}CH_2$, which is polymerized to form polystyrene. Toluene and the xylenes are useful as components in motor fuel and as chemical intermediates. Naphthalene still has some use in mothballs and is also important in the production of dyes and agricultural chemicals. The cresols are the main components in the wood preservative known as creosote.

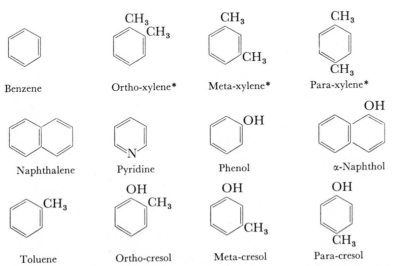

* The terms ortho, meta, and para are used as prefixes on names of aromatic substances to indicate the relative positions of two substituted groups.

Figure 24.2 Some aromatic substances found in coal tar.

Halogen-Containing Organic Compounds

In discussing the chemical reactions of olefins we have mentioned that a common product of addition reactions is a compound that can be thought of as a hydrocarbon in which one or more hydrogen atoms in the molecule have been replaced by halogen atoms. The halogen-containing derivatives of the hydrocarbons, though they rarely occur in nature, are very important to the synthetic organic chemist and also include some of the more common organic substances. Some of the typical reactions in which these substances are used will be discussed in the next section.

The geometries of the organic halogen derivatives are the same as those of the parent hydrocarbons. The bond angles within the substituted molecule are about the same as in the hydrocarbon. The C—X bond length is typically different from the C—H bond length and will vary with the halogen substituent. Since substitution on the hydrocarbon can usually occur at several nonequivalent positions, there are often several isomers of a halogen derivative with a given molecular formula. For instance, monochlorinated pentane could have any of the following isomeric structures:

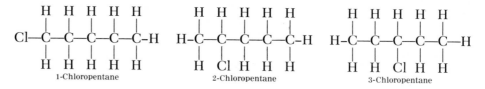

1-Chloropentane   2-Chloropentane   3-Chloropentane

Since pentane is only one of three isomers with the formula C_5H_{12}, it is clear that the number of possible isomers of a derivative with the formula $C_5H_{11}Cl$ is substantial.

Several of the relatively simple organic halides are well known because of their common uses. Carbon tetrachloride, CCl_4, is a solvent used in dry cleaning and since it is nonflammable, has had application as a fire extinguisher. This latter use is less frequent in recent years, since CCl_4 at high temperatures tends to partially oxidize to form phosgene, $COCl_2$, a poisonous gas used extensively in World War I. Chloroform, $CHCl_3$, is a low-boiling liquid, best known as an anesthetic. Since all the chlorinated hydrocarbons are toxic, chloroform has been essentially replaced in this use by other effective but less toxic materials. Freon, CCl_2F_2, which boils at $-28°C$, is chemically inert and is widely used as a refrigerant in home refrigerators and freezers. Paradichlorobenzene, $C_6H_4Cl_2$, is a solid aromatic organic halide which is poisonous to insects, particularly moth larvae. The materials known as DDT and chlordane are somewhat more complex chlorinated derivatives of organic hydrocarbons and are very effective insecticides.

Many related compounds such as $(CClF_2)_2$ are called "Freons."

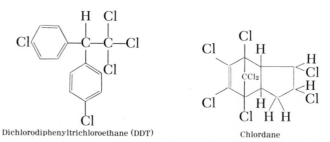

Dichlorodiphenyltrichloroethane (DDT)   Chlordane

Oxygen-Containing Organic Compounds

When one extends the discussion of organic compounds to include those which contain oxygen, the number and kinds of substances which can be prepared increases enormously. Whereas hydrogen and the halogen atoms form only one covalent bond, oxygen typically forms two and so can be present in organic molecules in other than the substitutional role to which the halogens are limited.

If the bonds to the oxygen atom in an organic molecule are all single, the substance is either an **ether** or an **alcohol**. A typical representative of each of these classes follows:

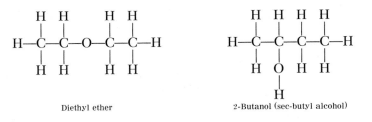

Diethyl ether   2-Butanol (sec-butyl alcohol)

These two substances are isomeric and differ in that in the ether the oxygen is between two carbon atoms and in the alcohol it is between a carbon atom and a hydrogen atom. Alcohols are hydroxy derivatives of hydrocarbons. (If the substitution of hydroxyl group for hydrogen occurs on an aromatic hydrogen, the product is called a phenol.)

Ethers are not highly reactive chemically, resembling the saturated hydrocarbon with the same number of atoms in the chain. They are mainly used as solvents, although diethyl ether is still used to some extent as an anesthetic. There is extensive industrial application of two cyclic ethers, ethylene oxide, which adds water to form ethylene glycol CH_2OHCH_2OH (permanent antifreeze), and tetrahydrofuran (made from corn cobs and oat hulls and used in nylon manufacture).

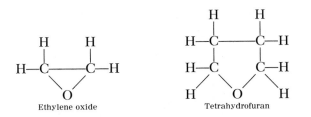

Ethylene oxide   Tetrahydrofuran

Alcohols and phenols are an important class of substances both from the synthetic and industrial viewpoints. The hydroxyl group, like the halogen group, affords a molecule a reactive center by which it can participate readily in chemical reactions. Some typical alcohol reactions will be considered in Section 24.2.

Alcohols can be prepared from natural sources or by synthetic means. The most important alcohol is ethyl alcohol, or ethanol, C_2H_5OH, which is commonly made by fermentation of sugar or starch solutions in the presence of natural or prepared yeasts and in the absence of air. The reaction may be schematically represented as

625

$$C_6H_{12}O_6 \longrightarrow 2\ C_2H_5OH + 2\ CO_2$$
$$\underset{\text{A sugar}}{\phantom{C_6H_{12}O_6}} \qquad\qquad \underset{\text{Ethanol}}{}$$

In addition to its use in intoxicating beverages, ethyl alcohol is an important solvent, a good antiseptic, and a reagent in many industrial processes. Methanol, CH_3OH, and isopropyl alcohol, $CH_3CHOHCH_3$, with physical and chemical properties similar to those of ethanol, are both prepared synthetically. Methanol is still used to some extent as an antifreeze and is the raw material for the manufacture of formaldehyde. Isopropyl alcohol is an antiseptic and the main component of rubbing alcohol.

The most important phenol is the parent compound, C_6H_5OH, from which the class gets its name. Phenols in general are solids at room temperature, are weak acids, and have marked antiseptic properties. Phenol (carbolic acid) was one of the first antiseptics, but because of its irritant properties is no longer used to disinfect wounds. The main use of phenol at present is in the manufacture of Bakelite resins.

If the oxygen atom in an organic molecule is double bonded to carbon, the substance is called an **aldehyde** or a **ketone**. Aldehydes and ketones have the general structures

in which R is an aliphatic group such as methyl, CH_3, or ethyl, C_2H_5, or an aromatic group such as phenyl, C_6H_5. Since aldehydes and ketones have similar structures, they resemble each other in many of their physical and chemical properties. The carbonyl group, $C{=}O$, is unsaturated and undergoes many addition reactions in both aldehydes and ketones. The carbonyl group may be converted in both classes of compounds by reduction reactions to either the hydroxyl group, to form alcohols, or the methylene group, CH_2, to form saturated compounds.

Most aldehydes and ketones are liquids at room temperature, with a solubility in water intermediate between that of the parent hydrocarbon and the related alcohol. Those of relatively high molecular weight have pleasant odors and are found in nature in flowers and spices. The lower members in both families have penetrating, rather sharp odors, and are prepared synthetically.

One of the common methods for the production of aldehydes is by the oxidation of alcohols. Formaldehyde, CH_2O, the first member of the aldehydes, is made commercially by the reaction of methanol vapor with air in the presence of a silver or copper catalyst:

$$2\ CH_3OH + O_2 \xrightarrow[250°C]{Ag} 2\ CH_2O + 2\ H_2O$$

The reaction is not carried to completion. The gaseous mixture is absorbed in water and marketed as formalin, a 40 per cent solution of formaldehyde containing some methanol, familiar to zoology students as a preservative.

Another important aldehyde is acetaldehyde, CH_3CHO, which is used mainly as a chemical intermediate. Acetaldehyde is produced commercially by an interesting reaction involving the hydration of acetylene:

$$H—C{\equiv}C—H + H_2O \xrightarrow[\text{HgSO}_4]{40\% \text{ H}_2\text{SO}_4} [CH_2{=}CHOH] \longrightarrow CH_3CHO$$

$$\underset{\substack{\text{Vinyl alcohol} \\ \text{(not isolable)}}}{} \qquad \underset{\text{Acetaldehyde}}{}$$

Among the ketones, acetone, CH_3COCH_3, is the simplest and best known. It is a volatile liquid widely used as a solvent in industrial and academic laboratories. It is used as a raw material for the plastic substance known as Lucite or Plexiglas. It can be produced by fermentation of sugar or starch solutions by the proper choice of fermenting agents and is also made commercially by the oxidation of isopropyl alcohol.

Another class of oxygen-containing compounds is that of the **organic acids**. If a molecule contains both a carbonyl and a hydroxyl group on the same carbon atom, the compound will behave as a weak acid in water solution.

$$\underset{\text{A carboxylic acid}}{\begin{matrix} R—C{=}O \\ | \\ O—H \end{matrix}} \rightleftharpoons \left(\begin{matrix} R—C{=}O \\ | \\ O \end{matrix}\right)^- + H^+$$

The organic acids of molecular weight below about 150 are typically liquids at room temperature, with sharp, penetrating odors. Those with R groups containing up to three carbon atoms are, because of the presence of the polar carbonyl and hydroxyl groups, completely miscible with water.

The carboxylic acids constitute one of the most important classes of organic compounds. Animal fats and other natural products afford a source of a wide variety of acids, many of which are of considerable commercial importance. Unlike most of the substances we have already mentioned, some of the acids and their derivatives are nontoxic, at least in moderate amounts, to humans, and are present in many foods.

The carboxyl group, containing both the carbonyl and hydroxyl groups, possesses many of the chemical properties of ketones and alcohols as well as properties of its own. The fact that the carboxyl group acts as a weak acid allows for rapid reaction of organic acids with aqueous alkaline solutions to form salts.

$$\begin{matrix} R—C{=}O \\ | \\ O—H \end{matrix} + OH^- \longrightarrow \left(\begin{matrix} R—C{=}O \\ | \\ O \end{matrix}\right)^- + H_2O$$

The carbonyl group of organic acids undergoes several reactions. Probably the most important of these involves reaction with an alcohol,

$$\underset{\text{An acid}}{\begin{matrix} R—C{=}O \\ | \\ O—H \end{matrix}} + \underset{\text{An alcohol}}{R'—O—H} \rightleftharpoons \left[\begin{matrix} O—R' \\ | \\ R—C—O—H \\ | \\ O—H \end{matrix}\right] \rightleftharpoons \underset{\text{An ester}}{\begin{matrix} O—R' \\ | \\ R—C{=}O \end{matrix}} + H_2O$$

Esters have much more pleasant odors than organic acids.

with water being eliminated from the unstable 1,1-dihydroxy intermediate, to form the class of substances called **esters**. The reaction is a common one and typically results in an equilibrium mixture. By removing water, the reaction can be driven to the right to produce the ester.

Fats consist of the esters of long-chain carboxylic, or fatty, acids. The alcohol involved is usually glycerol, $HOCH_2CHOHCH_2OH$, while the acid may be one of many, depending on the natural source. The carbon chain may be saturated or unsaturated and is only rarely branched. A typical fatty acid is palmitic acid, $CH_3(CH_2)_{14}COOH$, found in palm oil. A very important industrial reaction of fats is that of hydrolysis, or saponification, carried out by boiling fats with sodium hydroxide solution to produce the substances known as **soaps**:

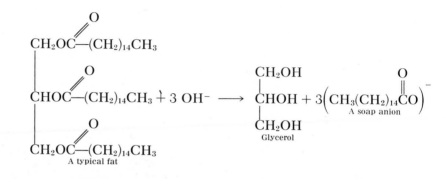

A soap is the sodium salt of a fatty acid. It obtains its useful cleaning properties by combining a long-chain hydrocarbon, which has good solvent action on other hydrocarbons, with the polar carboxy group and its high water solubility. Other substances of similar structure also have good cleaning properties and in recent years have replaced soaps to some extent. The difficulty with a soap is that in hard water, which typically contains calcium ions, the soap precipitates as a calcium salt. Much better solubility properties are obtained if one recovers the fatty acid from the reaction, reduces it to the alcohol with hydrogen, and then forms an ester by reacting the alcohol with concentrated sulfuric acid, as in the following reaction:

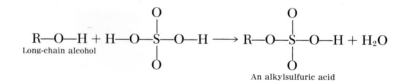

By neutralizing the remaining acid hydrogen with sodium hydroxide, one obtains a substance which belongs to that class of materials commonly known as **detergents**. These materials, mainly because of their improved solubility characteristics, are now widely used for laundry purposes, particularly in hard-water areas.

Detergents are a major problem in water pollution because they are so stable.

Many of the fatty acids are produced by the saponification reaction just described. Those with low molecular weight are, however, not present in fats and are prepared either by synthetic means or from other natural sources. Acetic acid, CH_3COOH, is the most widely known of the carboxylic acids and is the active component of vinegar, made for centuries by fermentation of apple or other ciders. Acetic acid is an important industrial chemical and is made synthetically by air oxidation of ethanol or by the oxidation of acetaldehyde.

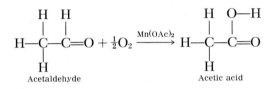

$$\text{Acetaldehyde} \qquad\qquad \text{Acetic acid}$$

24.2 SOME COMMON REACTIONS OF ORGANIC COMPOUNDS

Substitution Reactions

Perhaps the most important class of organic reactions is that involving substitution of one group for another in an organic molecule. Such reactions are extremely useful in organic synthesis, both in industry and in academic laboratories, and occur, by various mechanisms, in essentially all the different kinds of organic substances.

Nucleophilic Substitution Reactions. When the substitution reaction involves an electron-rich group displacing another electron-rich group from a saturated carbon atom, it is called a **nucleophilic substitution** reaction. An example of such a reaction is

$$CH_3CH_2CH_2Cl + OH^- \longrightarrow CH_3CH_2CH_2OH + Cl^-$$

In this reaction, which occurs readily when propyl chloride is dissolved in a solution of sodium hydroxide in aqueous ethanol, the electron-rich hydroxide ion displaces an electron-rich chloride ion from the propyl chloride molecule. This kind of reaction typically occurs between alkyl halides and negative ions or other Lewis bases. It is reversible and is therefore carried out under concentration conditions that favor the formation of the desired product; in the example given, the presence of excess hydroxide ion tends to drive the reaction to the right. The substitution reaction can be carried out with many different anions and can be used to prepare a wide variety of organic substances. In Table 24.1 are listed some of the possible reactions of this type, first in general and then with specific examples.

TABLE 24.1 SOME TYPICAL NUCLEOPHILIC SUBSTITUTION REACTIONS*

$R\text{—}X + X'^- \rightarrow RX' + X^-$ an alkyl halide	$C_2H_5Cl + I^- \rightarrow C_2H_5I + Cl^-$ ethyl iodide
$R\text{—}X + OR'^- \rightarrow ROR' + X^-$ an ether	$CH_3CH_2CH_2Br + NaOCH_2CH_3$ $\rightarrow CH_3CH_2OCH_2CH_2CH_3 + NaBr$ ethyl n-propyl ether
$R\text{—}X + 2\,NH_3 \rightarrow RNH_2 + NH_4X$ an amine	$CH_3CH_2Br + 2\,NH_3 \rightarrow CH_3CH_2NH_2 + NH_4Br$ ethyl amine
$R\text{—}X + HC\equiv C^- \rightarrow RC\equiv CH + X^-$ a substituted acetylene	$CH_3I + HC\equiv CNa \rightarrow CH_3C\equiv CH + NaI$ methyl acetylene

* The sodium derivatives in these reactions are prepared by reaction of metallic sodium with either ethyl alcohol or acetylene, producing sodium ethoxide, $NaOCH_2CH_3$, or sodium acetylide, $NaCCH$, respectively.

Recall the discussion of S_N1, S_N2 reactions in Chapter 19.

Nucleophilic substitution reactions have been studied extensively from kinetic and mechanistic viewpoints. Many of these reactions are first order, depending only on the concentration of the alkyl halide; these are called S_N1 reactions (substitution, nucleophilic, first order). In other cases these reactions are found to be second order overall, with rates which depend on both the concentration of the alkyl halide and that of the substituting group. Such reactions are described by the term S_N2 (substitution, nucleophilic, second order). The mechanisms of the two kinds of reactions necessarily differ. In an S_N1 reaction, the following mechanism seems likely:

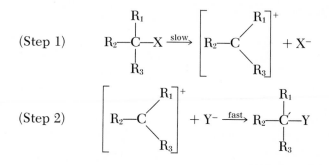

The positive ion formed in the first step is called a **carbonium** ion. Since the carbon atom at the center of the carbonium ion has only six electrons, the ion will tend to be planar, with approximately equal angles between the bonds to the three R groups (sp^2 hybridization). The product is formed as fast as the carbonium ion is produced, and hence the reaction is first order. The electrons on the Y^- group are attracted to the positive carbon nucleus, which explains why the name nucleophilic is assigned to this reaction.

In S_N2 reactions, the attacking Y^- group is thought to form an activated complex with the alkyl halide, approaching the molecule from the side opposite the X group (you will remember that the bond geometry around the central carbon atom in the halide will be tetrahedral). The mechanism would be written as

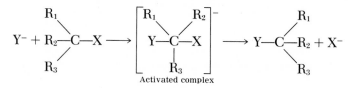

and would produce a second order reaction.

Nucleophilic substitution reactions may be caused to vary in their mechanism by changing the reaction conditions. As one might expect, the S_N1 reaction is favored in highly polar solvents, in which the ionization of the alkyl halide would tend to be greatest. It is also favored by relatively weak nucleophiles, which would not tend to form activated complexes readily.

Substitution Reactions Involving Free Radicals. Substitution reactions of the saturated hydrocarbons proceed by a quite different mechanism from that already discussed. A mixture of a paraffinic hydrocarbon and chlorine shows no reaction if kept in the dark. On exposure to ultraviolet light the system reacts with explosive violence, producing molecules in which chlorine atoms have replaced hydrogen. The final product will in general be a complex mixture of halogenated hydrocarbons, which can be separated into its many components only with great difficulty. The reaction between propane and chlorine is illustrated schematically as follows:

$$CH_3CH_2CH_3 + n\ Cl_2 \longrightarrow \begin{matrix} CH_3CH_2CH_2Cl + CH_3CHClCH_2Cl \\ + \qquad\qquad + \\ CH_3CHClCH_3\ + CH_2ClCH_2CH_2Cl \\ + \qquad\qquad + \\ \text{Other polyhalogenated propanes} \end{matrix} + n\ HCl$$

The halogenation of saturated hydrocarbons, like the reaction between gaseous hydrogen and a halogen (see Chapter 14), is considered to proceed via a mechanism involving free radicals. Once the halogen molecule is dissociated by a photon of ultraviolet light, many many molecules of hydrocarbon can be halogenated before another halogen molecule need dissociate. Such a reaction, as you may recall, is described as a chain reaction. The mechanism for the reaction between methane and chlorine is as follows:

initiation step:	$Cl_2 \xrightarrow{\text{UV photon}} 2\ Cl\cdot$
chain propagation steps:	$Cl\cdot + CH_4 \longrightarrow CH_3\cdot + HCl$
	$CH_3\cdot + Cl_2 \longrightarrow CH_3Cl + Cl\cdot$
termination:	$Cl\cdot + Cl\cdot \longrightarrow Cl_2$
	$CH_3\cdot + CH_3\cdot \longrightarrow C_2H_6$
	$CH_3\cdot + Cl\cdot \longrightarrow CH_3Cl$

In the reaction mixture, the concentrations of the very reactive chlorine atom and the methyl free radical $CH_3\cdot$ are both very low, so that perhaps as many as a million methyl chloride molecules are produced before a chain is terminated. Here we have stopped the halogenation at CH_3Cl, but it could clearly continue, by the same kind of mechanism, to produce CH_2Cl_2, $CHCl_3$, and CCl_4.

Elimination Reactions

Sometimes in nucleophilic substitution reactions, side reactions occur, producing other products. If, for instance, one is attempting to form an alcohol by the reaction

$$CH_3CH_2Br + OH^- \longrightarrow CH_3CH_2OH + Br^-$$

the following reaction, producing an olefin, may also occur:

$$CH_3CH_2Br + OH^- \longrightarrow CH_2{=}CH_2 + H_2O + Br^-$$

In this case the elements in hydrogen bromide are freed from the parent alkyl halide and react to form water and the halide ion. This kind of reaction, in which a small group is removed from a larger molecule, is called an **elimination** reaction. If the elements removed are hydrogen and a halogen, we speak of it as a **dehydrohalogenation** reaction. As you might expect, one can frequently adjust the reaction conditions to favor either the substitution or the elimination reaction. Ordinarily substitution will be more likely in an aqueous solution of potassium hydroxide; the elimination reaction is carried out in alcoholic KOH.

Alcohols can also be caused to undergo elimination reactions to form olefinic substances. In such reactions the elements in water are removed and the reaction is called a **dehydration**. The ease with which dehydration occurs depends on the number of carbon atoms attached to the carbon atom bonded to the hydroxyl group; the more carbon atoms, the easier is the dehydration reaction. A tertiary alcohol, like tertiary butyl alcohol, may dehydrate on being distilled, whereas a primary alcohol, like ethyl alcohol, may require much more drastic dehydrating conditions.

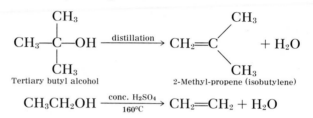

$$CH_3CH_2OH \xrightarrow[160°C]{conc.\ H_2SO_4} CH_2{=}CH_2 + H_2O$$

(Carbon atoms are classified as being primary, secondary, or tertiary, depending on whether they are bonded to one, two, or three other carbon atoms, respectively. In a tertiary alcohol the hydroxyl group is bonded to a tertiary carbon atom; secondary and primary alcohols are analogously defined. This approach to naming compounds has been superseded by the modern system, but is still applied to given carbon atoms and to many commonly encountered organic substances.)

Elimination reactions, like nucleophilic substitution reactions, may follow first order or second order kinetics. The mechanism of the first order elimination reaction, like the S_N1 reaction, is thought to involve the carbonium ion intermediate. This mechanism allows one to see a relation between nucleophilic substitution, described above, and elimination reactions which is not at once apparent. For both of these reactions, when they are first order, the first step in the accepted mechanism is the formation of the positively charged carbonium ion:

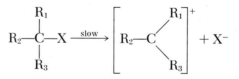

As soon as the carbonium ion forms, the bonds to the central carbon atom assume a planar configuration with all bond angles equal. At that point several possible reactions can occur:

1. Recombination of the carbonium ion with X^- to form the original halide.

2. Combination of the carbonium ion with a nucleophilic group present ·in the system, resulting in a nucleophilic substitution reaction.

3. Elimination of a hydrogen ion, H^+, from one of the carbon atoms adjacent to the central carbon atom, resulting in a dehydrohalogenation reaction. For example, assuming R_2 is a CH_3 group,

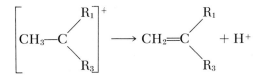

The reaction that is actually observed in a given case depends to some extent on the conditions in the reacting system. Since each of the possible reactions proceeds to an equilibrium state, and each has an equilibrium constant of roughly the same magnitude, it is often possible to obtain the desired product by proper choice of the reagent concentrations. Since dehydrohalogenation and dehydration reactions both produce a hydrogen ion, which reacts with hydroxide ion to form water, those reactions are favored in strongly alkaline alcoholic solutions, in which water concentration is kept low and hydroxide ion concentration is high. Nucleophilic substitution reactions would ordinarily be carried out in water solution in which there is a high concentration of the substituting negative ion.

Reaction via the carbonium ion intermediate occurs most readily when the carbon atom bonded to the nucleophilic group X is tertiary, that is, when it is attached to three other carbon atoms. Such substances form relatively stable carbonium ions and exhibit first order kinetics in substitution and elimination reactions. On the other hand, substances in which X is bonded to a primary carbon atom, as in RCH_2X, follow second order kinetics in such reactions. The mechanism for second order elimination reactions, like that for S_N2 reactions, involves in the first step a collision between the nucleophilic group and the starting compound.

24.3 PETROLEUM AND RUBBER: TWO MATERIALS OF IMPORTANCE IN THE CHEMICAL INDUSTRY

Although chemists, like other people interested in research, often work simply to learn more about their science, a great deal of research is done for the purpose of financial profit. Attempts by the chemical industry to improve and develop products obtained from natural sources have resulted in increased knowledge, both applied and theoretical, and have been responsible in large measure for the enormous growth of the industry since 1940. One of the naturally occurring materials of greatest interest to the industrial chemist has been the substance known as petroleum.

Petroleum

Petroleum, or oil, is found in underground pools in many regions of the earth. Its origin lies in plants and animals which lived on the earth and in the sea many millions of years ago. The residues from these organisms accumulated in certain regions, possibly as a result of geologic conditions, became buried, and were subjected to high pressures and reducing conditions over long periods of time. The resulting material, petroleum, is a complex mixture of hydrocarbons, containing paraffinic chain and ring molecules, aromatic molecules, and small amounts of oxygen- and sulfur-containing substances.

Petroleum has been known for thousands of years, being observed as surface seepages, particularly as oil films on streams and ponds. It had no known use, except as "medicine oil," until about the mid-nineteenth century when the first oil wells were drilled in this country and Rumania. Crude oil, on being distilled, yielded a fraction known as kerosene, which had immediate commercial importance as a lamp fuel; for many years kerosene was the main product of the petroleum industry.

With the development of the automobile, a lower-boiling fraction obtained in the distillation of petroleum, straight-run gasoline, became of dominant value. Since a 50-gallon barrel of crude oil yielded only about 7 or 8 gallons of gasoline, it became a matter of great commercial significance to increase the size of the gasoline fraction. Chemical research on this problem has been very successful and has resulted in (1) more and better gasolines, (2) a knowledge of what substances make up petroleum, (3) a knowledge of what substances make a good gasoline, and (4) much basic knowledge in organic chemistry, including much of the material presented in the previous section of this chapter.

As a result of petroleum research, chemists now are able to manipulate almost at will the end products of the petroleum refining process to meet many different kinds of demands, from home heating gases to jet fuels to road asphalts, in the way which makes best use of the starting material. A by-product of this research has been the petrochemical industry, which is rapidly assuming a major role as a supplier of the many organic chemicals which can now be produced from petroleum.

Prior to 1900, gasoline was of little value.

TABLE 24.2   FRACTIONS OBTAINED ON DISTILLATION OF PETROLEUM

FRACTION	BOILING RANGE (°C)	CARBON ATOM CONTENT	DIRECT USE
Gas	below 20	C_1—C_4	gas heating
Petroleum ether	20–60	C_5—C_6	industrial solvent
Light naphtha	60–100	C_6—C_7	industrial solvent
Straight-run gasoline	40–200	C_5—C_{10}	motor vehicle fuel
Kerosene	175–325	C_{11}—C_{18}	jet fuel
Gas oil	275–500		diesel fuel
Lubricating oil	above 400	C_{15}—C_{40}	lubricant
Asphalt	nonvolatile		roofing and road construction

Following a rough distillation of the crude oil into the fractions indicated in Table 24.2, the higher molecular weight fractions are carried through a controlled pyrolysis or *cracking* process, in which they are heated to about 500°C, often under catalytic conditions (see Section 24.1). In this process the molecules suffer a rupture of a carbon-carbon bond, yielding olefins and paraffins of lower molecular weight than the original fraction. The product contains a substantial fraction of substances which boil in the same range as gasoline, thus increasing significantly the yield of gasoline from crude oil. The lighter olefins, particularly ethylene and propylene, have in recent years found a market as raw materials in the plastics industry.

In order to further increase gasoline yield and quality, there are several other procedures used to treat both the original distillate fractions and the products from the cracking step. Since it has been found that branched-chain molecules perform better as motor fuels than do those containing straight chains, the middle distillation fractions are often subjected to a *reforming* or isomerization process in which the fraction is passed over a solid catalyst such as aluminum chloride at about 200°C. With normal pentane, for example, one would obtain on isomerization

Both these substances would be superior to the unbranched pentane as automobile fuel. Isomerization reactions also undoubtedly occur to some extent during cracking and so make an additional contribution in that step to the fuel quality of the product.

The light fractions, containing C_3 and C_4 hydrocarbons from the distillation and cracking operations, can be converted to useful gasolines by **polymerization** and **alkylation** processes, which are selective reversals of cracking. In polymerization reactions, gaseous olefins combine with each other to yield branched olefins, which can be used directly as fuel or hydrogenated to the saturated hydrocarbons:

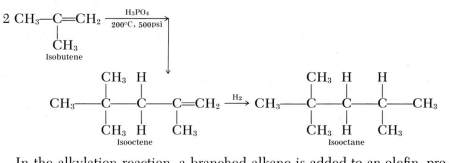

In the alkylation reaction, a branched alkane is added to an olefin, producing another branched alkane of more suitable molecular weight. The following reaction is illustrative and commercially important:

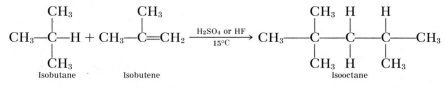

(Gasolines are rated in quality according to their resistance to "knock," or detonation, instead of smooth burning, in the automobile engine. Isooctane, produced in the foregoing reactions, is very resistant to knock, and has been

635

assigned an *octane number* of 100. Normal heptane, a straight-chain hydro-carbon prone to knocking, is given an octane number of 0. Commercial gaso-lines are assigned octane numbers on the basis of their performances against mixtures of these two reference hydrocarbons.)

As a final process for the production and improvement of gasolines we should mention **aromatization**. In this process straight-chain hydrocarbons are cyclized and dehydrogenated to form aromatic substances. The reaction is carried out at about 500°C in the presence of molybdenum oxide catalyst on alumina. The reaction as applied to normal heptane is indicated schematically as follows:

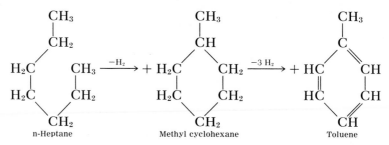

Rubber

Another naturally occurring organic material of chemical importance is the substance we know as rubber. The source is a latex or sap, produced by the Hevea or rubber tree, which is grown mainly in Ceylon and Indonesia. The latex is collected and treated with acetic acid to precipitate the raw rub-ber, which coagulates, is pressed dry, and is rolled into sheets for shipment.

Raw rubber is a gummy, light-colored substance, which becomes sticky when warmed and brittle when cold. In 1834 Charles Goodyear made the discovery that if natural rubber is heated with sulfur a reaction occurs, which produces a new material with the properties we normally associate with rubber, namely elasticity, flexibility, and resistance to abrasion. Ordinarily commercial rubber is prepared by this process, which is called vulcanization; rubber also usually contains certain additives, particularly carbon blacks, which are both less expensive than rubber and also greatly improve its re-sistance to wear and tear.

Soon after its discovery vulcanized rubber found significant uses in ar-ticles requiring good water resistance, such as raincoats and overshoes, and in tires for wagon wheels. With the development of the automobile, demand for tires resulted in an enormous increase in the rubber industry and provided great incentive for improving the quality of tire rubber. Again, chemical re-search on the problem has resulted in rubber with vastly improved properties as tire material and has also furnished us with knowledge of the chemical structure of rubber. This knowledge, applied to the production of synthetic rubbers, has led to a variety of rubbers tailored for special purposes.

It is now known that natural rubber is a polymeric substance, containing in its molecules many units of a monomer called isoprene, 2-methyl-1,3-buta-diene, attached to one another in the following manner:

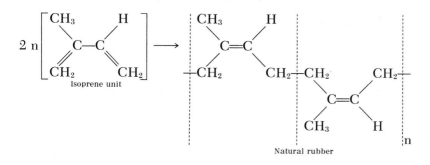

Natural rubber polymer molecules contain long chains of isoprene units, linked as shown, with a molecular weight average of about 400,000. During the vulcanization process the long polymer chains are cross-linked by reaction with fairly short chains of sulfur atoms (one to about six atoms long); it appears that in the process H_2S is eliminated and relatively little of the unsaturation in the rubber is lost. Ordinary rubber is produced by vulcanization with about 3 per cent sulfur by weight; hard rubber results when the amount of sulfur is increased to about 30 per cent.

Recently it has become possible to carry out the polymerization of isoprene synthetically to produce a product essentially identical to natural rubber. There is now substantial production of this synthetic "natural" rubber, which promises ultimately to make the rubber industry independent of the plantation-produced material. Though it only takes a moment to read these sentences, the determination of the true structure of natural rubber and its subsequent production by synthetic means was a problem that took many years and the work of many people for its solution.

Although the detailed evaluation of the structure of the rubber polymer was an extremely difficult problem, the nature of the monomer has been known for many years. It was found that isoprene could be polymerized to a rubberlike material even before 1900 (the substance produced was not natural rubber, since in natural rubber only the cis configuration shown in the foregoing diagram exists, whereas normal polymerization of isoprene yields chains with several different configurations). The knowledge of the nature of the monomer gave some incentive to production of synthetic rubbers, which were less expensive and more available than the natural product. In Germany during the first World War, since isoprene could not easily be produced, chemists turned to the following monomer,

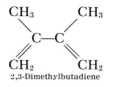

2,3-Dimethylbutadiene

and, by its polymerization, produced significant amounts of the first synthetic rubber, called methyl rubber. By vulcanization they were able to manufacture a reasonably satisfactory hard rubber, but soft methyl rubber, such as that needed in tire tubes, was completely inadequate to its task.

Since about 1930 several synthetic rubbers have been produced commercially. One of the most important of these is neoprene, a polymer of the

monomer chloroprene,

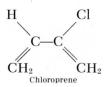

Chloroprene

This substance, when polymerized with due regard to the structure of the product, yields a rubber which for many purposes is superior to natural rubber. It is highly resistant to organic solvents, water, and oxidation. Although it is excellent as a tire rubber, it is somewhat more expensive than other satisfactory synthetic rubbers.

During World War II supplies of natural rubber from Malaya and the East Indies were cut off, making it imperative that the United States produce synthetic rubber in large volume on short notice. A copolymer of 1,3-butadiene, $CH_2{=}CHCH{=}CH_2$, and styrene, $C_6H_5C{=}CH_2$, was found to perform satisfactorily, and in 1945 this synthetic rubber (Buna S, GRS, or SBR are some of its common names) was produced in this country in the amount of 700,000 tons. This material is still used to some extent as a tire rubber, but is usually mixed with natural rubber or other synthetic rubbers to improve performance. Very recently, polymerized butadiene, having the cis structure present in the polymer double bonds in natural rubber, has been manufactured commercially, and promises to become an important rubber either in its own right or in blends with natural or styrene-butadiene rubbers.

24.4 NATURAL PRODUCTS

The organic substances we have so far considered contain relatively simple molecules, which can be derived for the most part from petroleum or coal. Many, perhaps most, organic compounds are obtained from the other natural source of such substances, namely, living plants and animals. The organic compounds which are found in living organisms are called *natural products* and include some of the most complex substances known. These have long been of interest to organic chemists and biochemists, who have studied their reactions both inside and outside the living system and have expended great effort in determining their molecular structures. In this section we shall be able to discuss but a few examples of this very important area of research.

Glucose: A Typical Carbohydrate and Sugar

A very common natural product is the substance known as glucose. Glucose is a member of that class of compound known as sugars; it is present in syrup, in some fruits, and is the sugar dissolved in blood. Glucose is obtained from the hydrolysis of both starch and cellulose. Because of its very wide occurrence in both the free and combined forms, it is perhaps the most abundant of all the organic substances.

Glucose also belongs to the **carbohydrate** family, which includes all the sugars. The molecular formula of glucose is $C_6H_{12}O_6$, which, like the formulas of all the carbohydrates, can be written as $C_x(H_2O)_y$. The carbohydrates are among the most important of the foods of man, being used by the body as a source of energy and as an intermediate in the formation of other substances. The chemical bonding in glucose is

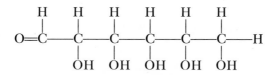

The glucose molecule is typical of the simple sugars. It contains a carbon chain, with one carbonyl group and hydroxyl groups on each carbon atom except the one in the carbonyl group. Because of the tetrahedral symmetry about the saturated carbon atom, the carbon chain in glucose is bent; this means that although there is free rotation around all the single C—C bonds, the relative positions of hydrogen atoms and hydroxyl groups on the chain are of significance, and the two structures below actually represent different substances:

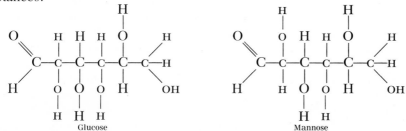

(In these drawings, the large and small symbols denote atoms above and below the plane of the paper.) Both substances are simple sugars, but, without breaking bonds, cannot be converted one to the other. The two substances are called **stereoisomers** of the basic structure (there are 16 in all).

The problem of finding which of the sixteen configurations of the stereoisomers of glucose actually belonged to naturally occurring glucose was solved by the great German chemist, Emil Fischer, in the twenty-year period beginning about 1885. By degrading longer-chain sugars to shorter ones of known configurations, synthesizing longer-chain sugars from shorter, and studying the relations between various derivatives formed from his sugar products, Fischer was finally able to determine the actual configuration of glucose and the other naturally occurring six-carbon sugars. For this work Fischer was awarded the Nobel Prize in chemistry in 1902.

Glucose and most other sugars in solution have the rather remarkable property of being able to rotate the plane of a beam of polarized light as it passes through the solution. These substances are said to be **optically active**. Optical activity is found in any organic substance in which there are carbon atoms bonded to four different groups; such carbon atoms are said to be **asymmetric**. In the aldohexoses there are four asymmetric carbon atoms, atoms to which are attached four nonidentical groups. Since each of the stereoisomers of glucose has a characteristic ability to rotate the plane of polarized light, Fischer was able to make use of their optical activities in his identifications of the various isomers.

Fischer's problem of determining the structures of the sugars was com-

Recall the discussion of optical isomerism, Chapter 19.

plicated by the fact that under ordinary circumstances sugars such as glucose have a cyclic structure. The most stable form of the glucose molecule is known to have the following configuration:

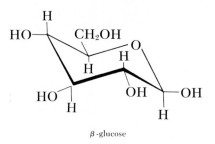

β-glucose

The ring is six-membered, containing one oxygen atom. From above it would appear as a hexagon, with carbon ring atoms alternately up and down, in a so-called chair configuration. The bulky OH and CH₂OH groups are oriented away from the ring center, more or less in the plane of the molecule, whereas the H atoms are in positions more directly above and below the ring; this structure, as might be expected, minimizes intramolecular strains caused by large groups in close proximity. When the ring is formed, the relative positions of the H and OH groups at the ring position at the extreme right in the drawing are sometimes interchanged, giving rise to an alternate, slightly less stable structure for the molecule, which is given the name α-glucose. The aqueous solution of glucose actually contains mainly cyclic molecules, 64 per cent β and 36 per cent α forms, along with a small amount of the open-chain form. In view of the enormous complexity of the properties of the aldohexoses, the determination of the configurations of stereoisomers of glucose by Fischer must even now be considered one of the major triumphs of the organic chemist.

Many sugars contain more than one simple sugar group per molecule. Ordinary sucrose, cane sugar, is a **disaccharide**, and on acid hydrolysis yields two simple sugars, of which one is glucose and the other, another common sugar, is called fructose. The molecular structure of sucrose is as follows:

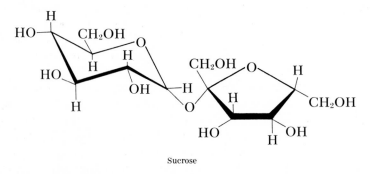

The carbon atoms of the ring are omitted for clarity.

Sucrose

The substances known as starch and cellulose both yield only glucose when subjected to acid hydrolysis. Starch is found in many plants as the carbohydrate stored in roots and seeds. It is present in large amounts in corn, potatoes, and wheat, and is one of the main sources of energy in our foods. Cellulose is the substance which makes up the cell membranes of most plants.

Ordinary wood is about 50 per cent cellulose; dry leaves, about 10 per cent. Cotton fiber, which contains about 98 per cent cellulose, is the best source of the pure material.

Both starch and cellulose are polymers of glucose, with molecular weights of about a million. The glucose units are linked through oxygen bridges, which may be considered to have formed in a condensation polymerization reaction in which one molecule of water is removed for each glucose molecule entering the chain. Cellulose molecules are unbranched; the chains are strongly hydrogen bonded, giving the material its high resistance to water and alcohols. Starch molecules are occasionally branched and cross-linked, perhaps once in 25 glucose units. The structures of starch and cellulose are similar and are indicated in the following formulas:

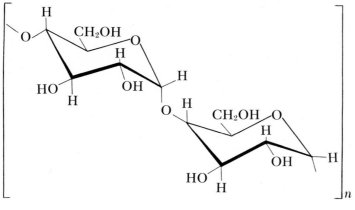

Starch

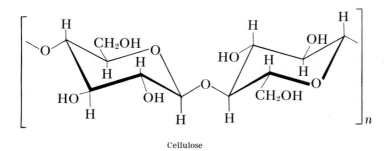

Cellulose

The essential difference between the cellulose and starch molecules appears to be that in the former the cyclic glucose units are in the β form, while in starch the α form exists. Starch can be readily hydrolyzed in the human body to the disaccharide called maltose by an enzyme called diastase. Cellulose, perhaps because of the more stable β configuration in its rings, is not affected by human enzymes and so is not useful as a food for man.

Amino Acids and Proteins

Although the structures of the carbohydrates are far from simple, some of the most complex substances belong to a different class of natural products,

641

called the **proteins**. It is in this area that much of the recent structural work has been done, with results that are truly impressive.

Proteins, like the polysaccharides starch and cellulose, are polymeric substances derived from a fairly large group of monomer units, called **amino acids**. The amino acids that are obtained from proteins are all α-amino acids, meaning that the amino, NH_2, group is on the carbon atom adjacent to the carboxyl group. The general formula for an α-amino acid can be written as

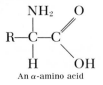

An α-amino acid

As you can see, unless R = H, the central carbon atom in the α-amino acid is asymmetric, making these substances optically active. The R group in the acids may be one of twenty-four groups. The names, abbreviations, and R groups present in these acids are given in Table 24.3.

Proteins constitute one of the major foods of man. They are present in most foods but are more abundant in lean meats and vegetables such as beans and peas than in fats or starchy foods such as potatoes and corn. In the animal body the proteins are hydrolyzed to α-amino acids, which are used to synthesize the many kinds of body proteins. The body cannot synthesize all these α-amino acids and must obtain about fourteen of them from the food which is eaten.

As we have noted, proteins are polymers of the α-amino acids. In a given protein, however, there are typically several amino acids, linked together in a specific order in a branched-chain structure. The linkage between amino acid molecules is from the acid carboxyl group on one molecule to the amino group on the next, with one water molecule being removed from the system per link formed; formally, then, the proteins may be considered to be condensation polymers of the amino acids. The bonding in the polymer chain is as indicated:

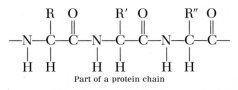

Part of a protein chain

The bonds between acid units are sometimes called **peptide bonds**, and the protein itself may be called a polypeptide. In the chain just illustrated, R, R', and R'' would not ordinarily be the same groups. The molecular weights of proteins are similar to those of many commercial polymers. The albumin obtained from egg white has a molecular weight of about 44,000, whereas that of urease, obtained from soy beans, is about 450,000.

In view of the nature of proteins, it is clear that unambiguous structure determinations for these substances are extremely difficult. One must first obtain a sample of the pure protein. This is accomplished by precipitating the protein selectively at its pH of minimum solubility, removing residual salt by placing the protein in a semipermeable membrane and washing with water, thereby ultimately increasing the purity of the protein within the mem-

TABLE 24.3 α-Amino Acids Obtained from Proteins

Name	Abbreviation	R—	Name	Abbreviation	R
Glycine	Gly	H—	Tyrosine	Tyr	HO—C₆H₄—CH₂—
Alanine	Ala	CH_3—			
Valine	Val	$(CH_3)_2CH$—	Thyroxine	Thy	(diiodo-hydroxyphenyl ether structure) CH₂—
Leucine	Leu	$(CH_3)_2CHCH_2$—			
Isoleucine	Ileu	$CH_3CH_2CH(CH_3)$—	Tryptophan	Try	(indole) CH₂—
Phenylalanine	Phe	$PhCH_2$—			
Serine	Ser	$HOCH_2$—			
Threonine	Thr	CH_3CHOH—			
Cysteine	CySH	$HSCH_2$—	Histidine	His	(imidazole) CH₂—
Methionine	Met	$CH_3SCH_2CH_2$—			
Asparagine	Asp-NH₂	H_2NCOCH_2—			
Glutamine	Glu-NH₂	$H_2NCOCH_2CH_2$—	Proline	Pro	Acid molecules —COOH (pyrrolidine)
Lysine	Lys	$H_2NCH_2CH_2CH_2CH_2$—			
δ-Hydroxylysine	Lys-OH	$H_2NCH_2CH(OH)CH_2CH_2$—			
Aspartic Acid	Asp	$HOOCCH_2$—			
Glutamic Acid	Glu	$HOOCCH_2CH_2$—	Hydroxyproline	Hypro	HO—(pyrrolidine)—COOH
Cystine	CySSCy	$-SCH_2$(an acid dimer)			
Arginine	Arg	$H_2NCNHCH_2CH_2CH_2$— $\parallel$ NH			

brane to the point at which it can be crystallized. Some of the protein is then completely hydrolyzed and the resultant amino acids determined as to kind and amount. This can be accomplished by thin-layer or paper chromatography, by ion exchange resins, or by other means. The protein is then partially hydrolyzed, into fragments containing from two to about seven acid units. These are analyzed as to type and relative amount, as were the amino acid units, thereby establishing the sequences of acid units in the protein fragments. The final protein structure is obtained by noting how the fragments overlap one another, and how, therefore, the amino acid sequence must exist in the whole molecule. Several proteins have been analyzed by this method, one of the most complex being insulin, the structure of which was reported by F. Sanger in 1955 (Nobel Prize, 1958). The amino acid sequence in this substance is as follows:

This structure is easier to write than to pronounce.

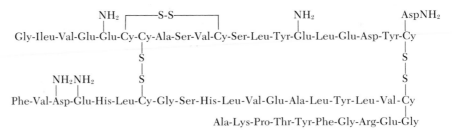

In addition to work on the problem of the amino acid sequence in proteins, there has been considerable research on the geometric arrangement of the protein chains in space. Within the protein chain there are many polar groups, some of which can interact through hydrogen bonding. It appears that in many proteins these interactions cause the chains to take on the form of a helix, like that of a coiled spring. In such a structure it is possible for hydrogen bonding to occur between a nitrogen atom at one point on the coil and an oxygen atom on a carbonyl group on the coil one turn down. In fibrous proteins such as hair and silk, several helices may intertwine, much in the manner in which the strands are twisted in a sisal rope, to give the fiber its strength and elasticity.

Some proteins, like insulin and egg albumin, yield only α-amino acids on hydrolysis. Other proteins are more complex, and can be resolved into two fractions, one containing a simple protein and the other a nonprotein group. Among the most interesting and challenging of the so-called *conjugated* proteins are the *nucleoproteins*, which can be resolved to produce a *nucleic acid* portion and a protein portion. Nucleoproteins are the main component of cell chromosomes; the nucleic acids are thought to be the substances which control cell reproduction and allow a species to reproduce its kind.

Nucleic acids are polymers with molecular weights of the order of one million. On complete hydrolysis they yield several kinds of substances, including one of two sugars, phosphoric acid, and a group of nitrogen ring compounds called nitrogen bases.

Partial hydrolysis of nucleic acids produces phosphoric acid plus fragments containing a sugar molecule bonded to a nitrogen base. This has led to the belief that the nucleic acid molecule consists of long chains of sugar–nitrogen base groups linked together by phosphate groups. When the sugar present is ribose, the nucleic acid is called RNA (ribonucleic acid). When

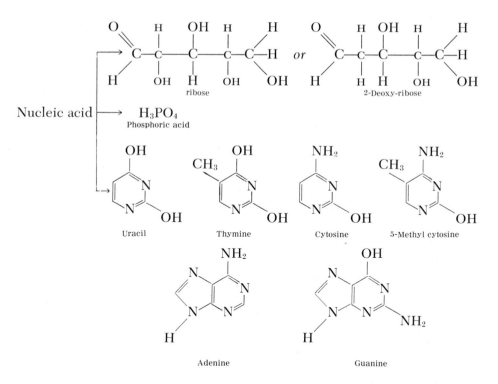

the sugar is 2-deoxy-ribose, the acid is given the name DNA (deoxyribo-nucleic acid). In a DNA molecule there are about 500 units.

All DNA molecules produce on hydrolysis equal numbers of moles of adenine and thymine, and equal numbers of moles of guanine and cytosine plus methyl cytosine. This has led to the belief that in the molecule adenine and thymine groups are paired and so are guanine and cytosine groups. If one draws scale models of these substances, he finds that hydrogen bonding can readily occur between molecules in the same pair, but not between molecules in different pairs. (Figure 24.3)

On the basis of this evidence and x-ray data on the DNA crystal, Watson

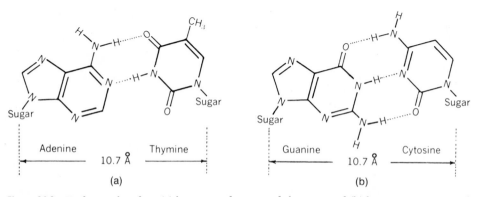

Figure 24.3 Hydrogen bonding (a) between adenine and thymine and (b) between guanine and cytosine. (From Noller, C. R.: *Chemistry of Organic Compounds*, W. B. Saunders Co., Philadelphia, 1965, 3rd edition.)

and Crick (Nobel Prize, 1962) proposed that the DNA polymer consists of a two-stranded helix in which the two strands are held together by bonding between the pairs of substances. The sugar and phosphate units can be readily accommodated on the outside of the helix, producing a relatively compact molecule. The general structure is as indicated in Figure 24.4. The structure suggested by Watson and Crick has been examined by others and appears to be substantially correct. One important feature of the structure is that it leads to a rather simple explanation of the mechanism by which DNA can be reproduced by the organism. The double helix is thought to separate, and the individual strands to attract and bind to themselves in proper sequence the proper nitrogen base–sugar groups, creating in the process another strand essentially identical to the one which split off.

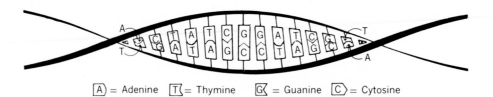

[A] = Adenine   [T] = Thymine   [G] = Guanine   [C] = Cytosine

Figure 24.4 Representation of the double-stranded spiral structure of a hypothetical deoxyribonucleic acid. (From Noller, C. R.: *Chemistry of Organic Compounds*, W. B. Saunders Co., Philadelpha, 1965, 3rd edition.)

In recent years there has been a tremendous amount of research on the manner in which RNA and DNA are able to establish the arrangement of amino acids in proteins synthesized by the living organism. It is believed that within RNA the sequence in which the component nitrogen bases are arranged constitutes a code by which the cell is able to establish the sequence in which the amino acids within a given protein are to be linked. This so-called genetic code has now been established for most of the naturally occurring amino acids. In 1968, Nirenberg, Holley, and Khorana, three American biochemists, were awarded the Nobel Prize in medicine and physiology for their contribution to the solution of this problem in molecular biology.

PROBLEMS

24.1   Define each of the following terms. For each term give the formula of an organic substance which meets the definition and so would be included within the group of substances covered by the term.

     a. Paraffin   d. Cycloparaffin
     b. Hydrocarbon   e. Aromatic substance
     c. Olefin   f. Alkene

24.2   Draw as many isomeric structures as you can for molecules with the formula C_5H_{12}.

24.3   Classify each of the following as a paraffin, an olefin, an acetylene, or an aromatic hydrocarbon.

     a. C_3H_6 b. C_6H_{14} c. C_5H_8 d. C_7H_8 e. C_4H_{10}

24.4 Classify each of the following molecules as belonging to an alcohol, an ether, an aldehyde, a ketone, or an acid.

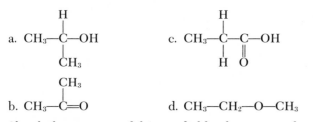

24.5 Sketch the structure of the paradichlorobenzene molecule, $C_6H_4Cl_2$.

24.6 Sketch the molecules of each of the following substances.

a. Methyl ethyl ether
b. Normal butyl alcohol
c. 2-Propanol
d. Acetaldehyde
e. Acetone
f. Glycerol
g. 1,2-Dichloroethane
h. 2-Methyl-3-ethylpentane

24.7 Consider the following possible reactions of 1-iodopropane.

a. $C_3H_7I + OH^- \longrightarrow C_3H_7OH + I^-$
b. $C_3H_7I + OH^- \longrightarrow C_3H_6 + H_2O + I^-$

What kind of reaction is reaction a? Reaction b? What reaction conditions would tend to favor reaction a? Reaction b?

24.8 The following would be expected to be an S_N2 reaction:

$$CH_3CH_2Cl + OH^- \longrightarrow CH_3CH_2OH + Cl^-$$

A certain reaction mixture containing CH_3CH_2Cl and OH^-, both initially at a concentration of 1.0 M, is 50% reacted after one hour. How long did it take for the first 20% of reaction to occur? (Recall the discussion of second order reactions, Chapter 14).

24.9 What does petroleum consist of? What problems must be solved if one is to obtain a maximum amount of usable gasoline from a given amount of petroleum?

24.10 What does natural rubber consist of? Sketch a portion of a molecule that would be present in vulcanized natural rubber.

24.11 Sketch the simplest optically active molecule you can think of.

24.12 Distinguish between a protein and a nucleic acid.

24.13 What are the components of the nucleic acids called DNA? Why are the nucleic acids of great biological interest?

24.14 Define and give an example of a substance falling within each of the following classes of organic compounds.

a. Ketone
b. Ether
c. Carboxylic acid
d. Ester
e. Fat
f. Alkyl halide
g. Aldehyde
h. Phenol

24.15 Define each of the following terms.

a. Carbohydrate
b. Protein
c. Asymmetric carbon atom
d. Stereoisomer
e. Optical activity

24.16 How many isomers are there with the molecular formula C_4H_9Cl? Sketch each of the isomeric molecules.

647

24.17 Draw structural formulas to correspond to the following molecular formulas.

    a. C_3H_6 b. $C_3H_4Cl_2$ c. C_7H_7Cl d. $C_6H_3Cl_3$

24.18 For each of the molecular formulas, indicate whether the compound could be an ether, an alcohol, an aldehyde, a ketone, or an acid. Sketch the structures of the molecules which you believe to be consistent with your answers.

    a. CH_4O b. C_2H_6O c. $C_2H_4O_2$ d. C_3H_6O e. CH_2O

24.19 C_8H_8 is called cyclooctatetraene. Assuming the carbon skeleton of the molecule forms a regular plane octagon, how many dibromo derivatives of cyclooctatetraene would you expect to exist? (Actually cyclooctatetraene is not planar and does not behave like a typical aromatic substance.)

24.20 Butane, C_4H_{10}, is a straight-chain saturated hydrocarbon. If each of the following substances contains a four-carbon unbranched chain, sketch its molecular structure.

    a. Butyraldehyde   d. 1-Butene
    b. Butyric acid   e. 1-Butanol
    c. 2-Butanol   f. 2-Butanone (a ketone)

24.21 What is a polymer? Polypropylene is an addition polymer of propylene. Sketch a portion of the polypropylene molecule. What fraction by weight of polypropylene would be carbon?

24.22 Tertiary butyl alcohol, $(CH_3)_3COH$, undergoes nucleophilic substitution or elimination reactions via S_N1 kinetics. Sketch the reaction intermediate for these reactions. What would be the product of nucleophilic substitution with I^-? What would be the products in elimination?

24.23 Name four processes used in the petroleum industry to improve the amount and quality of gasoline produced. Give an example of a reaction which might occur in each process.

24.24 If neoprene has a molecular structure similar to that of natural rubber, sketch a portion of a neoprene molecule.

24.25 Sketch a portion of the polybutadiene rubber recently produced with the molecular structure similar to that of natural rubber. Why would not simply polymerizing butadiene produce this material?

24.26 How many asymmetric carbon atoms are there in each of the following α-amino acids?

    a. Glycine   c. Asparagine
    b. Valine   d. Arginine

24.27 Sketch that portion of the insulin molecule containing the following chain of amino acid groups:

$$\text{-Ser-Leu-Tyr-Glu-}$$
$$|$$
$$\text{NH}_2$$

°24.28 Suggest how one might distinguish among ortho-, meta-, and para-dichlorobenzene by counting the isomers formed when one more chlorine atom is substituted for hydrogen.

°24.29 The general formula for a paraffin is C_nH_{2n+2}. Derive the general formula for an olefin. For an acetylene. What conditions must be satisfied if these formulas are to be valid?

°24.30 How many isomers would you expect for a paraffin with the molecular formula C_8H_{18}?

°24.31 Under what conditions would you expect the following reaction to proceed to the right?

$$C_2H_5OH + Br^- \longrightarrow C_2H_5Br + OH^-$$

°24.32 Write equations for the reactions, including conditions, by which one could prepare the following from propyl chloride, $CH_3CH_2CH_2Cl$.

a. An alcohol   b. An ether   c. An olefin   d. An amine

The following molecular formulas of substances named in these problems may prove to be useful.

methyl ethyl ether: $CH_3OC_2H_5$
normal butyl alcohol:
   $CH_3CH_2CH_2CH_2OH$
2-propanol: $CH_3CHOHCH_3$
acetaldehyde: CH_3CHO
acetone: CH_3COCH_3

glycerol: $CH_2OHCH_2OHCH_2OH$
1,2-dichloroethane: CH_2ClCH_2Cl
2-methyl,3-ethylpentane:
   $CH_3CH(CH_3)CH(C_2H_5)CH_2CH_3$
propylene: $CH_3CH{=}CH_2$

SUGGESTED READINGS

Perhaps the most important objective of the general chemistry course is to motivate the student to learn more about topics which have aroused his curiosity and interest. Instructors may also wish to spend more time on subjects such as thermodynamics, coordination chemistry, or organic chemistry which receive limited coverage in a general text. To fulfill either or both of these needs, we suggest the following paperbacks and short texts as particularly worthwhile.

"The Chemical Bond", J. J. Lagowski, Houghton Miflin Co. (1966). A fascinating historical account of the development of bonding theory.

"Electronic Structure and Chemical Bonding", Donald K. Sebera, Blaisdell Publishing Co. (1964). A classic; "written for the student" in the best sense of the phrase.

"Chemical Energy", Lawrence E. Strong and Wilmer J. Stratton, Reinhold Publishing Co. (1965). An elementary introduction to chemical thermodynamics with emphasis on chemistry.

"Why Do Chemical Reactions Occur?", J. Arthur Cambell, Prentice Hall Inc. (1965). A highly original (and eminently readable) approach that integrates thermodynamics and kinetics.

"Inorganic Complex Compounds", R. Kent Murmann, Reinhold Publishing Co. (1964). A survey of coordination chemistry by an authority in the field.

"Elementary Organic Chemistry, A Brief Course", H. O. Van Orden and G. L. Lee, W. B. Saunders Co., (1969). Particularly suitable as a supplementary text for those courses in which a half semester or more is devoted to organic chemistry.

Mathematics

EXPONENTIAL NOTATION

In chemistry one frequently has to deal with numbers that are either very large or very small. In a gram of lead metal there are about

$$2,910,000,000,000,000,000,000 \text{ atoms}$$

and each lead atom weighs about $0.000\,000\,000\,000\,000\,000\,000\,000\,344$ gram. Rather than write out so many zeroes when such numbers arise, scientists employ a shorthand notation to express these numbers. The notation is easy to learn and very convenient to use.

To represent the number 100, one can write its equivalent, 10×10, and then abbreviate the product by the expression 10^2. Similarly, 1,000,000, which is equal to $10 \times 10 \times 10 \times 10 \times 10 \times 10$, can be written as 10^6. The numbers 2 and 6, the superscripts in 10^2 and 10^6, are called **exponentials** and indicate how many times 10 is multiplied by itself to obtain the desired number.

Any number greater than unity can be written as the product of a number between 1 and 10 times 10 raised to some positive integral exponential. For example,

$$1325 = 1.325 \times (10 \times 10 \times 10) = 1.325 \times 10^3$$

$$26 = 2.6 \times 10^1$$

$$2,910,000,000,000,000,000,000 = 2.91 \times 10^{21}$$

For numbers which are less than unity, exponential notation can also be used.

$$0.01 = \frac{1}{100} = \frac{1}{10 \times 10} = \frac{1}{10^2} = 10^{-2}$$

The negative exponential indicates the number of times $\frac{1}{10}$, or 0.10, is to be multiplied by itself to obtain the desired number.

Any number less than unity can be expressed as the product of a number between 1 and 10, times 10 raised to some negative integral exponential. Some examples would be

$$0.0065 = 6.5 \times \frac{1}{10 \times 10 \times 10} = 6.5 \times 10^{-3}$$

$$0.1546 = 1.546 \times 10^{-1}$$

$$0.000\,000\,000\,000\,000\,000\,000\,344 = 3.44 \times 10^{-22}$$

A major advantage of expressing very large or very small numbers in exponential notation is that it greatly simplifies the operations of multiplication and division. To **multiply**, one **adds exponents**; to **divide**, one **subtracts exponents**. Thus we have

$$10^3 \times 10^4 = 10^7; \qquad 10^3 \times 10^{-5} = 10^{-2}; \qquad 10^x \times 10^y = 10^{x+y}$$

$$10^5 \ / \ 10^3 = 10^2; \qquad 10^4 \ / \ 10^{-2} = 10^6; \qquad 10^x \ / \ 10^y = 10^{x-y}$$

The product of two numbers written in exponential form is obtained by taking the product of the number terms and multiplying by the product of the exponential terms. The following examples are illustrative.

$$(1.64 \times 10^2) \times (2.31 \times 10^8) = (1.64 \times 2.31) \times (10^2 \times 10^8) = 3.79 \times 10^{2+8}$$

$$= 3.79 \times 10^{10}$$

$$(3.1 \times 10^6) \times (4.2 \times 10^{-11}) = (3.1 \times 4.2) \times (10^6 \times 10^{-11}) = 13 \times 10^{-5}$$

$$= 1.3 \times 10^{-4}$$

The quotient of numbers written in exponential notation is equal to the quotient of the number terms times the quotient of the exponential terms. Consider the following examples:

$$\frac{4.29 \times 10^5}{2.59 \times 10^2} = \frac{4.29}{2.59} \times \frac{10^5}{10^2} = 1.66 \times 10^{5-2} = 1.66 \times 10^3$$

$$\frac{(1.2 \times 10^{-3}) \times (4.7 \times 10^7)}{(6.0 \times 10^{15}) \times (3.1 \times 10^{-40})} = \frac{1.2 \times 4.7}{6.0 \times 3.1} \times \frac{10^{-3} \times 10^7}{10^{15} \times 10^{-40}} = 3.0 \times 10^{28}$$

To solve the last example without using exponential notation would be very tedious, and in all probability would result in a decimal error, at least.

It is, of course, not necessary to have the number multiplying the exponential term lie between 1 and 10, since clearly

$$1.65 \times 10^7 = 16.5 \times 10^6 = 165 \times 10^5$$

and in a given circumstance it might be most convenient to report the result by the last expression in the group.

Sometimes it is necessary to calculate the square or cube root of a number given in exponential notation. Since $(10^{2x})^{\frac{1}{2}}$ equals 10^x, it is most convenient to obtain a square root if the exponential is an even, rather than an odd, integer.

$$(72.4 \times 10^7)^{\frac{1}{2}} = (7.24)^{\frac{1}{2}} \times (10^8)^{\frac{1}{2}} = 2.69 \times 10^4$$

Similarly, a cube root is most easily obtained if the exponential is made a multiple of 3 by appropriate manipulation of the number.

$$(72.4 \times 10^7)^{\frac{1}{3}} = (724)^{\frac{1}{3}} \times (10^6)^{\frac{1}{3}} = 8.98 \times 10^2$$

Occasionally, it is necessary to add or subtract numbers written in exponential form. For example, we might wish to add 4×10^4 and 2×10^5. To do this, we rewrite one of the numbers so that the two exponents are the same.

$$4 \times 10^4 + 2 \times 10^5 = 0.4 \times 10^5 + 2 \times 10^5 = 2.4 \times 10^5$$

or
$$= 4 \times 10^4 + 20 \times 10^4 = 24 \times 10^4$$

The two answers, 2.4×10^5 and 24×10^4, are, of course, identical. To subtract 3×10^{-6} from 2×10^{-5}, we proceed in a similar manner.

$$2 \times 10^{-5} - 3 \times 10^{-6} = 20 \times 10^{-6} - 3 \times 10^{-6} = 17 \times 10^{-6}$$

or
$$= 2 \times 10^{-5} - 0.3 \times 10^{-5} = 1.7 \times 10^{-5}$$

COMMON AND NATURAL LOGARITHMS

In making calculations one frequently finds it convenient, or necessary, to employ logarithms. The ordinary, or common, logarithm of a number is equal to the exponential to which 10 must be raised to produce the number.

$$\log x = y \qquad \text{means } 10^y = x$$

$$\log 100 = 2 \qquad \text{means } 10^2 = 100$$

$$\log 0.001 = -3 \qquad \text{means } 10^{-3} = 0.001$$

Logarithms are useful for obtaining products of numbers, for taking their quotient, and for raising numbers to powers or extracting roots. The operations which must be carried out to accomplish such calculations are as follows.

Multiplication.   The logarithm of the product of two numbers is equal to the sum of the logarithms of the two numbers.

$$\log 6 = \log 2 + \log 3$$

$$\log 2 \times 10^5 = \log 2 + \log 10^5$$

Division.   The logarithm of the quotient of two numbers is equal to the difference between the logarithms of dividend and divider.

$$\log 3 = \log 6 - \log 2$$

$$\log \frac{2 \times 10^5}{6} = \log 2 + \log 10^5 - \log 6$$

Raising to a Power or Extracting a Root.   The logarithm of any number raised to a power is equal to the product of the exponent times the logarithm of the number.

$$\log 2^4 = 4 \log 2$$

$$\log 2^{\frac{1}{3}} = \frac{1}{3} \log 2$$

The logarithm of a number between 1 and 10 can be found directly from a table of logarithms such as that given on p. 656 or from a slide rule. Looking at the table, we note that

$$\log 1 = 0.0000; \quad 10^0 = 1$$

$$\log 2 = 0.3010; \quad 10^{0.3010} = 2$$

$$\log 3 = 0.4771; \quad 10^{0.4771} = 3$$

$$\log 6 = 0.7781; \quad 10^{0.7781} = 6$$

To find the logarithm of 6.02 in the table, we first locate the number 6.0 in the column at the far left and then follow across horizontally until we come to the column headed "2". We then read off the logarithm of 6.02 as 0.7796. To estimate the logarithm of 6.023, we note that the logarithm of 6.02 is 0.7796, while that of the next largest number, 6.03, is 0.7803. Since the number 6.023 is $\frac{3}{10}$ of the way between 6.02 and 6.03, we take $\frac{3}{10}$ of the difference between the two logs and add this to the log of 6.02.

$$\log 6.023 = \log 6.02 + 0.3 \, (\log 6.03 - \log 6.02)$$

$$= 0.7796 + 0.3 \, (0.0007) = 0.7798$$

With a four-place table of logarithms, this process of interpolation is necessary whenever a number is expressed to more than three digits.

The logarithm of a number less than 1 or greater than 10 can be obtained readily by writing the number in exponential form and applying the rules for multiplication.

$$\log 60.2 = \log (6.02 \times 10^1) = \log 6.02 + \log 10^1$$

$$= 0.7796 + 1 = 1.7796$$

$$\log 6.02 \times 10^{-5} = \log 6.02 + \log 10^{-5} = 0.7796 + (-5) = -4.2204$$

If the logarithm of a number is given, one can find the number by the inverse of the process used to obtain the logarithm. If the logarithm lies between 0 and 1, the number can be found directly from the table. Thus, to find the number whose logarithm is 0.8000, we locate 0.8000 in the body of the table, finding it in the horizontal row headed 6.3, under the vertical column headed "1". We deduce that the number whose logarithm is 0.8000 is 6.31. This fact is sometimes expressed by saying that the **antilogarithm** of 0.8000 is 6.31.

If the logarithm we are working with is greater than 1, let us say 2.8000, we express it as a sum:

$$2.8000 = 2 + 0.8000$$

We locate 0.8000 in the table, finding, as before, that its antilogarithm is 6.31. Since the antilogarithm of 2 is 10^2 ($\log 10^2 = 2$), we have

$$\text{antilog } 2.8000 = \text{antilog } 0.8000 \times \text{antilog } 2$$

$$= 6.31 \times 10^2$$

Similarly,

$$\text{antilog } 6.3010 = \text{antilog } 0.3010 \times \text{antilog } 6$$

$$= 2.00 \times 10^6$$

If the logarithm is less than zero, the process of finding the corresponding number is slightly more complicated. Suppose, for example, that we desire to find the antilogarithm of -5.6990. Since the logarithms in the table are expressed as positive fractions between 0 and 1, we must express -5.6990 as a fraction minus a whole number.

$$-5.6990 = 0.3010 - 6$$

$$\text{antilog of } -5.6990 = \text{antilog } 0.3010 \times \text{antilog of } -6$$

$$= 2.00 \times 10^{-6}$$

Similarly, if we wish to obtain the number whose logarithm is -2.3161:

$$-2.3161 = 0.6839 - 3$$

$$\text{antilog of } -2.3161 = \text{antilog } 0.6839 \times \text{antilog of } -3$$

$$= 4.83 \times 10^{-3}$$

In chemical calculations one usually uses a slide rule rather than logarithms, since slide rule calculations can be carried out very rapidly and with a precision appropriate to most chemical data. The student must, however, be familiar with the properties of logarithms since they are involved in certain very common chemical expressions.

Calculations with logarithms are carried out in most cases with the kinds of logarithms we have been discussing, in which 10 is the base, or the number which is raised to an exponential. Logarithms to the base 10 are called **ordinary** or **common logarithms**. It is possible, however, to set up a system of logarithms based on any arbitrary number, by using rules completely analogous to those used for the base 10. Let us consider the relations obtaining in a system of logarithms based on an arbitrarily chosen number, which we shall call N:

$$\log_N x = y \qquad \text{means } N^y = x$$

$$\log_2 8 = 3 \qquad \text{means } 2^3 = 8$$

$$\log_2 0.25 = -2 \qquad \text{means } 2^{-2} = 0.25$$

Note that the base of the logarithm is written as a subscript.

Many natural laws are found to be exponential in character. Radioactive decay, the temperature dependence of equilibrium constants and of reaction rate constants, and the distribution function for molecular speeds all involve equations which include exponential terms. The exponential terms in these equations are most conveniently expressed in terms of a number called e, raised to a power, rather than 10 raised to a power. The number e can be shown to have the value 2.71828. . . . Sometimes the natural law will be expressed in logarithmic rather than exponential form, with the system of logarithms being based on e rather than 10. Such logarithms are called **natural logarithms**, because the relations from which logarithms can be calculated are most simply expressed in terms of the base e; natural logarithms are assigned the symbol ln, to distinguish them from common logarithms, usually written as log with no subscript.

APPENDIX 1

TABLE OF LOGARITHMS

	0	1	2	3	4	5	6	7	8	9
1.0	.0000	.0043	.0086	.0128	.0170	.0212	.0253	.0294	.0334	.0374
1.1	.0414	.0453	.0492	.0531	.0569	.0607	.0645	.0682	.0719	.0755
1.2	.0792	.0828	.0864	.0899	.0934	.0969	.1004	.1038	.1072	.1106
1.3	.1139	.1173	.1206	.1239	.1271	.1303	.1335	.1367	.1399	.1430
1.4	.1461	.1492	.1523	.1553	.1584	.1614	.1644	.1673	.1703	.1732
1.5	.1761	.1790	.1818	.1847	.1875	.1903	.1931	.1959	.1987	.2014
1.6	.2041	.2068	.2095	.2122	.2148	.2175	.2201	.2227	.2253	.2279
1.7	.2304	.2330	.2355	.2380	.2405	.2430	.2455	.2480	.2504	.2529
1.8	.2553	.2577	.2601	.2625	.2648	.2672	.2695	.2718	.2742	.2765
1.9	.2788	.2810	.2833	.2856	.2878	.2900	.2923	.2945	.2967	.2989
2.0	.3010	.3032	.3054	.3075	.3096	.3118	.3139	.3160	.3181	.3201
2.1	.3222	.3243	.3263	.3284	.3304	.3324	.3345	.3365	.3385	.3404
2.2	.3424	.3444	.3464	.3483	.3502	.3522	.3541	.3560	.3579	.3598
2.3	.3617	.3636	.3655	.3674	.3692	.3711	.3729	.3747	.3766	.3784
2.4	.3802	.3820	.3838	.3856	.3874	.3892	.3909	.3927	.3945	.3962
2.5	.3979	.3997	.4014	.4031	.4048	.4065	.4082	.4099	.4116	.4133
2.6	.4150	.4166	.4183	.4200	.4216	.4232	.4249	.4265	.4281	.4298
2.7	.4314	.4330	.4346	.4362	.4378	.4393	.4409	.4425	.4440	.4456
2.8	.4472	.4487	.4502	.4518	.4533	.4548	.4564	.4579	.4594	.4609
2.9	.4624	.4639	.4654	.4669	.4683	.4698	.4713	.4728	.4742	.4757
3.0	.4771	.4786	.4800	.4814	.4829	.4843	.4857	.4871	.4886	.4900
3.1	.4914	.4928	.4942	.4955	.4969	.4983	.4997	.5011	.5024	.5038
3.2	.5051	.5065	.5079	.5092	.5105	.5119	.5132	.5145	.5159	.5172
3.3	.5185	.5198	.5211	.5224	.5237	.5250	.5263	.5276	.5289	.5302
3.4	.5315	.5328	.5340	.5353	.5366	.5378	.5391	.5403	.5416	.5428
3.5	.5441	.5453	.5465	.5478	.5490	.5502	.5514	.5527	.5539	.5551
3.6	.5563	.5575	.5587	.5599	.5611	.5623	.5635	.5647	.5658	.5670
3.7	.5682	.5694	.5705	.5717	.5729	.5740	.5752	.5763	.5775	.5786
3.8	.5798	.5809	.5821	.5832	.5843	.5855	.5866	.5877	.5888	.5899
3.9	.5911	.5922	.5933	.5944	.5955	.5966	.5977	.5988	.5999	.6010
4.0	.6021	.6031	.6042	.6053	.6064	.6075	.6085	.6096	.6107	.6117
4.1	.6128	.6138	.6149	.6160	.6170	.6180	.6191	.6201	.6212	.6222
4.2	.6232	.6243	.6253	.6263	.6274	.6284	.6294	.6304	.6314	.6325
4.3	.6335	.6345	.6355	.6365	.6375	.6385	.6395	.6405	.6415	.6425
4.4	.6435	.6444	.6454	.6464	.6474	.6484	.6493	.6503	.6513	.6522
4.5	.6532	.6542	.6551	.6561	.6571	.6580	.6590	.6599	.6609	.6618
4.6	.6628	.6637	.6646	.6656	.6665	.6675	.6684	.6693	.6702	.6712
4.7	.6721	.6730	.6739	.6749	.6758	.6767	.6776	.6785	.6794	.6803
4.8	.6812	.6821	.6830	.6839	.6848	.6857	.6866	.6875	.6884	.6893
4.9	.6902	.6911	.6920	.6938	.6937	.6946	.6955	.6964	.6972	.6981
5.0	.6990	.6998	.7007	.7016	.7024	.7033	.7042	.7050	.7059	.7067
5.1	.7076	.7084	.7093	.7101	.7110	.7118	.7126	.7135	.7143	.7152
5.2	.7160	.7168	.7177	.7185	.7193	.7202	.7210	.7218	.7226	.7235
5.3	.7243	.7251	.7259	.7267	.7275	.7284	.7292	.7300	.7308	.7316
5.4	.7324	.7332	.7340	.7348	.7356	.7364	.7372	.7380	.7388	.7396
5.5	.7404	.7412	.7419	.7427	.7435	.7443	.7451	.7459	.7466	.7474
5.6	.7482	.7490	.7497	.7505	.7513	.7520	.7528	.7536	.7543	.7551
5.7	.7559	.7566	.7574	.7582	.7589	.7597	.7604	.7612	.7619	.7627
5.8	.7634	.7642	.7649	.7657	.7664	.7672	.7679	.7686	.7694	.7701
5.9	.7709	.7716	.7723	.7731	.7738	.7745	.7752	.7760	.7767	.7774

TABLE OF LOGARITHMS—*Continued*

	0	1	2	3	4	5	6	7	8	9
6.0	.7782	.7789	.7796	.7803	.7810	.7818	.7825	.7832	.7839	.7846
6.1	.7853	.7860	.7868	.7875	.7882	.7889	.7896	.7903	.7910	.7917
6.2	.7924	.7931	.7938	.7945	.7952	.7959	.7966	.7973	.7980	.7987
6.3	.7993	.8000	.8007	.8014	.8021	.8028	.8035	.8041	.8048	.8055
6.4	.8062	.8069	.8075	.8082	.8089	.8096	.8102	.8109	.8116	.8122
6.5	.8129	.8136	.8142	.8149	.8156	.8162	.8169	.8176	.8182	.8189
6.6	.8195	.8202	.8209	.8215	.8222	.8228	.8235	.8241	.8248	.8254
6.7	.8261	.8267	.8274	.8280	.8287	.8293	.8299	.8306	.8312	.8319
6.8	.8325	.8331	.8338	.8344	.8351	.8357	.8363	.8370	.8376	.8382
6.9	.8388	.8395	.8401	.8407	.8414	.8420	.8426	.8432	.8439	.8445
7.0	.8451	.8457	.8463	.8470	.8476	.8482	.8488	.8494	.8500	.8506
7.1	.8513	.8519	.8525	.8531	.8537	.8543	.8549	.8555	.8561	.8567
7.2	.8573	.8579	.8585	.8591	.8597	.8603	.8609	.8615	.8621	.8627
7.3	.8633	.8639	.8645	.8651	.8657	.8663	.8669	.8675	.8681	.8686
7.4	.8692	.8698	.8704	.8710	.8716	.8722	.8727	.8733	.8739	.8745
7.5	.8751	.8756	.8762	.8768	.8774	.8779	.8785	.8791	.8797	.8802
7.6	.8808	.8814	.8820	.8825	.8831	.8837	.8842	.8848	.8854	.8859
7.7	.8865	.8871	.8876	.8882	.8887	.8893	.8899	.8904	.8910	.8915
7.8	.8921	.8927	.8932	.8938	.8943	.8949	.8954	.8960	.8965	.8971
7.9	.8976	.8982	.8987	.8993	.8998	.9004	.9009	.9015	.9020	.9026
8.0	.9031	.9036	.9042	.9047	.9053	.9058	.9063	.9069	.9074	.9079
8.1	.9085	.9090	.9096	.9101	.9106	.9112	.9117	.9122	.9128	.9133
8.2	.9138	.9143	.9149	.9154	.9159	.9165	.9170	.9175	.9180	.9186
8.3	.9191	.9196	.9201	.9206	.9212	.9217	.9222	.9227	.9232	.9238
8.4	.9243	.9248	.9253	.9258	.9263	.9269	.9274	.9279	.9284	.9289
8.5	.9294	.9299	.9304	.9309	.9315	.9320	.9325	.9330	.9335	.9340
8.6	.9345	.9350	.9355	.9360	.9365	.9370	.9375	.9380	.9385	.9390
8.7	.9395	.9400	.9405	.9410	.9415	.9420	.9425	.9430	.9435	.9440
8.8	.9445	.9450	.9455	.9460	.9465	.9469	.9474	.9479	.9484	.9489
8.9	.9494	.9499	.9504	.9509	.9513	.9518	.9523	.9528	.9533	.9538
9.0	.9542	.9547	.9552	.9557	.9562	.9566	.9571	.9576	.9581	.9586
9.1	.9590	.9595	.9600	.9605	.9609	.9614	.9619	.9624	.9628	.9633
9.2	.9638	.9643	.9647	.9652	.9657	.9661	.9666	.9671	.9675	.9680
9.3	.9685	.9689	.9694	.9699	.9703	.9708	.9713	.9717	.9722	.9727
9.4	.9731	.9736	.9741	.9745	.9750	.9754	.9759	.9763	.9768	.9773
9.5	.9777	.9782	.9786	.9791	.9795	.9800	.9805	.9809	.9814	.9818
9.6	.9823	.9827	.9832	.9836	.9841	.9845	.9850	.9854	.9859	.9863
9.7	.9868	.9872	.9877	.9881	.9886	.9890	.9894	.9899	.9903	.9908
9.8	.9912	.9917	.9921	.9926	.9930	.9934	.9939	.9943	.9948	.9952
9.9	.9956	.9961	.9965	.9969	.9974	.9978	.9983	.9987	.9991	.9996

$$P = C\ e^{-\Delta H/RT} \qquad \text{means} \qquad P = C \times 2.718^{-\Delta H/RT}$$

$$\ln P = \ln C - \Delta H/RT \quad \text{means} \quad \log_e P = \log_e C - \Delta H/RT$$

Clearly the logarithm of an exponential of e is most easily determined as a natural logarithm, and since natural laws typically involve exponentials based on e, natural logarithms are often used in writing equations for such laws.

Natural logarithms are readily converted to base 10 logarithms. Suppose, for example, that $\log_e x = A$. Then $e^A = x$. Taking base 10 logarithms, we have $A \log_{10} e = \log_{10} x$. But, from the table of logarithms, we find that the common logarithm of e (2.718) is 0.4343:

$$\log_{10} x = A\ (0.4343) = \log_e \times (0.4343)$$

This relationship is most often written in the form

$$\log_{10} x = \frac{\log_e x}{2.303}$$

This equation is commonly used to (1) find a number whose natural logarithm is known:

$$\log_e x = 1.500;\ x = ?$$

$$\log_{10} x = 1.500/2.303 = 0.6513$$

$$x = 4.48$$

(2) find a number expressed as a power of e.

$$x = 3.10\ e^{-1.600};\ x = ?$$

$$\log_e x = \log_e 3.10 - 1.600$$

$$\log_{10} x = \log_{10} 3.10 - 1.600/2.303$$

$$= 0.4914 - 0.6947 = -0.2033 = 0.7967 - 1$$

$$x = 6.261 \times 10^{-1}$$

(3) find the natural logarithm of a number.

$$x = 2.50 \times 10^{-6};\ \log_e x = ?$$

$$\log_{10} x = 0.3979 - 6 = -5.6021$$

$$\log_e x = 2.303 \times -5.6021 = -12.90$$

SIGNIFICANT FIGURES

Some numbers are, by their nature, exact. *Two* horses, *five* apples, or *sixteen* pennies indicate exactly how many items are described. Relations between dimensions in the same measuring system are also expressed in terms of exact numbers:

$$1 \text{ yard} = 3 \text{ feet} = 36 \text{ inches}$$

$$1000 \text{ millimeters} = 100 \text{ centimeters} = 1 \text{ meter}$$

These numbers arise by definition of the relationship between the various dimensional units and can all be considered to be perfectly exact.

When one relates dimensions in different measuring systems, the numbers involved are no longer all exact. Depending on the precision with which we wish to work, we can say

$$1 \text{ inch} = 2.540 \text{ centimeters}$$

or $\qquad 1 \text{ inch} = 2.54 \text{ centimeters}$

or $\qquad 1 \text{ inch} = 2.5 \text{ centimeters}$

In the foregoing equations the 1 may be taken to be exact. By stating that 1 in equals 2.54 cm, we imply that 1 in is closer in length to 2.54 cm than it is to either 2.53 cm or to 2.55 cm. To say that 1 in equals 2.540 cm means that it lies closer to 2.540 cm than to either 2.539 cm or to 2.541 cm. To use the first equation in a calculation means that we wish to work with a precision which is greater than that which is implied by the second or third equations. Somewhat loosely, we describe the precision of the numbers 2.540, 2.54, and 2.5, by saying that they contain four, three, and two **significant figures** respectively.

A similar situation arises when one obtains experimental data. Depending on the balance used, one might report that a sample of iron oxide weighed

1.6459 g,　or 1.646 g,　or 1.65 g,　or 1.6 g,

and, depending on the circumstances in the experiment, any one of these masses might be the most appropriate to use in a subsequent calculation. The implied precision of the masses varies widely, however, and we could indicate their precision by stating the number of significant figures in each number; these would range from five for the first mass to two for the last.

Obtaining the number of significant figures in a number often simply involves counting the number of digits in that number, as in the examples just given. However, if a number contains zeroes which *only* serve to locate a decimal point, the zeroes *do not* contribute to the precision of the number and *are not* counted. For example, if the masses in the example were reported in kilograms, rather than in grams, they would be

0.0016549 kg,　or 0.001646 kg,　or 0.00165 kg,　or 0.0016 kg

and those masses would clearly have the same precision as when reported in grams. The number of significant figures would remain, five, four, and so on, respectively, as before. The general rule in this regard is that the precision in the measurement of a quantity is not changed by the dimensions in which it is expressed.

659

An unambiguous method for determining number of significant figures is simply to express the number under consideration in exponential notation and count the digits in the number multiplying the exponential:

$$0.000000495 \text{ cm} = 4.95 \times 10^{-7} \text{ cm} \text{ (3 significant figures)}$$

$$0.00006 \text{ amp} = 6 \times 10^{-5} \text{ amp} \text{ (1 significant figure)}$$

Zeroes at the end of a number may either be significant or merely serve to fix a decimal point. Given the measurements

$$3100 \text{ cm} \quad \text{and} \quad 16.500 \text{ g}$$

the number of significant figures in 3100 is ambiguous and might be two, three, or four, depending on whether the zeroes serve to fix the decimal point or are experimentally meaningful. In 16.500 g there are *five* significant figures, since the decimal point is present in the number, and the purpose of the zeroes *must* be to describe its precision. Again, if exponential notation were used, the decimal point would be of necessity present, and no ambiguity would arise.

The importance of significant figures lies in calculations based on experimental data. The precision of an experimental result depends on the precision of the data used to obtain it. The calculation operation cannot by itself improve the precision of an experimental result. An illustration might be helpful in revealing some of the common pitfalls which arise in calculations involving experimental data.

Let us assume a student is given the problem of measuring the density of an unknown liquid. He performs the experiment by weighing an empty graduated cylinder, pouring in the liquid and measuring its volume, and then weighing the cylinder with the liquid in it. He obtains the following data by reading the balance and the markings on the graduated cylinder as best he can:

mass of cylinder plus liquid	262.1 g
mass of empty cylinder	128.4 g
mass of liquid in cylinder	133.7 g
volume of liquid	91.3 ml

Since density of liquid $= \dfrac{\text{mass of liquid}}{\text{volume of liquid}}$, the student obtains, by long division,

$$\text{density} = \frac{133.7 \text{ g}}{91.3 \text{ ml}} = 1.46_{440...} \text{ g/ml}$$

with as many digits in the quotient as he cares to determine. The obvious implication of the calculation is that, by using data of only moderate precision, one can find a density with very high precision. Any such implication is *incorrect*. The data used in the density calculation were found to *four* and *three* significant figures respectively, or, perhaps more meaningfully, the mass was measured with a precision of about 1 part in 1337 and the volume with a precision of about 1 part in 913. The calculated density *cannot* have a precision greater than that of the less precise of these numbers. This means that the liquid density found in the experiment should be reported as 1.46 g/ml, with an implied precision of 1 part in 146, rather than as 1.464 g/ml, which (improperly) has an implied precision of 1 part in 1464, greater than 1 part in 913 precision obtained in the measurement of the liquid volume.

As a general rule, in *multiplication* or *division* of experimentally obtained quantities we shall retain in the result a *number of significant figures equal to that in the least precise piece of data entering the calculation*. Since 91.3 ml is the least precise quantity entering the density calculation in the example, the rule would require that the density be reported to three significant figures, the same result we obtained by more careful consideration of the precision in the experiment.

In addition or subtraction of experimental quantities we retain the general principle that the calculation operation cannot improve the precision of an experimental result. If, for example, we dissolve two substances, A and B, in some water, and wish to report the mass of the final solution, we might obtain the following data:

		Implied precision
mass of water	994.2 g	± 0.1 g
mass of substance A	6.4545 g	± 0.0001 g
mass of substance B	29. g	± 1 g
sum of the masses	1029.$_{6545}$ g	± 1.1 g
properly reported mass	1030 g	$(\pm 1.1$ g$)$

To the right of the measured masses we have indicated the implied precision of the measurement. The likely error in the sum of the masses is clearly equal to the sum of the likely errors in each individual mass. The likely error in the total mass is about ± 1 g, caused mainly by the poor precision in the measurement of the mass of substance B. The reported total mass should not have a precision greater than the likely error and would be best reported as 1030 g, since the implied precision of that result, ± 1 g, is about equal to the likely error. Carrying the result to eight significant figures would be nonsense in view of the likely error, and rounding off to three significant figures would imply a greater likely error than is actually present. Clearly it is *not true* that the number of significant figures in a sum must not be larger than the smallest number of significant figures in any element entering the sum; in the example, the number of significant figures in the sum is clearly equal to four, whereas the number of significant figures in the least precise element in the sum, 29 g, is only two.

PROPORTIONALITY CONSTANTS

Many of the natural laws that we deal with in general chemistry are expressed in terms of mathematical equations which relate two or more variables. For example, we write the following equations to describe:

1. The relationship between the rate of decomposition of N_2O_5 and its concentration.

$$\text{rate} = k \,(\text{conc. } N_2O_5) \tag{1}$$

2. The relationship between the equilibrium concentration of molecular fluorine [F_2], and atomic fluorine [F] in the system $2\ F(g) \rightleftharpoons F_2(g)$.

$$\frac{[F_2]}{[F]^2} = K \tag{2}$$

3. The relationship between the pressure of a gas (p), its volume (V), number of moles (n), and absolute temperature (T).

$$\frac{p\ V}{n\ T} = R \tag{3}$$

In each of these equations, there is a **proportionality constant**. The quantity k in Equation 1 is referred to as a rate constant; K in Equation 2 is known as an equilibrium constant, while R in Equation 3 is the gas-law constant. In order to make effective use of equations such as these, it is important to understand where proportionality constants come from and precisely what they mean. The following general principles may be helpful in this regard.

1. *Once the form of the equation involving a proportionality constant has been established, its numerical magnitude can be determined by measuring simultaneously the values of the variables in the equation.*

Having established by experiment that Equation 1 is valid, i.e., that the rate of decomposition of N_2O_5 is directly proportional to its concentration, k can be calculated from the measured value of the rate at a particular concentration. Thus, when we find that the rate of decomposition of N_2O_5 at 70°C is 0.100 mole/lit hr when the concentration of N_2O_5 is 0.500 mole/lit, we can calculate k at this temperature.

$$k = \frac{\text{rate}}{\text{conc. } N_2O_5} = \frac{0.100 \text{ mole/lit hr}}{0.500 \text{ mole/lit}} = 0.200 \text{ hr}^{-1}$$

Again, knowing that one mole of a gas at 273°K and one atmosphere pressure occupies 22.4 liters, we can calculate R.

$$R = \frac{(1 \text{ atm}) (22.4 \text{ lit})}{(1 \text{ mole}) (273°K)} = 0.0821 \frac{\text{lit atm}}{\text{mole °K}}$$

2. *The magnitude of the proportionality constant is the same for all values of the variables that appear in the equation.*

Consider the following experimental data for the $F_2 - F$ equilibrium system at 1400°C.

[F_2]	[F]	$[F_2]/[F]^2 = K$
0.100	0.100	$0.100/(0.100)^2 = 10.0$
0.400	0.200	$0.400/(0.200)^2 = 10.0$
0.900	0.300	$0.900/(0.300)^2 = 10.0$

Clearly, the ratio $[F_2]/[F]^2$ is constant, regardless of the individual values of [F_2] or [F]. Stated another way, if one chooses a particular value for [F], that of [F_2] must become such that the ratio $[F_2]/[F]^2$ is 10.0. This allows us to calculate one of the variables in the equation, given the value of the other variable. Suppose, for example, that in a particular system, [F] = 0.0120. Then:

$$\frac{[F_2]}{(0.0120)^2} = 10.0; \quad [F_2] = (1.44 \times 10^{-4}) (10.0) = 1.44 \times 10^{-3}$$

Again, if, in another system, $[F_2] = 0.200$, it follows that:

$$\frac{(0.200)}{[F]^2} = 10.0; \qquad [F]^2 = 0.0200; \qquad [F] = 0.141$$

3. *The magnitude of the proportionality constant may and often does depend upon other variables which are not specified in the equation.*

Both the rate constant k in Equation 1 and the equilibrium constant K in Equation 2 are dependent upon temperature. Thus, k is 0.200 hr^{-1} at 70°C and 0.600 hr^{-1} at 80°C. The equilibrium constant for the reaction $2\,F(g) \rightleftharpoons F_2(g)$ changes from 10 at 1400°C to 5 at 1500°C. Clearly, if one is to perform calculations involving either of these "constants" it is important that the temperature be specified.

Rate constants and equilibrium constants can, of course, be written for a great many systems other than the two we have dealt with. The equilibrium constant for the association of chlorine atoms at 1400°C has a quite different value from that for fluorine atoms at the same temperature.

$$2\,Cl(g) \rightleftharpoons Cl_2(g); \qquad K = \frac{[Cl_2]}{[Cl]^2} = 10^{14} \text{ at } 1400°C$$

The Ideal Gas Law, Equation 3, is a more general relationship than Equations 1 or 2. To a good degree of approximation, it applies to all gases over a considerable range of temperatures.

4. *Proportionality constants ordinarily have units; their magnitude will, of course, depend upon the units in which they are expressed.*

We have expressed the gas law constant R in the units of lit atm/mole °K. For certain purposes, it is useful to know its magnitude in cal/mole °K.

$$R = 0.0821 \frac{\text{lit atm}}{\text{mole °K}} \times \frac{24.2 \text{ cal}}{1 \text{ lit atm}} = 1.99 \frac{\text{cal}}{\text{mole °K}}$$

As a further illustration, the rate constant for the decomposition of N_2O_5 could be expressed in min^{-1} as well as hr^{-1}.

$$k = \frac{0.200}{hr} \times \frac{1 \text{ hr}}{60 \text{ min}} = 3.33 \times 10^{-3} \text{ min}^{-1}$$

LINEAR FUNCTIONS

A variable y is said to be a linear function of another variable x if a plot of y vs. x is a straight line. A simple example of a linear function is the relationship described in the previous section between the rate of decomposition of N_2O_5 and its concentration: rate $= k$ (conc. N_2O_5). A plot of the rate of reaction vs. the concentration of N_2O_5 is a straight line, passing through the origin, with a slope equal to the rate constant k.

Another example of a linear relationship is that between °F and °C: °F $= 1.8$ °C $+ 32$°. In this case, the straight line relating the variables °F and °C, does not pass through the origin. Instead, the intercept on the vertical axis (often referred to simply as the "intercept") is 32°. That is, °F $= 32$° when °C $= 0$°. The slope of the line is 1.8; when the centigrade temperature increases by 1°, the Fahrenheit temperature increases by 1.8°.

663

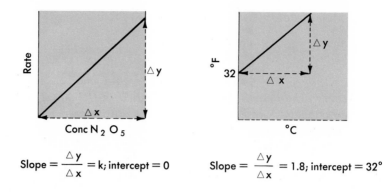

$$\text{Slope} = \frac{\triangle y}{\triangle x} = k; \text{intercept} = 0 \qquad \text{Slope} = \frac{\triangle y}{\triangle x} = 1.8; \text{intercept} = 32°$$

In general, y is a linear function of x if the equation relating them is of the form

$$y = a x + b \qquad (1)$$

In this equation, the constant *a* represents the slope of the line while *b* gives the value of the intercept on the y axis.

Frequently, we have occasion to work with experimental data which corresponds to a linear relationship between two variables. Ordinarily, the constants *a* and *b* are not known in advance. They can be determined by plotting the data, drawing the "best" straight line through the points, and estimating the slope and intercept from the graph. To illustrate how this is done, consider the following data:

x	1.00	2.00	3.00	4.00
y	15.5	26.2	36.0	46.5

When these data are plotted, we obtain the four points shown in the figure below. The "best" straight line through the points is drawn (note that the line does not pass exactly through each of the points). The intercept is obtained by extrapolating this line until it intersects the y axis; it appears to be approximately 6. We obtain the slope by choosing two points on the line and dividing the difference in their y values (Δy) by the difference in their x values (Δx).

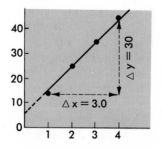

$$\text{slope} = \frac{\Delta y}{\Delta x} = \frac{46 - 16}{4.0 - 1.0} = 10$$

We deduce that the linear equation relating y and x is

$$y = 10\,x + 6$$

Frequently, a nonlinear relationship between two variables can be transformed into a linear function by an appropriate change in variables. Consider, for example, the equilibrium $2\,F(g) \rightleftharpoons F_2(g)$, where, as pointed out previously,

$$\frac{[F_2]}{[F]^2} = K$$

Clearly, since $[F_2] = K[F]^2$, a plot of $[F_2]$ vs. $[F]$ will not be linear. However, if we plot $[F_2]$ vs. $[F]^2$ rather than $[F]$, we obtain a straight line with a slope equal to the equilibrium constant K.

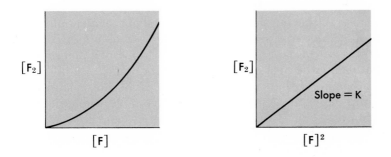

A somewhat more complex relationship which can be transformed into a linear function by an appropriate change of variables is that between the vapor pressure of a liquid (P) in mm Hg and the absolute temperature (T) in °K.

$$P = C\,e^{-\Delta H_{vap}/RT}$$

where $\Delta H_{vap} = \Delta H$ of vaporization (cal/mole), R = gas law constant (cal/mole °K), e = base of natural logarithms, C = constant. Taking the natural logarithm of both sides, we obtain

$$\ln P = \ln C - \frac{\Delta H_{vap}}{RT}$$

or
$$\log_{10} P = -\frac{\Delta H_{vap}}{2.30\,RT} + \log_{10} C \qquad (2)$$

Equation 2 is formally similar to the general equation

$$y = a\,x + b$$

provided we take as our variables

$$y = \log_{10} P, \qquad x = 1/T$$

In other words, a plot of $\log_{10} P$ vs. $1/T$ should give a straight line with a slope of $-\Delta H_{vap}/2.30\,R$, and an intercept of $\log_{10} C$.

665

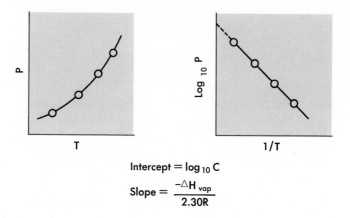

Intercept = $\log_{10} C$

Slope = $\dfrac{-\triangle H_{vap}}{2.30R}$

There are two distinct advantages to be gained by transforming a non-linear to a linear function by means of a change in variables.

1. It frequently happens that a quantity of physical significance can be obtained, either from the slope or the intercept of the linear plot. For example, by taking the slope of a plot of $[F_2]$ vs. $[F]^2$, we obtain directly the value of the equilibrium constant K. Again, from the slope of a plot of $\log_{10} P$ vs. $1/T$, we can calculate the heat of vaporization of a liquid.

2. A linear equation facilitates the process of interpolation (or extrapolation) required to estimate the value of one variable when that of the other is known. Suppose, for example, that by plotting $\log_{10} P$ vs. $1/T$, we have obtained from the observed slope and intercept, the following equation:

$$\log_{10} P = \frac{-2000}{T} + 7.00$$

This equation can now be used to calculate the vapor pressure at a given temperature. At 400°K,

$$\log_{10} P = \frac{-2000}{400} + 7.00 = +2.00$$

$$P = 100 \text{ mm Hg}$$

In principle, we could obtain this information from the curve drawn at the left above (P vs. T), by reading off the pressure corresponding to a given temperature. In practice, this is difficult to do accurately, particularly if only a few data points are available to fix the position of the curve.

PROBLEMS

1. Express the following numbers in exponential notation.

a. 262.4	d. 0.000 000 4659	g. 2000
b. 0.0039	e. 3294.5	h. 2000.0
c. 42,000,000,000	f. 20010	i. 0.000 0400

Ans: b. 3.9×10^{-3} f. 2.0010×10^4
g. 2×10^3

2. How many significant figures are there in each of the numbers in Problem 1?

Ans: b. Two   c. Probably two
f. Probably five   i. Three

3. Carry out the indicated operations, using exponential notation where useful and observing rules on significant figures.

a. 1594×0.0029
b. $1.40 \times 10.45 \div 16$
c. $23.961 \text{ cm} \times 10 \text{ mm}/1 \text{ cm}$
d. $52.6541 \text{ g} \div 17.0 \text{ ml}$
e. $13.6 \text{ ft} \times 12 \text{ in}/1 \text{ ft}$

f. $16.2 \text{ g} + 39.65 \text{ g} + 2.4792 \text{ g}$
g. $155.4 \text{ g} + 19 \text{ g} + 245.2 \text{ g}$
h. $16.54 \text{ g} + 208 \text{ g} - 203.9 \text{ g}$
i. $3.69 \text{ m} \times 100 \text{ cm}/1 \text{ m} \times 1 \text{ in}/2.540 \text{ cm}$
j. $2325 \text{ ml} + 16.8 \text{ ml} - 22.04 \text{ ml}$

Ans: a. 46
c. 2.3961×10^2 mm or 239.61 mm
f. 58.3 g
h. 21 g

4. Evaluate each of the following expressions, giving your result in exponential notation.

a. $\log_{10} 435$
b. $\log_{10} 2.785$
c. $\log_{10} 0.022$
d. $\ln 56.1$

e. $\ln 350$
f. $\ln 0.004$
g. $\ln 0.000\,068$
h. $\log_4 64$

i. $\log_{10} x = 2.43$; find x
j. $\log_{10} y = -5.6$; find y
k. $\ln z = 4.60$; find z
l. $\ln q = -24.2$; find q

Ans: a. 2.64   c. -1.66 e. 5.86
i. 2.70×10^2 l. 3×10^{-11}

5. Evaluate each of the following expressions, giving the result in exponential notation.

a. e^{16}
b. $e^{-2.88}$
c. $10^{4.3}$

d. $10^{-2.66}$
e. $\log_{10} e^{4.4}$
f. $\ln(6.3 \times 10^4)$

g. $85.1 \times e^{-8.3}$
h. $3.1 \ln 0.063$
i. $1.6 \times 10^3 \times e^{3.66}$

Ans: b. 5.6×10^{-2}
f. 11

6. a. The rate of a certain reaction is given by the expression

$$\text{rate} = k \, (\text{conc. A})^2 \, (\text{conc. B})$$

If the rate is 2.0×10^{-2} mole/lit sec when the concentrations of A and B are 0.100 and 0.050 mole/lit respectively, find k. Using this value of k, calculate the rate of reaction when the concentrations of A and B are both 0.0100 mole/liter.

b. Determine the value of the gas law constant in ergs/mole °K (use Appendix 4 for the appropriate conversion factors); ergs/molecule °K (1 mole = 6.02×10^{23} molecules); ml mm Hg/mole °K (760 mm Hg = 1 atm).

c. The average kinetic energy of a gas molecule is directly proportional to the absolute temperature.

$$\tfrac{1}{2} m u^2 = \text{constant} \times T; \qquad (m = \text{mass}, \, u = \text{velocity})$$

Calculate the ratio of the velocity of a CO_2 molecule to that of a Cl_2 molecule at 25°C; the ratio of the velocity of a CO_2 molecule at 25°C to that at -50°C; the temperature at which an O_2 molecule has the same velocity as a CH_4 molecule at 25°C.

Ans: b. 8.31×10^7 erg/mole °K, 1.38×10^{-16} erg/molecule °K
6.24×10^4 ml mm Hg/mole °K

c. 1.27, 1.16, 323°C

7. a. The following data are obtained for the variables x and y.

x	1.0	2.0	3.0	4.0
y	-0.8	2.3	5.6	8.8

Determine the constants of the linear equation relating y and x; estimate the values of y when $x = 2.5, 0.0$.

b. In an equilibrium involving two species A and B, the following data are obtained.

[A]	1.00	2.00	3.00	4.00
[B]	0.100	0.141	0.173	0.200

Determine the value of n in the equilibrium constant expression

$$K = \frac{[B]}{[A]^n}$$

Calculate K.

c. For each of the following functions, it is possible to obtain a linear relation by a proper choice of variables. State what these variables are.

$$y = a\, x^3$$
$$y = ax + b\, x^2$$
$$y = a/x + b$$
$$\log y^2 = a\, x^2$$
$$y = 6/(a - x) + 3$$

d. Show that if

$$k = A\, e^{-Ea/RT}$$

then

$$\log k = \frac{-Ea}{2.3\, RT} + \log A$$

and $\quad \log \dfrac{k_2}{k_1} = \dfrac{Ea\,(T_2 - T_1)}{2.3\, R\, T_2\, T_1}$, $\quad$ where the subscripts $_2$ and $_1$ refer to two different temperatures.

Ans: b. $n = \frac{1}{2}$, $K = 0.10$
c. y vs. x^3; y/x vs. x; yx vs. x; log y vs. x^2; $y(a - x)$ vs. x

Nomenclature of Inorganic Compounds

The composition of a compound may be specified by giving either its formula or its name. Throughout this text, we have discussed at some length how one can arrive at the chemical formulas of inorganic compounds. Here, we consider a related problem, that of developing a system of nomenclature for these compounds. In the interest of clarity and simplicity, we shall restrict our discussion to a relatively small number of rules which will suffice to name the great majority of inorganic compounds encountered in an introductory course in chemistry.

IONIC COMPOUNDS

The names of ionic compounds are derived from those of the ions of which they are composed. We shall first consider the nomenclature of individual ions and then the names of the compounds they form.

Positive Ions

Monatomic positive ions take the names of the metal from which they are derived:

$$Na^+ \text{ sodium} \qquad Ca^{2+} \text{ calcium} \qquad Al^{3+} \text{ aluminum}$$

When a metal forms more than one ion, it is necessary to distinguish between these ions. The accepted practice today is to indicate the oxidation number of the ion by a Roman numeral in parentheses immediately following the name of the metal:

$$Fe^{2+} \text{ iron(II)} \qquad Cu^+ \text{ copper(I)} \qquad Sn^{2+} \text{ tin(II)}$$
$$Fe^{3+} \text{ iron(III)} \qquad Cu^{2+} \text{ copper(II)} \qquad Sn^{4+} \text{ tin(IV)}$$

An earlier method, still widely used, adds to the stem of the Latin name of the metal the suffixes *-ous* or *-ic*, representing the lower and higher oxidation states respectively:

$$Fe^{2+} \text{ ferrous} \qquad Cu^+ \text{ cuprous} \qquad Sn^{2+} \text{ stannous}$$
$$Fe^{3+} \text{ ferric} \qquad Cu^{2+} \text{ cupric} \qquad Sn^{4+} \text{ stannic}$$

669

The only polyatomic cations to be considered here are:

NH_4^+ ammonium Hg_2^{2+} mercury(I) or mercurous

Negative Ions

Monatomic negative ions are named by adding the suffix *-ide* to the stem of the name of the nonmetal from which they are derived:

N^{3-}	nitride	O^{2-}	oxide	F^-	fluoride	H^-	hydride
		S^{2-}	sulfide	Cl^-	chloride		
		Se^{2-}	selenide	Br^-	bromide		
		Te^{2-}	telluride	I^-	iodide		

The nomenclature of polyatomic anions is more complex. The names of some of the more common oxyanions are:

OH^- hydroxide	NO_2^- nitrite	ClO_2^- chlorite
O_2^{2-} peroxide	SO_4^{2-} sulfate	ClO^- hypochlorite
CO_3^{2-} carbonate	SO_3^{2-} sulfite	MnO_4^- permanganate
PO_4^{3-} phosphate	ClO_4^- perchlorate	CrO_4^{2-} chromate
NO_3^- nitrate	ClO_3^- chlorate	$Cr_2O_7^{2-}$ dichromate

It will be noted that when a nonmetal such as nitrogen or sulfur forms two oxyanions in different oxidation states, the suffixes *-ate* and *-ite* are used to distinguish between the higher and lower states respectively. With elements such as chlorine which form more than two oxyanions, the prefixes *per-* (highest oxidation state) and *hypo-* (lowest oxidation state) are used as well.

Oxyanions that contain hydrogen as well as nonmetal and oxygen atoms are properly named as illustrated in the following examples:

HCO_3^- hydrogen carbonate	HPO_4^{2-} hydrogen phosphate
HSO_4^- hydrogen sulfate	$H_2PO_4^-$ dihydrogen phosphate

Compounds

The name of the positive ion is given first, followed by the name of the negative ion. Examples are

$CaCl_2$	calcium chloride
$FeBr_2$	iron(II) bromide
$(NH_4)_2SO_4$	ammonium sulfate
$Fe(ClO_4)_3$	iron(III) perchlorate
$NaHCO_3$	sodium hydrogen carbonate

In practice, compounds containing metal atoms, regardless of the type of bonding involved, are ordinarily named as if they were ionic. For example, the compounds $AlCl_3$ and $SnCl_4$, in both of which the bonding is primarily covalent, are named as follows:

$AlCl_3$ aluminum chloride $SnCl_4$ tin(IV) chloride

BINARY COMPOUNDS OF THE NONMETALS

When a pair of nonmetals form only one compound, that compound may be named quite simply. The name of the element whose symbol appears first in the formula is written first. The second portion of the name is formed by adding the suffix -ide to the stem of the name of the second nonmetal. Examples include

HCl   hydrogen chloride
H_2S hydrogen sulfide
NF_3 nitrogen fluoride

If more than one binary compound is formed by a pair of nonmetals, as is most often the case, the Greek prefixes, di = two, tri = three, tetra = four, penta = five, hexa = six, and so on, are used to designate the number of atoms of each element. Thus, for the oxides of nitrogen we have

*N_2O_5 dinitrogen pentoxide
*N_2O_4 dinitrogen tetroxide
NO_2 nitrogen dioxide
N_2O_3 dinitrogen trioxide
NO   nitrogen oxide
N_2O dinitrogen oxide

A great many of the best-known binary compounds of the nonmetals have acquired common names which are widely and, in some cases, exclusively used. These include

H_2O water
H_2O_2 hydrogen peroxide
NH_3 ammonia
N_2H_4 hydrazine

PH_3 phosphine
AsH_3 arsine
NO   nitric oxide
N_2O nitrous oxide

OXYACIDS

The names of some of the more common oxygen acids are listed as follows:

H_2CO_3 carbonic acid
H_3BO_3 boric acid
HNO_3 nitric acid
HNO_2 nitrous acid
H_2SO_4 sulfuric acid

H_2SO_3 sulfurous acid
$HClO_4$ perchloric acid
$HClO_3$ chloric acid
$HClO_2$ chlorous acid
HClO hypochlorous acid

It is of interest to compare the names of these oxyacids to those of the corresponding oxyanions listed previously. Note that oxyanions whose names end in -ate are derived from acids whose names end in -ic. Compare, for example, CO_3^{2-} (carbonate) and H_2CO_3 (carbonic acid); NO_3^- (nitrate) and HNO_3 (nitric acid); ClO_4^- (perchlorate) and $HClO_4$ (perchloric acid). Oxy-

---

* Note that in this case the a is dropped from the prefixes penta and tetra in the interests of euphony.

anions whose names end in *-ite* are derived from acids whose names end in *-ous*. Thus we have NO_2^- (nitr*ite*) and HNO_2 (nitr*ous* acid); ClO^- (hypochlor*ite*) and $HClO$ (hypochlor*ous* acid).

COORDINATION COMPOUNDS

The nomenclature of compounds containing complex ions in which a metal atom is held by coordinate covalent bonds to two or more ligands is perhaps more involved than that of any other type of inorganic compound. Several rules are required, the more pertinent of which are as follows:

1. As in simple ionic compounds, the cation is named first, followed by the anion.

2. If there is more than one ligand of a particular type attached to the central atom, Greek prefixes are used to indicate the number of these ligands. Where the name of the ligand itself is complex (e.g., ethylenediamine), the number of such ligands is indicated by the prefixes *bis-* or *tris-* instead of *di-* or *tri-* and the name of the ligand is enclosed in parentheses.

3. In naming a complex ion, the names of anionic ligands are written first, followed by those of neutral ligands, and finally that of the central metal atom. This is exactly the reverse of the order in which the groups are listed in the formula of the complex ion: the symbol of the central atom is written first, followed by the formulas of neutral ligands and then those of negatively charged ligands. In writing the formula of a coordination compound, the formula of the complex ion is often set off by brackets.

4. The names of anionic ligands are modified by substituting the suffix *-o* for the usual ending. Thus we have

Cl^- chloro	CO_3^{2-} carbonato
OH^- hydroxo	CN^- cyano

The names of neutral ligands are ordinarily not changed. Two important exceptions are:

$$H_2O \text{ aquo} \qquad NH_3 \text{ ammine}$$

5. The oxidation number of the central metal atom is indicated by a Roman numeral following the name of the metal. If the complex is an anion, the suffix *-ate* is added, often to the Latin stem of the name of the metal. Examples are

$[Co(NH_3)_6]Cl_3$	hexamminecobalt(III) chloride
$[Co(en)_3](NO_3)_3$	tris(ethylenediamine)cobalt(III) nitrate
$[Cr(NH_3)_4Cl_2]Cl$	dichlorotetramminechromium(III) chloride
$[Pt(H_2O)_3Cl]Br$	chlorotriaquoplatinum(II) bromide
$K_3[Fe(CN)_6]$	potassium hexacyanoferrate(III)
$K_4[Fe(CN)_6]$	potassium hexacyanoferrate(II)

The last two compounds are often referred to as potassium ferricyanide and potassium ferrocyanide respectively.

Atomic and Ionic Radii

Element	Atomic Number	Atomic Radius in Å	Ionic Radius in Å	Element	Atomic Number	Atomic Radius in Å	Ionic Radius in Å
H	1	0.37	(−1) 2.08	Ag	47	1.44	(+1) 1.26
He	2	0.93		Cd	48	1.49	(+2) 0.97
Li	3	1.52	(+1) 0.60	In	49	1.62	(+3) 0.81
Be	4	1.11	(+2) 0.31	Sn	50	1.40	
B	5	0.88		Sb	51	1.41	
C	6	0.77		Te	52	1.37	(−2) 2.21
N	7	0.70		I	53	1.33	(−1) 2.16
O	8	0.66	(−2) 1.40	Xe	54	1.90	
F	9	0.64	(−1) 1.36	Cs	55	2.62	(+1) 1.69
Ne	10	1.12		Ba	56	2.17	(+2) 1.35
Na	11	1.86	(+1) 0.95	La	57	1.87	(+3) 1.15
Mg	12	1.60	(+2) 0.65	Ce	58	1.82	(+3) 1.01
Al	13	1.43	(+3) 0.50	Pr	59	1.82	(+3) 1.00
Si	14	1.17		Nd	60	1.82	(+3) 0.99
P	15	1.10		Pm	61		
S	16	1.04	(−2) 1.84	Sm	62		
Cl	17	0.99	(−1) 1.81	Eu	63	2.04	(+2) 0.97
Ar	18	1.54		Gd	64	1.79	(+3) 0.96
K	19	2.31	(+1) 1.33	Tb	65	1.77	(+3) 0.95
Ca	20	1.97	(+2) 0.99	Dy	66	1.77	(+3) 0.94
Sc	21	1.60	(+3) 0.81	Ho	67	1.76	(+3) 0.93
Ti	22	1.46		Er	68	1.75	(+3) 0.92
V	23	1.31		Tm	69	1.74	(+3) 0.91
Cr	24	1.25	(+3) 0.64	Yb	70	1.93	(+3) 0.89
Mn	25	1.29	(+2) 0.80	Lu	71	1.74	(+3) 0.89
Fe	26	1.26	(+2) 0.75	Hf	72	1.57	
Co	27	1.25	(+2) 0.72	Ta	73	1.43	
Ni	28	1.24	(+2) 0.69	W	74	1.37	
Cu	29	1.28	(+1) 0.96	Re	75	1.37	
Zn	30	1.33	(+2) 0.74	Os	76	1.34	
Ga	31	1.22	(+3) 0.62	Ir	77	1.35	
Ge	32	1.22		Pt	78	1.38	
As	33	1.21		Au	79	1.44	(+1) 1.37
Se	34	1.17	(−2) 1.98	Hg	80	1.55	(+2) 1.10
Br	35	1.14	(−1) 1.95	Tl	81	1.71	(+3) 0.95
Kr	36	1.69		Pb	82	1.75	
Rb	37	2.44	(+1) 1.48	Bi	83	1.46	
Sr	38	2.15	(+2) 1.13	Po	84	1.65	
Y	39	1.80	(+3) 0.93	At	85		
Zr	40	1.57		Rn	86	2.2	
Nb	41	1.43		Fr	87		
Mo	42	1.36		Ra	88	2.20	
Tc	43			Ac	89	2.0	
Ru	44	1.33		Th	90	1.80	
Rh	45	1.34		Pa	91		
Pd	46	1.38		U	92	1.4	

APPENDIX 4

Conversion Factors and Constants

Acceleration of gravity	$g = 980.67$ cm/sec^2
Ampere	1 amp = 1 coulomb/sec
Angstrom	1 Å = 1×10^{-8} cm
	1 Å = 1×10^{-1} mμ
Atmosphere	1 atm = 760 mm Hg = 33.9 ft water
	1 atm = 14.70 lb/in^2
	1 atm = 1.013×10^6 dynes/cm^2
	1 atm = 1033 g/cm^2
Atomic mass unit	1 amu = 931 MeV
Avogadro's number	$N = 6.023 \times 10^{23}$
Boltzmann's constant	$k = 1.3805 \times 10^{-16}$ ergs/°K
British thermal unit	1 BTU = 252 cal
Calorie	1 cal = 4.184 joules
	1 cal = 0.04129 lit atm
Centimeter	1 cm = 1×10^8 Å
	1 cm = 0.3937 in
Centimeter/second	1 cm/sec = 0.02237 mile/hr
Cubic centimeter	1 cm^3 = 0.06102 in^3
Cubic inch	1 in^3 = 16.387 cm^3
Density	D water(1) = 1.000 g/cm^3 at 4°C
	D mercury = 13.6 g/cm^3 at 0°C
Electronic charge	$e^- = 4.80 \times 10^{-10}$ esu
	$e^- = 1.60 \times 10^{-19}$ coulomb
Electron volts per atom	1 eV/atom = 23.05 kcal/mole
Erg	1 erg = 2.389×10^{-8} cal
	1 erg = 1×10^{-7} joule
Faraday	1 faraday = 96500 coulombs = 23070 cal/V
	1 faraday = 6.023×10^{23} e$^-$
Gas constant	$R = 0.08205$ lit atm/mole °K
	$R = 1.987$ cal/(mole °K)
	$R = 8.314 \times 10^7$ ergs/(mole °K)
Gram	1 g mass = 2.15×10^{10} kcal
Grams per cubic centimeter	1 g/cm^3 = 62.43 lb/ft^3
Gram molecular volume	$V_0 = 22.413$ lit/mole at 0°C and 1 atm
Ice point	$T_0 = 273.15$°K
Inch	1 in = 2.540 cm
Joule	1 j = 0.2390 cal
Kilogram	1 kg = 2.205 lb
Liter	1 lit = 1000 cm^3
	1 lit = 1.0567 quarts
Liter-atmosphere	1 lit atm = 24.22 cal
Natural logarithms	$\ln x = 2.3026 \log_{10} x$
Pi	$\pi = 3.1416$
Planck's constant	$h = 6.626 \times 10^{-27}$ erg sec
Pound	1 lb = 453.6 g
Quart	1 qt = 0.9463 lit
Speed of light	$c = 2.998 \times 10^{10}$ cm/sec

ANSWERS TO SELECTED PROBLEMS

CHAPTER 1

1.15 127°B   **1.16** 0.821 atm   **1.17** 1.943 g/ml
1.19 63 g succinic acid, no tartaric acid   **1.20** 28° in a bath at 50°C
1.21   65 g tartaric acid, 14 g succinic acid
             53 g tartaric acid, 9 g succinic acid
             43 g tartaric acid, 5 g succinic acid
             35 g tartaric acid, 2 g succinic acid
             29 g tartaric acid, 0 g succinic acid
1.22 0.089 ml

CHAPTER 2

2.15 b. wt Pb/wt 0 = 6.46, 9.71, 12.9 = $1 : \frac{3}{2} : 2$
2.17 29 g   **2.18** 71.1 g   **2.19** 69.72 g   **2.20** hyperbola   **2.21** hyperbola
2.23 8, 15, 1×10^4 **2.24 a.** 128 g   **b.** 1.96 GAW   **c.** 19.6 g
2.25 a. 1.99×10^{-23} g   **b.** 6.6×10^{22} **c.** 1.56×10^{-10} g
2.26 29.4, 58.8, 88.2; Co or Ni
2.27 Specific heat = 0.11 cal/g°C; AW ≈ 55; GEW = 18.6 g; GAW = 55.8 g

CHAPTER 3

3.11 a. 7.12 g Sr   **b.** 0.585 g Ni   **c.** 1.18 g Co
3.12 b. $Tl_4V_2O_7$ **3.13** 45.6%   **3.14** $GaBr_3$ **3.15** $C_6H_4Cl_2$ **3.16** $CoCO_3$, Co_3O_4
3.17 a. 265 g   **b.** 0.494 mole   **c.** 2.05×10^{23} ions
3.18 2 $C_8H_{18}(l)$ + 25 $O_2(g)$ → 16 $CO_2(g)$ + 18 $H_2O(l)$
3.19 a. 11.0   **c.** 8.46×10^{18} **3.20 a.** 11.0 g   **d.** 123 g
3.21 a. 9.73 g   **b.** 7.7 g   **3.22 a.** 47.1 g   **b.** 76.4%
3.23 15 g of ?   **3.24** 67%   **3.26 a.** 0.605 g SO_2 **3.27** 1.65 mole

CHAPTER 4

4.13 a. −144 kcal   **b.** −288 kcal   **4.14 a.** −143 kcal   **b.** C_8H_{18} **4.15** −48.2 kcal
4.16 −94.5 kcal   **4.18 a.** −53.0 kcal   **b.** −31.1 kcal   **c.** −47.0 kcal

4.19 a. 230 cal/°C   **b.** −7.55 kcal/g   **4.20** −1.9 kcal/mole
4.21 a. −10.7 kcal, +0.9 kcal, −11.6 kcal   **b.** −10.7 kcal   **c.** −11.6 kcal
4.22 a. −52.7 kcal   **b.** −30.5 kcal   **c.** −46.4 kcal
4.23 a. −325 cal   **b.** 0   **c.** −445 cal   **4.24** None
4.25 a. −44 kcal   **b.** −20 kcal   **c.** −118 kcal

CHAPTER 5

5.19 3.33×10^{-3} mm Hg   **5.20** 2.96 atm   **5.21** 190 lb   **5.22** 1.60 g/lit
5.23 160   **5.24** $P_{H_2} = 1.5$ atm   **5.25** 0.0112 mole H_2; AW = 40   **5.26** 64%
5.27 29   **5.28** 1.004   **5.29 a.** T, M   **b.** T   **c.** M, T, V   **5.30** 460°C
5.31 MW = 44.010   **5.32** 56.5
5.33 a. $CCl_4 > SO_2 > O_2 > H_2$ (order of boiling points or heats of vaporization)
    b. $CCl_4 > SO_2 > O_2 > H_2$ (order of molecular volumes)
5.34 $KE = \frac{1}{2} Mu^2 = \frac{M}{2}\left(\frac{3RT}{M}\right) = \frac{3RT}{2}$

per °C: $\Delta KE = \frac{3R\Delta T}{2} = \frac{3R}{2} = \frac{3}{2}$ (2.0 cal/mole) = 3 cal/mole

5.35 672°   **5.36** 338.0°K, 64.8°C
5.37 a. Speed, energy, momentum unchanged, pressure 5 times as great.
    b. Speed increased by factor of $(2)^{\frac{1}{2}}$, energy increased by factor of 2, momentum increased by factor of $(2)^{\frac{1}{2}}$, pressure increased by factor of 2.
    c. Speed, energy, momentum unchanged, pressure doubles.
    d. Speed increased by factor of 4, energy unchanged, momentum $\frac{1}{4}$ original, p unchanged.

CHAPTER 6

6.14 a, d, e
6.15 Extrapolated values: mp = 310°C, bp = 380°C, ion. energy = 8 eV
6.16 b. 2.6 g/ml   **c.** 777°C   **d.** $SrCl_2$, SrO   **e.** 4.5 g/ml   **f.** 867°C
6.17 a. La_2S_3 **b.** BiI_3 **c.** KrF_4 **d.** K_2MoO_4 **e.** BaO_2 **f.** $Ca(TcO_4)_2$
6.18 Hint: Recall the Law of Dulong and Petit.
6.19 Hint: Look for a nonmetal with a GEW of about 18 g.

CHAPTER 7

7.21 a. +14, 16n, 14p, 14e   **b.** +6, 7n, 6p, 6e   **c.** +13, 13p, 10e
    d. +24, 28n, 24p, 21e
7.22 $\epsilon = 3.3 \times 10^{-6}$ erg, less than binding energy.
7.23 S: $1s^2\ 2s^2\ 2p^6\ 3s^2\ 3p^4$; $1s^2\ 2s^2\ 2p^6\ 3s^2\ 3p^3\ 4s^1$
    Na^+: $1s^2\ 2s^2\ 2p^6$; $1s^2\ 2s^2\ 2p^5\ 3s^1$
    Zn^{2+}: $1s^2\ 2s^2\ 2p^6\ 3s^2\ 3p^6\ 3d^{10}$; $1s^2\ 2s^2\ 2p^6\ 3s^2\ 3p^6\ 3d^9\ 4s^1$
7.24 Mg^{2+} no unpaired electrons, Fe^{3+} five, Zr^{4+} none

7.25 a. Two electrons in same orbital would have parallel spins.

    b. Single electrons in 2p orbitals would have opposed spins.

    c. Violates Hund's rule.

7.26 F^- ion

n	l	m_l	m_s
1	0	0	$\pm\frac{1}{2}$
2	0	0	$\pm\frac{1}{2}$
2	1	-1	$\pm\frac{1}{2}$
2	1	0	$\pm\frac{1}{2}$
2	1	$+1$	$\pm\frac{1}{2}$

7.27 $n = 4$, $l = 3$. m_l can be $-3, -2, -1, 0, +1, +2, +3$ (7 orbitals). $m_s = \pm\frac{1}{2}$

7.28 a. N   **b.** H, He, Li, Be   **c.** 1A or 3A metals   **7.29** 188 eV, 0.17 eV

7.30 Lyman: 1215 Å   **7.31** 2.2×10^4 Å, 1.4×10^5 Å   **7.32** 42

7.33 $\lambda = 1.55$ Å   **7.34 a.** 3,9   **b.** 27   **c.** $1s^3\ 2s^3\ 2p^1$; $1s^3\ 2s^3\ 2p^8$ **d.** 19

CHAPTER 8

8.21 a. MgH_2 **b.** Sc_2S_3 **c.** KAt   **d.** BaO   **e.** Li_2Te **f.** La_2O_3

8.22 a. 38, 39   **b.** 34, 35   **c.** 36   **d.** 3, 11, 19   **e.** 55, 87

8.23 a. $2\ Na(s) + S(s) \rightarrow Na_2S(s)$ **b.** $2\ K(s) + I_2(s) \rightarrow 2\ KI(s)$

    c. $4\ Al(s) + 3\ O_2(g) \rightarrow 2\ Al_2O_3(s)$ **d.** $Ba(s) + F_2(g) \rightarrow BaF_2(s)$

    e. $2\ Sr(s) + O_2(g) \rightarrow 2\ SrO(s)$

8.24 a. NH_4Br **b.** $(NH_4)_2SO_4$ **c.** $MgCO_3$ **d.** $Sc(C_2H_3O_2)_3$ **e.** $Al(NO_3)_3$

    f. Li_3PO_4 **g.** $Ca(OH)_2$ **h.** $NiCl_2$ **i.** $ZnSO_4$

8.25 a. 60%   **8.26** N—Cl < C—H < S—O < As—F < Mg—O

8.27 a. N≡N shorter, stronger **b.** O=O shorter, stronger **c.** H—Cl shorter, stronger

8.28 a. :C̈l—Äs—C̈l :   **d.** H—C≡N :   **e.**

8.29 a. :Ö—N—Ö :   **c.** :Ö—Br—Ö :   **e.** :Ö—S—Ö :

8.30 SeO_2, O_3, NO_3^-

8.31 a. O_3 **b.** CO   **c.** PH_4^+ **d.** C_6H_6 **e.** PBr_5 **8.32** I_3^-, SF_5^-, BeF_3^-

8.33 a. Tetrahedral   **b.** Triangular pyramid   **c.** Linear   **d.** Nonlinear

    e. Equilateral triangle   **f.** Octahedron

8.34 c, d   **8.35 a.** sp^3 **b.** sp^3 **c.** sp^3d^2 **d.** sp   **e.** sp^3 **f.** sp^2

8.36 a. sp^3d^2 **b.** sp^3d **c.** sp^3d^2 **d.** sp^3d^2 **e.** sp^3

8.37 a. $\frac{3}{2}$, 1   **b.** 2, 2   **c.** 3, 0   **d.** $\frac{1}{2}$, 1   **8.38** $(2)^{\frac{1}{2}} - 1$

8.39 H_2N—$(CH_2)_4$—NH_2; $(CH_3)_2$—N—N—$(CH_3)_2$; and so forth.

8.42 Triangular bipyramid (2 triangular pyramids joined at base). sp^3d

8.43 12. Octahedral, with electron pair at one apex (square pyramid considering atoms alone).

8.44 Pentagonal bipyramid   **8.45** Pentagonal bipyramid (14 e^-) if electron pair included.

CHAPTER 9

9.9 CaO, NaBr, LiF   **9.10 a.** $MgCO_3$ **b.** $SrCl_2 \cdot 6H_2O$ **c.** MgO

9.11 a. LiCl   **b.** AsH_3 **c.** SiO_2 **d.** HCl   **e.** $SiCl_4$ **f.** HCN

9.12 a. Ionic bonds   **b.** Dispersion forces   **c.** Covalent bonds   **d.** H bonds
   e. Dispersion forces, H bonds   **f.** Dispersion forces
9.13 a. Weak intermolecular forces   **b.** H bonds   **c.** Dipole forces
   d. Open crystal structure (see text)
9.14 a. S   **b.** O   **c.** S   **d.** P, C
9.15 a. Multiple bonds   **b.** 3 vs. 4   **c.** Layer structure
   d. Lower density of graphite
9.16 a. Mobile electrons   **b.** Mobile electrons   **c.** Electrons easily lost
9.17 a, b. Fewer bonds to break   **c.** Nonpolar
9.18 MW = 57; polymerization through H bonding   **9.21** appr. 2.6×10^{15} g

CHAPTER 10

10.17 a. 42.9 ml   **b.** 13500 ml   **c.** 17.3 ml
10.18 a. Increase   **b.** Increase   **c.** Increase   **d.** Increase
10.19 a. 768 mm Hg   **b.** 920 mm Hg
10.20 a. 0.0131 mole   **b.** 7.62 lit   **c.** 240 mm Hg   **d.** 183 mm Hg
10.21 290 mm Hg   **10.22 a.** 13.0 kcal
10.23 a. Sublimation occurs   **b.** Melting, vaporization   **c.** Condensation to liquid
10.24 4.07 Å; 10.6 g/ml   **10.25** CA; CA_3 **10.26** Decrease in density; raised
10.27 Na^+ **10.28** 0.476, 0.320, 0.260   **10.29** 0.66 g/ml
10.30 ΔE_{vap} = 9.9 kcal/mole
10.31 Energy to create 1 cm² of surface = 1.74×10^{-9} kcal; energy to bring one molecule
   to surface = 7.5×10^{-24} kcal; area occupied by one molecule = 4.3×10^{-15} cm².
10.32 a. 210°C   **b.** 6800 cal/mole   **c.** 310 mm Hg   **10.33** fcc   **10.34** AB_3C

CHAPTER 11

11.18 a. Same conc. H^+ **c.** dilute 4.8 lit conc. NH_3 to 12 lit   **11.19** 50 g
11.20 a. Dissolve 70.1 g NaCl in enough water to form one liter of solution
   b. Add 70.1 g NaCl to 1000 g of water
   c. Dissolve 53 g $NiSO_4 \cdot 6H_2O$ in enough water to form one liter of solution
   d. Add 11 g KOH to 26 g CH_3OH
11.21 a. X C_2H_5OH = 0.143, m = 9.25   **b.** X NaBr = 0.0596, m = 3.52
11.22 0.29, 0.24 lit; 31, 0.37; 0.76, 1.7; 50, 0.23 lit
11.23 Add 1.1 g KCl to 50 ml water; add 25 ml 0.20M KCl to 25 ml 0.40M KCl; dilute
   38 ml 0.40M KCl to 50 ml; evaporate 75 ml 0.20M KCl to 50 ml.
11.24 0.0232, 1.32, 1.24
11.25 a. Benzene   **b.** Biphenyl   **c.** Ar   **d.** Propane   **e.** Anthracene
11.26 1.1 g/lit   **11.27** 91.4 mm Hg   **11.29** 33 g   **11.30 b.** 100.046°C, −0.17°C
11.31 12 qt   **11.32** 114   **11.33 a.** 104   **b.** C_8H_{10} **11.34** 4%
11.35 −18.0°C   **11.36** 0.01 m sugar, 0.02 m urea, 0.01 m Li_2SO_4, 0.01 m AlF_3
11.37 0.352 atm   **11.39** appr. 65 ft (at 20°C)   **11.40** 557

CHAPTER 12

12.13 a. $K_c = [NO_2]^2/[NO]^2 \times [O_2]$ **b.** $K_c = [CS_2] \times [H_2]^4/[CH_4] \times [H_2S]^2$
   c. $K_c = [CO_2]/[CH_4] \times [O_2]^2$ **d.** $K_c = [C_2H_2]$

12.14 a. 1.01 **b.** 4.9×10^3 **12.15 a.** ← **b.** → **c.** ←

12.16 $[I_2] = 0.0053$, $[H_2] = 0.105$, $[HI] = 0.189$; 13 g I_2

12.17 $[I_2] = 0.011$, $[H_2] = 0.111$, $[HI] = 0.278$

12.18 a. 7.8×10^{-3} **b.** $[N_2] = 0.086$; $[H_2] = 0.26$; $[NH_3] = 0.028$

12.19 a. 9 **b.** K_c constant, amt. CO_2 constant, conc. CO_2 decreases

    c. 0.67 mole H_2; 0.83 mole CO_2

12.20 a. 0.15 **b.** 0.0017 **12.21 a.** 0.31 **b.** 1.76 **c.** 12 g/lit; $P_{NO_2} = 7.6$ atm,

                                                          $P_{N_2O_4} = 0.52$ atm, M = 49

12.23 3.1 g NH_3 **12.24** 2.8×10^{-3} **12.25** 0.24 mole SO_3 **12.26** 220

CHAPTER 13

13.14 $\Delta H = -38.1$ kcal, $\Delta G = -37.9$ kcal

13.15 G of HgO > G (Hg + $\frac{1}{2}$ O_2) at 500°C

13.16 $\Delta S° = -6.0$ cal/°K

13.17 a. $\Delta G° = +75.0$ kcal **b.** $\Delta G° = -99.9$ kcal

13.18 a. N.G. **b.** O.K.

13.19 Na(s) < NaCl(s) < Br_2(l) < Br_2(g) < N_2O_4(g)

13.20 a. $\Delta H = +0.6$ kcal, $\Delta G° = -8.8$ kcal, $\Delta S° = 11.7$ cal/°K

13.21 a. $K_p = 1.2 \times 10^{-8}$ **b.** $K_p = 10^{44}$

13.22 a. $K_c = 7 \times 10^{-6}$ **b.** $K_c = 10^{44}$ **13.23 a.** 0.22 atm **b.** 0.23 atm

13.24 a. 0.28 **b.** 210 mm Hg **c.** 60°C **d.** 3.3 atm **e.** Liquid **13.25** 1×10^4

13.26 a. $\frac{1}{16}$ **b.** 1 $SnCl_4$, 4 $SnCl_3Me$, 6 $SnCl_2Me_2$, 4 $SnClMe_3$, 1 $SnMe_4$

13.28 About 3400°K **13.29 a.** CuO **b.** Cu_2O

CHAPTER 14

14.15 a. 0.160 mole/lit sec **b.** 0.240 mole/lit sec **c.** 0.080 mole/lit sec

    d. 4.80 mole/lit min

14.16 c. 0.055 min^{-1} **14.17** 0.40 (mole/lit)$^{-1}$ sec^{-1}

14.18 a. 1.5×10^{-3} **b.** 1.9×10^{-3} **c.** 2.3×10^{-3} **d.** 3.7×10^{-3}

14.19 0.012 min^{-1} **14.20 a.** 6.0×10^{-2}, 6.0×10^{-7} **b.** 7.8 sec **c.** 3.0 sec

14.21 a. 0.012 mole/lit **c.** 31 sec **14.22 a.** 28 min **b.** 48 min **14.23** 36,100 cal

14.24 a. 1.4×10^{-6}, 2.7×10^{-6} **14.25** rate = $2kK(\text{conc. } O_3)^2/(\text{conc. } O_2)$

14.26 $\dfrac{\Delta \text{conc. } CH_3}{\Delta t} = k_1(\text{conc. } CH_3CHO) - k_3(\text{conc. } CH_3)^2 = 0$

$\dfrac{\Delta \text{conc. } CH_4}{\Delta t} = k_2(\text{conc. } CH_3)(\text{conc. } CH_3CHO) = k_2\left(\dfrac{k_1}{k_3}\right)^{\frac{1}{2}} (\text{conc. } CH_3CHO)^{\frac{3}{2}}$

14.27 1st order as conc. A → O

14.28 b. $dA/dt = -kA^3$; $-dA/A^3 = k\, dt$; $\frac{1}{2}A^2 - \frac{1}{2}A_0^2 = kt$

14.29

t	0	10	20	30	40	
p_A	200	159	126	100	80	k = 0.023 min^{-1}
$\log_{10} p_A$	2.301	2.201	2.100	2.000	1.903	

679

14.30 a. $-\Delta$conc. A$/\Delta t = k_1$(conc. A) $+ k_2$(conc. A)$^2 = 0.020$(conc. A)(1 + conc. A)
c. "Apparent order" increases from about 1.2 to about 1.5 as second order reaction becomes more important at higher concentrations.

14.31 $\dfrac{\Delta \text{conc. O}_2}{\Delta t} = k_1$(conc. NO$_2$)(conc. NO$_3$)

(conc. NO$_2$)(conc. NO$_3$) $= K_2$(conc. N$_2$O$_5$°) $= K_2 K_1$(conc. N$_2$O$_5$)

$\dfrac{\Delta \text{conc. O}_2}{\Delta t} = k_1 K_2 K_1$(conc. N$_2O_5$)

CHAPTER 15

15.17 a. $+5, -2$ **c.** $0, +1, -2$ **15.18 a.** N   **b.** O   **c.** NH$_3$ **d.** O$_2$
15.19 b. GeF$_4$ **c.** NO$_3^-$, NO$_2^-$ **15.20** H$_2$O, H$_2$O$_2$, BaO, BaO$_2$, Ba(OH)$_2$, BaH$_2$
15.21 a. LiH(s) $\rightarrow$ Li(s) $+ \frac{1}{2}$ H$_2$(g)   **c.** BaO$_2$(s) $\rightarrow$ BaO(s) $+ \frac{1}{2}$ O$_2$(g)
b. NH$_3$(g) $\rightarrow \frac{1}{2}$ N$_2$(g) $+ \frac{3}{2}$ H$_2$(g)   **d.** SO$_3$(g) $\rightarrow$ SO$_2$(g) $+ \frac{1}{2}$ O$_2$(g)
15.22 4×10^9 **15.23** Increase in p increases rate, does not change position of equilibrium; increase in T increases rate, shifts equilibrium to left.
15.24 a. mole% SO$_2 = 40$, O$_2 = 20$ **b.** n $= 0.0125$ **c.** [SO$_3$] $= 0.0050$, [SO$_2$] $= 0.0050$, [O$_2$] $= 0.0025$

15.25 7700 Å
15.26 b. Allow Na to oxidize in moist air   **c.** Expose Ba to air; add BaO$_2$ to water.
15.27 b. 3 TiO$_2$(s) $+ 4$ Al(s) $\rightarrow$ 3 Ti(s) $+ 2$ Al$_2$O$_3$(s)
c. Na$_2$O$_2$(s) $+ 2$ H$_2$O $\rightarrow$ 2 NaOH(s) $+$ H$_2$O$_2$
g. CaH$_2$(s) $+$ O$_2$(g) $\rightarrow$ CaO(s) $+$ H$_2$O(l)
15.29 600°C; K$_c = 5 \times 10^3$ **15.30** ΔH $= -51$ kcal; ΔG° $= -51$ kcal   **15.31** Fe
15.32 18   **15.33** Could be a mixture of Cs$_2$O and CsO$_2$

CHAPTER 16

16.17 c. no reaction   **d.** Hg$_2^{2+}$ $+ 2$ Cl$^-$ $\rightarrow$ Hg$_2$Cl$_2$(s)   **i.** Zn^{2+} $+$ S^{2-} $\rightarrow$ ZnS(s) and Ba^{2+} $+$ SO$_4^{2-}$ $\rightarrow$ BaSO$_4$(s)
16.18 0.345 mole Fe(OH)$_2$; 0.111 mole Fe^{2+}, 0.912 mole Cl$^-$, 0.690 mole Na$^+$
16.19 0.0746 mole K$^+$, OH$^-$; 0.135 mole Na$^+$, Cl$^-$; M K$^+ = 0.0901$, M Na$^+ = 0.163$
16.20 a. 9×10^{-10} **16.21 a.** 1.4×10^{-4} **16.22 a.** Yes   **b.** Yes   **c.** No   **d.** No
16.23 a. 3×10^{-6} **b.** 0.1%   **c.** 3×10^{-5} **16.24 a.** AgCl   **b.** 1.6×10^{-5}
16.25 2.5×10^{-11} **16.26** 41.4%   **16.27** 67.7%
16.28 b. Add SO$_4^{2-}$ to remove Pb^{2+}; add S^{2-} to remove Ni^{2+}.
16.30 a. Dissolve RbCl in water, titrate with AgNO$_3$, filter and evaporate.
16.31 a. Mg^{2+} $+ 2$ NaZ(s) $\rightarrow$ MgZ$_2$(s) $+ 2$ Na$^+$
16.32 a. 0.74 g Ca(OH)$_2$ **b.** 0.37 g Ca(OH)$_2$, 0.53 g Na$_2$CO$_3$ **16.33** 6.6 ft
16.34 a. 3 ml   **b.** 3.6 ml   **16.35** 51% NaBr

CHAPTER 17

17.21 a. 7   **c.** -1 **d.** 5.6   **17.22 a.** 10^{-6} **c.** 3×10^{-2}
17.23 a. [H$^+$] $= 1.0 \times 10^{-2}$, [OH$^-$] $= 1.0 \times 10^{-12}$, pH $= 2.0$
e. [OH$^-$] $= 0.1$, [H$^+$] $= 10^{-13}$, pH $= 13.0$

17.24 a. Dilute to 100 cc   **d.** Neutralize   **17.25 a.** Cl_2O_7, $HClO_4$, ClO_4^-

17.26 a. $HBr + H_2O \rightarrow H_3O^+ + Br^-$ **e.** $NH_4^+ + H_2O \rightarrow H_3O^+ + NH_3$

17.27 a. $Ca(OH)_2(s) \rightarrow Ca^{2+} + 2\ OH^-$ **d.** $CH_3NH_2 + H_2O \rightarrow CH_3NH_3^+ + OH^-$

17.28 b. $Cr(H_2O)_6^{3+} + H_2O \rightarrow Cr(H_2O)_5(OH)^{2+} + H_3O^+$

    c. $CO_3^{2-} + H_2O \rightarrow HCO_3^- + OH^-$

17.29 1.6×10^{-5} **17.30** $K_a = 1 \times 10^{-5}$

17.31 a. pH = 2.8; 1.2% ionized; $[H^+] = 1.5 \times 10^{-3}$; $[OH^-] = 6.7 \times 10^{-12}$

    d. pH = 7.0; 1.8×10^{-7}% ionized; $[H^+] = 1.0 \times 10^{-7} = [OH^-]$

17.32 1.0×10^{-4} **17.35** $Cu^{2+} + 4\ Cl^- \rightarrow CuCl_4^{2-}$

17.36 Multiply expression for K_1 by expression for K_2.

17.37 $[H_2O] = 55$; $[H_2S] = 10^{-1}$; $[H^+] = 10^{-4}$; $[HS^-] = 10^{-4}$; $[OH^-] = 10^{-10}$; $[S^{2-}] = 10^{-15}$

17.38 HZ < HY < HX. K_a for HY = 10^{-5}, HZ = 10^{-9} **17.39** 0.2M

17.40 a. $K = \dfrac{[HB] \times [A^-]}{[HA] \times [B^-]} = \dfrac{[HB]}{[H^+][B^-]} \times \dfrac{[H^+][A^-]}{[HA]} = \dfrac{K_a\text{ of HA}}{K_a\text{ of HB}}$

    b. $K = \dfrac{[A^-]}{[HA] \times [OH^-]} = \dfrac{[A^-][H^+]}{[HA]} \times \dfrac{1}{[H^+][OH^-]} = K_a/K_{H_2O}$

CHAPTER 18

18.19 a. $H^+ + OH^- \rightarrow H_2O$ **b.** $HC_2H_3O_2 + OH^- \rightarrow H_2O + C_2H_3O_2^-$

    c. $NH_3 + H^+ \rightarrow NH_4^+$

18.20 a. 2.2×10^3 **b.** 2.2   **18.21** 0.183 M   **18.22** 59.0%

18.24 Yellow, yellow, green, blue   **18.26 a.** 9.3   **b.** 9.1

18.27 c. Dissolve $CoCl_2$ in water, precipitate with OH^-. Titrate $Co(OH)_2$ with HBr; evaporate.

18.28 a. Precipitate CO_3^{2-}, SO_4^{2-} with Ba^{2+}. Dissolve $BaCO_3$ in acid.

    d. Precipitate Cu^{2+} with H_2S in strongly acidic solution; add base to precipitate ZnS.

18.29 a. 0.03 M   **b.** 10 M

18.30 a. no   **b.** Add 0.9 mole OH^- to reduce $[H^+]$ to 0.1 M

18.31 7, yes   **18.32** 10^{-29}, 10^{-43} **18.33 a.** 5.3   **b.** 6.4   **c.** 9.2   **d.** 12.0   **e.** 13.0

18.34 0.05 mole

CHAPTER 19

19.16 a. +3   **b.** 0   **c.** −1   **19.17 a.** Linear   **b.** Square, planar   **c.** Tetrahedral

19.18 a.

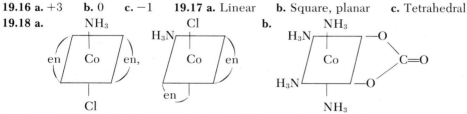

19.19 cis-$[Cr(NH_3)_4Cl_2]Br$, trans-$[Cr(NH_3)_4Cl_2]Br$, cis-$[Cr(NH_3)_4ClBr]Cl$, trans-$[Cr(NH_3)_4ClBr]Cl$

19.20 a. 36   **b.** 82   **c.** 36   **d.** 35   **19.21 a.** $5d^{10}\ 6s^2\ 6p^4$ **b.** $5d^8\ 6s^2\ 6p^6$

19.22 a. $5s^25p^2$ **b.** $5d^{10}6s^26p^4$ **c.** $3s^23p^63d^4$

19.23 Of the electrons in the 3d level of Ni^{2+}, six would fill the lower three orbitals; the other two electrons would go unpaired into the two higher orbitals.

19.24 a. 1×10^{-9} **b.** 3×10^{-5}

19.25 a. precipitate Ag^+ with HCl; add NH_3 to give complex with Ni^{2+}, ppt. with Al^{3+}. **b.** precipitate $SO_4{}^{2-}$ with Ba^{2+}; ppt. Cl^-, Br^- with Ag^+; dissolve AgCl in dilute NH_3.

19.26 a. HCl, EDTA   **b.** NH_3, $Na_2S_2O_3$ **c.** $Na_2S_2O_3$, NaCN   **19.27 a.** 0.1 M

19.28 a. $Zn^{2+} + 4\ OH^- \rightarrow Zn(OH)_4{}^{2-}$ **b.** $Al(OH)_3(s) + OH^- \rightarrow Al(OH)_4{}^-$ **c.** $Cu^{2+} + 4\ NH_3 \rightarrow Cu(NH_3)_4{}^{2+}$

19.29 Study kinetics in a solvent containing a small amount of water.

19.30 $Ni^{2+} + 2\ NH_3 + 2\ H_2O \rightarrow Ni(OH)_2(s) + 2\ NH_4{}^+$
$Ni(OH)_2(s) + 6\ NH_3 \rightarrow Ni(NH_3)_6{}^{2+} + 2\ OH^-$
$Al^{3+} + 3\ NH_3 + 3\ H_2O \rightarrow Al(OH)_3(s) + 3\ NH_4{}^+$
$Ni(NH_3)_6{}^{2+} + 3\ C_4H_8N_2O_2 \rightarrow Ni(C_4H_8N_2O_2)_3{}^{2+} + 6\ NH_3$
$Al(OH)_3(s) + OH^- \rightarrow Al(OH)_4{}^-$
$Al(OH)_4{}^- + H^+ \rightarrow Al(OH)_3(s) + H_2O$
$Al(OH)_3(s) + 3\ H^+ \rightarrow Al^{3+} + 3\ H_2O$

19.31 $[Ag^+] = 2.5 \times 10^{-9}$; $[Ag(NH_3)_2{}^+] = [Cl^-] = 6.3 \times 10^{-2}$; $[H^+] = 2.4 \times 10^{-12}$; $[OH^-] = 4.2 \times 10^{-3}$

19.32 $[Cu^{2+}] = 3 \times 10^{-12}$; $[OH^-] = 1.8 \times 10^{-4}$; $[Cu(NH_3)_4{}^{2+}] = 15$

19.34 $[Co(NH_3)_4Cl_2]Cl \cdot H_2O$

CHAPTER 20

20.14 a. Prevent recombination Na, Cl_2.   **b.** Low temperature

20.15 Maine (why?)

20.16 a. $2\ Cl^- + 2\ H_2O \rightarrow Cl_2(g) + H_2(g) + 2\ OH^-$ **b.** $Cu^{2+} + 2\ Br^- \rightarrow Cu(s) + Br_2$ **c.** $H_2O \rightarrow H_2(g) + \frac{1}{2}\ O_2(g)$ **d.** $2\ Cl^- + 2\ H^+ \rightarrow Cl_2(g) + H_2(g)$

20.17 a. Electrolyze water soln. KI   **b.** Electrolyze water soln. $CuSO_4$ **e.** Electrolyze water soln. KBr, evaporate.

20.18 a. Electrolyze to obtain soln. of NaOH; neutralize with HBr, evaporate. **c.** Electrolyze water solution $Cu(NO_3)_2$.

20.19 a. 0.209   **b.** 0.105 GAW Zn, 0.209 GAW Br   **c.** 6.83 g Zn, 16.7 g Br

20.20 a. 8.38 g/hr   **20.21** 11.7 hr   **20.22** 4.5 hr   **20.23** 41.4 g

20.24 a. 6   **b.** 12

20.25 a. $Cu(s) + 4\ H^+ + 2\ NO_3{}^- \rightarrow Cu^{2+} + 2\ NO_2(g) + 2\ H_2O$ **c.** $MnO_2(s) + 4\ H^+ + 2\ Cl^- \rightarrow Cl_2(g) + Mn^{2+} + 2\ H_2O$

20.26 a. $Cr_2O_7{}^{2-} + 6\ Fe^{2+} + 14\ H^+ \rightarrow 2\ Cr^{3+} + 6\ Fe^{3+} + 7\ H_2O$ **c.** $5\ PbO_2(s) + 2\ Mn^{2+} + 4\ H^+ \rightarrow 2\ MnO_4{}^- + 5\ Pb^{2+} + 2\ H_2O$

20.27 a. $2\ Cr(OH)_3(s) + 4\ OH^- + 3\ ClO^- \rightarrow 2\ CrO_4{}^{2-} + 5\ H_2O + 3\ Cl^-$ **c.** $3\ Cl_2(g) + 6\ OH^- \rightarrow 5\ Cl^- + ClO_3{}^- + 3\ H_2O$

20.28 a. $2\ NO_3{}^- + 6\ Cl^- + 8\ H^+ \rightarrow 2\ NO(g) + 3\ Cl_2(g) + 4\ H_2O$ **b.** $2\ MnO_4{}^- + 5\ S^{2-} + 16\ H^+ \rightarrow 2\ Mn^{2+} + 5\ S(s) + 8\ H_2O$

20.29 19.1 g   **20.30** 5.70 lit   **20.32 a.** 21.4 g

20.33 $10\ SO_4{}^{2-} + 54\ H^+ + 44\ I^- \rightarrow 5\ SO_3{}^{2-} + 3\ S(s) + 2\ H_2S(g) + 22\ I_2 + 25\ H_2O$

20.34 31.3% $CuCl_2$, 39.2% NaCl, 29.5% $NaNO_3$

CHAPTER 21

21.16 a. $2\ Ag^+ + Cu(s) \rightarrow 2\ Ag(s) + Cu^{2+}$ **c.** 0.46 V

21.17 a. No, yes   **b.** No, no   **c.** Yes, no   **d.** Yes, no

21.18 a. 0.25 V   **b.** 1.64 V   **c.** 0.83 V   **21.19 a.** 2.12 V   **b.** 0.01 V   **c.** 1.14 V

21.20 c. $2\ PbSO_4(s) + 2\ H_2O \rightarrow Pb(s) + PbO_2(s) + 4\ H^+ + 2\ SO_4^{2-}$

21.21 Negative

21.22 a. forms $Zn(NH_3)_4^{2+}$ **b.** H_2SO_4 replaced by water   **c.** Cell operated above 100°C, where vp H_2O exceeds 1 atm   **d.** Reagents less expensive   **21.23** 1.61 V

21.24 a. I_2 **b.** Ni(s)   **c.** Yes, no

21.25 a. Yes   **b.** Yes   **c.** No   **d.** No   **e.** No

21.26 a. $2\ Ag^+ + H_2(g) \rightarrow 2\ Ag(s) + 2\ H^+$

    b. $Cr_2O_7^{2-} + 6\ Fe^{2+} + 14\ H^+ \rightarrow 2\ Cr^{3+} + 6\ Fe^{3+} + 7\ H_2O$

    c. $2\ Fe^{3+} + 2\ I^- \rightarrow 2\ Fe^{2+} + I_2$ **d.** No reaction

    e. $Cl_2 + 2\ Br^- \rightarrow 2\ Cl^- + Br_2$

    f. $NO_3^- + 3\ Fe^{2+} + 4\ H^+ \rightarrow NO(g) + 3\ Fe^{3+} + 2\ H_2O$

21.27 1.35 V   **21.28** 0.04 M   **21.29** Yes   **21.30 a.** 1.25 V   **b.** 0.89 V

21.32 1.8×10^{-5}

21.33 $E = E° - \dfrac{0.059}{n} \log_{10} Q$, where Q = concentration quotient. At equilibrium,

$E = 0$, $Q = K$. $E° = \dfrac{0.059}{n} \log_{10} K$

21.34 a. $[Mn^{2+}] = 0.0026$, $[Cl^-] = 2$, $[H^+] = 2$ **b.** 63 cc

21.35 $E = E° - \dfrac{0.059}{2} \log_{10}$ (conc. H^+)^2. Setting $E° = 0.00$, conc. $H^+ = 10^{-14}$, $E = 0.83$ V

CHAPTER 22

22.18 a. O, R   **b.** R   **c.** O   **d.** O, R   **e.** O   **f.** R

22.19 a. Fe^{3+} **b.** Fe^{2+} **c.** ClO_4^- **d.** F_2

22.20 a. HCl   **b.** Neither   **c.** HCl   **d.** HCl, H_2O **e.** Neither

22.21 a. React Zn with HCl, evaporate.   **b.** React Co with dilute HNO_3, evaporate.

    d. Add Ca to water, evaporate.

22.22 Zn, Ni, Pb   **22.23 a.** No   **b.** No   **c.** Could get Sn(s)

22.24 a. $Sn^{2+} + \frac{1}{2} O_2(g) + H_2O \rightarrow SnO_2(s) + 2\ H^+$ **b.** $Sn(s) + Sn^{4+} \rightarrow 2\ Sn^{2+}$

22.25 a, b, d   **22.26** 0.36 V

22.27 a. H^+ involved in reduction   **b.** S(s) produced by reaction with O_2

22.28 a. Fe reacts directly with H^+.

    b. Increase in conductivity of soln. facilitates cell.

    c. Supplies O_2.

    d. Tin acts as cathode.

    e. Nickel acts as cathode.

22.29 a, b. Electrolyze water soln. KCl.

    c. Acidify solution obtained in **a**.

    d. Electrolyze hot soln. of KCl.

    e. Heat $KClO_3$ slightly above melting point.

22.30 a. 0.059 V   **b.** 0.071 V   **c.** 0.079 V   **d.** 0.094 V

22.31 a. $MnO_4^- \rightarrow Mn^{2+}$ **b.** $CrO_4^{2-} \rightarrow Cr_2O_7^{2-}$ **c.** NO_3^- more powerful oxid. agent than SO_4^{2-}.   **d.** $S_2O_3^{2-} + 2\ H^+ \rightarrow S(s) + SO_2(g) + H_2O$

22.32 22.9%   **22.33** 0.142 M

22.34 a. Oxidize I⁻ with Br₂ **b.** Precipitate as sulfides, oxidize CuS with HNO₃
c. Oxidize Cr³⁺ to Cr₂O₇²⁻ with ClO₃⁻

22.35 a. $2\ AgBr(s) + C_6H_6O_2 \rightarrow 2\ Ag(s) + 2\ Br^- + C_6H_4O_2 + 2\ H^+$

22.36 a. 0.600 N **b.** 0.500 N

22.37 E° = +1.28 V. Note that since Fe(OH)₂ is insoluble, one cannot get E° by adding 22.18a and 22.18b. In acidic soln., E° = +1.67 V.

22.38 Ag⁺ + I⁻; Ag⁺ + Fe²⁺; Fe²⁺ + MnO₄⁻; I⁻ + MnO₄⁻ **22.39** 4 M

CHAPTER 23

23.23 a. $^{224}_{88}Ra \rightarrow\ ^{4}_{2}He +\ ^{220}_{86}Rn$ **b.** $^{12}_{5}B \rightarrow\ ^{12}_{6}C +\ ^{0}_{-1}e$

23.24 8 alpha, 5 beta **23.25 a.** 0.105 g **b.** 45.2 hr **c.** 26.1 hr

23.26 40 min, 60 min **23.27** 6.4 min **23.28** 4.5×10^9 yr; 1.6 g; 2.4×10^{23}

23.29 340 yr **23.30** $^{63}_{29}Cu +\ ^{1}_{1}H \rightarrow\ ^{1}_{0}n +\ ^{63}_{30}Zn$ **23.31 a.** electron **b.** positron

23.32 a. $^{228}_{90}Th \rightarrow\ ^{4}_{2}He +\ ^{224}_{88}Ra$ **b.** $^{110}_{49}In \rightarrow\ ^{0}_{1}e +\ ^{110}_{48}Cd$

c. $^{112}_{49}In +\ ^{0}_{-1}e \rightarrow\ ^{112}_{48}Cd$

f. $^{239}_{94}Pu +\ ^{1}_{0}n \rightarrow\ ^{145}_{57}La + 4\ ^{1}_{0}n +\ ^{91}_{37}Rb$

23.33 a. −0.959 MeV **b.** -9.27×10^{17} ergs **c.** -2.00×10^6 kcal

23.35 -1.38×10^8 kcal **23.36** 4.64 MeV **23.37** Approximately 12

CHAPTER 24

24.16 4

24.17 a. 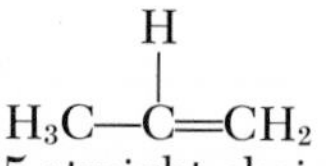 $H_3C-C=CH_2$ or

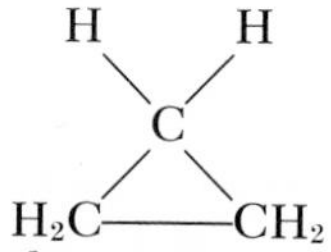

b. 5 straight chain and 2 cyclic isomers
c. Ortho-, meta-, and para-chlorotoluene
d. 3 trichlorobenzenes

24.18 a. Alcohol **b.** Alcohol, ether **c.** Acid
d. Ketone, aldehyde, unsaturated ether **e.** Aldehyde

24.19 4

24.20 butyraldehyde:

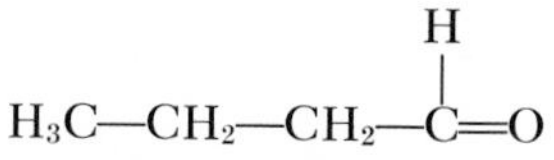

2-butanol:

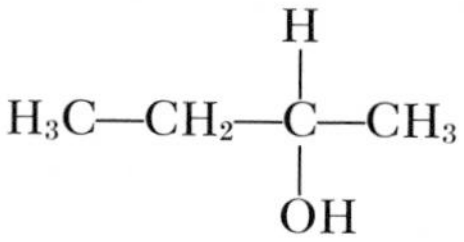

1-butene:

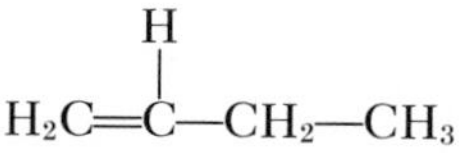

24.21

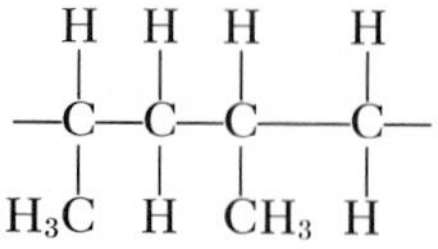

% C = 85.7

24.22 $(CH_3)_3C^+$; $(CH_3)_3CI$; $(CH_3)_2$—C=CH$_2$

24.23 Cracking, reforming, polymerization, aromatization.

24.24

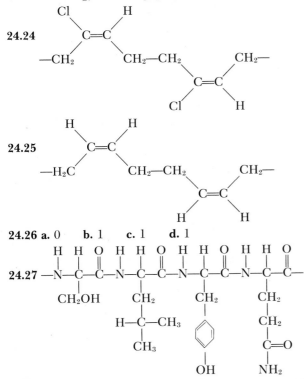

24.25

24.26 a. 0   **b.** 1   **c.** 1   **d.** 1

24.27

24.28 Ortho → 2 isomers, meta → 3 isomers, para → 1 isomer

24.29 C_nH_{2n}, C_nH_{2n-2}

24.30 18   **24.31** Acidic solution, high conc. Br$^-$

24.32 a. $CH_3CH_2CH_2Cl + OH^- \xrightarrow{H_2O} CH_3CH_2CH_2OH + Cl^-$

    b. $CH_3CH_2CH_2Cl + OCH_3^- \xrightarrow{CH_3OH} CH_3CH_2CH_2$—O—$CH_3 + Cl^-$

    c. $CH_3CH_2CH_2Cl + OH^- \xrightarrow{alcohol} CH_3CH{=}CH_2 + H_2O + Cl^-$

    d. $CH_3CH_2CH_2Cl + 2\ NH_3 \xrightarrow{sealed\ tube} CH_3CH_2CH_2NH_2 + NH_4^+ + Cl^-$

INDEX

Page numbers in *italics* refer to illustrations; those followed by (t) refer to tables.

Absolute temperature scale, 11, 86
Absorption spectrum, infrared, 20
Acetaldehyde, 626
  decomposition of, 357, 370
  oxidation of, 628
Acetic acid, 628
  colligative properties of, 433
  in buffer systems, 466
  ionization constant for, 436
  titration with sodium hydroxide, 464
Acetone, 626
Acetylene (molecule), 620
  bonding in, 184, 195, 201
    valence bond model of, *201*
  hydration of, 627
Acetylenes (class), 618
Acid(s). See also names of specific acids and
    Acid-base reactions.
  amino, 641
  Arrhenius concept of, 447
  as electrolytes, 433
  as solvents, solubility equilibria in, 473
    oxidizing agents in, 572
  Brönsted-Lowry, 447, 450(t)
  carboxylic, 627
  conjugate, 441
  equilibrium constant for, 434
  fatty, 628
  Lewis, 451
  neutralization of, 457
  organic, 627
  reactions of, with metals, 561
    with zinc, *434*
  strength of, 436
    determination of, 433, 449
    electronegativity and, 445
  strong, 433
    reaction with strong base, 457, 463
    reaction with weak base, 459, 464
  tests for, 425
  volatile, 470, 470(t)
  water solutions of, 424-456
    formation of, 430

Acid(s) *(Continued)*
  water solutions of, properties of, 425
  weak, 433
    ionization constants of, 434-439, 443(t)
    reaction with strong base, 457, 464
Acid-base indicator, 425, 462
  universal, 429
Acid-base reactions, 457-480
  Brönsted-Lowry concept of, 447
  concentrations of reactants in, 461
  direction of, 451
  end point in, 462
  equivalence point in, 463
  in inorganic synthesis, 469
  in ion separation, 472
  in qualitative analysis, 471-480
  in quantitative analysis, 460-469
  indicators in, 425, 429(t), 462
  Lewis concept of, 451
  Solvay process, 474
Acid-base titrations, 460
Acidity, and oxidizing strength, 572
Actinides, 119-121
Activation analysis, 601
Activation energy, 360
  catalysts and, *364*
Activity coefficient, 409
Actual yield, 51
Adsorption, chemical, 363
Air, 80
  composition of, 96
  ozone layer of, 230
Alcohol(s), 625
  dehydration of, 632
  reaction with organic acids, 627
  water solutions of, *274, 277*
Aldehydes, 626
Aldohexoses, 639
Alizarine yellow, 429
Alkali metals, atomic radii of, 165(t)
  electron configurations of, 163(t)
  reaction of oxygen with, 389
Alkaline earths, 119

Alkanes, 619
Alkenes, 619
Alkylation, 635
Alkynes, 620
Allotropy, 230
Alpha rays, 587
Aluminum, as reducing agent, 392
    oxidation of, 562
    production of, 513
    reaction with hydrochloric acid, 97
Aluminum chloride, 214
    in petroleum distillation, 635
Aluminum(III) ion, analysis for, 503
Aluminum oxide, electrolysis of, 513
Aluminum salts, reaction with water, 432
Aluminum sulfate, reaction with barium
    bromide, 401
Amino acids, 641
    from proteins, 643(t)
Ammonia, anhydrous, preparation of, 470
    as complexing agent, 504
    bonding in, 193, 194, 217
    conversion to ammonium salts, 469
    decomposition of, 364
    in Solvay process, 476
    ionization of, 441
    preparation of, 385
        Haber process for, 363, 386(t)
    reactions of, with copper(II) sulfate, 483
        with hydrogen chloride, 449
        with sodium hydride, 449
        with water, 440, 447
    titration of, with hydrochloric acid, 464
Ammonium chloride, 432
Ammonium ion, 432
    bonding in, 186
    test for, 471
Ammonium salts, preparation of, 469
Ampere, 519
Amphiprotic compound, 448
Amphoteric compound, 504
Analysis. See also *Qualitative analysis* and
        Quantitative analysis.
    activation, 601
    gravimetric, 412
    radioactive techniques in, 601
    volumetric, 576
Analytical balance, 6
Angstrom, 3
Angular momentum, 148
Anhydrides, acid, 430
    basic, 440
Anions, 28, 172
    hydrolysis of, 446
    in electrolytic cells, 512
Anisotropy, 27
Anode, 512
Antibonding orbital, 203
Aqua regia, 576
Aquocomplex, 483
Argon, 89
Aristotle, 24
Aromatization, 636
Arrhenius, S., 292, 360
    acid-base concepts, 447
Arsenic, Marsh test for, 385

Arsenic(III), sulfide, 408
Arsine, 384
Astatine, 120
Atmosphere, as unit of measurement, 11
    ozone layer of, 230
Atmospheric pressure, 80
Atom(s), 26, 127-169
    absolute mass of, 37
    electron clouds in, *147*
    electronic structure of, 137-162
    energy of, 133, 138
        radiant, 129
    foreign, 255
    interstitial, 255
    isotopes of, 585
    noble gas structure in, 186
    quantum state of, 134
    radii of, 175(t), 673(t). See also *Atomic radii.*
    size of, 37, 164
    theory of, 24
Atomic bomb, 602
Atomic mass unit, 605
Atomic number, 129
Atomic orbitals, *154, 159*
    combination of, 228
    covalent bonding and, 178
    hybrid, 197-201
    overlap of, 201
    population of, 149
    valence-bond approach, 201
Atomic radii, 175(t), 673(t)
    and complexing tendency, 483
    and electron configuration, 164, 165(t)
    and hydrogen bonding, 218
Atomic spectra, 134
    and quantum numbers, 151
    regularities in, 137
Atomic theory, 24
Atomic weight, 30, 38, 113(t)
    carbon-12 scale, 31
    determination of, 34, 41
Aufbau principle, 153
Avogadro, A., 95
Avogadro's Law, 89
Avogadro's number, 37, 46, 49

Bakelite, molecular structure of, *645*
Balance, analytical, *6*
Balmer, J. J., 137
Balmer series, for hydrogen, 137
Band theory, 227
Barium, analysis for, 412
    electronic structure of, *225*
Barium bromide, reaction with aluminum
    sulfate, 401
Barium chloride, anhydrous, formation of,
    213
Barium chloride dihydrate, 213
Barium hydroxide, reaction with nickel sul-
    fate, 402
Barium oxide, physical properties of, 213
Barium peroxide, 390
Barium sulfate, 401, 406
    preparation of, 415

Barometer, 81
Base(s). See also *Acid-base reactions.*
  Arrhenius concept of, 447
  Brönsted-Lowry, 447, 450(t), 485
  conjugate, 441
  Lewis, 451, 485
    in substitution reactions, 629
  neutralization of, 457
  nitrogen, 644
  strength of, 450
  strong, reaction with weak acid, 457, 464
    titration with strong acid, 463
  tests for, 426
  volatile, preparation of, 470
  water solutions of, 439-456
    properties of, 425
  weak, ionization constants of, 441, 443(t)
    reaction with strong acid, 459, 464
Battery, lead storage, 73, 537
  solar, 259
Bauxite, 514
Becquerel, H., 586
Benzene, 620, *621*
  as solvent, 214
  critical temperature of, *246*
  phase diagram for, *263*
  physical properties of, 249(t)
  reaction with bromine, 621
  resonance forms of, 622
  solubility of hydrocarbons in, 279(t)
  triple point of, 260
  vaporization of, 245
Berthollet, C. L., 25
Berthollide(s), 25
Beryllium, 202
  electronic structure of, 152, 205, 229, *229*
  hybrid bonds in, 197
Beryllium fluoride, bonding in, 195
  polarity in, 195
Berzelius, J. J., 30
Beta-emission, 587
Beta particle, 588
Beta rays, 587
Betatron, 597
Binding energy, in nuclear reactions, 606, 608(t)
Biochemical reactions, 469
Bjerrum, 293
Bohr, N., 130, 138
Bohr radius, 140
Bohr theory, 139
Boiling, 265
Boiling point, 17, 246
  critical temperature and, 247
  heat of vaporization and, 248
  of lighter elements, *112*, 113(t)
  of nonpolar substances, 218
  of solutions, 286
  pressure and, 246
Boltzmann, L., 100
Bomb calorimeter, 63
Bombardment, nuclear, 595, 597(t)
Bond(s), 28, 170-211
  and enthalpy changes, 71
  angle of, 192
  covalent, 176-185. See also *Covalent bond.*

Bond(s) *(Continued)*
  electron-pair, 176
  electron-sea model, 225
  energies of, 184(t)
    and electron distribution, 204
    in covalent bonds, 182
    in molecular orbitals, 204
  hybrid, 189, 197-201
  hydrogen, 217
    in ice, 218
  in nonmetal hydrides, 443
  in nonmetals, 199
  in organic compounds, 616
  in oxyacids, 444
  ionic, 170-175
  metallic, 224
  multiple, 184, 194
  orbitals, hybrid, 197
    molecular, 203
  peptide, 642
  pi, 201
  randomization of, in chemical reactions, 333
  resonance energy in, 190
  resonance hybrid, 189
  sigma, 200
Boric acid, 436
Boron, electronic structure of, 152
  hybrid bonds in, 198
Boron trifluoride, as Lewis acid, 452
  bonding in, 193, 198
Boyle, R., 24, 84
Boyle's Law, 83
Bragg, W., 129, 250
Bragg equation, 250
Bromide ion, analysis for, 414
Bromine, physical properties of, 215, 249(t)
  preparation of, 565
  reaction with benzene, 621
  reaction with hydrogen, 387
Bromobenzene, 621
Bromthymol blue, 462
Brönsted, J. N., 447
Brönsted-Lowry concept, 447
  of acids, 447, 450(t)
  of bases, 485
Buffers, 466
Building-up principle, 153
1,3-Butadiene, 620
Butane, 183(t), 617
2-Butanol, 625
1-Butene, 619
n-Butyl alcohol, 278

Calcium, reaction with water, 562
Calcium carbide, reaction with water, 620
Calcium carbonate, decomposition of, 213, 299, 318
Calcium chloride, 476
  drying agent, *33*
  solutions of, 292
Calcium fluoride, 29
Calcium hydride, 382
Calcium hydroxide, 418
  decomposition of, 213
  solubility of, 439

689

Calcium sulfate, 417
"Caloric," 8
Calorie, 11, 57
Calorimeter, 63
Calvin, M., 601
Camphor, as solvent, 291
Cannizzaro, S., 30, 33, 95
    determination of atomic weight, 41
Carbohydrate, 638
Carbolic acid, 626
Carbon, 27
    allotropic forms of, 233
    amorphous forms of, 234
    analysis for, 44
    bonds in, 200, 206, 233
    electron configuration in, 152
    electronic structure of, 152, 221, 233
    organic compounds of, 615-646
        bonding in, 183, 199, 620
    oxides of, 393
    properties of, 616
Carbon-12, 31, 38
Carbon-14, 594
Carbon dioxide, 27, 393
    analysis for, 472
    bonding in, 194
    in carbon-hydrogen analysis, 44
    in Solvay process, 475
    preparation of, 470
    solid, 259
Carbon-hydrogen analysis, absorption train
    for, 44
Carbon monoxide, 350, 394
    as reducing agent, 392
    molecular orbitals in, 207
Carbon tetrachloride, 624
    as solvent, 214, 277
    bonding in, 196
Carbonate ion, analysis for, 472
    as chelating agent, 485
    as test for acid, 425
Carbonic acid, 431
    preparation of, 470
    reaction with sodium hydroxide, 459
Carbonium ion, 630
Carbonyls, metal, 483
Carboxylic acid(s), 627
Cassiterite, 318, 392
Catalysis, 362
Catalyst, 362
    and activation energy, 364
Cathode, 127, 512
Cathode ray(s), 26, 127
Cathode ray tube, 128
Cathodic protection, against corrosion, 571
Cation(s), 28, 512
    formation of, 224
    hydrolysis of, 446
    metal, as oxidizing agents, 562
        oxidation of, 567
    monatomic, 172
Cell, electrolytic, 511-531. See also specific
    names.
Cellulose, 640, 641
Celsius scale, 10
Centigrade scale, 10

Cesium, electronic structure of, 225
Cesium chloride, arrangement of ions in, 254
    preparation of, 416
Cesium nitrate, 469
Chain reactions, 369
    nuclear, 603
Charge density, of ionic compounds, 405
Charcoal, 234
Charles' Law, 85, 87
Chelates, 485
Chelating agents, 484
    in qualitative analysis, 503
Chemical changes, 12
Chemical Equilibrium, Law of, 304
Chemical symbols, 21
Chemistry, basic concepts of, 1-23
Chlordane, 624
Chlorides, atom ratios in, 115
Chloride ion, 490
    analysis for, 411, 413
Chlorine, as oxidizing agent, 564
    bonding in, 206
    dissociation of, 387
    physical properties of, 215
    production of, 513
    reactions of, 565
        with hydrocarbons, 631
        with hydrogen, 387
Chlorine water, 565
Chlorobenzene, infrared spectrum of, 20
Chloroform, 624
    as solvent, 277
    chlorination of, 370
    molecule, bonding in, 196
Chloropentane, 624
Chlorophyll, 75, 486
Chloroprene, 638
Chromate ion, anticorrosive nature of, 569
Chromatography, 15
Chromium, oxidation states of, 573
Chromium(III), complex ions of, 499
Chromium(III) salts, analysis for, 577
Cis isomer, 488
Clausius, R. J. E., 100
Clausius-Clapeyron equation, 245
Cloud, of electrons, 27
Cloud chamber, Wilson, 589, 590
Cloud seeding, 265
Cobalt-60, 600
Cobalt(III), complex ions of, 490, 494, 499
Cobalt(II) iodide, preparation of, 561
Colligative properties, 283-294
    molecular weight and, 290
Collision theory, of reaction rate, 364
Combining Volumes, Law of, 94
Combustion, 70
Common ion effect, 406
Complex, activated, 361
Complex ions, 481-510
    analysis for, 502
    bonding in, 493
    charges of, 482
    composition of, 483
    coordination number and geometry of, 482,
        487(t), 491
    crystal field theory of, 495

Complex Ions (Continued)
  dissociation constants of, 502
  electrostatic forces in, 495
  formation rate of, 498
  in substitution reactions, 498
  in water solutions, 483
  ligands in, 482
  noble gas structure, 491
  separation of, 502
  solubility of, 482
  stability of, 505
  tests involving, 503
Complexing agent, ammonia as, 504
Complexing tendency, atomic radius and, 483
Compound(s), 20, 25
  amphiprotic, 448
  amphoteric, 504
  chemical formulas of, 42-55, 123
  chemical properties of, periodic nature of, 114
  coordination, 481-508, 672
  inorganic, analysis of, 44
    nomenclature, 669
    preparation of, 415, 469
  interhalogen, 210
  ionic, 28, 212. See also *Ionic compounds.*
  macromolecular, 221
  molecular, 214
  nonstoichiometric, 25
  organic, 615-649. See also *Organic compounds.*
Compressibility, of liquids, 239
Compression, of a gas, 79
Concentration(s), equilibrium, 306
Concentration cell, 551
Conductance, electrical, 259
Conductivity, electrical, 214, 223(t)
  in impure crystals, 256
  of metals, 223
  of water solutions, 283, 433
  valence electrons and, 226
  thermal, 223(t)
Conductors, 214
Conjugate acids and bases, 441
Conservation of Mass, Law of, 48
Constant(s), proportionality, 100, 352, 661
  tables of, 674
Constant composition, law of, 25
Contraction, lanthanide, 166
Conversion factors, table of, 674
  mass-energy, 605(t)
  mass units, 7(t)
Coordinating groups, in complex ions, 484
Coordination complexes, theory of, 490
Coordination compounds, 481-508, 672
Coordination number, 482
Copper, crystal structure of, 252
  electroplating with, 518
  oxides of, 390
  reaction with silver nitrate, 563
Copper(I), complex ions of, 493
Copper(II), complex ions of, 481, 498, 507
  oxide of, 391
Copper(II), salts of, complex ion formation, 498
Copper(II) chloride, electrolysis of, 516, 543

Copper(II) ferrocyanide, 289
Copper ion, 493
  analysis for, 503
Copper oxide, 391
Copper sulfate, anhydrous, 214
  water solution of, electrolysis of, 517
Copper(II) sulfate, anhydrous, 481
  reaction with ammonia, 483
Copper sulfate pentahydrate, 214
Corrosion, cathodic protection, 571
  impurities and, 570
  of iron, 568
Coulomb, 519
Covalent bond(s), 176-185
  and oxidation number, 378
  atomic orbital approach, 178
  coordinate, 186
  electronegativity and, 180, 181(t)
  energy of, 182
  in ice, 218
  in Lewis acids, 452
  in macromolecular substances, 222
  internuclear distances in, 182
  molecular orbital approach, 179
  noble gas structure in, 185
  polarity of, 179
  stability of, 201
  symmetrical, 180
Cracking, 618, 634
Cresol, 623
Crick, F., 646
Critical mass, 603
Critical reagent, 51
Critical temperature, and boiling point, 247
  in liquid-vapor equilibria, 245
Cryolite, 514
Crystal(s), defects in, 25, 255, 256
  and conductivity, 214
  density of, 20
  ionic, 29, 212
    conductivity of, 256
    structure of, 253
  lattice structure of, 29, 171
    diffusion in, 256
  macromolecular, 257
  nonstoichiometric, 257
Crystal field theory, 495
Crystal planes, movement of, 227
Crystallization, fractional, 14
Curie, I., 595
Curie, M., 30
Curie, P., 586
Cycloparaffin(s), 616
Cyclopropane, 617
Cyclotron, 597

Dalton, J., 24, 95
Dalton's Law, 95
Davisson, C. J., 141
DDT, 624
de Broglie, L., 141
de Broglie relation, 141
Debye, P. J., 293
Debye-Hückel equations, 293
Debye-Hückel theory, 424

Debye unit, 197
Decay, radioactive, 128, 598
 rate of, 591
Defects, crystal, 214, 255
Dehydration, 632
Dehydrohalogenation, 632
Deionization, 418
Deliquescence, 291
Democritus, 24
Density, 19, 113(t), 249(t)
 of gases, 91
 of lighter elements, *112*, 113(t)
 of liquids, 238
 of solids, 248
Deoxyribonucleic acid, 645
Detergent, 628
Dextrorotatory isomers, 488
Diamagnetism, 202
Diamond, bonding in, 221(t)
 crystal structure of, *233*
 properties of, 221(t)
Diazomethane, decomposition of, 373
Dibromobenzene, 621
Dichlorodiammineplatinum(II), 482, 488
Dichlorodiphenyltrichlorethane, 624
Dichlorotetramminecobalt(III) chloride, 490
Diethyl ether, 625
Diffraction, x-ray, 249, *250*
Diffusion, of gases, 102, *103*
 of liquids, 239
 of solids, 256
 rate of, 80
Dihydroxodiaquozinc(II), 482
Diisopropyl, 617
2,3-Dimethylbutadiene, 637
2,2-Dimethylbutane, 617
2,3-Dimethylbutane, 617
Dimethylglyoxime, 503
2,2-Dimethylpropane, 635
Dinitrogen tetroxide, 191
Dipoles, in molecular compounds, 216
 molecules as, 196
 temporary, in dispersion forces, 219
Dipole moments, 197, 197(t)
Disaccharide, 640
Dispersion forces, 218
Disproportionation, 564
Dissociation energy, 215
Distillation, 12
 of petroleum, 635
Distribution function, Maxwellian, *106, 366*
DNA, 645
Döbereiner, J., 115
Downs cell, 512
Dry cell, 536
Dry Ice, 259
 in cloud seeding, 265
Ductility, 224
Dulong, P. L., 31
Dulong and Petit, Law of, 31, 32(t)
Dumas, J. B. A., 30
Dyne, 11

ϵ, 100
E, 67

Δ E, 67
E°, 541
E_a, 360
EDTA, 507
Efflorescence, 214
Effusion, rate of, 103
Eicosane, 617
Einstein, A., 136
Einstein's equation, 604
Ekasilicon, 116
Electric current, amplification of, 258
Electrical energy, 73
 from chemical reactions, *326*
Electrode potential, standard, 539
Electrolysis, 128, 511-531
 Faraday's law of, 519
 of ionic compounds, 512
 of water solutions, 515
Electrolyte, 283
 colligative properties of, 291
 freezing points of, 292(t)
 interionic forces in, 409
 strong acids as, 433
Electrolytic cells, 511-531
Electron(s), 26, 127, 132
 arrangement of, 147
 charge on, 128
 density of, in molecular orbitals, 203
 emission of, 224
 energy of, 133, 159
 mass of, 128
 odd number of, 191
 orbits of, radii of, 140
 paired, 149, 176
 quantum numbers of, 147
  and energies of, 150
 transfer of, in chemical reactions, 376
 unpaired, 153
 valence, 165
  and chemical bonds, 224
  in metallic bonding, 225
  mobility of, 226
Electron cloud, 27, *147*
 and dipole forces, 219
 geometric structure of, 158
Electron configuration(s), 155, 156(t)
 and atomic radii, 164
 and chemical properties, 163
 and ionization energy, 160
 and periodicity, 163
 and physical properties, 212-237
 noble gas, 172(t)
Electron density, 158, 203
Electron-pair bond, 176
Electron-sea model, 225
Electron spin, 153
Electron trap, 579
Electron volt, 11
Electronegativity, 180
 and acid strength, 445
 and valence state, 181
Electronegativity difference, hydrogen bonds
  and, 217
 ionic character and, *181*
Electronic structure, atomic orbital approach,
  178

Electronic structure *(Continued)*
  atomic radii and, 164, 165(t)
  Heitler-London model, 177
  molecular orbital approach, 179
  of atoms, 147
  of ionic compounds, 171
  of ions, 155
  of Lewis acids, 452
  of nonmetal hydrides, 443
  of oxyacids, 444
  valence bond approach, 178
Electroplating, 518
Electrostatic repulsion, 192
Element (s), 20
  allotropic forms of, 230
  anisotropic, 27
  chemical properties of, and electron con-
      figuration, 163
    periodic nature of, 114
  chemical symbols for, 21
  classification of, 111-126
  discovery of, 21
  electronegative, 181
  electropositive, 121
  family of, 116
  isotopes of, 27, 585
  "octaves" of, 115
  oxidation states of, 379, *381*
  periodic character of, 111-114
  periodic table of, 116-121, 118(t)
  physical properties of, 112, 212-237
  transuranium, 597
  "triads" of, 115
Elimination reactions, 631
Emission, beta, 587
  neutron, 603
  thermionic, 224
Empirical formula, 42-46
End point, in acid-base titrations, 462
Endothermic reaction, 58
Energy, activation, 360
  atomic, 133, 138
  binding, in nuclear stability, 606, 608(t)
  bond, 182, 184(t), 204
  Einstein equation for, 604
  electrical, 73
    from chemical reactions, *326*
  electron, 133
    and quantum number, 159
  free, 323, 329(t)
  Heitler-London plot, *177*
  in chemical reactions, 162
  internal, 67
  ionization, *112*
  of formation, 329(t)
  lattice, 171
  light, 74
  mass and, 604
  measurement of, 11, 56
  mechanical, 73
  molecular, Maxwellian distribution func-
      tion for, 106, *366*
  nuclear, 604

Energy *(Continued)*
  of dissociation, 215
  of sublimation, 215
  quantized, 133
  radiant, 129
  resonance, 190
  rotational, 133
Energy changes, in chemical reactions,
      56-78, *362*
  in nuclear reactions 586
Enthalpy, 68, 69
  and chemical bonds, 71
Entropy, 330, 334(t)
Entropy unit, 333
Enzymes, 363
  in chemical reactions, 469
Equation(s), chemical, 42-55. See also names
      of specific equations.
  balancing, 47-51
  interpretation of, 48
  ideal gas, 87
  net ionic, 400
  oxidation-reduction, balancing, 523
Equilibrium, physical, 238
Equilibrium constant, 303
  for oxidation reduction reactions, 547
  for weak acids, 434
  temperature and, 317
  volume and, *314*, 314(t)
Equilibrium reactions, 57, 299-322
Equivalence point, 463
Erg, 11, 74
Escaping tendency, 286
Ester, 627
Ethane, 616
  bond energy of, 183
  structural geometry of, 193, *194*
Ethanol, 625
Ethene. See *Ethylene.*
Ether, 625
Ethyl alcohol, 625
  in thermometers, 10
  solubility of, 277
Ethylene, 28, 38, 619
  bonding in, 184, 195, 200
  hydrogenation of, 358
  valence bond model of, *201*
Ethylene glycol, 625
Ethylenediamine, 485
Ethylenediaminetetracetic acid, 507
Evaporation, 240, *288*
Exclusion principle, Pauli, 149
Exothermic reaction, 58
Expansion, of liquids, 239
  work of, 67
Exponentials, 651
Extensive properties, 19

Faraday, 519
Faraday, M., 621
Faraday's Law, of electrolysis, 519
Fahrenheit scale, 10

Fats, 628
Fatty acid, 628
Ferrocyanide ion, 495
Fischer, E., 639
Fission, 601, 604
Flask, volumetric, 5
Fluoride(s), binary, orbitals in, 200(t)
Fluoride ion, oxidation of, 515
Fluorine, atomic structure of, 585
    electronic structure of, 182, 185
    physical properties of, 215
    preparation of, electrolytic, 515
    reactions of, 377
        with hydrogen, 377, 387
        with sodium, 170
Formaldehyde, preparation of, 626
Formalin, 626
Formation, free energy of, 329(t)
Formula(s), chemical, 42-55, 123
Formula unit, 46
Fractional crystallization, 14
Fractional distillation, 12
Free energy, 327
    of formation, 329(t)
Freezing point, 261
    of electrolytes, 292(t)
    of solutions, 286
Frenkel defect, 255
Freon, 624
Fuel cell, 538
Fundamental particles, 131
Fusion, heat of, 248, 249(t), 260
        and molecular weight, 215
    nuclear, 610

γ, 240
$\triangle$G, 327
$\triangle$G°$_f$, 328
Gadolinium, 522
Gamma rays, 588
    in nuclear reactions, 597
Gas(es), compression of, 79
    concentration of molecules in, 241
    density of, 91
    diffusion of, 102, 103
    heat capacity of, 110
    ideal gas equation, 87
    inert, 119
    kinetic theory of, 99
    mixtures of, 95
        spontaneous formation, 333
    molecular energy of, 144
    molecular movement in, 100
    molecular weight of, 92
    noble, 27, 115, 119
        election configuration of, 172(t)
        solubility of, 279(t)
    partial pressure of, 95
    physical behavior of, 79-109
    pressure of, 80, 82, 100
    reactions of, 353(t)
    real, 98
    solubility of, 279
        partial pressure and, 282
    solutions of, 80

Gas(es) (Continued)
    translational motion of, 102
    volume of, 80, 94
Gas constant, 89, 89(t)
Gasoline, combustion of, 73
Gassendi, 24
Gay-Lussac, J. L., 94
Gay-Lussac's Law, 85, 87
Geiger-Müller counter, 589, 590
Germanium, 122
    as semiconductor, 257
    conductivity of, 223
    properties of, 116(t)
Germanium chloride, 122
Germer, L. H., 141
Gibbs, J. W., 325
Gibbs-Helmholtz equation, 335, 392
Gillespie, R. J., 192
Glycerol, 628
Goodyear, C., 636
Graham, T., 104
Graham's Law, 102
Gram, standard, 7
Gram atomic weight, 31
Gram equivalent weight, 31, 519
Gram molecular weight, 38
Graphite, crystal structure of, 233
Gravimetric analysis, 412
Gravity, 7
Gravity cell, 534
Guldberg, C. M., 305

h, 74
H, 68
H$_f$, 60
$\triangle$H$_f$, 61, 69
$\triangle$H$_{fus}$, 260
$\triangle$H$_{subl}$, 260
$\triangle$H$_{vap}$, 244, 260
Haber, F., 385
Haber process, 385
    temperature and pressure in, 386(t)
Hahn, O., 602
Half-life, 592
Halides, alkyl, in substitution reactions, 629
    hydrogen, 124(t)
        as strong acids, 433
        stability of, 388
    organic, 624
Hall, C., 513
Hall process, 513
    current efficiency in, 522
Halogen(s), 119
    hydrogen compounds of, 386
    in organic compounds, 624
    physical properties of, 215(t)
Halogenation, 631
Heat, 8, 68, 226
    and free energy change, 330
    flow of, 69
    from nuclear fission, 604
    in chemical reactions, 56-60
    measurement of, 57, 63, 64
    of formation, 60 62(t)
    of fusion, 248, 249(t), 260

Heat (*Continued*)
  of fusion, and molecular weight, 215
  of sublimation, 260
  of vaporization, 249(t)
    and boiling point, 248
    and molecular weight, 215
    experimental determination of, 244
    of liquids, 239, 240(t)
Heat capacity, 64
  of gases, 110
  of metals, 229
  of water, 218
Heat engine, 73
Heitler, W. A., 177
Heitler-London energy plot, *177*
Helium, as eluting agent, 16
  critical temperature of, 246
  electronic structure of, 152
  mass spectrum of, 35
  solubility of, *291*
Hematite, 392
Hemoglobin, 486
Henry's Law, 281, 285
Heptane, 636
Heroult, P. L. T., 514
Hess's Law, 60
Hevea tree, 636
Hexacyanoferrate ion, 494
Hexane, 617
  combustion of, 70
Hexaquoiron(II) ion, 494
Higgins, 24
Hückel, 293
Hund, F., 179
Hund's rule, 153
Hydration, water of, 213
Hydrides, 124
  interstitial, 383
  metallic, 383
  molecular, 384
  nonmetal, bonding in, 443
  reaction with water, 443
  saline, 382
Hydriodic acid, oxidation of, 567
Hydrocarbons, 616
  aliphatic, 277
  aromatic, 620
  oxidation of, 231, 370
  physical properties of, 215
  saturated, 616
  solubility of, 277
    in benzene, 279(t)
  unsaturated, 618
  volatility of, 215
Hydrochloric acid, 449
  as oxidizing agent, 561
  formation of, 430, 447
  reactions of, with ammonia, 464
    with sodium hydroxide, 457, 463
    with sulfides, 472
Hydrofluoric acid, 176, 436, 443
    as Brönsted Lowry acid, 448
  formation of, 433
Hydrogen, analysis for, 44
  atomic energies of, 138, 138(t)
  atomic spectrum of, *135*
  Balmer series for, 137(t)

Hydrogen (*Continued*)
  Bohr theory of, 138
  covalent bonding of, 176, 185
  electron behavior in, 145
  electron density in, *177*
  halides of, 124(t)
    as strong acids, 433
    stability of, 388
  in acidic solutions, 430
  ion, as oxidizing agent, 561
  ionization energy of, 145
  periodic classification of, 124
  reactions of, with fluorine, 377
    with halides, 443
    with halogens, 386
    with metals, 383
    with nitrogen, 384
    with nonmetals, 379, 383, 384(t)
Hydrogen bomb, 610
Hydrogen bonds, 217
  electronegativity differences and, 217
  in ice, 218
Hydrogen bromide, preparation of, 471
Hydrogen carbonate ion, 417, 440
  amphiprotic nature of, 448
  analysis for, 472
Hydrogen chloride, bonding in, 71
  formation of, 369, 387
  preparation of, *471*
  reaction with ammonia, 449
  reaction with water, 430, 447
Hydrogen cyanide, bonding in, 218
Hydrogen fluoride, electronic structure of, 179
  hydrogen bonding in, 217
  polarity of, 195
  reaction with water, 433
  stability of, 308
  structure of, 176
Hydrogen iodide, chemical equilibrium of, 300, 303(t)
  decomposition of, 364
  preparation of, 470
Hydrogen peroxide, in determining acid strength, 449
  reaction with iodide ion, 468
Hydrogen sulfate ion, 431
Hydrogen sulfide, analysis for, 472
  preparation of, 470
Hydrolysis, of salts, 446
  of fats, 628
Hydrolysis constant, 443
Hydronium ion, 426
Hydroquinone, 580
Hydroxide(s), formation of, 439
  reaction with acids, 457
  reaction with water, 439
Hydroxide ion, in water solutions, 426
Hydroxylamine, 580
Hypochlorite ion, 566
  electronic structure for, 187
Hypochlorous acid, 444

Ice, geometrical structure of, 218, *219*
Ideal Gas Law, 88, 100, 102

Impurities, and changes in state, 264
  removal from solids, 14
Indicators, acid-base, 425, 462
  color change intervals of, 429(t)
  in precipitation reactions, 413
  universal, 429
Inert gases, 119
Infrared light, 75
Inorganic chemistry, 1
Inorganic compounds, nomenclature of, 669
  preparation of, 469
  quantitative analysis of, 45
    precipitation reactions in, 415
Inorganic synthesis, acid-base reactions in, 469
  precipitation reactions in, 416
Insulin, *644*
Intensive properties, 19
Interhalogen compounds, 210
Intermolecular forces, in liquids, 239
  in molecular substances, 216
Internal energy, 67
International Bureau of Weights and Measures, 2
Interstitial particles, in crystals, 255
Iodide ion, analysis for, 414
  oxidation of, 567
  reaction with hydrogen peroxide, 468
Iodine, physical properties of, 215
  reaction with hydrogen, 387
  sublimation of, 259
    solid, 299
Iodine chloride, crystal structure of, 216, *217*
Iodine heptafluoride, 211
Iodine pentafluoride, 210
Ion(s), 28
  aquocomplex, 483
  charge of, and melting point, 213(t)
  common ion effect, 406
  complex, 481-510. See also *Complex ions.*
  concentrations of, equilibrium, 407
    Nernst equation and, 553
  electron configurations of, 155
  electronic structure of, 171
  forces between, and solubility product, 409
  foreign, 255
  formation of, 376
  in solutions, 293
  internuclear distances of, 174
  interstitial, 255
  metal, 562, 567
    reaction with water, 432
    separation of, 503
  monatomic, 171
  negative, 28, 172
    hydrolysis of, 446
    in electrolytic cells, 512
  polyatomic, 174(t)
    decomposition of, 212
    electron-pair repulsion in, 192
  positive, 28, 172, 512
    as oxidizing agents, 562
    formation of, 224
    hydrolysis of, 446
    oxidation of, 567

Ion(s) (*Continued*)
  positive, quantitative analysis for, 471
  radii of, 175(t), 673(t)
  resonance structures, 189
  separation of, 471
  size of, 174
    and melting point, 213(t)
Ion-exchange process, 418
Ionic bond, 170-175
Ionic character, electronegativity difference and, *181*
  partial, 180
Ionic compounds, 28, 212
  as electrical conductors, 214
  charge density, and solubility, 405
  charges of ions in, 171
  decomposition of, 212
  electrolysis of, 512
  formation of, 170, 376, 382
  insoluble, 404, 415
  melting point of, 212
  nomenclature for, 669
  solubility of, 214, 279, 403
Ionic crystals, defects in, 255
  electrical conductivity of, 256
  structure of, 253
Ionic equations, 400
Ionic radii, 175(t) 673(t)
Ionic substance, 212
  movement of crystal planes in, *227*
Ionization, radiation and, 589
Ionization constants, of acids and bases, 443(t)
  of water, 427
  uses of, 437
Ionization energy, *112*
  and electron configuration, 160
  of isoelectronic series, *162*
Iron, complex ions of, 484, 494
  corrosion of, 568
    impurities and, 570
  electronic structure of, 155
Iron carbide, 570
Iron(III) chloride, electrolysis of, 523
Iron(III) hydroxide, 568
Iron(III) ions, analysis for, 503
Iron(III) oxide, 392
Iron(II) salts, in photographic methods, 579
  oxidation of, 568
Iron(III) salts, analysis for, 507
Isobutane, 617, 635
Isobutene, 635
Isoelectronic series, 161, *162*
Isohexane, 617
Isomer, 618
  geometrical, 488, 619
Isooctane, 635
Isooctene, 635
Isoprene, polymerization of, 637
Isopropyl alcohol, 626
Isotope(s), 27, 585
  decay of, 598
  fissionable, 602
  neutron-to-proton ratio in, 599
  radioactive series of, 588
  uses of, 600

Joliot, F., 595

k, 353
K_a, 434-439, 443(t)
K_b, 441
K_c, 303-310
K_h, 443
K_p, 340
K_{sp}, 407
K_w, 427
K-electron capture, 600
Kekule, F. A., 621
Kelvin, W. T., 86
Kelvin scale, 11, 86
Kerosene, 634
Ketone, 626
Kinetic theory, 349-375
    of gases, 99

Lability, of complex ions, 498
Lampblack, 234
Lanthanide contraction, 166
Lanthanides, 119, 121
    atomic radii of, 166(t)
Lattice, crystal, 29, 251
    defects in, 255
    diffusion within, 256
    energy in, 171
Lawrence, E. O., 597
Lead, 223
Lead bromide, 326
    formation of, 330
Lead chloride, solubility of, 403
Lead chromate, 415
Lead ion, analysis for, 415
Le Chatelier's Principle, 316
Leclanché cell, 536
Leucippus, 24
Levorotatory isomers, 489
Lewis, G. N., 185, 451
Lewis acid, 451
Lewis base, 451, 485
    in substitution reactions, 629
Lewis structures, 185
Libby, W. F., 594
Ligand field theory, 498
Ligands, 482
Light, 74, 129, 226
    in nonspontaneous reactions, 75
    infrared, 75
    photoelectric effect, 224
    ultraviolet, 75
        as source of ozone, 230
    wavelengths of, 75
Light quanta, 136
Lime, 417
Lime-soda method, 417
Limestone, 476
Limiting density method, 110
Linear functions, 663
Liquids, 238-247
    boiling point of, 246
    compressibility of, 239

Liquids (Continued)
    density of, 18, 238
    diffusion of, 239
    expansion of, 239
    freezing point of, 261
    hydrogen-bonded, as solvents, 279
    intermolecular forces in, 239
    organic, solubility of, 277
    particle spacing in, 239
    physical properties of, 248(t)
    pressure of, 81
    structure of, 238
    surface to volume ratio, 240
Liter-atmosphere, 11
Lithium, 228
    atomic spectrum of, 135
    band structure of, 229
    conductivity of, 229
    electronic structure of, 152
    molecular orbitals in, 205
Lithium aluminum hydride, 383
Lithium chlorate, solubility of, 403
Lithium chloride, crystal structure of, 254
Lithium hydride, 382
Lithium iodide, 176
Lithium oxide, 439
Litmus, 425
Logarithmic Rate Law, 591
Logarithms, 653, 657(t)
London, F., 177
London forces, 218
Lowry, T. M., 447
Luminescence, radioactive, 589
Luster, 224, 226
Lye, 426, 439
Lyman series, 138(t)

Macromolecular crystals, 257
Macromolecular substances, 221
Magnesium, as acid strength indicator, 425
    isotopes of, 131(t), 132(t)
    reaction with oxygen, 58
Magnesium chloride, as test for base, 426
Magnesium hydroxide, 439
Magnesium oxide, physical properties of, 213
Magnesium salts, as test for base, 426
Magnesium sulfate, physical properties of, 292(t)
    solubility of, 406
Magnetic orbital, 148
Malleability, 224
Maltase, 363
Maltose, hydrolysis of, 363
Manganese, oxidation states of, 573
    oxides of, 390
Manganese dioxide, 257
Manometer, 82, 82, 242
Marsh test, for arsenic, 385
Mass, and energy, 604
    atomic, absolute, 37
    Conservation of, 48
    conversion factors for, 7(t)
    critical, 603
    measurement of, 5
    nuclear, 607(t)

Mass *(Continued)*
 of electrons, 128
 of molecules, 38
Mass Action, Law of, 305
Mass decrement, 608
Mass number, 131
Mass spectrometer, 34, *34*
Matter, particulate nature of, 24
Maxwell, J. C., 100, 106, 365
Maxwellian distribution function, *106, 366*
McLeod gauge, 108
McMillan, 597
Mechanical energy, 73
Mechanics, 133
Medicine, radioactive isotopes in, 600
Meitner, L., 602
Melting point, 17, 261
 interionic forces and, 212
 molecular weight and, 215
 of impure solids, 17
 of ionic compounds, 212
 of lighter elements, *112*, 113(t)
 of macromolecular substances, 221
 of mixtures, *18*
 of nonpolar substances, 218
 solubility and, 278
Mendeleev, D. I., 34, 115
 predicted properties of germanium, 116(t)
Mendelevium, 598
Mercuric sulfide, 578
Mercury, in barometers, 81
 in thermometers, 9
 physical properties of, 248(t), 249(t)
Mercury(II) chloride, 214
Mercury(II) oxide, 391
Mercury(II) sulfide, 403
Metals, alkali, atomic radii of, 165(t)
  electron configurations of, 163(t)
  reaction of oxygen with, 389
 electrical conductivity of, 223
 electropositive nature of, 119
 heat capacity of, 229
 hydrogen compounds of, 379, 383
 movement of crystal planes in, 227
 oxidation of, 568
  impurities and, 570
 oxygen compounds of, stability of, 391
 physical properties of, 222-229
 reaction with acids, 561, 575
 reaction with hydrogen, 383
 reaction with water, 562, 568
 transition, 117
  crystal defects in, 256
  electron configuration of, 155
  hydrogen compounds of, 383
  in complex ion formation, 483, 484(t)
  oxidation states of, 380
  oxygen compounds of, 390
Metal ions, separation of, 503
 as oxidizing agents, 562
 oxidation of, 567
Metallic bonds, 224
Metalloids, 120
Methane, 616
 bond angles in, 193
 bond energy of, 183

Methane *(Continued)*
 oxidation of, 620
 reaction with chlorine, 631
 structure of, *194*
Methanol, 626
Methyl alcohol, 277
Methyl amine, 440
2-Methylbutane, 635
Methyl cyclohexane, 636
2-Methylpentane, 617
2-Methylpropane, 617
Methyl red, 463
Methyl rubber, 637
Metric system, 2, 7
Milk of magnesia, 439
Millikan, R. A., 128
Misch metal, 121
Miscibility, 277
Mixture, melting point of, *18*
Mohr titration, 413
Molality, 274
Molarity, 275
 and normality, 461
Mole, 46, 49
Mole fraction, 273
Molecular formula, 42, 46
Molecular orbital(s), 201
 formation of, *203*, 228
 types of, 203
Molecular orbital method, 179
Molecular speeds, 102, 105
Molecular substances, intermolecular forces
  in, 215
 physical properties of, 214-220
 volatility of, 215
Molecular weight, 38
 and colligative properties, 290
 and Graham's Law, 104
 and molecular formula, 46
 and physical properties, 218
 and rate of effusion, 104
 and volatility, 215
 limiting density method, 110
 of gases, 92
 vapor density method, 93, *93*
Molecule(s), 27
 as dipoles, 196
 bond angles in, 192
 collisions of, 100
 diatomic, 27, 206(t)
 dipole moments of, 197
 electron density in, *180*
 electron-pair repulsion in, 192
 in gases, 100, 144
 Lewis structures for, 186
 mass of, 38
 models of, 28
 multiple bonds in, 184, 194
 odd-electron, 191
 orbitals in, 201, *203, 228*
 paramagnetic, 191
 polar, 180, 195, 216
 resonance structures of, 189
 size of, 220
 speed of, 105
 valence-bond model, 201

Momentum, 101
  angular, 148
Mond process, 483
Morely, E. W., 30
Moseley, H., 129
Motion, Newton's Laws of, 101
  translational, 102
  translational energy of, 133
Mulliken, R. S., 179
Multiple Proportions, Law of, 25

n-p junction, in semiconductors, 258
Naphthalene, 623
  solubility of, 279
Neohexane, 617
Neon, mass spectrum of, 35
Neoprene, 637
Nernst equation, 550
Net ionic equation, 400
Neutralization reactions, 457
Neutron, 131, 597
Neutron emission, 603
Newlands, 115
Newton, I., 24, 133
Newton's Law, 6, 101
Nickel(II), complex ions of, 492, 494
Nickel carbonyl, 483
Nickel chloride, reaction with sodium hy-
  droxide, 400
Nickel hydroxide, 400
Nickel oxide, crystal defects in, 256, 257
Nickel sulfate, reaction with barium hydrox-
  ide, 402
Nitrates, preparation of, 416
Nitrate ion, as oxidizing agent, 572
  bonding in, 195
  resonance forms, 190
Nitric acid, 430
  Ostwald process for, 386
  reaction, with metals, 575
Nitric oxide, electronic structure of, 191
  molecular orbitals in, 207
Nitrogen, atmospheric, 393
  bonding in, 184
  hydrogen compounds of, 385
  oxidation states of, 573
  oxides of, 393
Nitrogen dioxide, 350
  electronic structure of, 191
Nitrous oxide, decomposition of, 363
Nobelium, 598
Noble gases, 27, 115, 119
  electron configurations of, 172(t)
  solubility of, 279(t)
Noble gas structure, in covalent bonding, 185
  monatomic ions with, 171
Nonelectrolyte(s), 283
  colligative properties of, 284-290
Nonequilibrium phase behavior, 264
Nonmetals, as oxidizing agents, 564
  atomic radii of, 182
  binary compounds of, 671
  bonding in, 199, 443

Nonmetals (Continued)
  covalent radii of, 182
  hydrogen compounds of, 379, 383, 384(t),
    443
  in periodic table, 120
  molecular orbitals in, 206
  oxidation states of, 380
  oxygen compounds of, 383
  stability of, 393
Nonpolar substances, intermolecular forces
  in, 218
  boiling points of, 216(t)
Normality, 276
  molarity and, 461
Nuclear chemistry, 1
Nuclear fission, 601
  as energy source, 604
Nuclear fusion, 610
Nuclear reactions, 585-614
  binding energy in, 606
  bombardment, 595
  chemical changes in, 585
  energy changes in, 586
  gamma rays in, 597
  stability of, 606
Nuclear reactors, 604
Nucleic acid, 644
Nucleoprotein, 644
Nucleus, atomic, 26, 128
  binding energy and, 606
  mass of, 129
  stability of, 146
  structure of, 131

"Octaves," of elements, 115
Octet rule, 185, 190
Octyl alcohol, 278
Olefin, 618
Onsager, 293
Optical activity, 639
Optical isomer, 488
Orbital(s), 148
  antibonding, 203
  atomic 154, 159
    combination of, 228
    hybrid, 197-201
    overlap of, 201
    population of, 149
  bonding, 203
  covalent bonding and, 179
  magnetic, 148
  molecular, 201, 203, 228
Orbital diagram, 153, 158
Order number, 130
Organic acids, 627
Organic chemistry, 1, 615-649
Organic compounds, 615-649
  bonding in, 616
  common reactions of, 629-633
  halogen-containing, 624
  natural, 638
  optical activity in, 639
  oxygen-containing, 625
  quantitative analysis of, 44

Organic Compounds (Continued)
  solubility of, 277
  synthesis of, 629
Organic matter, age of, 594
Organic synthesis, substitution reactions in, 629
Osmosis, 288, 288
Osmotic pressure, measurement of, 289
Ostwald, W., 26
Ostwald process, 386
Overvoltage, 543
Oxalate ion, 485
Oxidation, 376-382, 512. See also Oxidation-reduction reactions.
  in water purification, 231
Oxidation number, 377
  and acid strength, 445
  covalent bonding and, 378
  in balancing oxidation-reduction equations, 526
Oxidation potential, standard, 539
Oxidation-reduction reactions, 377, 511-531.
  direction of, 549
  equations for, balancing, 523
  equilibrium constant for, 547
  extent of, 549
  in photography, 578
  in qualitative analysis, 560
  nonspontaneous, 511
  oxidation number and, 526
  oxidizing agents in, 560-584
  simultaneous, 564
  spontaneity of, 532-559
Oxidation state, 378, 381
Oxide(s), metal, 389, 391
  nonmetal, 383, 393
  reaction with water, 439
  reduction by hydrogen, 33
Oxidizing agent(s), 382, 560-584
  in acid solutions, 572
  metal ions as, 562
Oxyacid, 671
  as strong acid, 433
  electronic structure of, 430, 444
Oxyanion, 431
  as oxidizing agent, 572
  in qualitative analysis, 577
  reduction of, 573
Oxygen, 28
  allotropes of, 230
  atomic weight of, 30
  boiling point of, 247
  compounds of, 389
  critical temperature of, 246
  electronic structure of, 191
  in organic compounds, 625
  isotopes of, 131(t), 132(t)
  paramagnetism in, 191
  physical properties of, 231(t)
  reactions of, 566
    with magnesium, 58
    with metals, 389
    with nonmetals, 383, 393
Oxyhemoglobin, 486
Ozone, atmospheric, 230
  physical properties of, 231(t)

Packing, crystal, 252
Paired electrons, 149, 176
Palmitic acid, 628
Paradichlorobenzene, 624
Paraffin(s), 616
Paraffin wax, 216
Paramagnetism, 191
Partial ionic character, 180
Partial pressures, 95
Particle(s), wave properties of, 141
Particle in a box model, 144
Pauli exclusion principle, 149
Pauling, L., 178
Pentane, 277
  in thermometers, 10
Peptide bonds, 642
Percentage yield, 51
Perchloric acid, 426, 430
Periodic law, 114
Periodic table, 116-121, 118(t)
  limitations of, 124
Periodicity, 114
  and electron configuration, 163
Permanganate ion, as oxidizing agent, 576
Permeability, 288
Peroxides, 389
Petit, A. T., 31
Petroleum, 633
pH, 428
pH meter, 554
Phase diagram, 262
Phenol, 623, 625
Phenol-resins, 626
Phenolphthalein, 426, 463
Phosgene, 624
Phosphoric acid, 462
Phosphorus, allotropic forms of, 234
  bonding in, 199, 235
  electronegativity values, 182
  electronic structure of, 235
  physical properties of, 234
Photoelectric effect, 141, 224
Photography, 578
Photon, 74, 135
  in nuclear reactions, 597
Photosynthesis, 75
  radioactive analysis of, 601
Physical changes, 12
Physical chemistry, 2
Pi bonds, 201
Pitchblende, 586
Planck's constant, 136
Planck's equation, 74
Platinum, as catalyst, 363, 395
Platinum(II), complex ions of, 482
Plato, 24
Polarity, and physical properties, 216
  in covalent bonding, 179
  of molecules 180, 195
Polarization, 220
Polonium, 586
Polyene, 620
Polyethylene, 619
Polymer(s), 619
  molecular weight of, 291
  of sulfur, 232

Polymerization, 619, 635
Polypeptide, 642
Polypropylene, 619
Polystyrene, 619
Polyvinyl chloride, 619
Positron, 595, 600
Potassium, 113
    atomic spectrum of, *135*
Potassium chloride, 292(t)
Potassium chromate, 413
Potassium ferrocyanide, 503
Potassium fluoride, electrolysis of, 515
Potassium hexacyanoferrate, 503
Potassium hydroxide, 632
Potassium nitrate, 409
Potassium permanganate, as oxidizing agent,
    576
    preparation of, 572
Potassium uranyl sulfate, 586
Potentials, of electrolytic cells, 539
Potentiometer, *555*
Powell, 192
Precipitation reactions, 400-424
    and inorganic preparations, 415
    indicators in, 413
    in qualitative analysis, 414
    in quantitative analysis, 411
    in water softening, 416
    ion-exchange in, 418
    solubility equilibria in, 406
Pressure, and boiling point, 246
    and solubility, 281
    atmospheric, 80
    effect on state, 261
    measurement of, 11, 82
    osmotic, *289*
    partial, 95
    sublimation, 259
    vapor, 259
Priestly, J. 20, 391
Propane, 183(t), 617
    reaction with chlorine, 631
Propene, 619
Proportionality constant, 100, 352, 661
Proteins, 641
Proton, 131
    hydrated, 426
    transfer of, in acidic solutions, 430
        in Brönsted-Lowry acids, 451
Proust, J., 25
Pure substances, 12
    heats of formation, 60
    identification of, 16
    physical properties of, 18
Pycnometer, 19
Pyridine, 440, *623*
Pyrolusite, 572
Pyrolysis, 634

Qualitative analysis, acid-base reactions in,
    457-480, 471
    chelating agents in, 503
    for complex ions, 502

Qualitative Analysis *(Continued)*
    oxidation-reduction reactions in, 560
    oxyanions in, 577
    precipitation reactions in, 414
Quantitative analysis, 42, 506
    acid-base reactions in, 460-469
    carbon-hydrogen, 44
    EDTA titrations in, 507
    gravimetric, 412
    of organic compounds, 44
    potassium permanganate in, 576
    precipitation reactions in, 411
    volumetric, 412
Quantum, 133
    of light, 136
Quantum mechanics, 133
Quantum number, 133, 143, 147
    allowed sets of, 150(t)
    and atomic spectra, 151
    energy and, 159
    principal, 148
Quantum state, 134
Quantum theory, 133
Quartz, 221, *222*

Rad, 590
Radiant energy, atomic, 129
    conversion to electrical energy, 259
Radiation, 587
    exposure to, 591(t)
    interaction with matter, 589
    ionization effects of, 589
Radicals, 369
    free, 630
Radioactive decay, 128
    of artificially produced isotopes, 598
    rate of 591
Radioactive isotopes, 600
Radioactive series, 588
Radioactivity, 128
    artificial, 595
    natural, 586
Radium, 586
Rain, 265
Randomization, of chemical bonds, 333
Raoult's Law, 284
    molecular interpretation of, *285*
Rare earths, 119, 121
Rate constant, for chemical reactions, 353
    temperature-dependence of, 359
Rate law, logarithmic, 591
Rays, cathode, 127
Reactions, acid-base, 457-480. See also
    Acid-base reactions.
    actual yield of, 51
    alkylation, 635
    bimolecular, 367
    biochemical, 469
    catalysts in, 362
    chain, 369
        nuclear, 603
    collision theory of, 364
    dehydration, 632
    dehydrohalogenation, 632

Reactions (*Continued*)
  direction of, 306
  electrochemical, 73
  electron transfer in, 376
  elimination, 631
  endothermic, 58
  energy changes in, 56-78, 162, 170, *362*
  enzymes in, 363, 469
  equilibrium, 57, 299-322. See also *Equilibrium reactions.*
  exothermic, 58
  extent of, 308
  free energy in, 323, 327
  gas phase, 353(t), 354
  half-life of, 356
  mechanical energy and, 73
  neutralization, 457
  nonspontaneous, 75
  nuclear, 585-614. See also *Nuclear reactions.*
  nucleophilic, 500
  of organic compounds, 629-633
  order of, 353
  oxidation, 376-382, 512
  oxidation-reduction, 511-531. See also *Oxidation-reduction reactions.*
  percentage yield of, 51
  polymerization, 635
  precipitation, 400-424. See also *Precipitation reactions.*
  radioactive analysis of, 601
  rates of, 349-375
  reduction, 376, 382, 512
  "scrambling," 333
  spontaneity of, 58, 323-348
  substitution, 500, 629. See also *Substitution reactions.*
  synthesis, 629
  theoretical yield of, 51
Reagent(s), critical, 51
  in metal-ion titrations, 507
Rectifier, 258
Redox reactions. See *Oxidation-reduction reactions.*
Reducing agent, 382
Reduction, 376-382, 512
Reduction potential, standard, 539, 542(t)
Reflectance, 226
Repulsion, electrostatic, 192
Resistance, 259
Resonance, 189
Ribonucleic acid, 645
Richards, T. W., 30
RNA, 645
Roentgen, W., 129
Rotational energy, 133
Rubber, 636
Rubidium iodide, 174, *176*
Rust, 568
Rutherford, E., 26, 128

△S, 332
Salts, as corrosion inhibitors, 569
  electrolysis of, 544
  formation of, 627

Salts (*Continued*)
  hydrated, 214
  hydrolysis of, 446
  preparation of, acid-base reactions in, 469
  water solutions of, 446, 544
Salt-bridge cell, 535
Salt water, distillation of, *13*
Sanger, F., 644
Saponification, 628
Schottky defect, 256
Schrödinger, W., 146
Schrödinger wave equation, 146
"Scrambling," 333
Seaborg, G., 597
Semiconductors, 257
Sidgwick, 192, 491
Sigma bonds, 200
Sigma orbitals, 203
Significant figures, 659
Silica gel, in chromatography, 15
Silicon, conductivity of, 223, 257
  semiconductors from, *258*
Silicon dioxide, crystal structure of, 221, *222*
Silver, complex ions of, 487, 501
  halides of, 124(t)
  reaction with nitric acid, 575
Silver acetate, saturated solution of, 406
Silver bromide, 578
Silver chloride, 409
Silver fluoride, 173
Silver ion, as oxidizing agent, 562
Silver nitrate, 413
  reaction with copper, 563
Silver oxide, 338, 391
  free energy change in, 328, 328(t)
Silver sulfide, 256
Simplest formula, 42-46
Slater, J. C , 178
Soaps, 628
Sodium, atomic spectrum of, *135*
  electronic structure of, 155
  ionization energy of, 160, *161*
  production of, by electrolysis, 512
  reactions of, with fluorine, 170
    with water, 562
Sodium acetate, 440
  reaction with acetic acid, 466
  water solution of, 272
Sodium bromide, reaction with sulfuric acid, 574
Sodium carbonate, 440
  in water softening, 417
  preparation of, 476
Sodium chloride, 400
  arrangement of ions in, *254*
  as ionic conductor, 214
  crystal structure of, *29*
  electrolysis of, 512, *513*, 516, 543
  in solutions, 291
  in Solvay process, 475
  physical properties of, 212, 249(t)
  separation from water, 12
  solid, forces in, 217
Sodium fluoride, formation of, 170, 376
  ionic structure of, *171*
Sodium hydride, reaction with ammonia, 449

Sodium hydrogen carbonate, 432, 440
  preparation of, 475
Sodium dihydrogen phosphate, 432
Sodium hydrogen sulfate, 432
Sodium hydrogen sulfite, 432
Sodium hydroxide, 426, 439
  preparation of, 517
  reactions of, with acetic acid, 464
    with carbonic acid, 459
    with hydrochloric acid, 457, 463
    with nickel chloride, 400
Sodium ion, electronic structure of, 155
Sodium peroxide, 389
  as oxidizing agent, 578
Sodium sulfide, water solution of, 446
Sodium stearate, 416
Sodium thiosulfate, 580
Solar battery, 259
Solids, 248-267
  crystal structure in, 249
  density of, 248
  diffusion of, 256
  intermolecular forces in, 248
  macromolecular, 221
  melting point of, 261
  nonpolar, 279
  solubility of, 278
Solubility, and fractional crystallization, 14
  melting point and, 278
  of hydrocarbons, 277
  of ionic compounds, 214, 279, 403
  of macromolecular substances, 222
  pressure and, 279, 281
  principles of, 276-280
  rules for, 404(t)
  temperature and, 280
Solubility equilibria, in water solutions, 406
  in acid solutions, 473
Solubility product, 407, 410(t)
Solute, 271
Solutions, 269-298. See also *Water solutions*.
  acidic, 271(t)
    equilibria in, 473
  basic, 271(t)
  boiling point of, 286
  colligative properties of, 283-294
  concentration of, 271
    and equilibria, 311
  dilute, 271
  freezing point of, 286
  gaseous, 93
  ideal, 285
  interstitial, 270
  ionic atmosphere in, 293
  liquid-liquid, 277
  liquid-vapor, 279
  neutral, 427
  phases of, 269
  saturated, 272
  solid, 269
  solid-liquid, 278
  structure of, 270
  substitutional, 270
  supersaturated, 272
  vapor pressure in, 284
Solvay process, 474

Solvent(s), 271
  in fractional crystallization, 15
  nonpolar, and ionic compounds, 214
  vapor pressure of, 284
Sörenson, P. L., 428
Sound, 106
Specific heat, 31
Spectrometer, 34
Spectrum(a), atomic, 134
  continuous, 134
  infrared absorption, 20
  mass, 35
  quantum numbers and, 151
  regularities in, 137
  x-ray, 129
Spin, electron, 153
Spontaneity, of chemical reactions, 323-348,
  545
Standard electrode potential, 539
Starch, 640, *641*
Stas, J. S., 30
State, changes in, 12, 79, 238-267
  impurities and, 264
  internal energy and, 67
  nonequilibrium behavior, 264
  phase diagram of, 262
  pressure and, 261
Stereoisomer, 639
Storage battery, 73, 537
Strassman, 602
Strontium-90, 603
Styrene, 623
Sublimation, 259
  energy of, 215
  heat of, 260
Sublimation pressure, 259
Substitution reactions, 500, 629
  in complex ion formation, 498
  Lewis bases in, 629
  nucleophilic, 629
Succinic acid, *14*
Sucrose, *640*
  hydrolysis of, 358
Suction pump, *81*
Sugar, 638
Sulfate ion, 187
  analysis for, 472
Sulfide(s), 472
Sulfide ion, hydrolysis of, 446
  oxidation of, 567
Sulfite ion, bonding in, 189, 195
Sulfur, allotropic forms of, 230
  atomic weight of, 30
  monoclinic, 231
  oxidation states of, 573
  oxides of, 395
  "plastic," 232
  rhombic, 231
Sulfur dioxide, electronic structure of, 189
Sulfur hexafluoride, bonding in, *194*, 199
Sulfur trioxide, 430
  as Lewis acid, 452
  bonding in, 189
Sulfuric acid, 239
  as oxidizing agent, 561
  formation of, 430

Sulfuric acid (*Continued*)
in preparing weak acids, 470
preparation of, 395
reaction with sodium bromide, 574
Superoxides, 389
Surface tension, 240, 240(t)
intermolecular forces and, 218
Synthesis, inorganic, acid-base reactions in, 469
precipitation reactions in, 416

Tartaric acid, 14
Temperature, absolute, 11, 86
and equilibrium constant, 317
and free energy change, 330
and reaction rate, 359
and solubility, 280
and vapor pressure, 244
Kelvin, 86
Temperature scales, 9
Tetracyanonickelate(II) ion, 494
Tetraethyl lead, 370
Tetrahydrofuran, 625
Tetramminecopper(II) ion, 481, 483, 502
Tetramminecopper(II) sulfate, 481
Tetramminezinc(II) ion, 493
Theoretical yield, 51
Thermionic emission, 224
Thermistor, 259
Thermit process, 392
Thermochemistry, Laws of, 59
Thermodynamics, 65-73
chemical, 323-348
First Law of, 65, 331
Second Law of, 332
Third Law of, 335
Thermometer, 8
Thin-layer chromatography, 15
Thomson, J. J., 26, 127
Tin, allotropic forms of, 230
Tin(II) chloride, oxidation of, 567
Titration, acid-base, 460
chloride, 413
EDTA, 507
equivalence point in, 463
indicators in, 425, 462
metal ion, reagents for, 507
Mohr, 413
oxidation-reduction, 576
Toluene, 623, 636
Torr, 11
Torricelli, E., 11, 81
Trans isomer, 488
Transistor, 258
Transition elements, 119
Transition metals, 117
crystal defects in, 256
electronic structure of, 155
in complex ion formation, 483, 484(t)
oxidation states of, 380
reaction with hydrogen, 383
reaction with oxygen, 390
Translational energy, 133
Translational motion, 102

Transuranium elements, 597
"Triads," of elements, 115
Triperoxychromate ion, 578
Triple point, 260
Trouton's rule, 248

Ultraviolet light, 75
and ozone formation, 230
Uncertainty principle, 133
Unit cell, in crystal structure, 251
Uranium-235, preparation of, 602
Uranium-238, radioactive series for, 588(t)
Uranium salts, 586

Vacancies, in crystals, 255
Valence-bond theory, 201
Valence electron, 165
and chemical bonds, 224
in metallic bonding, 225
mobility of, 226
Valence state, 181
Vanadium pentoxide, as catalyst, 363, 395
Van der Waals equation, 99
Van der Waals forces, 218
Vapor, particle spacing in, *239*
Vapor density, in molecular weight determination, 93, *93*
Vapor-liquid equilibrium, 240
Vapor phase chromatography, 16
Vapor phase equilibrium, 242
Vapor pressure, 241, 259
and boiling point, 248
measurement of, *241*
of solvents, 284
temperature and, 244
volume and, 242
Vapor-solid equilibrium, 259
Vaporization, heat of, 239, 260
boiling point and, 248
experimental determination of, 244
molecular weight and, 215
Vervey, 257
Vibrational energy, 133
Viscosity, of water, 218
Volatility, and distillation, 12
and polarity, 216
molecular weight and, 215
of acids, 470
of molecular substances, 215
of nonpolar substances, 218
Voltage, 550
standard, 541
Voltaic cells, 532-559
commercial, 535
direction of current in, 550
zinc-copper, 533
Volume(s), 5(t), 5
Law of Combining, 94
of gases, 94
Volumetric analysis, 412
Von Laue, M., 250
Vulcanization, 636

Waage, P., 305
Water, 28, 42
  amphiprotic nature of, 448
  as Brönsted acid and base, 448
  as oxidizing agent, 562
  bonding in, 193, 217, 220
  deionization of, 418
  dissociation of, 326
  entropy of, *335*
  hard, 416
  heat capacity of, 218
  impurities in, 421
  of hydration, 213
  physical properties of, 218, 248(t), 249(t),
    261
  polarity of, 195
  purification of, by oxidation, 231
  reactions of, with aluminum salts, 432
    with hydrides, 443
    with hydroxides, 439
    with metal ions, 432
    with metals, 562
    with oxides, 439
    with salts, 446
    with sodium, 562
  sublimation of, 259
  vapor pressure of, temperature and, *244*
  vaporization of, 336
Water gas, 318, 395
Water softening, 416
Water solutions, acidic, 430
  basic, 439-456
  complex ions in, 483
  conductivity of, 283, 433
  electrolysis of, 515
  electrolytic properties of, 283
  equilibria in, 427
  ion concentrations in, 427

Water solutions *(Continued)*
  ionic reactions in, 401
  of metals, 568
  of salts, 544
  oxidizing agents in, 560-583
  pH of, 428
  properties of, 286, 425
Watson, J., 646
Wave function(s), 142, *142*
Weight, 6
Werner, A., 491
Wilson cloud chamber, 589
Wöhler, F., 513
Work, 70
  and chemical reactions, 57, 66, 325
  of expansion, 67

X-ray(s), in nuclear reactions, 597
X-ray diffraction, 249, *250*
X-ray spectra, 129
Xenon difluoride, 200
Xylene, 623

Zeolite, 419, *419*
Zinc, as acid strength indicator, 425
  as reducing agent, 575
  complex ions of, 492
  reaction with acids, *434*
  reaction with nitric acid, 575
Zinc salts, reaction with water, 432
Zinc sulfate, 432
  anticorrosive nature of, 569
Zinc sulfide, 589